CONFOCAL AND TWO-PHOTON MICROSCOPY

CONFOCAL AND TWO-PHOTON MICROSCOPY
FOUNDATIONS, APPLICATIONS, AND ADVANCES

Edited by

Alberto Diaspro
National Institute for the Physics of Matter
Department of Physics
University of Genoa
Genoa, Italy

WILEY-LISS

A JOHN WILEY & SONS, INC., PUBLICATION

This book is printed on acid-free paper. ♾

Published simultaneously in Canada.

For ordering and customer service information please call 1-800-CALL-WILEY.

Library of Congress Cataloging-in-Publication Data:

Confocal and two-photon microscopy : foundations, applications, and advances / edited by Alberto Diaspro.
p. ; cm.
Includes bibliographical references and index.
ISBN 0-471-40920-0 (cloth : alk. paper)
1. Confocal microscopy. 2. Fluorescence microscopy. 3. Three-dimensional imaging- -Methodology. I. Diaspro, Alberto, 1959-
[DNLM: 1. Microscopy, Confocal- -methods. 2. Imaging, Three-Dimensional- -methods. 3. Microscopy, Fluorescence- -methods. QH 224 C747 2002]
QH224 .C64 2000
502′.8′2- -dc21

Printed in the United States of America.

10 9 8 7 6 5 4 3 2 1

This book is warmly dedicated with enormous love to my wife Teresa, to my daughter Claudia, and to Pippi, a white and lovely rabbit. A particular dedication is for Bruno and Franca Bianco, who are, with Francesca, Pietro, and Elisabetta, a sweet and special family for me. This book is also dedicated to the memories of Ivan Krekule, who acted like a father, and of Mario Arace, my grandfather, who purchased my first oscilloscope.

CONTENTS

FOREWORD

The last few decades have been an exciting time for optical microscopy. New imaging concepts together with the availability of lasers and the use of computers have completely transformed this field. Not the least of these new developments is the emergence of first confocal and later two-photon or multiphoton microscopy. Both techniques not only add increased resolution to optical microscopy, but are also capable of imaging an optical section inside an extended specimen independently of out-of-focus regions. This phenomenon results in what is also known as three-dimensional imaging, with the basic image a three-dimensional (3D) dataset in a computer.

These developments have been of tremendous benefit to fields where the 3D organization of an object is sought, and have found wide application in materials and food science, and especially the biomedical sciences. In the latter field, the ability to create over time 3D image sequences of cells and tissues in the native, watery environment is particularly important. Studies of, for instance, cell division and cell and tissue development are facilitated by this tool.

The incoherence of emitted fluorescence makes it the contrast mode of choice for image formation in sectioning microscopy. Fluorescent labeling techniques are particularly valuable because they are very powerful and are rapidly evolving. Moreover, recently developed green fluorescent protein (GFP) techniques provide, in combination with optical sectioning, access to 3D organization and function in living biological specimens.

In view of the importance of 3D imaging and its wide application, an up-to-date account of the present state of the field is of great value. The present book treats both the basics as well as the various aspects of governing resolution, contrast, and 3D image formation and processing. In addition, a number of relevant applications are presented, illustrating both the power of 3D imaging and the ways in which it can be applied.

My expectation is that 3D microscopy still holds for us a number of significant developments – possibly even surprises – with respect to resolution and specificity of contrast formation.

G.J. Brakenhoff

PREFACE

"It was morning, and the new sun sparkled gold across the ripples of a gentle sea. [. . .] More than anything else, Jonathan Livingston Seagull loved to fly. [. . .] No limits, Jonathan? he thought, and he smiled. His race to learn had begun."

—From *Jonathan Livingston Seagull: A Story*, Richard Bach, 1972, Pan Books Ltd, London, UK

Since I was a student in Electronic Engineering, I was fascinated by microscopy. In 1983, I began my first work in microscopy, using quantitative phase contrast and digital holography. The leitmotif of my interest was and still is understanding the complex relationship between structure and function, as well as an obsession for resolution. I decided to pursue a book on two microscopical techniques that are unique in allowing for structural studies on functioning systems, mainly biological ones, at a low level of structural and functional perturbation. This book was conceived in the spirit of assembling information useful to begin or evolve in the use of three- and four-dimensional optical microscopy.

Tony Wilson and Colin Sheppard recently reminded me of the historical paper by Lukosz, published in November 1966 in the *Journal of the Optical Society of America*. Entitled "Optical Systems with Resolving Powers Exceeding the Classical Limit," the paper's abstract reads: "A new theorem on the ultimate limit of performance of optical systems is established: Not the band-width of the transferred spatial frequencies but only the number of degrees of freedom of the optical message transmitted by a given optical system is invariant. It is therefore possible to extend the bandwidth by reducing the object area." As commented by Wilson in his chapter, this theorem forms the basis for confocal microscopy as well scanning probe techniques.

Although created more than three decades ago, only in the past decade has the confocal microscope become a practical and widely disseminated instrument. The advent of two-photon excitation microscopy and new developments in the architecture of scanning heads, as well as imaging technology and software, have favored the growth in applications for biological research, chemical analysis, materials testing, and clinical methods. This book intends to fill the need for a substantive, focused, and up-to-date treatment integrating the applications and technological advances relevant across biology, medicine, physics, engineering, and chemistry. It offers crucial information about fluorescent molecules, imaging software, and new commercial trends. With fundamental principles presented alongside advanced topics, both potential new users and seasoned researchers should find it useful.

For this book, as is common in science, contributions draw upon a range of disciplines, from optics to biophysics, histochemistry to biochemistry and so on, in a truly multidisciplinary scientific arena. I hope that its publication will stimulate and encourage further interdisciplinary activities in this field, where a lot of work can and has to be done.

Color versions of many of the figures in this book are available at the following online address: ftp://ftp.wiley.com/public/sci_tech_med/confocal_microscopy

ALBERTO DIASPRO
Genoa, Italy

ACKNOWLEDGMENTS

First, I thank all the authors who agreed to contribute and without whom this project could not have been realized.

During my career, I have had the opportunity of interacting with scientists who have introduced me to different microscopic techniques. Among them, in no particular order, are Bruno Bianco, Francesco Beltrame, Miguel Aguilar, Cesare Ascoli, Carlo Frediani, Ranieri Rolandi, Pio Benedetti, Ivan Krekule, Roger Johnston, Gary Salzman, Ignacio Tinoco, Jr., Enrico Gratton, Weyming Yu, and David Piston. I would also like to acknowledge the debt I owe to previous publications in this field, including *Introduction to Fourier Optics* by Goodman, *Principles of Optics* by Born and Wolf, two papers in *Nature* by Agard and Sedat, and by Fred Brakenhoff and his colleagues, a *Cytometry* paper by Ted Young, *Classical Electrodynamics* by Jackson, *QED* by Feynman, the mythical book by Wilson and Sheppard, papers by Quate on acoustical microscopy and by Binnig, Rohrer, Betzig, and van Hulst on scanning probe techniques, the bible entitled *Digital Image Processing* by Ken Castleman, *Biophysics* by Volkenstein, *Physical Biochemistry* by van Holde, *Biophysical Chemistry* by Cantor and Schimmel, and, in particular, *Handbook of Biological Confocal Microscopy* by Jim Pawley.

Finally, in random order, I am very sincerely grateful to: Cristiana Ricci, Stefano Schiappacasse, Kathleen O'Keefe, Enrico Gratton, Weyming Yu, David Piston, Christian Soeller, Marc Cannell, Osamu Nakamura, Satoshi Kawata, Dixon Chung, Joseph Lakowicz, Henryk Malak, Hans Gerritsen, Hans van der Voort, Peter Drent, Peter So, Tiziana Parasassi, Cesare Usai, Enrico Ambi, Salvatore Cannistraro, Alessandra Gliozzi, Mauro Robello, Bruno Samorì, Giulio Peirone, Giacomo Pasini, Ranieri Rolandi, Umberto Maria Marinari, Alberto Blasi, Bruno Bianco, Kees van Ord, Marteen Balzar, Stanley Schwartz, Ammasi Periasamy, Tony Wilson, Colin Sheppard, John Johnston, Ignacio Tinoco Jr., Carlos Bustamante, David Dunlap, Ken Castleman, George Ruben, Alessandro Chiabrera, Francesco Beltrame, Cesare Ascoli, Fred Brakenhoff, Jim Pawley, Giuliano Mazzini, Beppe Starace, Raffaele De Vita, Miguel Aguilar, Alvin Wald, Luna Han, Paola Ramoino, Mario Bertero, Stanley Zietz, Eligio Patrone, Giuseppe Aceti, Ted Young, Marco Raimondo, Pio Benedetti, Marco Fato, Paolo Gualtieri, Valtere Evangelista, Michele Bertolotto, Heinz Gundlach, Giuseppe Cittadini, Emilio Di Maria, Giovanna Castellaro, Swamy Laxminarayan, Carlo Pellicciari, Isabel Freitas, Gianni Melioli, Paola Barboro, and Antonio Borsellino.

CONTRIBUTORS

Luis Bagatolli, Departamento de Quimica Biologica, Cuidad Universitaria, Cordoba, Argentina

Keith M. Berland, Department of Physics, Emory University, Atlanta, Georgia, U.S.A.

Mario Bertero, INFM, DISI. University of Genoa, Genoa, Italy

Patrizia Boccacci, INFM, DISI, University of Genoa, Genoa, Italy

Christof Buehler, Paul Scherrer Institut, Villigan, Switzerland

Slivano Capitani, Dipartimento di Morfologia ed Embriologia, Sezione di Anatomia Umana Normale, Università di Ferrara, Ferrara, Italy

Walter Carrington, University of Massachusetts Medical Center, Shrewsbury Campus, Shrewsbury, Massachusetts, U.S.A.

Kenneth Castleman, Advanced Digital Imaging Research, LLC, League City, Texas, U.S.A.

Alberto Diaspro, National Institute for the Physics of Matter, Department of Physics, University of Genoa, Genoa, Italy

Chen Y. Dong, Department of Physics, National Laboratory of Taiwan, Taipei, Taiwan

Mary Dickinson, Biological Imaging Center, Beckman Institute, California Institute of Technology, Pasadena, California, U.S.A.

Scott E. Fraser, Biological Imaging Center, Beckman Institute, California Institute of Technology, Pasadena, California, U.S.A.

P. L. T. M. Frederix, Utrecht University, Molecular Biophysics, Utrecht, The Netherlands

Katsamusa Fujita, Department of Pathology and Cell Regulation, Kyoto Prefectural University of Medicine, Kyoto, Japan

H. C. Gerritsen, Utrecht University, Molecular Biophysics, Utrecht, The Netherlands

John Girkin, Institute of Photonics, University of Strathclyde, Wolfson Centre, Glasgow, U.K.

Enrico Gratton, Laboratory for Fluorescence Dynamics, University of Illinois at Urbana-Champaign, Urbana, Illinois, U.S.A.

Kees de Grauw, Utrecht University, Molecular Biophysics, Utrecht, The Netherlands

Thomas Hancewicz, Unilever Edgewater Laboratory, Edgewater, New Jersey, U.S.A.

Lily Hsu, Department of Mechanical Engineering, Massachusetts Institute of Technology, Cambridge, Massachusetts, U.S.A.

Jiří Janáček, Institute of Physiology, Academy of Sciences of the Czech Republic, Prague, Czech Republic

James E. N. Jonkman, Cell Biology and Biophysics Programme, European Molecular Biology Laboratory, Heidelberg, Germany

Peter Kaplan, Unilever Edgewater Laboratory, Edgewater, New Jersey, U.S.A.

Raymond Keller, Department of Biology, University of Virginia, Charlottesville, Virginia, U.S.A.

Ki H. Kim, Department of Mechanical Engineering, Massachusetts Institute of Technology, Cambridge, Massachusetts, U.S.A.

Karsten König, Institute of Anatomy II, Friedrich Schiller University Jena, Jena, Germany

Michael Kozubek, Faculty of Informatics, Laboratory of Optical Microscopy, Masaryk University, Brno, Czech Republic

Ivan Krekule, Institute of Physiology, FGV Academy of Sciences of the Czech Republic, Prague, Czech Republic

Lucie Kubínová, Institute of Physiology, FGV Academy of Sciences of the Czech Republic, Prague, Czech Republic

Moshe Levi, Department of Internal Medicine, University of Texas Southwestern Medical Center and VA Medical Center, Dallas, Texas, U.S.A.

Alberto M. Martelli, Instituto di Citomorfologia Normale e Patologica del C.N.R., Dipartimento Di Scienze Anatomiche Umane e Fisiopatologia dell'Apparato Locomotore, Università di Bologna, Bologna, Italy

Barry R. Masters, Department of Opthalmology, University of Bern, Bern, Switzerland

Jim McNally, Fluorescence Imaging Facility, Laboratory of Receptor Biology and Gene Expression, National Cancer Institute, Bethesda, Maryland, U.S.A.

Luca M. Neri, Dipartimento di Morfologia ed Embriologia, Sezione di Anatomia Umana Normale, Università di Ferrara, Ferrara, Italy

Colten Noakes, W.M. Keck Center for Cellular Imaging, Department of Biology, University of Virginia, Charlottesville, Virginia, U.S.A.

Stamatis Pagakis, Confocal and Image Analysis Laboratory, Membrane Biology, National Institute for Medical Research, London, U.K.

Tiziana Parasassi, Istituto di Medicina Sperimentale, CNR, Roma, Italy

Ammasi Periasamy, Keck Center for Cellular Imaging (KCCI), University of Virginia, Charlottesville, Virginia, U.S.A.

Roy Pike, Physics Department, King's College London, London, U.K.

Colin Sheppard, School of Physics, The University of Sydney, Australia

Irini Skaliora, University Laboratory of Physiology, University of Oxford, Oxford, U.K.

Paul Skoglund, Department of Biology, University of Virginia, Charlottesville, Virginia, U.S.A.

Carolyn L. Smith, National Institute for Neurological Disorders and Stroke, Light Imaging Facility, Bethesda, Maryland, U.S.A.

Peter T. C. So, Department Mechanical Engineering, Massachusetts Institute of Technology, Cambridge, Massachusetts, U.S.A.

Ernst H. K. Stelzer, Cell Biology and Biophysics Programme, European Molecular Biology Laboratory, Heidelberg, Germany

Ann E. Sutherland, Department of Cell Biology, University of Virginia Health Sciences Center, Charlottesville, Virginia, U.S.A.

Tetsuro Takamatsu, Department of Pathology and Cell Regulation, Kyoto Prefectural University of Medicine, Kyoto, Japan

Uday K. Tirlapur, Institute of Anatomy, Friedrich-Schiller-University, Jena, Germany

Peter Török, Department of Engineering Science, University of Oxford, Oxford, U.K.

Fulvio Ursini, Dipartimento di Chimica Biologica, Università di Padova, Padova, Italy

Tony Wilson, Department of Engineering Science, University of Oxford, Oxford, U.K.

David L. Wokosin, Centre for Biophotonics, Strathclyde Institute for Biomedical Sciences, University of Strathclyde, Glasgow, U.K.

Ralf Wolleschensky, Carl Zeiss Jena GmbH, Advanced Imaging Microscopy Division, Jena, Germany

Chris Xu, Bell Laboratories, Lucent Technologies, Holmdel, New Jersey, U.S.A.

Weiming Yu, Department of Cell and Molecular Biology, Northwestern University, School of Medicine, Chicago, Illinois, U.S.A.

Hubert K. Zajicek, Department of Internal Medicine, University of Texas Southwestern Medical Center and VA Medical Center, Dallas, Texas, U.S.A.

CHAPTER 1

The Generalized Microscope*

COLIN J. R. SHEPPARD

School of Physics, and Australian Key Centre for Microscopy and Microanalysis, The University of Sydney, NSW 2006, Australia

1 INTRODUCTION

Two-photon fluorescence microscopy and second-harmonic-generation microscopy (Gannaway and Sheppard, 1978) are two examples of a range of different modes of nonlinear optical microscopy, as discussed by Sheppard and Kompfner (1978). In fact, microscope imaging modes can be based on a wide range of different physical principles. In particular, in principle, virtually any form of spectroscopy can be the basis of building up an image by measuring the spatial variations in the signal (Sheppard, 1987a). However, there are also fundamentally different types of microscope: we can have conventional (non-scanning), scanning, and confocal microscopes. This chapter discusses the different ways of constructing a microscope, together with the basic principles and properties of these various systems.

2 THE MICROSCOPE

According to the Oxford English Dictionary, a microscope is "an instrument magnifying objects by means of lenses so as to reveal details invisible to the naked eye." Thus a microscope forms an image of an object. But what does a microscope image? Of course, the first microscopes were of the optical variety. Actually, an ordinary optical microscope images variations in the reflectivity (or transmissivity) of the object. However, in general, a microscope can image variations in a wide range of different physical properties of an object. Thus a microscope, as well as forming a magnified image, can render visible variations in a variety of different characteristics of the sample.

* This chapter is based upon an invited presentation at the Symposium of the Australian Society for Electron Microscopy, University of Sydney, 1996.

Confocal and Two-Photon Microscopy: Foundations, Applications, and Advances, Edited by Alberto Diaspro.
ISBN 0-471-40920-0

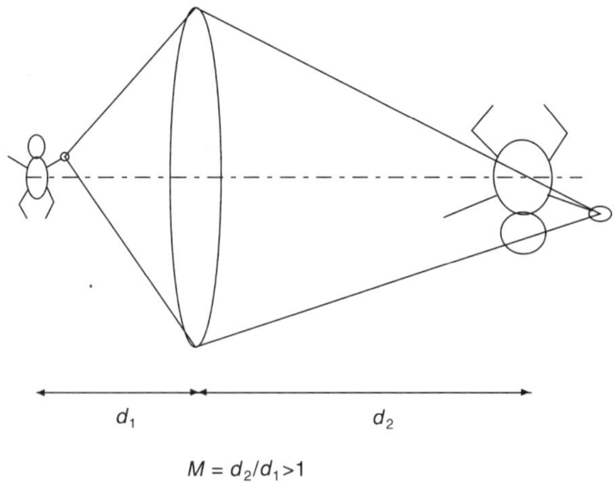

FIGURE 1 A microscope produces a magnified image of an object. A distorted three-dimensional image is formed.

In a simple magnifying system, a lens forms a real magnified and inverted image of an object, as illustrated in Figure 1. If the distance of the lens from the object is d_1 and the distance of the image from the lens is d_2, then the linear magnification M^2 is given by

$$M = \frac{d_2}{d_1}, \tag{1}$$

which for a magnifying system is greater than unity. Sometimes the magnification is specified as negative to specify that the image is inverted.

As shown in Figure 1, the microscope produces a three-dimensional (3D) image of a three-dimensional object. A point of the object closer to the lens appears further from the lens in the image. The longitudinal magnification is equal to M^2. However, it can be shown that it is impossible to devise an instrument that can produce a perfect 3D image of a 3D object (Maxwell, 1851). The reason is that to have a perfect transverse image, the system must satisfy Abbé's sine condition – but to have a perfect axial image, the system must satisfy the Herschel condition. These two conditions are not mutually compatible. Actually we find that the longitudinal magnification is not exactly equal to M^2, and is not even constant in space (Sheppard and Gu, 1993). Different depths of the object are imaged differently: the microscope objective only produces a good image for one plane of focus, and other planes will exhibit spherical aberration. However, all of these problems can be overcome by bringing the part of the object under observation into focus, without refocusing the eye or other detector. A 3D image can thus be generated by scanning the object in the axial direction, and recording a stack of images at different depths in a computer.

3 THE SCANNING MICROSCOPE

Thus an image can be formed by a lens. On the other hand, an image can be generated by a scanning system, as illustrated in Figure 2. A probe of light is scanned over the object in a raster, and the reflected (or transmitted) light detected by a photodetector. If the signal from

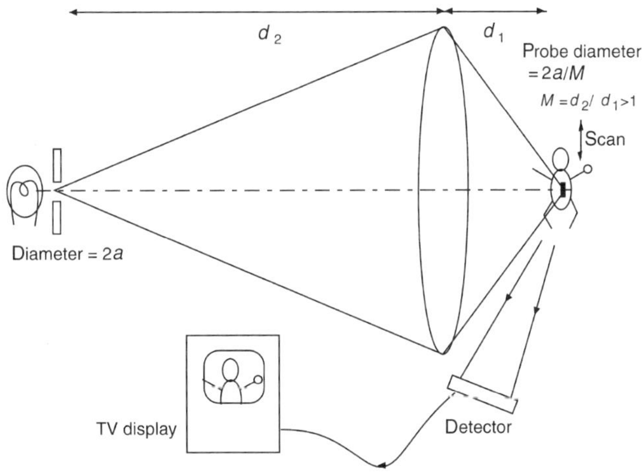

FIGURE 2 A scanning microscope. A demagnified probe is scanned relative to the sample.

the photodetector is used to modulate the brightness of a TV display, which is scanned in synchronism with the probe of light, an image of the object is formed. If the probe of light is produced by imaging a source of light (for example from a small aperture placed in front of a lamp), then the size of the probe (neglecting diffraction effects) is a factor M smaller than that of the aperture, where now d_2 is the distance of the lens from the aperture, and d_1 is the distance of the probe from the lens (Figure 2). If $d_2 > d_1$, then $M > 1$, and the probe is smaller than the aperture. The size of the probe of light limits the resolution of the system – that is, the smallest detail in the object that can be seen in the image. It should be noted that the magnification of the image is given simply by the ratio of the distance scanned in the image to the distance scanned by the probe. It is thus unrelated to M, and the magnification can be altered without changing the lens and is not in any way related to the resolution.

It has been shown that the imaging properties of scanning and conventional microscopes are identical under analogous conditions, this being known as the principle of equivalence (Cowley, 1969; Zeitler and Thomson, 1970; Welford, 1972; Barnett, 1973; Engel, 1974; Kermisch, 1977; Sheppard and Wilson, 1986a,b). This property is based on the principle of reciprocity, which is a very general physical law that holds even with diffraction, multiple scattering, and stray light. The principle of reciprocity states that the field measured by a detector at point 2 due to a particular source at point 1 is equal to the field measured by the detector when placed at point 1 due to the source placed at point 2. This principle is generally true even in the presence of absorption or multiple scattering, except if there are magnetic or polarization effects involved. The principle of equivalence also breaks down if there is inelastic scattering In electron microscopy, since then a loss in energy for the conventional system is equivalent to an increase in energy for the scanning system (or vice versa). Of course it also breaks down in fluorescence microscopy, since energy is lost in the excitation process (an anti-Stokes process would be reciprocal to a Stokes process, but is very inefficient).

Scanning systems exhibit a number of important advantages over conventional systems. Broadly, these are based on two classes of property. First, in a scanning system, the image is in the form of an electronic signal. As a result, this is advantageous for quantitative measurements, as well as for image processing, including image enhancement, image restoration, and image analysis. Second, in a scanning system, the object is illuminated by a focused spot, which extends the range of imaging modes available. Confocal microscopy, for example, is a powerful imaging mode, which is only available by using some form of scanning. In OBIC (optical-beam-induced current) imaging, the image signal results from the current induced in a semiconductor sample, thus allowing electronic defects to be imaged (Sheppard, 1987b).

Differential phase contrast (Dekkers and De Lang, 1977; Hamilton and Sheppard, 1984) is another useful imaging mode, which allows visualization of phase variations in the object (Figure 3). A split detector, consisting of two semicircular elements, gives two signals, which are subtracted by an amplifier. In the absence of an object, the net signal is zero, but in the presence of a phase gradient, the beam is deflected so that there is a net output, which to a first approximation is proportional to the phase gradient. This approach can be used either in transmission, or in reflection, where the output is related to the surface slope of the object. Actually, the signal also depends on the transmissivity (or reflectivity), which can be measured by adding, rather than subtracting, the outputs from the two elements. The phase gradient can thus be calculated by dividing the difference signal by the sum signal, and the absolute phase by integrating this quotient.

Various different types of microscope are illustrated in Figure 4. We begin by considering image formation in a conventional brightfield microscope as illustrated schematically in Figure 4*a*. This simplified diagram shows a microscope with critical illumination: a large-area incoherent source is focused by a condenser onto the specimen such that a comparatively large area of the specimen is illuminated, corresponding to the whole field of view of the objective. Information from each illuminated point in the specimen is simultaneously transmitted by the objective lens to form the primary image. Of course, in a practical microscope, this image is further magnified by the eyepiece. However, the important property to note is that it is the objective that is responsible for forming the image, with the condenser playing only a secondary role in determining the resolution of the system.

In fact, the aperture of the condenser acts to control the spatial coherence of the overall imaging process, which can vary between fully coherent (in which case field amplitudes add according to their respective phases) and fully incoherent (when intensities add) (Hopkins, 1953). The aberrations of the condenser are not important, so that the source and condenser together behave simply as a partially coherent effective source. For this reason, in the widely used Köhler illumination, in which the specimen plane is made conjugate with the field diaphragm rather than the light source, imaging is in principle identical to critical illumination. If the numerical aperture of the condenser is made small

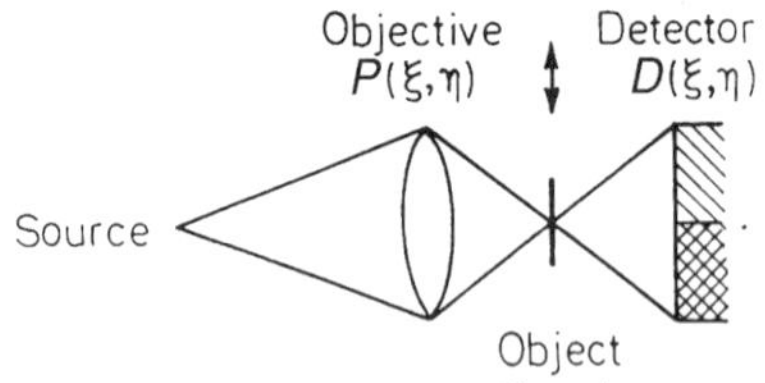

FIGURE 3 Differential phase contrast using a split detector. Reproduced with permission from Hamilton, D. K. and Sheppard, C. J. R., *J. Microsc.* **133**, 27–39 (1984).

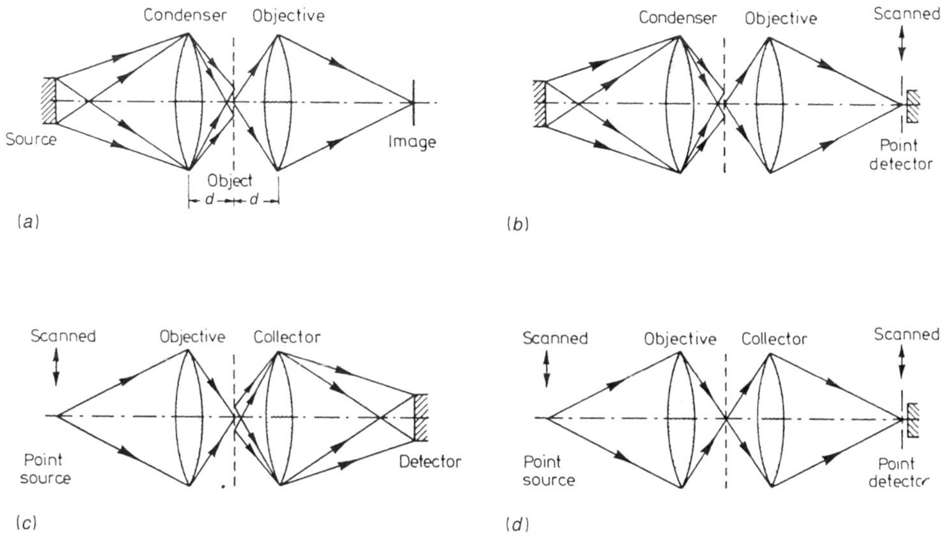

FIGURE 4 Different types of microscope. (*a*) A conventional microscope. (*b*) The image of a conventional microscope is measured point by point. (*c*) A scanning optical microscope. (*d*) A confocal scanning optical microscope. Reproduced with permission from Sheppard, C. J. R., In: *Scanned Image Microscopy* (ed. E. A. Ash). London. Academic Press, 1980, pp. 201–225.

compared with that of the objective, imaging becomes coherent. Coherent imaging exhibits fringing on edges in the image (Watrasiewitz, 1965), and phase information in the object (resulting from variations in optical thickness) can be imaged as intensity changes by defocusing the system. Incoherent imaging, on the other hand, is obtained if the numerical aperture of the condenser is made large compared with that of the objective, and although this depends on the form of the specimen, it generally results in superior resolution (Hopkins and Barham, 1950). It should be noted, however, that in practice it is impossible to obtain truly incoherent brightfield imaging when using a high-aperture objective, and that in practice the condenser aperture diaphragm is adjusted to give a compromise between resolution and contrast in the image, thus resulting in partially coherent imaging. For the particular case, commonly adopted in practice, where the numerical apertures of condenser and objective are equal (so-called matched or full illumination), imaging, although partially coherent, looks generally similar to incoherent imaging.

In Figure 4*b* the image of a conventional microscope is measured point by point by a detector in front of which is placed a small aperture. This method is often used for microdensitometry or micrometrology. If the aperture is now scanned in a raster, the output of the detector gives an image, in electrical form, which is identical to the optical image observed in Figure 4*a*, providing of course that the aperture is small enough that its finite size does not degrade the image resolution, and that the bandwidth of the electrical detection circuit is high enough to transmit the signal variations occurring during scanning. This arrangement is identical in principle to the technique of video microscopy, where the image on a TV camera is electronically scanned. Because, in video microscopy, contrast can be electronically enhanced, it is often preferable to use full illumination in this case, thus improving the resolution as compared with a microscope with visual inspection.

In Figure 4*c*, a scanning optical microscope is illustrated. A single point in the specimen is illuminated by the focused and demagnified point source, while a large-area detector, in conjunction with a collector lens, collects light from the full field of view of the specimen. It will be observed that the ray paths shown in Figures 4*b* and 4*c* are identical but reversed and, because of the optical principles of reciprocity and equivalence, the microscopes illustrated in Figures 4*b* and 4*c* have identical imaging properties. In the scanning optical microscope shown in Figure 4*c*, the resolution of the system is primarily limited by the properties of the first lens, called the objective or projector. The second lens, the collector (analogous to the condenser of Figure 4*a* or 4*b*), controls only the degree of spatial coherence in the imaging process, and its aberrations are unimportant. Because this optical arrangement is analogous to critical illumination, it is called critical detection (Figure 5*c*). An alternative arrangement, in which the detector is placed in the back focal plane of the collector lens, can be termed Köhler detection: it behaves identically in principle but has the advantage that spatial sensitivity variations in the detector are not so important (Figure 5*b*). Indeed, as the collector and detector together act as an effective detector, the collector may in principle be dispensed with altogether and the detector placed in that plane (Figure 5*a*). It is seen that in all three of these arrangements, the reflector is merely measuring the total power of the emitted radiation.

Finally in Figure 4*d*, we combine the arrangements of Figures 4*b* and 4*c* to give a confocal scanning optical microscope, in which a point source illuminates just a small region of the object, and a confocal point detector detects light from this illuminated region. If the point source and detector are scanned in unison, a two-dimensional image is generated. However, this system now behaves very differently from the previous ones. The confocal microscope is thus not a special case of the general partially coherent imaging

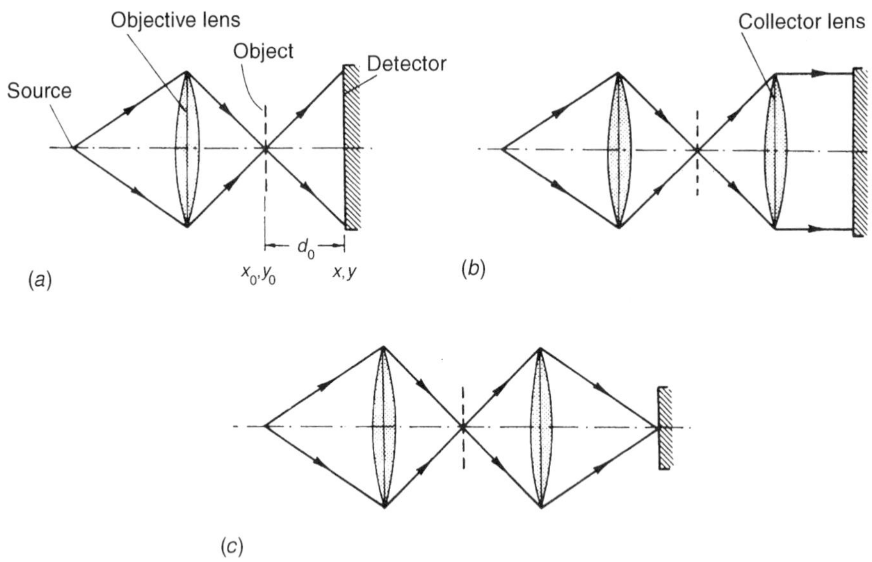

FIGURE 5 Different arrangements for the scanning optical microscope. (*a*) As in the scanning transmission electron microscope. (*b*) Köhler detection. (*c*) Critical detection. Reproduced with permission from Sheppard, C. J. R. and Choudhury, A., *Optica Acta* **24**, 1051–1073 (1977).

system as analysed by Hopkins (1953). From the symmetry of Figure 4*d*, it is clear that the two lenses play an equal part in the imaging process. We might expect that this would result in an improvement in resolution, which is indeed the case. In fact, the confocal system behaves as a coherent imaging system, but with a sharper effective point spread function than in a conventional coherent microscope.

Although Figure 4 is drawn for a transmission geometry, in practice most confocal systems operate in the reflection or epi-illumination mode, in which the same objective lens is used for both illumination and detection. For a specimen placed in the focal plane, the properties of confocal transmission and reflection systems are identical. However, once the object is moved from the focal plane, certain differences arise: in particular, in confocal reflection, a strong optical sectioning effect occurs that allows a single section through a thick object to be imaged. This is, of course, the major advantage of the confocal microscope arrangement. Figure 4 could also equally well apply to fluorescence imaging. However, in this case, because of the incoherent nature of fluorescence emissions after excitation by coherent light, imaging is always incoherent.

The imaging properties of any lens or imaging system may be described in terms of the 3D intensity point spread function (PSF), $I(x, y, z)$, which is the intensity variation in the image formed by the lens of an ideal point object (or, in practice, of a subresolution object such as a 50 nm diameter gold particle or fluorescent microsphere). While, in conventional fluorescence microscopy, the image formation is determined by the PSF of the imaging objective, in a confocal scanning optical microscope, the image formation is determined by the PSFs of both the illuminating and the collecting lenses. This is because both lenses now play an equal role in image formation – the former in defining the size of the illuminating spot and the latter in focusing this to the confocal pinhole. The confocal system's PSF is thus the product of the individual PSFs of the two lenses. For a confocal epi-illumination arrangement, these two PSFs are, of course, identical (neglecting a Stokes shift in fluorescence), since the same objective is employed for both illumination and collection. Thus the PSF for an epi-illumination confocal system is equal to the square of the PSF of the objective lens.

Confocal microscopes can be achieved either by using a point detector, in practice performed by placing a pinhole in front of a photodetector, or by using a coherent detector (Figure 6). Such a coherent detector, sensitive to the amplitude of the incident radiation, is not directly available for light, but is available for acoustic radiation, as in a scanning acoustic microscope. For light, a confocal effect can also be achieved by using an

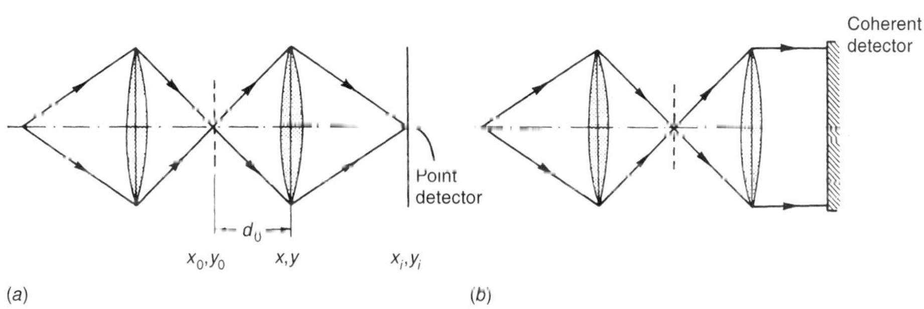

FIGURE 6 Different arrangements for the confocal microscope. (*a*) Using a point detector. (*b*) Using a coherent detector, as in a scanning acoustic microscope. Reproduced with permission from Sheppard, C. J. R. and Choudhury, A., *Optica Acta* **24**, 1051–1073 (1977).

interferometric method to synthesize a coherent detector. Thus, in interference microscopes using the high aperture condenser and objective lenses, an optical sectioning effect arises similar to that in the confocal microscope. Another way of producing a coherent detector is to use a single-mode optical fiber.

4 THE GENERALIZED MICROSCOPE

We can generalize the conventional microscope to a system in which the specimen is subjected to some stimulus, and the emitted radiation, which can be, for example, light or electrons, is focused to produce a magnified image of the object (Figure 7).

The conventional microscope is the simplest example of this generalized microscope. In fact, we do not need to use visible light, but can use any electromagnetic radiation, over a broad range of different available wavelengths. In Figure 8, we show the electromagnetic spectrum, illustrating how the wavelength in air varies with frequency. Instead of visible light, shorter wavelengths allow greater resolution to be achieved. Ultraviolet (UV) light is commonly used to excite fluorescence. X rays have also been used successfully to image biological samples.

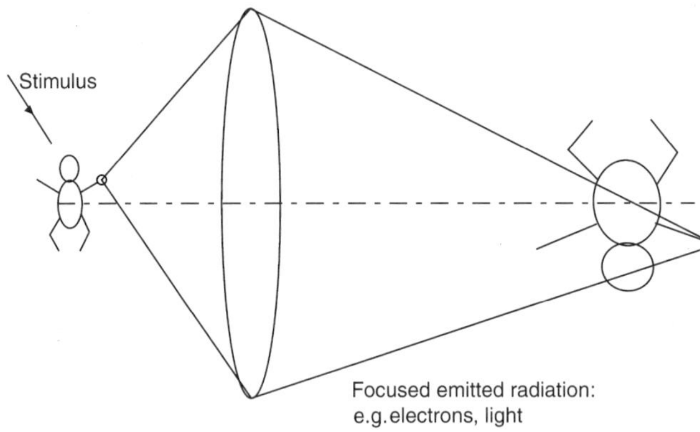

FIGURE 7 The generalized microscope, in which the object is subjected to some stimulus, and emitted radiation focused to an image.

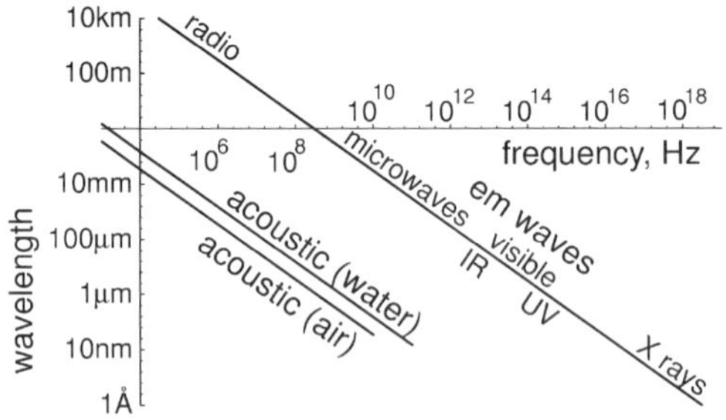

FIGURE 8 The electromagnetic spectrum, showing wavelength as a function of frequency. The behaviour of acoustic waves in air and water is also shown.

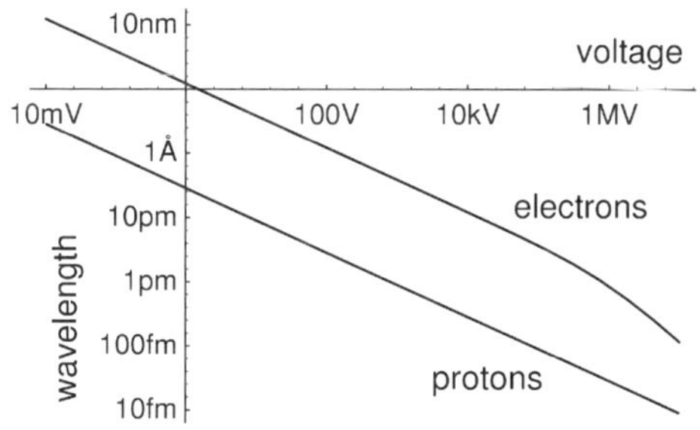

FIGURE 9 The wavelength of electrons and protons, as a function of the accelerating voltage.

In the longer-wavelength region, infrared (IR) radiation is used for observation of semiconductors or for molecular spectroscopic imaging. Imaging of emitted mid-infrared radiation can show up variations in temperature, such as hotspots in semiconductors. Microwaves have also been used, but resolution is limited by the long wavelength. For this reason, microwaves (and mid-IR) have been used for near-field microscopy.

The other important class of radiation that can be used for illumination is matter waves. The most common of these, of course, are electrons, the wavelength of which, as a function of acceleration voltage, is shown in Figure 9. The bend in the curve at very high voltages is caused by relativistic effects. The wavelength of the electrons is very small: even for 100 V electrons, the wavelength is only 0.1 nm. But aberrations caused by the electron lenses limit resolution so that at present atomic resolution is only just achievable, even with much higher voltages.

Figure 10 shows the overall layout of a transmission electron microscope (TEM). A beam of electrons is produced by an electron gun. The specimen is illuminated via a condenser lens (or lenses). Usually these are magnetic lenses in which a current in a coil produces a magnetic field that focuses the electrons. The electrons that are transmitted through the specimen are then focused by a series of lenses to form a final image that can be viewed directly on a phosphor screen. Figure 10 shows how an X-ray detector can be used to characterize the elements in the specimen. The figure also shows an electron-energy-loss spectrometer with a detector that can measure the energy of the electrons in a particular region of the image.

5 THE GENERALIZED SCANNING MICROSCOPE

Figure 11 illustrates how a generalized scanning microscope is constructed. The specimen is illuminated by a probe of radiation, which can be, for example, electrons or light. The probe is scanned relative to the specimen. Then any output signal can be detected an used to generate an image on a TV display. For example, if the specimen is illuminated with a probe of electrons, the signal could be secondary electrons (low energy electrons that are produced by the interaction of an incident high-energy electron with the specimen), backscattered electrons, X rays, or luminescent light. If the specimen is a semiconductor device, then the incident electrons can cause a current to flow in the device, which can be measured to produce a signal that depends on the electrical properties of the device, such as the diffusion length of the charge carriers. This imaging mode is called the EBIC (electron-beam-induced current) mode. Figure 12 shows a schematic diagram of a

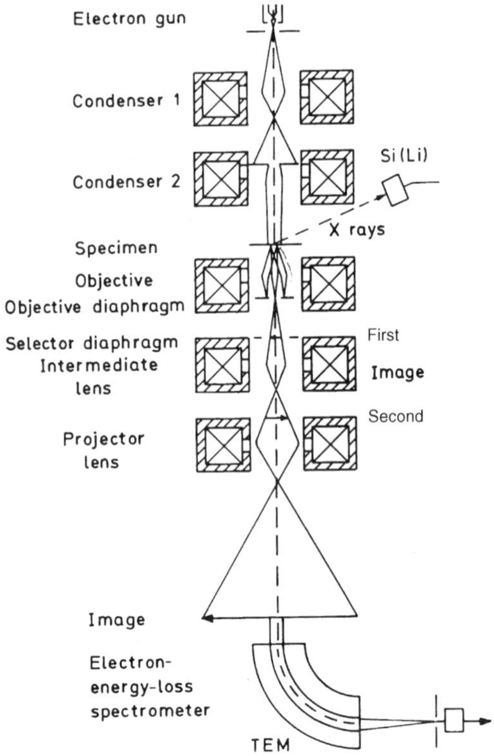

FIGURE 10 The transmission electron microscope. Reproduced with permission from Reimer, L., *Transmission Electron Microscopy.* Berlin: Springer-Verlag, 1983.

scanning electron microscope (SEM). As in a TEM, lenses are used to focus the electrons, but Figure 12 shows the use of electrostatic lenses, which focus electrons with an electrostatic field. The electron beam is scanned in a raster by deflector coils, and a cathode-ray tube (CRT) display is scanned in unison. The figure shows the secondary electrons being detected using an electron multiplier for amplification: a phosphor coupled to a photomultiplier is more common. Again, in modern instruments, a variety of signal can be detected, and used to form an image.

Instead of electrons, other particle beams can be used for illumination. Figure 9 shows the wavelength of a beam of protons as a function of the accelerating voltage. Because of the larger mass of protons, relativistic effects are not apparent for protons in the curve in Figure 9. Figure 13 shows a schematic diagram of the analogous scanning optical microscope (SOM). In this, the object is scanned mechanically under the focused laser spot. Again an image can be formed directly on a TV display; nowadays, however, it is usually stored in a computer. Figure 13 shows a transmission microscope, but a reflection form is similar in principle. Any signal can be used to form an image. Figure 14 shows a scanning second-harmonic-generation (SHG) microscope. A second-harmonic signal is generated by the interaction of the focused laser spot with the sample.

Any radiation from the electromagnetic spectrum, including UV and IR light, can be used to illuminate the specimen. Figure 15 shows a scanning X-ray microscope. Or instead

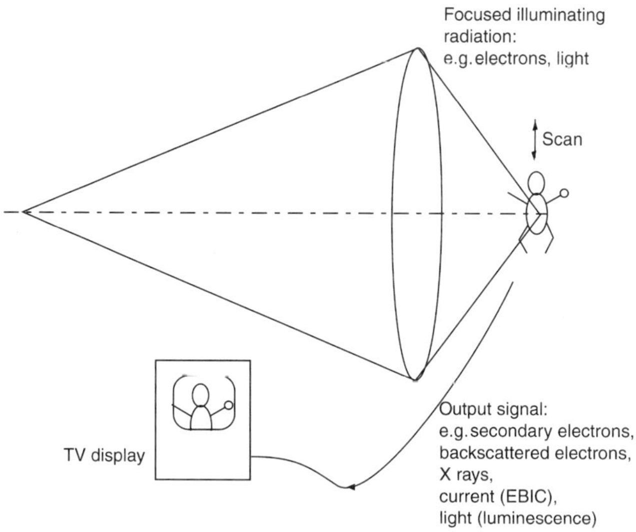

FIGURE 11 The generalized scanning microscope, in which a focused probe of radiation gives rise to an output signal that is used to generate an image.

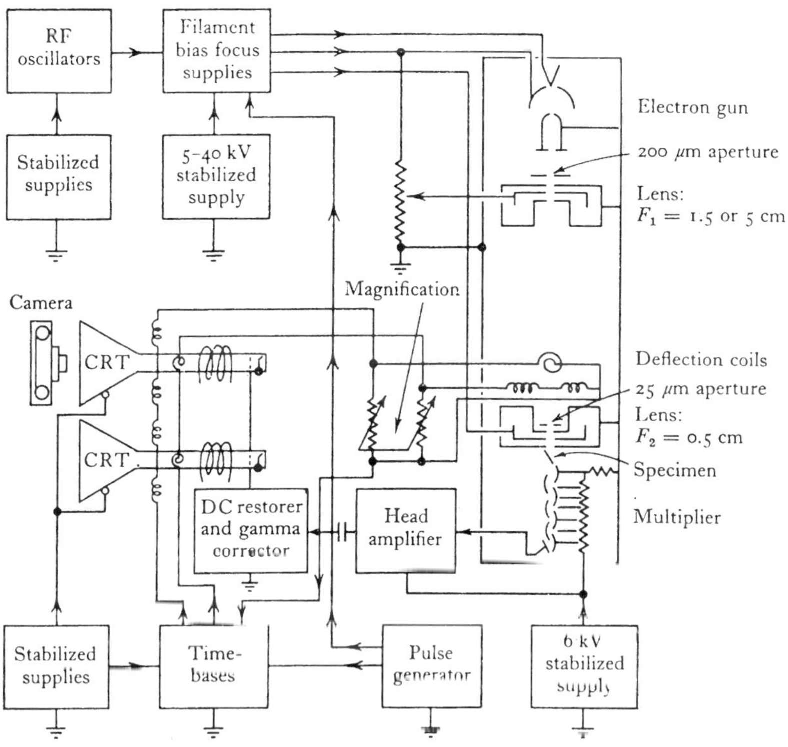

FIGURE 12 The scanning electron microscope. Reproduced with permission from Oatley, C. W., *The Scanning Electron Microscope*. Cambridge: Cambridge University Press, 1972.

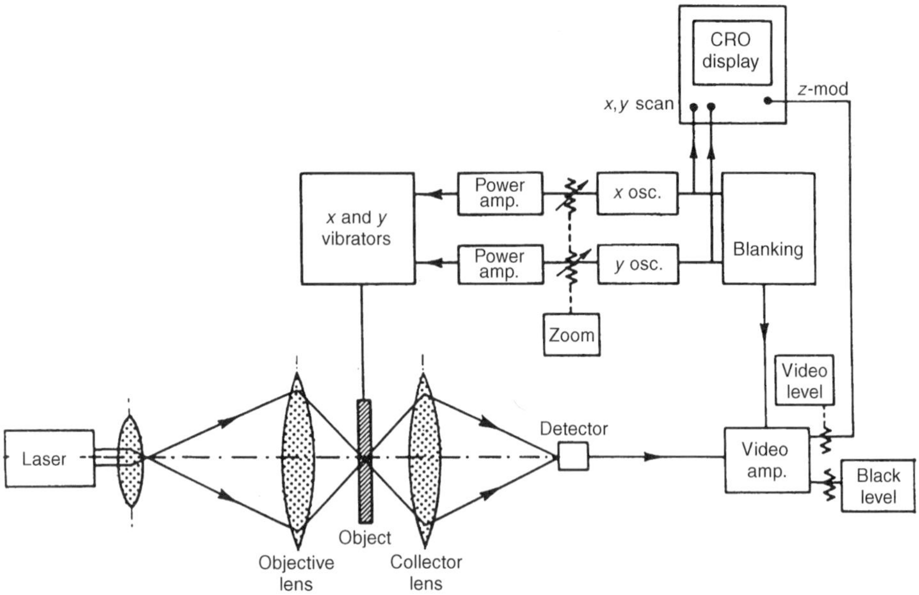

FIGURE 13 The scanning optical microscope. Reproduced with permission from Sheppard, C. J. R., *The Physics Teacher* **16**, 648–651 (1978).

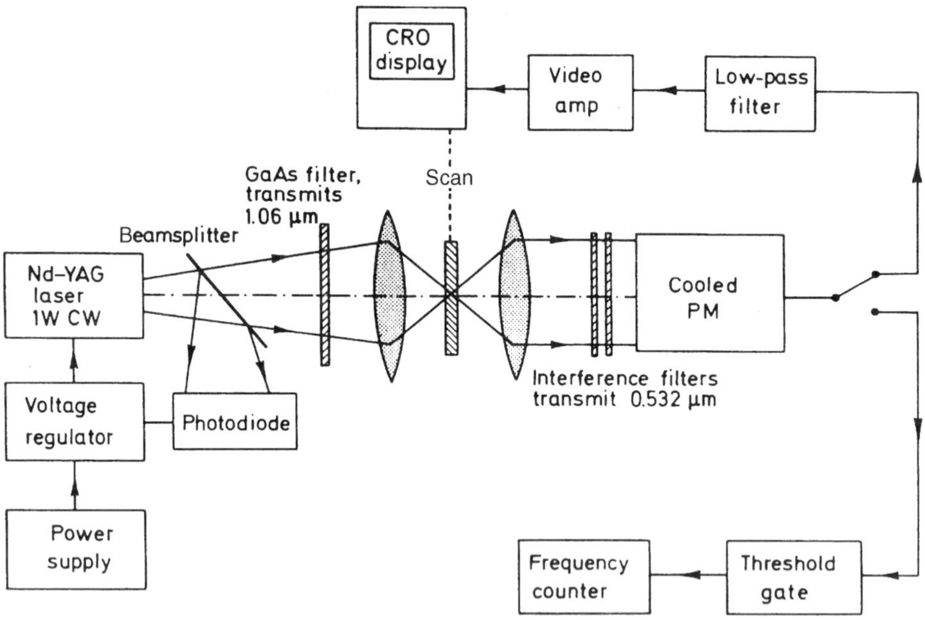

FIGURE 14 The scanning second-harmonic-generation (SHG) microscope. Reproduced with permission from Sheppard, C. J. R., Kompfner, R., Gannaway, J. N., and Walsh, D., *IEEE J. Quant. Electron.* **QE-13**, 100D (1977).

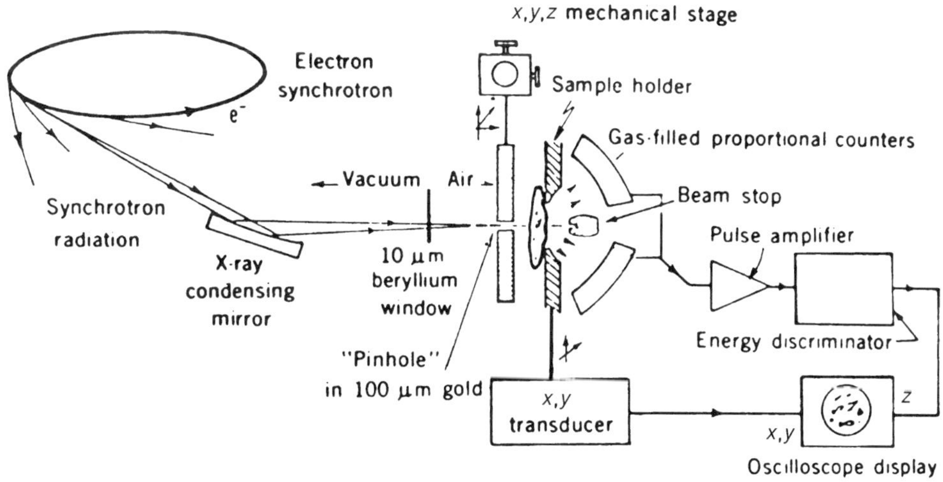

FIGURE 15 The scanning X-ray microscope. Reproduced with permission from Horowitz, P. *Ann. NY Acad. Sci.* **306**, 203 (1978).

of electromagnetic waves, acoustic radiation can be used, in a scanning acoustic microscope (SAM). In Figure 16, ultrasonic waves are focused into a water immersion medium. Mechanical scanning is again used to build up an image. Figure 8 shows the wavelength of the acoustic wave as a function of frequency. By using microwave frequencies, wavelengths, and therefore resolution, in the submicrometer range can be generated. The SAM

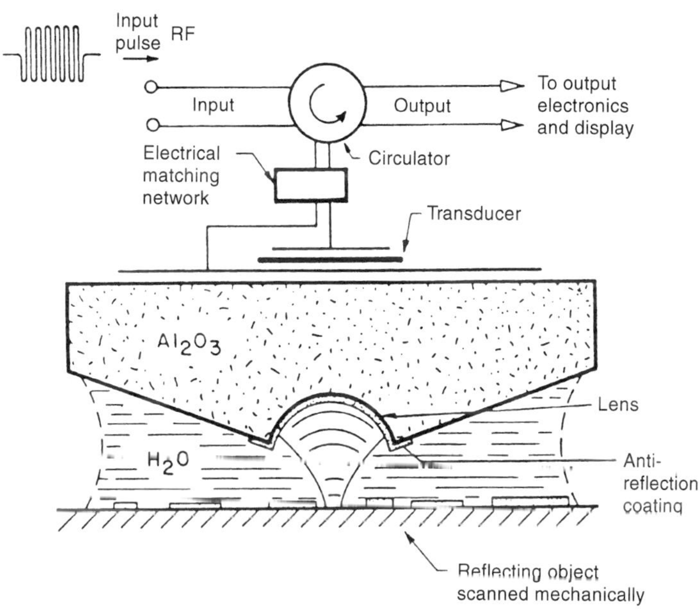

FIGURE 16 The scanning acoustic microscope. Reproduced with permission from Wickramasinghe, H. K., In: *Advances in Optical and Electron Microscopy*, Vol 11 (ed. T. Mulvey and C. J. R. Sheppard). London: Academic Press, 1989, pp. 153–182.

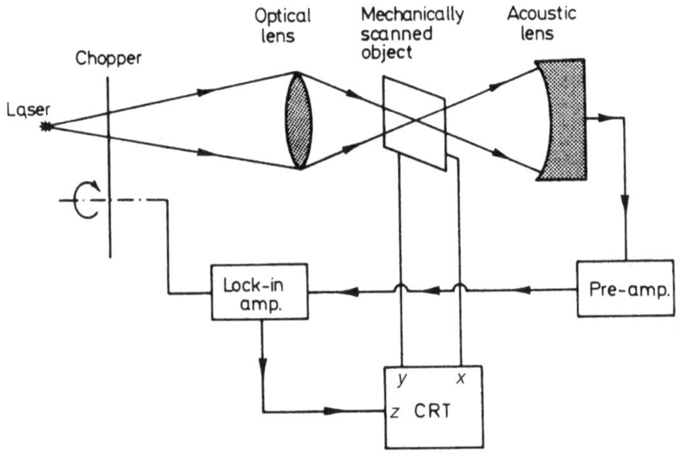

FIGURE 17 The scanning photoacoustic microscope. Reproduced with permission from Cox, I. J. and Sheppard, C. J. R., *J. Acoust. Soc. Am.* **76**, 513–515 (1984).

acts as a confocal microscope because the detector is sensitive to the amplitude of the acoustic wave.

Figure 17 shows a schematic diagram of a scanning photo-acoustic microscope (SPAM). A modulated laser beam produces periodic heating, and hence expansion of the sample. This generates an acoustic signal that is used to form an image. Figure 17 shows a confocal system: the sample is illuminated with a focused light spot, and the acoustic wave is also focused and detected with a coherent detector. A wide range of different signals can be detected, each imaging different physical properties of the sample. Figure 18 shows the generation of acoustic and thermal waves in the sample and in the ambient air. In addition, emitted IR radiation (photothermal microscope), or direct

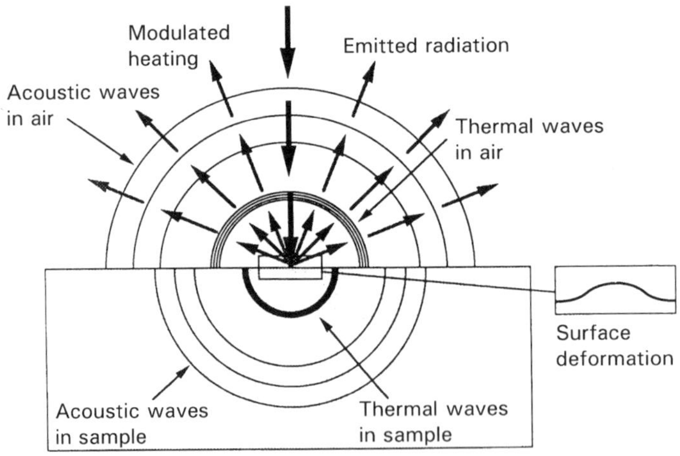

FIGURE 18 Physical processes connected with thermal waves. Reproduced with permission from Hartikainen, J., Jaarinen, J., and Luukkala, M., In: *Advances in Optical and Electron Microscopy,* Vol. 12 (eds. T. Mulvey and C. J. R. Sheppard). London: Academic Press, 1991, pp. 313–359.

measurement of expansion with an interferometer (photodisplacement microscope) measures the local temperature of the illuminated object.

A scanning microscope is more versatile than a conventional (nonscanning) microscope, since the emitted radiation need not be refocused. In this way, a scanning microscope can be constructed based on any form of spectroscopy. In fact, the signal need not come from any radiation, but instead from the measurement of any property that will change with scanning. If the emitted radiation can be focused, this can be used to construct a type of confocal microscope.

Finally, Figure 19 shows different arrangements for a near-field microscope. In a near-field microscope, a very small tip or aperture is used to attain resolution greater than the classical limit set by the wavelength of the radiation. As Figure 19 shows, many different designs of near-field microscope have been proposed, but it is now evident that these again fall into the categories described earlier. The sample can be uniformly illuminated, and a signal detected using a near-field probe (*a*). Or the sample is illuminated using the near-

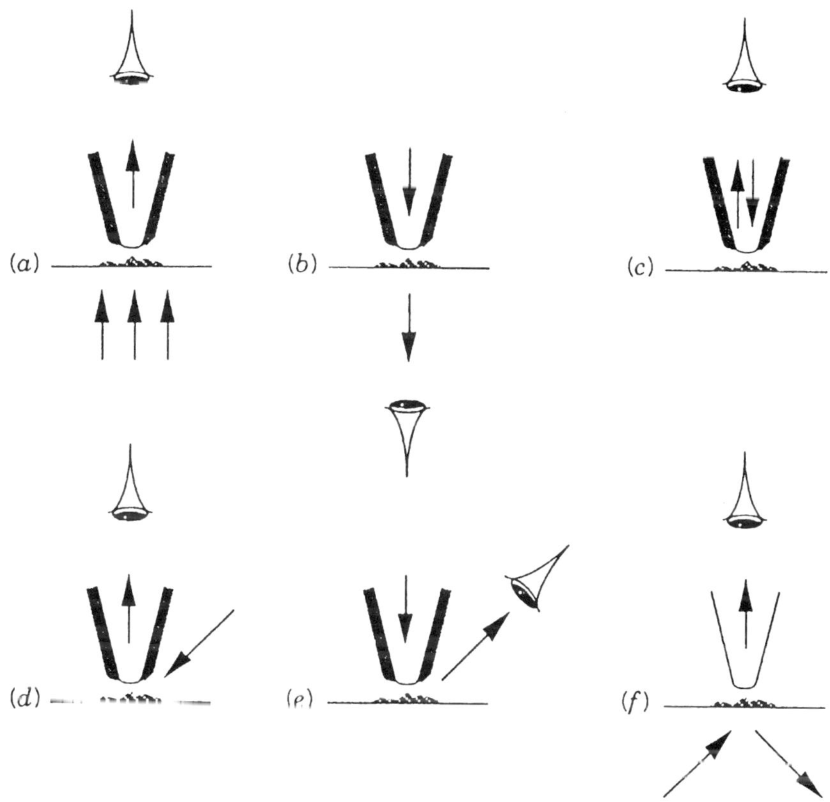

FIGURE 19 Different arrangements for the scanning near-field microscope. The adjectives used to describe each mode are (*a*) collection, (*b*) illumination, (*c*) collection/illumination, (*d*) oblique collection, (*e*) oblique illumination, and (*f*) dark-field. Reproduced with permission from Paesler, M. A. and Moyer, P. J., *Near-Field Optics: Theory, Instrumentation, and Applications*. New York: Wiley, 1996.

field probe (*b*). Or a confocal arrangement can be used, with a near-field probe used for both illumination and collection (*c*).

6 SUMMARY

We have described how a generalized conventional (nonscanning) microscope can be constructed, using variety of different stimuli, and focused image signals. Alternatively, generalized scanning microscopes are constructed using focused illumination and a detected signal. Or, by using focused illumination, and by focusing the emitted radiation on to a point detector, we can construct a generalized confocal microscope.

REFERENCES

Barnett, M. E. (1973). The reciprocity theorem and the equivalence of conventional and scanning transmission microscopes. *Optik* **38**, 585–588.

Cowley, J. M. (1969). Image contrast in a transmission scanning electron microscope. *Appl. Phys. Lett.* **15**, 58.

Dekkers, N. H. and De Lang, H. (1977). A detection method for producing phase and amplitude images simultaneously in a STEM. *Philips Tech. Rev.* **37**, 1.

Engel, A. (1974). Principle of reciprocity and its application to conventional and scanning dark field electron microscopy. *Optik* **41**, 117–126.

Gannaway, J. and Sheppard, C. J. R. (1978). Second harmonic imaging in the scanning optical microscope. *Opt. Quant. Electron* **10**, 435–439.

Hamilton, D. K. and Sheppard, C. J. R. (1984). Differential phase contrast in scanning optical microscopy. *J. Microsc.* **133**, 27–39.

Hopkins, H. H. (1953). On the diffraction theory of optical images. *Proc. R. Soc. Lond.* **A217**, 408–432.

Hopkins, H. H. and Barham, P. M. (1950). The influence of the condenser on microscopic resolution. *Proc. Phys. Soc.* **B63**, 737.

Kermisch, D. (1977). Principle of equivalence between scanning and conventional optical imaging systems. *J. Opt. Soc. Am.* **67**, 1357–1360.

Maxwell, J. C. (1851). On the general laws of optical instruments. *Q. J. Pure Appl. Math.* **2**, 271–285.

Sheppard, C. J. R. (1987a). Scanning optical microscopy. In: *Advances in Optical and Electron Microscopy*, Vol. 10 (eds. R. Barer and V. E. Cosslett). London: Academic Press, 1–98.

Sheppard, C. J. R. (1987b). Scanning optical microscopy of semiconductor materials and devices. *Scanning Microsc.* **3**, 15–24.

Sheppard, C. J. R. and Gu, M. (1993). Imaging by a high aperture optical system. *J. Mod. Opt.* **40**, 1631–1651.

Sheppard, C. J. R. and Kompfner, R. (1978). Resonant scanning optical microscope. *Appl. Opt.* **17**, 2879–2882.

Sheppard, C. J. R. and Wilson, T. (1986a). On the equivalence of scanning and conventional microscopes. *Optik* **78**, 39–43.

Sheppard, C. J. R. and Wilson, T. (1986b). Reciprocity and equivalence in scanning microscopes. *J. Opt. Soc. Am.* **A3**, 755–756.

Watrasiewitz, B. M. (1965). Theoretical calculations of images of straight edges in partially coherent illumination. *Optica Acta* **12**, 391–400.

Welford, W. T. (1972). On the relationship between the modes of image formation in scanning microscopy and conventional microscopy. *J. Microsc.* **96**, 105–107.

Zeitler, E. and Thomson, M. G. R. (1970). Scanning transmission electron microscopy, I and II. *Optik* **31**, 258–280, 359–366.

CHAPTER 2

Confocal Microscopy: Basic Principles and Architectures

TONY WILSON

Department of Engineering Science, University of Oxford, Parks Road, Oxford OX1 3PJ, U.K.

1 INTRODUCTION

There have been a variety of reasons for the growth of interest in scanning *optical* microscopy during the last few years. Amongst these, we must certainly include the ready availability of laser light sources and the tremendous increase in powerful low-cost computer image- and data-processing systems. However, these reasons are insufficient in themselves. It is probably fair to say that if these instruments could not operate in the *confocal* mode, they would not attract a fraction of their present popularity. In essence, the confocal system offers, amongst many other things, the chance to image thick biological tissue in three dimensions. These instruments operate well in the brightfield and fluorescence modes.

In this chapter, we will introduce the confocal principle and describe typical optical implementations. We begin, however, by considering the conventional optical microscope, and shall show that, by choosing a scanning approach, we have the freedom to modify the optical system in simple ways to permit many novel imaging modes, of which the confocal arrangement is one of the most powerful.

In essence, the conventional microscope is a parallel-processing system that images the entire object field simultaneously. This is quite a severe requirement for the optical components. We can relax this requirement if we no longer try to image the whole object at once. The limit of this relaxation is to require an image of only *one* object point at a time. In this case, all that we ask of the optics is that it should provide a good image of one point. The price that we have to pay is that we must *scan* in order to build up an image of the entire field. The answer to the question as to whether this price is worth paying will, to some extent, depend on the application in question

A typical arrangement of a scanning confocal optical microscope is shown in Figure 1, where the system is built around a host conventional microscope. The essential components are some form of mechanism for scanning the light beam (usually from a laser)

Confocal and Two-Photon Microscopy: Foundations, Applications, and Advances, Edited by Alberto Diaspro.
ISBN 0-471-40920-0

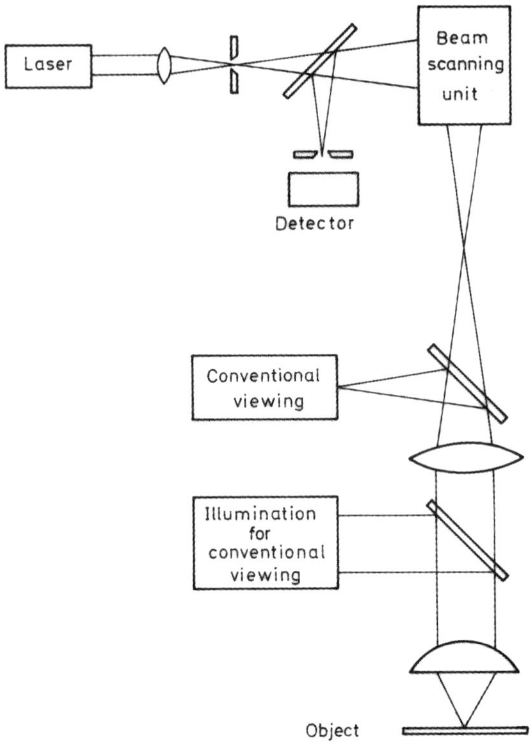

FIGURE 1 Schematic diagram of a confocal microscope.

relative to the specimen and appropriate photodetectors to collect the reflected or fluorescent light (Wilson and Sheppard, 1984). Most of the early systems were analogue in nature, but it is now universal, thanks to the serial nature of the image formation, to use a computer both to drive the microscope and to collect, process, and display the image.

In the beam-scanning confocal configuration of Figure 1, the scanning is typically achieved by using vibrating galvanometer-type mirrors or acousto-optic beam deflectors. The use of the latter gives the possibility of TV-rate scanning, whereas vibrating mirrors are often relatively slow when imaging an extended region of the specimen. We should note that other approaches to scanning may be implemented, such as specimen-scanning and lens-scanning. These methods, although not generally available commercially, do have advantage in certain specialized applications (Wilson and Sheppard, 1984; Wilson, 1990). Since this chapter is necessarily introductory, much of the material may be found in other sources, such as Wilson and Sheppard (1984), Wilson (1990), Pawley (1995), Corle and Kino (1996), Gu (1996), and Masters (1996), where a detailed list of references may be found.

2 IMAGE FORMATION IN SCANNING MICROSCOPES

We shall base our discussion on a very simple model of image formation in conventional microscopes. Using this model, we can derive an equivalent form of scanning microscope that will allow us to suggest the modification which leads to the confocal optical system.

Figure 2*a* illustrates the key elements of the optical system of a conventional optical microscope. In essence, the object is illuminated by a patch of light from an extended source via a condenser lens. The illuminated patch of the object is then imaged by the objective lens into the image plane, where it is finally viewed by eye. In this case, the resolution results primarily from the objective lens. The condenser lens plays a much less important role: its aberrations are unimportant, and it serves to determine the coherence of the imaging. A scanning microscope could be built by simply scanning a point detector through the image plane to measure the image intensity of each object point; see Figure 2*b*. In this way, we would build up an image point by point, which could be stored in a computer and displayed on a computer screen. An analogous optical system is shown in Figure 2*c*, which uses a point source and a large-area incoherent detector. Here the first lens, the objective lens, probes the specimen, point by point, with a finely focused spot of light, while the second lens merely serves to collect the transmitted light. This collector lens is analogous to the condenser lens in conventional microscopy, and thus plays a minor role in determining the resolution.

In order to make the second lens contribute significantly to the resolution, we need to make it into an *imaging* lens rather than a *collector* lens. We achieve this by using a point detector rather than a large-area one. This is the form of the confocal scanning microscope; see Figure 2*d*. Here light from the point source probes a very small region of the object, and the point detector ensures that only light from that same small area is detected. In the figure, an image is built up by scanning the source and detector in synchronism. In practice, it will usually prove more convenient to scan the light beam or the object rather than the detector, and also to operate the system in reflection.

In this symmetric confocal configuration, we see that both lenses play equal roles in the imaging. The use of the point detector has caused the second lens to *image* the region of the object probed by the first lens. We might expect, therefore, that as two lenses are

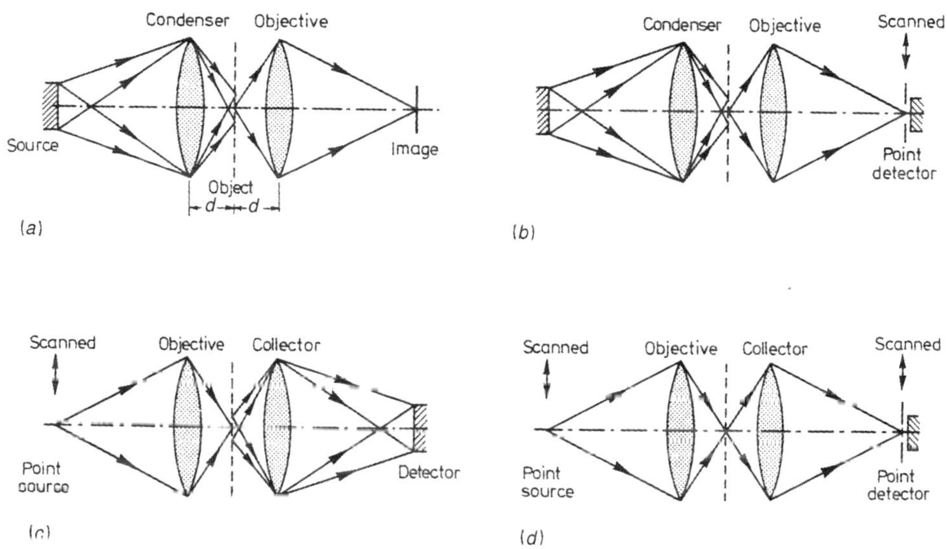

FIGURE 2 The optical arrangement of various forms of scanning optical microscopes. (*a*) Conventional microscope. (*b*) A form of conventional scanning microscope. (*c*) A form of conventional scanning microscope. (*d*) The confocal optical system.

employed simultaneously to image the object, the resolution will be improved. This prediction is borne our both in theory and in practice. Figure 3*a* shows the form of the image of a single point object in a conventional and confocal microscope. It is clear that the confocal image is significantly sharper than the conventional together with relatively low side-lobe level. This suggests that superior lateral resolution should be possible in confocal microscopy. This is confirmed in Figure 3*b*, where the image of two points is shown. The two points are clearly distinguished in the confocal case.

This improvement may at first seem implausible. However, it can be simply explained by a principle given by Lukosz (1966), which states, in essence, that resolution can be increased at the expense of field of view. The field of view can then be increased by scanning. One way of taking advantage of Lukosz's principle is to place an extremely small aperture extremely close to the object. The resolution is now determined by the size of the hole rather than the radiation. In the confocal microscope, we do not use a physical aperture in the focal plane. Instead, we use the back-projected image of a point detector in conjunction with the focused point source.

The confocal principle was first described by Minsky (1961). His motivation was the desire to obtain an image of a slice of a specimen without the distracting presence of out-of-focus information from surrounding planes. The confocal optical systems fulfils this requirement at the same time as providing enhanced lateral resolution.

This optical-sectioning or depth-discrimination property has become the major motivation for using confocal microscopes, and is the basis of many of the novel imaging modes

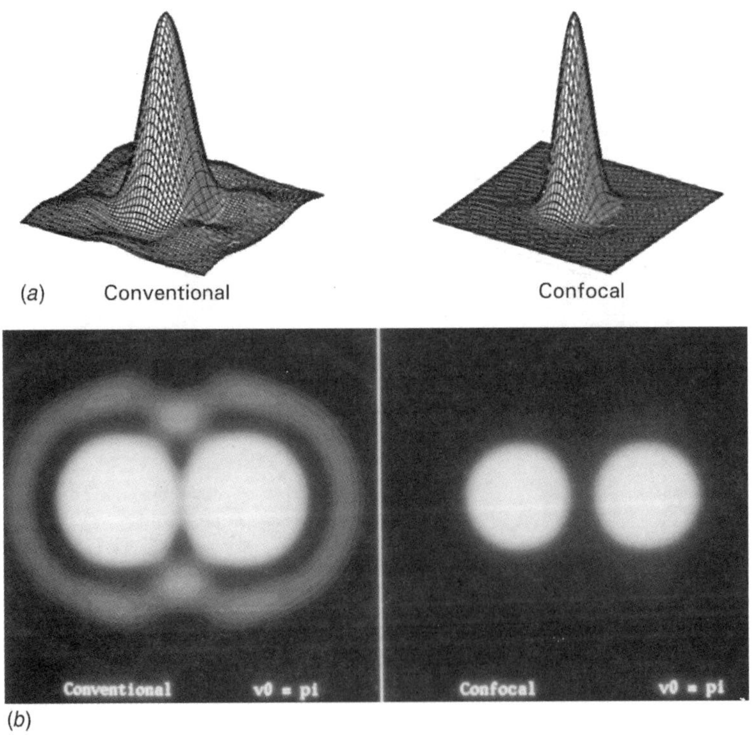

FIGURE 3 The point spread functions of the conventional and confocal microscopes are shown in (*a*), together with comparison images of two-point objects in (*b*). In both cases, the superior resolution of the confocal system is apparent.

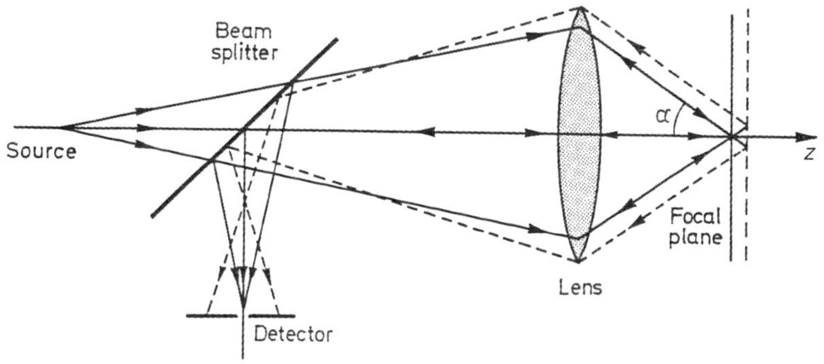

FIGURE 4 The origin of the optical-sectioning or depth-discrimination property of the confocal optical system.

of these instruments. It may be understood very easily from Figure 4, where we show a reflection-mode confocal microscope and consider the imaging of a specimen with a rough surface. The full lines show the optical path when an object feature lies in the focal plane of the lens. At a later scan position, the object surface is supposed to be located in the plane of the vertical dashed line. In this case, simple ray tracing shows that the light reflected back to the detector pinhole arrives as a defocused blur, only the central portion of which is detected and contributes to the image. In this way, the system discriminates against features that do not lie within the focal region of the lens. A very simple method of both demonstrating the effect and giving a measure of its strength is to scan a perfect reflector axially through focus and measure the detected signal strength. Figure 5 shows a typical response. These responses are frequently termed the $V(z)$, by analogy with a similar technique in scanning acoustic microscopy, although the correspondence is not perfect. A simple paraxial theory models this response as

$$I(u) = \left(\frac{\sin \frac{1}{2} u}{\frac{1}{2} u} \right)^2, \tag{1}$$

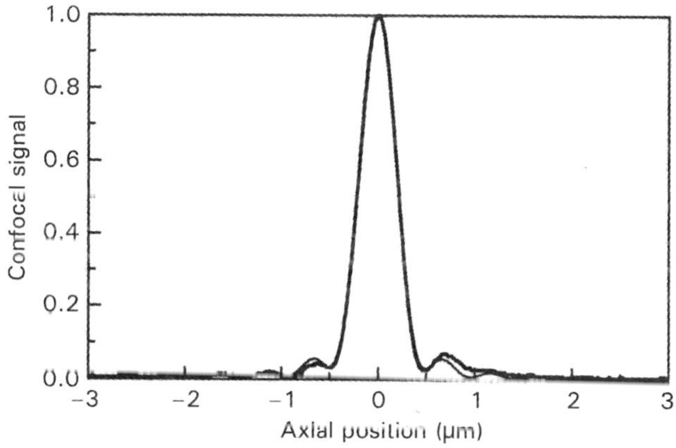

FIGURE 5 The variation in detected signal as a plane reflector is scanned axially through focus. The measurement was taken with a 1.3 numerical-aperture objective and 633 nm radiation.

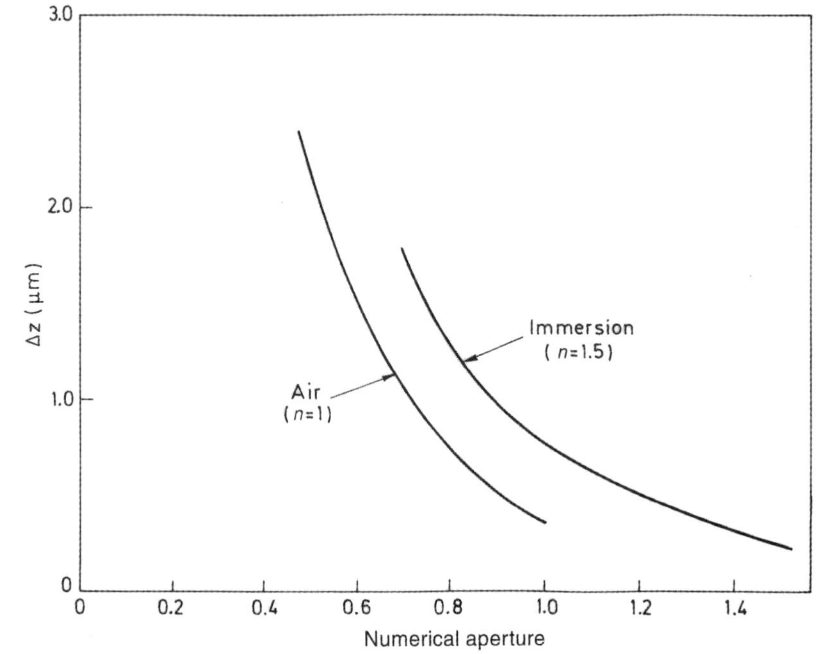

FIGURE 6 The optical sectioning width as a function of numerical aperture. The curves are for red light (0.6328 μm wavelength). Δz is the full-width at the half-intensity points of the curves of $I(u)$ against u.

where u is a normalized axial coordinate, which is related to the real axial distance z via

$$u = \frac{8\pi}{\lambda} z \sin^2 \tfrac{1}{2}\alpha, \qquad (2)$$

where λ is the wavelength and $\sin\,\alpha$ the numerical aperture. As a measure of the strength of the sectioning, we can choose the full-width at half-intensity of the $I(u)$ curves. Figure 6 shows this value as a function of numerical aperture for the specific case of imaging with red light from a helium–neon laser. These curves were obtained using a high-aperture theory, which is more reliable than Eq. (1) at the highest values of numerical aperture.

3 APPLICATIONS OF DEPTH DISCRIMINATION

Since this property is one of the major reasons for the popularity of confocal microscopes, it is worthwhile, at this point, to review briefly some of the novel imaging techniques that have become available with confocal microscopy.

Figure 7 illustrates the essential effect: Figure 7*a* shows a conventional image of a planar microcircuit that has deliberately been mounted with its normal at an angle to the optical axis. We see that only one portion of the circuit, running diagonally, is in focus. Figure 7*b* shows the corresponding confocal image: here the discrimination against detail outside the focal plane is clear. The areas that were out of focus in Figure 7*a* have been rejected. Furthermore, the confocal image appears to be in focus throughout the visible band, which illustrates that the sectioning property is stronger than the depth of focus.

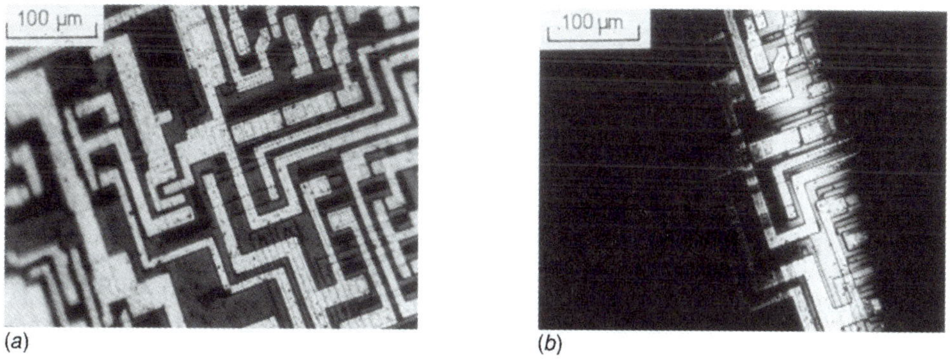

FIGURE 7 (*a*) Conventional scanning microscope image of a tilted microcircuit: the parts of the object outside the focal plane appear blurred. (*b*) Confocal image of the same microcircuit: only the part of the specimen within the focal region is imaged strongly.

This suggests that if we try to image a thick translucent specimen, we can arrange, by the choice of our focal position, to image detail from one specific volume region. In essence, we can section the specimen optically without having to resort to mechanical means. Figure 8 shows an idealized schematic of the process. The portion of the beehive-shaped object that we see is determined by the focus position. In this way, it is possible to take a *through-focus* series and obtain data about the three-dimensional structure of the specimen. If we represent the volume image by $I(x, y, z)$, then by focusing at a position $z = z_1$, we obtain, ideally, the image $I(x, y, z_1)$. This, of course, is not true in practice, because the optical section is not infinitely thin.

It is clear that the confocal microscope allows us to form high-resolution images with a depth of focus sufficiently small that all the detail that is imaged appears in focus. This suggests immediately that we can extend the depth of focus of the microscope by adding together (integrating) the images taken at different focal settings *without* sacrificing the lateral resolution. Mathematically, this *extended-focus* image is given by

$$I_{EF}(x, y) = \int I(x, y, z)\, dz \tag{3}$$

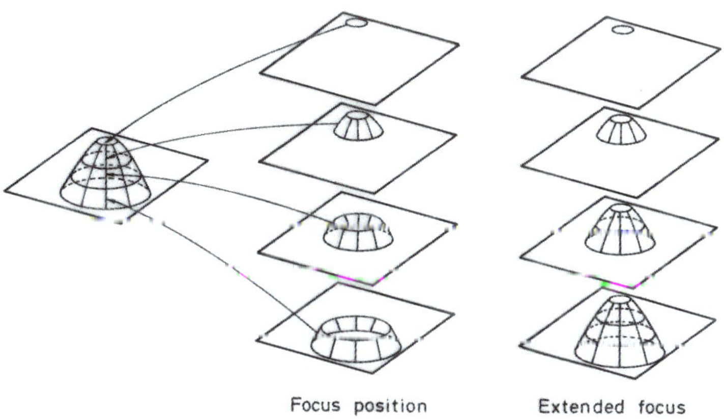

FIGURE 8 An idealization of the optical-sectioning property, showing the ability to obtain a through-focus series of images, which may then be used to reconstruct the original volume object at high resolution.

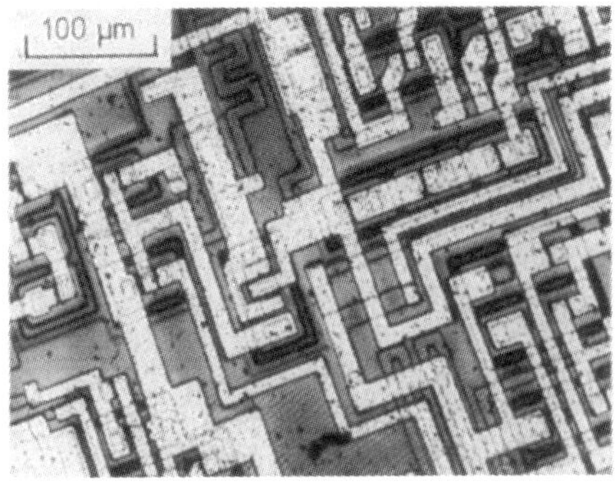

FIGURE 9 The extended-focus image of the specimen of Figure 8.

Figure 9 shows the result of scanning the specimen of Figure 7 axially by an amount such that each portion of its surface passes through focus.

As an alternative to the extended-focus method, we can form an *autofocus* image by scanning the object axially and, instead of integrating, selecting the focus at each picture point by recording the maximum in the detected signal. Mathematically, this might be written as

$$I_{AF}(x,\ y) = I(x, y, z_{\max}), \tag{4}$$

where $z_{\max}$ corresponds to the focus setting giving the maximum signals. The images obtained are somewhat similar to those with the extended-focus method and, again, substantial increases in depth of focus may be obtained. We can go one step further in this case and turn the microscope into a noncontacting surface profilometer. Here we simply display $z_{\max}$.

It is clear by now that the confocal method gives us a convenient tool for studying three-dimensional structures in general. We essentially record the image as a series of slices and play it back in any desired fashion. Naturally, in practice it is not as simple as this, but we can, for example, display the data as an x–z image rather than an x–y image. This is somewhat similar to viewing the specimen from the side. As another example, we might choose to recombine the data as stereo pairs by introducing a slight lateral offset to each image slice as we add them up. If we do this twice, with an offset to the left in one case and the right in another, we obtain, very simply, stereo pairs. Mathematically, we form images of the form

$$\int I(x \pm \gamma z, y, z)\, dz, \tag{5}$$

where γ is a constant. In practice, it may not be necessary to introduce offsets in both directions to obtain an adequate stereo view.

All that we have said so far on these techniques has been by way of simplified introduction. In particular, we have not presented any fluorescence images, because these will be dealt with adequately later. The key point is that, in both brightfield and fluorescence modes, the confocal principle permits the imaging of specimens in three dimensions. The situation is, of course, more involved than we have implied. A thorough knowledge of the image-formation process, together with the effects of lens aberrations and absorption, is necessary before accurate data manipulation can take place.

4 FLUORESCENCE MICROSCOPY

We now turn our attention to confocal fluorescence microscopy, since this is the imaging mode that is most usually employed in biological applications. Although we have introduced the confocal microscope in terms of brightfield imaging, the comments we have made concerning the origin of the optical sectioning and so on, carry over directly to the fluorescence case, although the numerical values describing the strength of the optical sectioning are, of course, different and we shall return to this point later.

We begin by referring back to the conventional scanning microscope system of Figure 2*c* and imagine that suitable filters or dichroics are present such that only the fluorescence radiation is detected. We see immediately that the resolution results essentially from the primary radiation and *not* from the *longer-wavelength* fluorescence: the collector essentially just collects the fluorescence radiation onto a detector. In the nonscanning traditional conventional microscope of Figure 2*a* the opposite is the case. Here the primary radiation merely excites the fluorescence, which is then imaged by the objective lens. The resolution is now determined by the longer-wavelength fluorescence radiation. This leads us to conclude that we should expect superior imaging in a *scanning* embodiment of a fluorescence microscope than the usual wide-field configuration. Naturally the imaging is improved further by introducing a confocal pinhole to obtain optical sectioning.

If we assume that the fluorescence in the object destroys the coherence of the illuminating radiation and produces an incoherent fluorescent field proportional to the intensity of the incident radiation, $I(v,\ u)$, then we may write the effective intensity point spread function, which describes image formation in the incoherent *confocal* fluorescence microscope, as

$$I(v,\ u)I\left(\frac{v}{\beta},\ \frac{u}{\beta}\right) \tag{6}$$

where the optical coordinates u and v are defined relative to the *primary* radiation and $\beta = \lambda_2/\lambda_1$ is the ratio of the fluorescence to the primary wavelength. We note that $v = (2\pi/\lambda)rn \sin \alpha$, where r denotes the actual radial distance.

This suggests that the imaging performance depends on the value of β. In order to illustrate this, we show in Figure 10 the variation in detected signal strength as a perfect fluorescent sheet is scanned through focus. This serves to characterize the strength of the optical sectioning in fluorescence microscopy. We note that the half-width of these curves is essentially proportional to β, and so for optimum sectioning the wavelength ratio should be as close to unity as possible.

We have just discussed what we might call single-photon fluorescence microscopy in the sense that a fluorophore is excited by a single photon of a particular wavelength. It then returns to the ground state and emits a photon at the (slightly longer) fluorescence wavelength. It is this radiation that is detected via the confocal pinhole. Recently, however, much interest has centred on two-photon excitation fluorescence microscopy (Denk et al., 1990; Göppert-Mayer, 1931). This process relies on the simultaneous absorption of two, longer-wavelength photons, following which a single fluorescence photon is emitted. The excitation wavelength is typically twice that used in the single-photon case. The beauty of the two-photon approach lies in the quadratic dependence of the fluorescence intensity on the intensity of the illumination. This leads to fluorescence emission, which is always confined to the region of focus. In other words, the system possesses an inherent optical-

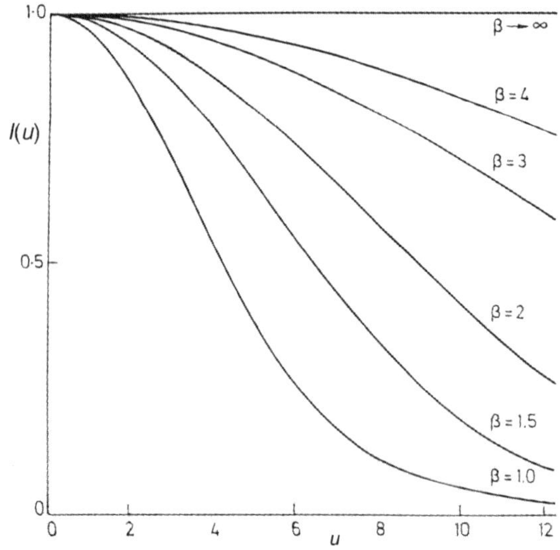

FIGURE 10 The detected signal as a perfect planar object is scanned axially through focus for a variety of fluorescent wavelengths. We note that if we measure the sectioning by the half-width of these curves, the strength of the sectioning is essentially proportional to β.

sectioning property. Other benefits of two-photon fluorescence over the single-photon case include the use of red or infrared lasers to excite ultraviolet dyes, confinement of photobleaching to the focal region (the region of excitation), and the reduced effects of scattering and greater penetration. However, it should also be remembered that, compared with single-photon excitation, the fluorescence yield of many fluorescent dyes under two-photon excitation is relatively low (Xu and Webb, 1996). The practical advantages and drawbacks of two-photon excitation microscopy will be dealt with in other chapters in this book, and so here we shall merely compare the *optical* performance of the single- and two-photon modalities.

In order to make this comparison, we shall assume that the *emission* wavelength λ_{em} is the same irrespective of the mode of excitation. Since the wavelength required for single-photon excitation is generally shorter than the emission wavelength, we may write it as $\gamma\lambda_{em}$ where $\gamma < 1$. Since two-photon excitation requires the simultaneous absorption of two photons of half the energy, we shall assume that the excitation wavelength in this case may be written as $2\gamma\lambda_{em}$, which has been shown to be a reasonable approximation for many dyes (Xu et al., 1996).

If we introduce optical coordinates u and v normalized in terms of λ_{em}, we may now write the effective point spread functions in the single-photon confocal and two-photon cases as

$$I_{1p,\,\mathrm{conf}} = I\left(\frac{v}{\gamma}, \frac{u}{\gamma}\right) I(v, u) \tag{7}$$

and

$$I_{2p} = I^2\left(\frac{v}{2\gamma}, \frac{u}{2\gamma}\right). \tag{8}$$

We note, that although a pinhole is not usually employed in two-photon microscopy, it is perfectly possible to include one if necessary. In this confocal two-photon geometry, the effective point spread function becomes

$$I_{2p,\ \mathrm{conf}} = I^2\left(\frac{v}{2\gamma}, \frac{u}{2\gamma}\right) I(v, u). \tag{9}$$

If we now look at (7) and (8) in the $\gamma = 1$ limit, we find

$$I_{1p,\ \mathrm{conf}} = I^2(v, u) \tag{10}$$

and

$$I_{2p} = I^2(\tfrac{1}{2}v, \tfrac{1}{2}u). \tag{11}$$

We now see that because of the longer excitation wavelength used in two-photon microscopy, the effective point spread function is *twice* as large as that of the single-photon confocal in *both* the *lateral* and *axial* directions. The situation is somewhat improved in the confocal two-photon case, but it is worth remembering that the advantages of two-photon excitation microscopy are accompanied by a reduction in optical performance as compared with the single-photon case.

5 OPTICAL ARCHITECTURES

In addition to the generic architecture of Figure 1, a number of alternative implementations of confocal microscopes have recently been developed in order to reduce the alignment tolerances of the architecture of Figure 1, as well as to increase image acquisition speed. Since great care is required to ensure that the illumination and detector pinholes lie in equivalent positions in the optical system, a reciprocal-geometry confocal system has been developed that uses a single pinhole to both launch light into the microscope and detect the returning confocal image signal. A development on this is to replace the physical pinholes with a single-mode optical fiber. The optical system of such a reciprocal-geometry microscope is shown in Figure 11.

However these systems are essentially developments of the traditional confocal architecture. We shall now describe two recent attempts to develop new confocal architectures. The first system is based on the pinhole source/detector concept whereas the second approach describes a simple modification to a conventional microscope so as to permit optically sectioned images to be obtained.

5.1 The Aperture Mask System

The question of system alignment and image-acquisition speed have already been answered to a certain extent by Petrán and his colleagues, who originally developed the tandem scanning (confocal) microscope (Petrán et al., 1968). This is shown in Figure 12*a* together with a later one-sided disc embodiment in Figure 12*b* (see, e.g., Corle and Kino,

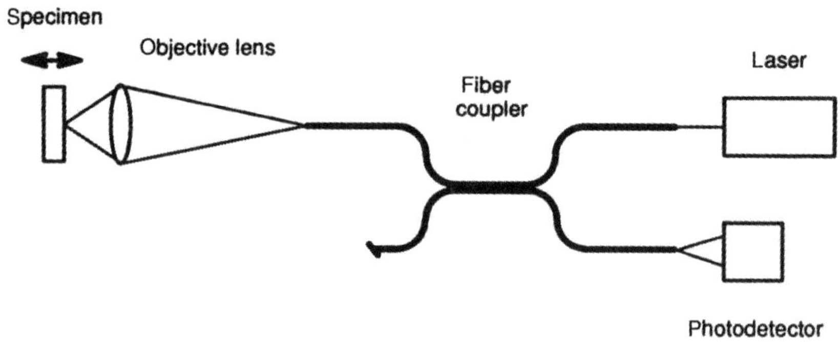

FIGURE 11 The optical system of a reciprocal-geometry confocal microscope using a single-mode optical fiber to both launch light into the microscope system and detect the confocal image signal.

1996). The main component of the system is a disc containing many pinholes. Each pinhole acts as both the illumination and the detection pinhole, and so the system acts rather like a large number of parallel, reciprocal-geometry, confocal microscopes, each imaging a specific point on the object. However, since it is important that no light from a particular confocal system enters an adjacent system (i.e., there should be no crosstalk between adjacent confocal systems), the pinholes in the disc must be placed far apart–typically ten pinhole diameters–which has two immediate consequences. First, only a small amount–typically 1%–of the available light is used for imaging, and secondly, the wide spacing of the pinholes means that the object is only sparsely probed. In order to probe–and hence image–the whole object, it is necessary to arrange the pinhole apertures in a series of Archimedean spirals and to rotate the disc. These systems have been developed and are capable to producing high-quality images without the need to use laser illumination in real time with both TV-rate and higher imaging speeds.

In order to make greater use of the available light in this approach, we must, inevitably, place the pinholes closer together. This, of course, leads to crosstalk between the neighboring confocal systems, and hence a method must be devised to prevent this. In order to do this, we replace the Nipkow disc of the tandem scanning microscope with an aperture mask consisting of many pinholes placed as close together as possible. This aperture mask has the property that any of its pinholes can be opened and closed independently of the others in any desired time sequence. This might be achieved, for example, by using a liquid crystal spatial light modulator. Since we require there to be no crosstalk between the many parallel confocal systems, it is necessary to use a sequence of openings and closings of each pinhole that is completely uncorrelated with the openings and closings of all the other pinholes. There are many such orthonormal sequences in the literature. However, they all require the use of both positive and negative numbers, and unfortunately we cannot have negative intensity of light! The pinhole is either open, which corresponds to 1, or closed, which corresponds to 0. There is no position that can correspond to -1. The way out of the dilemma is to not obtain the confocal signal directly. In order to use a particular orthonormal sequence, $b_i(t)$, of plus and minus 1s, for the ith pinhole we must add a constant shift to the desired sequence in order to make a sequence of positive numbers that can be encoded in terms of pinhole opening and closing. Thus, we encode each of the pinhole openings and closings as $\frac{1}{2}[1 + b_i(t)]$, which will correspond to

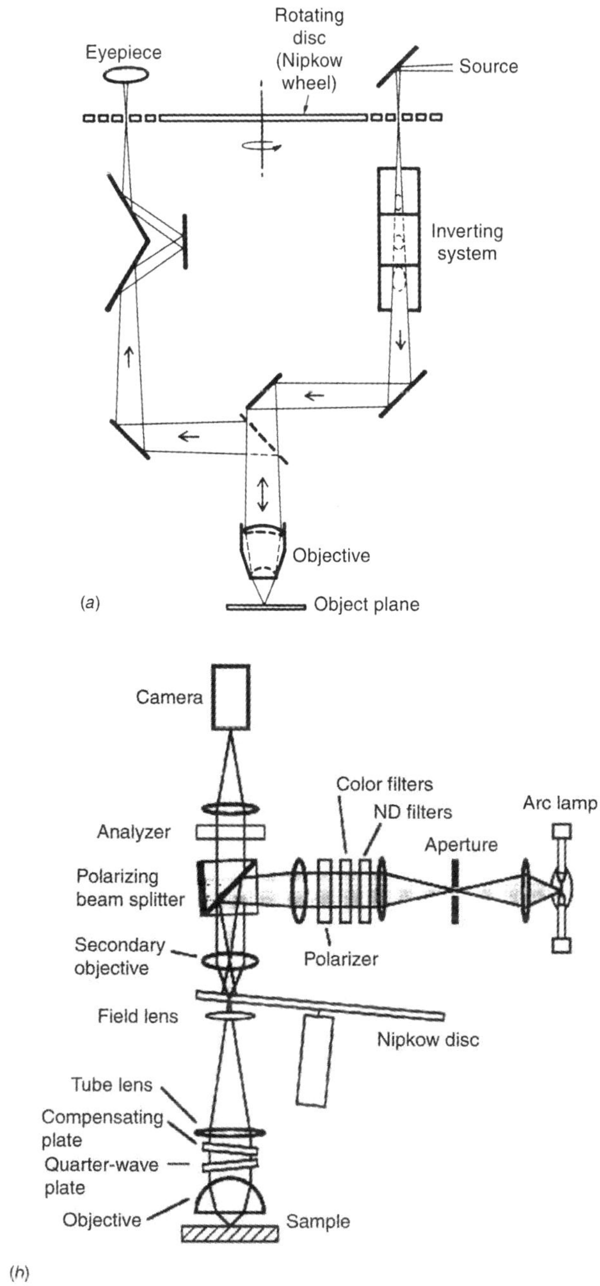

FIGURE 12 The optical system of the original tandem scanning microscope (*a*), together with that of a later, one-sided-disc, implementation (*b*). The microscope essentially comprises many confocal systems, each of which images a specific point on the object. All object points are imaged by rotating the disc.

open (1) when $b_i(t) = 1$ and to closed (0) when $b_i(t) = -1$. The effect of adding the constant offset to the desired sequence is to produce a composite image that will be partly confocal owing to the $b_i(t)$ terms and partly conventional owing to the constant term. The method of operation is now clear. We first take an image with the pinholes encoded as we have just discussed, and so obtain a composite conventional plus confocal image. We then switch all the pinholes to the open state to obtain a conventional image. It is then a simple matter to subtract the two images in real time in a computer to produce the confocal image.

Although this approach may be implemented using a liquid crystal spatial light modulator, it is cheaper and simpler merely to impress the correlation codes photolithographically on a disc and to rotate the disc so that the transmissivity at any picture point varies according to the desired orthonormal sequence. A blank sector may be used to provide the conventional image. If this approach is adopted, then all that is required is to replace the single-sided Nipkow disc of the tandem scanning microscope of Figure 10 with a suitably encoded aperture disc, as in Figure 13 (Juškaitis et al., 1996; Wilson et al., 1996). Figure 14 shows a through-focus series of images of a fly's eye. Figures 15 and 16 show three-dimensional representations of the fly's eye and a portion of a transistor, respectively, obtained with this white-light real-time confocal system.

5.2 The Use of Structured Illumination to Achieve Optical Sectioning

The motivation for the development that we have just described was to produce a light-efficient, real-time, three-dimensional imaging system. The approach was to start with the traditional confocal microscope design – point source and point detector – and to engineer a massively parallel, light-efficient, three-dimensional imaging system (see also Liang et al., 1997; Hanley et al., 1999).

An alternative approach is to realize that the conventional light microscope already possesses many desirable properties, such as real-time image capture, standard illumination, and ease of alignment. However, it does not produce optically sectioned images in the sense usually understood in confocal microscopy. In order to see how this deficiency may be corrected via a simple modification of the illumination system, let us begin by looking at the theory of image formation in a conventional fluorescence microscope, and ask in

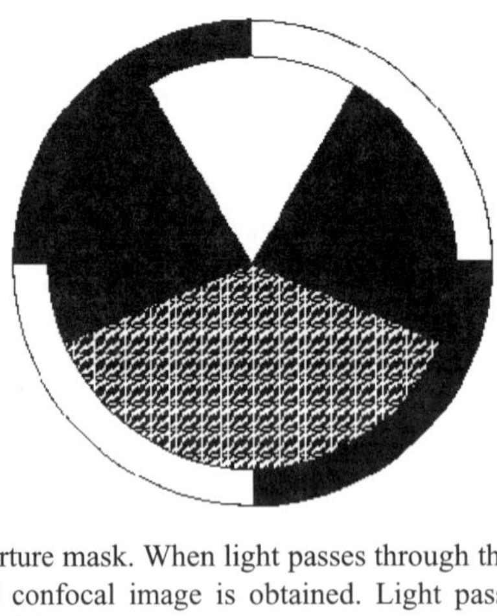

FIGURE 13 A typical aperture mask. When light passes through the encoded region of the disc, a composite conventional and confocal image is obtained. Light passing through the unobstructed sector provides a conventional image.

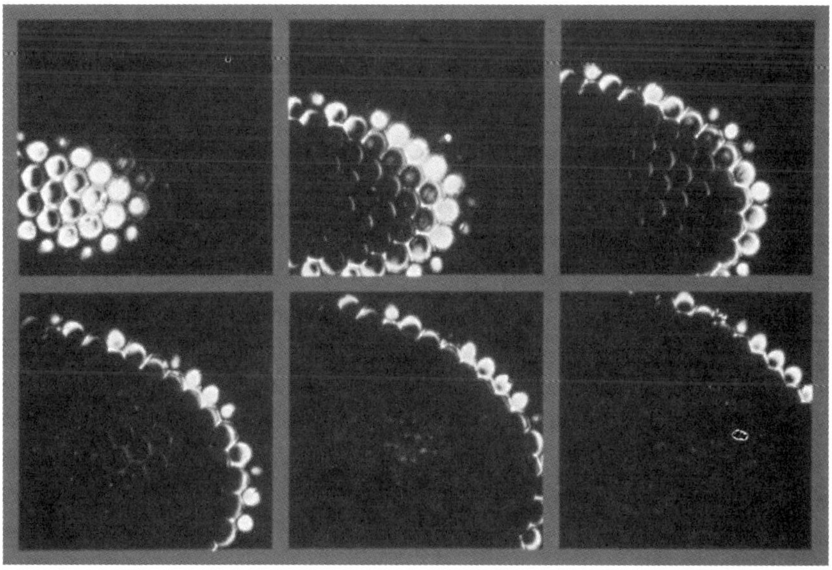

FIGURE 14 See ftp site for color figure. A through-focus series of images of a portion of a fly's eye. Each optical section was recorded in real time using standard nonlaser illumination.

what way the image changes as the microscope is defocused. We know that in a confocal microscope, the image signal from all object features attenuates with defocus and that this does not happen in a conventional microscope. However, when we look closely at the image-formation process, we find that it is only the zero-spatial-frequency (constant) component that does not change with defocus – see Figure 17 – all other spatial frequencies actually do attenuate with defocus to a greater or lesser extent. Figure 18 shows the image of a single-spatial-frequency one-dimensional bar pattern object for increasing degrees of defocus. When the specimen is imaged in focus, a good image of the bar pattern is

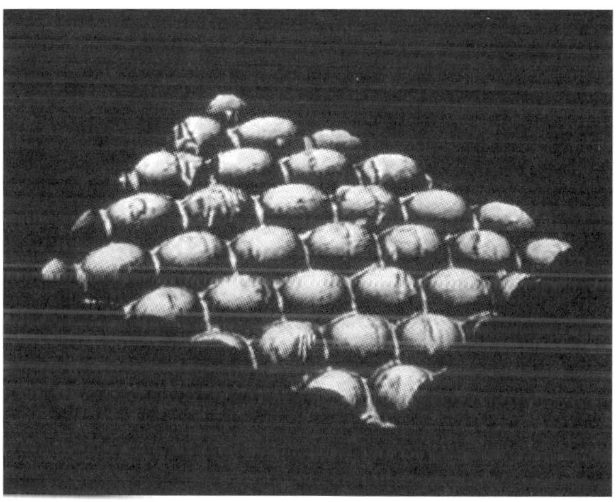

FIGURE 15 A computer-generated three-dimensional representation of the fly's eye.

FIGURE 16 See ftp site for color figure. A computer-generated false-color image of a portion of a transistor. The color represents surface height and the depth range from pale yellow to blue is 6 mm.

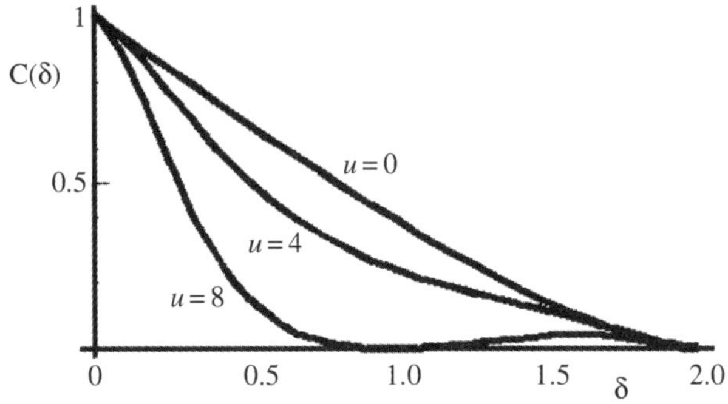

FIGURE 17 The transfer function for a conventional fluorescence microscope for a number of values of defocus (measured in optical units). We notice that the response for all nonzero spatial frequencies decays with defocus.

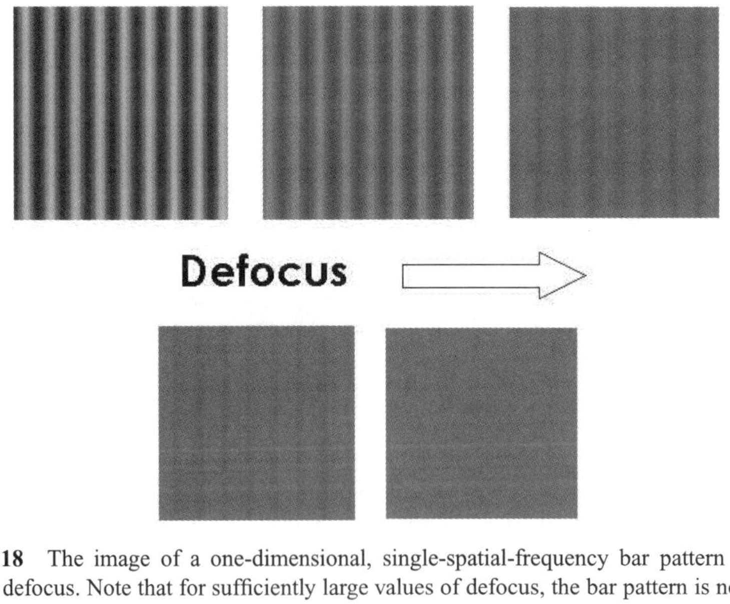

FIGURE 18 The image of a one-dimensional, single-spatial-frequency bar pattern for varying degrees of defocus. Note that for sufficiently large values of defocus, the bar pattern is not imaged at all.

obtained. However, with increasing defocus, the image becomes progressively poorer and weaker until it eventually disappears, leaving a uniform gray level. This suggests a simple way to perform optical sectioning in a conventional microscope.

If we simply modify the illumination path of the microscope so as to project onto the object the image of a one-dimensional, single-spatial-frequency fringe pattern, then the image that we see through the microscope will consist of a sharp image of those parts of the object where the fringe pattern is in focus, together with an out-of-focus blurred image of the rest of the object. In order to obtain an optically sectioned image, it is necessary to remove the blurred out-of-focus portion as well as the fringe pattern from the in-focus optical section. There are many ways to do this – one of the simplest involves simple processing of three images taken at three different spatial positions of the fringe pattern. The out-of-focus regions remain fairly constant between these images, and the relative spatial shift of the fringe pattern allows the three images to be combined in such a way as to remove the fringes. This permits us to retrieve both an optically sectioned image as well as a conventional image in real time (Neil et al., 1997).

Since the approach involves processing three conventional microscope images, the image formation is fundamentally different from that of the confocal microscope. However, the depth-discrimination or optical-sectioning strength is very similar, and this approach, which requires very minimal modifications to the instrument, has been used to produce high-quality three-dimensional images of volume objects that are directly comparable to those obtained with confocal microscopes. A schematic of the optical arrangement is shown in Figure 19*a*, together with experimentally obtained axial responses in Figure 19*b*. It will be noted that these response are substantially similar to those obtained in the true confocal case. As examples of images, we show in Figure 20 two images of a lily pollen grain. The first is an autofocus image of greatly extended depth of field constructed from a through-focus series of images. The second is a conventional image taken at mid-focus. A further example is shown in Figure 21, which is an autofocus image of a spiracle of a head louse. The dramatic increase in depth of field is again clear when compared with a mid-focus conventional image. We note that these images were taken using a standard microscope illuminator as light source. Indeed, the system is so light-efficient that good-quality optical sections have been obtained or transistor specimens simply using a candle as light source!

Imaging using fluorescence light is also possible using this technique. However, an alternative approach that does not require a physical grid is possible if a laser is used as the light source. In this system, the laser illumination is split into two beams, which are allowed to interfere at an angle in the fluorescent specimen. This has the effect of directly "writing" a one-dimensional fringe pattern in the specimen. Spatial shift of the fringe pattern is achieved by varying the phase of one of the interfering beams. As before, three images are taken, from which both the optically sectioned image and conventional image may be obtained (Neil et al., 1998). The beauty of this approach is that no imaging optics are required at the illuminating wavelength, and system alignment is trivial.

6 SUMMARY

We have discussed the origin of the optical-sectioning property in the confocal microscope in order to introduce the range of imaging modes that this unique form of microscope

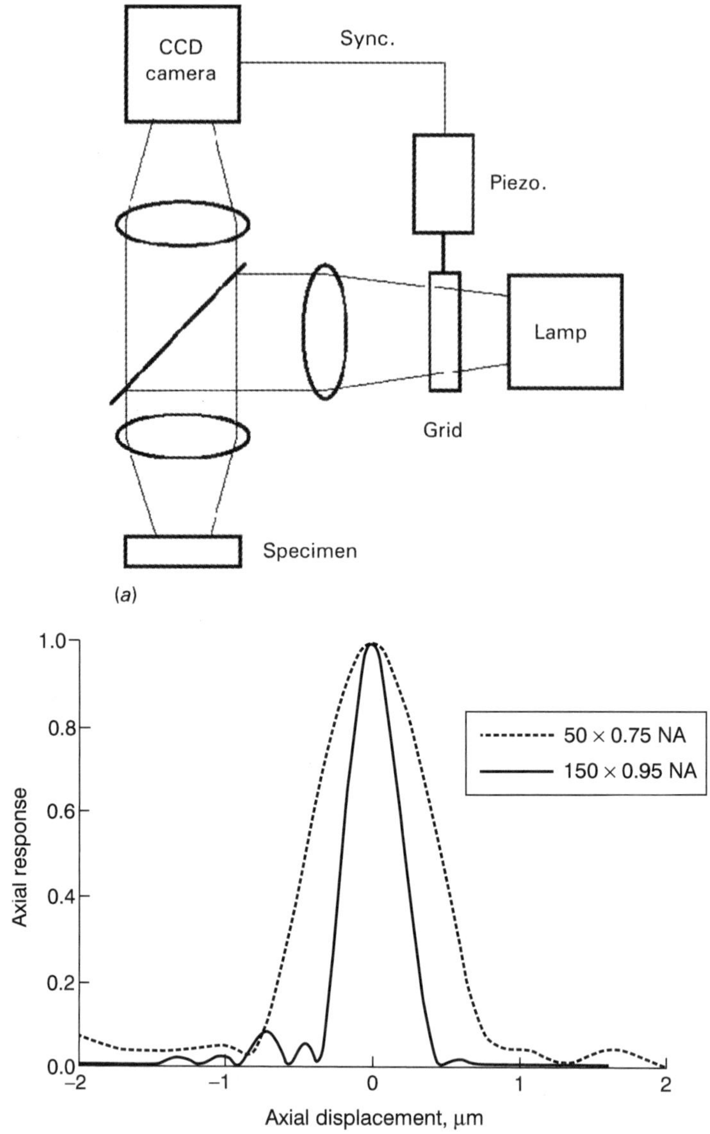

FIGURE 19 (*a*) The optical system of the structured-illumination microscope. (*b*) Experimentally obtained axial responses, which confirm the optical-sectioning ability of the instrument.

leads to. A range of optical architectures have also been described. By far the most universal is that shown in Figure 1, where a confocal module is integrated around a conventional optical microscope. Other, more recent, real-time implementations have also been described. A number of practical aspects of confocal microscopy have not been discussed. This is because they are readily available elsewhere, such as advice on the correct choice of detector pinhole size, or because they are still the focus of active research, such as the problem of overcoming the inevitable resolution-impairing aberrations, which are ubiquitous when focusing deep into index-mismatched specimens.

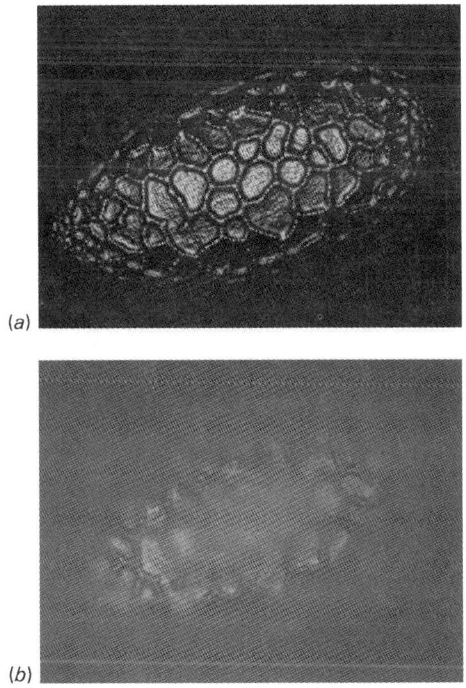

FIGURE 20 Two images of a grain of lily pollen. (*a*) An autofocus image obtained by processing a series of through-focus images in order to obtain an image with vastly extended depth of field. (*b*) A conventional image taken at a mid-focus setting for comparison.

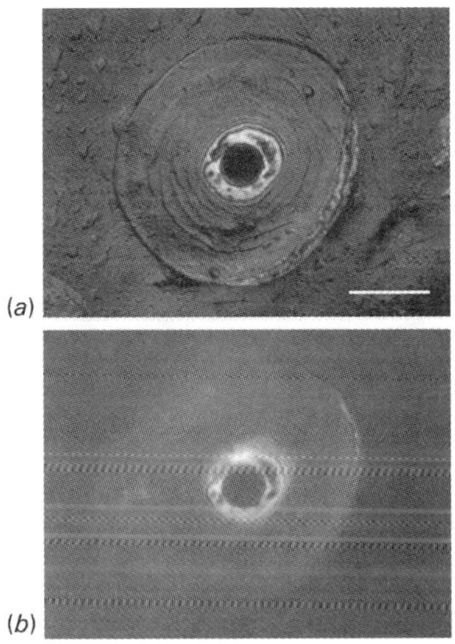

FIGURE 21 Two images of the region around the spiracle of a head louse. As in Figure 20, (*a*) is an autofocus image while (*b*) shows a mid-focus conventional image. The scale bar depicts a length of 10 μm.

REFERENCES

Corle, T. R. and Kino, G. K. (1996). *Confocal Scanning Optical Microscopy and Related Imaging Systems*. New York: Academic Press.

Denk, W., Strikler, J. H., and Webb, W. W. (1990). *Science* **248**, 73.

Göppert-Meyer, M. (1931). *Annln. Phys.* **9**, 273.

Gu, M. (1996). *Principles of Three-Dimensional Imaging in Confocal Microscopes*. Singapore: World Scientific.

Hanley, Q. S., Verveer, P. J., Gemkow, M. J., Arndt-Jovin, D., and Jovin, T. M. (1999). *J. Microsc.* **196**, 317.

Juškaitis, R., Wilson, T., Neil, M. A. A., and Kozubek, M. (1996). *Nature* **383**, 804.

Liang, M. H., Stehr, R. L., and Krause, A. W. (1997). *Opt. Lett.* **22**, 751.

Lukosz, W. (1966). *J. Opt. Soc. Am.* **56**, 1463.

Masters, B. R. (1996). *Confocal Microscopy.* Bellingham, WA: SPIE Press.

Minsky, M. (1961). Microscopy Apparatus. United States Patent 3,013,467, Dec. 19, 1961 (Filed Nov. 7, 1957).

Neil, M. A. A., Juskaitis, R., and Wilson, T. (1997). *Opt. Lett.* **22**, 1905.

Neil, M. A. A., Juskaitis, R., and Wilson, T. (1998). *Opt. Commun.* **153**, 1.

Pawley, J. B. (Ed.) (1995). *Handbook of Biological Confocal Microscopy.* New York: Plenum Press.

Petrán, M., Hadravsky, M., Egger, M. D., and Galambos, R. (1968). *J. Opt. Soc. Am.* **58**, 661.

Wilson, T. (Ed.) (1990). *Confocal Microscopy.* London: Academic Press.

Wilson, T. and Sheppard, C. J. R. (1984). *Theory and Practice of Scanning Optical Microscopy.* London: Academic Press.

Xu, C. and Webb, W. W. (1996). *J. Opt. Soc. Am.* **B13**, 481.

Xu, C., Williams, R. M., Zipfel, W., and Webb, W. W. (1996). *Bioimaging* **4**, 198.

CHAPTER 3

Two-Photon Excitation Fluorescence Microscopy

ALBERTO DIASPRO[1] and COLIN J. R. SHEPPARD[2]

[1] National Institute for the Physics of Matter (INFM) and Department of Physics, University of Genoa, 16146 Genova, Italy
[2] School of Physics and Australian Key Centre for Microscopy and Microanalysis, The University of Sydney, NSW 2006, Australia

1 INTRODUCTION

Two-photon excitation (TPE) microscopy and other nonlinear optical techniques constitute one of the most promising and fastest-growing areas in biological and medical imaging at the optical-resolution level (Sheppard, 1989; Pawley, 1995; Hell, 1996; Diaspro, 1999a,b; Diaspro and Robello, 2000; So et al., 2000; König, 2000); see Chapter 1 of this book (Sheppard, 2002). In particular, TPE belongs to a class of imaging techniques that allow observation of biological specimens in conditions very close to their natural environment, increasing the possibility of understanding the complex and delicate relationship existing between structure and function (Cantor and Schimmel, 1980; Beltrame et al., 1985; Chance, 1989; Manders et al., 1992; Carrington et al., 1995). Moreover, while conventional fluorescence microscopy is appropriate for passive observation, the ability of TPE to interact with the specimen in a very small region of the order of 0.1 femtoliter makes it suitable for both active and interactive experimentation methods. One of the keywords for this emerging technique is the term "scanning," as used earlier for acoustic microscopy (Lemons and Quate, 1970) and confocal microscopy (Sheppard, 1977; Sheppard and Choudhury, 1977; Brakenhoff et al., 1979; Sheppard and Wilson, 1980; Wilson and Sheppard, 1984; Minsky, 1988; Webb, 1996).

TPE microscopy is a comparatively new form of scanning far-field fluorescence optical microscopy. The three-dimensional (3D) resolving properties of the optical microscope are well known and consolidated in their modern forms, namely computational wide field and confocal laser scanning optical-sectioning microscopy (Wilson and Sheppard, 1984; Agard et al., 1989; Wilson, 1990; Kriete, 1992; Shotton, 1993; Pawley, 1995; Masters, 1996; Sheppard and Shotton, 1997; Benedetti, 1998; Pike, this book, Chapter 24); see Chapters 1, 2 (Wilson, 2002), and 5 (Jonkman and Stelzer, 2002). Among the most important and

Confocal and Two-Photon Microscopy: Foundations, Applications, and Advances, Edited by Alberto Diaspro.
ISBN 0-471-40920-0

well-established developments in the field of microscopic techniques are 3D applications of confocal microscopy that, coupled to fluorescence mechanism of contrast, allow understanding of the intriguing 3D properties of biological systems (Cox, 1984; Carlsson et al., 1985; Brakenhoff et al., 1989; Sheppard, 1989; White et al., 1987; Wilson, 1989).

In the ideal confocal microscope, the illumination is focused into a diffraction-limited spot scanned on the specimen, confining both the illuminated focal region and the detected area of the emitted light; see Chapters 1, 2, and 5. The light emitted due to reflection or fluorescence is imaged by the objective lens into the image plane, where a circular aperture (pinhole) is placed in front of a light detector. This pinhole is responsible for rejection of axial out-of-focus light and of lateral overlapping diffraction patterns. Thus, the axial and the lateral resolution of the light microscope are improved by a factor of approximately 1.4, resulting in a volume selectivity 2.7 times better than in the wide-field case (Brakenhoff et al., 1979; Wilson and Sheppard, 1984); see Chapters 5 and 6 (Török and Sheppard, 2002). The physical suppression of the contributions from out-of-focus layers to image formation produces the so-called optical-sectioning effect. This is consolidated by the fact that the application of image-processing algorithms can greatly improve the situation (Cox and Sheppard, 1984; Agard et al., 1989; Kriete, 1992; Carrington et al., 1995; Castleman, 1996; Bertero and Boccacci, 1998; Schrader et al., 1996; Diaspro et al., 2000); see Chapter 13 (Periasamy et al., 2002). Unfortunately, one of the major drawbacks of confocal microscopy is that during excitation of fluorescent molecules using photons of appropriate energy, the whole thickness of the specimen within a hourglass-shaped region is affected by every scan. This means that even though out-of-focus fluorescence is not detected, it is generated. The problem with this effect is that many interesting and important biological investigations involving microscopic three-dimensional imaging are defeated by photobleaching of the fluorescent label and by phototoxicity. The situation becomes particularly serious when there is a need for three-dimensional and temporal imaging coupled to the use of fluorochromes that require excitation in the ultraviolet regime (Stelzer et al., 1994; Denk, 1996). In this context, the advent of two-photon excitation laser scanning microscopy has allowed the possibility of alleviating these problems, opening new avenues to the application of microscopic techniques to the study of biological systems while providing attractive and effective advantages over classical confocal microscopy (Denk et al., 1990; Potter, 1996; Centonze and White, 1998; Gu and Sheppard, 1995; Piston, 1999; Squirrel et al., 1999; Diaspro and Robello, 2000; So et al., 2000); see Chapter 2. TPE fluorescence microscopy is not only revolutionary in its ability to provide optical sections, together with other practical advantages, but also in its elegance and the effectiveness of its application of quantum physics (Loudon, 1983; Feynman, 1985; Shih et al., 1998). This nonlinear microscopy, together with those based on three-photon excited fluorescence (Hell et al., 1996), second-harmonic generation (Gannaway and Sheppard, 1978; Gauderon et al., 1998; Campagnola et al., 1999), and third-harmonic generation (Mueller et al., 1998; Squier et al., 1998), has been shown to be a valuable technique for biomedical imaging (Denk, 1996; König, 2000; So et al., 2000; Straub et al., 2000; Periasamy, 2001).

2 CHRONOLOGICAL STEPS

The story dates back to 1931 when the Nobel laureate Maria Göppert-Mayer (Fig. 1) predicted the phenomenon of two-photon absorption in her doctoral dissertation (Göppert-Mayer, 1931): one atom or molecule can simultaneously absorb two photons in the same

FIGURE 1 Maria Göppert-Mayer.

quantum event. The keyword "simultaneous" here means within a temporal window of 10^{-16} s (0.1 fs). This should give an idea of the rarity of the event. As reported by Denk and Svoboda (1997), in bright daylight a molecule of a good one- or two-photon absorber absorbs a photon through one-photon interaction about once a second, and a photon pair by two-photon simultaneous interaction every 10 million years. This should also explain why it was only around 1961, after the development of laser sources (Wise, 1999; Gratton and van de Ven, 1995), that Kaiser and Garret (1961) reported two-photon excitation of fluorescence in $CaF_2:Eu^{2+}$ and Singh and Bradley (1964) estimated the three-photon absorption cross-section for naphthalene crystals. These two results followed the experimental observations by Franken et al. (1961) of second-harmonic generation in a quartz crystal using a ruby laser. Later, Rentzepis et al. (1970) produced photographs of three-photon excited fluorescence in organic dyes. The nonlinear harmonic generation is function of the molecular structure (Birks, 1970; Cantor and Schimmel, 1980), and, as simply and effectively described by Wilson and Sheppard (1984), allows the probing of the non-centrosymmetric geometry exhibited by several biological samples and crystals. In fact, Hellwarth and Christensen (1974) exploited optical nonlinearity for second-harmonic generation in ZnSe polycrystals at a microscopic level. The original idea of incorporating nonlinear optical effects, such as higher-order harmonic generation and two- or multi-photon excitation of fluorescence, was first suggested and demonstrated by Sheppard, Kompfner, Gannaway, and Choudhury of the Oxford group, who outlined how such a nonlinear optical phenomenon could be exploited in microscopy when combined with the evolving scanning optical microscope (Sheppard et al., 1977; Gannaway and Sheppard, 1978; Sheppard and Kompfner, 1978); see Figure 2. It was the Oxford group who realized the optical-sectioning ability due to the confinement of the event at the focal plane of the objective, because the image intensity had a quadratic dependence on the illumination power. However, for many years, the application of two-photon absorption was mainly limited to spectroscopic studies (Friedrich and McClain, 1980; Friedrich, 1982; Birge, 1986). Only the development of mode-locked lasers, providing high-peak-power femto-

Resonant scanning optical microscope

C. J. R. Sheppard and R. Kompfner

The enhancement of nonlinear interactions in the scanning optical microscope by placing the object in a resonant cavity is discussed. The effect of resonance of the primary radiation is also considered. Practical geometries for the resonant scanning optical microscope are given.

In the scanning optical microscope[1,2] nonlinear interactions are expected to occur between the object and a highly focused beam of light, which we hope will open new ways of studying matter in microscopic detail hitherto not available. Nonlinear interactions[3-5] include the generation of sum frequencies, Raman scattering, two-photon fluorescence, and others. We feel that the method will be of particular interest in studying biological materials, some of which have large second-harmonic generation coefficients,[6-9] and the wide range of these coefficients should give very strong contrast in the images formed. Furthermore, frequency mixing should give information concerning the chemical structure of the object.[10]

Objects suitable for microscopy are usually no thicker than a few microns, and hence the ever-present dispersion of the refractive index of optically transparent materials which would cause the fundamental light wave and the harmonic wave excited by it to get out of step need not concern us. This means, however, that the intensity of the signal is low. We have estimated[11] that a fundamental power of 0.1 W, focused down to a diffraction limited spot, will give a second-harmonic power of about 10^{-9} W from a resonably nonlinear substance such as KDP or a protein. This may be detected with a SNR of the order of a thousand, which is adequate to produce a picture in a reasonable time.

However, heating in the object sets a limit to the amount of power that can be used because some fraction of it will be absorbed and transformed into heat. We distinguish between two kinds of scanning: slow and fast. In the slow-scan case, where the probe is effectively stationary on the object, we estimate that the permissible optical beam power is about 35 mW. In the fast-scan case, the whole specimen reaches an equilibrium temperature, and the power is limited to about 100 mW.[11] It is clear that heat dissipation is a serious problem for the harmonic scanning optical microscope, and ways of improving either the efficiency of conversion of fundamental to harmonic power, for example by beam pulsing, or the thermal conductance of the object supports are of importance.

One way to enhance the efficiency of the generation of the signal is to place the volume element in which the signal is generated within a resonator[11] tuned to the frequency of the signal and in such a position that maximum power can be generated and abstracted from the resonator, as shown Fig. 1.

If a resonator is to function, it has to have certain geometric and physical properties which we shall now describe by way of a particular example. In order to be used in a scanning microscope it must have a region of maximum power concentration—a focus. In order to be effective in improving the signal generation efficiency it must have as high a Q as is practically possible. And in order to obtain the highest signal extraction efficiency it should be critically coupled or matched to the detecting device or system. All these criteria are satisfied by an open resonator of the almost concentric type, that is, by a resonator bounded by spherical reflectors having equal radii R spaced a distance slightly less than $2R$ apart. One reflector can serve as the output coupler by allowing a certain fraction of the power circulating inside the resonator to be transmitted through it. On the assumption that the transverse amplitude distribution inside a resonator is Gaussian, that the smallest spot radius (where the amplitude is e^{-1} of the maximum) is w_0, and that the refractive index of the object equals that of the medium in the resonator, the distance l between the centers of curvature of the reflectors is approximately equal to

$$l = (4\pi^2 w_0^4)/(\lambda^2 d), \quad (1)$$

where λ is the wavelength in the medium and d the distance between the reflectors.

When this work was done both authors were with Oxford University, Department of Engineering Science, Oxford, England.

Received 1 November 1976; revised manuscript received 22 September 1977.

0003-6935/78/0915-2879$0.50/0.

© 1978 Optical Society of America.

15 September 1978 / Vol. 17, No. 18 / APPLIED OPTICS 2879

FIGURE 2 First page of the pioneering paper by Sheppard and Kompfner (1978).

second pulses with repetition rate of around 100 MHz (Spence et al., 1991; Gosnell and Taylor, 1991; Gratton and van de Ven, 1995; Fisher et al., 1997; Wise, 1999) made TPE laser scanning microscopy a practical technique and allowed its widespread use. It was the excellent work of Denk and colleagues in the W.W. Webb Laboratories at Cornell University that revolutionized fluorescence microscopic imaging (Denk et al., 1990). They coupled the potential of two-photon excited fluorescence imaging in a scanning microscope with utilization of ultrafast pulsed lasers. In summary, we can say that the success of two-photon excitation microscopy has been due to developments in two, still-evolving, areas: laser scanning microscopy and mode-locked lasers.

3 TWO-PHOTON EXCITATION OF FLUORESCENT MOLECULES

3.1 Basic Aspects

In confocal microscopy, an extremely small volume of the sample is imaged point by point (Wilson and Sheppard, 1984); see Chapters 1 and 2. This focal volume is of the order of $0.02\,\mu m^3$. In general, in the fluorescence image formation mode (see Chapter 2), the absorption of one photon is used to excite fluorescing molecules to a higher-energy state, from which emission of one photon of fluorescence occurs. Due to the characteristics of the processes involved, the fluorescence emission is shifted towards a longer wavelength than that used for excitation. This shift is in the range of 50–200 nm, which is typical of the fluorescent probes in use in fluorescence imaging of biological specimens. For example, a fluorescent molecule that absorbs one photon at 350 nm, in the ultraviolet (UV) region, exhibits fluorescence at 450 nm, around blue–green wavelengths. Unfortunately, at UV or blue excitation wavelengths, photobleaching, and more generally phototoxicity, strongly affect experimental results. This is alleviated in the case of TPE (Brakenhoff et al., 1996; Patterson and Piston, 2000). The TPE of fluorescent molecules is related to the simultaneous absorption of two lower-energy photons, which cooperatively provide the energy needed to prime the fluorescence process.

The relationship between photon energy E and wavelength λ is given by

$$E = hc/\lambda, \tag{1}$$

where $h = 6.6 \times 10^{-34}$ J s is Planck's constant and $c = 3 \times 10^8$ m s^{-1} is the speed of light in vacuum. This energy should be such that the molecule is brought to an excited state, from which it decays to an intermediate state, giving off a photon of light having an energy lower than that needed to prime excitation. In TPE, the energies of the two photons involved in the excitation process of the same fluorescent molecule can combine only if they are able to interact with the same fluorophore at the same time. The timescale for simultaneity is the timescale of molecular energy fluctuations at photon energy scales, as determined by Heisenberg's uncertainty principle, namely 10^{-16} s (Louisell, 1973). In principle, these two photons do not necessarily have to be identical. What it is relevant for TPE is that their wavelengths λ_A and λ_B are such that

$$\lambda_{1-\text{photon}} \approx \left(\frac{1}{\lambda_A} + \frac{1}{\lambda_B}\right)^{-1}, \tag{2}$$

where $\lambda_{1-\text{photon}}$ is the wavelength needed to prime fluorescence emission in conventional or one-photon mode according to the energy requirement imposed by Eq. (1). This situation, compared with the conventional one-photon excitation process, is illustrated in Figure 3 using a Jablonski-like diagram. It is worth noting that for practical reasons the experimental choice is usually such that (Diaspro, 2001)

$$\lambda_A = \lambda_B \approx 2\lambda_{1-\text{photon}}; \tag{3}$$

see Chapter 10 (Girkin and Wokosin, 2001). As it will be more evident from the following, the TPE process requires high photon flux densities, which can be obtained by tight focusing of a laser beam. In a TPE process, it is crucial to combine sharp spatial focusing

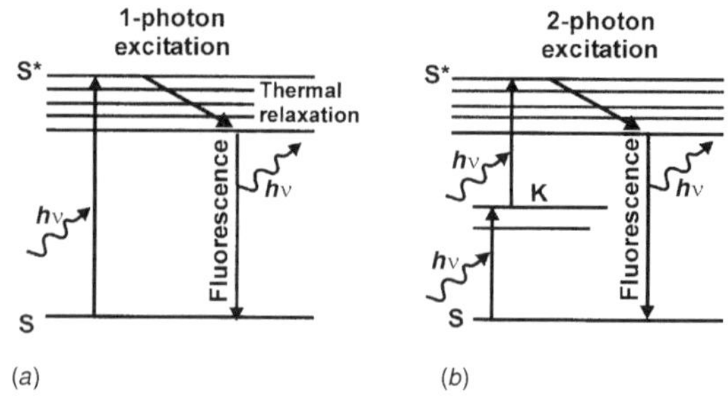

FIGURE 3 Simplified Jablonski diagram sketching the excitation process in one and two photon absorption. A general situation is depicted. S* is the excited state. K is a virtual state that is only permitted for a very short time and not an allowed energy state. [After Soeller and Cannell (1999).]

with temporal confinement of the excitation beam. The process can be extended to n photons, requiring higher photon densities temporally and spatially confined.

An elegant treatment in terms of quantum theory for two-photon transition has been proposed by Nakamura (1999) using perturbation theory: describing the process by a time-dependent Schrödinger equation, where the Hamiltonian contains electric-dipole interactions terms; the first-order solution is related to one-photon excitation, while higher-order solutions are related to n-photon excitations.

Now, let us discuss TPE on the basis of the following simple assumption: the probability of a molecule undergoing n-photon absorption is proportional to the probability of finding n photons within the volume it occupies at any moment in time (Louisell, 1973; Andrews, 1985) or to the probability of finding them within the interval of time that the molecule spends in a virtual state (Moscatelli, 1986). Here we shall refer to the first case, discussed earlier by Andrews (1985).

Rule zero is related to the fact that we are dealing with molecules that are endowed with a suitable set of energy levels such that all possible n-photon transitions are possible. We consider the relationship between the mean number of photons, m, at any time within a molecular volume and the intensity, I, of the laser beam (i.e., energy per unit area per unit time). According to the previous statement, we are interested in finding the probability p_n that n photons are in the molecular volume. Consider a cube of side s through which the laser beam passes and the beam width much larger than s. The mean energy in this cubic box, for a beam of wavelength λ, is

$$E_m = mhc/\lambda. \tag{4}$$

Since the cross-sectional area is s^2, and the time needed for each photon to cross the box is s/c, then

$$I = \frac{E_m}{s^2 \cdot s/c} = \frac{mhc^2}{\lambda \cdot s^3}. \tag{5}$$

Considering $V = s^3 = V_m/N_a$, where for a molecule the mean volume occupied is the molar volume V_m divided by Avogadro's constant $N_a = 6.022 \times 10^{23}$ mol^{-1}, we obtain

$$m = \frac{IV_m\lambda}{N_a hc^2}. \tag{6}$$

As an example, considering a wavelength of 780 nm delivered at peak intensities of the order of GW cm^{-2} into a molecular volume of the order of 10^{-4} m^3 mol^{-1}, the resulting value for m is of the order of 10^{-5}.

In order to find p_n, we can use a Poisson distribution (Louisell, 1973), resulting in

$$p_n = \frac{m^n}{n!} e^{-m}. \tag{7}$$

Here, since m is small, one can expand the exponential in a Taylor series and consider only the first term, thus setting the exponential to unity. The resulting probability for TPE, $n = 2$, is therefore given by

$$p_2 = \tfrac{1}{2} m^2 \propto \Upsilon I^2, \tag{8}$$

where Υ is a proportionality factor. From the relationship (8), the dependence of TPE on I^2 should be evident.

Since we have shown that TPE is a process that has a quadratic dependence on the instantaneous intensity of the excitation beam, we can introduce the molecular cross-section, as the propensity to absorb in a TPE event photons having a certain energy or wavelength, and refer the fluorescence emission as function of the temporal characteristics of the light, $I(t)$, to this cross-section. The fluorescence intensity $I_f(t)$ can be considered to be proportional to the molecular cross-section $\delta_2(\lambda)$ and to $I(t)^2$:

$$I_f(t) \approx \delta_2 I(t)^2 \approx \delta_2 P(t)^2 \left[\pi \frac{(NA)^2}{hc\lambda}\right]^2, \tag{9}$$

where $P(t)$ is the laser power.

The time-averaged two-photon fluorescence intensity per molecule within an arbitrary time interval T, $\langle I(t) \rangle$, can be defined as

$$\langle I_f(t) \rangle = \frac{1}{T}\int_0^T I_f(t)^2 \, dt = \delta_2 \left[\pi \frac{(NA)^2}{hc\lambda}\right]^2 \frac{1}{T}\int_0^T P(t)^2 \, dt, \tag{10}$$

in the case of continuous-wave (CW) laser excitation. For a pulsed laser, $T = 1/f_P$, where f_P is the pulse-repetition rate.

For a CW laser beam, where $P(t) = P_{\text{ave}}$, (10) becomes

$$\langle I_{f,\,cw}(t) \rangle = \delta_2 P_{\text{ave}}^2 \left[\pi \frac{(NA)^2}{hc\lambda}\right]^2. \tag{11}$$

Now, for a pulsed laser beam with pulse width τ_p, repetition rate f_p, and average power $P_{\text{ave}} = DP_{\text{peak}}(t)$, where $D = \tau_p f_p$, the $P(t)$ profile, in a simple but effective approximation, can be described as

$$P(t) = \begin{cases} \dfrac{P_{\text{ave}}}{\tau_p f_p} & \text{for} \quad 0 < t < \tau_p, \\ 0 & \text{for} \quad \tau_p < t < \dfrac{1}{f_P}. \end{cases} \tag{12}$$

Equation (10) becomes (So et al., 2001)

$$\langle I_{f,\,p}(t)\rangle = \delta_2 \frac{P^2_{\text{ave}}}{\tau_p^2 f_P^2}\left[\pi\frac{(NA)^2}{hc\lambda}\right]^2 \frac{1}{T}\int_0^{\tau_p} dt = \delta_2 \frac{P^2_{\text{ave}}}{\tau_p f_P}\left[\pi\frac{(NA)^2}{hc\lambda}\right]^2. \tag{13}$$

The conclusion that can be drawn here is that CW and pulsed lasers operate at the same excitation efficiency if the average power of the CW laser is kept higher by a factor of $(\tau f_P)^{-1/2}$. This means that 10 W delivered by a CW laser, allowing the same efficiency as conventional excitation performed at approximately 0.1 mW, becomes nearly equivalent to 30 mW for a pulsed laser.

This leads to the commonly used relationship given below that applies to the practical situation of a train of beam pulses focused through an high-numerical-aperture objective, with a duration τ_p and a repetition rate f_p. In this case, the probability n_a that a certain fluorophore simultaneously absorbs two photons during a single pulse, in the paraxial approximation, is given by (Denk et al., 1990)

$$n_a \propto \frac{\delta_2 P^2_{\text{ave}}}{\tau_p f_p^2}\left(\frac{NA^2}{2\eta c\lambda}\right)^2, \tag{14}$$

where $\eta = h/2\pi$, P_{ave} is the time-averaged power of the beam, and λ is the excitation wavelength. If we define 1 GM (Göppert-Mayer) $= 10^{-58}$ m^4 s, then, for a δ_2 of approximately 10 GM per photon (Chapter 4: Xu, 2002), focusing through an objective of numerical aperture 1.2–1.4, an average incident laser power of about 1–50 mW, operating in a wavelength range of 640–1000 nm with 100 fs pulsewidth and 100 MHz repetition rate, would saturate the fluorescence output at the same order as the saturational levels found for one-photon excitation. It follows that for optimal fluorescence generation, the desirable repetition time of pulses should be of the order of a typical excited-state lifetime, which is a few nanoseconds for many of the fluorescent molecules used in fluorescence imaging. In practice, this is what sets the desired repetition rate at around 100 MHz. The other condition for the validity of Eq. (14) is that the probability for each fluorophore to be excited during a single pulse has to be less than unity. The reason for this lies in the observation that during the pulse (of 10^{-13} s duration and with a typical excited-state lifetime in the 10^{-9} s range), the molecule has insufficient time to relax to the ground state. This can be considered a prerequisite for absorption of another photon pair. Therefore, whenever n_a approaches unity, saturation effects start to occur. The use of Eq. (14) allows one to choose optical and laser parameters that maximize excitation efficiency without saturation. Figure 4 depicts (not to scale) the practical time-scale condition typically used.

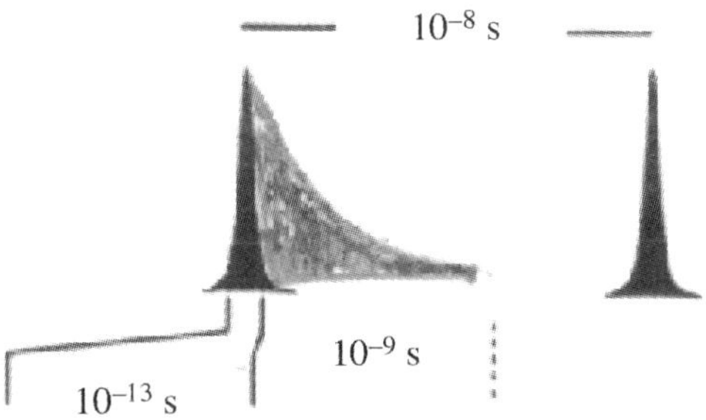

FIGURE 4 Relevant timescales for fluorescence, pulse width, and repetition rate (not to scale).

The optical parameter for enhancing the process is the numerical aperture of the lens, NA.

In order to have a realistic example, we can estimate n_a, accordingly to Eq. (14), for a common fluorescent molecule such as fluorescein, which possesses a two-photon cross-section of 38 GM at 780 nm (So et al., 2001). For our example, we use NA = 1.4, a repetition rate of 100 MHz, and a pulsewidth of 100 fs, and values of $P_{ave} = 1,\ 10,\ 20,$ and 50 mW. From Eq. (14), we obtain

$$n_a = 38 \times 10^{-58} \frac{P_{ave}^2}{100 \times 10^{-15}(100 \times 10^8)^2} \left[\frac{(1.4)^2}{2 \times 1.06 \times 10^{-34} \times 3 \times 10^8 \times 780 \times 10^{-9}} \right]^2$$
$$\approx 5.83 P_{ave}^2.$$

The results for $P_{ave} = 1,\ 10,\ 20,$ and 50 mW are 5.83×10^{-3}, 0.583, 1.16, and 2.9, respectively. It is evident that saturation begins to occur at 10 mW (Koester et al., 1999; So et al., 2001). The related rate of photon emission per molecule, at 10 mW of excitation, assuming no photobleaching (Patterson and Piston, 2000; So et al., 2001), is given by n_a multiplied by the repetition rate of the pulses, and is approximately 3×10^7 photons s^{-1}.

3.2 TPE Cross-Sections (see Chapter 4)

The quantum-mechanical selection rules for TPE differ from those for one-photon excitation (Loudon, 1983; Birge, 1979). Notwithstanding this, several common fluorescent molecules can be used. The main problem that arises here is that knowledge of one-photon cross-sections does not permit any quantitative prediction of the two-photon ones. This means that cross-sections for TPE or higher orders of excitation have to be measured. A "rule of thumb" that can help (but still does not allow optimal TPE conditions to be predicted) states that one may expect the TPE cross-section to peak at twice the wavelength with respect to one photon excitation conditions, especially in symmetrical molecules. However, the cross-section parameter has been measured for a wide range of dyes (Xu et al., 1995; Albota et al., 1998a); see Chapter 4. It is worth noting that, with the increasing use of TPE microscopy, new ad hoc organic molecules, engineered to have large two-photon absorption cross-sections have recently been developed (Albota et al., 1998b). Table 1 summarizes the properties of some common fluorescent molecules under one and two-photon excitation (Xu et al., 1995; So et al., 2000); see Chapter 4. Figure 5 shows a

TABLE 1 Properties of some common fluorescent molecules[a]

Fluorescent molecule	One-photon absorption peak λ (nm)	One-photon emission peak (nm)	Two-photon excitation peak (nm)	Two-photon action cross-section (GM)	Two-photon absorption cross-section (GM)
Rhodamine B	520	572	840		210
Fluorescein	498	518	782		38
Indo-1 free	345	475		4.5 at 700 nm	12
Indo-1 Ca	335	400		1.2 at 700 nm	2.1
Fura-2 free	362	512		11 at 700 nm	
Fura-2 Ca	335	505		12 at 700 nm	
Bis_MSB	340	420		6.0 at 691 nm	6.3
DiI C-18	500	570		95 at 700 nm	
Coumarin 307	395	488	776	19	
Cascade blue	376, 399	423	750	2.1	
Lucifer yellow	430	533	860	0.95	
BODIPY	480	520	920	17	
NADH	340	430		0.01 at 700 nm	
Dansylhydrazine	336	531		0.72 at 700 nm	
DAPI	345	455		0.16 at 700 nm	

[a] So et al. (2000) and Chapter 4.

sampling of measured (Xu et al., 1995) cross-sections for the wavelength ranges of the most popular mode-locked lasers (Gratton and van de Ven, 1995); see Chapter 10. In addition to extrinsic fluorescent molecules, TPE fluorescence from NAD(P)H, flavoproteins (Piston et al., 1995; So et al., 2000), and tryptophan and tyrosine in proteins (Lakowicz and Gryczynski, 1992) has been measured. Special mention should also be made of imaging of green fluorescent protein (GFP), which is an important molecular marker for gene expression (Chalfie et al., 1994; Chalfie and Kain, 1998; Potter, 1996). Cross-sections are around 6 GM (at 800 nm) and 7 GM (at 960 nm) in the case of wild-type and S65T GFP respectively. For comparison, one should consider that the cross-section for NADH, at the excitation maximum of 700 nm, is approximately 0.02 GM (So et al., 2000).

3.3 Optical Consequences (see Chapters 5 and 6)

For microscopy, the two-photon effect has the important consequence of limiting the excitation region to within a sub-femtoliter volume. The 3D confinement of the two-photon excitation volume can be understood on the basis of optical diffraction theory (Born and Wolf, 1999). Using excitation light with wavelength λ, the intensity distribution at the focal region of an objective with numerical aperture $\mathrm{NA} = \sin\alpha$ is described, in the paraxial regime, by (Sheppard and Gu, 1990; Born and Wolf, 1999)

$$I(u, v) = \left| 2 \int_0^1 J_0(v\rho) e^{-iu\rho^2/2} \rho \, d\rho \right|^2, \tag{15}$$

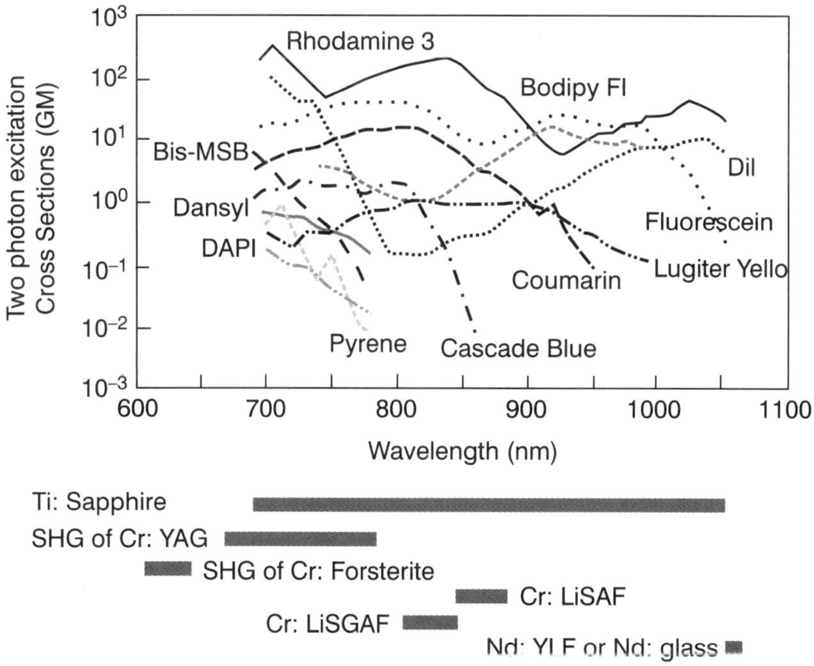

FIGURE 5 Sample of measured cross-sections for common fluorescent molecules compared with the tuning range of the most commonly utilized laser sources. [After Xu et al. (1995).]

where J_0 is the zeroth-order Bessel function, ρ is a radial coordinate in the pupil plane, and

$$u = \frac{8\pi z \sin^2 \frac{1}{2}\alpha}{\lambda}, \qquad v = \frac{2\pi r \sin \alpha}{\lambda}$$

are dimensionless axial and radial coordinates, respectively, normalized to the wavelength (Wilson and Sheppard, 1984). Now, the intensity of fluorescence distribution within the focal region has a $I(u, v)$ behavior for the one-photon case and $I^2(\frac{1}{2}u, \frac{1}{2}v)$ for the TPE case, as shown earlier. The arguments of $I^2(\frac{1}{2}u, \frac{1}{2}v)$ take into account the fact that in the latter case one utilizes wavelengths that are approximately twice those used for one photon excitation. These distributions are called the point spread functions (PSFs) of the microscope (Born and Wolf, 1999); see Chapters 5, 11 (Castleman, 2002), and 12 (Boccacci and Bertero). In comparison with the one-photon PSF, the TPE PSF is axially confined (Nakamura, 1993; Gu and Sheppard, 1995); see Chapter 5. In fact, considering the integral over v, keeping u constant, its behavior is constant along z for the one-photon case and has a half-bell shape for TPE. This behavior, discussed further in Chapters 2, 5, and 6, explains the discrimination property in TPE.

In general, two-photon microscopy has a radial resolution comparable to that of one-photon conventional microscopes thanks to a better signal-to-noise ratio and an effective and narrow depth of focus that makes it well suited for 3D optical sectioning. Figure 6 shows the behavior of PSFs for the one- and two-photon excitation cases, along with the overall behavior of fluorescence (Cannell and Soeller, 1997). Table 2 gives the values of the 3D intensity PSF in the transversal and axial directions (Gu and Sheppard, 1995).

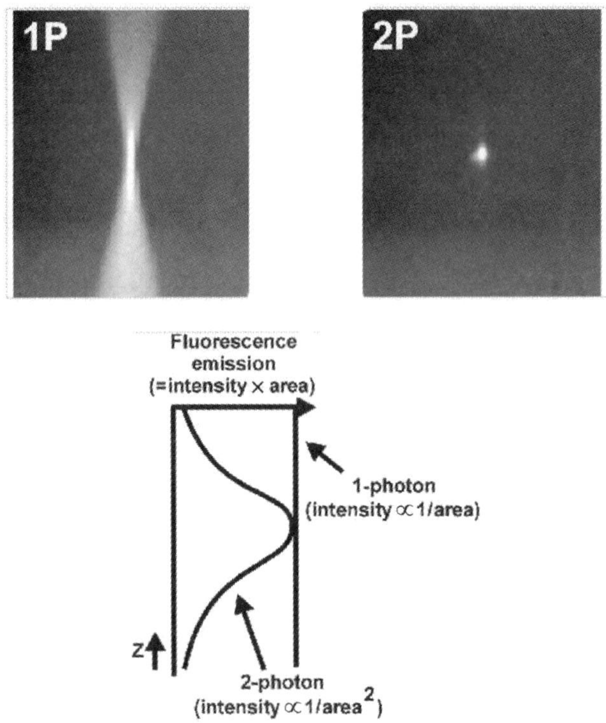

FIGURE 6 One-photon versus two-photon fluorescence emission from a solution of fluorescein. The behavior of the fluorescence intensity is strongly related to that of the PSF. The region of highest illumination intensity occurs at the waist of the excitation cone (Bianco and Diaspro, 1989; Born and Wolf, 1999). In the case of no saturation of fluorescent molecules, the local intensity dependence on the area is also sketched. [After Soeller and Cannell (1999).]

Comparison of the 3D intensity PSFs for confocal one-photon and two-photon imaging reveals that the resolution is almost the same in the two cases; see Chapters 5 and 6.

Now, the most interesting aspect, also predicted by Eq. (14), is that the excitation power falls off as the square of the distance from the lens focal point, within the approximation of a conical illumination geometry. In practice, this means that the quadratic relationship between the excitation power and the fluorescence intensity results in the TPE falling off as the fourth power of the distance from the focal point of the objective. This implies that those regions away from the focal volume of the objective lens, directly related to the numerical aperture of the objective itself, do not suffer photobleaching or phototoxicity effects and do not contribute to the signal detected when a TPE scheme is used. Because

TABLE 2 Values of the half-width of the 3D intensity point spread function in the transverse and axial directions, $v_{1/2}$ and $u_{1/2}$, respectively

	Conventional 1P	Confocal 1P	Conventional 2P	Confocal 2P
$v_{1/2}$	1.62	1.17	2.34	1.34
$u_{1/2}$	5.56	4.01	8.02	4.62

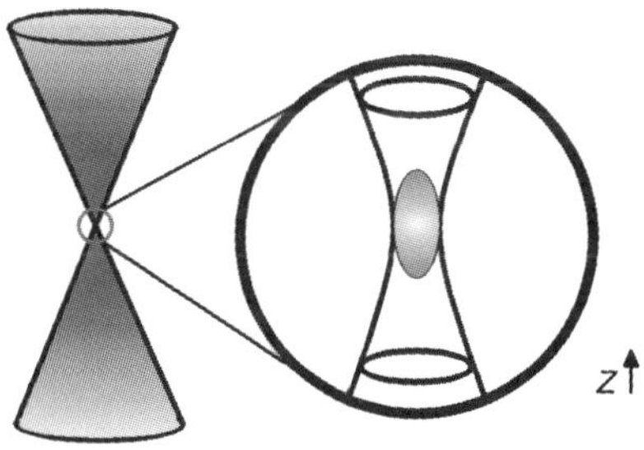

FIGURE 7 Sketch illustrating fluorescence confinement due to two-photon excitation [After Cannell and Soeller (1997).]

they are simply not involved in the excitation process, a confocal-like effect is obtained without the necessity for a confocal pinhole. Figure 7 illustrates the situation. Figure 8 shows the spatial extension of the fluorescence emission from a solution containing fluorescent molecules and subjected to one- and two-photon excitation regimes. Consequently, photodamage and photobleaching effects are extremely localized, as shown in Figure 9. In TPE, over 80% of the total intensity of fluorescence comes from a 700–1000 nm-thick region about the focal point for objectives with numerical apertures in the range from 1.2 to 1.4 (Brakenhoff et al., 1979; Wilson and Sheppard, 1984); see Chapters 2, 5, and 6. This also implies a reduction in background that allows compensation of the reduction in spatial resolution due to the wavelength. The use of infrared wavelengths

FIGURE 8 See ftp site for color figure. Spatial extension of the fluorescence emission from a solution containing fluorescent molecules under one- and two-photon excitation. In the case of TEP, the excitation phenomenon, small spots in the lower part of the cuvette, takes place only within a diffraction-limited volume. This volume can be roughly quantified using the resolution parameters of the system. In the case of conventional or one-photon excitation, the double-cone geometry (the upper bright area in the cuvette) is evident. The main implications of this behavior are an inherent confocal effect and consequent optical sectioning property, and photobleaching and phototoxicity confined to a sub-femtoliter volume. (Picture courtesy of John Girkin from Biorad web homepage.)

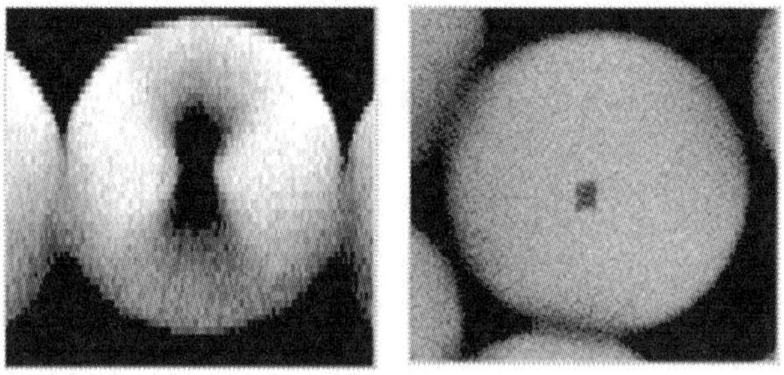

FIGURE 9 Three-dimensional projections (*y–z* plane cut) of a large fluorescent sphere 22 μm in diameter, where photobleaching has been induced in a central *x–y* section in single-photon confocal (*a*) and TPE (*b*) mode using 488 nm and 720 nm excitation wavelengths, respectively. In both cases, a 3 μs dwell time has been used for bleaching and for imaging and a 0.5 μm axial step has been performed, producing 128 optical slices. Diaspro (2001).

instead of UV or visible ones also allows deeper penetration than in the conventional case (So et al., 2001); see Chapters 9 (König and Tirlapur, 2002) and 13. In fact, Rayleigh scattering produced by small particles is proportional to the inverse fourth power of the wavelength. Thus, the longer wavelengths used in TPE, or in general in multiphoton excitation, will be scattered less than the ultraviolet–visible wavelengths used for conventional excitation. It is worth noting that when considering deep imaging in thick samples, optical aberrations should be taken into account (de Grauw et al., 2002; Centonze and White, 1998); see Chapter 7 (de Grauw et al., 2002). With TPE, deeper targets within a thick sample can be reached. Of course, for the fluorescence light, on the way back, scattering can be overcome by acquiring the emitted fluorescence using a large-area detector and collecting not only ballistic photons (Soeller and Cannel, 1999; Buehler et al., 1999); see Chapter 10.

4 TPE MICROSCOPE

4.1 General Considerations

It is evident from the previous sections that the main components of a TPE microscope are the following: a beam scanning system, a high-numerical-aperture objective, and a high-peak-power laser at moderate average power (femtosecond- or picosecond-pulsed at relatively high repetition rate) emitting infrared or near-infrared wavelengths (Denk et al., 1995; König et al., 1996; So et al., 1996; Soeller and Cannell, 1996, 1999; Wokosin and White, 1997; Centonze and White, 1998; Potter et al., 1996; Wolleschensky et al., 1998; Diaspro et al., 1999a; Tan et al., 1999; Mainen et al., 1999; Majewska et al., 2000; Wier et al., 2000; Diaspro, 2001); see Chapter 10. Moreover, two other important and critical issues are those related to a high-sensitivity detection system and high-throughput microscope pathway. In a typical TPE microscope, images are built by raster-scanning the *x–y* mirrors of a galvanometrically driven scanner. The scanning speed is mainly determined by the mechanical properties of the device, and the time needed for single-line scanning is of the order of milliseconds. In TPE set-ups, particular attention should be given to the surfaces of the mirrors and to the way they are mounted on the scanners in order to get the best reflection efficiency and scanning stability. Enhanced silver coating of the mirrors is most commonly used to optimize reflectivity of infrared excitation wavelengths (Wokosin and White, 1997). The excitation light should reach the microscope objective, passing through the minimum number of optical components and possibly along the shortest path. Typically, high-numerical-aperture objectives, with high infrared transmission, are used to maximize the TPE efficiency. As also described in Chapter 10, the signal detection efficiency can be further enhanced by using an additional reflector in the condenser assembly. While the *x–y* scanners provide lateral focal-point scanning, axial scanning can be achieved by means of different positioning devices, the most popular being a belt-driven system using a DC motor and a single-objective piezo nano-positioner. Acquisition and visualization is generally completely computer-controlled by dedicated software. It is usually possible to switch simply between the one-photon and two-photon modes.

There are two popular approaches to TPE microscopy: one called the descanned mode, uses the same optical pathway and mechanism employed in confocal laser scanning microscopy, while the second, called the non-descanned mode, optimizes the optical pathway, minimizing the number of optical elements encountered and increasing the detector area. In the first approach, pinholes are removed or set to their maximum aperture and the emission signal is captured using the galvanometric scanning mirrors. This is why it is called the descanned mode. In the second approach, the confocal architecture is modified in order to increase the collection efficiency: pinholes are removed and the emitted radiation is collected using dichroic mirrors on the emission path or external detectors without passing through the galvanometric scanning mirrors. Figure 10 illustrates these two approaches, also showing the conventional confocal scheme with a pinhole along with the descanned pathway.

A high-sensitivity detection system is another critical component (Wokosin et al., 1998). The fluorescence emitted is collected by the objective and transferred to the detection system through a dichroic mirror along the emission path. Typically, because of the high excitation intensity, an additional barrier filter is needed to avoid mixing of the excitation and emission light at the detection system, which is placed differently depending on the acquisition scheme being used. Photodetectors that can be used include photo-multiplier tubes, avalanche photodiodes, and charge-coupled-device (CCD) cameras (Denk et al., 1995); see Chapter 10. The first of these are the most commonly used. This is due to their low cost, good sensitivity in the blue–green spectral region, high dynamic range, large size of the sensitive area, and single-photon counting-mode availability (Hamamatsu Photonics, 1999). Avalanche photodiodes are excellent in terms of sensitivity. Unfortunately they are expensive and the small active photosensitive area (< 1 mm) could introduce drawbacks in the detection scheme. CCD cameras are used in video-rate multifocal imaging; see Chapters 10 and 23 (Fujita and Takamatsu, 2002).

In order to obtain better spatial resolution, it is also possible to retain the confocal pinhole (Cannell and Soeller, 1997, Gauderon et al., 1999; Periasamy et al., 1999); see Chapters 6 and 10. Unfortunately, in some practical experimental situations, the low efficiency of the TPE fluorescence process may rule out such a solution. However when pinhole insertion is possible, the major advantage is that the axial resolution can be improved by approximately 40%. In Chapter 6, Török and Sheppard discuss the dependence of the point spread function on pinhole size, both theoretically and experimentally. Figure 11 shows a typical architecture based on a commercial scanning head and

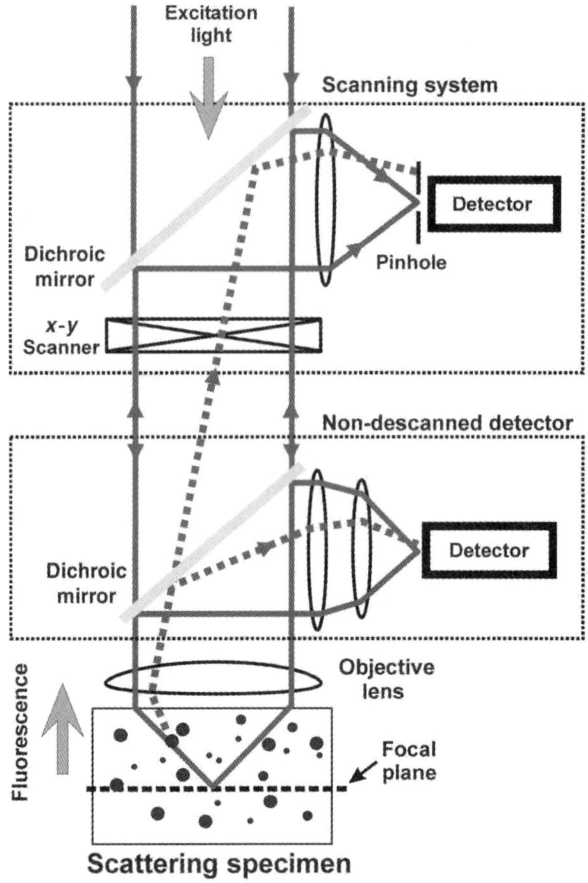

FIGURE 10 Principles of scanned and non-descanned detection. See the text for details. [After Soeller and Cannell (1999).]

operating in descanned mode. In Figure 12, a comparison between a conventional and TPE image is shown, demonstrating the multifluorescence excitation capability of TPE: by illuminating a specimen with only a single wavelength, it is possible to excite three or more fluorescent molecules simultaneously. The effect of the confocal pinhole is demonstrated in Figure 13. It can be seen that the resolution, particularly in the axial direction, is improved by using a confocal pinhole. Because the chromosomes are dispersed in 3D, the specimen is particularly suited for 3D imaging. Table 3 summarizes the characteristics of one-photon confocal imaging compared with TPE descanned imaging.

4.2 Laser Excitation Beam (see Gratton and van de Ven, 1995, and Chapters 8 and 10)

Following the discussion about the relationship (14) and referring elsewhere for details (Wokosin et al., 1996; Gratton and van de Ven, 1995; Chapter 10), the parameters that are most relevant in the selection of the laser source are average power, pulsewidth and repetition rate, and wavelength. Simplicity in use and in maintenance should also be

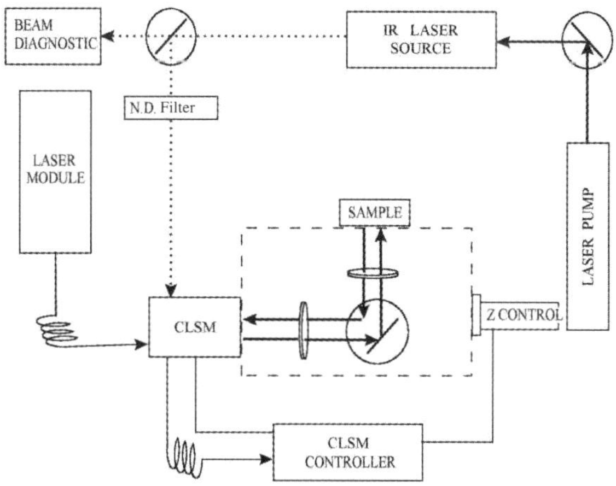

FIGURE 11 Typical components of a TPE architecture based on a confocal laser scanning head. See the text for details.

considered. Unfortunately, cost is still a restrictive parameter. The features of an infrared pulsed laser are high average power, an 80 MHz high repetition rate, and a 100 fs short pulsewidth. At present, the use of short pulses and small duty cycles is mandatory to allow image acquisition in a reasonable time with power levels that are biologically tolerable (Denk et al., 1994; Denk, 1996; Koester et al., 1999; König et al., 1996, 1998; König, 2000); see Chapter 9.

The laser beam can be coupled directly to the scanning head or an optical fiber can be used to launch the excitation within the scanning head. Direct coupling allows the maximum flexibility, together with the minimum level of "extra" problems at the excitation stage. In fact, even if the fiber coupling solution seems better in terms of practicality and safety at first glance, some practical problems can arise related to the fiber coupling solution itself, especially when managing a multi-user facility. For example, owing to high-order nonlinearities in the fiber, there is a limit to how much power can be put through the fiber (60–100 mW). Moreover, because the fiber coupler needs some devices for maintaining pulsewidth and beam shape, it is not totally "turnkey," demanding further control and alignment procedures, especially when wavelength changes are required.

Pulsewidth is another very delicate issue. In fact, because it is not very easy to measure it at the focal volume within the sample, little can be definitely said on it (Hanninen and Hell, 1996; Guild et al., 1997); see Chapter 8 (Wolleschensky et al., 2002). However, pulsewidth is especially relevant for two principal reasons: the average power delivered to the sample at proper peak power necessary to prime the fluorescence process; and the two- (or n-) photon efficiency related to pulse width as the $(n-1)$th inverse power.

Although users do not perform measurement of the pulsewidth at the sample when they use two-photon microscopy, which would require a specific procedure that, even if not too complex for a researcher in the field, could be irksome for the majority of users, it is a reasonable approximation to assume that at the focal volume a 1.5–2 times temporal pulse broadening occurs using high-quality optics; see Chapters 8 and 10. As an example, for a measured laser pulsewidth of about 100 fs, an estimate at the sample is about 150–180 fs

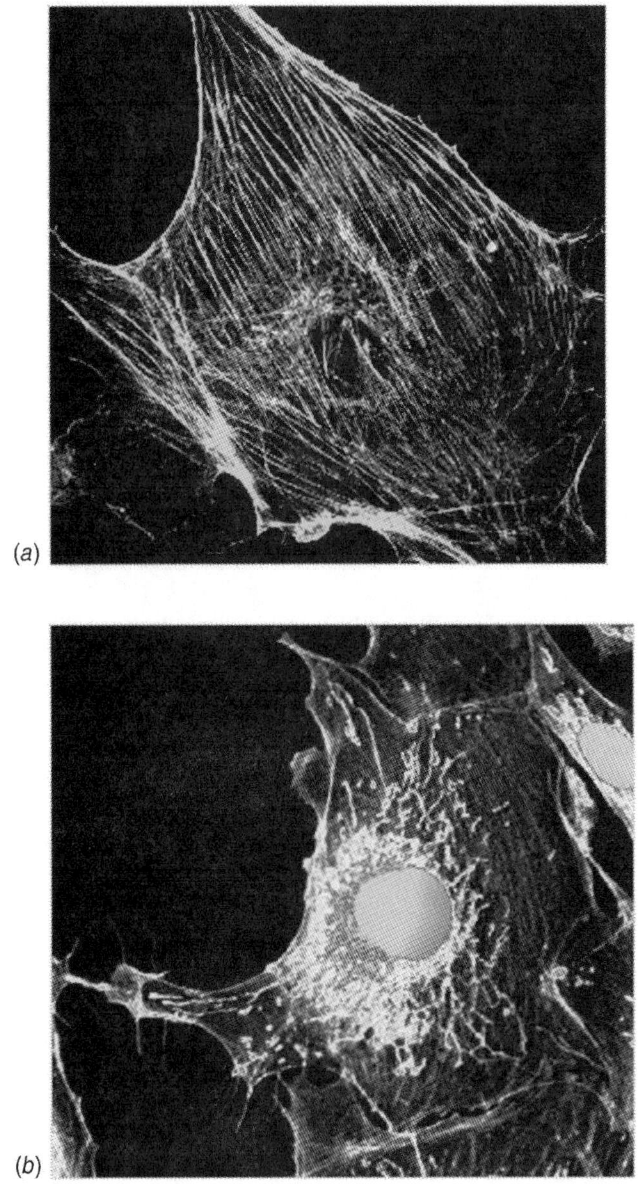

FIGURE 12 See color insert. Bovine pulmonary artery endothelial cells (F-147780, Molecular Probes, Oregon) marked with Mito Tracker Red, Bodipy FL phallacidin, and DAPI specific for mitochondria, F-actin, and DNA respectively. Confocal imaging (*a*) has been performed at 488 and 543 nm, TPE imaging (*b*) was performed at 720 nm. In the confocal image, DAPI could not be excited owing to the lack of an appropriate UV source.

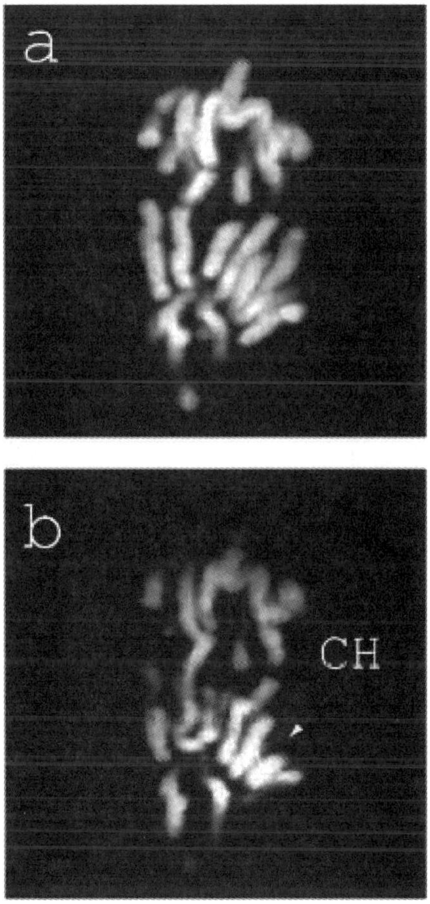

FIGURE 13 Optical sections of two groups of DAPI-stained onion root chromosomes in a three-dimensional volume imaged by two-photon excited fluorescence. (*a*) Confocal pinhole almost fully open. (*b*) optimized confocal pinhole size. CH: chromosomes in the focal plane. [After Gauderon et al. (1999).]

under favorable experimental conditions, including sample characteristics. Sample properties are mentioned because for thick samples the role played by thickness in pulsewidth broadening is not so obvious (Centonze et al., 1998; Gu et al., 1999; So et al., 2001); see Chapter 7.

Table 4 gives data for the most popular laser systems available on the market.

4.3 A Practical Example

Although commercial systems for TPE microscopy are available, a TPE microscope can also be constructed from separately supplied components (So et al., 1996; Cannell and Soeller, 1997; Wier et al., 2000) or by adapting an existing confocal microscope (Denk

TABLE 3 Characteristics of TPE descanned imaging and one-photon confocal imaging

	TPE	Confocal
Excitation source	Laser, IR, fs–ps pulsed, 80–100 MHz repetition rate, tunable 680–1050 nm	Laser visible–UV CW (365, 488, 514, 543, 568, 633, 647 nm)
Excitation/emission separation	Wide	Close
Detectors	PMT (typical), APD, CCD	PMT (typical), APD, CCD
Volume selectivity	Intrinsic (fraction of femtoliter)	Pinhole required
Image formation	Beam scanning, rotating-disc–multilens, sample scanning	Beam scanning, rotating-disc–multilens, sample scanning
Deep imaging	> 400 μm (problems related to pulse-shape modifications and scattering)	Yes, but limited compared with TPE (problems related to shorter-wavelength scattering)
Spatial resolution	Less than that of confocal, because of the focusing of IR radiation, compensated by the higher signal-to-noise ratio; better resolution if using pinhole at the cost of a reduction in collected intensity	Diffraction-limited, depending on pinhole size
Real time imaging	Possible	Possible
Signal to noise ratio	High (especially in non-descanned mode)	Good
Fluorophores	All those available for conventional excitation plus some designed specifically for TPE	Selected fluorophores depending on laser lines in use
Photobleaching	Only in the focus volume, defined through resolution parameters	Within the whole double cone of excitation defined by the lens characteristics
Contrast mechanisms	Fluorescence, second- and third-harmonic generation, lifetime	Fluorescence, reflection, transmission
Critical costs	Laser care	—

TABLE 4 Commercially available laser systems

	Tuning range	Wavelength (nm)	Pulsewidth	Average power (mW) at 5 W pump	Size
Spectra Physics, Tsunami	Wide	680–1050	25 fs, < 100–130 fs, up to 100 ps	100–800	Large
Coherent, Mira	Wide	680–1000	< 100 fs, 5–10 ps	100–700	Large
Spectra Physics, Mai Tai	Limited	750–850 780–920	< 100 fs, 25 fs	750	Compact
Coherent, Vitesse	Limited	700–760 760–860	100 fs	< 1000	Compact

et al., 1990; Potter et al., 1996; Mainen et al., 1999; Diaspro et al., 1999b; Majewska et al., 2000); see Chapter 10.

The basic designs are similar and the critical components have been outlined above. There are several advantages in modifying an existing confocal system into a TPE microscope. Important and practical advantages are flexibility in choosing components, tailoring the system to particular requirements, understanding and controlling its operation, and price. Of course, there are some disadvantages, such as the scattered customer support due to different companies being involved in different sections of the design and some difficulties in building or adapting the microscope. Table 5 summarizes some characteristics of commercially available systems, and in the appendix of this book there are short notes from the main companies involved in the confocal and TPE microscopy market. Here, as an example, we report some steps related to the conversion of a compact commercial scanning head into a TPE microscope (Diaspro et al., 1999b; Diaspro, 2001). A panoramic view of the experimental set-up is given in Figure 14 and a schematic drawing is shown in Figure 15. The core of the architecture is a mode-locked Ti : sapphire infrared pulsed laser (Tsunami 3960, Spectra Physics Inc., Mountain View, California), pumped by a high-power (5 W at 532 nm) solid state laser (Millennia V, Spectra Physics Inc.). The Ti : sapphire laser output can be tuned across a range from 680 nm to 1100 nm, allowing excitation of a variety of fluorescent molecules. A dichroic mirror, optimized for high-power ultrashort infrared pulses (CVI, U.S.A.), is used to bring the Tsunami beam to the scanning head, and a neutral-density rotating wheel (Melles Griot, U.S.A.) allows control of the output power into the scanning head. For an average power of 40 mW at the entrance of the scanning head, the average power before the microscope objective is measured at about 15–16 mW and that at the sample is estimated as between 11 and 12 mW (Diaspro et al., 2001). For a measured laser pulsewidth of about 80 fs at the laser output window, the estimate at the sample is about 150 fs in the most favorable situation. The laser beam can be aligned using the conventional laser source of the scanning head by marking some reference positions inside the head itself. The beam overfills the back aperture of the objective, ensuring that the full aperture of the objective is used and the light is focused into a diffraction-limited spot (So et al., 1996); see Chapter 10.

The scanning and acquisition system for microscopic imaging is based on a commercial single-pinhole scanning head (Nikon PCM2000; Nikon Instruments, Florence) mounted on the lateral port of a common inverted microscope (Nikon Eclipse TE300). The Nikon PCM2000 has a simple and compact light path that makes it very appropriate for conversion to a two-photon microscope, as can be seen in Figure 16 (Diaspro et al., 1999b; Diaspro, 2001). The scanning head operates in the "open pinhole" condition when operating as a TPE microscope.

A dichroic mirror (first) has been substituted in the original scanning head to allow excitation from 680 nm to 1050 nm (Chroma Inc., Brattleboro, Vermont). The substituted dichroic mirror reflects very efficiently (> 95%) from 680 nm to 1050 nm. The 50% cutoff is at around 640 nm. The mirror has its maximum transmission (> 90%) from 410 nm to 620 nm. The galvanometer mirrors are silver coated on fused silica, and have a high damage threshold. The minimum pixel dwell time is 3 μs, related to the mechanical response of the scanners. A series of custom made emission filters, which block infrared radiation (> 650 nm) to an optical density of 6–7 within 50 mW of beam power incident on the filters themselves, have been utilized: E650P, HQ 460/50, HQ 535/50, HQ 485/30 and HQ 405/30 (Chroma Inc.) (Stanley, 2001). The E650P filter has been initially tested to check its blocking performances with respect to the infrared reflections coming from stray

TABLE 5 Commercially available TPE microscopy systems

	Company	Size	Pulsewidth regime	Wavelength range (nm)	Average power (mW)	Laser coupling	Acquisition	Other features
LSM 510 NLO	Zeiss	Compact/normal	fs	700–900	50	Direct-box/fiber	Descanned	Simultaneous, confocal
MRC 1024 MP	BIORAD	Normal/large	fs	690–1000	Not reported	Direct-box	Descanned/non-descanned	None relevant
Radiance 2000 MP	BIORAD	Compact/normal	fs	690–1000	Not reported	Direct-box	Descanned/non-descanned	Faster scanning (> 750 Hz)
RTS2000 MP	BIORAD	Large	fs	690–1000	Not reported	Direct-box	Descanned/non-descanned (?)	130 frames s^{-1} video rate
TCS SP2	Leica	Normal/large	ps	720–900	Not reported (120 maximum at the sample)	Fiber	Descanned/non-descanned (in preparation)	Spectral capability

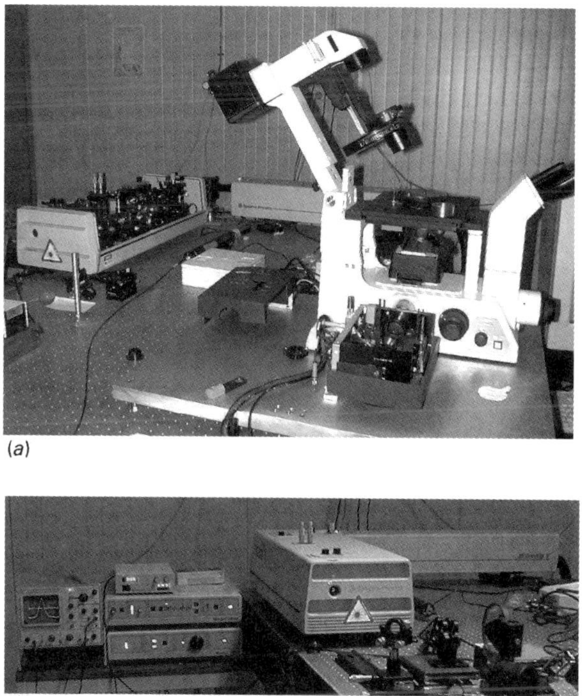

FIGURE 14 (*a*) The TPE microscope operating in the Laboratory of Physics of Biosystems (LFB, Department of Physics, University of Genoa, Italy), within the strategic framework of a national project of the National Institute of the Physics of Matter (INFM, Istituto Nazionale Fisica della Materia) (Diaspro et al., 1999b). This microscope, constructed in 1999, is part of a multipurpose architecture, also including lifetime imaging and fluorescence correlation spectroscopy modules. On the left the open Tsunami cavity is visible. The microscope and the PCM2000 scanning head mounted on its lateral port are visible in the center, while on the right is the video unit. (*b*) The Tsunami laser output and its electronics, along with the oscilloscope monitoring the pulsing status and the wavelength spectrum. Beam diagnostics is performed using an ultrafast laser spectrum analyzer RE201 (Ist-Rees, U.K.) and a thermopile detector power meter AN2/10A-P (Ophir, Israel). Diaspro (2001).

rays within the scanning head or from the sample, and constitutes the base for the other HQ filters.

One-photon and two-photon modes can be simply obtained, and sample positioning is retained as shown in Figure 17. More specifically and very simply, the single-mode optical fiber used for one-photon excitation-beam delivery is unplugged and the optical path in air is available for direct coupling of the infrared laser beam (Diaspro, 2001). At the emission stage, a high-throughput optical fiber delivers the emitted fluorescence from the scanning head to the PCM2000 control unit, where photomultiplier tubes (R928, Hamamatsu, Japan) are physically plugged in. This solution was particularly useful in the first stages of the adaptation, because it allowed operation with the scanning box open. Axial scanning for confocal and TPE three-dimensional imaging is actuated by means of two different positioning devices, depending on the experimental circumstances and axial accuracy needed: a belt-driven system using a DC motor (RFZ-A, Nikon, Japan) and a single-

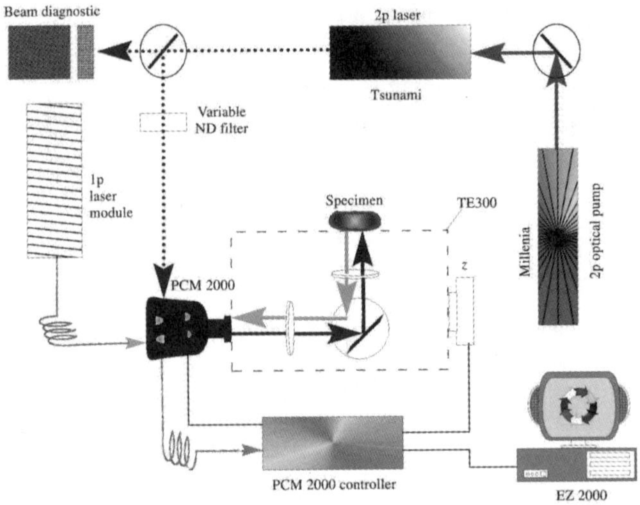

FIGURE 15 Schematic drawing of the system shown in Figure 14*a* and described in the text as example.

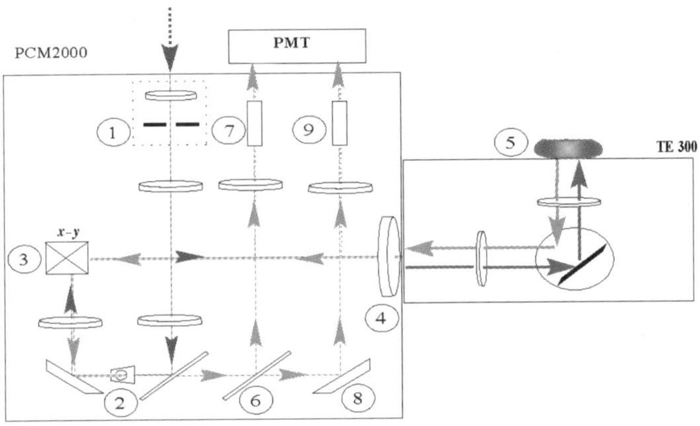

FIGURE 16 Optical scheme of the Nikon PCM2000 confocal scanning head. The excitation beam enters into the PCM2000 scanning head through an optical coupler (1) in order to reach the sample on the *x–y–z* stage (5). The beam passes through the pinhole holder (2) kept in open position, the galvanometric mirrors (3), and the scanning lens (4). Fluorescence generated from the sample (5) is delivered to the photomultiplier tube (PMT) through acquisition channels directed by two selectable mirrors (6,8) via optical fiber (7,9) coupling. One-photon and two-photon modes can be simply obtained by switching from the single-mode optical fiber (one-photon), coupled to a module containing conventional laser sources (Ar-ion, He–Ne green), to the TPOC, allowing delivery of the Tsunami laser beam (two-photon). Axial scanning for confocal and TPE three-dimensional imaging is actuated by means of two different positioning devices, depending on the experimental circumstances and axial accuracy needed: a belt-driven system using a DC motor (RFZ-A, Nikon, Japan) and a single-objective piezo nano-positioner (PIFOC P-721-17, Physik Instrumente, Germany). The piezoelectric axial positioner allows an axial resolution of 10 nm within a motion range of 1000 nm at 100 nm steps, and utilizes a linear variable differential transformer (LVDT) integrated feedback sensor. Acquisition and visualization are completely computer controlled by a dedicated software, EZ2000 (Coord, The Netherlands; http://www.coord.nl).

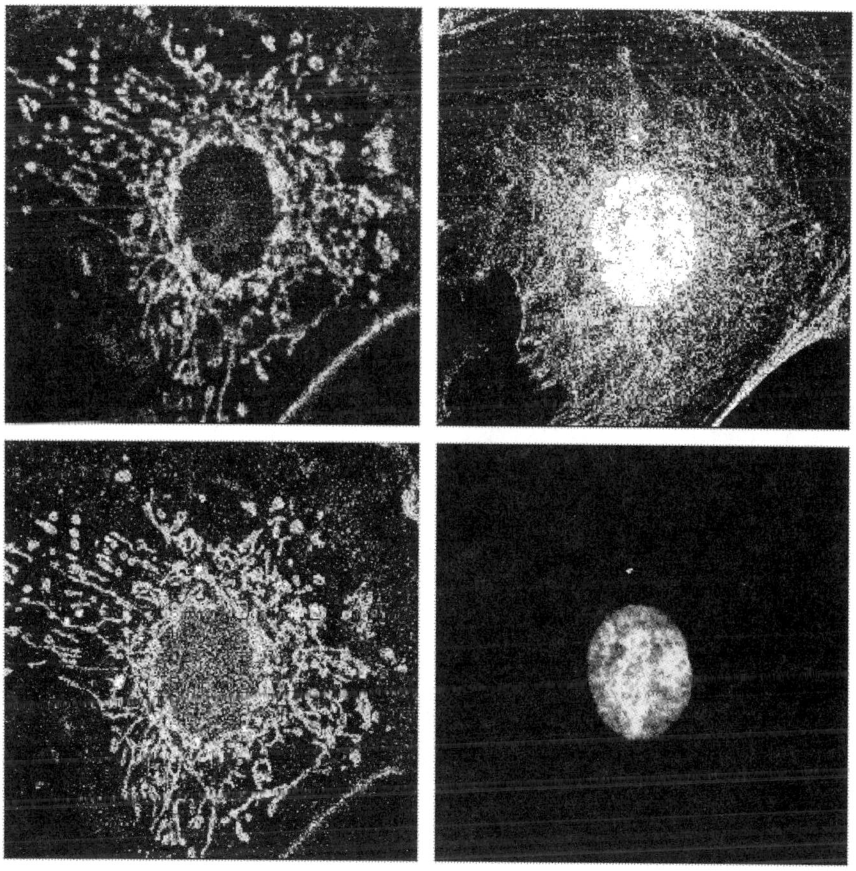

FIGURE 17 See color insert. In the first row, a confocal (left) and a TPE (right) image are shown, demonstrating the maintenance of positioning when switching from one mode to another. The second row displays the signal detected after insertion of specific emission filters for red–green (left) and blue (right). TPE allows triple fluorescence imaging using one excitation wavelength. Bovine pulmonary artery endothelial cells (Molecular Probes, G-18015) have been loaded with DAPI (DNA), mitotracker red (mitochondria) and bodipy (actin).

objective piezo nano-positioner (PIFOC P-721-17, Physik Instrumente, Germany). Acquisition and visualization are completely computer-controlled and came with the commercial confocal scanning head.

Point spread function (PSF) measurements were performed for a planachromatic Nikon 100×, 1.4 NA immersion oil objective with enhanced transmission in the infrared region (Diaspro et al., 1999a and 1999b). Blue fluorescent carboxylate modified microspheres 0.1 μm diameter (F-8797, Molecular Probes, Oregon) were used. A drop of water-diluted microspheres was spread between two coverslips of nominal thickness 0.17 mm. It is worth noting that these microspheres constitute a good compromise between the use of subresolution point scatterers and acceptable fluorescence emission. Axial scanning was in steps of 100 nm. The x–y scan step was 35 nm. The measured full width at half maximum (FWHM) lateral and axial resolutions were 210 ± 40 nm and 700 ± 50 nm, respectively. Intensity profiles along with the x y–z directions of experimental data and

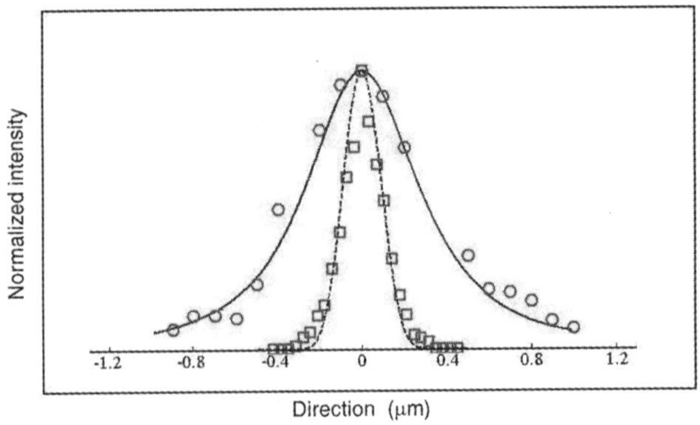

FIGURE 18 The radial (□) and axial (○) intensity profiles of the TPE point spread function (Diaspro et al., 1999b).

theoretical expectations are shown in Figure 18. In order to be sure of operating in TPE regime, the quadratic behavior of the fluorescence intensity with respect to excitation power has to be checked; see also Figure 19. Moreover, during any fluorescence acquisition, a simple and effective test for TPE conditions can be performed by delivering continuous radiation instead of pulsed at the same average power. This can be accomplished by interrupting the pumping at the Ti : sapphire laser for a while and switching off the pulse control. When the pump is activated, if there are not too many vibrations, it is possible to get a continuous beam that is not appropriate for TPE in terms of incident power on the sample. In this case, there is no fluorescence signal to be imaged. On restoring the pulse at any moment during the scanning, fluorescence becomes visible,

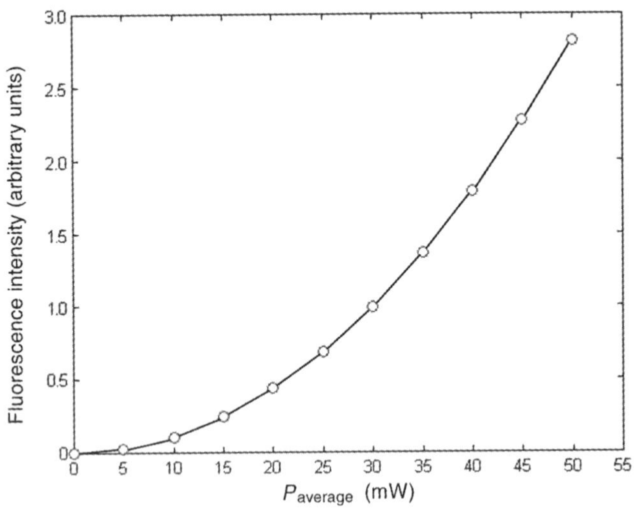

FIGURE 19 The quadratic dependence on the excitation intensity, which forms the basis for the application of TPE for 3D microscopy, has been checked for a solution of fluorescein (Merck, Germany) in water.

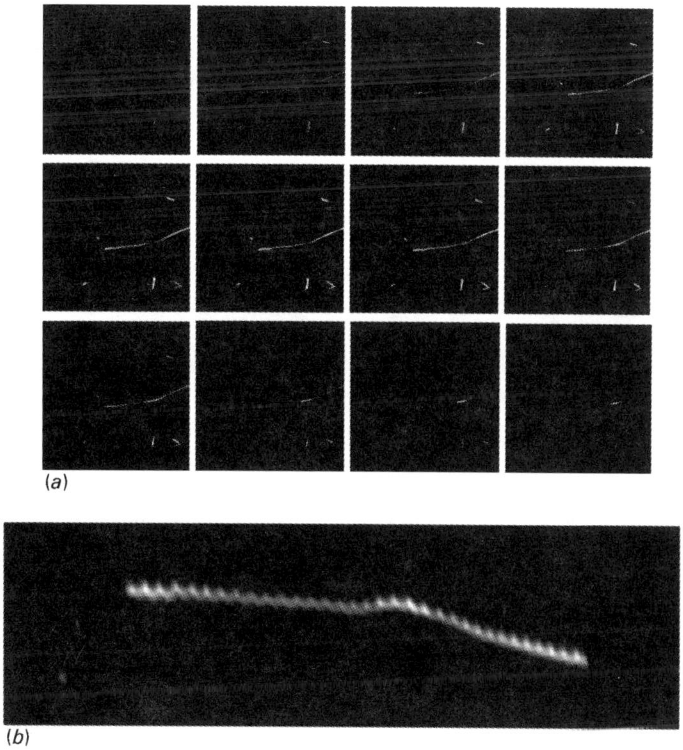

FIGURE 20 The 3D assembly of the mature sperm head of the octopus *Eledone cirrhosa*, loaded with DAPI, from 12 optical sections (*a*), visualized in a 3D fashion (*b*). [After Diaspro et al. (1999b).]

confirming the TPE imaging condition. The three-dimensional imaging capabilities are shown in Figure 20.

5 CONCLUSIONS

The development of TPE microscopy has allowed intrinsic three-dimensional resolution, the elimination of background fluorescence, and the attractive possibility of exciting UV-excitable fluorescent molecules and thus increasing sample penetration. In fact, TPE implies that, unlike conventional or one-photon excitation, fluorescence can be achieved at a wavelength shorter than the excitation wavelength: two 700 nm photons combine to produce fluorescence at, say, 450 nm. It is the exploitation of this phenomenon that eliminates the necessity for UV radiation to excite fluorescence. The elimination of UV excitation improves specimen viability, too (König et al., 1997, 1999). Overall photobleaching is considerably reduced, since the infrared pulsed laser light excitation occurs only at the focal plane, and multiple excitation of fluorescent molecules can be more easily and efficiently accomplished than in conventional excitation. The excitation of the fluorescent molecules bound to a biological system under study takes place mainly in an excitation volume of the order or smaller than 1 femtoliter. This means that 3D optical sectioning is achieved without the need for a confocal scheme. This allows the use of different schemes for fluorescence acquisition that do not require descanning of the signal; this is very useful, especially when the fluorescence intensity is particularly low (Denk et al., 1995; So et al., 1996); see Chapter 10. At the same time, scattering effects at the excitation wavelengths are greatly reduced, allowing deeper penetration into specimens than normally occurs in the case of visible or ultraviolet excitation. This implies that when seeking to obtain 3D information, more imaging scans are permitted, thus allowing 4D (3D–*t*) studies on samples kept under physiological conditions. Two-photon excitation microscopy has become an essential imaging strategy for imaging of thick samples and live cells.

From a fluorescence acquisition perspective, ballistic and scattered emitted photons can be allowed to contribute to the total acquired signal: they have been produced only in a very specific volume. What is invaluable for imaging of cells – in particular of live cells – is the fact that the weak endogenous absorption and highly localized spatial confinement of the TPE process dramatically reduces phototoxicity stress. Some work has been directed at elucidating the possible induced photodamage and irreversible sample modifications due to the use of high laser intensities; see Chapters 9 and 10. To the best of our knowledge, the situation is still favorable when compared with the damage induced by means of conventional fluorescence excitation. Notwithstanding this, some care has to be taken, and some experimental parameters need to be carefully controlled, such as power, dwell time and pulsewidth (Diaspro, 2001); see Chapters 8 and 10.

In order to summarize some other advantages of TPE, we recall the following further aspects:

(i) Repeated scanning on the specimen in confocal microscopy, particularly with UV light, induces rapid photoisomerization and high background autofluorescence. TPE microscopy reduces these complications, providing better penetration at infrared wavelengths and thus prolonging cell viability during image acquisition.

(ii) The conventional excitation technique requires special UV optics for the UV excitation probe, while TPE uses conventional microscope optics.

(iii) In one-photon excitation, the emission wavelength is close to the excitation wavelength (in fact, about 50–200 nm longer). In TPE, the fluorescence emission occurs at a wavelength substantially shorter than the excitation wavelength.

Most of the outstanding applications of TPE microscopy are described in this book. However, it is also worth noting the usefulness of the TPE scheme for spectroscopic and lifetime studies (So et al., 1996; Sytsma et al., 1998), for optical data storage and microfabrication (Cumpston et al., 1999), and for detection of single molecules (So et al., 2000). Moreover, a very interesting application has arisen in the study of impurities affecting the growth of protein crystals (Caylor et al., 1999). Mention should also be made of imaging in the field of plant biology; see Chapter 21 (Tirlapur and König, 2002). Finally, there are the applications of TPE in active devices such as those used for nanosurgery (König, 2000) and photodynamic therapy (So et al., 2000).

At present, the main disadvantage of TPE fluorescence microscopy results from the cost of femtosecond laser sources. As has been pointed out by other researchers, once the technology becomes less expensive and simpler, there will not be a confocal microscope that is not also a two-photon microscope.

REFERENCES

Agard, D. A., Hiraoka, Y., Shaw, P. J., and Sedat, J. W. (1989). Fluorescence microscopy in three-dimensions. *Meth. Cell. Biol.* **30**, 353–378.

Albota, M. A., Xu, C., and Webb, W. W. (1998a). Two-photon fluorescence excitation cross sections of biomolecular probes from 690 to 960 nm. *Appl. Opt.* **37**, 7352–7356.

Albota, M. A., Beljonne, D., Brédas, J. L., Ehrlich, J. E., Fu, J.-Y., Heikal, A. A., Hess, S. E., Kogej, T., Levin, M. D., Marder, S. R., McCord-Maughon, D., Perry, J. W., Rockel, H., Rumi, M., Subramaniam, G., Webb, W., Wu, X.-L., and Xu, C. (1998b). Design of organic molecules with large two-photon absorption cross sections. *Science* **281**, 1653–1656.

Andrews, D. L. (1985). A simple statistical treatment of multiphoton absorption. *Am. J. Phys.* **53**, 1001–1002.

Beltrame, F., Bianco, B., Castellaro, G., and Diaspro, A. (1985). Fluorescence, absorption, phase-contrast, holographic and acoustical cytometries of living cells. In: *Interactions between Electromagnetic Fields and Cells* (ed. A. Chiabrera and H. P. Schwan). New York: Plenum Press, pp. 483–498.

Benedetti, P. (1998). From the histophotometer to the confocal microscope: the evolution of analytical microscopy. *Eur. J. Histochem.* **42**, 11–17.

Bertero, M. and Boccacci, P. (1998). *Introduction to Inverse Problems in Imaging*. Bristol: IOP Publishing.

Bianco, B. and Diaspro, A. (1989). Analysis of the three dimensional cell imaging obtained with optical microscopy techniques based on defocusing. *Cell Biophys.* **15**, 189–200.

Birge, R. R. (1979). A theoretical analysis of the two-photon properties of linear polyenes and the visual chromophores. *J. Chem. Phys.* **70**, 165–169.

Birge, R. R. (1986). Two-photon spectroscopy of protein-bound fluorophores. *Acc. Chem. Res.* **19**, 138–146.

Birks, J. B. (1970). *Photophysics of Aromatic Molecules*. London: Wiley-Interscience.

Boccacci, P. and Bertero, M. (2002). Image-restoration methods: basics and algorithms. In: *Confocal and Two-Photon Microscopy: Foundations, Applications and Advances* (ed. A. Diaspro). New York: Wiley-Liss.

Born, M. and Wolf, E. (1999). *Principles of Optics*, 7th (expanded) edn. Cambridge: Cambridge University Press.

Brakenhoff, G. J., Blom P., and Barends, P. (1979). Confocal scanning light microscopy with high aperture immersion lenses. *J. Microsc.* **117**, 219–232.

Brakenhoff, G. J., van Spronsen, E. A., van der Voort, H. T., and Nanninga, N. (1989). Three-dimensional confocal fluorescence microscopy. *Meth. Cell. Biol.* **30**, 379–398.

Brakenhoff, G. J., Muller, M., and Ghauharali, R. I. (1996). Analysis of efficiency of two-photon versus single-photon absorption for fluorescence generation in biological objects. *J. Microsc.* **183**, 140–144.

Buehler, C., Kim, K. H., Dong, C. Y., Masters, B. R., and So, P. T. C. (1999). Innovations in two-photon deep tissue microscopy. *IEEE Eng. Med. Biol. Mag.* **18**, 23–30.

Campagnola, P., Wei, M.-d., Lewis, A., and Loew, L. (1999). High-resolution nonlinear optical imaging of live cells by second harmonic generation. *Biophys. J.* **77**, 3341–3351.

Cannell, M. B. and Soeller, C. (1997). High resolution imaging using confocal and two-photon molecular excitation microscopy. *Proc. R. Microsc. Soc.* **32**, 3–8.

Cantor, C. R. and Schimmel, P. R. (1980). *Biophysical Chemistry*, Part II: *Techniques for the Study of Biological Structure and Function*. New York: W. H. Freeman.

Carlsson, K., Danielsson, P. E., Lenz, R., Liljeborg, A., Majlof, L., and Aslund, N. (1985). Three-dimensional microscopy using a confocal laser scanning microscope. *Opt. Lett.* **10**, 53–55.

Carrington, W. A., Lynch, R. M., Moore, E. D. W., Isenberg, G., Fogarty, K. E., and Fay, F. S. (1995). Super resolution in three-dimensional images of fluorescence in cells with minimal light exposure. *Science* **268**, 1483–1487.

Carrington, W. (2002). Imaging live cells in 3D using wide-field microscopy with image restoration. In: *Confocal and Two-Photon Microscopy: Foundations, Applications and Advances* (ed. A. Diaspro). New York: Wiley-Liss.

Castleman, K. R. (1996). *Digital Image Processing*. Englewood Cliffs, NJ: Prentice-Hall.

Castleman, K. R. (2002). Sampling, resolution and digital image processing in the spatial and Fourier domains: basic principles. In: *Confocal and Two-Photon Microscopy: Foundations, Applications and Advances*. (ed. A. Diaspro). New York: Wiley-Liss.

Caylor, C. L., Dobrianov, I., Kimmer, C., Thorne, R. E., Zipfel, W., and Webb, W. W. (1999). Two-photon fluorescence imaging of impurity distributions in protein crystals. *Phys. Rev.* **E59**, 3831–3834.

Centonze, V. E. and White, J. G. (1998). Multiphoton excitation provides optical sections from deeper within scattering specimens than confocal imaging. *Biophys. J.* **75**, 2015–2024.

Chalfie, M. and Kain, S. (Eds.) (1998). *Green Fluorescent Protein. Properties, Applications, and Protocols*. New York: Wiley-Liss.

Chalfie, M., Tu, Y., Euskirchen, G., Ward, W. W., and Prasher, D. C. (1994). Green fluorescent protein as a marker for gene expression. *Science* **263**, 802–805.

Chance, B. (1989). *Cell Structure and Function by Microspectrofluorometry.* New York: Academic Press.

Cox, I. J. (1984). Scanning optical fluorescence microscopy. *J. Microsc.* **133**, 149–153.

Cox, I. J. and Sheppard, C. J. R. (1983). Digital image processing of confocal images. *Image Vision Comput.* **1**, 52–56.

Cumpston, B. H. et al. (1999). Two-photon polymerization initiators for three-dimensional optical storage and microfabrication. *Nature* **398**, 51–54.

de Grauw, C. J., Frederix, P. L. T. M., and Gerritsen, H. C. (2002). Aberrations and penetration in in-depth confocal and two-photon microscopy. In: *Confocal and Two-Photon Microscopy: Foundations, Applications, and Advances* (ed. A. Diaspro). New York: Wiley-Liss.

Denk, W. (1996). Two-photon excitation in functional biological imaging. *J. Biomed. Opt.* **1**, 296–304.

Denk, W., Strickler, J. H., and Webb, W. W. (1990). Two-photon laser scanning fluorescence microscopy. *Science* **248**, 73–76.

Denk, W., Delaney, K. R., Gelperin, A., Kleinfeld, D., Strowbridge, B. W., Tank, D. W., and Yuste, R. (1994). Anatomical and functional imaging of neurons using 2-photon laser scanning microscopy. *J. Neurosci. Meth.* **54**, 151–162.

Denk, W., Piston, D., and Webb, W. W. (1995). Two-photon molecular excitation in laser scanning microscopy. In: *Handbook of Confocal Microscopy* (ed. J. B. Pawley), New York: Plenum, pp. 445–457.

Denk, W. and Svoboda, K. (1997). Photon upmanship: why multiphoton imaging is more than a gimmick. *Neuron* **18**, 351–357.

Diaspro, A. (Guest Ed.) (1999a). Two-photon microscopy. *Microsc. Res. Tech.* **47**, 163–212.

Diaspro, A. (Guest Ed.) (1999b). Two-photon excitation microscopy. *IEEE Eng. Med. Biol. Mag.* **18** (5), 16–99.

Diaspro, A. (2001). Building a two-photon microscope using a laser scanning confocal architecture. In: *Methods in Cellular Imaging* (ed. A. Periasamy). New York: Oxford University Press, pp. 162–179.

Diaspro, A. and Robello, M. (2000). Two-photon excitation of fluorescence for three-dimensional optical imaging of biological structures. *J. Photochem. Photobiol.* **B55**, 1–8.

Diaspro, A., Annunziata, S., Raimondo, M., and Robello, M. (1999a). Three-dimensional optical behavior of a confocal microscope with single illumination and detection pinhole through imaging of subresolution beads. *Microsc. Res. Tech.* **45**, 130–131.

Diaspro, A., Corosu, M., Ramoino, P., and Robello, M. (1999b). Adapting a compact confocal microscope system to a two-photon excitation fluorescence imaging architecture. *Microsc. Res. Tech.* **47**, 196–205.

Diaspro, A., Annunziata, S., and Robello, M. (2000). Single-pinhole confocal imaging of sub-resolution sparse objects using experimental point spread function and image restoration. *Micros. Res. Tech.* **51**, 464–468.

Diaspro, A., Chirico, G., Cannone, F., Beretta, S., Federici, F., Robello, M., and Olivini, F. (2001). Two-photon microscopy and spectroscopy based on a compact confocal scanning head. *J. Biomed. Opt.* (in press).

Feynman, R. P. (1985). *QED: The Strange Theory of Light and Matter.* Princeton, NJ: Princeton University Press.

Fisher, W. G., Watcher, E. A., Armas, M., and Seaton, C. (1997). Titanium : sapphire laser as an excitation source in two-photon spectroscopy. *Appl. Spectrosc.* **51**, 218–226.

Franken, P. A., Hill, A. E., Peters, C. W., and Weinreich, G. (1961). Generation of optical harmonics. *Phys. Rev. Lett.* **7**, 118–119.

Friedrich, D. M. (1982). Two-photon molecular spectroscopy. *J. Chem. Educ.* **59**, 472–483.

Friedrich, D. M. and McClain, W. M. (1980). Two-photon molecular electronic spectroscopy. *Annu. Rev. Phys. Chem.* **31**, 559–577.

Fujita, K. and Takamatsu, T. (2002). Real-time in situ calcium imaging with single- and two-photon confocal microscopy. In: *Confocal and Two-Photon Microscopy: Foundations, Applications, and Advances* (ed. A. Diaspro). New York: Wiley-Liss.

Gannaway, J. N. and Sheppard, C. J. R. (1978). Second harmonic imaging in the scanning optical microscope. *Opt. Quant. Electron.* **10**, 435–439.

Gauderon, R., Lukins, R. B., and Sheppard, C. J. R. (1999). Effects of a confocal pinhole in two-photon microscopy. *Microsc. Res. Tech.* **47**, 210–214.

Girkin, J. M. and Wokosin, D. L. (2002). Practical multiphoton microscopy. In: *Confocal and Two-Photon Microscopy: Foundations, Applications, and Advances* (ed. A. Diaspro). New York: Wiley-Liss.

Göppert-Mayer, M. (1931). Über Elementarakte mit zwei Quantensprüngen. *Annln. Phys.* **9**, 273–295.

Gosnell, T. R. and Taylor, A. J. (Eds.) (1991). *Selected Papers on Ultrafast Laser Technology.* Bellingham, WA: SPIE Press.

Gratton, E. and van de Ven, M. J. (1995). Laser sources for confocal microscopy. In: *Handbook of Confocal Microscopy* (ed. J. B. Pawley). New York: Plenum, pp. 69–97.

Gu, M. and Sheppard, C. J. R. (1995). Comparison of three-dimensional imaging properties between two-photon and single-photon fluorescence microscopy. *J. Microsc.* **177**, 128–137.

Guild, J. B., Xu, C., and Webb, W. W. (1997). Measurement of group delay dispersion of high numerical aperture objective lenses using two-photon excited fluorescence. *App. Opt.* **36**, 397–401.

Hamamatsu Photonics (1999). *Photomultiplier Tubes: Basics and Applications.* 2nd edn. Japan: Hamamatsu Photonics K. K.

Hanninen, P. E. and Hell, S. W. (1994). Femtosecond pulse broadening in the focal region of a two-photon fluorescence microscope. *Bioimaging* **2**, 117–121.

Hell, S. W. (Guest Ed.) (1996). Nonlinear optical microscopy. *Bioimaging* **4**, 121–172.

Hell, S. W., Bahlmann, K., Schrader, M., Soini, A., Malak, H., Gryczynski, I., and Lakowicz, J. R. (1996). Three-photon excitation in fluorescence microscopy. *J. Biomed. Opt.* **1**, 71–74.

Hellwarth, R. and Chistensen, P. (1974). Nonlinear optical microscopic examination of structures in polycrystalline ZnSe. *Opt. Commun.* **12**, 318–322.

Jonkman, J. E. N. and Stelzer, E. H. K. (2002). Resolution and contrast in confocal and two-photon microscopy. In: *Confocal and Two-Photon Microscopy: Foundations, Applications, and Advances* (ed. A. Diaspro). New York: Wiley-Liss.

Kaiser, W. and Garrett, C. G. B. (1961). Two-photon excitation in $CaF_2 : Eu^{2+}$. *Phys. Rev. Lett.* **7**, 229–231.

Koester, H. J., Baur, D., Uhl, R., and Hell, S. W. (1999). Ca^{2+} fluorescence imaging with pico- and femtosecond two-photon excitation: signal and photodamage. *Biophys. J.* **77**, 2226–2236.

König, K. (2000). Multiphoton microscopy in life sciences. *J. Microsc.* **200**, 83–104.

König, K., Liang, H., Berns, M. W., and Tromberg, B. J. (1996). Cell damage in near infrared multimode optical traps as a result of multiphoton absorption. *Opt. Lett.* **21**, 1090–1092.

König, K., So, P. T. C., Mantulin, W. W., and Gratton, E. (1997). Cellular response to near-red femtosecond laser pulses in two-photon microscopes. *Opt. Lett.* **22**, pp. 135–136.

König, K., Boehme, S., Leclerc, N., and Ahuja, R. (1998). Time-gated autofluorescence microscopy of motile green microalga in an optical trap. *Cell. Mol. Biol.* **44**, 763–770.

König, K., Becker, T. W., Fischer, P., Riemann, I., Halbhuber, K. J. (1999). Pulse-length dependence of cellular response to intense near-infrared laser pulses in multiphoton microscopes. *Opt. Lett.* **24**, 113–115.

König, K. and Tirlapur, U. K. (2002). Cellular and subcellular perturbations during multiphoton microscopy. In: *Confocal and Two-Photon Microscopy: Foundations, Applications, and Advances* (ed. A. Diaspro). New York: Wiley-Liss.

Kriete, A. (Ed.) (1992). *Visualization in Biomedical Microscopies*. Weinheim: VCH.

Lakowicz, J. R. and Gryczynski, I. (1992). Tryptophan fluorescence intensity and anisotropy decays of human serum albumin resulting from one-photon and two-photon excitation. *Biophys. Chem.* **45**, 1–6.

Lemons, R. A. and Quate, C. F. (1975). Acoustic microscopy: biomedical applications. *Science* **188**, 905–911.

Loudon, R. (1983). *The Quantum Theory of Light*. Oxford: Oxford University Press.

Louisell, W. H. (1973). *Quantum Statistical Properties of Radiation*. New York: Wiley.

Mainen, Z. F., Malectic-Savic, M., Shi, S. H., Hayashi, Y., Malinow, R., and Svoboda, K. (1999). Two-photon imaging in living brain slices. *Methods* **18**, 231–239.

Majewska, A., Yiu, G., and Yuste, R. (2000). A custom-made two-photon microscope and deconvolution system. *Pflugers Arch.–Eur. J. Physiol.* **441**, 398–408.

Manders, E. M. M., Stap, J., Brakenhoff, G. J., van Diel, and Aten, J. A. (1992). Dynamics of three-dimensional replication patterns during the S′-phase analyzed by double labelling of DNA and confocal microscopy. *J. Cell. Sci.* **103**, 857–862.

Masters, B. R. (1996). *Selected Papers on Confocal Microscopy.* Bellingham, WA: SPIE Press.

Minsky, M. (1988). Memoir of inventing the confocal scanning microscope. *Scanning* **10**, 128–138.

Moscatelli, F. A. (1986). A simple conceptual model for two-photon absorption. *Am. J. Phys.* **54**, 52–54.

Mueller, M., Squier, J., Wilson, K. R., and Brakenhoff, G. J. (1998). 3D microscopy of transparent objects using third-harmonic generation. *J. Microsc.* **191**, 266–274.

Nakamura, O. (1993). Three-dimensional imaging characteristics of laser scan fluorescence microscopy: two-photon excitation vs. single-photon excitation. *Optik* **93**, 39–42.

Nakamura, O. (1999). Fundamentals of two-photon microscopy. *Microsc. Res. Tech.* **47**, 165–171.

Patterson, G. H. and Piston, D. W. (2000). Photobleaching in two-photon excitation microscopy. *Biophys. J.* **78**, 2159–2162.

Pawley, J. B. (Ed.) (1995). *Handbook of Biological Confocal Microscopy.* New York: Plenum Press.

Periasamy, A., Skoglund, P., Noakes, C., and Keller, R. (1999). An evaluation of two-photon excitation versus confocal and digital deconvolution fluorescence microscopy imaging in *Xenopus* morphogenesis. *Micr. Res. Tech.* **47**, 172–181.

Periasamy, A. (Ed.) (2001). *Methods in Cellular Imaging.* New York: Oxford University Press.

Periasamy, A., Noakes, C., Skoglund, P., Keller, R., and Sutherland, A. E. (2002). Two-photon excitation fluorescence microscopy imaging in *Xenopus* and transgenic mouse embryos. In: *Confocal and Two-Photon Microscopy: Foundations, Applications, and Advances* (ed. A. Diaspro). New York: Wiley-Liss.

Pike, R. (2002). Superresolution in fluorescence confocal microscopy and in DVD optical storage. In: *Confocal and Two-Photon Microscopy: Foundations, Applications, and Advances* (ed. A. Diaspro). New York: Wiley-Liss.

Piston, D. W. (1999). Imaging living cells and tissues by two-photon excitation microscopy. *Trends Cell Biol.* **9**, 66–69.

Piston, D. W., Masters, B. R., and Webb, W. W. (1995). Three-dimensionally resolved NAD(P)H cellular metabolic redox imaging of the in situ cornea with two-photon excitation laser scanning microscopy. *J. Microsc.* **178**, 20–7.

Potter, S. M. (1996). Vital imaging: two-photons are better than one. *Curr. Biol.* **6**, 1596–1598.

Potter, S. M., Wang, C. M., Garrity, P. A., and Fraser, S. E. (1996). Intravital imaging of green fluorescent protein using two-photon laser-scanning microscopy. *Gene* **173**, 25–31.

Rentzepis, P. M., Mitschele, C. J., and Saxman, A. C. (1970). Measurement of ultrashort laser pulses by three-photon fluorescence. *Appl. Phys. Lett.* **17**, 122–124.

Schrader, M., Hell, S. W., and van der Voort, H. T. M. (1996). Potential of confocal microscope to resolve in the 50–100 nm range. *Appl. Phys. Lett.* **69**, 3644–3646.

Saloma, C., Saloma-Palmes, C., and Kondoh, H. (1998). Site-specific confocal fluorescence imaging of biological microstructures in a turbid medium. *Phys. Med. Biol.* **43**, 1741.

Sheppard, C. J. R. (1977). The use of lenses with annular aperture scanning optical microscopy. *Optik* **48**, 329–334.

Sheppard, C. J. R. (1989). Axial resolution of confocal fluorescence microscopy. *J. Microsc.* **154**, 237–241.

Sheppard, C. J. R. (2002). The generalized microscope. In: *Confocal and Two-Photon Microscopy Foundations, Applications, and Advances* (ed. A. Diaspro). New York: Wiley-Liss.

Sheppard, C. J. R. and Choudhury, A. (1977). Image formation in the scanning microscope. *Optica Acta* **24**, 1051–1073.

Sheppard, C. J. R. and Gu, M. (1990). Image formation in two-photon fluorescence microscopy. *Optik* **86**, 104–106.

Sheppard, C. J. R. and Kompfner, R. (1978). Resonant scanning optical microscope. *Appl. Opt.* **17**, 2879–2882.

Sheppard, C. J. R. and Shotton, D. M. (1997). *Confocal Laser Scanning Microscopy.* Oxford: BIOS.

Sheppard, C. J. R. and Wilson, T. (1980). Image formation in confocal scanning microscopes. *Optik* **55**, 331–342.

Sheppard, C. J. R., Kompfner, R., Gannaway, J., and Walsh, D. (1977). The scanning harmonic optical microscope. IEEE/OSA Conference on Laser Engineering and Applications, Washington, DC.

Shih, Y. H., Strekalov, D. V., Pittman, T. D., and Rubin, M. H. (1998). Why two-photon but not two photons? *Fortschr. Phys.* **46**, 627–641.

Shotton (Ed.) (1993). *Electronic Light Microscopy. Techniques in Modern Biomedical Microscopy*, Vol. 1. New York: Wiley-Liss.

Singh, S. and Bradley, L. T. (1964). Three-photon absorption in naphthalene crystals by laser excitation. *Phys. Rev. Lett.* **12**, 162–164.

So, P. T. C., Berland, K. M., French, T., Dong, C. Y., and Gratton, E. (1996). Two photon fluorescence microscopy: time resolved and intensity imaging. In: *Fluorescence Imaging Spectroscopy and Microscopy* (ed. X. F. Wang and B. Herman). New York: Wiley, pp. 351–373.

So, P. T. C., Dong, C. Y., Masters, B. R., and Berland, K. M. (2000). Two-photon excitation fluorescence microscopy. *Annu. Rev. Biomed. Eng.* **2**, 399–429.

So, P. T. C., Kim, K. H., Buehler, C., Masters, B. R., Hsu, L., and Dong, C. Y. (2001). Basic principles of multi-photon excitation microscopy. In: *Methods in Cellular Imaging* (ed. A. Periasamy). New York: Oxford University Press, pp. 152–161.

Soeller, C. and Cannell, M. B. (1996). Construction of a two-photon microscope and optimisation of illumination pulse duration. *Pfluegers Arch. Eur. J. Physiol.* **432**, 555–561.

Soeller, C. and Cannell, M. B. (1999). Two-photon microscopy: imaging in scattering samples and three-dimensionally resolved flash photolysis. *Microsc. Res. Tech.* **47**, 182–195.

Spence, D. E., Kean, P. N., and Sibbett, W. (1991). 60-fsec pulse generation from a self-mode-locked Ti : sapphire laser. *Opt. Lett.* **16**, 42–45.

Squier, J. A., Muller, M., Brakenhoff, G. J., and Wilson, K. R. (1998). Third harmonic generation microscopy. *Opt. Express* **3**, 315–324.

Squirrel, J. M., Wokosin, D. L., White, J. G., and Barister, B. D. (1999). Long-term two-photon fluorescence imaging of mammalian embryos without compromising viability. *Nature Biotechnol.* **17**, 763–767.

Stanley, M. (2001). Improvements in optical filter design. *Proc. SPIE* **4262**, 52–61.

Stelzer, E. H. K., Hell, S., Lindek, S., Pick, R., Storz, C., Stricker, R., Ritter, G., and Salmon, N. (1994). Non-linear absorption extends confocal fluorescence microscopy into the ultra-violet regime and confines the illumination volume. *Opt. Commun.* **104**, 223–228.

Straub, M., Lodemann, P., Holroyd, P., Jahn, R., and Hell, S. W. (2000). Live cell imaging by multifocal multiphoton microscopy. *Eur. J. Cell Biol.* **79**, 726–734.

Sytsma, J., Vroom, J. M., De Grauw, C. J., and Gerritsen, H. C. (1998). Time-gated fluorescence lifetime imaging and microvolume spectroscopy using two-photon excitation. *J. Microsc.* **191**, 39–51.

Tan, Y. P., Llano, I., Hopt, A., Wurriehausen, F., and Neher, E. (1999). Fast scanning and efficient photodetection in a simple two-photon microscope. *J. Neurosci. Meth.* **92**, 123–135.

Tirlapur, U. K. and König, K. (2002). Two-photon near-infrared femtosecond laser scanning microscopy in plant biology. In: *Confocal and Two-Photon Microscopy: Foundations, Applications, and Advances* (ed. A. Diaspro). New York: Wiley-Liss.

Török, P. and Sheppard, C. J. R. (2002). The role of pinhole size in high-aperture two- and three-photon microscopy. In: *Confocal and Two-Photon Microscopy: Foundations, Applications, and Advances* (ed. A. Diaspro). New York: Wiley-Liss.

Webb, R. H. (1996). Confocal optical microscopy. *Rep. Prog. Phys.* **59**, 427–471.

White, J. G., Amos, W. B., and Fordham, M. (1987). An evaluation of confocal versus conventional imaging of biological structures by fluorescence light microscopy. *J. Cell Biol.* **105**, 41–48.

Wier, W. G., Balke, C. W., Michael, J. A., and Mauban, J. R. (2000). A custom confocal and two-photon digital laser scanning microscope. *Am. J. Physiol.* **278**, H2150–H2156.

Wilson, T. (1989). Optical sectioning in confocal fluorescent microscope. *J. Microsc.* **154**, 143–156.

Wilson, T. (Ed.) (1990). *Confocal Microscopy.* London: Academic Press.

Wilson, T. (2002). Confocal microscopy: basic principles and architectures. In: *Confocal and Two-Photon Microscopy: Foundations, Applications, and Advances* (ed. A. Diaspro). New York: Wiley-Liss.

Wilson, T. and Sheppard, C. J. R. (1984). *Theory and Practice of Scanning Optical Microscopy.* London: Academic Press.

Wise, F. (1999). Lasers for two-photon microscopy. In: *Imaging: A Laboratory Manual* (ed. R. Yuste, F. Lanni, and A. Konnerth). Cold Spring Harbor, NY: Cold Spring Harbor Press, pp. 18.1–18. 9.

Wokosin, D. L. and White, J. G. (1997). Optimization of the design of a multiple-photon excitation laser scanning fluorescence imaging system. *Proc. SPIE* **2984**, 25–29.

Wokosin, D. L., Centonze, V. E., White, J., Armstrong, D., Robertson, G., and Ferguson, A. I. (1996). All-solid-state ultrafast lasers facilitate multiphoton excitation fluorescence imaging. *IEEE J. Select. Top. Quant. Electron.* **2**, 1051–1065.

Wokosin, D. L., Amos, W. B., and White, J. G. (1998). Detection sensitivity enhancements for fluorescence imaging with multiphoton excitation microscopy. *Proc. IEEE Eng. Med. Biol. Soc.* **20**, 1707–1714.

Wolleschensky, R., Feurer, T., Sauerbrey, R., and Simon, U. (1998). Characterization and optimization of a laser scanning microscope in the femtosecond regime. *Appl. Phys.* **B67**, 87–94.

Wolleschensky, R., Dickinson, M. E., and Fraser, S. E. (2002). Group-velocity dispersion and fiber delivery in multiphoton laser scanning microscopy. In: *Confocal and Two-Photon Microscopy: Foundations, Applications, and Advances* (ed. A. Diaspro). New York: Wiley-Liss.

Xu, C. (2002). Cross-sections of fluorescence molecules in multiphoton microscopy. In: *Confocal and Two-Photon Microscopy: Foundations, Applications, and Advances* (ed. A. Diaspro). New York: Wiley-Liss.

Xu, C., Guild, J., Webb, W. W., and Denk, W. (1995). Determination of absolute two-photon excitation cross sections by in situ second-order autocorrelation. *Opt. Lett.* **20**, 2372–2374.

CHAPTER 4

Cross-Sections of Fluorescence Molecules in Multiphoton Microscopy

CHRIS XU

Bell Laboratories, Lucent Technologies, Holmdel, NJ 07733, U.S.A.

1 INTRODUCTION

Maria Göppert-Mayer first predicted that an atom or a molecule could absorb two photons simultaneously in the same quantum event (Göppert-Mayer 1931). The first experimental evidence of this phenomenon came 30 years later when Kaiser and Garret (1961) demonstrated two-photon excitation (2PE) in a $CaF_2:Eu^{2+}$ crystal. Several techniques have been used to determine 2PE cross-sections of various materials in the past two decades (Hermann and Ducuing, 1972a; Twarowski and Kliger, 1977; Sheik-Bahae et al., 1990) (Kang et al., 1997; Kaatz and Shelton, 1999). Direct measurement of two-photon absorption cross-sections is usually difficult because only a small fraction of photons are absorbed via the two-photon process (Sengupta et al., 2000). Two-photon fluorescence excitation offers an alternative approach to determining 2PE cross-sections, provided that the material is fluorescent and its fluorescence quantum efficiency η is known. Although fluorescence measurements provide excellent sensitivity, accurate determination of absolute two-photon cross-sections is hampered by the strong dependence of absorption rates on the temporal and spatial coherence of the excitation light (Swofford and McClain, 1975; Birge, 1983).

Denk et al. (1990) first succeeded in applying two-photon excited fluorescence to laser scanning microscopy. Because the excitation rate is proportional to the square of the illumination intensity, two-photon laser scanning microscopy (2PLSM) provides intrinsic three-dimensional (3D) resolution without background fluorescence as compared with conventional confocal fluorescence microscopy. 2PLSM has since become a powerful tool for intrinsic 3D fluorescence imaging (Williams et al., 1994; Yuste and Denk, 1995; Heikal et al., 2000). This discovery has stimulated new interest in the physics of 2PE. Previously, two-photon spectroscopy had mainly been used to study the electronic structure of excited states in molecules (McClain and Harris, 1977), where it can access different excited states because its selection rules are different from those for one-photon absorption. Less effort

Confocal and Two-Photon Microscopy: Foundations, Applications, and Advances, Edited by Alberto Diaspro.
ISBN 0-471-40920-0

had been devoted to accurate quantitative studies of common fluorophores widely used in 2PLSM, leading to substantial disagreement amongst published values of two-photon cross-sections. In this chapter, we review our published data with absolute values of 2PE cross-sections for 25 commonly used fluorophores in the Ti : sapphire tuning range from 690 nm to 1050 nm. Detailed methods of obtaining these spectra will be summarized and compared (Xu et al., 1995; Xu and Webb, 1996). These data, especially with their wide spectral range, should also be valuable to molecular spectroscopists, particularly as a guide for predicting 2PE spectra.

In this chapter we also review our measurements of three-photon excited (3PE) fluorescence from ultraviolet (UV) fluorophores, including the Ca^{2+} indicators fura-2 and indo-1, the DNA stain DAPI, and the fluorescent reagent dansylhydrazine by excitation at $\lambda \approx 1.0\,\mu m$. We compare the one-, two-, and three-photon fluorescence excitation cross-sections of the fluorophores measured and discuss the resulting features of nonlinear microscopy for biological imaging (Xu et al., 1996).

2 ESTIMATE OF MULTIPHOTON CROSS-SECTIONS

The essence of the theory of multiphoton processes can be represented in perturbation theory of one form or another. It is sufficient to use the time-dependent semiclassical interaction to derive the multiphoton transition amplitude; for virtually all processes in the laser field, this gives completely equivalent results to those obtained from a more accurate description using quantum field theory. Details of the rigorous derivation can be found elsewhere (Faisal, 1987). Here, for the purpose of an order-of-magnitude estimation, a greatly simplified approach will be presented to describe multiphoton excitation (MPE) processes, and only the lowest-order dipole transition will be considered.

Intuitively, multiphoton processes require two or more photons to interact *simultaneously* with the molecule. For example, in a two-photon process, the "first" photon excites the molecule to an intermediate state, and the "second" photon further excites the molecule to its final state. Because the intermediate state can be a superposition of molecular states instead of an eigenstate of the molecule, it is usually referred to as the virtual intermediate state. Although the two-photon transition strength consists of contributions from all eigenstates as possible intermediate states, the single-intermediate-state (SIS) approximation (Birge, 1983) can be used to give an order of magnitude estimation. Figure 1 shows a schematic illustration of the SIS approximation. Therefore,

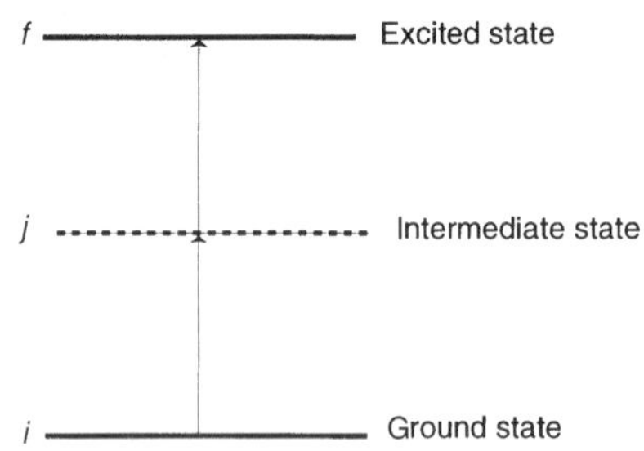

FIGURE 1 Schematic illustration of the single-intermediate-state (SIS) approximation. *i*, *j*, and *f* represent the initial (ground) state, the intermediate state, and the final (excited) state, respectively.

TABLE 1 Order-of-magnitude estimation of MPE cross-sections: *j*, *k*, and *l* are the intermediate states, and τ_j, τ_k, and τ_l represent the corresponding lifetimes of the intermediate states

*n*PE	Approximate expression	Estimated cross-section
2PE	$\sigma_2 = \sigma_{ij}\sigma_{jf}\tau_j$	10^{-49} cm^2 (photons/cm^2 s)$^{-1}$
3PE	$\sigma_3 = \sigma_{ij}\sigma_{jk}\sigma_{kf}\tau_j\tau_k$	10^{-82} cm^2 (photons/cm^2 s)$^{-2}$
4PE	$\sigma_3 = \sigma_{ij}\sigma_{jk]}\sigma_{kl}\sigma_{lf}\tau_j\tau_k\tau_l$	10^{-115} cm^2 (photons/cm^2 s)$^{-3}$

by simple examination of the absorption rates, the two-photon absorption cross-section σ_2 can be obtained:

$$\sigma_2 = \sigma_{ij}\sigma_{jf}\tau_j, \tag{1}$$

where σ_{ij} and σ_{jf} represent the appropriate one-photon absorption cross-sections and τ_j is the intermediate-state lifetime, which determines the timescale for photon coincidence (simultaneously arrival). τ_j can be estimated from the uncertainty principle; that is, τ_j must be short enough in order not to violate energy conservation. Thus,

$$\tau_j \approx \frac{1}{\Delta\omega} = \frac{1}{|\omega_{ij} - \omega|}, \tag{2}$$

where ω_{ij} and ω are the transition frequency and the incident photon frequency, respectively. For an electronic transition (ω_{if}) in the visible frequency range, and further assuming that the intermediate state and the final state are close in energy, we have $\tau_j \approx 10^{-15}$–10^{-16} s. The one-photon absorption cross-section of a molecule can be estimated from its dipole transition length (typically $\sigma_1 \approx 10^{-16}$–$10^{-17}$ cm^2 for a transition length of 10^{-8} cm). Hence, the estimated two-photon absorption cross-sections should be approximately 10^{-49} cm^4 s/photon: Eq. (1). The estimations for three- and four-photon excitation cross-sections can be obtained by parallel analysis. They are listed in Table 1.

It is important to note that the above estimates are only valid for simultaneous MPE; that is, virtual states with lifetimes of the order of 10^{-15} s serve as the intermediate states. However, in a sequential MPE process, real states with lifetimes of approximately 10^{-9}–10^{-12} s serve as the intermediate states. Consequently, the cross-section for a sequential MPE process can be orders of magnitude larger than the above estimate. Although the enhanced MPE cross-sections may seem desirable for some applications, there are many limitations in such multistep MPE, which we choose not to discuss in this chapter.

3 THEORETICAL ANALYSIS OF MPE

The crucial problem in MPE cross-section measurements is that the absorption rates depend strongly on the high-order (≥ 2) spatial and temporal coherence of the excitation light. Knowing the relationship between the experimentally measured fluorescence power and the excitation power, is essential in MPE cross-section measurements. In this section, we first examine the relationship in general and discuss the relevant parameters in any MPE cross-section measurements. Detailed analysis of several special cases will follow.

Because n-photon excitation is an nth-order process, the number of photons absorbed per molecule per unit time via n-photon excitation is proportional to the n-photon absorption cross-section σ_n and to the nth-power of the incident intensity I (McClain and Harris, 1977). In a particular experiment, the total number of photons absorbed per unit time N_{abs} is also a function of the dye concentration C and the illuminated sample volume V:

$$N_{\text{abs}}(t) = \int_V d\mathbf{r}\ \sigma_n c(\mathbf{r}, t) I^n(\mathbf{r}, t). \tag{3}$$

In the absence of ground-state depletion and photobleaching, $C(\mathbf{r}, t)$ can be assumed to be constant. Moreover, in the case where one can separate the time and space dependences of the excitation intensity, we have

$$N_{\text{abs}}(t) = C\sigma_n I_0^n(t) \int_V d\mathbf{r}\ S^n(\mathbf{r}), \tag{4}$$

where $S(\mathbf{r})$ and $I_0(t)$ describe the spatial and temporal distributions of the incident light, that is, $I(\mathbf{r}, t) = S(\mathbf{r}) I_0(t)$. We chose $S(\mathbf{r})$ to be a dimensionless spatial distribution function and $I_0(t)$ to be the intensity at the geometric focal point.

Assuming no stimulated emission and no self-quenching, the number of fluorescence photons collected per unit time, F, is given by

$$F(t) = \frac{1}{n}\phi\eta N_{\text{abs}}, \tag{5}$$

where η and ϕ are the fluorescence quantum efficiency of the dye and the fluorescence collection efficiency of the measurement system, respectively. Here the factor $1/n$ simply reflects the fact that n photons are needed for each excitation event. In practice, only the time-averaged fluorescence photon flux $\langle F(t)\rangle$ is measured:

$$\langle F(t)\rangle = \frac{1}{n}\phi\eta C\sigma_n \langle I_0^n(t)\rangle \int_V d\mathbf{r}\ S^n(\mathbf{r}). \tag{6}$$

We note that $\langle F(t)\rangle$ is proportional to $\langle I_0^n(t)\rangle$, not to $I_0(t)\rangle n$. Because most detectors only give a signal that is proportional to $\langle I_0(t)\rangle$, we rewrite Eq. (6) in terms of the average intensity,

$$\langle F(t)\rangle = \frac{1}{n} g^{(n)} \phi\eta C\sigma_n \langle I_0(t)\rangle^n \int_V d\mathbf{r}\, S^n(\mathbf{r}), \tag{7a}$$

$$g^{(n)} = \frac{\langle I_0^n(t)\rangle}{\langle I_0(t)\rangle^n}, \tag{7b}$$

where $g^{(n)}$ is a measure of the nth-order temporal coherence of the excitation source.

All of the relevant quantities in a MPE cross-section measurement are presented in Eq. (7). Experimental determination of σ_n usually involves the characterization of three parameters: the spatial distribution of the incident light $\int_V S^n(\mathbf{r})\ d\mathbf{r}$, the nth-order temporal coherence $g^{(n)}$, and the fluorescence collection efficiency of the system ϕ.

3.1 Spatial Considerations

The spatial distributions of two limiting cases, namely the GL focus and the diffraction-limited focus, will be discussed in detail. In practice, only these two profiles have been used in quantitative measurements in order to avoid the complex nature of the beam profiles that result from the partial filling of the back aperture of the focusing lens. We shall only consider two- and three-photon cross-section measurements in the following sections. The general formula for measuring the n-photon cross-section for higher n can be obtained by parallel analysis.

3.1.1 Excitation with a Gaussian–Lorentzian (GL) Beam A laser that operates in TEM_{00} mode has a GL spatial profile. However, diffraction by a finite aperture distorts the wavefront. A detailed theory of a GL beam propagating through such finite apertures has been developed (Dickson, 1970). The resulting spatial profile after the aperture is complex, and depends on the ratio of the GL beam waist w_0 and the aperture size. In practice, such difficulties can be circumvented by keeping the diameter of the incident beam much smaller than that of the back aperture of the focusing lens. The consequences of diffraction can be neglected, and therefore a perfect GL profile can be obtained after the lens.

The intensity distribution of a GL beam is of the form

$$I(\rho, z, t) = \frac{2P(t)}{\pi w^2(z)} \exp\left[-\frac{2\rho^2}{w^2(z)}\right], \tag{8a}$$

$$w(z) = w_0 \left[1 + \left(\frac{z}{z_R}\right)^2\right]^{1/2}, \tag{8b}$$

$$z_R = \frac{n_0 \pi w_0^2}{\lambda}, \tag{8c}$$

where z is the distance along the optical axis, ρ is the distance from the optical axis, λ is the wavelength of the excitation light in vacuum, n_0 is the refractive index of the sample medium, and $P(t)$ is the incident power. z_R, frequently referred to as the Rayleigh length of the beam, characterizes the focal depth of the Gaussian beam. We can easily obtain expressions for $S(\mathbf{r})$ and $I_0(t)$ from Eq. (8):

$$I_0(t) = \frac{2P(t)}{\pi w_0^2}, \tag{9a}$$

$$S(\mathbf{r}) = \frac{w_0^2 \exp[-2\rho^2/w^2(z)]}{w^2(z)}. \tag{9b}$$

Assuming the sample thickness to be much greater than the Rayleigh length, an analytical expression can be readily obtained for the integral of $S^n(\mathbf{r})$ (for $n \geq 2$):

$$\int_{V\to\infty} S^n(\mathbf{r})\, d\mathbf{r} = \frac{(2n-5)!!}{2n(2n-4)!!} \frac{n_0 \pi^3 w_0^4}{\lambda}, \tag{10}$$

where !! denotes the double factorial, that is, $2n!! = 2 \cdot 4 \cdot 6 \cdots 2n$, $(2n-1)!! = 1 \cdot 3 \cdot 5 \cdots (2n-1)$, $(-1)!! = 1$, $0!! = 1$.

3.1.2 A Plane Wave Focused by Diffraction-Limited Optics We first consider the three-dimensional intensity distribution at the focal spot of a diffraction-limited objective lens with uniform illumination. In practice, this is achieved by expanding the incident laser beam so that the beam diameter is much larger than the back aperture of the focusing lens. Properties of the lens are then determined by its point spread function (PSF). The dimensionless distance from the optical axis, v and that from the in-focus plane, u, are given by

$$v = \frac{2\pi(NA)\rho}{\lambda}, \tag{11a}$$

$$u = \frac{2\pi(\mathrm{NA})^2 z}{n_0\lambda}, \tag{11b}$$

where the numerical aperture $\mathrm{NA} = n \sin\theta$ and θ is the half-angle of collection for the lens. We use the paraxial form of the normalized intensity PSF, $h^2(u, v)$, for a diffraction-limited lens (Wilson and Sheppard, 1984):

$$I_0(t) = \frac{\pi \mathrm{NA}^2}{\lambda^2} P(t), \tag{12a}$$

$$S(\mathbf{r}) = h^2(u, v) = \left| 2\int_0^1 J_0(v\theta) \exp[-\tfrac{1}{2} iu\theta^2]\, \theta\, d\theta \right|^2. \tag{12b}$$

The paraxial approximation is satisfactory for $\sin\theta < 0.7$, or NA< 1.0 with an oil-immersion objective (Sheppard and Matthews, 1987). After changing ρ and z to their optical units v and u using Eq. (11), the integral of $S(\mathbf{r})$ becomes:

$$\int_{V\to\infty} S^n(\mathbf{r})\, d\mathbf{r} = a_n \left(\frac{n_0 \lambda^3}{8\pi^3 \mathrm{NA}^4} \right), \tag{13}$$

where

$$a_n = \int_0^\infty 2\pi v\, dv \int_{-\infty}^\infty du\, [h(u, v)]^{2n}.$$

Here we have again assumed that the sample thickness is much greater than the focal depth of the lens. It is interesting to note that, in the paraxial approximation, the volume integrals are proportional to NA^{-4}, regardless of the order of photon processes. Because analytical expressions for a_n cannot be obtained, numerical integrations of $[h(u, v)]^{2n}$ over the entire space have been performed for selected values of n. Results are listed in Table 2.

TABLE 2 Numerical integrations of a_n under the paraxial approximation[a]

n	2	3	4	5	6	7	8	10	12	14
a_n	70	28.1	18.3	13.2	10.1	8.06	6.62	4.76	3.64	2.89

[a] Because numerical integrations can only be done over a limited space, the estimated uncertainty of the integral is $<4\%$ for $n = 2$ and negligible for $n > 2$. Absolute values of a_n are expected to be different in the high-NA limit, where the paraxial approximation is no longer valid. However, numerical calculations for NA $= 1.2$ showed that the *relative* values of a_n ($n = 2, 3, 4, 5, 6$) are approximately the same as those listed in this table.

3.1.3 Two-Photon Excitation

In the thick-sample limit, the expression for experimentally detected 2PE fluorescence in a GL focus can be obtained using Eqs. (7) and (10):

$$\langle F(t)\rangle = \frac{1}{2} g^{(2)} \phi \eta C \sigma_2 n_0 \frac{\pi \langle P(t)\rangle^2}{\lambda}. \tag{14}$$

A similar equation can also be obtained for a diffraction-limited focus using Eqs. (7) and (13):

$$\langle F(t)\rangle \approx \frac{1}{2} g^{(2)} \phi \eta C \sigma_2 n_0 \frac{8.8 \langle P(t)\rangle^2}{\pi \lambda}. \tag{15}$$

Although the collection efficiency ϕ is dependent on the NA of the collecting lens, we note that the total fluorescence generation is independent of the NA of the focusing lens (or the beam-waist size of the GL beam) in thick samples. Intuitively, one sees that the effect of the increased intensity by tighter focusing is just compensated by the shrinking excitation volume. Thus, the amount of fluorescence summed over the entire space remains unchanged for 2PE. Working with thick samples in 2PE measurements takes advantage of this interesting property: the generated fluorescence power is insensitive to the size of the focal spot, assuming no aberration. Hence, a small variation of the laser beam size (with excitation power monitored after the objective lens) or sample thickness would not affect the results of measurements.

One possible concern when using thick samples is lens aberration. Although microscope objectives are highly corrected, index mismatch between the immersion media and the sample introduces aberration that generally reduces the amount of two-photon fluorescence (Hell et al., 1993). However, such aberration effects are negligible when low-NA lenses are used.

3.1.4 Three-Photon Excitation

In the thick-sample limit, the expression for experimentally detected 3PE fluorescence in a GL focus can be obtained using Eqs. (7) and (10):

$$\langle F(t)\rangle = \frac{1}{3} g^{(3)} \phi \eta C \sigma_3 n_0 \frac{2 \langle P(t)\rangle^3}{3 \lambda w_0^2}. \tag{16}$$

A similar equation can also be obtained for a diffraction-limited focus using Eqs. (7) and (13):

$$\langle F(t)\rangle \approx \frac{1}{3} g^{(3)} \phi \eta C \sigma_3 n_0 \frac{3.5\mathrm{NA}^2 \langle P(t)\rangle^3}{\lambda^3}. \quad (17)$$

It is clear that the amount of 3PE fluorescence depends on the size of the focal spot (i.e., the NA of the lens or the beam-waist size). Thus, unlike 2PE experiments with thick samples, quantitative measurements of 3PE fluorescence require characterization of the focal-spot size. However, working with thick samples still provides the advantage of less sensitivity to the beam-waist measurements, because $\langle F(t)\rangle \propto w_0^{-6}$ or NA^6 in the thin-sample limit, in which case a small deviation of the focal-spot size can cause a much larger error in the measured cross-section.

3.2 Temporal Considerations

In addition to the characterization of the spatial profiles of the excitation beam, experimental determination of MPE cross-sections is difficult mainly because the multiphoton absorption rates depends crucially on the temporal profiles of the excitation beam, that is on $g^{(n)}$ as defined by Eq. (7b). An effective and simple experimental approach to solve this problem is to compare the generated fluorescence of the specimen with some known two- or three-photon references provided that reliable two- or three-photon standards in the wavelength range of interest exist (Kennedy and Lytle, 1986; Jones and Callis, 1988). Such a reference technique also takes into account the spatial distribution if the same excitation geometry is used for both the standard and the specimen to be measured. However, several measurement methods can be used in the absence of such multiphoton standards. They will be briefly reviewed in the following sections.

3.2.1 Measurements with a Single-Mode Continuous-Wave (CW) Laser An ideal single-mode CW laser has $g^{(n)} = 1$. Measuring MPE cross-sections using such lasers only requires knowledge of the value of the average excitation intensity. However, single-mode CW excitation usually requires many orders of magnitude more average power than pulsed excitation to obtain the same rate of excitation, because of the much smaller value of $g^{(n)}$. Using the same average power, the excitation rate achieved with a single-mode CW laser is only approximately 10^{-5} and 10^{-10} for two- and three-photon excitation, respectively, of that using a mode-locked femtosecond Ti : sapphire laser with 80 MHz repetition rate and 100 fs pulsewidth. Nevertheless, CW excitation is feasible for dyes with relatively large cross-section in 2PE. The prospect of CW 3PE is not good except for molecules with extraordinarily large three-photon cross-sections (such as those reported by He et al., 1995). For example, with an average power of 100 mW at the specimen of 1 mM concentration and using a three-photon fluorescence action cross-section $\eta\sigma_3$ of 10^{-82} cm^6 (s/photon)^2, Eq. (17) predicts that the detected fluorescence rate is only 0.2 photon s^{-1}, assuming NA $= 1.3$, $\lambda = 1000$ nm, and a high detection efficiency ϕ of 1%.

It is important to stress the "true CW" nature of a single-mode CW laser. A multimode CW laser may retain enough RMS noise (such as caused by mode beating) to prevent its use as a stable CW source.

3.2.2 Measurements with a Pulsed Laser Measurements performed with a pulsed laser require careful evaluation of the temporal coherence factor $g^{(n)}$. The focused intensity obtained from a mode-locked laser is a periodic function in time,

$$I_0(t) = I_0\left(t + \frac{m}{f}\right), \qquad m = 1, 2, 3, \ldots, \tag{18}$$

where f is the pulse-repetition rate. Let $t = 0$ be at the peak of one excitation pulse. Because of the periodic nature of the pulse train, one only needs to calculate $g^{(n)}$ for one cycle. Defining τ as the excitation pulsewidth (full-width at half-maximum, FWHM) and $f\tau$ as the duty cycle, $g^{(n)}$ can be expressed in terms of the ratio of the dimensionless quantity $g_p^{(n)}$, which depends only on the shape of the laser pulse and the duty cycle:

$$g^{(n)} = \frac{g_p^{(n)}}{(f\tau)^{n-1}}, \tag{19a}$$

$$g_p^{(n)} = \tau^{n-1} \int_{-1/2f}^{1/2f} I_0^n(t)\, dt \Big/ \left[\int_{-1/2f}^{1/2f} I_0(t)\, dt\right]^n. \tag{19b}$$

For pulses with a Gaussian temporal profile, one finds $g_p^{(2)} = 0.66$ and $g_p^{(3)} = 0.51$, whereas for a hyperbolic-secant square pulse, one finds $g_p^{(2)} = 0.59$ and $g_p^{(3)} = 0.41$. Intuitively, $g^{(n)}$ represents the n-photon excitation efficiency of the excitation light at a fixed time-averaged intensity. We note that $g^{(n)}\langle I_0(t)\rangle$ is generally not the commonly referred *peak* intensity or *pulse* intensity I_p, which should be given by $I_p = \langle I_0(t)\rangle / f\tau$.

Mode-locked lasers with their high peak intensities and broad tunability offer tremendous advantages in multiphoton fluorescence excitation. However, determination of MPE cross-sections depends crucially on the estimation of the temporal coherence at zero time delay (Loudon, 1983), $g^{(n)} = \langle I_0^n\rangle / \langle I_0\rangle^n$, of the excitation light *inside the sample*. Although the pulsewidth and the laser repetition rate can be routinely measured in any laser laboratory, the determination of $g_p^{(n)}$ in the femtosecond time region is difficult. In addition, external measurements of the pulse shapes do not provide reliable estimates of $g^{(n)}$ *inside the sample*, because ultrashort pulses are distorted by the optical nonlinearities and group delay dispersion (GDD) (e.g., a GDD of $-2000\ \text{fs}^2$ at 730 nm for a Zeiss Neofluo 10× 0.3 NA lens; Guild et al., 1997). To circumvent the difficulties of determining the pulse shape inside the sample while still taking full advantage of pulsed excitation, we shall briefly review a technique that we have developed in obtaining 2PE cross-section (Xu et al., 1995).

3.2.3 Autocorrelation Techniques The method described here takes advantage of the fact that the fluorescence from the test chromophore itself can be used not only to measure the temporal autocorrelation of the pulsed excitation light but also to gauge the amount of excitation and, by the combination of both, the absolute 2PE cross-section. The sample is illuminated through focusing optics at the output of the Michelson interferometer that generates a spatial superposition of two halves of the excitation beam shifted in time by a variable delay (Fig. 2). The set-up resembles the collinear autocorrelator used routinely for pulse-duration measurements, but with the sample fluorophore serving as the quadratic medium. While the autocorrelation function, measured externally or in situ,

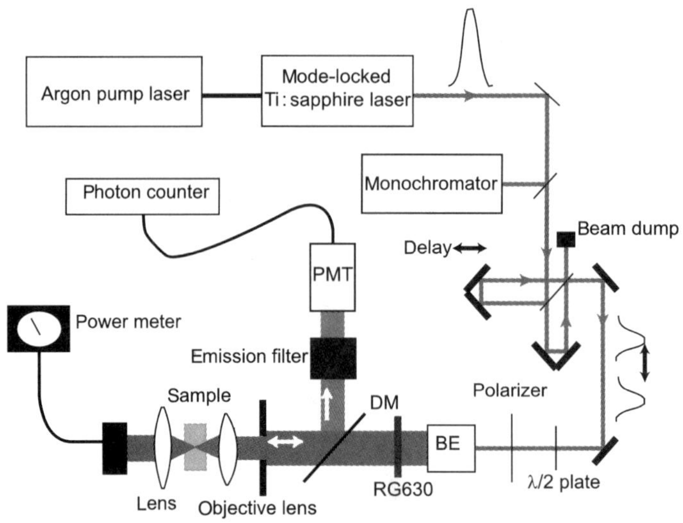

FIGURE 2 Experimental apparatus for measuring 2PE cross-sections using the autocorrelation technique. A long-pass filter (> 630 nm) eliminates residual argon-pump light and Ti : sapphire fluorescence. The beam expander (BE) expands the beam to about 25 mm in diameter ($1/e^2$), which is sufficient to overfill the back aperture (10 mm diameter) of the objective (Zeiss Neofluoar 0.3 NA 10×). A long-pass dichroic mirror (DM) with reflectivity of greater than 95% for $\lambda < 610$ nm separates the fluorescence from the excitation light. The incident power at the sample is measured by recollimating the transmitted beam onto a calibrated power meter. To ensure that the total generated fluorescence does not depend on the NA, a sample is used that is significantly thicker (100 μm) than the focal depth (approximately 40 μm). Fluorescence is detected by a photomultiplier tube (PMT) (R1924, Hamamatsu) connected to a photon counter, and is recorded as a function of the interferometer delay.

does not determine the temporal profile of the pulse uniquely, it contains enough information for one to calculate $g^{(2)}$ without any assumptions about the pulse shape other than that all the pulse energy is contained within a finite temporal interval.

For simplicity, we shall again only consider a single excitation pulse. In the relevant case of two beams with equal intensities that overlap perfectly in space but are temporally shifted by a delay τ_d, the fluorescence F recorded as a function of τ_d is simply the second-order autocorrelation function of the excitation pulse (Diels, 1990):

$$\begin{aligned} F(\tau_d) = \frac{1}{2} B\sigma_2 C\eta\phi \int_{-\infty}^{\infty} dt \, \{ & P^2(t) + P^2(t+\tau_d) + 4P(t)P(t+\tau_d) \\ & + 2P(t)P(t+\tau_d)\cos 2[\omega\tau_d + \varphi(t+\tau_d) - \varphi(t)] \\ & + 4[P(t)P(t+\tau_d)]^{1/2}[P(t) + P(t+\tau_d)]\cos[\omega\tau_d + \varphi(t+\tau_d) - \varphi(t)]\}, \end{aligned} \qquad (20)$$

where $\int_V S^2(\mathbf{r})\, d\mathbf{r} = B$. The original and delayed beams are represented by $P(t)$ and $P(t+\tau_d)$, respectively. Integration of Eq. (20) with respect to τ_d over a range $2T$ much greater than the pulsewidth eliminates the oscillatory terms, after the integrals of t and τ_d

are exchanged. Since $P(t)P(t+\tau_d) \approx 0$ for $|\tau_d| \geq T$, the integration region can be extended from $-\infty$ to ∞. We then obtain

$$\int_{-T}^{T} F(\tau_d)\, d\tau_d = \frac{1}{2} B\sigma_2 C\eta\phi \left\{ 4\left[\int_{-\infty}^{\infty} P(t)\, dt\right]^2 + 4T \int P^2(t)\, dt \right\}. \tag{21}$$

The amount of 2PE fluorescence detected at $|\tau_d| \geq T$, F_∞, is simply twice the amount generated by a single pulse of one of the half-beams:

$$F_\infty = B\sigma_2 C\eta\phi \int_{-\infty}^{\infty} P^2(t)\, dt. \tag{22}$$

Combining Eqs. (21) and (22) yields

$$\sigma_2 = \frac{\int_{-T}^{T} F(\tau_d)\, d\tau_d - 2TF_\infty}{2BC\eta\phi\left[\int_{-\infty}^{\infty} P(t)\, dt\right]^2}. \tag{23}$$

Once C, η, and ϕ have been independently determined (e.g., by one-photon fluorescence calibration), the calibrated cross-section σ_2 can be obtained by recording the autocorrelation function $F(\tau_d)$ and the incident pulse energy $\int_{-\infty}^{\infty} P(t)\, dt$. Therefore, 2PE cross-section measurements by this technique do not require any prior knowledge about the temporal profile of the excitation source. Note that Eqs. (21) and (22) yield

$$g^{(2)} = 2F_\infty \left[f \int_{-T}^{T} F(\tau_d)\, d\tau_d - 2f\, TF_\infty \right]^{-1}. \tag{24}$$

Although only absolute measurement of 2PE cross-sections has been used to demonstrate this new method, it is relatively simple to extend the above analysis to 3PE, in which case a third-order correlation trace needs to be obtained in addition to the second-order autocorrelation function. We note that such technique may be most valuable in measuring the cross-sections of higher ($>$ 2nd) order processes, where absolute measurement with CW lasers becomes difficult because of the much weaker fluorescence excitation.

3.3 System Collection Efficiency ϕ and Fluorescence Quantum Efficiency η

The system collection efficiency ϕ is a function of the collection efficiency of the collection lens, the transmission of the optics, and the detector quantum efficiency. The most convenient approach to determine ϕ is to use one-photon excited (1PE) fluorescence. Direct comparison with 1PE fluorescence also gives the absolute values of 2PE cross-sections, assuming equal emission quantum efficiencies. However, care must be taken to ensure that the collection efficiencies for 1PE and 2PE fluorescence are the same. For example, the use of short-path-length sample cuvettes ensures uniform fluorescence collection efficiency throughout the sample, which is necessary because 1PE occurs throughout the entire sample thickness (i.e., it is not localized to the focal region as in 2PE). In general, it is assumed that 1PE, 2PE, and 3PE fluorescence emission spectra are the same. As discussed later, we observe no exceptions to this assumption for the fluorophores discussed in this chapter.

One can also determine ϕ directly from the fluorescence emission spectrum, the measured NA of the collection lens, and the transmission of the optics. The detector response can be obtained from the manufacturer's catalog or can be independently calibrated using a known-intensity source (e.g., with a laser and calibrated neutral-density filters). In our experiments, the collection efficiencies determined by the above two methods were quite consistent.

A measurement of the n-photon excited fluorescence quantum efficiency η is very difficult, because direct measurement of n-photon absorption (usually very small) is required. It is usually assumed that fluorescence quantum efficiencies are the same for 1PE, 2PE, and 3PE, and are constant over the entire spectral range.

4 TWO- AND THREE-PHOTON EXCITATION SPECTRA

Figure 3 summarizes our measurements of two-photon fluorescence excitation spectra $\sigma_2(\lambda)$ of several fluorophores and Ca^{2+} indicators in the spectral range of 690–1050 nm. For comparison, the tuning ranges of several mode-locked laser sources are also plotted. 2PE cross-section measurements of another eight dyes, including fluorescein, rhodamine B, DiI, and excitation ratiometric Ca^{2+} indicator fura-2, are shown in Figure 4, with their corresponding one-photon absorption spectra. These spectra show that nonlinear fluorescence microscopy based on 2PE can be expected to work well with most existing fluorophores and many available laser sources. For nonratiometric Ca^{2+} indicators, such as Ca-crimson, Ca-orange and Ca-green-1, the 2PE spectra of the Ca^{2+}-free and Ca^{2+}-bound forms are indistinguishable. The ratios of fluorescence intensity for the Ca^{2+}-free to that for the Ca^{2+}-bound forms of these indicators are also comparable to their reported one-photon values. Thus, the binding of Ca^{2+} does not change the shapes of the 2PE spectra for these Ca^{2+} indicators. Similar properties were observed in the case of fluorescein–dextran molecules, where the binding of fluorescein to dextran does not change its 2PE spectrum (Albota et al., 1998a).

Green fluorescent proteins (GFPs) have attracted tremendous interest as biological reporters for gene expression (Chalfie et al., 1994). Images of cultured cells using 2PE GFP fluorescence have been reported (Potter et al., 1996). Wild-type GFP has a high UV absorption at 400 nm and relatively low visible absorption at 480 nm. To facilitate the use of GFPs with visible excitation, mutant S65T (replacement of serine 65 with threonine) was engineered to enhance the visible absorption band (Cubitt et al., 1995). The measured 2PE spectra of GFP wild-type and the S65T mutant are similar to the corresponding one-photon spectra (Fig. 5), indicating that wild-type GFP can be two-photon excited using the same wavelength needed to excite UV and visible fluorophores (Figs. 3 and 4), while only long-wavelength visible fluorophores can be simultaneously excited with the S65T mutant. The 2PE spectra of another GFP mutant, EGFP (S65T and F64L) (Cormack et al., 1996; Chen et al. 2000), are similar to that of GFPS65T. However, the 2PE cross-section at 960 nm is approximately 102 GM.

Two-photon excited cellular autofluorescence from intrinsic chromophores such as NAD(P)H has been used to study the cellular metabolic state (Piston et al., 1995). Photodamage of cells can also be detected by monitoring autofluorescence changes (König et al., 1996). However, for most fluorescence imaging applications, autofluores-

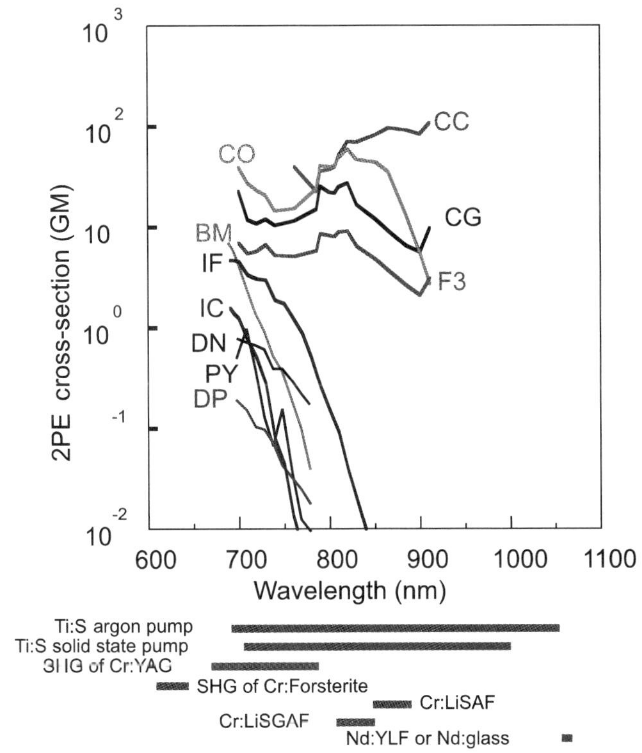

FIGURE 3 See ftp site for color figure. Two-photon fluorescence excitation spectra of fluorophores. For BM, data are for the two-photon absorption cross-sections. For all the other fluorophores, data are for the two-photon action cross-section (i.e., the product of the fluorescence emission quantum efficiency and the two-photon absorption cross-section). 1 GM $\equiv 10^{-50}$ cm^4 s/photon. Spectra are excited with linearly polarized light using a mode-locked Ti : sapphire laser. For comparison, the tuning ranges of conveniently available mode-locked laser sources are also plotted. The fluorophores illustrated are as follows: BM $\equiv$ *p*-bis(*o*-methylstyryl)benzene; DP (DAPI not DNA-bound) $\equiv$ 4′,6-diamidino-2-phenylindole, dihydrochloride; DN (dansyl) $\equiv$ 5-dimethylaminonaphthalene-1-sulfonylhydrazine; PY $\equiv$ 1,2-bis-(1-pyrenedecanoyl)-*sn*-glycero-3-phosphocholine; IC $\equiv$ indo-1 with Ca^{2+}; IF $\equiv$ indo-1 without Ca^{2+}; CG $\equiv$ calcium green-1 with Ca^{2+}; CO $\equiv$ calcium orange with Ca^{2+}; CC $\equiv$ calcium crimson with Ca^{2+}; F3 $\equiv$ fluo-3 with Ca^{2+}. All of the samples were purchased from either Eastman Kodak or Molecular Probes. Note that the *y* axis is a logarithmic scale.

cence background is undesirable since it limits the detection sensitivity for fluorescent probes. Thus, knowledge of the 2PE spectra of NADH and flavins should help to optimize 2PE fluorescence imaging, and may also provide insight into the photodamage mechanisms of living biological preparations in the near IR (König et al., 1995). Figure 6 shows the 2PE spectra of NADH and FMN. The 2PE spectrum of NADPH is nearly identical to that of NADH. The 2PE spectrum of FAD is similar to that of FMN, although the action cross-sections of FAD are approximately 9 times smaller than those of FMN (presumably owing to quenching of fluorescence by the adenine).

One general feature conserved in all measured 2PE spectra is that the 2PE peak absorption wavelengths never appear redshifted (but are frequently blueshifted) relative to twice the one-photon absorption peaks. For example, large blueshifts were observed for

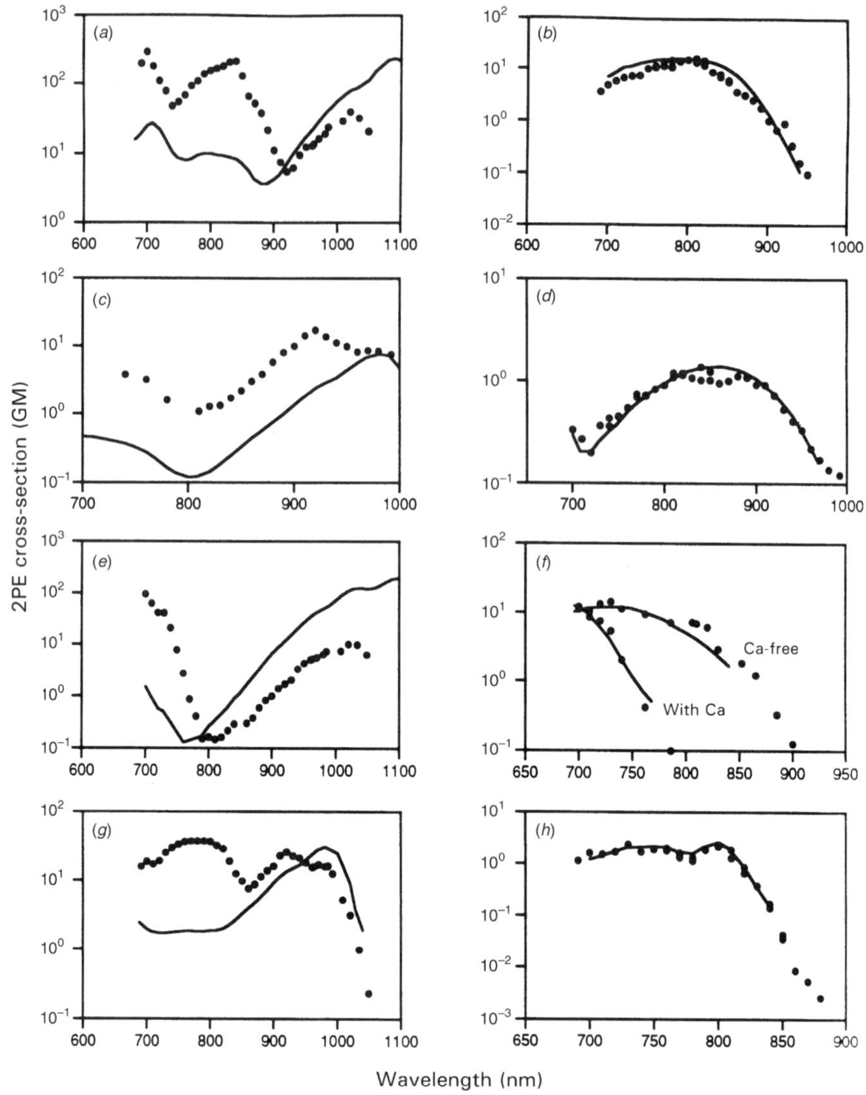

FIGURE 4 Comparison of one-photon (curves) and two-photon (filled circles) fluorescence excitation spectra. The *y* axis values represent two-photon action cross-sections, except for rhodamine B (*a*) and fluorescein (*g*), where values of the two-photon absorption cross-sections are given. For the one-photon results, the *x*-axis values represent twice the one-photon transition wavelengths and the *y*-axis values are arbitrarily scaled. The dyes illustrated are as follows: (*a*) rhodamine B in methanol; (*b*) coumarin 307 in methanol; (*c*) Bodipy (4,4-difluoro-1,3,5,7,8-pentamethyl-4-bora-3*a*,4*a*-diazaindacene-2,6-disulfonic acid, disodium salt) in water; (*d*) lucifer yellow CH, ammonium salt, in water; (*e*) DiI (1,1-dioctadecyl-3,3,3′,3′-tetramethylindocarbocyanine perchlorate) in methanol; (*f*) fura-2; (*g*) fluorescein in water, pH $\approx$ 13; (*h*) cascade blue hydrazide, trisodium salt, in water. Note that the *y*-axis is a logarithmic scale.

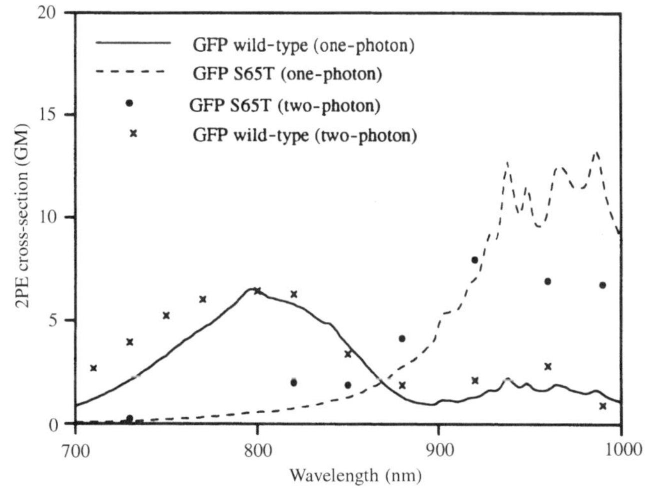

FIGURE 5 Two-photon action cross-sections of GFPs. The y-axis values represent two-photon action cross-sections for GFPs. For the one-photon results, the x-axis values represent twice the one-photon transition wavelengths and the y-axis values are arbitrarily scaled. GFPs were provided by George Patterson and David W. Piston, Vanderbilt University.

rhodamine B, DiI, fluorescein, and several Ca^{2+} indicators (Figs. 3 and 4). Although these interesting spectral features are still not understood quantitatively, there are several desirable consequences of such significant blueshifts. First, the resolution of 2PLSM using these fluorophores is considerably higher than predicted by assuming 2PE peak wavelengths at twice the one-photon absorption peak wavelengths. Second, the blueshifts allow conveniently available mode-locked laser sources to excite popular dyes, as

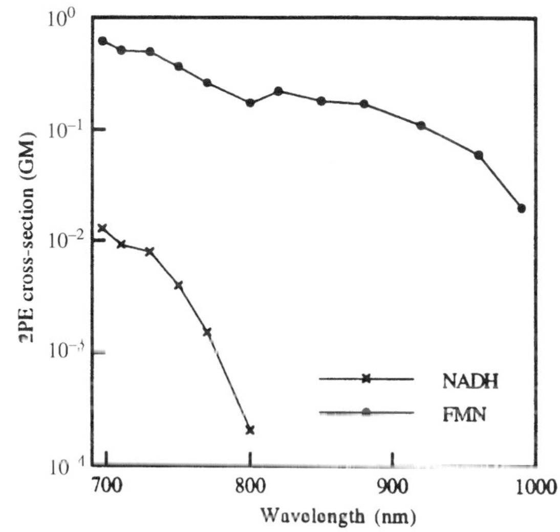

FIGURE 6 Two-photon action cross-sections of the native fluorophores NADH and FMN. NADH and FMN were purchased from Sigma. Note that the y-axis is a logarithmic scale.

TABLE 3 One-, two-, and three-photon excitation cross-sections σ. Cross-sections were measured in 10^{-4} M solution. η is the fluorescence quantum efficiency. The estimated uncertainties are 30% for the two-photon cross-sections and a factor of about 3 for the three-photon cross-sections

	σ_1 at λ nm (10^{-16} cm^2)	$\eta\sigma_2$ at 700 nm (10^{-50} cm^4 s/photon)	$\eta\sigma_3$ at 1.0 μm (10^{-83} cm^6 (s/photon)^2)
DAPI[a]	1.3 (345 nm)	0.16	0.25
Dansyl	0.17 (336 nm)	1	0.3
Fura-2 with Ca^{2+}	1.2 (335 nm)	12	30
Fura-2, free	1.0 (362 nm)	11	20
Indo-1, with Ca^{2+}	1.3 (340 nm)	1.5	6
Indo-1, free	1.3 (345 nm)	3.5	2

[a] DAPI not bound to DNA. The fluorescence quantum efficiency of DAPI is expected to go up by 20-fold upon binding to DNA (Haugland, 1992–94).

indicated in Figures 3 and 4. Finally, conventional dyes possessing different one-photon absorption spectra can be excited by 2PE at a single wavelength. This capability, combined with the large spectral separation between the excitation light and the fluorescence light, greatly simplifies experiments requiring multiple fluorophores.

We have observed and measured 3PE fluorescence from UV fluorophores such as the Ca^{2+} indicators fura-2 and indo-1, the DNA stain DAPI, and the fluorescent reagent dansylhydrazine by excitation at $\lambda \approx 1.0$ μm. These fluorophores are normally excited at approximately 330 nm to 360 nm, and exhibit no linear absorption at 500 nm or 1.0 μm. The three-photon energy of the 1.0 μm radiation falls into the UV absorption band, and therefore three-photon fluorescence is expected. The approximate three-photon cross-sections σ_3 are listed in Table 3. For comparison, the corresponding one- and two-photon cross-sections are also given.

To investigate the photophysics of 3PE of fluorophores, we have measured several 3PE spectra $\sigma_3(\lambda)$ from 960 nm to 1050 nm for comparison with 1PE spectra; see Figure 7. The qualitative agreement shown in Figure 7 suggests that these 3PE spectra parallel the corresponding 1PE spectra – as expected, because the same initial excited states can be reached via one- or three-photon excitation without violating any selection rules.

5 POWER DEPENDENCE OF 2PE AND 3PE FLUORESCENCE

The power-squared dependence of 2PE fluorescence for most of the dyes listed in this chapter was tested at one or several excitation wavelengths (Xu and Webb, 1996). Data for rhodamine B and fluorescein are summarized in Figure 8. Other results are listed in Table 4. In all cases, nearly perfect power-squared dependence of 2PE fluorescence was recorded.

Theoretically, 2PE fluorescence should obey the square-law dependence at low excitation intensity. However, because of various factors, such as stimulated emission (Hermann and Ducuing, 1972b), excited-state absorption (Bradley et al., 1972), saturation, intensity-dependent 2PE cross-section, lack of corrections for 1PE, etc., significant deviations from the square-law dependence of 2PE fluorescence have frequently been

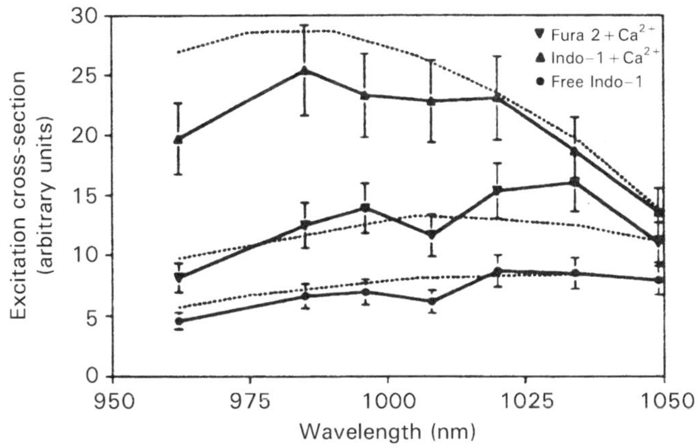

FIGURE 7 Comparison of one-photon (dashed lines) and three-photon (solid lines) fluorescence excitation of indo-1 and fura-2. For the one-photon results, the *x*-axis values represent three times the one-photon transition wavelengths. The one-photon results are normalized to the corresponding three-photon data at 1050 nm.

observed. Hermann and Ducuing (1972a,b) found that rhodamine B showed a linear dependence at 694 nm at peak intensity levels above 2×10^{25} photons $cm^{-2}\,s^{-1}$) using nanosecond pulses. They suggested that the deviations were caused by stimulated emission from the fluorescence state. Bradley et al. (1972) found that rhodamine B at 1064 nm and rhodamine 6G at 694 nm had similar deviations from a power-squared dependence at a peak intensity level of about 10^{28} photons ($cm^{-2}\,s^{-1}$) using picosecond pulses. They suggested stimulated emission and excited-state absorption as the cause for rhodamine 6G and excited-state absorption for rhodamine B. Figure 8*a* shows the intensity-squared dependence for rhodamine B when excited at 702 nm for peak intensity levels from 10^{27} to 10^{28} photons $cm^{-2}\,s^{-1}$ using 100 fs pulses. Despite the fact that a two orders of magnitude higher peak intensity than that in Hermann's experiment was used, the fluorescence obeys an intensity-squared dependence. Assuming that the deviation from square-law dependence observed by Hermann and Ducuing was indeed caused by stimulated emission alone (as they suggested), the discrepancy between the two experiments can be explained as follows. The proposed stimulated emission is a one-photon process and is proportional to the total number of photons incident on the excited fluorophore. Although a femtosecond light source provides a very high peak intensity, the average intensity is quite low. Therefore, very little stimulated emission is expected in our case. However, if pulses with

TABLE 4 Dependence of 2PE fluorescence on excitation peak intensity

Dye	Excitation wavelength (nm)	Peak intensity (photons $cm^{-2}\,s^{-1}$)	Slope of logarithmic plot
DiI	702	2.8×10^{27}–1.2×10^{28}	2.01 ± 0.03
Bodipy®	770	8.6×10^{27}–2.6×10^{28}	2.02 ± 0.03
Lucifer yellow	870	4.2×10^{27}–4.6×10^{28}	1.98 ± 0.03
Rhodamine B	1000	4.3×10^{28}–3.9×10^{29}	2.01 ± 0.03
Indo-1 (high-Ca)	754	2.9×10^{28}–5.5×10^{29}	2.00 ± 0.03

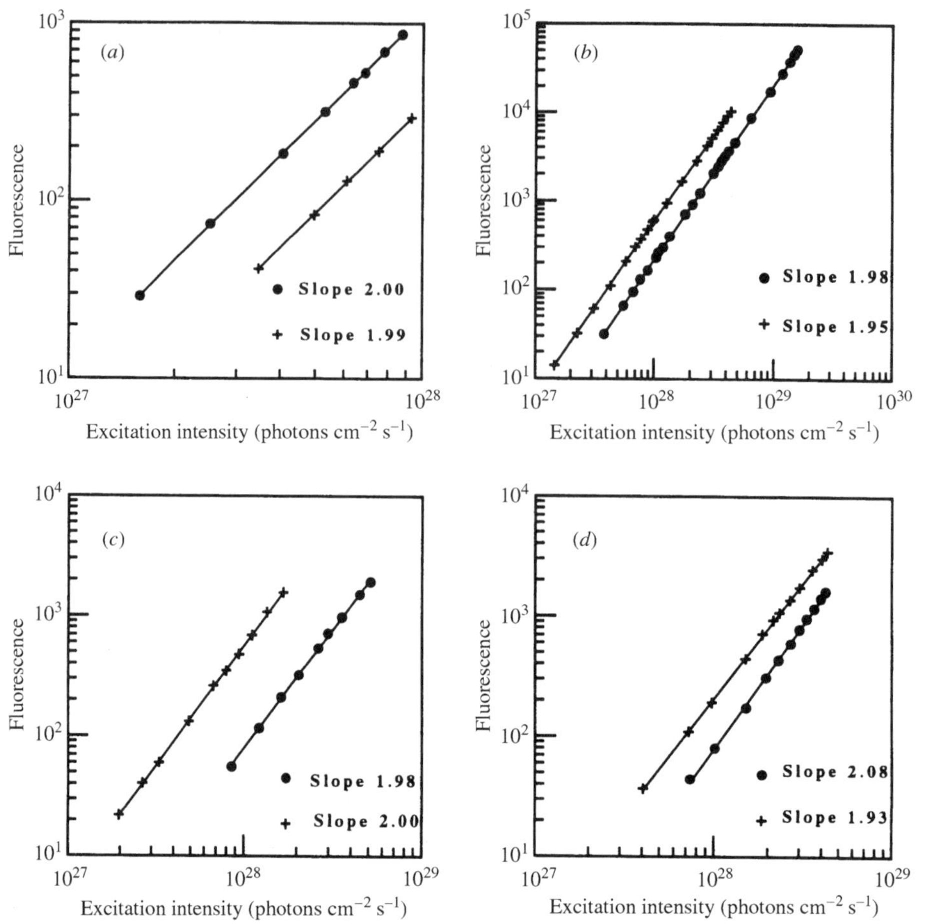

FIGURE 8 Logarithmic plots of the dependence of relative two-photon induced fluorescence on pulse intensity for a 10^{-4} M solution of rhodamine B in methanol (filled circles) and a 10^{-4} M solution of fluorescein in water (crosses). The excitation wavelengths λ were as follows: (*a*) 702 nm; (*b*) 770 nm; (*c*) 825 nm; (*d*) 870 nm. The fitted slopes are shown on each graph. The excitation pulsewidth is about 100 fs (assuming a sech^2 pulse shape). The ordinate scales of fluorescence are in unrelated arbitrary units. The estimated uncertainties of the fitted slopes are ± 0.03.

much longer durations are used, as in the experiments performed by Bradley et al. (ps and ns) and by Hermann and Ducuing (ns), stimulated emission could occur with significant probability at high intensity levels.

Three-photon excited fluorescence should ideally obey a power-cubed dependence. Figure 9 shows the power dependence for several fluorophores at relatively low intensity range. Using a high-NA water-immersion objective lens (NA = 1.2), no significant deviations from the cubic dependence were observed at I_p up to 6×10^{30} photons cm^{-2} s^{-1} for fura-2 and DAPI at 992 nm. However, dramatic decreases in the apparent power exponent were observed at photon flux densities higher than 10^{31} photons cm^{-2} s^{-1} for both fluorophores. Although photobleaching may account for this deviation, other factors, such as field-induced changes in the electronic properties of the fluorophores, may become

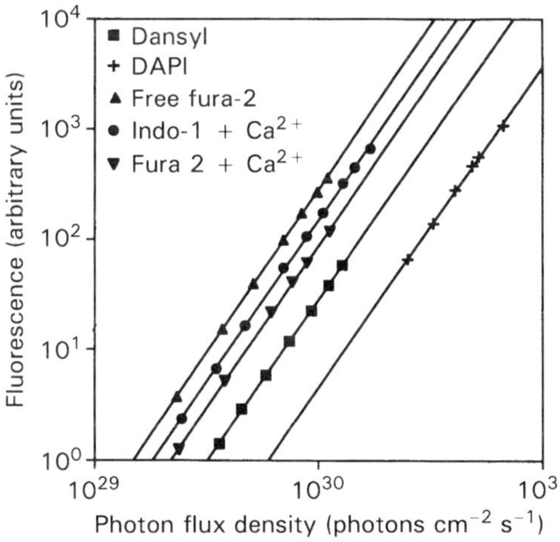

FIGURE 9 Logarithmic plots of the dependence of relative 3PE fluorescence on the peak incident photon flux density at the geometric focal points in five fluorophore solutions. The concentrations are 10^{-4} M and the excitation wavelength is 1.0 μm. The excitation pulsewidth is about 110 fs (assuming a sech^2 pulse shape). The solid lines represent the best power law fits to the experimental data, with slopes of 2.92, 2.96, 2.93, 2.94 and 2.91 for dansyl, free fura-2, indo-1+Ca^{2+}, fura-2 + Ca^{2+}, and DAPI, respectively. The estimated uncertainties in the slopes are ±3%.

important at very high intensity levels (Birge, 1983). The instantaneous electric field E at the focal spot is approximately 3×10^7 V cm^{-1} at $I_p = 10^{31}$ photons cm^{-2} s^{-1}, which is probably sufficient to modify the photophysical properties of the dye molecules.

A measurable combination of n-photon and $(n+1)$-photon excited fluorescence is expected within certain wavelength and intensity ranges. For example, thermally assisted 1PE could become significant at low intensity levels if the wavelength of attempted 2PE overlaps the absorption band for 1PE. This causes a severe departure from the square-law dependence. We have observed a nearly linear power dependence with rhodamine B at 705 nm using a single-mode CW laser at an excitation intensity of around 10^{24} photons cm^{-2}s^{-1}. A significant deviation from square-law dependence is found for rhodamine B at intensity levels of approximately 10^{24}–10^{25} photons cm^{-2} s^{-1} for $\lambda < 730$ nm, which were also observed by Hermann and Ducuing at 694 nm.

Mixture of fluorescence resulting from different-order photon excitations can also be obtained in attempting 3PE. For example, as an excitation ratiometric calcium indicator, the one-photon absorption maximum of fura-2 shifts from 362 nm in the Ca-free form to 335 nm on binding to Ca^{2+} (Grynkiewicz et al., 1985), and the corresponding shifts are also apparent in its two-photon fluorescence excitation spectra (Fig. 4). This absorption change makes the ratio between 3PE and 2PE fluorescence of the two species very different from each other. At $I_p \approx 10^{30}$ photons cm^{-2} s^{-1} (which is normally used in nonlinear laser microscopy; Williams et al., 1994), it is found that for the Ca-bound species, 2PE dominates at $\lambda < 830$ nm and 3PE at $\lambda > 900$ nm; while for the Ca-free species, 2PE dominates up to 910 nm and 3PE only at $\lambda > 960$ nm. At intermediate wavelengths, a mixture of two- and three-photon excitation was observed. Table 5 lists the measured power exponents for fura-2 at several wavelengths, which shows the gradual transition from 2PE to 3PE. Although the three-dimensional imaging capability is

TABLE 5 Slopes in the logarithmic plot of fluorescence output versus incident intensity at selected wavelengths. Slopes were measured at $I_p \approx 10^{30}$ photons $cm^{-2}s^{-1}$. A slope between 2.0 and 3.0 indicates that a mixture of 2PE and 3PE fluorescence was generated. The uncertainties in the slopes are about 6%

	838 nm	861 nm	880 nm	900 nm	922 nm	970 nm
Fura-2, $+Ca^{2+}$	2.2	2.4	2.6	2.7	2.9	3.0
Fura-2, free	2.0	2.0	2.0	2.1	2.3	2.9

preserved in this wavelength range, quantitative imaging, as is required in calcium measurements using ratiometric indicators, would be complicated by the calibration dependence on incident intensities.

Experimentally, intersystem crossing, photobleaching, and saturation cause severe deviation from the expected power-law dependence of the excited fluorescence. Photobleaching may account for the nonquadratic dependence of 2PE fluorescence in some earlier reports (Fischer et al., 1995).

6 1PE, 2PE, AND 3PE FLUORESCENCE EMISSION SPECTRA COMPARED

In most cases reported in the literature, 1PE and 2PE fluorescence emission spectra were the same (Curley et al., 1992). However, there are cases where difference between 1PE and 2PE fluorescence emission spectra were observed (Parma and Omenetto, 1978). Figure 10 shows the emission spectra of fluorescein under 1PE and 2PE. For all the fluorophores described in this chapter, 1PE and 2PE fluorescence emission spectra are identical. In general, 2PE fluorescence emission spectra are independent of excitation wavelengths (Fig. 11), supporting the assumption that the fluorescence quantum efficiency is a constant regardless of the excitation wavelength. There has been no reported difference between

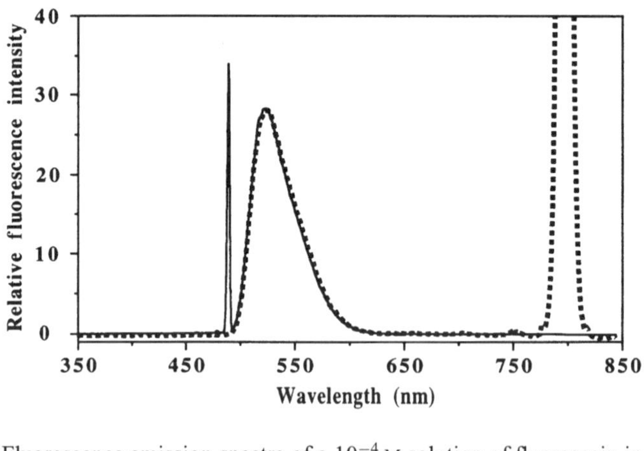

FIGURE 10 Fluorescence emission spectra of a 10^{-4} M solution of fluorescein in water excited by one photon (488 nm) (solid line) and by two photons (800 nm) (dashed line). The peaks at 488 nm and 800 nm are residual excitation light. The instrument wavelength resolution is 1.0 nm.

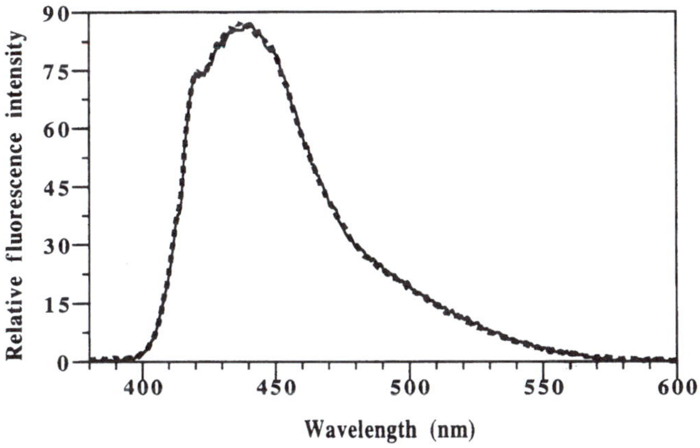

FIGURE 11 Two-photon excited fluorescence emission spectra of a 10^{-4} M solution of cascade blue in water excited at 750 nm (solid line) and at 800 nm (dashed line).

1PE and 3PE fluorescence emission spectra, suggesting that quite generally the same fluorescent states are reached, regardless of the excitation mode.

7 DISCUSSION

The maximum fluorescence output available for image formation is obtained using ultrashort (of order 100 fs) pulsed excitation. However, fluorescence saturates at the limit of one transition per pulse per fluorophore. For an n-photon process, fluorescence output per excitation pulse, neglecting ground-state depletion, is proportional to $\alpha \equiv \sigma_n I^n_{\text{peak}}\tau$, where I_{peak} is the peak intensity at the geometric focal point (Xu and Webb, 1996), σ_n is the n-photon cross-section, and τ is the pulsewidth. When $\alpha \ll 1$, α rigorously represents the excitation probability per molecule per pulse. For a square pulse in time, saturation occurs when $\alpha \approx 1$. With a high-NA (NA = 1.3) objective lens and a mode-locked Ti : sapphire laser providing 100 fs pulses at 80 MHz repetition rate and 1.0 μm excitation wavelength, the approximate estimated saturation powers for two-, three-, and four-photon processes are, respectively, 30 mW, 150 mW, and 300 mW using the excitation cross-sections estimated in Section 2. Although these incident power levels are easily available from femtosecond laser sources such as the mode-locked Ti : sapphire laser, solvent dielectric breakdown may occur before saturation is reached in three- and four-photon excitation.

The three dimensional resolution of two-photon laser scanning microscopy results from the quadratic power dependence of the 2PE fluorescence. However, such a quadratic relation no longer holds under the saturation condition (Xu and Webb, 1997). Thus, the spatial localization of 2PE begins to degrade when $\alpha \approx 1$, and will become nonlocalized when $\alpha \gg 1$. More detailed analysis of saturation under MPE can be found in Xu and Webb (1997). It is apparent that the axial resolution degrades significantly faster than does the radial resolution. Intuitively, this can be understood because the fluorescence power decreases much faster in the radial direction (exponentially) than in the axial direction (power-law). In the extreme hypothetical case where the edge of the fluorescence distribution is infinitely sharp (i.e., for a "tophat" distribution), there should be no decrease in resolution, regardless of the saturation level.

To increase the probability of simultaneous multiphoton incidence, a pulsed laser with high peak intensity is generally used as the excitation source. For example, a modelocked Ti : sapphire laser with 100 MHz repetition rate and 100 fs pulses provides enhancement factors of 10^5 and 10^{10} for 2PE and 3PE, respectively, when compared with a CW laser. However, such a high degree of temporal focus, combined with the tight spatial focus (typically $< 1\ \mu m^2$), results in a peak intensity that is many orders of magnitude higher than in normal 1PE. Although peak intensities at the focal point are generally less than $3 \times 10^{11}\ S\ cm^{-2}$ in routine 2PLSM, other applications can necessitate the use of far higher intensities. Uncaging of biological effector molecules in three-dimensionally specified volumes, for example, can require as much as $10^{12}\ W\ cm^{-2}$ (Brown et al., 1999). Note that the intensities used in 1PE are typically less than $10^6\ W\ cm^{-2}$. The high peak intensities used in MPE opens a range of new laser–matter interactions that have previously been neglected in 1PE. Although dielectric breakdown of pure water is not expected until peak intensities above $10^{13}\ W\ cm^{-2}$ in the femtosecond region (Smith, 1978; Du et al., 1994; Stuart et al., 1995), the photophysical responses of cellular materials and biological buffers to very intense radiation fields are yet to be explored in detail (Shear et al., 1997). In fact, knowledge of high-intensity damage phenomena is scarce even in nonbiological systems when 100 fs pulsed excitation is used. Nonetheless, there have been interesting applications in the area of micromachining using intense femtosecond pulsed sources (Perry et al., 1999).

Laser–matter interactions are strongly intensity-dependent. At relatively low intensities (generally $< 10^{12}\ W\ cm^{-2}$), linear and perturbative nonlinear phenomena dominate. At intensities significantly above $10^{12}\ W\ cm^{-2}$, nonperturbative nonlinear phenomena become important. It is worth noting that the intensities required for two-photon uncaging and three- or even four-photon imaging may lie between the perturbative and nonperturbative regions – an issue that can cause considerable complications in both the theoretical and experimental investigations of MPE. The relative contributions of various-order processes depends on the incident intensity. As the intensity increases, so does the relative contribution of the higher-order processes. Thus, mechanisms of photobleaching and photodamage can change fundamentally at high intensities. Higher-order processes may contribute significantly in the high-intensity region, a phenomenon that has been observed in our measurements of serotonin photobleaching (Shear et al., 1997). The possibility of higher-order excitation complicates quantitative experiments on possible photodamage during two-photon imaging.

8 CONCLUSIONS

We have reviewed two-photon fluorescence excitation spectra of the most common biological fluorophores in the tuning range of a mode-locked Ti : sapphire laser. These spectra show that nonlinear fluorescence microscopy based on 2PE works well with most existing fluorophores and conveniently available laser sources. We have also measured three-photon fluorescence excitation cross-sections. Although three-photon microscopy requires 5–10 times higher incident power than comparable two-photon laser scanning microscopy to obtain a comparable fluorescence excitation rate, it does provide alternative wavelength windows to probe biological specimens.

The range of the measured cross-section values lies around the estimate based on perturbation theory. Although such two-photon cross-sections have proven to be large enough for imaging, the required high peak power necessitates the use of ultrashort excitation pulses, and therefore complex and costly laser systems. A much larger two-photon cross-section is clearly desirable for two-photon laser scanning microscopy and some other fields. Significant efforts have already been devoted to the design and synthesis of dyes with very large two-photon cross-sections (Albota et al., 1998b; Ventelon et al., 1999). Although challenging, such endeavors will certainly be richly rewarded.

ACKNOWLEDGMENT

The author thanks Professor Watt W. Webb for providing a wonderful research environment at Cornell University, where all the cross-section work was done.

REFERENCES

Albota, M. A., Xu, C., et al. (1998). Two-photon excitation cross sections of biomolecular probes from 690 to 980 nm. *Appl. Opt.* **37**, 7352–7356.

Albota, M. A., Beljonne, D., Brédas, J.-L., Ehrlich, J. E., Fu, J.-Y., Heikal, A. A., Hess, S. E., Kogej, T., Levin, M. D., Marder, S. R., McCord-Maughon, D., Perry, J. W., Röckel, H., Rumi, M., Subramaniam, G., Webb, W. W., Wu, X.-L., and Xu, C. (1998b). Design of organic molecules with large two-photon absorption cross sections. *Science* **281**, 1653–1656.

Birge, R. R. (1983). One-photon and two-photon excitation spectroscopy. In: *Ultrasensitive Laser Spectroscopy* (ed. D. S. Kliger). New York: Academic Press, pp. 109–174.

Bradley, D. J., Hutchinson, M. H. R., et al. (1972). Interaction of picosecond laser pulses with organic molecules I: Two-photon fluorescence quenching and singlet states excitation in rhodamine dyes. *Proc. R. Soc. Lond.* **A328**, 97–121.

Brown, E. B., Shear, J. B., et al. (1999). Photolysis of caged calcium in femtoliter volumes using two-photon excitation. *Biophys. J.* **76**, 489–499.

Chalfie, M., Tu, Y., et al. (1994). Green Fluorescence Protein as a marker for Gene Expression. *Science* **263**, 802–805.

Chen, B. E., L. B., Nimchinsky, E. A., Burbach, B., Fox, K., and Svoboda, K. (2000). Imaging high-resolution structure of GFP-expressing neurons in neocortex in vivo. *Learning and Memory* **7**, 433–441.

Cormack, B. P., Valdivia, R. H., et al. (1996). FACS-optimized mutants of the green fluorescent protein (GFP). *Gene* **173**, 33–38.

Cubitt, A. B., Heim, R., et al. (1995). Understanding, improving and using fluorescent proteins. *TIBS* **20**, 448–455.

Curley, P. F., Ferguson, A. I., et al. (1992). Application of a femtosecond self-sustaining mode-locked Ti : sapphire laser to the field of laser scanning confocal microscopy. *Opt. Quant. Electron.* **24**, 851–859.

Denk, W., Strickler, J. H., et al. (1990). Two-photon laser scanning fluorescence microscopy. *Science* **248**, 73–76.

Dickson, L. D. (1970). Characteristics of a propagating gaussian beam. *Appl. Opt.* **9**, 1854–1861.

Diels, J.-C. (1990). Femtosecond dye lasers. In: *Dye Laser Principles* (ed. F. J. Duarte and L. W. Hillman). San Diego: Academic Press, pp. 104–109.

Du, D., Liu, X., et al. (1994). Laser-induced breakdown by impact ionization in SiO_2 with pulsewidths from 7 ns to 150 fs. *Appl. Phys. Lett.* **64**, 3071–3073.

Faisal, F. H. M. (1987). *Theory of Multiphoton Processes*. New York: Plenum Press.

Fischer, A., Cremer, C., et al. (1995). Fluorescence of coumarins and xanthenes after two-photon absorption with a pulsed titanium–sapphire laser. *Appl. Opt.* **34**, 1989–2003.

Göppert-Mayer, M. (1931). Über Elementarakte mit zwei Quantensprüngen. *Annln. Phys.* **9**, 273–295.

Grynkiewicz, G., Poenie, M., et al. (1985). A new generation of Ca^{2+} indicators with greatly improved fluorescence properties. *J. Biol. Chem.* **260**, 3440–3450.

Guild, J., Xu, C., et al. (1997). Measurement of group delay dispersion of high numerical aperture objectives using two-photon excited fluorescence. *Appl. Opt.* **36**, 397–401.

Haughland, R. P. (Ed.) (1992–94). *Molecular Probes Handbook 1992–1994*, p. 222.

He, G. S., Bhawalkar, J. D., et al. (1995). Three-photon-absorption-induced fluorescence and optical limiting effects in an organic compound. *Opt. Lett.* **20**, 1524–1526.

Heikal, A. A., Hess, S. T., et al. (2000). Molecular spectroscopy and dynamics of intrinsically fluorescent proteins: coral red (dsRed) and yellow (Citrine). *Proc. Natl. Acad. Sci. USA* **97**, 11996–12001.

Hell, S., Reiner, G., et al. (1993). Aberrations in confocal microscopy induced by mismatches in refractive index. *J. Microsc.* **169**, 391–405.

Hermann, J. P. and Ducuing, J. (1972a). Absolute measurement of two-photon cross-sections. *Phys. Rev.* **A5**, 2557–2568.

Hermann, J. P. and Ducuing, J. (1972b). Dispersion of the two-photon cross-section in rhodamine dyes. *Opt. Comm.* **6**, 101–105.

Jones, R. D. and Callis, P. R. (1988). A power-square sensor for two-photon spectroscopy and dispersion of second-order coherence. *J. Appl. Phys.* **64**, 4301–4305.

Kaatz, P. and Shelton, D. P. (1999). Two-photon fluorescence cross-section measurements calibrated with hyper-Rayleigh scattering. *J. Opt. Soc. Am.* **B16**, 998–1006.

Kaiser, W. and Garrett, C. G. B. (1961). Two-photon excitation in $CaF_2:Eu^{2+}$. *Phys. Rev. Lett.* **7**, 229.

Kang, I., Krauss, T., et al. (1997). Sensitive measurement of nonlinear refraction and two-photon absorption by spectrally-resolved two-beam coupling. *Opt. Lett.* **22**, 1077–1079.

Kennedy, S. M. and Lytle, F. E. (1986). *p*-Bis(*o*-methylstyryl)benzene as a power-square sensor for two-photon absorption measurements between 537 and 694 nm. *Anal. Chem.* **58**, 2643–2647.

König, K., Liu, Y., et al. (1995). Autofluorescence spectroscopy of optically trapped cells. *Photochem. Photobiol.* **62**, 830–835.

König, K., So, P. T. C., et al. (1996). Two-photon excited lifetime imaging of autofluorescence in cells during UVA and NIR photostress. *J. Microsc.* **183**, 197.

Loudon, R. (1983). *The Quantum Theory of Light.* Oxford: Clarendon Press.

McClain, W. M. and Harris, R. A. (1977). Two-photon molecular spectroscopy in liquids and gases. In: *Excited States* (ed. E. C. Lim). New York: Academic Press, pp. 1–56.

Parma, L. and Omenetto, N. (1978). Fluorescence behavior of 7-hydroxycoumarine excited by one-photon and two-photon absorption by means of a tunable dye laser. *Chem. Phys. Lett.* **54**, 544–546.

Perry, M. D., Stuart, B. C., et al. (1999). Ultrashort-pulse laser machining of dielectric materials. *J. Appl. Phys.* **85**, 6803–6810.

Piston, D. W., Masters, B. R., et al. (1995). Three dimensionally resolved NAD(P)H cellular metabolic redox imaging of the in situ cornea with two-photon excitation laser scanning microscopy. *J. Microsc.* **178**, 20–27.

Potter, S. M., Wang, C. M., et al. (1996). Intravital imaging of green fluorescent protein using two-photon laser-scanning microscopy. *Gene* **173**, 25–31.

Sengupta, P. B. J., Banerjee, S., Philip, R., Kumar, G. R., and Maiti, S. (2000). Sensitive measurement of absolute two-photon absorption cross-sections. *J. Chem. Phys.* **112**, 9201–9205.

Shear, J. B., Xu, C., et al. (1997). Multiphoton-excited visible emission by serotonin solutions. *Photochem. Photobiol.* **65**, 931–936.

Sheik-Bahae, M., Said, A. A., et al. (1990). Sensitive measurement of optical nonlinearities using a single beam. *IEEE J. Quant. Electron.* **26**, 760–769.

Sheppard, C. J. R. and Matthews, H. J. (1987). Imaging in high-aperture optical systems. *J. Opt. Soc. Am.* **A4**, 1354–1360.

Smith, W. L. (1978). Laser-induced breakdown in optical materials. *Opt. Eng.* **17**, 489–503.

Stuart, B. C., Feit, M. D., et al. (1995). Laser-induced damage in dielectrics with nanosecond to subpicosecond pulses. *Phys. Rev. Lett.* **74**, 2248–2251.

Swofford, R. and McClain, W. M. (1975). The effect of spatial and temporal laser beam characteristics on two-photon absorption. *Chem. Phys. Lett.* **34**, 455–460.

Twarowski, A. J. and Kliger, D. S. (1977). Multiphoton absorption spectra using thermal blooming. *Chem. Phys.* **20**, 259.

Ventelon, L., Moreaux, L., et al. (1999). New quadrupolar fluorophores with high two-photon excited fluorescence. *Chem. Commun.* **20**, 2055–2056.

Williams, R. M., Piston, D. W., et al. (1994). Two-photon molecular excitation provides intrinsic 3-dimensional resolution for laser-based microscopy and microphotochemistry. *FASEB J.* **8**, 804–813.

Wilson, T. and Sheppard, C. J. R. (1984). *Theory and Practice of Scanning Optical Microscopy.* New York: Academic Press.

Xu, C. and Webb, W. W. (1996). Measurement of two-photon excitation cross-sections of molecular fluorophores with data from 690 nm to 1050 nm. *J. Opt. Soc. Am.* **B13**, 481–491.

Xu, C. and Webb, W. W. (1997). Multiphoton excitation of molecular fluorophores and nonlinear laser microscopy. In: *Topics in Fluorescence Spectroscopy*, Vol. 5 (ed. J. Lakowicz). New York: Plenum Press, pp. 471–540.

Xu, C., Guild, J., et al. (1995). Determination of absolute two-photon excitation cross-sections by in situ second-order autocorrelation. *Opt. Lett.* **20**, 2372–2374.

Xu, C., Zipfel, W., et al. (1996). Multiphoton fluorescence excitation: new spectral windows for biological nonlinear microscopy. *Proc. Natl. Acad. Sci. USA* **93**, 10763–10768.

Yuste, R. and Denk, W. (1995). Dendritic spines as basic function units of neuronal integration. *Nature* **375**, 682–684.

CHAPTER 5

Resolution and Contrast in Confocal and Two-Photon Microscopy

JAMES E. N. JONKMAN and ERNST H. K. STELZER

Cell Biology and Biophysics Programme, European Molecular Biology Laboratory, Meyerhofstrasse 1, D-69117 Heidelberg, Germany

1 GOALS OF LIGHT MICROSCOPY

All light microscopes, including conventional, confocal, and two-photon microscopes, are fundamentally limited in the resolution that they can achieve. The resolution, defined as the minimum separation of two point objects that provides a certain contrast between them, depends on the wavelength of the light (both incident and detected), on the numerical aperture (NA) of the optical arrangement, and on the specimen or the experiment that one wishes to perform. By this last limitation, we mean that any "real" specimen cannot be imaged at the theoretically achievable resolution of the instrument. For example, fueled by recent advances in fluorescent protein technology (Ellenberg et al., 1999; White and Stelzer, 1999; Zacharias et al., 2000), it has become easier to investigate active processes in living cells and tissues. However, such live specimens are optically thick and inhomogeneous: a focused beam of light cannot maintain an ideal profile while penetrating the sample. Furthermore, live cells are dynamic, and sensitive to photobleaching and thermal damage: these factors limit the duration for which they can be observed and the power of the incident light. The inherent properties of a specimen restrict the way in which it can be imaged and, consequently, limit the achievable resolution.

The goal of light microscopy is to provide a tool for visualizing a specimen, with a certain resolution and contrast, and on a particular timescale. These parameters can vary widely from one experiment to the next. In cell biology, some experiments are best accomplished by triple immunofluorescence labeling, fixing of the specimen, and subsequent high-resolution, three-dimensional imaging to determine the relative localization of the fluorescently labeled structures (Figure 1*a*). In such an experiment, the microscope should provide excellent resolution, but the speed with which it acquires the images is not critical. Study of the dynamics of proteins in vivo, however, requires that the microscope be capable of collecting images quickly (at a rate of, say, 5–10 images per second). In this case, we must allow that the quality of the images will not be as high as in

Confocal and Two-Photon Microscopy: Foundations, Applications, and Advances, Edited by Alberto Diaspro.
ISBN 0-471-40920-0

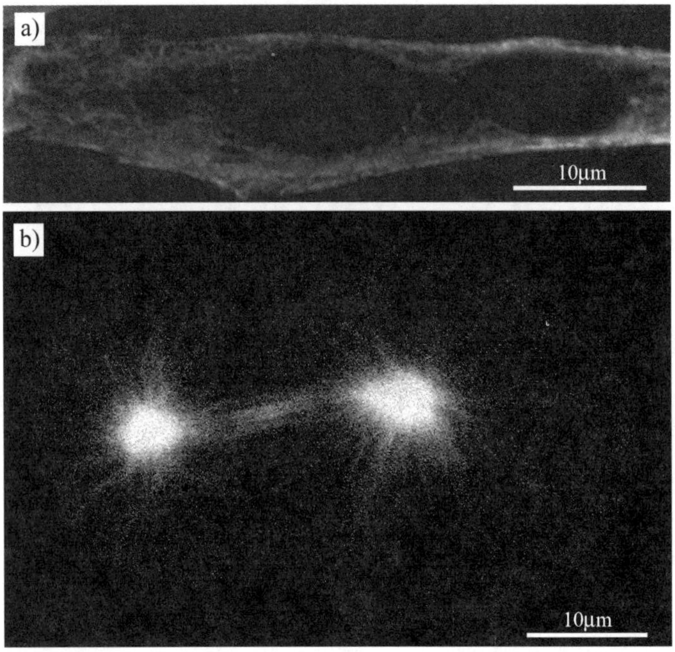

FIGURE 1 See color insert. (*a*) Confocal fluorescence image (slice 20 in a stack of 30 images) of an immunofluorescently labeled, fixed myoblast cell. In this experiment, the relative localization of three proteins (green: tubulin; red: actin; blue: myosin) is investigated at different time points during muscle tissue morphogenesis. Since the samples were fixed and sufficient time was available, each fluorophore distribution was recorded separately. This guaranteed the highest excitation efficiency and the optimal spectral separation. The acquisition time was 30 s per slice. (*b*) Confocal fluorescence image of GFP-labeled tubulin in a live *Caenorhabolitis elegans* embryo. Measuring protein dynamics during the first cell division requires a rapid sequence of images (about two images per second for several minutes).

the fixed-specimen case (Figure 1*b*). In terms of resolution, then, a light microscope should provide the best possible resolution within the constraints of a given experiment.

We begin this chapter by calculating and comparing the lateral and axial extents of the point spread functions of confocal, two-photon, and several other fluorescence microscope arrangements. We also explain why optical sectioning and not an increased axial resolution is the main advantage of both confocal and two-photon microscopy over conventional microscopy. Next, we describe how the properties of real specimens limit resolution from a practical point of view. Techniques for improving the resolution of confocal and two-photon microscope arrangements are considered. We conclude by putting confocal and two-photon resolution in perspective against the backdrop of live-cell microscopy.

2 COMPARISON OF MICROSCOPE PROPERTIES

2.1 Point Spread Functions

A useful tool for comparing the performance of optical microscopes is the point spread function (PSF). The amplitude PSF of a microscope is related to the strength of the electromagnetic field in the image plane when observing a pointlike light source (Born and Wolf, 1999, pp. 484–499). The amplitude PSF can be regarded as a kind of system response of the microscope. We cannot observe amplitude fluctuations directly; but we can observe the squared modulus of the amplitude PSF, namely the intensity PSF, which is the image formed by the point source in the microscope ocular or camera. Of course, real specimens are rarely point sources; but they can be regarded as a superposition of pointlike, subresolution objects. The extents of the intensity PSF in both the lateral and axial directions are related to the resolution of a microscope.

One can approximately measure the intensity PSF of a microscope by recording the image of a subresolution bead as the bead is scanned through the focus. However, aberrations, noise, and other nonlinear effects degrade the image, and hence it requires some effort to experimentally observe the calculated intensity PSFs presented in this chapter. For simplicity, the calculations presented here do not take into account polarization, although this can be done (Rohrbach and Stelzer, 2001; van der Voort and Brakenhoff, 1990; Hell et al., 1993). Comparing the calculated intensity PSFs allows one to evaluate the lateral and axial performances, as well as the optical-sectioning capabilities of conventional, confocal, two-photon, and other microscope arrangements.

The point spread function is a description of the system response of a microscope in real space. There are advantages to comparing the system responses in frequency space by calculating transfer functions rather than PSFs (Goodman, 1968, pp. 111–120; Frieden, 1967). However, coherent transfer functions are the Fourier transforms of amplitude PSFs; and optical transfer functions are the autocorrelation functions of coherent transfer functions. Hence, the two descriptions are mathematically equivalent.

In calculating the PSFs, several assumptions are made: (1) the optical system is linear – no light is absorbed or scattered in the system; (2) the optical system is invariant – the image of an object is independent of the object's position in the field of view. These assumptions lead to an appreciable error in the calculated PSFs for high-NA (>1.3) systems.

2.2 Calculating PSFs

A detailed treatment of how one calculates the three-dimensional distribution of light near the focus of a well-corrected lens is presented in Born and Wolf (1999, pp. 484–499). For a source whose vacuum wavelength is λ, the amplitude PSF is given by

$$h(u, v) = -i\frac{2\pi nA\sin^2\alpha}{\lambda}e^{iu/\sin^2\alpha}\int_0^1 J_0(v\rho)e^{-iu\rho^2/2}\rho\, d\rho, \tag{1}$$

where n is the refractive index of the medium in which the source is embedded, $n\sin\alpha$ is the numerical aperture of the objective lens, A is the area of the illumination or detection aperture, and J_0 is the zeroth-order Bessel function. With the optical axis in the z direction, and $r = \sqrt{x^2+y^2}$ is the lateral distance from the optical axis; v and u are normalized optical units perpendicular and parallel to the optical axis, respectively, and are given by

$$v = \frac{2\pi nr\sin\alpha}{\lambda}, \qquad u = \frac{2\pi nz\sin^2\alpha}{\lambda}$$

The intensity PSF $|h(u, v)|^2$ is calculated as the product of the amplitude PSF and its complex conjugate:

$$|h(u, v)|^2 = h(u, v)h^*(u, v). \tag{2}$$

This equation describes the three-dimensional intensity distribution of a beam of light focused by a lens, or equivalently the spatial absorption pattern of a uniform fluorophore solution in the vicinity of the focus. It can also describe the probability of a photon, emitted from a certain location in the object, hitting a point detector in an image plane of the microscope. In these two senses, the PSFs are referred to as the illumination intensity PSF and detection intensity PSF, respectively.

Where appropriate the calculations were performed using parameters suitable for exciting enhanced green fluorescent protein EGFP (Xu et al., 1996; Jakobs et al., 2000):

Single-photon illumination wavelength $\lambda_{\text{ill}} = 488\,\text{nm}$
Two-photon illumination wavelength $\lambda_{2p\text{-ill}} = 900\,\text{nm}$
Detection wavelength $\lambda_{\text{det}} = 520\,\text{nm}$
Numerical aperture NA $= 1.3$ or 0.95
Refractive index $n = 1.518$

The pinhole diameter for calculations involving confocal intensity PSFs was assumed to be smaller than one Airy unit.

2.3 Comparing PSFs

Calculated PSFs are shown in Figure 2 for various microscope arrangements. Important properties of the PSF are its lateral and axial extents and the heights of the higher-order maxima (side lobes) relative to its central peak. The lateral and axial resolution of a microscope is related to the lateral and axial extents of its intensity PSF. The full-width at half-maximum (FWHM) extents of the central peak are used to quantify the comparison (Table 1).

In a laser scanning fluorescence microscope (such as the confocal microscope), a focused beam excites the object point-by-point. The focused beam is described by the illumination intensity PSF shown in Figure 2*a*. In the conventional fluorescence microscope, the entrance aperture of the objective lens is uniformly illuminated, but pointwise (i.e., spatially resolved) detection is performed by the camera or the eye of the observer. The detection intensity PSF of a conventional fluorescence microscope is shown in Figure 2*b*. The FWHM extent of the central peak is 0.20 μm in the lateral direction and 0.84 μm in the axial direction. The illumination and detection intensity PSFs are both described by Eq. (2), using the wavelengths $\lambda_{\text{ill}} = 488\,\text{nm}$ for the illumination intensity PSF and $\lambda_{\text{det}} = 520\,\text{nm}$ for the detection intensity PSF. The extents of the detection intensity PSF are broader than those of the illumination intensity PSF by the factor $\lambda_{\text{ill}}/\lambda_{\text{det}}$ in both lateral and axial directions.

A confocal microscope uses both pointwise illumination and pointwise detection. In a confocal fluorescence microscope, the objective lens forms an image of the illumination pinhole and the detection pinhole in their common conjugate image plane, the object plane. Only the fluorophores that are in the volume shared by the illumination and

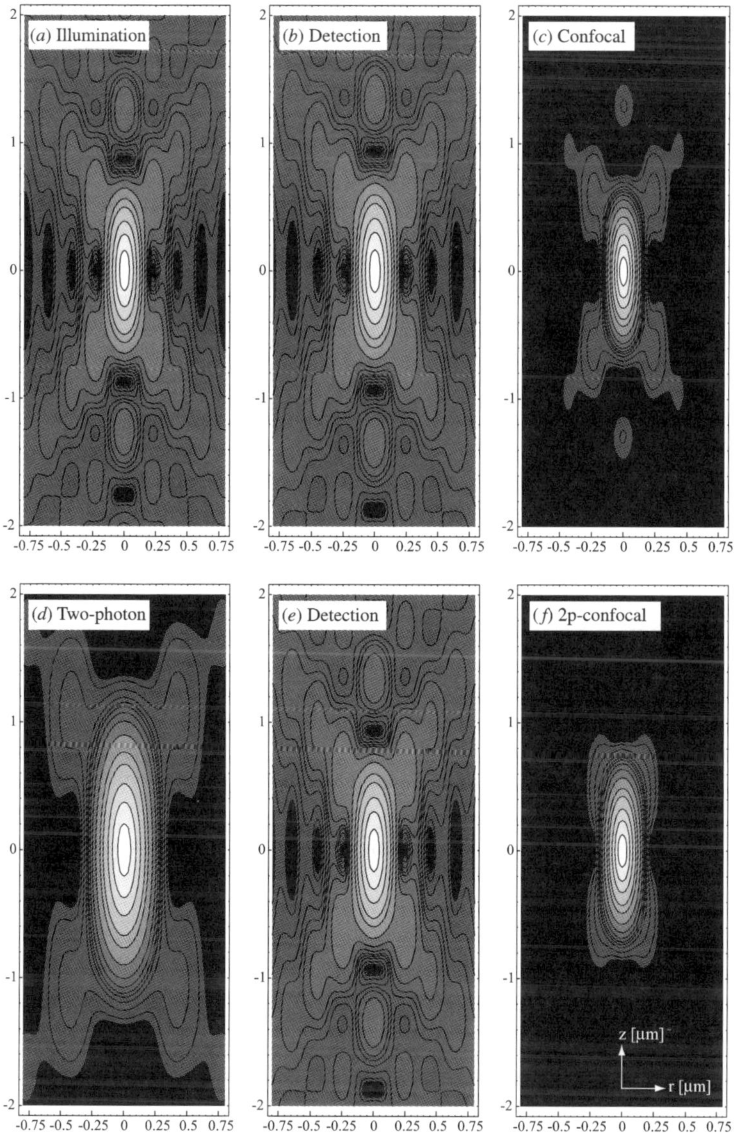

FIGURE 2 Calculated point spread functions for fluorescence microscope arrangements. Cylindrical symmetry and circular polarization are assumed. Effects due to aberration or low Fresnel numbers are not present. (*a*) Illumination PSF ($\lambda_{ill} = 488\,nm$). (*b*) Fluorescence-detection PSF ($\lambda_{det} = 520\,nm$). (*c*) Confocal fluorescence PSF, which is the product of the illumination and detection PSFs. (*d*) Two-photon PSF ($\lambda_{2p\text{-}ill} = 900\,nm$). (*e*) Fluorescence-detection PSF (same as (*b*)). (*f*) 2p-confocal PSF, which is the product of the two-photon PSF and the fluorescence detection PSF. All PSFs are calculated using NA = 1.3, $n = 1.518$. Contour lines are calculated for intensities of 0.9, 0.7, 0.5, 0.3, 0.2, 0.1, 0.05, 0.03, 0.02, 0.015, 0.01, 0.005, 0.002, and 0.001 (Born and Wolf, 1999, p. 490).

TABLE 1 FWHM extents of calculated PSFs; and resolutions according to Eqs. (6)–(11). The parameters are the same as those described in Figure 2

	FWHM extent (μm)		Resolution (μm)	
PSF	Lateral	Axial	Lateral	Axial
Illumination (ill)	0.19	0.78	0.23	0.88
Detection (det)	0.20	0.84	0.24	0.93
Confocal = ill × det	0.14	0.57	0.16	0.65
Two-photon = $(\text{ill})^2$	0.23	0.93	0.28	1.07
$2p$-confocal = $(\text{ill})^2$ × det	0.16	0.63	0.18	0.72

detection PSFs have a reasonable chance to be both excited and detected. Since these two events – excitation and detection – are independent, the confocal fluorescence intensity PSF is the product of the illumination intensity PSF and the detection intensity PSF:

$$|h_{\text{confocal}}(x, y, z)|^2 = |h_{\text{det}}(x, y, z)^2|h_{\text{ill}}(x, y, z)|^2. \tag{3}$$

The resulting confocal intensity PSF, shown in Figure 2*c*, is narrower in both the lateral and axial extents by about 30% (Table 1). (Assuming that the illumination and detection wavelengths are approximately the same, the FWHM extents of the intensity PSF are reduced by a factor of $1/\sqrt{2}$ compared with the conventional microscope PSF.) The side lobes of the confocal PSF are also reduced relative to the central peak, meaning that more of the intensity is concentrated at the center of the focal volume. Both of these factors contribute to an improvement in resolution for the confocal over the conventional fluorescence microscope.

In a two-photon microscope, the fluorescence emission is proportional to the probability of (nearly) simultaneously absorbing two low-energy photons. Since two independent events have to occur, this probability is proportional to the square of the excitation intensity. Uniform (i. e., nonconfocal) detection is performed using a large-area detector. The intensity PSF of a two-photon microscope is, therefore, the square of the illumination intensity PSF (Sheppard and Gu, 1990):

$$|h_{\text{two-photon}}(x, y, z)|^2 = [|h_{\text{ill}}(x, y, z)|^2]^2. \tag{4}$$

Comparing two-photon and single-photon absorption cross-sections for EGFP, we find that the peak of the two-photon absorption cross-section occurs at a little less than twice the wavelength of the single-photon absorption peak (Xu et al., 1996). Using $\lambda_{2p\text{-ill}} = 900\,\text{nm}$ for the two-photon illumination wavelength, $\lambda_{\text{ill}} = 488\,\text{nm}$ for the single-photon illumination wavelength, and $\lambda_{\text{det}} = 520\,\text{nm}$ for the fluorescence detection wavelength, we find that, for observing EGFP-labeled specimens, the extents of the two-photon intensity PSF are more than 60% greater than those of the confocal PSF, and 15% greater than those of a conventional fluorescence microscope.

The extent of the two-photon microscope intensity PSF can be improved by adding a point detector (i. e., a pinhole) to the detection path (Stelzer et al., 1994). The two-photon confocal (2p-confocal) intensity PSF (Figure 2*e*) is the product of the squared illumination intensity PSF and the detection intensity PSF:

$$|h_{2p\text{-confocal}}(x, y, z)|^2 = [|h_{\mathrm{ill}}(x, y, z)|^2]^2 |h_{\mathrm{det}}(x, y, z)|^2. \tag{5}$$

Again using EGFP as an example, the extents of the 2*p*-confocal PSF are still about 10% worse than those of a comparable single-photon confocal microscope.

To summarize, the components of the calculated intensity PSFs in the lateral ($z = 0$) and axial ($r = 0$) directions are plotted in Figure 3 for various microscope arrangements. Taking the extents of these PSFs as indications of resolution, one might conclude that the confocal microscope provides improved resolution, both laterally and axially, over the conventional fluorescence and 2p-confocal microscopes. Moreover, the confocal microscope has a significant resolution advantage over the two-photon microscope. In the lateral direction, these conclusions are valid under ideal imaging conditions: practical limitations are considered below. In the axial direction, the true advantage of confocal and two-photon microscopy over conventional fluorescence microscopy is that they provide an optical sectioning property in thick fluorescent specimens, as discussed in the following section.

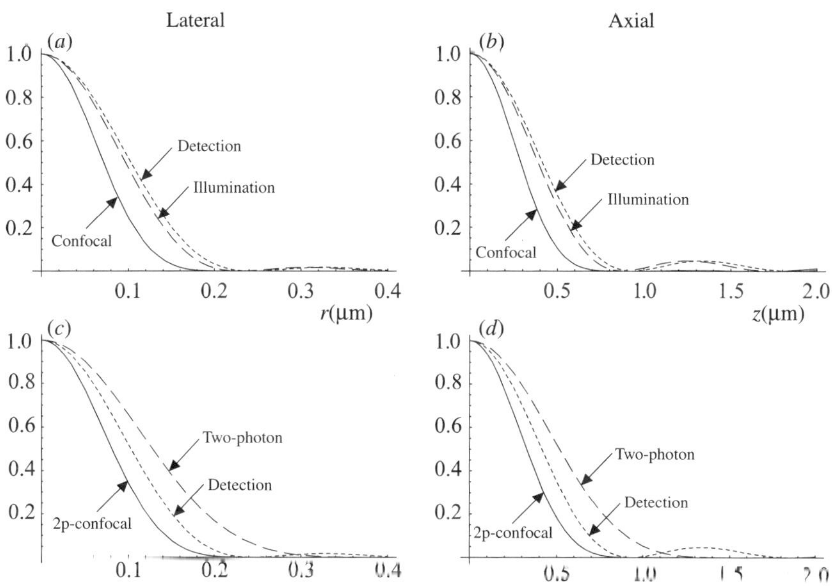

FIGURE 3 Orthogonal extents of the illumination, detection, and confocal PSFs calculated under exactly the same conditions as the contour plots: (*a*) lateral ($z = 0$) and (*b*) axial ($r = 0$) extents. Orthogonal extents of the two-photon, detection, and 2*p*-confocal PSFs in (*c*) lateral ($z = 0$) and (*d*) axial ($r = 0$) directions. Note the different scales for the lateral and axial plots. The plots show clearly that all three techniques are capable of distinguishing point objects along the optical axis. With such objects, every conventional microscope will perform better than a two-photon microscope.

3 DEFINITIONS OF RESOLUTION

3.1 Lateral Resolution

The lateral components of the intensity PSFs (i. e., the Airy patterns) for two pointlike objects of equal intensity in a conventional fluorescence microscope are shown in Figure 4 for various separation distances. The resolution of a microscope is not defined by, but rather is related to, the FWHM extent of its PSF in the component directions. Resolution is defined as the minimum separation between two point objects, such that they can still be "sufficiently" distinguished. There is some ambiguity as to what constitutes "sufficiently" separated PSFs. In the Sparrow criterion (Hecht, 1998, p. 465), the two objects are no longer resolved when the dip between their two Airy disc maxima just disappears (Fig. 4*c*). The well-known Rayleigh criterion (Hecht, 1998, p. 465) suggests that two point objects are resolved when the first minimum of one Airy disc is aligned with the central maximum of the second Airy disc (Fig. 4*b*). To quantify resolution, we use the term contrast, defined as the difference between the maximum intensity of the two objects and the minimum intensity found between them. Using this definition, the Sparrow criterion defines the resolution as the cutoff distance, that is, the separation at and below which zero contrast is achieved. The Rayleigh criterion corresponds to the separation at which a 26.4% contrast

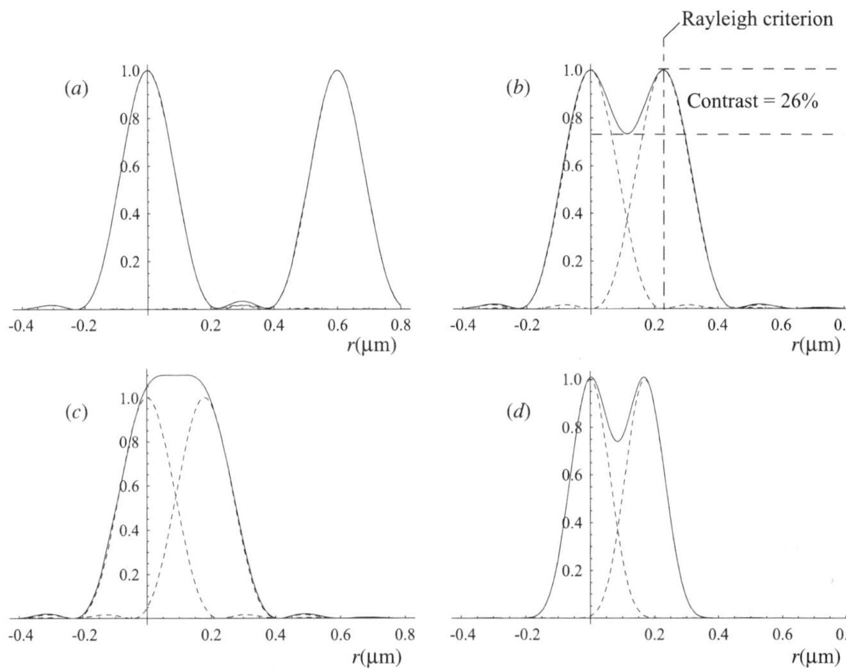

FIGURE 4 Lateral behavior of the single-photon illumination intensity PSF (i.e., the Airy function) for two point objects at various separation distances. (*a*) The point objects are well separated. (*b*) The point objects, separated by the distance specified by the Rayleigh criterion, provide a 26% contrast. (*c*) The point objects, separated by the distance specified by the Sparrow criterion, provide no contrast. At this cutoff distance, the objects are not resolved. (*d*) Lateral behavior of the confocal intensity PSFs for two point objects at the separation that produces a 26% contrast. Note that the two point objects can be closer together.

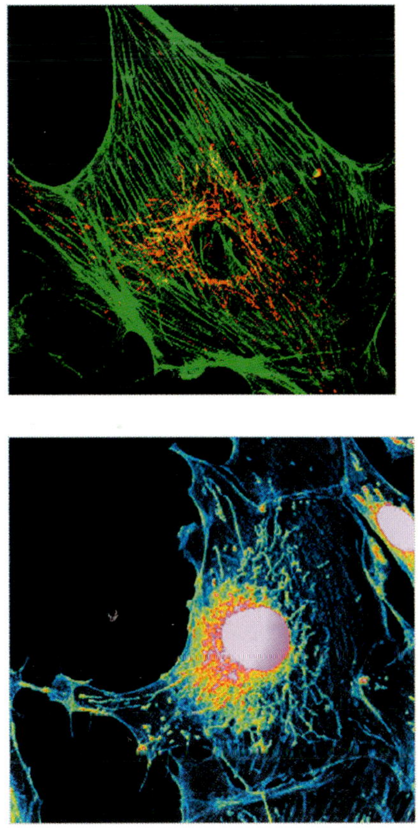

Figure 3.12 Bovine pulmonary artery endothelial cells (F-147780, Molecular Probes, Oregon) marked with Milo Tracker Red, Bodipy FL phallacidin, and DAPI specific for mitochondria, F-actin, and DNA respectively, Confocal imaging (*a*) has been performed at 488 and 543 nm, TPE imaging (*b*) was performed at 720 nm. In the confocal image, DAPI could not be excited owing to the lack of an appropriate UV source.

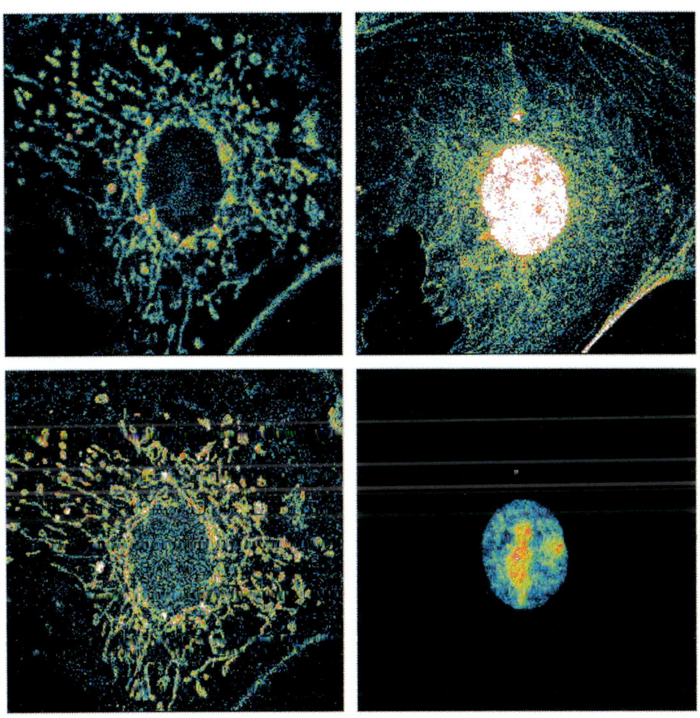

Figure 3.17 In the first row, a confocal (*left*) and a TPE (*right*) image are shown, demonstrating the maintenance of positioning when switching from one mode to another. The second row displays the signal detected after insertion of specific emission filters for red-green (*left*) and blue (*right*). TPE allows triple fluorescence imaging using one excitation wavelength. Bovine pulmonary artery endothelial cells (Molecular Probes, G-18015) have been loaded with DAPI (DNA), mitotracker red (mitochondria) and bodipy (actin).

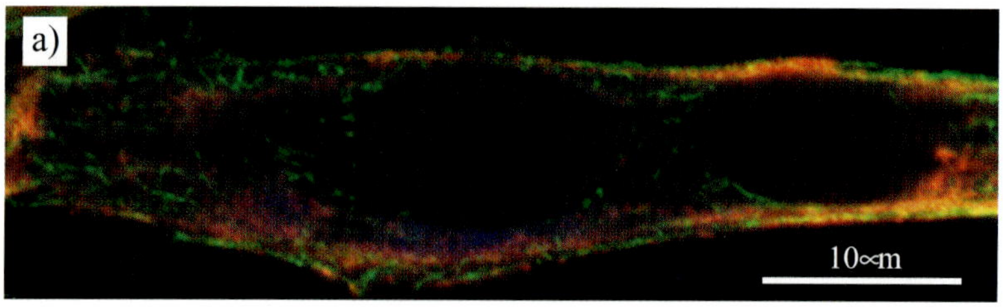

Figure 5.1 Confocal fluorescence image (slice 20 in a stack of 30 images) of an immunofluorescently labeled, fixed myoblast cell. In this experiment, the relative localization of three proteins (green: tubulin; red: actin; blue: myosin) is investigated at different time points during muscle tissue morphogenesis. Since the samples were fixed and sufficient time was available, each fluorophore distribution was recorded separately. This guaranteed the highest excitation efficiency and the optimal spectral separation. The acquisition time was 30 s per slice.

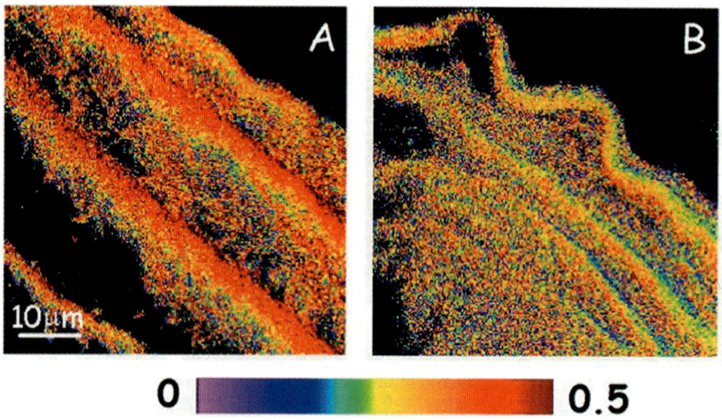

Figure 22.2 Pseudocolor representation of autofluorescence polarization values of matrix fibers in the control aorta ring (*A*) and after incubation with minimally oxidized LDL (*B*). The lumen is on the top right. (*C*) Normalized histograms of the pixel frequency versus autofluorescence polarization were obtained from images of the control (continuous black line) and of LDL-treated samples (dotted red line). Each histogram represents the average of three images.

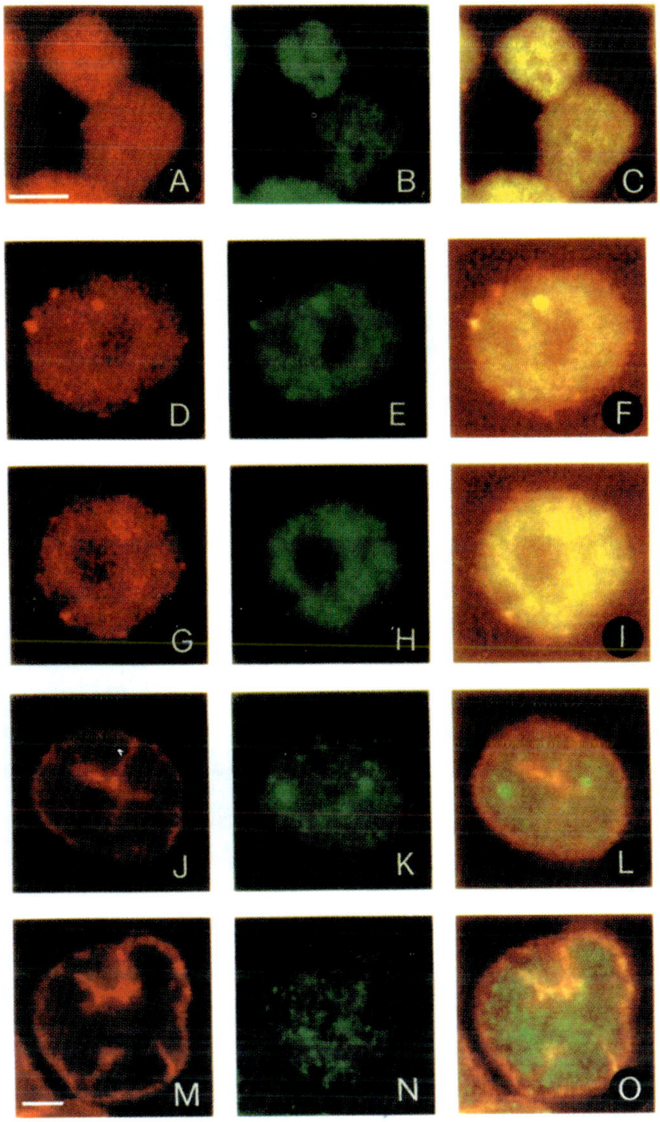

Figure 17.1 Double labeling immunofluorescence imaged by confocal microscopy to verify whether intranuclear lamin A colocalized with ribonucleoprotein particle (RNP) antigen known to be a constituent of the nuclear matrix. As shown in (*A*)–(*C*), we observed that in 37°C-stabilized, $(NH_4)SO_4$-extracted nuclear matrices, lamin A shared an almost-complete colocalization with RNP, as detectable from the superimposed yellow fluorescence. Next, we wanted to analyze in isolated matrices the distribution of both lamin A and B1 relative to either p80 coilin, a typical component of coiled bodies, or the transcription factor GATA-1, which is particularly expressed in erythroid cells and is a class of nuclear inclusions. *See page 354 for complete figure caption.*

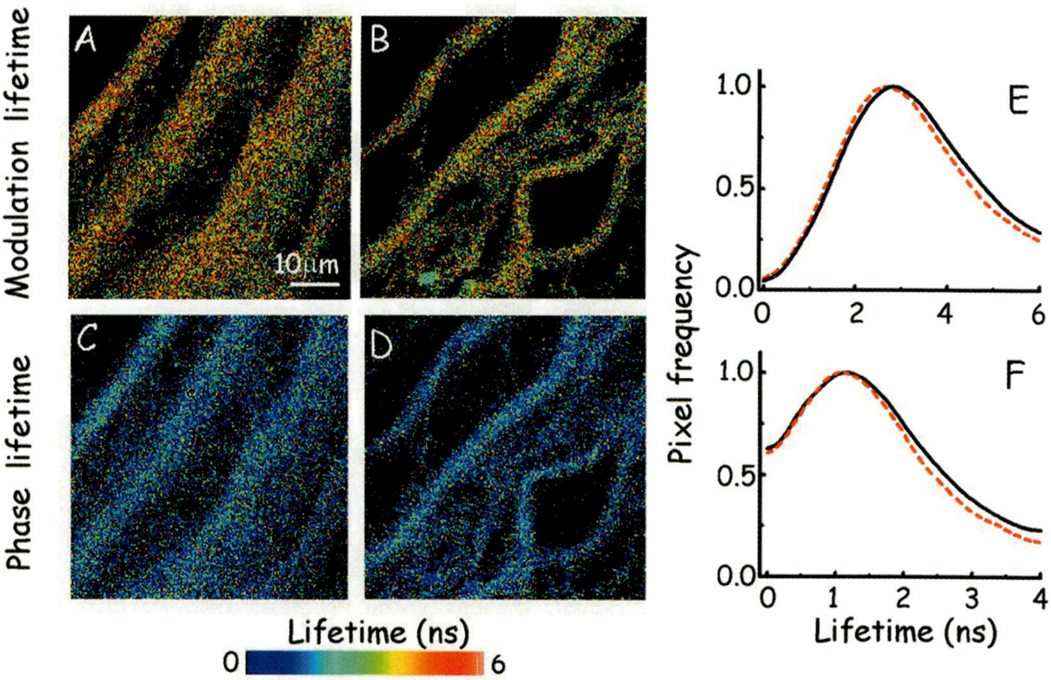

Figure 22.3 Pseudocolor representation of autofluorescence lifetime values (*A, B*: modulation lifetimes; *C, D*: phase lifetimes) of matrix fibers in the control aorta ring (*A, C*) and after incubation with minimally oxidized LDL (*B, D*). The lumen is on the top right. (*E, F*) Normalized histograms of the pixel frequency versus autofluoresence lifetime were obtained from images of the control (continuous black line) and of LDL-treated samples (dotted red line). Each histogram represents the average of four images.

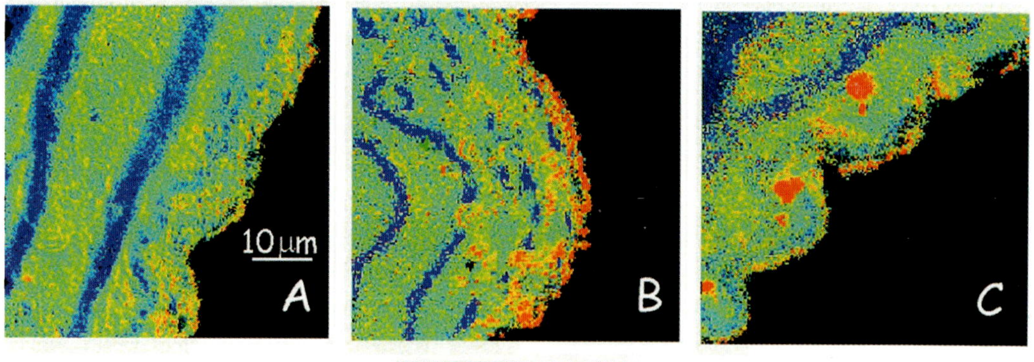

Figure 22.5 Pseudocolor representation of Laurdan GP values in rat aorta sections. In (*A*), a control samples is imaged; in (*B*) and (*c*), the aorta samples were incubated for 30 minutes at 37°C with minimally oxidized LDL. In all panels, the lumen is on the right.

is achieved. Essentially, any contrast between zero and 99% can be used to define a resolution; but a 26% contrast is reasonable for most cases. A complete treatment of the concept of contrast and resolution in fluorescence microscopy has been presented elsewhere (Stelzer, 1998).

Under ideal imaging conditions, the lateral separation of two point objects by the 26% criterion is approximately equal to the Rayleigh criterion. Therefore, we can define the lateral resolution for a conventional fluorescence microscope as the distance between the central maximum and first minimum of the Airy disc, which is given by

$$r_{xy} = \frac{1.22\lambda_{\mathrm{em}}}{2\mathrm{NA}} \approx \frac{0.6\lambda_{\mathrm{em}}}{\mathrm{NA}}. \tag{6}$$

This equation is convenient for calculating the lateral resolution of a conventional fluorescence microscope, ensuring a reasonable contrast. In practice, it is difficult to measure the location of an intensity minimum; hence, to determine the resolution experimentally, it is quite common to measure the FWHM extents of the microscope's PSF. As shown in Table 1, the FWHM values are approximately 17% smaller than the resolution calculated using the Rayleigh criterion.

For confocal fluorescence, the FWHM lateral extent of the PSF is about 30% narrower than that of the conventional fluorescence PSF. Again under ideal imaging conditions, the separation required for a reasonable contrast in a confocal microscope (Figure 4*d*) is approximately

$$r_{xy,\mathrm{confocal}} \approx \frac{0.4\lambda_{\mathrm{em}}}{\mathrm{NA}}. \tag{7}$$

For two-photon microscopy, the FWHM lateral extent of the PSF is broader than the conventional fluorescence PSF by about 15%:

$$r_{xy,\text{two-photon}} \approx \frac{0.7\lambda_{\mathrm{em}}}{\mathrm{NA}}. \tag{8}$$

In all of these microscope arrangements, the lateral resolution is proportional to the wavelength and inversely proportional to the NA of the objective lens.

3.2 Axial Resolution for Pointlike Objects

For pointlike objects, the same approach can be applied to determine a microscope's axial resolution. As in the lateral direction, a reasonable contrast between pointlike objects is achieved when they are separated by the distance between the central maximum and first minimum of the axial component of the PSF. In a conventional fluorescence microscope, this distance is given by (Born and Wolf, 1999, p. 491):

$$r_z = \frac{2\lambda_{\mathrm{em}} n}{\mathrm{NA}^2}, \tag{9}$$

where n is the refractive index of the specimen. The axial extents of the confocal and two-photon microscope PSFs are again 30% smaller and 15% greater, respectively, compared with the conventional fluorescence microscope PSF:

$$r_{z,\text{confocal}} \approx \frac{1.4\lambda_{\text{em}} n}{\text{NA}^2}, \tag{10}$$

$$r_{z,\text{two-photon}} \approx \frac{2.3\lambda_{\text{em}} n}{\text{NA}^2}. \tag{11}$$

The axial resolution is proportional to the wavelength and refractive index, and inversely proportional to the square of the NA of the objective lens. The axial resolutions based on these calculations are 10% greater than the axial FWHM extents of the corresponding PSFs (Table 1).

3.3 Optical Sectioning

The axial extent of the confocal fluorescence PSF is not much better than that of the conventional fluorescence microscope; therefore, for pointlike objects, the axial resolution in a confocal microscope is only slighter better than in a conventional instrument. For thick specimens, however, both the confocal and two-photon microscopes show a dramatic improvement in axial resolution due to the optical sectioning properties of these microscopes.

To understand optical sectioning, one must consider a microscope's intensity PSF, integrated over its lateral extent. For the illumination PSF, this integral as a function of depth is

$$E_{\text{ill,int}}(z) \propto \int_{r=0}^{r=\infty} |h_{\text{ill}}(r,z)|^2 2\pi r \, dr. \tag{12}$$

The integral is constant as a function of depth. This statement simply reflects conservation of energy. At each depth in the specimen, the distribution of the illumination may differ, but the total optical power remains the same. In other words, the laterally integrated intensity of the detection intensity PSF is a constant, meaning that a conventional fluorescence microscope has no optical-sectioning capability. However, the integrated intensity PSF for the confocal microscope is

$$E_{\text{confocal,int}}(z) \propto \int_{r=0}^{r=\infty} |h_{\text{ill}}(r,z)|^2 |h_{\text{det}}(r,z)|^2 2\pi r \, dr \tag{13}$$

In contrast to the integrated intensity PSF for the conventional fluorescence microscope, this function has a maximum in the focal plane (Figure 5*a*). This is the explanation for the depth-discrimination capability of a confocal microscope (Cox et al., 1982; Wijnaendts van Resandt et al., 1985). Similarly, the integrated intensity PSF for the two-photon microscope peaks at the focus: two-photon microscopes also exhibit optical-sectioning. However, because the illumination wavelength for two-photon is longer than its single-photon counterpart, the peak is not as sharp, and hence the optical-sectioning performance is poorer than that of a comparable confocal microscope.

The best practical illustration for optical sectioning is to record the intensity as one focuses through the cover slip into a thick layer of fluorophore dissolved in the immersion medium of the lens. The light distributions described by the confocal and two-photon

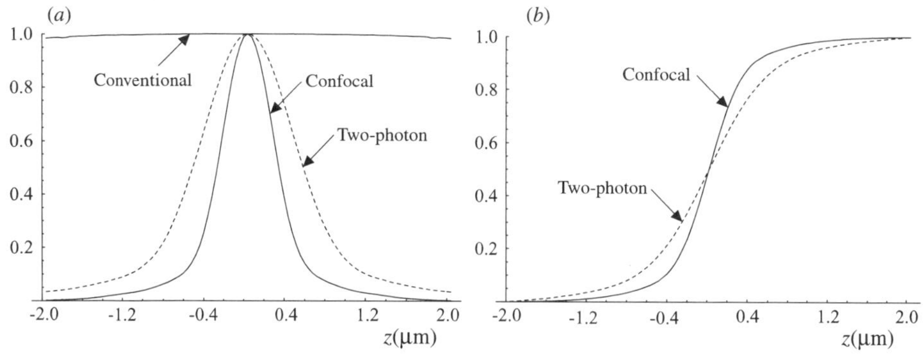

FIGURE 5 The sea resolution indicates the behavior of various microscope arrangements when observing thick specimens. Parameters for the calculations are the same as those used above. (*a*) The integrated intensity of the illumination PSF is constant as a function of depth. It resembles the behavior of conventional fluorescence microscopes. The integrated intensities of the confocal and two-photon PSFs peak at the focus. In contrast to conventional fluorescence microscopes, both techniques observe only the fluorophores in the neighborhood of the focus. This illustrates why confocal and two-photon microscopes can perform optical sectioning. (*b*) The sea response of the confocal microscope has a greater slope than the two-photon microscope's sea response. The confocal fluorescence microscope has a better resolution for axial edges, such as the one encountered when entering a thick object.

intensity PSFs, Eqs. (3) and (4), experience no fluorophores outside the layer; but they find fluorophores everywhere in their immediate environments once deep inside the thick layer. The confocal sea response

$$E_{\text{confocal, sea}}(z_0) \propto \int_{z=-\infty}^{z=z_0} \int_{r=0}^{r=\infty} |h_{\text{ill}}(r, z)|^2 |h_{\text{det}}(r, z)|^2 2\pi r \, dr \tag{14}$$

and the equivalent two-photon sea response are plotted in Figure 5*b*. The slopes of the sea-response curves can be used to characterize the axial resolutions of the confocal, two-photon, and other microscope arrangements for thick specimens.

3.4 Alternative Approaches

An alternative approach for estimating the obtainable resolution of pointlike objects in a microscope is derived from the uncertainty principle (Grill and Stelzer, 1999). This approach calculates the standard deviations of the intensity PSF for a conventional microscope, along the lateral and axial directions. Applying an additional factor, the respective resolutions are given by

$$\Delta x = \frac{4\sqrt{3}\lambda}{2\pi n(3 - 2\cos\alpha_{\max} - \cos 2\alpha_{\max})^{1/2}}, \tag{15}$$

$$\Delta z = \frac{4\sqrt{3}\lambda}{2\pi n(1 - \cos\alpha_{\max})}, \tag{16}$$

where α_{max}, the opening angle of the objective lens, is related to the numerical aperture by

$$\mathrm{NA} = n \sin \alpha_{max}. \tag{17}$$

These calculations have been extended to account for microscope arrangements in which binary annular apertures have been used. They are also particularly useful for computing the resolution of microscopes that employ large solid angles, such as the 4Pi microscope discussed below.

Another approach based on the evaluation of the Maxwell equations, which takes arbitrary distributions in the aperture plane as well as polarization and aberrations into account, is found in Rohrbach and Stelzer (2001).

4 PRACTICAL RESOLUTION LIMITS

Equations (7)–(12) all assume "ideal imaging conditions"; but these conditions can only be achieved for certain specimens and with considerable effort. Many factors contribute to reduce the contrast and, therefore, the resolution of microscope images.

4.1 The Diameter of the Pinhole

The effect of changing the pinhole diameter in a confocal microscope can be dramatic (Gu and Sheppard, 1993; Wilson, 1995; Gauderon et al., 1999). To demonstrate this effect, fluorescently labeled microtubules in fibroblast cells were imaged on a confocal microscope using various pinhole diameters. The images in Figure 6*a–d*, together with the graph in Figure 6*e*, clearly demonstrate that the optical-sectioning performance of a confocal microscope improves as the pinhole is reduced. In addition, there is a slight improvement in the lateral resolution as the pinhole approaches an ideal size: in Figure 6*f*, one sees a 10% reduction in the lateral width of the microtubules. Since the pinhole was still about 10% larger than necessary in order to be considered a point detector, the ideal 30% improvement in resolution was not observed. For many biological specimens, and especially for live specimens observed using EGFP, the number of fluorescent photons emitted by the specimen is too low to permit the pinhole to be closed down to its minimum diameter.

The resolution and optical-sectioning performance of a 2*p*-confocal microscope also improves as the pinhole diameter is reduced (Stelzer et al., 1994; Gu and Gan, 1996). In practice, however, confocal detection is rarely used with two-photon microscopy.

4.2 Pixelation

So far, PSFs and their components have been presented as smooth and continuous. However, real microscope images are recorded with a finite number of pixels in all three dimensions. The pixels will not generally fall on exactly the peak intensity of one point object, or on the intensity minimum that separates it from a neighboring point object. The peaks tend to be underestimated while the valleys tend to be overestimated. The result is that pixelation reduces the contrast and, consequently, the resolution of images recorded on the microscope. The often-quoted Nyquist sampling criterion says that one must sample at least 2.3 times for each resolution element (Webb and Dorey, 1995). However, the Nyquist

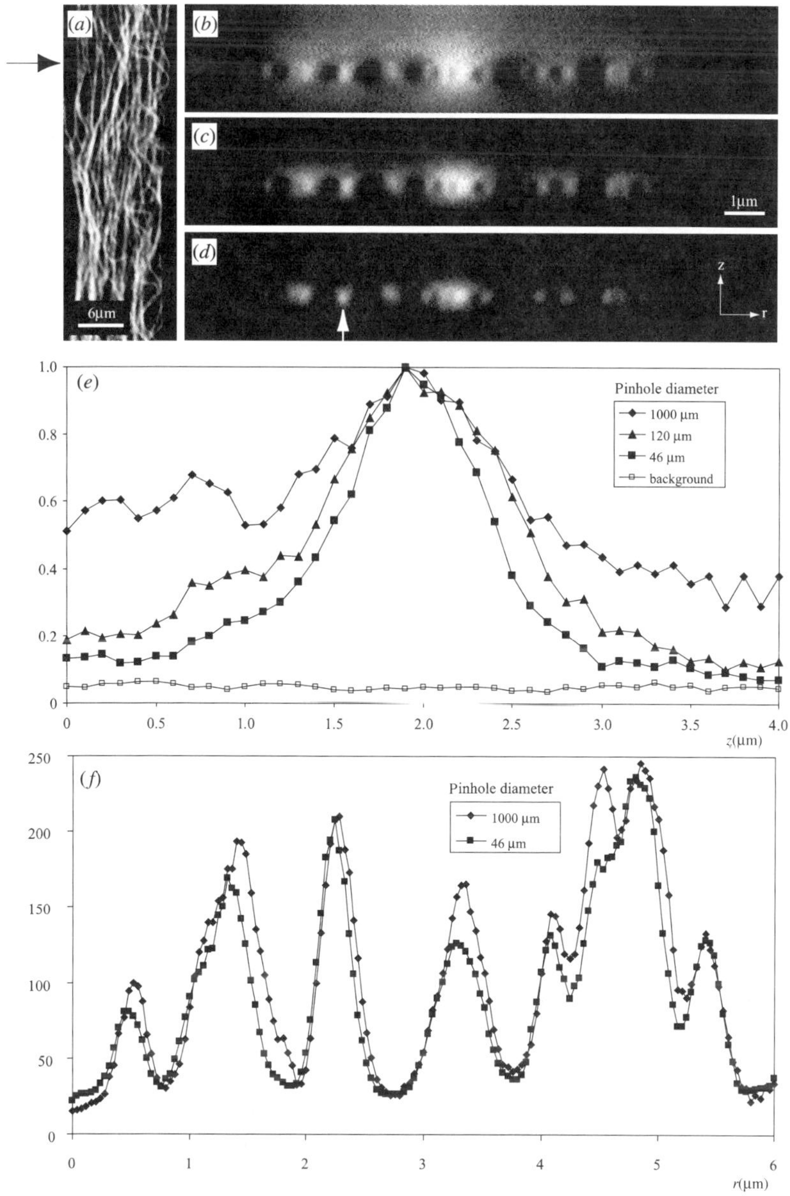

FIGURE 6 Fluorescently labeled microtubules in fibroblast cells were imaged on a confocal microscope using various pinhole diameters. (*a*) Confocal image in a plane normal to the optical axis (i.e., the *x*–*y* plane) with a pinhole diameter of 120 μm (i.e. about twice as large as it should be for high-resolution imaging). (*b*), (*c*), and (*d*) are confocal *x*–*z* images, along the line indicated by the black arrow next to (*a*), using pinhole diameters of 1000 μm, 120 μm, and 46 μm, respectively. The optical- sectioning performance clearly improves when the pinhole diameter is reduced. (*e*) Normalized intensities versus axial distance across the microtubule indicated by the white arrow in (*d*), for all three pinhole diameters. (*f*) Normalized intensities versus lateral distance across the microtubules, for the smallest and largest pinhole diameters. While the axial discrimination varies significantly only, a slight improvement in resolution is noticeable when the pinhole diameter is reduced.

criterion (like the Sparrow criterion) is a cutoff criterion: sampling at the Nyquist frequency does not guarantee a minimum contrast between features. Sampling at a rate of 8 pixels per central maximum of the Airy disc guarantees a contrast of at least 17% for objects that would have a contrast of 26% under ideal conditions (Stelzer, 1998).

4.3 Noise

Noise also affects the resolution of a recorded microscope image. Thin, fixed, immunofluorescently labeled specimens usually emit enough fluorescent photons to provide a high signal-to-noise ratio (SNR). However, many confocal and two-photon microscopy applications (including live-cell imaging with GFPs) involve imaging low concentrations of fluorophores, deep within specimens that readily absorb and scatter photons. In these situations, noise can strongly contribute to or even dominate the image. Noise creates an uncertainty in the quantification of the fluorescence intensity for a given pixel (Carlsson, 1991). This results in an uncertainty in the measured contrast. Since we have to accept an error in our estimation of the intensity, we have to accept that the contrast is, in general, underestimated in a noisy image. Of course, a reduction in contrast due to noise also means a reduction in resolution, as demonstrated in Figure 7. Particularly in the regions of the image marked with an arrow, features become visible with 16-frame averages that were not distinguished in a single image frame. For each type of specimen, one must find an appropriate balance between the scanning speed, the illumination intensity, and the resolution improvement provided by a reduced pinhole diameter (Gauderon and Sheppard, 1999).

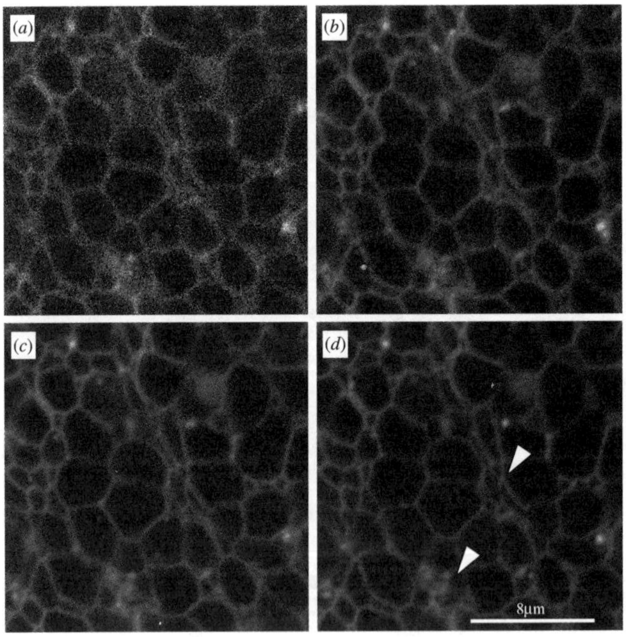

FIGURE 7 Confocal x–y images of an Alexa488-labelled *Drosophila* embryo, taken with line averages of 1, 4, 8, and 16 respectively. Noise reduces contrast, and hence resolution, in an image. Several features (indicated by the arrows) are only resolved when the number of averages is 8 or 16.

4.4 Precision

Resolution should not be confused with precision. Resolution has been defined as the distance between point objects in an image, such that they are discernable as separate objects with a given contrast (say, 26%). However, the location of a single point object can be measured to a precision much greater than the resolution of the microscope. This is accomplished by calculating the intensity-weighted "center of mass equivalent" of the object. Many applications take advantage of precision to make measurements in the tens-of- nanometers regime. Single-particle tracking in video sequences (Saxton and Jacobson, 1997) yields precisions of as high as 20 nm. The photonic force microscope (Florin et al., 1997) tracks the position of a single bead to within a precision of 10 nm, in order to determine the three-dimensional structure of molecules. Finally, surface topologies can be determined to within a fraction of a micrometer using confocal microscopes (Wijnaendts van Resandt et al., 1985).

5 IMPROVING THE RESOLUTION

The simplest way to improve the resolution of an optical microscope is to optimize the illumination wavelength and the numerical aperture of the objective used in the microscope. Both lateral and axial resolutions are proportional to the wavelength; hence, the resolution can be improved in both directions simultaneously by using a fluorophore that has shorter absorption and emission wavelengths. For single-photon (i. e., conventional or confocal) fluorescence, excitation wavelengths generally range from 350 nm to about 650 nm (Tsien and Waggoner, 1995). For two-photon microscopy, the wavelengths range from about 690 nm (for UV-excitable dyes) to 1100 nm (Fischer et al., 1995; Xu et al., 1996). Since the illumination wavelength determines the resolution of a two-photon microscope, the obtainable resolutions for two-photon microscopy are generally worse than that of a comparable confocal arrangement, as shown above for EGFP.

Of course, the specimen often limits the choice of fluorescence probe. Most live specimens are sensitive to UV (<350 nm) light exposure (Carlsson et al., 1992; Montag et al., 1991). For live-cell imaging of protein dynamics, there are several fluorescent proteins available whose absorption spectra range from 430 nm to 570 nm (Jakobs et al., 2000). For a single fluorescent label, one usually selects a GFP mutant that performs the best in terms of photostability, quantum efficiency, size, etc., as opposed to one that offers the best resolution (EGFP is currently the best choice in most cases). For double labeling, one is limited to combinations of spectrally separated GFP mutants, such as cyan and yellow fluorescent proteins (CFP and YFP), or EGFP and DsRed. In essence, the specimen and the nature of the experiment usually determine which wavelengths can be used.

The lateral resolution of a microscope is inversely proportional to the NA of the objective lens; and the axial resolution is inversely proportional to the square of the NA. Therefore, another way to improve the resolution (especially the axial resolution) is to use an objective lens with a higher numerical aperture. There are again limitations. One factor that limits the NA of the objective lens is the working distance of the lens: a NA = 1.4 oil immersion objective lens with a 63 times magnification (63×) generally has a working distance around 250 μm; while a NA = 0.95, 63×, water-immersion lens can have a working distance as large as 2.0 mm. A relatively long working distance is required for imaging below the specimen surface with confocal or two-photon microscopes.

Another limitation arises when imaging biological specimens. For thin specimens imaged close to the cover slip, the best choice of objective is usually an NA = 1.4 oil-immersion lens. However, for biological specimens in an aqueous medium, using an oil-immersion lens causes aberrations that severely degrade the imaging performance further into the specimen (Wilson and Carlini, 1989; Allred and Mills, 1989; Carlsson, 1991; Török et al., 1995; Rohrbach and Stelzer, 2001). Intensity PSFs for a confocal microscope arrangement that take into account a change in refractive index (immersion oil refractive index 1.518; sample refractive index 1.33) have been calculated and verified experimentally (Hell et al., 1993; Hell and Stelzer, 1995). They show that aberrations due to refractive-index mismatch become severe at a depth of 15 μm. These aberrations result in a factor-of-two increase in the axial FWHM extent of the PSF, as well as an axial focal shift of 2.5 μm from the nominal focal position and a 60% reduction in detected intensity (relative to the intensity detected at the cover slip). Similarly, in a two-photon microscope arrangement, experiments using oil-immersion objectives to image into biological specimens show a strong reduction in fluorescence intensity as a function of depth, as well as a decrease in axial resolution (Jacobsen et al., 1994; Gerritsen and de Grauw, 1999). A water-immersion lens with a lower NA performs better in biological specimens than the high NA oil-immersion objective.

5.1 Confocal Theta Microscopy

In a custom confocal or two-photon microscope, there are several techniques for providing enhanced resolution that require special modifications to the microscope. As shown in Figure 2, the PSFs for the confocal and two-photon fluorescence microscopes have a much greater axial than lateral extent. Confocal theta microscopy overcomes this elongation of the observation volume to achieve an almost-isotropic resolution (Stelzer and Lindek, 1994). The technique works by changing the spatial configuration of the illumination and detection volumes by using different axes for illumination and detection, as depicted in Figure 8*a*. One objective lens is used to illuminate the sample. A second objective lens, whose optical axis is at an angle θ to the illumination axis, is used to collect the emitted fluorescence. The PSF, given by

$$\begin{aligned} |h_{\text{confocal theta},90^\circ}(x, y, z)|^2 &= |h_{\text{ill}}(x, y, z)|^2 |h_{\text{det},90^\circ}(x, y, z)|^2 \\ &= |h_{\text{ill}}(x, y, z)|^2 |h_{\text{det}}(z, y, x)|^2, \end{aligned} \tag{18}$$

is almost isotropic, with both lateral and axial extents comparable to the lateral extent of the confocal fluorescence PSF of similar numerical aperture (Figure 8*b*). While an angle of $\theta = 90^\circ$ results in the smallest confocal volume (Lindek et al., 1994b), angles between 70° and 110° still result in acceptably small volumes (Lindek et al., 1994a).

There are several ways to realize such an optical arrangement. The most obvious is depicted in Figure 8*a*: two microscope objectives are positioned in such a way that an angle θ of nearly 90° between illumination and detection axis is achieved (Lindek et al., 1995). This arrangement is difficult to realize on a standard microscope base. A more practical solution, which can be adapted to any standard confocal microscope, is the single-lens theta microscope (SLTM) arrangement (Lindek et al., 1997; Swoger et al., 1998). In both arrangements, the specimen must be mounted on, for example, a glass capillary and scanned through the focus of the objective. Theta microscopy can also be

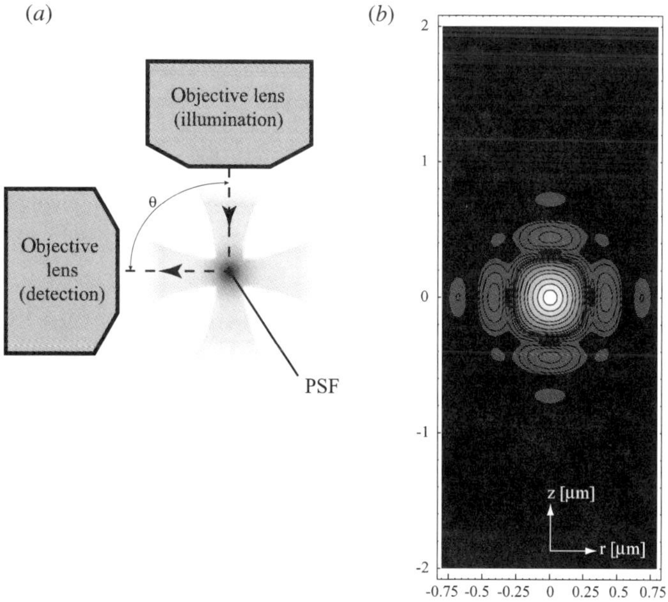

FIGURE 8 (*a*) In the theta microscope, the illumination and detection axes are tilted by an angle θ, close to 90°, relative to each other. (*b*) The PSF of the theta microscope is nearly isotropic. The calculation used NA = 0.95, $n = 1.518$, $\lambda_{ill} = 488$ nm, and $\lambda_{em} = 20$ nm.

realized in a nonconfocal arrangement using two-photon excitation (Lindek and Stelzer, 1999).

Theta microscopy is not intended for systems that require the resolution provided by a high-NA (1.3, oil-immersion) objective. It is, however excellently suited for optical systems that use low-NA, large-working-distance objective lenses (Stelzer et al., 1995). It should be noted that several two-photon applications have also been reported that use such low-NA objectives for penetrating deep into the specimen (Palmes-Saloma and Saloma, 2000; Periasamy et al., 1999; Masters and So, 2001). Ideal lenses for theta microscopy are the water-immersion, no-cover-slip 40×, NA = 0.8 and 63×, NA = 0.95 objective lenses from Zeiss (models 440090 and 1036-023, respectively). Although a water-immersion lens with a numerical aperture of 0.95 has a better axial resolution than an oil-immersion lens with NA = 1.4, the volume resolution of the water-immersion lens is still worse (Stelzer and Lindek, 1994; Sheppard, 1995). A detailed discussion of the resolution, efficiency, and working distance achieved in a single-lens theta microscope has been published elsewhere (Lindek et al., 1999).

5.2 4Pi Confocal Microscopy

Another technique for improving the resolution of a microscope is the use of 4Pi illumination together with confocal detection. In a 4Pi fluorescence microscope, a sample is illuminated coherently through two opposing objective lenses having a common focus (Hell and Stelzer, 1992a; Lindek et al., 1995). The use of two objective lenses essentially doubles the angular aperture (approaching the maximum 4π solid angle –

hence the name), and therefore provides an improvement in resolution, most notably in the axial direction. An axial FWHM extent (for the central peak of the PSF) of 75 nm has been confirmed experimentally for a 4Pi confocal arrangement (Hell et al., 1994a).

In 4Pi microscopy, coherent illumination wavefronts from the two objective lenses interfere in the focal volume. Depending on the phase difference between these wavefronts, the interference in the geometrical focal plane can be constructive, destructive, or something intermediate. The calculation of the 4Pi illumination intensity PSF is accomplished by adding two counterpropagating amplitude PSFs, both of which are multiplied by a phase function:

$$|h_{4\text{Pi,ill}}(x, y, z)|^2 = |\Phi_1(\alpha)h_{\text{ill}}(x, y, z) + \Phi_2(\beta)h_{\text{ill}}(x, y, -z)|^2, \tag{19}$$

where the phase functions are related to each other by

$$\Phi_1(\alpha) - \Phi_2(\beta) \propto e^{-\phi}. \tag{20}$$

The 4Pi illumination PSF for the case of constructive interference is shown in Figure 9*a*. Modulation of the PSF by interference produces a central main maximum whose FWHM extent is four to five times smaller than the FWHM extent of the conventional microscope's intensity PSF. Unfortunately, strong secondary maxima are also present along the optical axis (Fig. 9*e*): their influence has to be taken into account when calculating the total illumination volume.

In the 4Pi confocal microscope, illumination occurs coherently through two lenses, while the fluorescently emitted light is gathered incoherently using only one of the two lenses (Hell and Stelzer, 1992b). (In the literature, this is referred to as 4Pi(A) confocal microscopy – two microscope arrangements that involve coherent detection are called 4Pi(B) and 4Pi(C).) The 4Pi confocal PSF is calculated, as usual, by multiplying the illumination and detection intensity PSFs:

$$|h_{4\text{Pi(A)}}(x, y, z)|^2 = |h_{4\text{Pi,ill}}(x, y, z)|^2 |h_{\text{det}}(x, y, z)|^2. \tag{21}$$

The secondary maxima in the 4Pi illumination intensity PSF are substantially damped when multiplied by the confocal PSF (Figure 9*b*).

A significant improvement in 4Pi confocal microscopy has been realized by the use of two-photon excitation, which improves the 4Pi illumination PSF by damping the side lobes, as shown in Figure 9*c*. Combining 4Pi illumination with two-photon excitation and confocal detection results in an intensity PSF (Fig. 9*d*) whose FWHM extent is broader than that of the 4Pi confocal microscope; but whose axial side lobes are substantially reduced compared with the single-photon 4Pi confocal PSF (compare Figs. 9*e* and 9*f*). Such a microscope arrangement has been extensively investigated (Hell and Stelzer, 1992a; Hell et al., 1994b; Lindek et al., 1995; Gu and Sheppard, 1995; Hänninen et al., 1995). Further improvements have been the use of point deconvolution to computationally remove the axial side lobes, and active compensation of relative phase changes between the interfering wavefronts (Hell et al., 1996, 1997; Schrader et al., 1998). A two-photon 4Pi confocal microscope has achieved an axial resolution of 145 nm for observing fluorescently labeled actin fiber bundles in a fixed fibroblast cell, at depths of up to 5 μm in the cell (Schrader et al., 1998).

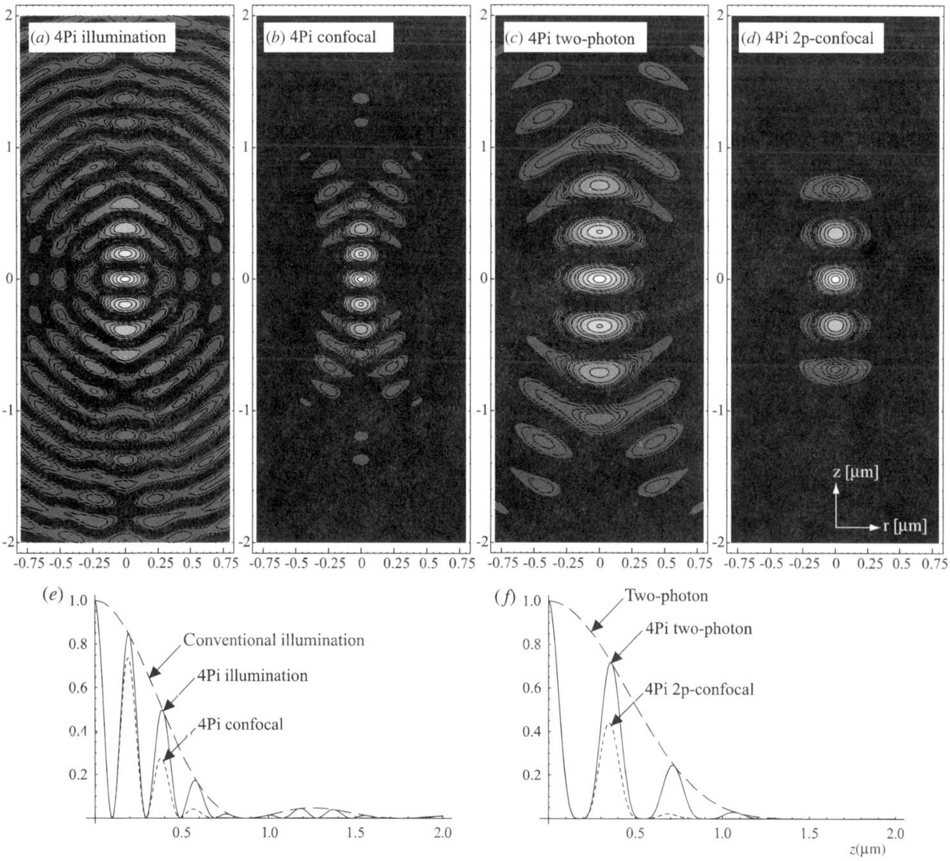

FIGURE 9 Calculated PSFs for various single- and two-photon 4Pi arrangements: (*a*) Single-photon 4Pi illumination; (*b*) single-photon 4Pi(A) confocal; (*c*) 4Pi two-photon illumination; (*d*) 4Pi(A) 2*p*-confocal. The PSFs were calculated using $\mathrm{NA} = 1.3$, $n = 1.518$, $\lambda_{\mathrm{ill}} = 488\,\mathrm{nm}$, $\lambda_{\mathrm{em}} = 520\,\mathrm{nm}$, and $\lambda_{\mathrm{ill},2p} = 900\,\mathrm{nm}$. (*e*, *f*) Axial components of the various PSFs, showing severe ringing. The ringing is damped somewhat by two-photon illumination. Note that the 4Pi arrangement requires high numerical apertures.

5.3 Stimulated Emission Depletion

Another promising technique for achieving enhanced three-dimensional resolution has recently been experimentally verified (Klar and Hell, 1999; Klar et al., 2000). The technique is to quench the fluorescence from the outer part of the focal spot using stimulated emission depletion (Hell and Wichmann, 1994). A visible illumination pulse excites fluorescence in a standard PSF volume. A second pulse of slightly longer wavelength follows a few picoseconds after the excitation pulse. Using an optical phase plate, the NIR pulse is shaped so that the PSF is intense in a ring around the focal point, and dark at its center. The NIR pulse stimulates emission of fluorescence in the focal ring immediately after excitation; the longer-lived fluorescence in the (nonstimulated) center of the focal volume is subsequently detected. Basically, this is a single-photon technique; but the power used for depletion is in the milliwatt range, close to the two-photon power requirements. An axial resolution of about 100 nm was achieved using stimulated emission depletion, corresponding to a five-fold improvement over a similar confocal microscope arrangement. For further comments on the potentials and limitations of this technique, see Weiss (2000).

5.4 Deconvolution

By measuring the PSF of a microscope, computational methods can be used to enhance both the lateral and axial resolutions of images taken on that microscope. This technique, called deconvolution, can be used for confocal and two-photon microscope images, as well as conventional fluorescence images. Deconvolution does not, however, provide optical sectioning for thick specimens. For a detailed treatment of deconvolution techniques, see Boccecci and Bertero, Chapter 12 of this book.

6 CONCLUSIONS AND OUTLOOK

Each microscope has its own advantages and disadvantages, and some may even find a niche that makes them applicable. For very thin specimens, conventional fluorescence microscopy using high-NA oil-immersion objective lenses can provide high-resolution images with little effort. DIC and other transmission contrasts can readily be used to supplement conventional fluorescence images for visualizing unlabeled structures in thin specimens. Several widefield techniques, most notably structured illumination (Neil et al., 1997; Frohn et al., 2000; Gustafsson, 2000), are emerging that provide enhanced resolution for thin specimens in a conventional fluorescence microscope arrangement.

Confocal microscopes provide optical sectioning for observing thicker specimens, without compromising resolution. Confocal microscope technology is well understood, well established, and reliable. A beam-scanning confocal microscope, which offers flexibility for zoom, magnification, and selection of arbitrary regions, provides an excellent base for live-specimen experiments such as fluorescence recovery after photobleaching (FRAP) (White and Stelzer, 1999) and fluorescence resonant energy transfer (FRET) (Bastiaens *et al.*, 1996). Multiple excitation wavelengths provide optimal spectral separation of multiple fluorophore signals (e. g., for the combination CFP–YFP; Ellenberg et al., 1999). Water-immersion objective lenses provide excellent resolution when imaging deep inside biological specimens.

Two-photon microscopes also provide optical sectioning for observing thick specimens. However, the resolution of a two-photon microscope for pointlike objects is worse than the resolution of a conventional microscope. The optical sectioning performance of a two-photon microscope is worse than that of a comparable confocal microscope. Considerably higher optical powers are needed for two-photon excitation of fluorophores; and the excitation efficiency (in terms of collected photons per bleached fluorophore) is worse than in the single-photon case (Brand *et al.*, 1997). The broad two-photon absorption cross-sections of many dyes do not permit selective excitation of fluorophores in a multiply labeled specimen. Because illumination is confined to the focal plane, two-photon microscopy results in reduced photobleaching outside of the focal volume; however, photobleaching in the focal plane is higher than in the confocal case (Patterson and Piston, 2000).

Despite these well-known drawbacks, the two-photon microscope seems to find a niche in imaging deep within scattering specimens (Centonze and White, 1998; Piston, 1999). Such applications benefit from two aspects of the two-photon microscope. First, there is reduced scattering in biological specimens at infrared excitation wavelengths, allowing for deeper penetration of the excitation beam. Second, the fact that all emitted fluorescent photons must stem from the confined illumination volume permits the use of large-area detectors, which collect photons more efficiently than point detectors. There remains, however, some controversy (Palmes-Saloma and Saloma, 2000). Two-photon microscopes operated with confocal detection cannot take advantage of this enhanced detection efficiency, and are hence not useful for imaging deep inside scattering specimens. Similarly, the recently proposed multifocal two-photon microscope (Straub et al., 2000), which uses a CCD camera (i. e., an array of point detectors), suffers from reduced detection efficiency compared with the standard two-photon microscope, and hence its applicability is not apparent. A second potential niche for two-photon microscopy, made possible by the three-dimensional confinement of the excitation, is experiments in which three-dimensional uncaging or photobleaching of fluorophores is observed (Piston et al., 1992; Piston, 1999).

In conclusion, we are currently of the opinion that most of the applications in modern cell and developmental biology can be easily handled with an excellent conventional and a state-of-the-art confocal laser scanning microscope in combination with FRET, FRAP, and fluorescent protein technologies.

ACKNOWLEDGMENTS

The authors wish to thank Valentina Greco, Stephan Grill, A. A. Hyman, Luisa Pieroni, Eric Karsenti, and Veronique Pizon for various specimens. Also, thanks to James H. Swoger, Alexander Rohrbach, and Jan Huisken for their critical reviews of the manuscript, and for assistance with the illustrations.

REFERENCES

Allred, D. B. and Mills, J. P. (1989). Effect of aberrations and apodization on the performance of coherent optical systems. 3: The near field. *Appl. Opt.* **28**, 673–681.

Bastiaens, P.I. H., Majoul, I. V., Verveer, P. J., Söling, H.-D., and Jovin, T. M. (1996). Imaging the intracellular trafficking and state of the AB_5 quaternary structure of cholera toxin. *EMBO J* **15**, 4246–4253.

Born, M. and Wolf, E. (1999). *Principles of Optics*, 7th (expanded) edn. Cambridge: Cambridge University Press.

Brand, L., Eggeling, C., Zander, C., Drexhage, K. H., and Seidel, C. A. M. (1997). Single-molecule identification of coumarin-120 by time-resolved fluorescence detection: comparison of one- and two-photon excitation in solution. *J. Phys. Chem. A* **101**, 4313–4321.

Carlsson, K. (1991). The influence of specimen refractive index, detector signal integration, and non-uniform scan speed on the imaging properties in confocal microscopy. *J. Microsc.* **163**, 167–178.

Carlsson, K., Mossberg, K., Helm, P. J., and Philip, H. (1992). Use of UV excitation in confocal laser scanning fluorescence microscopy. *Micron and Microscopica Acta* **23**, 413–428.

Centonze, V. E. and White, J. G. (1998). Multiphoton excitation provides optical sections from deeper within scattering specimens than confocal imaging. *Biophys. J.* **75**, 2015–2024.

Cox, I. J., Sheppard, C. J. R., and Wilson, T. (1982). Improvement in resolution by nearly confocal microscopy. *Appl. Opt.* **21**, 778–780.

Ellenberg, J., Lippincott-Schwartz, J., and Presley, J. F. (1999). Dual-colour imaging with GFP variants. *Trends Cell Biol.* **9**, 52–56.

Fischer, A., Cremer, C., and Stelzer, E. H. K. (1995). Fluorescence of coumarins and xanthenes after two-photon absorption with a pulsed titanium–sapphire laser. *Appl. Opt.* **34**, 1989–2003.

Florin, E.-L., Pralle, A., Hörber, J. K. H., and Stelzer, E. H. K. (1997). Photonic force microscope based on optical tweezers and two-photon excitation for biological applications. *J. Struct. Biol.* **119**, 202–211.

Frieden, B. R. (1967). Optical transfer of the three-dimensional object. *J. Opt. Soc. Am.* **57**, 56–66.

Frohn, J T., Knapp, H. F., and Stemmen, A. (2000). True optical resolution beyond the Rayleigh limit achieved by standing wave illumination. *Proc. Nat. Acad. Sci. USA* **97**, 7232–7236.

Gauderon, R. and Sheppard, C. J. R. (1999). Effect of a finite-size pinhole on noise performance in single-, two- and three-photon confocal fluorescence microscopy. *Appl. Opt.* **38**, 3562–3565.

Gauderon, R., Lukins, P. B., and Sheppard, C. J. (1999). Effect of a confocal pinhole in two-photon microscopy. *Microsc. Res. Tech.* **47**, 210–214.

Gerritsen, H. C. and de Grauw, C. J. (1999). Imaging of optically thick specimen using two-photon excitation microscopy. *Microsc. Res. Tech.* **47**, 206–209.

Goodman, J. W. (1968). *Introduction to Fourier Optics*. San Francisco: McGraw-Hill.

Grill, S. and Stelzer, E. H. K. (1999). Method to calculate lateral and axial gain factors of optical setups with a large solid angle. *J. Opt. Soc. Am.* **16**, 2658–2665.

Gu, M. and Gan, X. S. (1996). Effect of the detector size and the fluorescence wavelength on the resolution of three- and two-photon confocal microscopy. *Bioimaging* **4**, 129–137.

Gu, M. and Sheppard, C.J . R. (1993). Effects of a finite-sized pinhole on 3D image formation in confocal two-photon fluorescence microscopy. *J. Mod. Opt.* **40**, 2009–2024.

Gu, M. and Sheppard, C. J. R. (1995). Optical transfer function analysis for two-photon 4Pi confocal fluorescence microscopy. *Opt. Commun.* **114**, 45–49.

Gustafsson, M. G. L. (2000). Surpassing the lateral resolution limit by a factor of two using structured illumination microscopy. *J. Microsc.* **198**, 82–87.

Hänninen, P. E., Hell, S. W., Salo, J., Soini, E., and Cremer, C. (1995). Two-photon excitation 4Pi confocal microscope: enhanced axial resolution microscope for biological research. *Appl. Phys. Lett.* **66**, 1698–1700.

Hecht, E. (1998). *Optics*, 3rd edn. New York: Addison-Wesley.

Hell, S. and Stelzer, E. H. K. (1992a). Fundamental improvement of resolution with a 4Pi-confocal fluorescence microscope using two-photon excitation. *Opt. Commun.* **93**, 277–282.

Hell, S. and Stelzer, E. H. K. (1992b). Properties of a 4Pi confocal fluorescence microscope. *J. Opt. Soc. Am.* **A9**, 2159–2166.

Hell, S. W. and Stelzer, E. H. K. (1995). Lens aberrations in confocal fluorescence microscopy. In: *Handbook of Biological Confocal Microscopy* (ed. J. B. Pawley). New York: Plenum Press, pp. 347–354.

Hell, S. W. and Wichmann, J. (1994). Breaking the diffraction resolution limit by stimulated emission: stimulated-emission–depletion fluorescence microscopy. *Opt. Lett.* **19**, 780–782.

Hell, S., Reiner, G., Cremer, C., and Stelzer, E. H. K. (1993). Aberrations in confocal fluorescence microscopy induced by mismatches in refractive index. *J. Microsc.* **169**, 391–405.

Hell, S. W., Lindek, S., Cremer, C., and Stelzer, E. H. K. (1994a). Measurement of the 4Pi-confocal point spread function proves 75 nm axial resolution. *Appl. Phys. Lett.* **64**, 1335–1337.

Hell, S. W., Lindek, S., and Stelzer, E. H. K. (1994b). Enhancing the axial resolution in far-field light microscopy: two-photon 4Pi confocal fluorescence microscopy. *J. Mod. Opt.* **41**, 675–681.

Hell, S. W., Schrader, M., Hanninen, P. E., and Soini, E. (1996). Resolving fluorescence beads at 100–200 nm axial distance with a two-photon 4Pi-microscope operating in the near infrared. *Opt. Commun.* **128**, 394.

Hell, S. W., Schrader, M., and van der Voort, H. T. M. (1997). Far-field fluorescence microscopy with three-dimensional resolution in the 100-nm range. *J. Microsc.* **187**, 1–7.

Jacobsen, H., Hänninen, P. E., Soini, E., and Hell, S. W. (1994). Refractive-index-induced aberrations in two-photon confocal fluorescence microscopy. *J. Microsc.* **176**, 226–230.

Jakobs, S., Subramaniam, V., Schonle, A., Jovin, T. M., and Hell, S. W. (2000). EFGP and DsRed expressing cultures of *Escherichia coli* imaged by confocal, two-photon and fluorescence lifetime microscopy. *FEBS Lett.* **479**, 131–135.

Klar, T. A. and Hell, S. (1999). Subdiffraction resolution in far-field fluorescence microscopy. *Opt. Lett.* **24**, 954–956.

Klar, T. A., Jakobs, S., Dyba, M., Egner, A., and Hell, S. (2000). Fluorescence microscopy with diffraction resolution barrier broken by stimulated emission. *Proc. Nat. Acad. Sci. USA* **97**, 8206–8210.

Lindek, S. and Stelzer, E. H. K. (1999). Resolution improvement by nonconfocal theta microscopy. *Opt. Lett.* **24**, 1505–1507.

Lindek, S., Pick, R., and Stelzer, E. H. K. (1994a). Confocal theta microscope with three objective lenses. *Rev. Sci. Instr.* **65**, 3367–3372.

Lindek, S., Salmon, N., Cremer, C., and Stelzer, E. H. K. (1994b). Theta microscopy allows phase regulation in 4Pi(A)-confocal two-photon fluorescence microscopy. *Optik.* **98**, 15–20.

Lindek, S., Stelzer, E. H. K., and Hell, S. (1995). Two new high-resolution confocal fluorescence microscopies (4Pi, Theta) with one- and two-photon excitation. In: *Handbook of Biological Confocal Microscopy* (ed. J. B. Pawley). New York: Plenum Press, pp. 417–430.

Lindek, S., Stefany, T., and Stelzer, E. H. K. (1997). Single-lens theta microscopy – a new implementation of confocal theta microscopy. *J. Microsc.* **188**, 280–284.

Lindek, S., Swoger, J., and Stelzer, E. H. K. (1999). Single-lens theta microscopy: resolution, efficiency, and working distance. *J. Mod. Opt.* **46**, 843–858.

Masters, B. R. and So, P. T. C. (2001). Confocal microscopy and multi-photon excitation microscopy of human skin in vivo. *Opt. Express* **8**, 2–10.

Montag, M., Kukulies, J., Jörgens, R., Gundlach, H., Trendelenburg, M. F., and Spring, H. (1991). Working with the confocal scanning UV-laser microscope: specific DNA localization at high sensitivity and multiple-parameter fluorescence. *J. Microsc.* **163**, 201–210.

Neil, M. A. A., Juskaitis, R., and Wilson, T. (1997). Method of obtaining optical sectioning by using structured light in a conventional microscope. *Opt. Lett.* **22**, 1905–1907.

Palmes-Saloma, C. and Saloma, C. (2000). Long-depth imaging of specific gene expressions in whole-mount mouse embryos with single-photon excitation confocal fluorescence microscopy and FISH. *J. Struct. Biol.* **131**, 56–66.

Patterson, G. H. and Piston, D. W. (2000). Photobleaching in two-photon excitation microscopy. *Biophys. J.* **78**, 2159–2162.

Periasamy, A., Skoglund, P., Noakes, C., and Keller, R. (1999). An evaluation of two-photon excitation versus confocal and digital deconvolution fluorescence microscopy imaging in *Xenopus* morphogenesis. *Microsc. Res. Tech.* **47**, 172–181.

Piston, D. W. (1999). Imaging living cells and tissues by two photon excitation microscopy. *Trends Cell Biol.* **9**, 66–69.

Piston, D., Wu, E., and Webb, W. (1992). 3-dimensional diffusion measurements in cells by 2-photon excitation fluorescence photobleaching recovery. *Faseb J.* **6**, A34.

Rohrbach, A. and Stelzer, E. H. K. (2001). Optical trapping of dielectric particles in arbitrary fields. *J. Opt. Soc. Am. A.* **18**, 839–853.

Saxton, M. J. and Jacobson, K. (1997). Single-particle tracking: applications to membrane dynamics. *Annu. Rev. Biophys. Biomol. Struct.* **26**, 373–399.

Schrader, M., Bahlmann, K., Giese, G., and Hell, S. W. (1998). 4Pi-confocal imaging in fixed biological specimens. *Biophys. J.* **75**, 1659–1668.

Sheppard, C. J. R. and Gu, M. (1990). Image formation in two-photon fluorescence microscopy. *Optik.* **86**, 104–106.

Sheppard, C. J. R. (1995). Fundamental reduction of the observation volume in far-field light microscopy by detection orthogonal to the illumination asis: confocal theta microscopy – Comment. *Opt. Commun.* **119**, 639–695.

Stelzer, E. H. K. (1998). Contrast, resolution, pixelation, dynamic range and signal-to-noise ratio: fundamental limits to resolution in fluorescence light microscopy. *J. Microsc.* **189**, 15–24.

Stelzer, E. H. K. and Lindek, S. (1994). Fundamental reduction of the observation volume in far-field light microscopy by detection orthogonal to the illumination axis: confocal theta micrsocopy. *Opt. Commun.* **111**, 536–547.

Stelzer, E. H. K., Hell, S., Lindek, S., Stricker, R., Pick, R., Storz, C., Ritter, G., and Salmon, N. (1994). Nonlinear absorption extends confocal fluorescence microscopy into the ultra-violet regime and confines the illumination volume. *Opt. Commun.* **104**, 223–228.

Stelzer, E. H. K., Lindek, S., Albrecht, S., Pick, R., Ritter, G., Salmon, N. J., and Stricker, R. (1995). A new tool for the observation of embryos and other large specimens: confocal theta fluorescence microscopy. *J. Microsc.* **179**, 1–10.

Straub, M., Lodemann, P., Holroyd, P., Jahn, R., and Hell, S. W. (2000). Live cell imaging by multifocal multiphoton microscopy. *Eur. J. Cell Biol.* **79**, 726–734.

Swoger, J., Lindek, S., Stefany, T., Haar, F.-M., and Stelzer, E. H. K. (1998). A confocal fiber-coupled single-lens theta microscope (FC-SLTM). *Rev. Sci. Instr.* **69**, 2956–2963.

Török, P., Varga, P., and Booker, G. R. (1995). Electromagnetic diffraction of light focused through a planar interface between materials of mismatched refractive indices: structure of the electromagnetic field I. *J. Opt. Soc. Am.* **A12**, 2136–2144.

Tsien, R. Y. and Waggoner, A. (1995). Fluorophores for confocal microscopy. In: *Handbook of Biological Confocal Microscopy* (ed. J. B. Pawley). New York: Plenum Press, pp. 267–279.

van der Voort, H. T. M. and Brakenhoff, G. J. (1990). 3-D image formation in high-aperture fluorescence confocal microscopy: a numerical analysis. *J. Microsc.* **158**, 43–54.

Webb, R. H. and Dorey, C. K. (1995). The pixelated image. In: *Handbook of Biological Confocal Microscopy* (ed. J. B. Pawley). New York: Plenum Press, pp. 55–67.

Weiss, S. (2000). Shattering the diffraction limit of light: A revolution in fluorescence microscopy? *Proc. Natl. Acad. Sci.* **97**, 8747–8749.

White, J. and Stelzer, E. H. K. (1999). Photobleaching GFP reveals protein dynamics inside live cells. *Trends Cell Biol.* **9**, 61–65.

Wijnaendts van Resandt, R. W., Marsman, H.J. B., Kaplan, R., Davoust, J., Stelzer, E. H. K., and Stricker, R. (1985). Optical fluorescence microscopy in three dimensions: microtomoscopy. *J. Microsc.* **138**, 29–34.

Wilson, T. (1995). The role of the pinhole in confocal imaging system. In: *Handbook of Biological Confocal Microscopy* (ed. J. B. Pawley). New York: Plenum Press, pp. 167–182.

Wilson, T. and Carlini, A. R. (1989). The effect of aberrations on the axial response of confocal imaging systems. *J. Microsc.* **154**, 243–256.

Xu, C., Zipfel, W. R., Shear, J. B., Williams, R. M., and Webb, W. W. (1996). Multiphoton fluorescence excitation: new spectral windows for biological nonlinear microscopy. *Proc. Natl. Acad. Sci. USA* **93**, 10763–10768.

Zacharias, D. A., Baird, G. S., and Tsien, R. Y. (2000). Recent advances in technology for measuring and manipulating cell signals. *Curr. Opin. Neurobiol.* **10**, 416–421.

CHAPTER 6

The Role of Pinhole Size in High-Aperture Two- and Three-Photon Microscopy

PETER TÖRÖK[1] and COLIN J. R. SHEPPARD[2]

[1] Department of Engineering Science, University of Oxford, Parks Road, Oxford OX1 3PJ, U. K. and [2] School of Physics and Australian Key Centre for Microscopy and Microanalysis, The University of Sydney, NSW 2006, Australia

1 INTRODUCTION

In this chapter, we investigate the role of pinhole size in two- and three-photon fluorescence microscopy. It is frequently emphasized in the literature that multiphoton microscopes, which definition here means two- or three-photon fluorescence microscopes, possess an inherent optical sectioning property. In an early paper [1], one of us showed that the use of finite-sized pinholes in confocal microscopy results in reduced signal-to-noise ratio. In other words, in conventional scanning microscopes, a pinhole can be used in front of the detector to gain optical sectioning at the expense of signal-to-noise ratio. Actually, the signal-to-noise ratio depends strongly on the assumed noise properties. If noise originates from background fluorescence (i. e., from other parts of the specimen) or from autofluorescence from optical components, then an optimum pinhole size exists [2, 3]. In nonlinear microscopy, where optical sectioning is inherent in the optical system, the presence of a pinhole could appear to be counterintuitive or even detrimental owing to signal-to-noise considerations. One would then say that any work investigating the role of the pinhole size in multiphoton microscopy appears to be superfluous. However, as the use of a pinhole results in a contribution to resolution from the imaging effect of the shorter-wavelength fluorescent light, an improvement in resolution is not surprising.

In this chapter (which is chiefly based on a paper by Higdon et al. [4]), we consider the effect of polarization of the fluorescent emission on the imaging properties of multiphoton microscopes. The lateral and axial resolutions are calculated for one-, two-, and three-photon excitation of *p*-quaterphenyl for high- and low-aperture objective lenses, which case is used as a characteristic example. Significant improvement in lateral resolution is found to be achieved by employing a confocal pinhole. Subsequently, we average over the orientations of the dye molecule to model the image of a fluorescently labelled,

Confocal and Two-Photon Microscopy: Foundations, Applications, and Advances, Edited by Alberto Diaspro.
ISBN 0-471-40920-0

subresolution bead. This improvement increases with the order of the multiphoton process. Simultaneously, it is found that, when the size of the pinhole is reduced to achieve close to confocal resolution, the signal-to-noise ratio is not degraded by more than 30%. The degree of optical sectioning achieved is found to improve dramatically with the use of confocal detection. For two- and three-photon excitation, an improvement of 30% in the axial full-width at half-maximum is predicted.

2 BACKGROUND

Conventional and confocal fluorescence microscopes are routinely used, primarily in biological sciences, to investigate three-dimensional structures of tissues. The image contrast in these microscopes arises from collecting and measuring the intensity of the radiation emitted by the ensemble of those individual fluorochrome molecules with which the specimen has been labeled [5]. Therefore, image formation in fluorescence microscopes may be studied by determining how the microscope images a single fluorescent molecule. In the first part of this chapter, we shall develop a theory to do just that. Our theory considers the role of the polarization state of illumination as well as the finite size of the confocal pinhole. Whereas it would be flattering to assume that our writing is sufficiently clear that the theory is readily understood by all, the reality is that thorough understanding of Section 4 requires in-depth knowledge of physical optics. Undoubtedly some of our audience will not possess this knowledge. Therefore the numerical results section (5), which immediately follows the theory section (4), is written in such a way that it can be understood even if Section 4 is not read.

We shall begin the composition of the theory by considering a fluorescent microscope. In our model, a high-aperture lens is used to focus the incident plane wave into the specimen. This lens in a conventional fluorescence microscope corresponds to the condenser lens. Light emitted by the fluorescent molecules is collected by another high-aperture lens. This lens in a conventional fluorescence microscope corresponds to the objective lens. The objective lens images the volume emitting the fluorescent radiation, with the help of another lens (which we term the detector lens) onto the detector. In the confocal operation mode, we choose to place a small hole, the so-called pinhole, in front of the detector.

The theoretical model starts by determining the electric field that excites the fluorescent molecule in the focal region of a high-aperture, high-Fresnel-number, aplanatic microscope objective lens using the Debye–Wolf diffraction integral [6]. The fluorescent molecule will be treated as a harmonically oscillating electric dipole whose probability of excitation (strength) is proportional to the intensity of the illumination to the power of the order of the multiphoton process [7]. In general, the fluorescent emission is partially polarized [8, 9], and in this section the two limiting cases of polarized and unpolarized emission are considered. The emitted field passes back through the high-aperture lens and the associated detector lens to a finite-sized detector. The image signal is averaged over dipole orientations to model imaging of a fluorescent bead. The theory is used to describe the axial and lateral resolution of multiphoton fluorescent microscopes employing linearly and circularly polarized illumination and finite-sized detectors. The effect of the partial polarization of the fluorescent emission on the imaging properties is analyzed. The signal-to-noise ratio is calculated for finite-sized confocal pinholes, and is used to find an "optimum" pinhole size. The theory is also applicable to low-aperture lenses, and it is found that, as expected, the degree of partial polarization of the fluorescent emission may be ignored in such cases.

A brief introduction to those aspects of multiphoton fluorescence imaging that are considered to be important for our current discussions is given. This is followed by a detailed description of the theoretical considerations, and finally numerical results describing the imaging properties of one-, two-, and three-photon fluorescent microscopes are presented.

3 MULTIPHOTON FLUORESCENCE IMAGING

A fluorescent molecule is excited by absorbing the energy of a photon and subsequently radiating another photon as it returns to a lower-energy state. The wavelength of the emission for Stokes fluorescence is longer than that of the excitation (the emitted photon possesses less energy than the absorbed photon owing to energy lost between excitation and emission). Alternatively, the energy of two photons may be absorbed simultaneously by the fluorescent molecule, and, again, the energy is irradiated as a single photon. In this case, the energy of the emitted photon is limited by the combined energy of the two absorbed photons. Hence, the emission spectrum has a lower limit of approximately half the excitation wavelength. In theory, any number of photons, N, may be simultaneously absorbed and a single photon emitted. By similar consideration of the energy involved, the wavelength of the emission must be greater than $1/N$ of the illumination wavelength. When a single absorbed photon causes the emission of a single photon, the process is called one-photon (1p) fluorescence. When N simultaneously absorbed photons cause the emission of a single photon the process is called N-photon (Np), or multiphoton, fluorescence. Two-photon scanning optical microscopy was proposed in 1978 [10] and demonstrated successfully in 1990 [11]. Three-photon scanning microscopy was demonstrated in 1995 [12]. It was shown [11, 12] that 2p and 3p fluorescence microscopy have distinct advantages over the 1p case: for example, photobleaching is confined to the illumination volume of the specimen, and UV dyes can be excited with visible or near-infrared light. More importantly, perhaps, optical sectioning can be achieved without a confocal pinhole, so that fluorescent light can be detected even after scattering in a thick sample. It is not surprising that a great deal of work has been carried out on the theory of image formation of multiphoton fluorescent microscopy. Most notable is the paraxial fluorescence theory presented by Sheppard and Gu and their co-workers. These authors have analyzed the lateral and axial resolution of two-photon [13, 14] and three-photon [15] microscopes and the three-dimensional image formation for finite-sized [16] and infinitely small detectors [17, 18]. However, the theories presented are only applicable to low-aperture objective lenses, and no account is taken of the polarization state of the illumination or the polarization properties of the fluorescent dye. There are two possible ways to model the mechanism of fluorescent emission. The first is to consider the fluorescent emission as being randomly polarized. In previous publications [19–21], this led to mathematical formulas in which the point spread functions of the illumination and excitation optical paths were simply multiplied together. The other, more rigorous, approach is to consider the mechanism of fluorescence emission as a polarization-dependent process. Extensive attention has been given in the literature to the study of the relationship between absorbed and emitted polarization states since the theories of Perrin [8] and Soleillet [9] were presented. In general, fluorescent emission is partially polarized, and the number of parameters that determine the state of the polarization is extremely large [9]. The relationship between illumination and emission polarization direction also depends on a number of parameters, such as the fluorescent molecule's rotational mobility [22] and its individual characteristics [7]. Accounting for the polarization states in 2p and 3p fluorescence is a more involved process [23–25], but the conclusion is still that the fluorescent emission is partially polarized. As the state of the polarization varies with the order of the multiphoton process, the fluorescent molecule used, and many experimental conditions, it would be of little value to model any one specific case. Rather, in the current work, the limiting cases of polarized and nonpolarized emission will be considered, either of which can be used as an approximation to given experimental conditions.

Recently, the present authors derived a high-aperture vectorial theory for conventional and confocal one-photon fluorescence microscopes [26]. However, in this publication, we did not account for finite-sized pinholes. Here, the following fluorescent excitation and emission processes are considered. The fluorescent material is illuminated by the electric field produced by a high-aperture, high-Fresnel-number condenser lens. The probability of excitation is proportional to the intensity to the power N (i. e., the squared modulus of the electric field to the power 1, 2, or 3 for 1p, 2p, or 3p processes, respectively). Once the molecule is excited, it re-radiates as a harmonically oscillating electric dipole. The light emitted from the fluorescent dipole is incoherent with respect to the illumination. The direction of the dipole moment defines the polarization state of the emission. Three cases will be considered throughout this work:

- *Case I*: Induced dipole, unpolarized emission. The dipole is excited by the full incident electric field. The electric dipole moment of the emission is randomly oriented.
- *Case II*: Permanent dipole, unpolarized emission. The dipole is excited by only the component of the incident electric field parallel to the dipole moment. The electric dipole moment of the emission is randomly oriented.
- *Case III*: Induced dipole, polarized emission. The dipole is excited by the full incident electric field. The electric dipole moment of the emission coincides with the incident field.

It should be noted that, while Cases I and II are physically plausible, for example for dye in solution, Case III should be treated with care. Gryczynski et al. [25] obtained expressions for the angular displacement of the emission transition moment from the direction of the polarized excitation for 1p, 2p, and 3p fluorescence. In Case III, it is assumed that there is zero angular displacement between excitation and emission, which is never actually the case but, rather, a limiting model that is a good approximation to a highly polarized emission. A fourth polarized-emission case that is not considered in this work is that of fixed dipoles that are aligned in a given direction. In such a case, the direction of the fixed dipole alignment with respect to the incident polarization direction is the important factor in image formation. As it is not possible to consider all possible dipole orientations in this work, and because a theory for image formation of 1p fluorescence confocal and conventional microscopes with dipoles fixed in three orthogonal directions has already been developed [26], this fourth case is omitted.

4 THEORY

Practical multiphoton microscopes use either linearly or circularly polarized illumination. In order to study the effects of incident polarization on the image formation, we chose to analyze the general model in which a Babinet–Soleil compensator is introduced into the illumination optical system. This allows study of the microscope response to a range of incident illumination polarization states between linear and circular. The plane wave leaving the Babinet–Soleil compensator is then incident upon a high-aperture, high-Fresnel-number objective lens. The spherical wavefront produced by this lens may be represented as a coherent superposition of plane waves. The integral formula of Richards and Wolf [6] then gives the electric field in the focal region. From this distribution, it is possible to calculate the intensity (squared modulus of the electric field) at any position in the focal region and the unit vector of polarization at that point. The fluorescent molecule is excited, and subsequently radiates as a harmonically oscillating electric dipole. The relationship between the incident field and the dipole moment of the fluorescent emission is presented for Cases I, II, and III as discussed in Section 3. In the detection system, a lens identical to that used for the illumination recollimates the light. The collimated light is not analyzed by further polarizing components in this treatment, but is simply focused by a detector lens on to a finite-sized pinhole detector.

4.1 Fluorescent Dipole Moment

A schematic representation of the illumination system is shown in Figure 1. The illumination $\mathbf{E}_0$ is linearly polarized in the x direction with unit electric vector; that is,

$$\mathbf{E}_0 = \begin{pmatrix} 1 \\ 0 \\ 0 \end{pmatrix}. \tag{1}$$

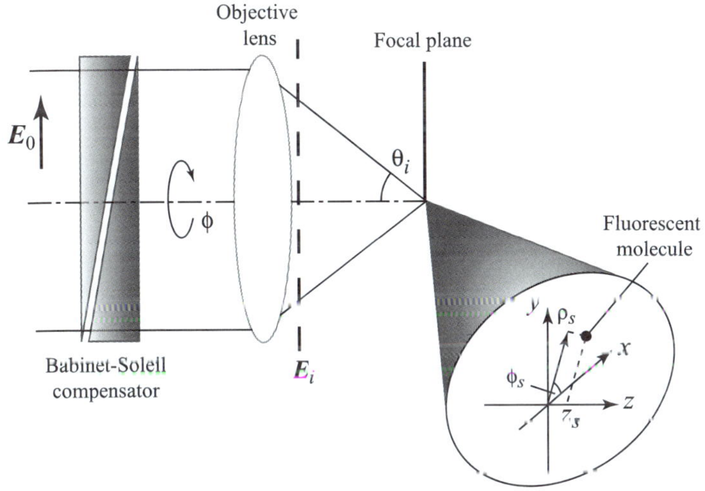

FIGURE 1 Schematic diagram of the illumination system.

The electric vector $\mathbf{E}_i$ of the field after propagating through the Babinet–Soleil compensator and the high-aperture lens is given by

$$\mathbf{E}_i = \mathbf{T}_f \cdot \mathbf{BS} \cdot \mathbf{E}_0, \tag{2}$$

where the matrix $\mathbf{T}_f$ is the transformation representing the focusing by the high-aperture objective lens, and may be written as [27]

$$\mathbf{T}_f = \sqrt{\cos\theta_i} \times \begin{pmatrix} \cos^2\phi\cos\theta_i + \sin^2\phi & \sin\phi\cos\phi(\cos\theta_i - 1) & -\sin\theta_i\cos\phi \\ \sin\phi\cos\phi(\cos\theta_i - 1) & \sin^2\phi\cos\theta_i + \cos^2\phi & -\sin\theta_i\sin\phi \\ \sin\theta_i\cos\phi & \sin\theta_i\sin\phi & \cos\theta_i \end{pmatrix}, \tag{3}$$

where the spherical polar coordinate system is defined such that $r > 0$, $0 \leqslant \theta_i < \pi$, and $0 \leqslant \phi < 2\pi$, and the $\sqrt{\cos\theta_i}$ term describes the apodization function of the aplanatic lens. **BS** represents the generalized Jones matrix of a Babinet–Soleil compensator [27]:

$$\mathbf{BS} = \begin{pmatrix} \cos(\frac{1}{2}\delta) + i\cos(2\gamma)\sin(\frac{1}{2}\delta) & i\sin(2\gamma)\sin(\frac{1}{2}\delta) & 0 \\ i\sin(2\gamma)\sin(\frac{1}{2}\delta) & \cos(\frac{1}{2}\delta) - i\cos(2\gamma)\sin(\frac{1}{2}\delta) & 0 \\ 0 & 0 & 1 \end{pmatrix}, \tag{4}$$

where δ is the relative retardation and γ is the orientation of the optical axis to the x direction. If the Babinet–Soleil compensator is taken to have its optic axis at 45° to the direction of incident polarization, the form of the matrix **BS** is greatly simplified, becoming

$$\mathbf{BS} = \begin{pmatrix} A & B & 0 \\ B & A & 0 \\ 0 & 0 & 1 \end{pmatrix}, \tag{5}$$

where the functions A and B are given by

$$A = \cos(\tfrac{1}{2}\delta), \qquad B = i\sin(\tfrac{1}{2}\delta). \tag{6}$$

The electric vector $\mathbf{E}_i$ is given by substitution of Eqs. (1), (3), and (5) into Eq. (2). This results in

$$\mathbf{E}_i(\theta_i, \phi) = \sqrt{\cos\theta_i} \times \begin{pmatrix} A[(1+\cos\theta_i) - (1-\cos\theta_i)\cos 2\phi] - B(1-\cos\theta_i)\sin 2\phi \\ B[(1+\cos\theta_i) + (1-\cos\theta_i)\cos 2\phi] - A(1-\cos\theta_i)\sin 2\phi \\ 2\sin\theta_i(A\cos\phi + B\sin 2\phi) \end{pmatrix}. \tag{7}$$

The electric field in the vicinity of the focus is readily given by the Debye–Wolf diffraction integral [6], with $\mathbf{E}_i$ taken as the strength vector:

$$\mathbf{e}_i = \int_0^{\alpha_i} \int_0^{2\pi} \mathbf{E}_i(\theta_i, \phi) \exp[ik_i\rho_s \sin\theta_i \cos(\phi - \phi_s)] \times \exp(ik_i z_s \cos\theta_i) \sin\theta_i \, d\theta_i \, d\phi, \tag{8}$$

where the cylindrical coordinates (ρ_s, ϕ_s, z_s) denote the position where the fluorescent dipole is located, k_i is the wavenumber of the illumination, and α_i is the convergence semi-angle of the objective lens. After performing the integration over ϕ, Eq. (8) gives

$$\mathbf{e}_i = \begin{pmatrix} e_{ix} \\ e_{iy} \\ e_{iz} \end{pmatrix} = \begin{pmatrix} AI_0 + I_2(A\cos 2\phi_s + B\sin 2\phi_s) \\ BI_0 + I_2(A\sin 2\phi_s - B\cos 2\phi_s) \\ -2iI_1(A\cos\phi_s + B\sin\phi_s) \end{pmatrix}, \tag{9}$$

where the functions I_n are given by

$$I_0 = \int_0^{\alpha_i} \sqrt{\cos\theta_i} \sin\theta_i (1 + \cos\theta_i) J_0(k_i\rho_s \sin\theta_i) \exp(ik_i z_s \cos\theta_i) \, d\theta_i \tag{10a}$$

$$I_1 = \int_0^{\alpha_i} \sqrt{\cos\theta_i} \sin^2\theta_i J_1(k_i\rho_s \sin\theta_i) \exp(ik_i z_s \cos\theta_i) \, d\theta_i \tag{10b}$$

$$I_2 = \int_0^{\alpha_i} \sqrt{\cos\theta_i} \sin\theta_i (1 - \cos\theta_i) J_2(k_i\rho_s \sin\theta_i) \exp(ik_i z_s \cos\theta_i) \, d\theta_i, \tag{10c}$$

where J_n is the nth-order Bessel function of the first kind. Equations (9) and (10), usually referred to as the point spread function, represent the general solution to the electric field in the focal region of a high-aperture lens illuminated by a plane wave of arbitrary polarization state. If the Babinet–Soleil compensator is set to zero retardance ($\delta = 0$), then Eq. (9) reduces to the corresponding equations of Richards and Wolf [6] for linearly polarized illumination.

For Cases I and II, the electric dipole moment of the emission is randomly oriented. The unit vector, and hence the dipole moment, oriented at angles θ_{dp} and ϕ_{dp} to the z and x axes, respectively, are given by

$$\mathbf{p}^I = \mathbf{p}^{II} = \begin{pmatrix} \cos\phi_{dp} \sin\theta_{dp} \\ \sin\phi_{dp} \sin\theta_{dp} \\ \cos\theta_{dp} \end{pmatrix}. \tag{11}$$

Integrating the intensity in the detector plane over all dipole orientation (θ_{dp}, ϕ_{dp}) will then give the response of an "average" dipole moment, corresponding to observation of a random cluster of dye molecules within the point spread function volume.

In Case III, the electric dipole moment of the emission, $\mathbf{p}^{III}$, is taken to be the unit vector in the direction of the incident electric field at the fluorescent molecule. The unit vector of polarization at position $\mathbf{r}_s = (\rho_s, \phi_s, z_s)$ in the object space is given by

$$\hat{\mathbf{e}}_i(\mathbf{r}_s) = \frac{\mathbf{e}_i(\mathbf{r}_s)}{|\mathbf{e}_i(\mathbf{r}_s)|}. \tag{12}$$

The Cartesian components of the electric dipole moment are thus given by

$$\mathbf{p}^{III} = \begin{pmatrix} p_x \\ p_y \\ p_z \end{pmatrix} = \hat{\mathbf{e}}_i(\mathbf{r}_s) = \frac{1}{|\mathbf{e}_i(\mathbf{r}_s)|}\begin{pmatrix} e_{ix} \\ e_{iy} \\ e_{iz} \end{pmatrix}. \tag{13}$$

The probability of excitation is given by the squared modulus of the component of the electric field in the direction of the dipole, raised to the power of the multiphoton process order. For Cases I and III, corresponding to an induced dipole, the incident field is in the direction of the dipole, and hence

$$P_N^I = P_N^{III} = |\mathbf{e}_i|^{2N}, \tag{14}$$

where N is the order of the process, that is, $N = 1$ for 1p, $N = 2$ for 2p, etc. It can be seen from (9) that

$$|\mathbf{e}_i|^2 = |I_0|^2 + |I_1|^2 + |I_2|^2 + 2\cos(\tfrac{1}{2}\delta)\cos(2\phi_s)(\Re\{I_0 I_2^*\} + |I_1|^2), \tag{15}$$

where $\Re$ denotes the real part of a complex number and * denotes the complex conjugate. For Case II, the probability of excitation depends only on the component of the incident electric field parallel to the permanent dipole moment, and hence

$$P_N^{II} = |\mathbf{p}^{II} \cdot \mathbf{e}_i|^{2N}. \tag{16}$$

Having expressions for the probability of excitation and the dipole moments for each case, all that remains is to consider the detection arm of the microscope.

4.2 The Detected Signal

The light emitted from the fluorescent molecule is collected by a high-aperture objective lens and imaged on to a finite-sized pinhole, as shown in Figure 2. The expression for the electric field in the detector plane due to a radiating electric dipole in the focal region is [28]

$$\mathbf{e}_d = \begin{pmatrix} p_x K_0^I - 2ip_z K_1^I \cos(\phi_d) + K_2^I(p_x\cos(2\phi_d) + p_y\sin(2\phi_d)) \\ p_y K_0^I - 2ip_z K_1^I \sin(\phi_d) + K_2^I(p_x\sin(2\phi_d) - p_y\cos(2\phi_d)) \\ p_z K_0^{II} - iK_1^{II}(p_x\cos(\phi_d) + p_y\sin(\phi_d)) \end{pmatrix}. \tag{17}$$

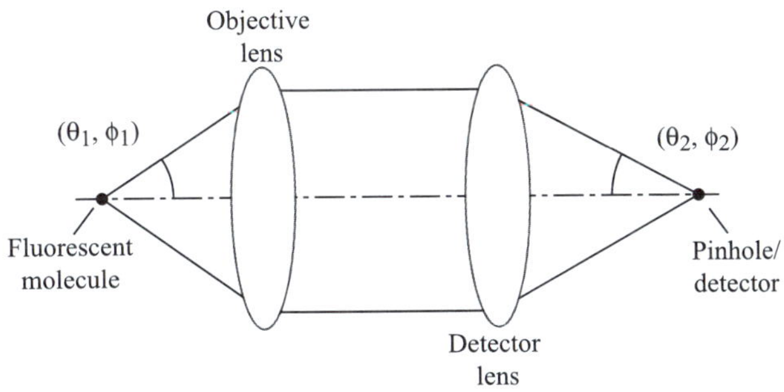

FIGURE 2 Schematic diagram of the detection system.

Here the functions K_n are given by

$$K_0^I = \int_0^{\alpha_2} \sqrt{\frac{\cos\theta_2}{\cos\theta_1}} \sin\theta_2 (1 + \cos\theta_1 \cos\theta_2) J_0(k_f \rho_d \sin\theta_2) \exp(ik_f z_s \cos\theta_1)\, d\theta_2, \tag{18a}$$

$$K_1^I = \int_0^{\alpha_2} \sqrt{\frac{\cos\theta_2}{\cos\theta_1}} \sin\theta_2 \sin\theta_1 \cos\theta_2 J_1(k_f \rho_d \sin\theta_2) \exp(ik_f z_s \cos\theta_1)\, d\theta_2, \tag{18b}$$

$$K_2^I = \int_0^{\alpha_2} \sqrt{\frac{\cos\theta_2}{\cos\theta_1}} \sin\theta_2 (1 - \cos\theta_1 \cos\theta_2) J_2(k_f \rho_d \sin\theta_2) \exp(ik_f z_s \cos\theta_1)\, d\theta_2, \tag{18c}$$

$$K_0^{II} = \int_0^{\alpha_2} \sqrt{\frac{\cos\theta_2}{\cos\theta_1}} \sin^2\theta_2 \sin\theta_1 J_0(k_f \rho_d \sin\theta_2) \exp(ik_f z_s \cos\theta_1)\, d\theta_2, \tag{18d}$$

$$K_1^{II} = \int_0^{\alpha_2} \sqrt{\frac{\cos\theta_2}{\cos\theta_1}} \sin^2\theta_2 \cos\theta_1 J_1(k_f \rho_d \sin\theta_2) \exp(ik_f z_s \cos\theta_1)\, d\theta_2, \tag{18e}$$

where k_f is the wavenumber of the fluorescent emission and (ρ_d, ϕ_d, z_d) is the position in the detector plane in cylindrical coordinates.

Here we have assumed that the polarization in the back focal plane of the objective is unchanged with axial scan of the object, which will be close to true for small defocus distances. The functions $K_n^{I,II}$ represent the electric and magnetic multipole components in the detected electric field distribution, respectively. If θ is small, K_n^I reduce to the same form as the Richards and Wolf integrals, and K_n^{II} becomes very small. We note that Eq. (18) is slightly different from the corresponding equation given by Higdon et al. [4].

The angular aperture of the detector lens, α_2, and the coordinate θ_2 are related to the angular aperture of the collimating objective lens, α_1, and the coordinate θ_1 via

$$\frac{n_1 \sin\alpha_1}{n_2 \sin\alpha_2} = \frac{n_1 \sin\theta_1}{n_2 \sin\theta_2} = \beta, \tag{19}$$

where β is the nominal magnification of the objective lens and n_1 and n_2 are the refractive indices in the object and image space, respectively

For Cases I and II, the integration must be performed over all possible dipole orientations, because the dipoles can freely rotate during the fluorescent lifetime; that is,

$$I_d^{I,II} = \int_0^{\pi} \int_0^{2\pi} P^{I,II} |\mathbf{e}_d^{I,II}|^2 \sin\theta_{dp}\, d\theta_{dp}\, d\phi_{dp}. \tag{20}$$

For Case III, the intensity distribution in the detector plane is given by

$$I_d^{III} = P^{III} |\mathbf{e}_d^{III}|^2. \tag{21}$$

Substituting the probability functions from Eqs. (14) and (16) gives

$$I_d^{I} = |\mathbf{e}_i|^{2N} \int_0^{\pi} \int_0^{2\pi} |\mathbf{e}_d^{I}|^2 \sin\theta_{dp}\, d\theta_{dp}\, d\phi_{dp}, \tag{22a}$$

$$I_d^{II} = \int_0^{\pi} \int_0^{2\pi} |\mathbf{p}^{II} \cdot \mathbf{e}_i|^{2N} |\mathbf{e}_d^{II}|^2 \sin\theta_{dp}\, d\theta_{dp}\, d\phi_{dp}, \tag{22b}$$

$$I_d^{III} = |\mathbf{e}_i|^{2N} |\mathbf{e}_d^{III}|^2. \tag{22c}$$

We note that for Cases I and II, the image separates into a product of illumination and detection factors as a result of the depolarization of the fluorescent emission. Another substitution for the electric dipole moment and electric fields $\mathbf{e}_i$ and $\mathbf{e}_d$ for Case I, after performing the integrations over θ_{dp} and ϕ_{dp}, leads to

$$I_d^{I} = \tfrac{8}{3}\pi |\mathbf{e}_i|^{2N} [|K_0^I|^2 + 2|K_1^I|^2 + |K_2^I|^2 + 2(|K_0^{II}|^2 + |K_1^{II}|^2)], \tag{23}$$

where Eq. (22) has been used.

For Case II, it is not possible to obtain a general solution for an arbitrary order of multiphoton process. However, the first three orders have been evaluated analytically to give

$$\begin{aligned} I_d^{II} = \tfrac{8}{15}\pi[&2(|K_0^I|^2 + |K_1^I|^2 + |K_2^I|^2 + |K_0^{II}|^2 + 2|K_1^{II}|^2)(|e_{ix}|^2 + |e_{iy}|^2) \\ &+ (|K_0^I|^2 + 6|K_1^I|^2 + |K_2^I|^2 + 6|K_0^{II}|^2 + 2|K_1^{II}|^2)|e_{iz}|^2 - 4f_1 + 2f_2] \end{aligned} \tag{24}$$

for a 1-photon process. For a 2-photon process, the expression becomes

$$\begin{aligned} I_d^{II} = \tfrac{8}{35}\pi |\mathbf{e}_i|^2 [&(3|K_0^I|^2 + 2|K_1^I|^2 + 3|K_2^I|^2 + 2|K_0^{II}|^2 + 6|K_1^{II}|^2)(|e_{ix}|^2 + |e_{iy}|^2) \\ &+ (|K_0^I|^2 + 10|K_1^I|^2 + |K_2^I|^2 + 10|K_0^{II}|^2 + 2|K_1^{II}|^2)|e_{iz}|^2 - 8f_1 + 4f_2], \end{aligned} \tag{25}$$

and for a 3-photon process,

$$\begin{aligned} I_d^{II} = \tfrac{8}{63}\pi |\mathbf{e}_i|^4 [&2(2|K_0^I|^2 + |K_1^I|^2 + 2|K_2^I|^2 + |K_0^{II}|^2 + 4|K_1^{II}|^2)(|e_{ix}|^2 + |e_{iy}|^2) \\ &+ (|K_0^I|^2 + 14|K_1^I|^2 + |K_2^I|^2 + 14|K_0^{II}|^2 + 2|K_1^{II}|^2)|e_{iz}|^2 - 12f_1 + 6f_2], \end{aligned} \tag{26}$$

where we have introduced

$$f_1 = \Re\{K_1^I(K_0^{I*} + K_2^{I*}) + 2K_0^{II}K_1^{II*}\}, \tag{27a}$$

$$f_2 = (\Re\{K_0^I K_2^{I*}\} + |K_1^{II}|^2)[(|e_{ix}|^2 - |e_{iy}|^2)\cos(2\phi_d) + 2\Re\{e_{ix}e_{iy}^*\}\sin(2\phi_d)]. \tag{27b}$$

Substituting for the electric dipole moment and electric fields $\mathbf{e}_i$ and $\mathbf{e}_d$ in Case III gives

$$\begin{aligned} I_d^{III} = {} & \tfrac{8}{15}\pi|\mathbf{e}_i|^{2(N-1)}[(|K_0^I|^2 + |K_2^I|^2 + 2|K_1^{II}|^2)(|e_{ix}|^2 + |e_{iy}|^2) \\ & + 4(|K_1^I|^2 + |K_0^{II}|^2)|e_{iz}|^2 - 4f_1 + 2f_2]. \end{aligned} \tag{28}$$

The detected signal of a fluorescence microscope employing a pinhole of radius R is given by integrating the power over the area of the detector, but since θ_2 is small, this is very close to integrating the intensity; that is,

$$I^{I,II,III} = \int_0^R \int_0^{2\pi} I_d^{I,II,III} r_p \, dr_p \, d\phi_p, \tag{29}$$

where (r_p, ϕ_p) is a coordinate system centered on the pinhole. It should be noted that the I_n in Eq. (10) are functions of $\mathbf{r}_s$, whereas the K_n in Eq. (18) are functions of $\mathbf{r}_d$. The vector relationship between the coordinates of the detector plane and those of the pinhole is given by [28]

$$\mathbf{r}_p = \mathbf{r}_d - \mathbf{r}_s\beta. \tag{30}$$

Equation (29) is used to compute numerical results with the appropriate substitution for each case.

5 NUMERICAL RESULTS

We trust that the reader arriving at this point has read at least Sections 2 and 3. Description of the multiphoton microscopes, discussion on the specific fluorescence cases considered in this section, and general background knowledge that we consider essential in understanding what follows are all described in those sections. The mathematical details in Section 4 are not necessary in order to understand the following numerical results.

The theory presented in the preceding section is generally applicable to any multiphoton fluorescence microscope. However, certain parameters must be chosen before numerical results can be presented. First, it is well known that, for example, a 3p fluorescence microscope exhibits better lateral and axial resolution than a 2p microscope when characteristic quantities are computed for the same excitation wavelength and for different fluorescence wavelengths [29]. Conversely, when the system has different excitation wavelengths and the same fluorescence wavelength, the 3p microscope exhibits worse lateral and axial resolution than a 2p microscope. This is due to the increase in the spatial distributions of the illuminating field due to the longer wavelength inherent with a higher-order process. For this reason, one has to decide whether the first or the second case is computed numerically, although the results can be easily scaled for different wavelength conditions. Gryczynski et al. [25] recently demonstrated multiphoton excitation of *p*-quaterphenyl (*p*-QT) dyes. In this study, the emission spectrum of the fluorescence was measured at different excitation wavelengths for one-, two-, and three-photon excitation. For *p*-QT, the same emission spectrum was observed for three-photon excitation at 850 nm, two-photon excitation at 586 nm, and one-photon excitation at 283 nm. The peak emission is at around 360 nm. All numerical results presented in this chapter will be calculated for these wavelengths. In previous publications on the theory of multiphoton fluorescence microscopes, numerical results were presented in terms of normalized optical coordinates. In two papers in particular [15, 18], results of similar theories may appear to be contradictory purely because of the choice of normalization wavelength employed in the two cases. To avoid such confusion in this work, and because it is not possible to express the results of high-aperture theory in universally applicable normalized optical coordinates, results will be presented for specific examples in terms of absolute dimensions. However, pinhole radii will also be expressed in terms of Airy units (A.u.), where 1 A. u. is the radial distance of the first zero of the Airy distribution in the detector plane of the 1.4 NA, 100× microscope objective lens used in the numerical simulations. Thus, for our assumed nominal magnification of 100×, 1 A. u. corresponds to a pinhole radius of 19.6 μm. It should be noted that true pinhole size is thus different for different commercial systems. Other papers have used another normalized pinhole radius, namely $v_d =$ 3.83 A.u.

5.1 Lateral and Axial Resolution

We will begin by considering the two limiting cases of linearly and circularly polarized illumination. It will be useful to clarify the physical meaning of the three polarization-dependent fluorescent emissions (Cases I, II, and III) described in Section 3. Cases I and II, the two limiting models of unpolarized emission, describe much the same process whatever the polarization state of the illumination. In Case I, the circularly polarized illumination is absorbed and a randomly oriented dipole emits. In Case II, the dipole moment of emission is fixed in a random direction and only the component of the illumination polarized parallel to the dipole moment is absorbed. In Cases I and II, averaging over all dipole directions leaves the fluorescent emission completely unpolarized. In Case III, all of the incident illumination is absorbed, and the dipole moment of the emission is aligned with the electric vector of the illuminating field. This statement is equally true for circularly as well as linearly polarized illumination. When we say all the illumination is absorbed, of course we really mean that a constant proportion is absorbed for all three cases, dependent on the excitation cross-section of the dye molecules.

In Figure 3, the images of a fluorescent molecule scanned in the focal plane for the specific example of unpolarized emission (Case I) in a microscope employing a 100×, 1.4 NA oil-immersion ($n_{\text{oil}} = 1.518$) objective lens for 1p, 2p, and 3p processes are presented. The left-hand images were computed for a 100 μm (5.1 A.u.) radius pinhole, which will later be shown to be close to equivalent to conventional imaging, whilst the right-hand images were calculated for a microscope employing a confocal pinhole of 1 μm (0.051 A.u.) radius. The $R = 100$ μm plots give an idea of the relative image size for the first three multiphoton processes in a conventional scanning fluorescence microscope imaging a single molecule. It is seen that the image of the single molecule increases dramatically in size, and hence the lateral resolution worsens, as the order of the multiphoton process increases. Such a worsening in lateral resolution, caused by the longer excitation wavelength required with higher-order processes as illustrated in Figure

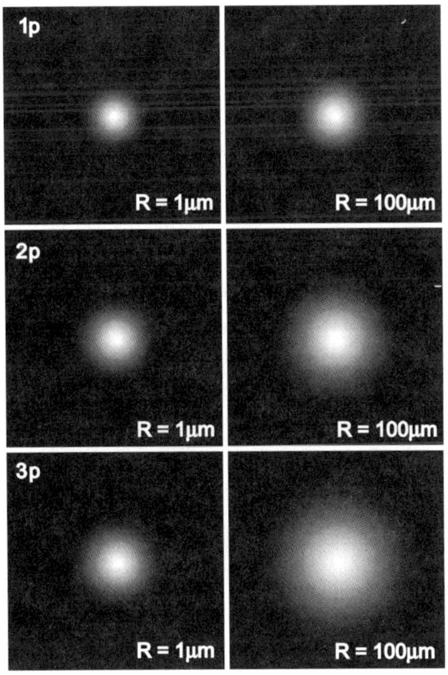

FIGURE 3 Image of a fluorescent molecule illuminated with circularly polarized light with unpolarized emission (Case I) for 1p, 2p, and 3p, and for pinhole radii of 1 and 100 μm. The image size is 0.8 μm × 0.8 μm.

3, is a high price to pay for the advantages that are inherent with the higher-order (2p and 3p) processes. However, when the 1 μm pinhole is employed, and hence the imaging is confocal, the process order has a much smaller effect on the lateral resolution, because the predominant factor in resolution is the emission, which is at the same wavelength for any order of process. Note that as the magnification is specified for a particular focal length of detector lens, the radius of the objective pupil, and $\sin\alpha_2$, are proportional to NA/β. The image size still increases with the number of photons – but by a significantly smaller amount. This is an important result, which suggests that, although a confocal pinhole is not required to achieve optical sectioning in 2p and 3p fluorescence imaging, considerably higher-resolution imaging will result if a confocal pinhole is employed. Each of the images in Figure 3 is circularly symmetric. This is due to the fact that the circularly polarized illumination results in a circularly symmetric electric field in the focal region when focused by the high-aperture lens and, in Case I, the emission is unpolarized. Cases II and III also result in circularly symmetric images: Case III because the emission, as well as the illumination, is circularly polarized, and Case II for exactly the same reasons as Case I.

To show how the limiting cases of the fluorescent emission affect the resolution properties of the microscope, Figure 4 is presented. In this figure, the axial and lateral half-width at half-maximum (HWHM) curves are plotted as functions of pinhole radius for 1p, 2p, and 3p, and for Cases I, II, and III. Figure 4 shows the transition between the two limiting cases of confocal and conventional imaging, which were illustrated in Figure 3, as the pinhole size is increased. The improvement in lateral resolution as the pinhole size is reduced is clearly seen to increase with the process order for Cases I, II, and III. In fact, the dependence of the lateral resolution on the pinhole radius is similar for Cases I, II, and III.

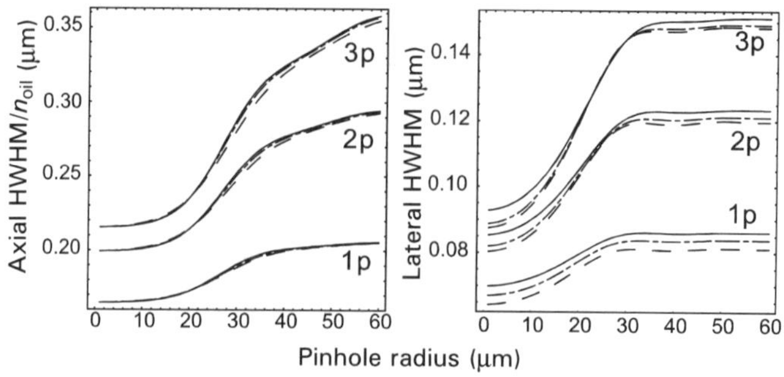

FIGURE 4 Axial and lateral half-width at half-maximum (HWHM) curves for 1p, 2p, and 3p as functions of pinhole radius for Cases I (continuous line), II (dash-dotted line), and III (dashed line) with circularly polarized illumination. $100\times$, 1.4 NA.

So, when circularly polarized illumination is employed, the state of polarization of the fluorescent emission does not greatly affect the imaging properties of the microscope. The most significant resolution difference (of less than 7%) is seen for the 1p process and is between Case III (polarized emission) and Case I (unpolarized with all illumination absorbed). Case III is seen always to exhibit the best lateral resolution and Case I the worst. Hence, the resolution of the microscope can be assumed to fall between these two limits, and is determined by the state of polarization, which depends on many parameters, as discussed in the introduction to this chapter. In general, the higher the degree of polarization of the fluorescent emission, the better the resolution. Figure 4 also shows that, for lateral resolution, pinhole radii greater than 30 μm (1.53 A.u.) correspond to conventional imaging, whereas pinhole radii smaller than 10 μm (0.51 A.u.) give close to confocal imaging for each of the multiphoton processes. It should be noted that these values are applicable only to the current numerical parameters. Although results have also been given for the reader's information in normalized units for the current numerical parameters, it is not possible to make general statements that are independent of lens aperture when considering fluorescence processes in high-aperture systems.

A measure of the axial resolution of the multiphoton fluorescence microscope imaging a single fluorescent molecule is also given in Figure 4 in terms of the axial HWHM as it is scanned along the optic axis. It is seen that the polarization state of the emission has little influence on the axial HWHM, and that reducing the size of the pinhole reduces the axial HWHM. This improvement in the axial resolution due to the introduction of a confocal pinhole is seen to increase with process order (as with lateral resolution). In the case of the axial resolution, close to confocal imaging is obtained for pinhole radii less than 20 μm (1.2 A.u.) and conventional imaging for greater than 60 μm (3.05 A. u.). Although it is known that two- and three-photon fluorescence imaging exhibits optical sectioning without the need to use a confocal point detector, it is important, particularly in light of the results presented in Figure 3, to determine whether any advantage results from the use of confocal detection. To investigate this, we introduce the integrated intensity I_{int}, as

$$I_{int}(z_s) = \int_0^\infty \int_0^{2\pi} I(r_s, \phi_s, z_s) r_s \, dr_s \, d\phi_s, \tag{31}$$

which is essentially the total energy in the image as a function of defocus. In the case of conventional single-photon fluorescence imaging, it is well known that this function is constant, whereas for a confocal geometry it decays, that is, the system exhibits optical sectioning. This is illustrated in Figure 5 for Case I and circularly polarized illumination.

Similar trends are to be expected for Cases II and III. The 2p and 3p cases illustrate the optical sectioning capability of a conventional multiphoton fluorescence microscope. However, as also shown in this figure, the introduction of a confocal pinhole leads to a further improvement in the optical sectioning for both 2p and 3p fluorescence scanning microscopes, with a decrease in the integrated intensity HWHM of about 30%. This result, together with those of Figure 3, suggest that considerable resolution advantage is likely to result from implementing confocal rather than conventional multiphoton fluorescence imaging.

5.2 Image Asymmetry

It has been shown previously [28] that when a point scatterer is imaged by linearly polarized incident light, the lateral asymmetry in the electric field in the focal region causes a laterally asymmetric image of the scatterer. This effect is now investigated for multiphoton fluorescence imaging as follows. A measure of the image asymmetry is introduced

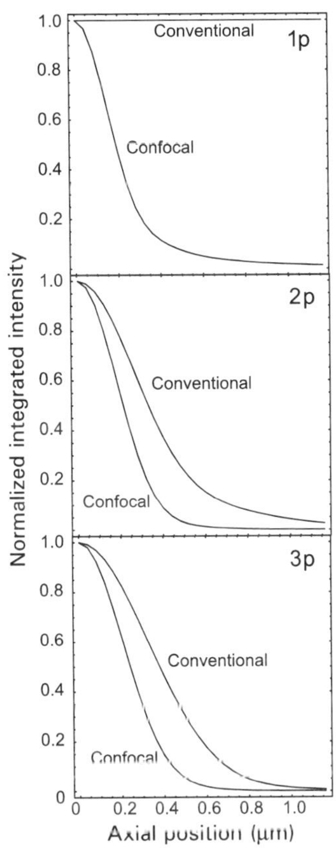

FIGURE 5 Normalized integrated intensity as a function of axial position for 1p, 2p, and 3p for conventional and confocal detection. 100×, 1.4 NA.

as the ratio of the HWHM of a scan along the x direction to the HWHM of a scan along the y direction. For circularly polarized illumination ($\delta = 90°$), the asymmetry ratio is unity, whereas if the Babinet–Soleil is set at any relative retardance other than 90° ($0° < \delta < 180°$), then the electric field in the focal region becomes asymmetric. In general, the incident polarization is elliptical, and for the special cases of $\delta = 0°$ and $\delta = 180°$, the incident illumination is linearly polarized in the x and y directions, respectively.

Figure 6 shows the asymmetry ratio (i. e., unity for a symmetric image) as a function of relative retardation setting δ for the case of polarized 1p, 2p, and 3-p fluorescent emission detected with a 100 μm pinhole. There is a smooth transition in the symmetry ratio, with maxima and minima corresponding to the two linear illumination cases as expected.

Figure 7 shows the equivalent images to those presented in Figure 3, but for linearly polarized illumination ($\delta = 0$), i.e., Case I, 100× 1.4 NA lens, 1p, 2p, and 3p, for $R = 100$ μm and $R = 1$ μm and a field of view of 0.8 μm × 0.8 μm. These lateral images are asymmetric, as expected, as a consequence of employing a high-aperture microscope objective lens and linear polarization. Although the form of the image is more complicated for linearly polarized light, the general dependences of image size on process order and pinhole size are the same as for circular polarization; that is, the improvement in lateral resolution due to the confocal pinhole increases with the order of the multiphoton fluorescence process. The symmetry properties in Figure 7 are seen to vary from image to image, with process order and pinhole radius.

As the image asymmetry with linearly polarized illumination is of practical importance, Figure 8 is presented. This figure shows the variation in image asymmetry as a function of pinhole radius, and it is seen that, in general, the asymmetry is smaller when the pinhole radius is reduced. The images in Figure 7 were plotted for the case of unpolarized fluorescent emission (Case I), and hence the asymmetry is a "mapping" of the asymmetry in the illuminating electric field. When fluorescent permanent dipoles are randomly oriented (Case II), or the emission is polarized (Case III), the asymmetry becomes a more complicated combination of multiple asymmetries in the high-aperture system.

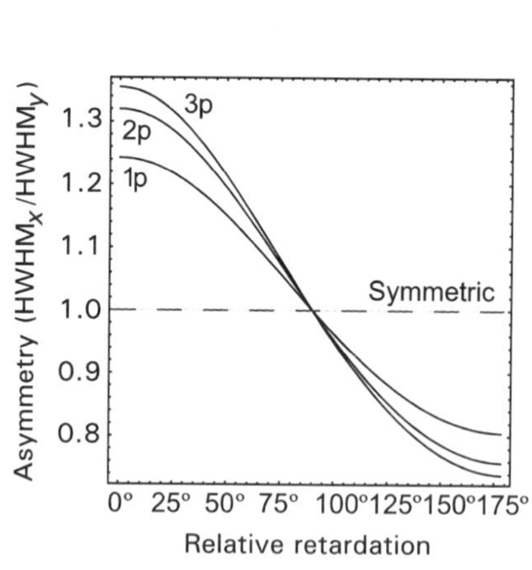

FIGURE 6 Asymmetry in the image of a fluorescent molecule for 1p, 2p, and 3p excitation as a function of the relative retardance setting of the Babinet–Soleil compensator for $R = 100$ μm for Case I.

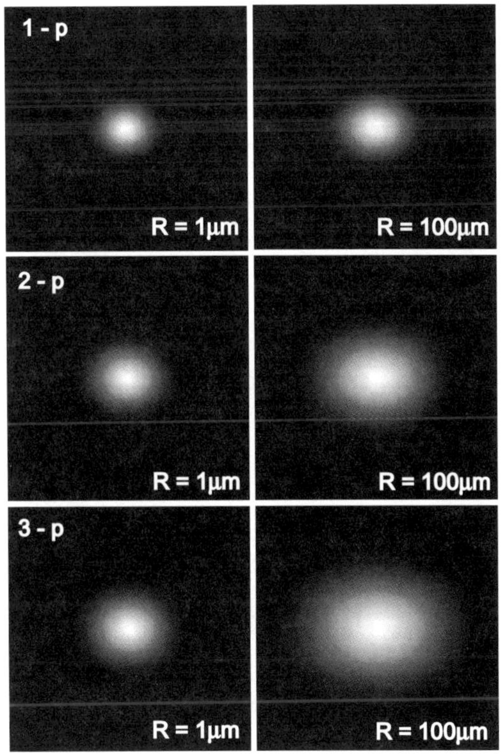

FIGURE 7 Image of a fluorescent molecule illuminated with linearly polarized light (horizontal direction) with unpolarized emission (Case I) for 1p, 2p, and 3p, and for pinhole radii of 1 and 100 μm. The image size is 0.8 μm × 0.8 μm.

5.3 Paraxial Approximation

All of the asymmetry effects are due to the combination of high-aperture objective lens and the linearly polarized illumination used for the preceding numerical results. Figure 9 shows axial and lateral HWHM curves calculated for a low-aperture, 10×, 0.15 NA, lens. Again the dashed and full lines in the lateral plot correspond to orthogonal scan directions. All of the Cases I, II, and III are plotted in this figure, and, although there is a slight widening of some of the curves, they cannot be distinguished from one another, indicating that the degree of polarization does not affect the axial or lateral resolution in low-aperture systems. The asymmetry in the image is also seen to be negligible for each case. It should also be noted that when plotted for circularly polarized illumination, almost identical results to those in Figure 9 were obtained.

These results may also be obtained from the less complicated paraxial theory for multiphoton fluorescence, since the polarization state of the emission and high-aperture effects may be ignored. However, in practice, such low-numerical-aperture lenses are seldom used except in developmental biology. Nevertheless, the paraxial theory is often used to obtain a rough estimate concerning the performance of a high-aperture optical system. Since our theory is a high-aperture description, it is easy to approximate our formulas paraxially. When such an approximation is made, one can compare the results of the approximation with those obtained in the high aperture case. The result of such a

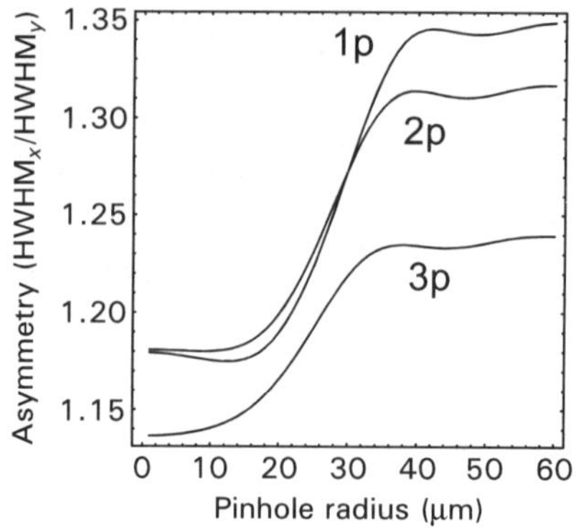

FIGURE 8 Asymmetry in the image of a fluorescent molecule for 1p, 2p, and 3p excitation as a function of pinhole size for linearly polarized illumination for Case I.

comparison could be used to evaluate the accuracy of the low-aperture approximation. When, for example, such a comparison is performed for linear illumination, Case I, and 3p configuration, we find that the paraxial approximation yields a lateral resolution value about 0.55 times higher than is obtained in the high-aperture case. This represents a deviation from the actual value by 55%!

It is also important to calculate how much the presence of a confocal pinhole would improve on the lateral and axial resolution as compared with the conventional case. The ratio improvement in axial and lateral resolution for an unpolarized emission (Case I) of a fluorescent molecule illuminated with circular polarization between the conventional and confocal imaging limits is given in Table 1 for 1p, 2p, and 3p. The improvements in both lateral and axial resolution are approximately the same for the two lenses, though there seems to be an increasing difference with increasing process order.

The results for the 10×, 0.15 NA lens were very close to those of the paraxial theory. Since the paraxial theory may be written in normalized coordinates, its results are simply

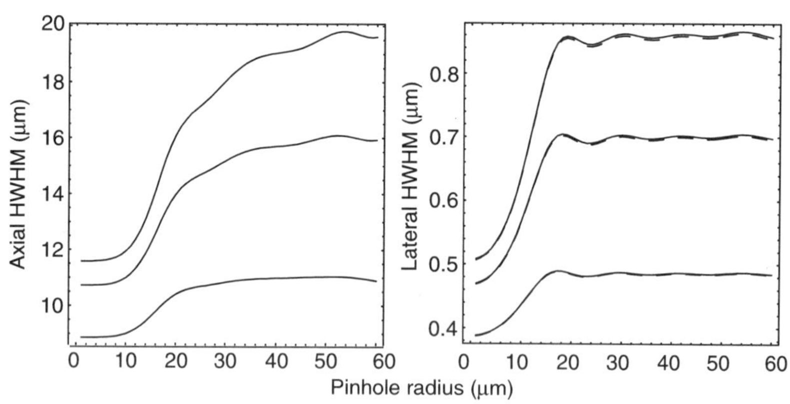

FIGURE 9 Axial and lateral HWHM curves for 1p, 2p, and 3p as functions of the pinhole radius for Cases I, II, and III with linearly polarized illumination. The corresponding curves fully overlap. The dashed curves indicate scans in the direction perpendicular and the full curves scans parallel to the direction of incident polarization, 10×, 0.15 NA.

TABLE 1 Ratio increase in resolution due to a confocal pinhole over conventional imaging ($HWHM_{conv}/HWHM_{conf}$) for 1- , 2-, and 3-photon fluorescence for low- and high-aperture lenses

		Resolution improvement		
Lens		1p	2p	3p
10×, 0.15	Lateral	1.26	1.52	1.73
	Axial	1.27	1.53	1.74
100×, 1.4	Lateral	1.26	1.49	1.69
	Axial	1.27	1.53	1.77

scaled according to the lens. Therefore, the difference in lateral resolution results from the paraxial theory are the same for any lens. Considering the data in Table 1, this means that the paraxial theory estimates the *ratio* of the resolution values for the high-aperture lens fairly accurately for the 1p, 2p, and 3p processes. On the other hand, we should not forget that the *actual* resolution values are very inaccurate when the paraxial theory is used to compute the high-aperture objective.

5.4 Signal-to-Noise Ratio

In the previous subsections, it has been shown that employing a confocal pinhole in 2p and 3p fluorescence microscopes leads to great improvements in the lateral and axial resolution. However, it is important to consider that in such microscopes, the light budget can be limited. It is therefore the aim in this section to determine an optimum pinhole radius in terms of maximizing the signal-to-noise ratio of the microscope. The approach is similar to that employed by Sheppard [30], but employing the high-aperture theory. That is, shot noise on the background is assumed to be proportional to the area of the detector, of radius R, and a measure of the signal-to-noise ratio is given by

$$\frac{S}{N}(R) = \frac{I(R)}{\sqrt{I(R) + \xi R^2}}, \tag{32}$$

where the constant ξ represents the noise level. This model is applicable for background emanating from autofluorescence of the optical components (cements and coatings), as is important in single-molecule fluorescence imaging. A typical example of how the signal-to noise ratio changes with pinhole radius is given in Figure 10. These curves are plotted for 3p fluorescence on a log–log scale. Figure 11 shows the optimum pinhole radius plotted as a function of the noise level ξ for the two limiting cases of polarized (Case III) and unpolarized (Case I) emission for 1p, 2p, and 3p fluorescence in the high-aperture system.

In the high-noise limit, when the shot noise on the signal becomes large, it is seen that the optimum pinhole radius varies between approximately 14 μm and 16 μm (0.71 0.81 A.u.), depending on the degree of polarization, and is independent of the order of the multiphoton process. In the low-noise limit, the optimum pinhole radius varies between

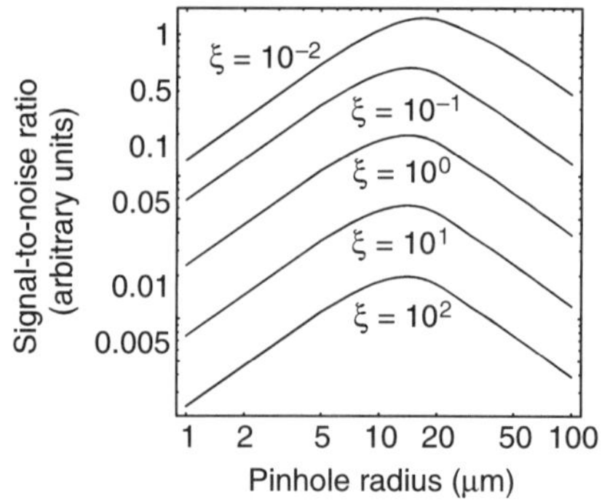

FIGURE 10 Signal-to-noise ratio as a function of pinhole radius for 3p fluorescence imaging and for Case I, 100×, 1.4 NA.

approximately 17 μm and 25 μm (0.86–1.27 A.u.) for unpolarized and polarized emission, respectively. In this case, there is seen to be a small variation in the curves corresponding to the 1p, 2p, and 3p processes, but in practice these differences may be ignored. These optimum pinhole radii correspond to imaging in the regime between the conventional and confocal limits; that is, employing the optimum pinhole already leads to a distinct improvement in the axial resolution over the conventional imaging case. However, to obtain the maximum possible improvement in resolution, an even smaller pinhole need be employed.

Figure 12 is plotted to determine the cost in terms of the signal-to-noise ratio of employing a pinhole that leads to close to true confocal imaging. For these calculations, a pinhole of radius 10 μm (0.051 A.u.) is used. Such a pinhole would lead to resolution properties only slightly worse than an ideal confocal microscope ($R \to 0$). Figure 12 shows the signal-to-noise ratio with the 10 μm pinhole as a percentage of the signal-to-noise ratio obtained by employing the optimum pinhole radius as a function of the noise level.

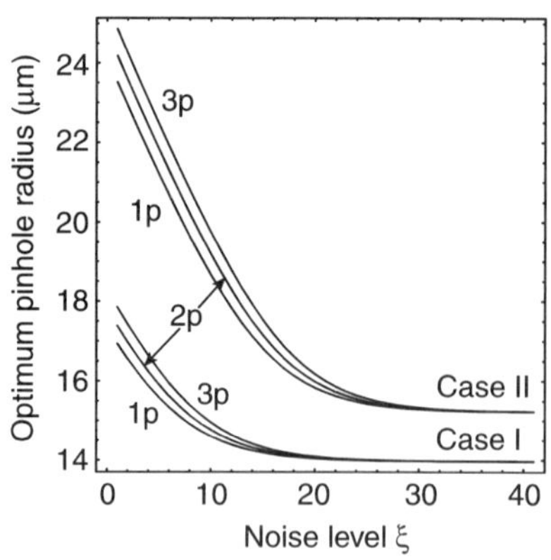

FIGURE 11 Pinhole radius corresponding to maximum signal-to-noise ratio as a function of noise level for fluorescence imaging for 1p, 2p, and 3p excitation, and for Cases I and II. 100×, 1.4 NA.

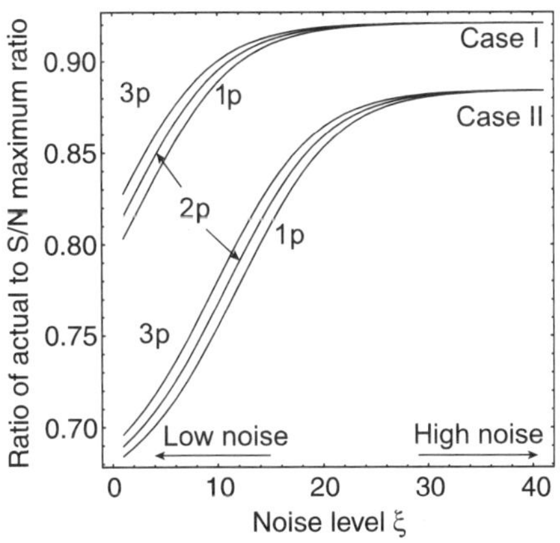

FIGURE 12 Signal-to-noise ratio for a confocal pinhole ($R = 10\,\mu m$) as a ratio of the maximum signal-to-noise ratio. Plotted as a function of noise level for fluorescence imaging for 1p, 2p, and 3p excitation, and for Cases I and II. 100×, 1.4 NA.

It is seen that in the high-noise limit the signal-to-noise ratio is reduced from its maximum value by only 8–12% (i. e. about 0.5 dB), depending on the state of polarization of the fluorescent emission. In the low-noise case, the decrease in signal-to-noise ratio is more significant at 17–32% of its maximum value (i. e. about 1–2 dB). Again, the reduction in signal-to-noise ratio is due mainly to the state of polarization rather than the order of the multiphoton process. However, it should be noted that 3p fluorescence, which benefits the most in terms of resolution from confocal imaging, actually suffers the least in terms of signal-to-noise ratio. Ultimately, even the largest loss in the signal-to-noise ratio mentioned above is a relatively small price to pay for the vast improvement in the lateral resolution of the 2p and 3p microscopes achieved by employing the confocal pinhole.

5.5 High Background Noise

In the previous subsection, the signal-to-noise ratio has been investigated based on a model in which noise power is represented by a term proportional to the area of the pinhole. This was shown experimentally [30] to be a form of noise present in a confocal reflection microscope. As explained earlier, such noise could be generated in the fluorescence mode by cements and coatings on optical components. Another potential source of noise is the background autofluorescence from neighboring regions of the sample [3, 31]. Assuming this to be featureless, it is equivalent to integrating over the three-dimensional point spread function. This has been analyzed for a 1p fluorescence microscope in the paraxial approximation [31] and also for 2p and 3p microscopes [32]. It is found that for 1p fluorescence, the noise power increases as the area of the pinhole for small pinhole sizes, but eventually increases linearly with pinhole radius as the noise power becomes proportional to the thickness of the optical section. A qualitatively similar result is expected for the high-aperture case. The resulting signal-to-noise ratio is shown in Figure 13, where for 1p fluorescence the signal-to-noise is normalized to unity at peak value, and for 2p and 3p fluorescence it is normalized to unity for large pinholes. Note that in the latter cases, the signal-to-noise ratio does not fall off to zero with large pinhole size, since the thickness in optical section is limited by the excitation process. The optimum

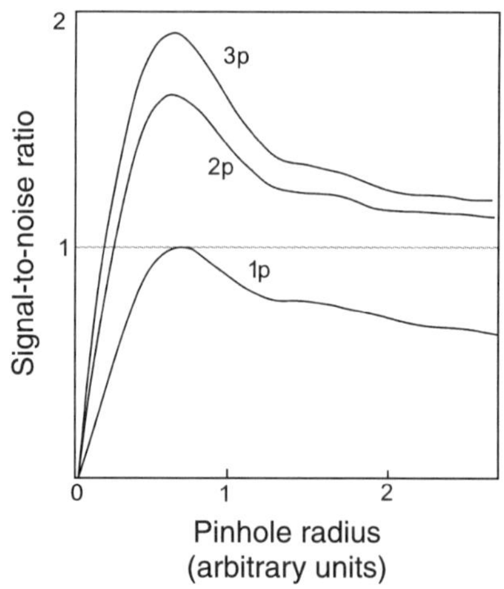

FIGURE 13 The signal-to-noise ratio in a case where the specimen exhibits strong background fluorescence, as a function of pinhole size. Results obtained from paraxial theory. For 2p and 3p, the curves are normalized to a signal-to-noise ratio of unity for large pinholes. For 1p, the curve is normalized to a peak signal-to-noise ratio of unity. After Gauderon and Sheppard [33].

pinhole sizes are found to be 0.60, 0.63 and 0.64 A. u. for 1p, 2p, and 3p processes, respectively, that is, 11.8, 12.3 and 12.5 μm for the 1.4 NA, 100× objective. The improvement in signal-to-noise ratio over using a large pinhole is then by a factor of 1.66 and 1.92 for 2p and 3p, respectively (2.2 and 2.8 dB).

Figure 14 shows images of a 210 nm (subresolution) diameter fluorescent bead imaged in 2p fluorescence with different pinhole sizes [33]. The images were recorded with a Leica TCS NT confocal microscope coupled to a Coherent Mira 900 femtosecond Ti : sapphire laser pumped by a Coherent Verdi solid state green laser. The bead was imaged using 120 fs pulses at 800 nm wavelength using a 63×, 1.3 NA oil-immersion objective lens. The images clearly show an improvement in resolution for small pinhole sizes.

6 CONCLUSIONS

In this chapter a theory for 1p, 2p, and 3p multiphoton polarization-dependent fluorescence processes has been presented. From analysis of the numerical results, the following conclusions are drawn. In conventional imaging, the lateral resolution is reduced significantly as the order of the multiphoton process increases. However, the improvement in lateral resolution due to a confocal pinhole increases significantly with increasing process order; that is, for confocal imaging, the lateral resolution is not degraded significantly as the process order increases. When employing circularly polarized illumination, although the state of polarization of the fluorescent emission does not affect the lateral resolution by more than 7% even for the 100×, 1.4 NA lens considered, polarized emission always produces slightly better resolution than unpolarized. The improvement in axial sectioning due to a confocal pinhole increases significantly with increasing process order, suggesting that confocal detection should be used whenever possible in multiphoton

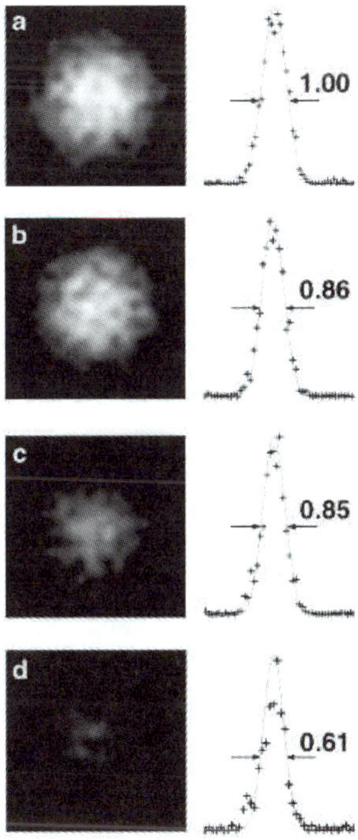

FIGURE 14 Series of 2p experimental fluorescence images and line scans of a single fluorescent bead of diameter 210 nm for four different sizes of the confocal pinhole: (*a*) 4 A.u.; (*b*) 2 A.u.; (*c*) 1 A.u.; (*d*) 0.5 A.u. The FWHM values on the line scans are quoted relative to case (*a*). Reproduced with permission from Gauderon and Sheppard [33].

imaging. When employing linearly polarized illumination, the degree of image asymmetry depends strongly on the degree of polarization of the emission. This image asymmetry is greatly reduced by the introduction of a confocal pinhole, and this reduction in asymmetry increases with the order of the multiphoton process. For the high-aperture system considered, with low noise, the optimum pinhole radius varies between 15 μm and 25 μm (0.75–1.28 A.u.) depending on the state of polarization. For high noise, the optimum pinhole radius is independent of the state of polarization at 15 μm (0.75 A. u.). The order of the multiphoton process has very little effect on the optimum pinhole size. To achieve true confocal imaging, the signal-to-noise ratio is reduced by between 8% and 30% from its maximum value, depending on the noise level and polarization of emission.

While we have stressed the advantages to be gained from using a confocal pinhole in multiphoton microscopy, in terms of resolution, asymmetry, and signal-to-noise ratio, we should finally point out that these conclusions are not applicable if the fluorescent light is scattered significantly on exciting from the sample. Furthermore, non-descanned non-confocal detectors may in principle offer increased signal levels.

ACKNOWLEDGMENT

Peter Török is an EPSRC Advanced Fellow at the University of Oxford.

REFERENCES

1. Cox, I. J. and Sheppard, C. J. R. Information capacity and resolution in an optical system. *J. Opt. Soc. Am.* **3**, 1152–1158 (1986).
2. Sheppard, C. J. R., Gan, X. S., Gu, M., and Roy, M., Signal-to-noise in confocal microscopes. In: *Handbook of Biological Confocal Microscopy* (ed. J. B. Pawley). New York: Plenum, 1995, pp. 363–372.
3. Sandison, D. R., Williams, R. M., Wells, K. S., Stricker, J., and Webb, W. W., Quantitative fluorescence confocal laser scanning microscopy (CLSM). In: *Handbook of Biological Confocal Microscopy* (ed. J. B. Pawley). New York: Plenum, 1995, pp. 417–430.
4. Higdon, P. D., Török, P., and Wilson, T. Imaging properties of high aperture multiphoton fluorescence scanning optical microscopes. *J. Microsc.* **193**, 127–141 (1999).
5. Xie, X. S. and Dunn, R. C. Probing single molecule dynamics. *Science* **265**, 361–364 (1994).
6. Richards, B. and Wolf, E. Electromagnetic diffraction in optical systems II. Structure of the image field in an aplanatic system. *Proc. R. Soc. London.* **A253**, 358–379 (1959).
7. Gryczynski, I., Malak, H., and Lakowicz, J. R. Multiphoton excitation of the DNA stains DAPI and Hoechst. *Bioimaging* **4**, 138–148 (1996).
8. Perrin, F. La fluorescence des solutions. *Ann. Phys. (Paris)* **12**, 169–275 (1929).
9. Soleillet, P. Sur les parame'tres caracterisant la polarization partielle de la lumie're dans les phenome'nes de fluorescence. *Ann. Phys. (Paris)* **12**, 23–86 (1929).
10. Sheppard, C. J. R. and Kompfner, R. Resonant scanning optical microscope. *Appl. Opt.* **18**, 2879–2882 (1978).
11. Denk, W., Strickler, J. H., and Webb, W. W. Two-photon fluorescence scanning microscopy. *Science* **248**, 73–76 (1990).
12. Hell, S. W., Bahlmann, K., Schrader, M., Soini, A., Malak, H., Gryczynski, I., and Lakowicz, J. R. Three-photon excitation in fluorescence microscopy. *J. Biomed. Opt.* **1**, 71–74 (1995).
13. Sheppard, C. J. R. and Gu, M. Image formation in 2-photon fluorescence microscopy. *Optik* **86**, 104–106 (1990).
14. Gu, M. and Sheppard, C. J. R. Comparison of three dimensional imaging properties between two-photon and single-photon fluorescence microscopy. *J. Microsc.* **177**, 128–137 (1995).
15. Sheppard, C. J. R. Image formation in three-photon fluorescence microscopy. *Bioimaging* **4**, 124–128 (1996).
16. Gu, M. and Sheppard, C. J. R. Effects of a finite-sized pinhole on 3D image-formation in confocal 2-photon fluorescence microscopy. *J. Mod. Opt.* **40**, 2009–2024 (1993).
17. Gu, M. and Sheppard, C. J. R. 3D image formation in confocal microscopy under ultra-short laser pulse illumination. *J. Mod. Opt.* **42**, 747–762 (1994)..
18. Gu, M. and Gan, X. S. Effect of the detector size and the fluorescence wavelength on the resolution of three- and two-photon confocal microscopy. *Bioimaging* **4**, 129–137 (1996).
19. van der Voort, H. T. M. and Brakenhoff, G. J. 3-D image formation in high-aperture fluorescence confocal microscopy – a numerical analysis. *J. Microsc.* **158**, 43–54 (1990).
20. Hell, S. W., Lindek, S., and Stelzer, E. H. K. Enhancing the axial resolution in far-field light-microscopy – 2-photon 4Pi confocal fluorescence microscopy. *J. Mod. Opt.* **41**, 675–681 (1994).
21. Visser, T. D. and Wiersma, S. H. Electromagnetic description of image formation in confocal fluorescence microscopy. *J. Opt. Soc. Am.* **A12**, 599–608 (1994).
22. Axelrod, D. Carbocyanine dye orientation in red cell membrane studied by microscopic fluorescence polarization. *Biophys. J.* **26**, 557–574 (1979).

23. Friedrich, D. M. Tensor patterns and polarization ratios for three-photon transitions in fluid media. *J. Chem. Phys.* **75**, 3258–3268 (1981).
24. Callis, P. R. On the theory of two-photon induced fluorescence anisotropy with applications to indoles. *J. Chem. Phys.* **99**, 27–37 (1993).
25. Gryczynski, I., Malak, H., and Lakowicz, J. R. Three photon excitation of *p*-quaterphenyl with a mode-locked femtosecond Ti : sapphire laser. *J. Fluoresc.* **6**, 139–145 (1996).
26. Sheppard, C. J. R. and Török, P. An electromagnetic theory of imaging in fluorescence microscopy, and imaging in polarization fluorescence microscopy. *Bioimaging* **5**, 205–218 (1997).
27. Török, P., Higdon, P. D., and Wilson, T. On the general properties of polarizing conventional and confocal microscopes. *Opt. Commun.* **148**, 300–315 (1998).
28. Török, P., Higdon, P. D., and Wilson, T. Theory for confocal and conventional microscopes imaging small dielectric scatterers. *J. Mod. Opt.* **45**, 1681–1698 (1998).
29. Lindek, S., Stelzer, E.H. K., and Hell, S. W., Two new high resolution confocal fluorescence microscopes (4Pi, theta) with single- and two-photon excitation. In: *Handbook of Biological Confocal Microscopy* (ed. J. B. Pawley). New York: Plenum, 1995, pp. 417–430.
30. Sheppard C. J. R. Stray light and noise in confocal microscopy. *Micron.* **22**, 239–243 (1991).
31. Gan, X. S. and Sheppard, C. J. R. Detectability: a new criterion for evaluation of the confocal microscope. *Scanning* **15**, 187–192 (1993).
32. Gauderon, R. and Sheppard, C. J. R. Effect of a finite-size pinhole on noise performance in single-, two-, and three-photon confocal fluorescence microscopy. *Appl. Opt.* **38**, 3562–3565 (1999).
33. Gauderon, R. and Sheppard, C. J. R. Effect of a confocal pinhole in two-photon microscopy. *Microsc. Res. Techn.* **47**, 210–214 (1999).

CHAPTER 7

Aberrations and Penetration in In-Depth Confocal and Two-Photon-Excitation Microscopy

C. J. de GRAUW, P. L. T. M. FREDERIX, and H. C. GERRITSEN

Ultrecht University, Molecular Biophysics, P.O. Box 80.000, 3508 TA Utrecht, The Netherlands

1 INTRODUCTION

Fluorescence laser scanning microscopy and conventional confocal microscopy (Pawley, 1995), as well as two-photon excitation microscopy (Denk et al., 1990), are now widely used in many fields. In laser scanning microscopy, laser light is used to excite molecules in a sample. The laser beam is focused by a microscope objective and scanned over an area in the sample from where the emission is recorded.

In conventional (single-photon) confocal laser scanning microscopy (CLSM), laser wavelengths ranging from the UV to the red are used to excite the fluorescent molecules. The fluorescence signal coming from the focal region is passed through a spatial filter that blocks the signal from outside the focus. Fluorescent molecules are excited by absorbing one photon, making the fluorescence yield linearly dependent on the excitation intensity.

In two-photon excitation microscopy, a wavelength of about twice the single-photon absorption wavelength is used for excitation. The simultaneous absorption of two photons, with about half the energy required for single-photon absorption, results in the excitation of the fluorescent molecules in the specimen. Two-photon excitation is a single-quantum effect, and depends on the square of the excitation intensity. It is a nonlinear process that depends on the third-order nonlinear susceptibility tensor (Göppert-Mayer, 1930), which implies that the signal is weak and requires a high intensity for excitation. This is realized in practice by employing high-numerical-aperture (NA) objectives that strongly focus the laser power. Generally, ultrashort (80 fs–1 ps) laser pulses are used for two-photon excitation. These pulses are usually provided by mode-locked solid state lasers. In this way, a high peak power is present in the focal region during each pulse, while the time-averaged power is comparatively low.

Confocal and Two-Photon Microscopy: Foundations, Applications, and Advances, Edited by Alberto Diaspro.
ISBN 0-471-40920-0

In single-photon confocal microscopy, the signal outside the focal plane is rejected by a pinhole. This requires a careful selection of the pinhole size in order to optimize the axial resolution without reducing the fluorescence signal substantially. Two-photon microscopy has an intrinsic axial resolution, even without a confocal pinhole in the detection path (Williams et al., 1994). This property arises from the quadratic dependence of the two-photon absorption process on the excitation intensity. The highest intensity is found in the focal area of the microscope objective. This, in combination with the quadratic dependence, leads to a localized volume with efficient two-photon absorption.

1.1 Aberrations and Penetration Depth

One of the advantages of two-photon excitation microscopy, compared with single-photon excitation microscopy, is its enhanced penetration depth. Two-photon excitation microscopy can yield images hundreds of micrometers deep into optically dense samples (Centonze and White, 1998; de Grauw et al., 1999; Daria et al., 1998). In contrast, confocal single-photon microscopy yields images much less deep in these samples. In two-photon microscopy, the optical properties of the microscope and the sample become of great importance for the image quality. Chromatic and spherical aberrations introduced by the optical elements in the microscope, aberrations resulting from focusing through a turbid sample, and a refractive-index mismatch between the objective and the sample introduce a distortion of the focus. These effects reduce the intensity in the focal region, and thus reduce the fluorescence signal (Carlsson, 1991; Hell et al. 1993a,b; Jacobsen et al., 1994; Visser et al., 1992). Apart from a reduced fluorescence signal, a distorted focus also influences the image quality. The focal region where excitation and emission takes place is described by the point spread function (PSF). The resolution of the recorded images depends directly on the dimensions of the PSF. A distorted PSF, therefore, yields a reduction of the imaging resolution.

The effects of chromatic and spherical aberrations on the PSF are identical for single-photon and two-photon excitation because they are mathematically identical. Apart from the wavelength dependence of the aberrations, the distortion of the focus in both types of microscopes leads to a similar decrease in excitation intensity. However, the fluorescence yield in the case of two-photon excitation is further reduced owing to the quadratic dependence of two-photon absorption on the incident intensity.

The reduction of intensity due to focusing through an immersion medium and a cover glass into a sample can be ascribed to refraction of the photons by the different materials and to scattering by components in the sample. Small objects in the sample may scatter the excitation and emission photons. Refraction takes place at interfaces with different refractive indices. Therefore, refraction may occur when there is a refractive-index mismatch between objective, immersion medium and sample. Refraction can also occur on objects in the sample that have a refractive index different from that of their environment.

All these effects distort the PSF, and therefore reduce the penetration into a sample and the resolution of the recorded images.

2 THE POINT SPREAD FUNCTION

The optical system projects a single point in the object plane as a blurred spot in the image plane. The way that a single point source is observed in the image is described by the PSF of the system (van der Voort and Brakenhoff, 1988; Wilson, 1990). Any object in the specimen is observed as a convolution of the original object with the PSF. The PSF is influenced by the optical properties of the system, such as the numerical aperture of the objective and optical aberrations. In an ideal scanning microscope, with no optical distortions, the light is focused into, and observed from, a diffraction-limited spot in the sample. Generally, the smaller the focus, the smaller the PSF, and the better the image resolution.

In a confocal single-photon microscope, the PSF is determined by the wavelength, the size of the pinhole, the properties of the objective, the sample, and the rest of the optical system. In a two-photon microscope, no pinhole is required, and often widefield detection is employed As a consequence, the PSF is directly related to the localized focal volume where the two-photon absorption occurs. Here, the quadratic dependence of the excitation on the incident intensity makes the fluorescence yield very sensitive to the size of the PSF.

2.1 Experimental Two-Photon PSF

The PSF of a system can be obtained by recording the three-dimensional (3D) image of a point source. Generally, small fluorescent beads are used for this purpose. As a compromise between fluorescence yield and the size of the bead, we used latex beads of 220 nm diameter, homogeneously filled with yellow green (FluoresBrite, PolySciences, Warrington, U.S.A.). A diluted drop of distilled water containing beads was dried onto a cover slip, to which they adhered. After rinsing, the cover slip was sealed on a microscope slide with water between the two glasses. 3D images of the beads attached to the coverslip were recorded on our home-built two-photon microscope (Sytsma et al., 1998) using a water-immersion objective (Nikon, 60 × /NA 1.2). As excitation source, we used a Ti : sapphire laser with a repetition rate of 82 MHz at 800 nm wavelength. The size of the images (*xyz*: 12 μm × 12 μm × 5.2 μm in 256 × 256 × 40 pixels) was chosen to amply oversample in all dimensions.

In order to obtain the optimal performance of the optical system, it is of crucial importance that all the optical elements are aligned well and that the entrance aperture of the objective is overfilled. The latter assures that the maximum NA of the objective is used and the highest resolution is obtained. Most water-immersion objectives correct for the presence of the cover glass that separates the immersion fluid from the specimen. It is of great importance that the correction match the thickness of the coverglass used. We employed a water-immersion objective with a variable cover glass thickness correction. The correction collar setting was optimized to give the smallest PSF at the glass surface.

Figure 1 shows lateral (*xy*) and axial (*xz*) slices through the center of the two-photon PSFs at the surface. The full-width at half-maximum (FWHM) serves as a good measure for the dimensions of the PSF. The measured FWHMs of the PSF are listed in Table 1. The theoretical values in Table 1 for a two-photon microscope were calculated using time-averaged electromagnetic energy distribution theory for the two-photon excitation processes (van der Voort and Brakenhoff, 1990). The experimental PSF for a single-photon confocal microscope was measured by Brakenhoff et al. (1986). For the single-photon confocal microscope, the theoretical values were derived from estimation of the illumination distribution (Brakenhoff et al., 1986; Born and Wolf, 1999).

3 REFRACTIVE INDEX MATCHING

In a laser scanning microscope, the excitation beam is focused in the sample by an objective. The imaging depth is varied by changing the distance between the objective and

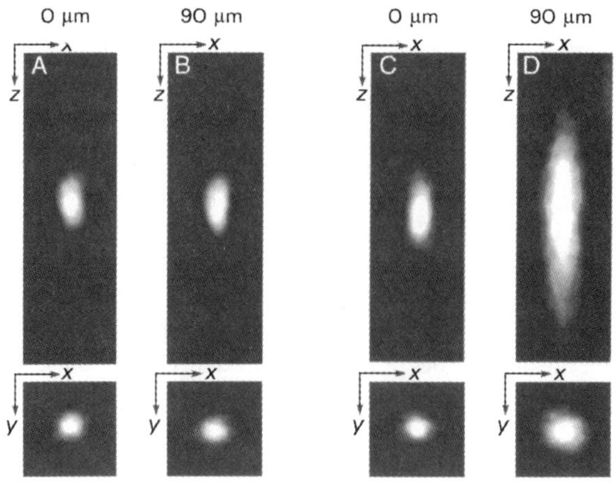

FIGURE 1 Images taken from a 3D data set of measured PSFs. Axial (*xz*, 1.5 μm × 5 μm) and lateral (*xy*, 1.5 μm × 1.5 μm) slices through the center of the PSF measured at the surface (0 μm depth) and at 90 μm depth in a water sample are shown. (*A*) and (*B*) are measured with a 60 × /NA 1.2 water-immersion objective; (*C*) and (*D*) are measured with a 40 × /NA 1.3 oil-immersion objective.

the coverslip (Figure 2). The space between the objective lens and the coverslip can be filled either with air or with an immersion fluid. The immersion fluid allows matching of the refractive index between objective lens and sample. In a perfectly matched system, the position of the focus in the sample is referred to as the nominal (apparent) focus position (NFP). A mismatched system introduces a difference between the nominal NFP and the actual focus position (AFP). The AFP and NFP may be related by using the low-NA approximation AFP = NFP × n_1/n_2 (Hell et al., 1993a,b). Here n_1 is the refractive index of the immersion fluid and n_2 that of the sample. A mismatch in the refractive index between the immersion fluid and the sample results in an incorrect scaling of the image along the optical axis (Visser et al. 1992; Hell et al., 1993a,b; Jacobsen et al., 1994).

A second effect of a mismatch in the refractive index is an increase in spherical aberrations. These aberrations broaden the PSF and have a serious effect on resolution and signal intensity. This broadening increases with depth, and is particularly important when imaging deep inside a sample. In single-photon confocal microscopy, the detected volume

TABLE 1 Dimensions (FWHM) of PSFs

	PSF dimensions (FWHM): lateral × axial (μm × μm)				
	Depth = 0 μm				Depth = 90 μm
	Two-photon		Single-photon confocal[a]		Two-photon
Objective	Measured	Theoretical	Measured	Theoretical	Measured
60 × /NA 1.2, water	0.27 × 0.67	0.22 × 0.73	0.22 × 0.73	0.23 × 0.76	0.29 × 0.62
40 × /NA 1.3, oil	0.27 × 0.79	0.20 × 0.57			0.44 × 2.2

[a] Brakenhoff et al. (1986).

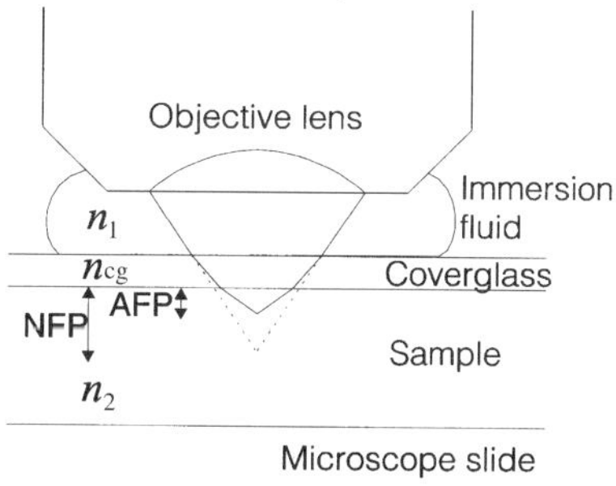

FIGURE 2 Schematic impression showing the immersion fluid between the objective and the coverglass and the difference between the nominal focus position (NFP) and the actual focus position (AFP). n_1, n_{cg}, and n_2 are the refractive indices of the immersion fluid, coverglass, and sample, respectively. In a perfectly matched system, $n_1 = n_2$, and the AFP is identical to NFP.

is limited by the pinhole size. Here, spherical aberrations result in a loss of signal passing through the pinhole (Carlsson, 1991). This effect increases when focusing deeper into the specimen. In two-photon excitation microscopy, an enlargement of the PSF reduces the local intensity and thus the total two-photon absorption. Consequently, the fluorescence signal is reduced (Hell et al., 1993a,b).

3.1 Axial Scaling

In the case of a mismatch between the refractive indices of the sample and the immersion fluid, a difference occurs between the AFP and NFP. The difference in axial scaling can be measured by comparing the observed depths of objects in the sample with different objectives. For the water-immersion objective (60 × /NA 1.2) and the oil-immersion objective (40 × /NA 1.3, $n_{oil} = 1.5118$) employed here, we determined this difference in a sample containing fluorescent beads in water at various depths. Here, the depths observed with the water-immersion objective served as the reference depths. Analysis of the observed depths showed that, as expected, the axial scaling effect is linear with depth. Linear regression of the data results in a scaling factor of 0.866 for the oil objective. The scaling factor is close to the geometrical approximation of $n_1/n_2 = 0.882$. This is in good agreement with both experimental and theoretical values found in the literature (Carlsson, 1991; Hell et al., 1993a,b; Jacobsen et al., 1994).

3.2 PSFs and Refractive-Index Mismatch

The effects on the PSF of a refractive-index mismatch between the objective and the specimen were investigated for a two-photon microscope using a fully matched and an unmatched system. PSFs were measured at various depths in a water sample with an oil-immersion objective (Nikon, 40×, NA 1.3, plane fluor, $n_{oil} = 1.5118$) and a water-immersion objective (Nikon, 60×, NA 1.2, plane apochromatic). To this end, 220 nm fluorescent beads were dried onto an object slide and on a coverslip. The coverslip was placed in reverse on the object slide with a drop of distilled water in between. A strip of glass served as a spacer on one side of the coverslip to give it a small fixed inclination of 0.3° (Figure 3). The small inclination was used to obtain bead images with a variable water spacing. The measurements at zero depth were carried out with the beads attached to the

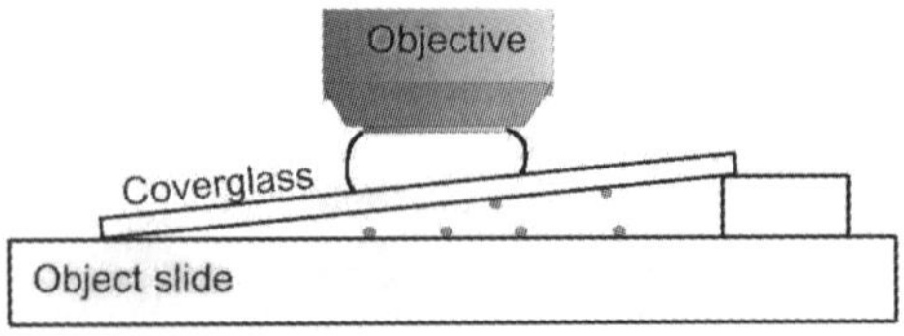

FIGURE 3 Configuration for measuring the PSF using small fluorescent beads at various depths in a water sample.

coverslip, and measurements at other depths with the beads attached to the object slide. Other imaging settings were identical to those of the measurements of the PSFs described above. 3D images of the beads were recorded at depths of 0, 15, 45 , 75 and 90 μm (AFP) in the water. The depths of the beads were determined by measuring the distance between the coverslip and the beads using the water objective. Since this is a fully matched situation, the AFP equals the NFP. These depths were used as reference values for the measurements with the oil-immersion objective.

Figure 1 shows lateral (*xy*) and axial (*xz*) slices through the centre of the two-photon PSFs. The dimensions of the PSFs are listed in Table 1. Reasonable agreement is found between the calculated and experimental values of the FWHMs, although all measured values are slightly larger than the theoretical values. The discrepancy is somewhat larger for the oil-immersion objective than for the water-immersion objective. The additional broadening can be attributed to spherical aberrations. Intensity profiles through the PSFs measured with the oil-immersion objective show a asymmetric axial (*z*) broadening of the PSF, which is characteristic of spherical aberrations (Entwistle, 1995; Wilson and Juškaitis, 1995).

In Figure 4, the FWHMs of the measured PSFs are given as functions of the imaging depth. As expected for a fully matched system, constant lateral and axial FWHMs are observed for the (matched) water-immersion objective. The mismatched system (oil-immersion objective) shows a clear broadening and elongation of the PSF.

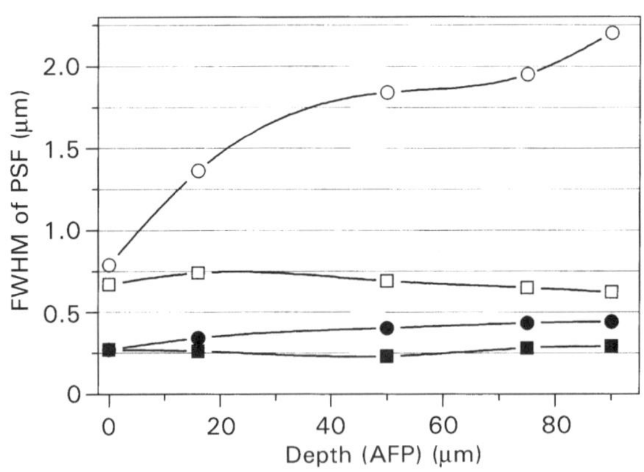

FIGURE 4 Dimensions of measured PSFs for the 60 × /NA 1.2 water-immersion objective (squares) and the 40 × /NA 1.3 oil-immersion objective (circles). Lateral (closed) and axial (open) dimensions are indicated. [Figure reprinted with permission from de Grauw et al. (1999).]

In the mismatched system, the relative axial elongation of the PSF between the surface and depth of 90 μm is 280%. The relative lateral broadening is about 160%. Clearly the PSF, and thus the resolution, are more affected in the axial than in the lateral dimension. In the two-photon excitation microscope, widefield detection is employed, making it insensitive to chromatic effects and other aberrations in the detection light path. The optical response of this system is mainly determined by the properties of the excitation light path. Therefore, the degradation of the PSF can be attributed to aberrations of the excitation path.

In a single-photon confocal geometry, the PSF is sensitive to both the excitation and the detection path. Theoretical values for the relative broadening of the confocal single-photon PSFs are given by Hell et al. (1993). The dimensions were calculated for a depth of 90 μm, for a confocal setup with an NA 1.3 oil-immersion objective at an excitation wavelength of 488 nm using an aqueous FITC solution. The authors find an increase in FWHM from 0.45 μm to 1.75 μm axially and from 0.16 μm to 0.255 μm laterally. The relative axial elongation of 278% and the lateral broadening of 163% are similar to the values found for the two-photon microscope. This confirms that effects of chromatic and spherical abberations on the PSF are identical for both types of microscopes.

3.3 Signal Strength and Refractive-Index Mismatch

In general, scattering and absorption by the sample and an enlargement of the PSF due to a refractive-index mismatch are the causes of signal loss in in-depth imaging experiments. The effects of a refractive-index mismatch on the observed fluorescence intensity when imaging deep in a sample can be easily determined. In a transparent sample of 50 μM rhodamine 6G in water, the fluorescence was measured as a function of depth. Since the sample contains only water and fluorophore, no significant scattering is present. Therefore, the signal attenuation at greater depth in the sample only arises from absorption and distortion of the PSF. The fluorescence yield as a function of depth is plotted in Figure 5. For the two-photon excitation experiments with the water-immersion objective (fully matched situation), no signal attenuation is observed at increasing depth. This is consistent with the depth-independent PSF, and implies negligible absorption. It also confirms that no significant scattering is present.

For the oil-immersion objective, a clear signal decrease is found in the two-photon excitation experiments. At 80 μm depth, the signal intensity has decreased to 20% of the value at the surface. Since absorption and scatting are negligible, this decrease can be attributed to enlargement of the PSF volume. The measured PSFs can be used to verify this. The integrated intensity of the PSFs measured at 0, 16, 49, 75, and 90 μm depth with the oil-immersion objective are plotted in Figure 5. It should be noted that all the PSFs were recorded with the same pixel dwell times and excitation intensities. The fact that the signals obtained from the fluorescent solution and contained in the PSF overlap confirms that the dependence of the signal with depth arises from the enlargement of the PSF volume only.

3.4 Absorption

The fluorescence signal for a two-photon excitation microscope with a matched refractive index between the objective and the sample (50 μM rhodamine 6G in water sample, water-immersion objective) shows no attenuation with increasing depth (Figure 5). This implies

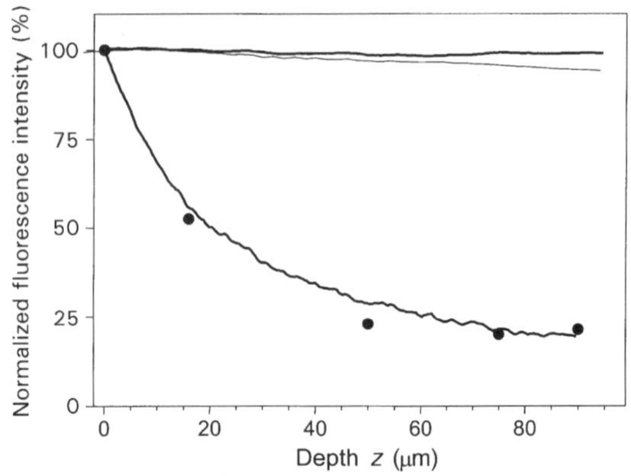

FIGURE 5 Fluorescence intensity as function of depth in a homogeneous fluorescent sample for the $60\times$/NA 1.2 water-immersion objective and the $40\times$/NA 1.3 oil-immersion objective. All intensities are normalized to the intensity at the coverglass-sample interface at 0 μm depth. Thick lines represent two-photon excitation, and the thin line single-photon excitation. The integrated intensities of the PSFs measured with the oil-immersion objectives, normalized to the value at the surface, are indicated by filled circles. [Figure partly reprinted with permission from de Grauw et al. (1999).]

that there is no absorption of excitation light outside the focal region. The fluorescence signal measured with confocal single-photon excitation in the same matched system is also plotted in Figure 5. The diameter of the pinhole was backprojected in the object plane, 0.33 μm. The signal shows a small decrease with increasing depth. The signal decrease is attributed to absorption of excitation light before it reaches the focus. The fact that no absorption is observed in two-photon excitation means that also reabsorption of emitted photons by the rhodamine molecules can be ruled out. Therefore, the signal decrease observed in single-photon excitation arises from absorption of the excitation light only.

4 SCATTERING AND REFRACTION

Apart from refraction due to refractive-index mismatch between the objective and the sample, light can also be refracted by objects in the sample. This refraction occurs at interfaces with objects with a different refractive index. Here, it is assumed that the objects are considerably larger than the wavelength. In the case of objects smaller than or the same size as the wavelength, scattering starts to play a role. Both refraction on large objects and scattering on small particles affect the size of the PSF, and may lead to signal reduction.

When a ray of light passes through a sample, it may be scattered by small objects. In biological samples, the scatterers may be vesicles, membranes, cell organelles, or bacteria. The size of the scattering particles, relative to the wavelength of the light λ, determines the scattering behavior. For particles substantially smaller than the wavelength of the incident light, the scattering shows an inverse fourth-power dependence on the wavelength (Rayleigh scattering). For particles of the same size or larger than the wavelength, the more general Mie scattering formalism has to be used. Here, the wavelength dependence is less strong, and the amount of scattering depends on the scattering efficiency (van de Hulst, 1957). For very large particles, scattering becomes normal refraction, which depends on the wavelength dependence of the refractive index. Typically, biological samples contain scatterers with a large variety in size. It can therefore be expected that the wavelength dependence of scattering in these samples results in a wavelength dependence of less than λ^4.

Focusing deep into a scattering sample results in an increased amount of scattered light. This results in a reduced intensity in the focus and an increased size of the PSF. As a consequence, the fluorescence yield decreases and the image resolution is reduced. The loss of signal due to PSF distortion is particularly important in two-photon excitation microscopy, where the absorption depends quadratically on the intensity.

The fluorescence intensity at a certain depth in a sample in the absence of a PSF enlargement can be described by a Boltzmann transport equation:

$$\frac{dI(z,\theta)}{dz} = -I\mu_s - I\mu_a + F(\theta), \tag{1}$$

where I is the intensity at a distance z from the coverglass. The scattering coefficient μ_s and the absorption coefficient μ_a describe the signal attenuation due to scattering and absorption. $F(\theta)$ is a function describing the angular dependence of the intensity. This function depends on the anisotropy, which is defined as the mean of the cosine of the scattering angle (Prahl et al., 1993). It accounts for the scattered photons that contribute to the intensity at depth z.

In the case of single- and two-photon excitation microscopy, the excitation light is strongly focused. As a consequence, scattering of a photon means that the photon has only a limited chance of reaching the focus. Therefore, as an approximation, we choose here to neglect the influence of anisotropy. In the case of widefield detection, as is often used in two-photon microscopy, scattered fluorescent photons may contribute to the signal. A photon scattered at a sufficiently small angle will still reach the detector. In that case, the anisotropy of the scattering plays a role.

In fluorescence microscopy, the excitation and emission wavelengths are different, and the coefficients are described by μ_{exc} and μ_{em}. As was seen in the previous section the reabsorption of emitted photons can be neglected. In the case of two-photon excitation microscopy, absorption of excitation light outside the focus can be neglected. Therefore, solutions of Eq. (1) that describe the signal attenuation on the way to and from the focus for the two types of microscopes are given by (Flock et al., 1992; van de Hulst, 1957).

$$I_{SPE}(z,\lambda) \propto I_0 \exp[-(\mu_{s,\text{exc}} + \mu_{s,\text{em}} + \mu_{a,\text{exc}})z], \tag{2}$$

$$I_{TPE}(z,\lambda) \propto [I_0 \exp(-\mu_{s,\text{exc}}z)]^2 \exp(-\mu_{s,\text{em}}z) = I_0^2 \exp[-(2\mu_{s,\text{exc}} + \mu_{s,\text{em}})z]. \tag{3}$$

where I_0 is the incident intensity and I is the collected intensity for single-photon excitation (SPE) and two-photon excitation (TPE). For two-photon excitation, the quadratic dependence of the fluorescence signal on the excitation intensity yields a factor of two before the scattering coefficient. For a low-NA objective, the distance z is approximately equal to the AFP. For a high-NA objective, the integrals of Eq. (2) and (3) over the light cone should be used.

In two-photon excitation, near-infrared wavelengths are typically used for excitation. The wavelength dependence of scattering implies that, compared with single-photon excitation, reduced scattering may be expected. Less scattering, together with the effect of reduced two-photon absorption outside the focal region, results in increased depth penetration. The use of widefield detection in two-photon excitation microscopy makes the imaging properties independent of the detection light path. This allows the detection of (multiple) scattered fluorescence light and improves the detection efficiency. This is particularly important in thick specimens and dense materials, where significant scattering of the emitted light takes place.

4.1 Scattering in Turbid Media

4.1.1 Intralipid Intralipid is a phospholipid suspension in water that is used as an intravenous nutrient. It consists of glycerol, lecithin, and soybean oil, which form vesicles in water. It has no strong absorption in the wavelength range from 400 nm to 1100 nm ($\mu_a \approx 10^{-5}$ mm^{-1}/ml of 10% intralipid), and has been used in many tissue-simulating scattering experiments (Moes et al.,1989; van Staveren et al., 1991). The diameter of the phospholipid vesicles shows a broad distribution, between about 10 nm and 500 nm, with an average of 100 nm (van Staveren et al., 1991). A solution of intralipid-20% (Pharmacia & Upjohn, Woerden, The Netherlands) was diluted four times to a 5% solution with water containing Oregon green 488 (Molecular Probes, Leiden, The Netherlands) to a final concentration of 50 μM Oregon green. The fluorescence signal was recorded as a function of depth in the sample using a confocal single-photon excitation microscope and a two-photon excitation microscope (Figure 6). The single-photon excitation wavelength was $\lambda_{\text{exc}} = 488$ nm, and the signal was recorded using a pinhole diameter, backprojected in the object plane, of 0.33 μm. The two-photon excitation wavelength was 800 nm and widefield detection was used. Both experiments were carried out using the 60 × /NA 1.2 water-immersion objective. The average distance travelled by a photon in the light cone of such an objective is greater than the perpendicular AFP distance. As an estimate the factor relating the average distance z_{av} to z was calculated for our objective. Taking into account the Gaussian intensity as a function of distance from the optical axis, and entrance pupil diameter, this resulted for our microscope in $z_{\text{av}} = 1.45z$.

As a reference, the fluorescence signal attenuation was measured in a solution containing only Oregon green and no intralipid. As in the rhodamine 6G experiment

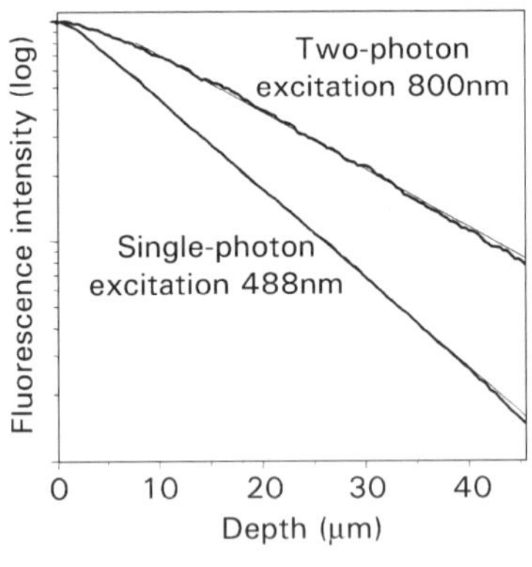

FIGURE 6 Measured fluorescence signal in a solution containing 5% intralipid and 50 μM Oregon green 488. The two lines represent the intensity (log) as a function of depth in the sample, measured with a confocal single-photon excitation microscope and with a two-photon excitation microscope. The curves are normalized to the intensity at the coverglass–solution interface.

(Figure 5), the attenuation with two-photon excitation was negligible, which with single-photon excitation, a small signal decrease was observed. This decrease is due to the absorption of the excitation light by the dye. Reabsorption of emitted photons can be neglected, as was seen in the previous section. The absorption coefficient for the single-photon excitation light, calculated from the measured signal attenuation, yields $\mu_a = 0.427\ \text{mm}^{-1}$ at 488 nm. The slopes of the curves in Figure 6 yield the total extinction coefficient $\mu_{\text{ext}} = \mu_s + \mu_a$, and can be used to obtain the scattering coefficient.

For single-photon excitation, both the excitation light and emission light are scattered, and only the excitation light is absorbed. The average scattering on the way to and from the focus, defined here as the combined effect of scattering of excitation and emission photons on the detected signal, allows one to make a comparison between single-photon-excitation and two-photon-excitation microscopes. For single-photon excitation, the average scattering is given by $\mu_{s.\text{av.SPE}} = \mu_{s.\text{em}} + \mu_{s.\text{exc}}$, while for two-photon excitation it is given by $\mu_{s.\text{av.TPE}} = 2\mu_{s.\text{em}} + \mu_{s.\text{exc}}$. The mean free path (MFP) is a measure of the average distance that a photon can travel before it is scattered or absorbed. The MFP, defined as the reciprocal of the extinction coefficient, can be derived from the slopes of the curves in Figure 6. In the case of two-photon excitation, this mean that the MFP describes the effective average distance including the quadratic dependence of the absorption on the excitation intensity. The slope should be corrected for the fact that the slope describes the average attenuation on the path z_{av} to and from the focus: MFP = slope $\times\ 2 \times 1.45$.

For single-photon excitation, the slope of the curve in Figure 6 yields an MFP of 31.5 μm. The average scattering coefficient can be written as $\mu_{s,\text{av,SPE}} = 1/\text{MFP} - \mu_a$, and yields $31.7\ \text{mm}^{-1}$.

For two-photon excitation, an MFP of 47.7 μm is found. Since there is no absorption in this case, the influence of scattering to and from the focus can be described by $\mu_{s,\text{av,TPE}} = 1/\text{MFP} = 21.1\ \text{mm}^{-1}$ (Table 2). These results are valid under the restriction that the refractive index of the intralipid solution is close to that of water and that there is no refractive-index mismatch.

Van Staveren et al. (1991) determined the scattering properties of intralipid for a collimated light beam. The scattering coefficient was found to be $\mu_s = 1.6\ \lambda^{-2.4}\ \text{mm}^{-1}/\%$intralipid, with the wavelength λ in μm, for concentrations from 0.4% to 1.7% intralipid. Linear extrapolation to a 5% intralipid concentration results in $\mu_s = 44.76\ \text{mm}^{-1}$ for 488 nm and $\mu_s = 13.67\ \text{mm}^{-1}$ for 800 nm excitation light. At the

TABLE 2 Overview of the scattering coefficients and MFP. In the two-photon excitation experiments, the specimen was excited with 800 nm; for single-photon excitation, this was 488 nm. The maximum emission of Oregon green is 524 nm (in the intralipid experiments)

	Intralipid-5%				Biofilm	
	Measurement		Literature[a]		Measurement	
	Two-photon	Single-photon confocal	800 nm + 524 nm	488 nm + 524 nm	Two-photon	Single-photon confocal
$\mu_{s,\text{av}}$ (mm^{-1})	21.1	31.7	25.7	41.3	20.6	87.2
MFP (μm)	47.4	31.1	38.9	24.2	48.5	11.5

[a] van Staveren et al. (1991).

wavelength of maximum emission for Oregon green (524 nm), $\mu_s = 37.75\ \text{mm}^{-1}$ is found. Using these values, combined absorption plus emission scattering coefficients of $\mu_{s,\text{SPE}} = 41.25\ \text{mm}^{-1}$ and $\mu_{s,\text{TPE}} = 25.72\ \text{mm}^{-1}$ are found for single- and two-photon excitation, respectively. The MFP is the reciprocal of the scattering coefficient, and amounts to 38.9 μm and 24.2 μm for single- and two-photon excitation, respectively. The difference between these and the measured MFP values may be partly due to the use of the average pathlength z_{av}. The larger MFP values in the measurement may also be the result of the high concentrations of the measured intralipid. At concentrations above 1.7%, the density, in combination with the size of the particles, cause a shadowing effect (van de Hulst, 1957). This shadowing results in a larger MFP for the photons in the measurements presented.

4.1.2 Biofilm Biofilms consist of several species of bacteria embedded in an extracellular protein matrix. A mixture of 10 different species of oral bacteria was grown on a coverslip in a chemostat system in mucin-based growth medium for 10 days to form a biofilm. It was shown that these biofilms can serve very well as a model system for natural dental plaque (Bradshaw et al., 1996).

Images of biofilm stained with 50 μM rhodamine 6G were measured at different depths. The intensity images reveal the complicated microstructure of the biofilm (Figure 7*B*). The average refractive index of this type of biofilm is very close to that of water (Gerritsen et al., 1999; Vroom et al., 1999). Therefore, the water-immersion objective was used in the experiments described here. The biofilm contains strongly scattering objects such as membranes, bacteria, and macromolecules.

The integrated fluorescence intensity of *xy* images is plotted in Figure 7*A* as a function of depth. From the attenuation curves, the MFP and scattering coefficients were determined as described above for intralipid. The single-photon absorption can be derived from Figure 5, and yields $\mu_a = 0.21\ \text{mm}^{-1}$. For single-photon excitation, we find MFP = 11.5 μm and $\mu_{s,\text{av},\text{SPE}} = 87.0\ \text{mm}^{-1}$. For two-photon excitation, we find MFP = 48.5 μm and $\mu_{s,\text{av},\text{TPE}} = 20.6\ \text{mm}^{-1}$.

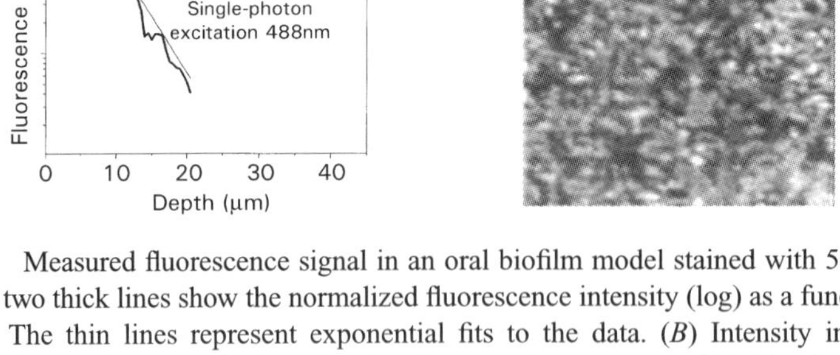

FIGURE 7 Measured fluorescence signal in an oral biofilm model stained with 50 μM rhodamine 6G. (*A*) The two thick lines show the normalized fluorescence intensity (log) as a function of depth in the sample. The thin lines represent exponential fits to the data. (*B*) Intensity image (*xy* plane) measured with two-photon excitation, showing the complex structure of biofilm. Image dimensions 18 μm × 18 μm (*xy*).

Here two-photon microscopy yields a much larger MFP, and thus better penetration, than the confocal microscope. For the intralipid solution, the MFP enhancement in the two-photon case was much less. This difference may be ascribed to a refractive-index mismatch due to the phospolipid particles in the solution.

In the two-photon microscope, we were able to record images of biofilm at depths exceeding 100 μm in the sample with good resolution and image contrast (images not shown). Using a confocal single-photon microscope, similar images could be recorded up to 30 μm deep. The main signal attenuation in biofilm arises from scattering. Therefore, it can be concluded that scattering limits imaging to a depth of two to three times the MFP.

4.2 Wavelength Dependence of Scattering

To investigate the wavelength dependence of scattering, images from the same region in a biofilm were recorded at various wavelengths in the confocal single-photon excitation microscope. As illumination source, various laser lines of an argon–krypton laser were used. In addition, the same sample was measured in the two-photon microscope at 800 nm in reflection mode.

The scattered intensity of the laser light was recorded from the unstained biofilm. Photons from the laser are scattered, and some of them are backscattered towards the objective. These photons, on their way back, may be scattered again. The resulting photons were measured as a function of depth of the focus using the water-immersion objective. The resulting intensity curves could be well fitted to a mono-exponential decay. The slope of these curves was employed to calculate the extinction coefficient $\mu_{ext} = 1/\mathrm{MFP}$. Since absorption is negligible, the extinction coefficient is a measure of the scattering coefficient at the wavelength applied.

In Figure 8, the extinction coefficient is plotted as a function of wavelength. The extinction coefficient at 800 nm (two-photon excitation) is a factor of 3.3 lower than the value found at 488 nm (single-photon excitation). This again illustrates the enhanced penetration depth with two-photon microscopy due to the reduced scattering that results from the longer excitation wavelength. The fact that the scattering coefficients found at 488 nm and 800 nm are slightly smaller than found in the biofilm measurements can be explained by variations in the density of bacteria in the biofilm.

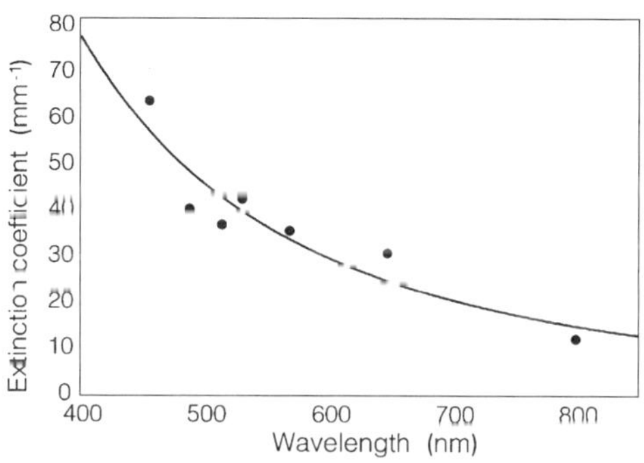

FIGURE 8 Wavelength dependence of extinction coefficient in oral biofilm.

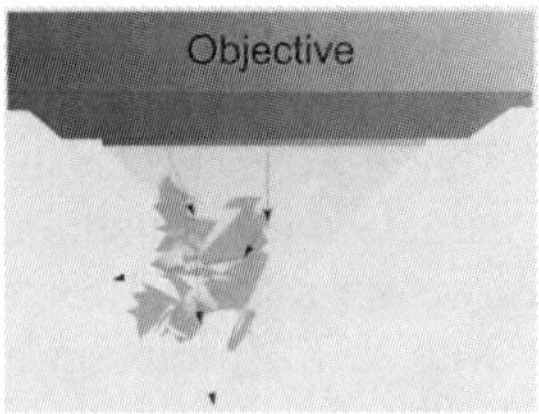

FIGURE 9 Impression of photons refracted by an object in the excitation light cone of an objective.

The data points plotted in Figure 8 were fitted to the function $y = a\lambda^{-b}$. From the figure, we find $b = 2.4 \pm 0.5$. For very small particles, Rayleigh scattering occurs, with a wavelength dependence of $b = 4$. Therefore, it can be concluded that the scattering objects are of intermediate size compared with the wavelength. The bacteria involved are known to have a size of the order of 0.5–1.5 μm (Holt, 1984).

4.3 Influence of Refraction

When objects in the sample are substantially larger than the wavelength, effects other than scattering occur. In the case where large objects with various refractive indices are present, the excitation and emission photons are refracted. As a consequence, parts of the excitation beam will not reach the focus (Figure 9). In both confocal single-photon and two-photon microscopy, this results in a decrease in signal. The signal decreases with the number of refractive-index boundaries that the beam has to pass. This is particularly detrimental in the case of two-photon excitation, because of the quadratic dependence of the fluorescence signal on the excitation intensity in the focus. Hence, this mechanism reduces the advantage of enhanced penetration in a two-photon microscope due to the longer excitation wavelength.

The mechanism of refraction was investigated using a sample of fresh cheese. Fresh cheese consists of large connected clusters of proteins and fat (lipids) interspersed with water. In the water, salts and polysaccharides are dissolved. Water has a refractive index $n = 1.33$ at room temperature, and the refractive index of oil is typically around 1.5. The various structures present in the sample result in large heterogeneities in the refractive index (Figure 10*B*). The measured signal extinction curves are presented in Figure 10*A*. It is clear from the curves that the two imaging methods result in very similar extinctions (slopes). The main difference results from the first few micrometers below the coverglass. This is where the photons are still relatively undisturbed by refractive-index boundaries. Below a depth of 3 μm, the extinction in the two microscopes is almost identical.

5 CONCLUSIONS

The recording of images deep in a sample is generally limited by attenuation of the signal and by image resolution. Refraction leads to a distortion of the point spread function (PSF) of the microscope, which results in a lower signal and decreased resolution.

Absorption of two-photon excitation light outside the focus was found to be negligible for solutions of 50 μM rhodamine 6G and Oregon green 488. For single-photon confocal microscopy, a small amount of absorption was found.

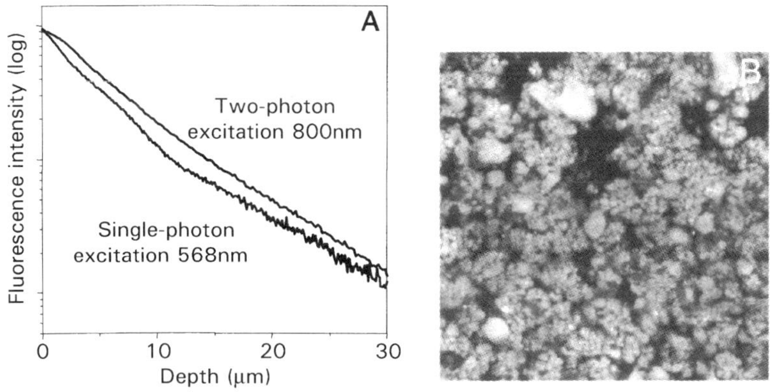

FIGURE 10 Measured fluorescence signal in a fresh cheese sample containing large lipophilic objects in a water environment. The sample was stained with Nile red, which is hydrophobic. (*A*) The lines represent the normalized fluorescence intensity (log) as a function of depth in the sample. (*B*) Intensity image (*xy* plane, 50 μm × 50 μm) measured with two-photon excitation, showing the structure of the sample.

A refractive-index mismatch between the sample and the immersion fluid results in enlargement of the PSF. This results in spreading of the excitation intensity over a larger volume and to a reduction in the fluorescence signal. This is particularly important for two-photon excitation, because of the quadratic dependence of the absorption on the excitation intensity.

Scattering reduces both the excitation and emission light. Scattering by particles in the sample depends on the wavelength of the light. The reduced scattering at longer wavelengths yields a larger penetration depth of two-photon excitation over single-photon excitation. It was shown that for a dense biofilm sample, where no refractive-index mismatch between the objective and the sample was present, the scattering coefficient for the two-photon excitation microscope was a factor of 3.3 smaller than that for the confocal single-photon microscope. Depending on the morphology and the density of the scattering particles, this factor can be different for various samples.

Refraction is particularly important in samples that contain structures that are considerably larger than the wavelength and have a different refractive index from their environment. Here, refraction of parts of the excitation beam occurs. It was shown for a sample containing large fat structures within a water environment that the extinction coefficient was equal for the two-photon and single-photon confocal microscopes.

ACKNOWLEDGMENTS

The authors would like to thank Unilever Research, Port Sunlight, U.K. and Dr. D. Bradshaw from the Centre of Applied Microbiology and Research, Salisbury, U.K. for their help and for the supply of biofilm.

REFERENCES

Born, M. and Wolf, E. (1999). *Principles of Optics*, 7th (expanded) edn. Cambridge: Cambridge University Press.

Bradshaw, D.J., Marsh, P.D., Schilling, K.M., and Cummins, D. (1996). A midified chemostat system to study the ecology of oral biofilms. *J. Appl. Bacteriol.* **80**, 124–130.

Brakenhoff, G.J., van der Voort H.T.M., van Spronsen, E.A., and Nanninga, N. (1986). Three-dimensional imaging by confocal scanning fluorescence microscopy. *Ann. NY Acad. Sci.* **483**, 405–415.

Carlsson, K. (1991). The influence of specimen refractive index, detector signal integration, and non-uniform scan speed on the imaging properties in confocal microscopy. *J. Microsc.* **163**, 167–178.

Centonze, V.E. and White, J.G. (1998). Multiphoton excitation provides optical sections from deeper within scattering specimens than confocal imaging. *Biophys. J.* **75**, 2015–2024.

Daria, V., Nakamura, O., Palmes-saloma, C., and Kawata, S. (1998). Enhanced depth penetration in imaging turbid biological samples by two-photon fluorescence microscopy. *Jpn. J. Appl. Phys.* **37**, L959–L961.

de Grauw, C.J., Vroom, J.M., van der Voort, H.T.M., and Gerritsen, H.C. (1999). Imaging properties in two-photon excitation microscopy and effects of refractive index mismatch in thick specimen. *Appl. Opt.* **38**, 5995–6003.

Denk, W., Stickler, J.H., and Webb, W.W. (1990). Two-photon laser scanning fluorescence microscopy. *Science* **248**, 73–76.

Entwistle, A. (1995). The effects of total internal reflection on the spread function along the axis in confocal microscopy. *J. Microsc.* **180**, 148–157.

Flock, S.T., Jacques, S.L., Wilson, B.C., Star, W.M., and van Gemert, M.J.C. (1992). Optical properties of intralipid: a phantom medium for light propagation studies. *Lasers Surg. Med.* **12**, 510–519.

Gerritsen, H.C., Vroom, J.M., and de Grauw, C.J. (1999). Combining two-photon excitation with fluorescence lifetime imaging. *IEEE Eng. Med. Biol.* **18**, 31–35.

Göppert-Mayer, M. (1930). Über elementarakte mit zwei quantensprüngen. *Annln. Phys.* **9**, 273–294.

Hell, S.W. (1995). Effect of the specimen refractive index on the imaging of a confocal fluorescence microscope employing high aperture oil immersion lenses. *Bioimaging* **3**, 39–47.

Hell, S.W., Reiner, G., Cremer, C., and Stelzer, E.H.K. (1993a). Refractive-index-induced aberrations in two-photon confocal microscopy. *J. Microsc.* **176**, 226–230.

Hell, S.W., Reiner, G., Cremer, C., and Stelzer, E.H.K. (1993b). Abberations in confocal fluorescence microscopy induced by mismatches in refractive index. *J. Microsc.* **169**, 391–405.

Holt, J.G. (1984). *Bergey's Manual of Systematic Bacteriology*, Vol. 1, Baltimore: Williams & Wilkins.

van de Hulst, H.C. (1981). *Light Scattering by Small Particles*. New York: Dover.

Kano, H., van der Voort, H.T.M., Schrader, M., van Kempen, G.M.P., and Hell, S.W. (1996). Avalanche photodiode detection with object scanning and image restoration provides 2–4 fold resolution increase in two-photon fluorescence microscopy. *Bioimaging* **4**, 187–197.

Moes, C.J.M., van Gemert, M.J.C., Star, W.M., and Prahl, S.A. (1989). Measurements and calculations of the energy fluence rate in a scattering and absorbing phantom at 633 nm. *Appl. Opt.* **28**, 2292–2296.

Pawley, J.B. (Ed.) (1995). *Handbook of Biological Confocal Microscopy*, 2nd edn. New York: Plenum Press.

Prahl, S.A., van Gemert, M.J.C., and Welch, A.J. (1993). Determining the optical properties of turbid media by using the added-doubling method. *Appl. Opt.* **32**, 559.

Sytsma, J., Vroom, J.M., de Grauw, C.J., and Gerritsen, H.C. (1998). Time-gated fluorescence lifetime imaging and microvolume spectroscopy using two-photon excitation. *J. Microsc.* **191**, 39–51.

van der Voort, H.T.M. and Brakenhoff, G.J. (1988). Determination of the 3-dimensional optical properties of a confocal scanning laser microscope. *Optik* **78**, 48–53.

van der Voort, H.T.M. and Brakenhoff, G.J. (1990). 3-D image formation in a high-aperture fluorescence confocal microscope: a numerical analysis. *J. Microsc.* **158**, 43–54.

van der Voort, H.T.M. and Strasters, K.C. (1995). Restoration of confocal images for quantitative image analysis. *J. Microsc.* **178**, 165–181.

van Staveren, H.J., Moes, C.J.M., van Marle, J., Prahl, S.A., and van Gemert, M.J.C. (1991). Light scattering in intralipid-10% in the wavelength range of 400–1100 nm. *Appl. Opt.* **30**, 4507–5414.

Visser, T.D., Oud, J.L., and Brakenhoff, G.J. (1992). Refractive index and axial distance measurements in 3D microscopy. *Optik* **90**, 17–19.

Vroom, J.M., de Grauw, C.J., Gerritsen, H.C., Bradshaw, D.J., Marsh, P.D., Watson, G.K., Birmingham, J.J., and Allison, C. (1999). Depth penetration and detection of pH gradients in biofilms by two-photon excitation microscopy. *Appl. Env. Microbiol.* **65**, 3502–3511.

Williams, R.M., Piston, D.W., and Webb, W.W. (1994). Two-photon molecular excitation provides intrinsic 3-dimensional resolution for laser-based microscopy and microphotochemistry. *FASEB J.* **8**, 804–813.

Wilson, T. (1990). *Confocal Microscopy.* London: Academic Press.

Wilson, T. and Juškaitis, R. (1995). The axial response of confocal microscopes with high numerical aperture objective lenses. *Bioimaging* **3**, 35–38.

CHAPTER 8

Group-Velocity Dispersion and Fiber Delivery in Multiphoton Laser Scanning Microscopy

RALF WOLLESCHENSKY,[1] MARY E. DICKINSON,[2] and SCOTT E. FRASER[2]

[1] Carl Zeiss Jena GmbH, Advanced Imaging Microscopy Division, Jena D-07745, Germany
[2] Biological Imaging Center, Beckman Institute, Caltech, Pasadena, CA 91125, U.S.A.

1 INTRODUCTION

Multiphoton laser scanning microscopy (MPLSM) has emerged as an important technical advance for live and deep tissue fluorescence imaging. In MPLSM, fluorescent molecules are excited by the simultaneous absorption of two or more near-infrared (NIR) photons. Multiphoton excitation has a quadratic dependence, producing excitation only at the focal plane; thus, out-of-focus fluorescence does not contribute to image background, and photodamage outside the plane of focus is greatly reduced. In practical terms, MPLSM makes it possible to acquire images with a high signal-to-noise ratio. In living samples, MPLSM can be used to collect images with minimal damage to the tissue, and the use of IR light makes it possible to image deeper in the specimen, owing to less scattering and absorption of the incident light.

Ultrafast lasers emitting femtosecond pulses are required for multiphoton excitation of fluorescent dyes. Such lasers are capable of producing high peak power to favor multiphoton excitation, using a low average power. The average power can be calculated by multiplying the pulse energy by the repetition rate. Although MPLSM offers advantages over confocal imaging because the use of IR light causes less photodamage to cells than visible-range photons, too much power from the IR laser can produce heat in the sample that can damage vital functions. For instance, power levels above 6 mW may disturb cell replication [1]. The peak power, which is the power at the plane of focus, can be defined as the pulse energy divided by the pulse length. High peak intensities (more than $10^8\,\mathrm{W\,cm^{-1}}$) are required for multiphoton excitation. Thus, in order to keep the average power low, one has to use the shortest possible pulses to maximize fluorochrome excitation and the amount of signal from the sample.

Confocal and Two-Photon Microscopy: Foundations, Applications, and Advances, Edited by Alberto Diaspro.
ISBN 0-471-40920-0

Many lasers are now available that can produce ultrashort pulses at high repetition rates. Titanium : sapphire lasers are most commonly used. These lasers are capable of producing pulses of order 100 fs length over a broad tunable wavelength range (690–1064 nm) with a high repetition rate of about 80 MHz. Similarly, solid state, doubled neodymium-doped yttrium lithium fluoride (Nd : YLF) lasers, emitting 1047 nm, 175 fs pulses at 120 MHz, have also be used for multiphoton excitation. Although these lasers are able to supply short pulses with a high duty cycle, how these lasers are routed into and through a microscope can greatly effect the efficiency of excitation at the sample. Optical materials, such as glass elements in a microscope, are normal dispersive media. This means that the refractive index of the material decreases as the wavelength increases. Since a short pulse consists of a broad spectral band, the pulselength becomes broader in such media. This is because the redshifted frequency components, with respect to the center wavelength, are traveling faster through such materials than the blueshifted frequency components. The temporal broadening due to this effect is referred to as group-velocity dispersion (GVD), which is described in more detail in [2–4]. Thus, 100 fs pulses coming out of the laser may be considerably longer by the time they get to the sample as a result of GVD.

Throughout this chapter, we shall examine the effects of GVD in multiphoton microscopy, and describe methods designed to compensate for this effect in imaging systems. The chapter will describe the theory related to GVD, will provide information about how to measure the pulselength in MPLSM systems, and will discuss appropriate objective lens parameters for MPLSM. In addition, we will discuss the design of a multiphoton imaging system in which the laser is routed into the microscope using 3 m fiber and a grating dispersion compensator (GDC).

2 GENERAL REMARKS ON ULTRASHORT LASER PULSES

As mentioned briefly above, ultrashort laser pulses have some unique features, in comparison with continuous-wave (CW) light, which makes it necessary to use proper optical set-ups in order to achieve the shortest possible pulses at the location of the specimen. The behavior of ultrashort pulses in an optical set-up such as a microscope and an optical fiber will be discussed in this section.

Ultrashort pulses have a certain spectral bandwidth $\Delta\lambda$ that determines the shortest possible pulselength. In general, the larger the spectral bandwidth, the shorter the pulselength can be. Typical values are in the range of $\Delta\lambda/\lambda \approx 0.01$–$0.1$, whereby λ is the center wavelength.

A convenient and sufficient way to describe the shape of ultrashort laser pulses in the time and frequency domains is to use Gaussian profiles. Gaussian pulses will be used throughout this chapter to describe the propagation of pulses through optical materials.

For a complete characterization of the pulses, it is sufficient to measure the full-width at half-maximum (FWHM) of the profiles in the time and frequency domains. The spectral bandwidth $\Delta\omega$ can be measured easily using a spectrum analyzer, but for the measurement of the pulselength τ, special techniques are required, since the time responses of conventional detectors are too slow to measure the temporal pulse shape of ultrashort lasers directly. These techniques are referred to as autocorrelation techniques. In typical autocorrelation measurements, a Michelson interferometer is used. Here, two replicas of the laser pulse are obtained, and by varying the length of one arm of the interferometer, the time delay between the two pulses can be adjusted. Combining the two pulses in a nonlinear medium yields, depending on the special geometry of the set-up, either the interferometric or the intensity autocorrelation [3, 4]. From the autocorrelation function, which is the response of the nonlinear medium as a function of the time delay between two pulses, the actual pulselength can be calculated. If a Gaussian pulse shape is assumed the FWHM of the autocorrelation function divided by 1.44 yields the pulselength τ.

Owing to the high numerical aperture, of the objective lenses used, in commercially available microscopic set-ups, autocorrelators cannot be used to measure the pulselength at the location of the specimen. A way around this is to use a Michelson interferometer in front of the microscope to generate the two replicas. The two replicas then travel through the microscope. At the location of the specimen, the two pulses combine and produce, for instance, two-photon fluorescence from a dye cuvette. This method described in [5, 6] can be used to measure the pulselength at the location of the specimen.

3 PROPAGATION OF ULTRASHORT LASER PULSES

As mentioned above, optical materials, such as glass elements in a microscope, are normal dispersive media and short pulses are subject to temporal broadening due to GVD. Although the pulse is broadened by this effect, the spectral bandwidth is not affected by dispersion. Instead, the frequency components become temporally ordered after passing through normal dispersive media, with faster-moving spectral components (longer-wavelength) appearing before slower-moving components (shorter-wavelength). The pulse in this case is referred to as a positively chirped pulse. In the case of a bandwidth-limited pulse, all spectral components arrive at the same time, and the chirp is zero. The chirp of a pulse, C, is expressed in terms of the time–bandwidth product:

$$\tau \, \Delta\omega = 4 \ln(2) \sqrt{1 + C^2}, \tag{1}$$

where $\Delta\omega$ and τ are the spectral bandwidth and the pulselength of the pulse, respectively.

The GVD of a material is described by the dispersion parameter β_2 and the length of the material, z. The dispersion parameter multiplied by the material length can be calculated from material constants, that is, in terms of the second derivative of the refractive index with respect to the wavelength λ:

$$\beta_2 z = \frac{\lambda_0^3 z}{2\pi c_0^2} \frac{d^2 n}{d\lambda^2}\bigg|_{\lambda_0}, \tag{2}$$

where c_0 is the speed of light and λ_0 the wavelength in vacuum and equals in a first approximation for laser pulses the center wavelength. Since the microscope consists of several different optical components, the dispersion parameter for the whole system is the sum over all individual parameters for each optical element. By using Eq. (2), it is possible to calculate the dispersion of the system using the material constants only.

Using the dispersion parameter, the length of the material, the chirp C, and the pulselength before the material, τ_0, the pulselength at the output of the material, τ, can be calculated using the following equation:

$$\tau = \tau_0 \sqrt{\left[1 + \frac{4 \ln(2)\, \beta_2 z C}{\tau_0^2}\right]^2 + \left[\frac{4 \ln(2)\, \beta_2 z}{\tau_0^2}\right]^2}. \tag{3}$$

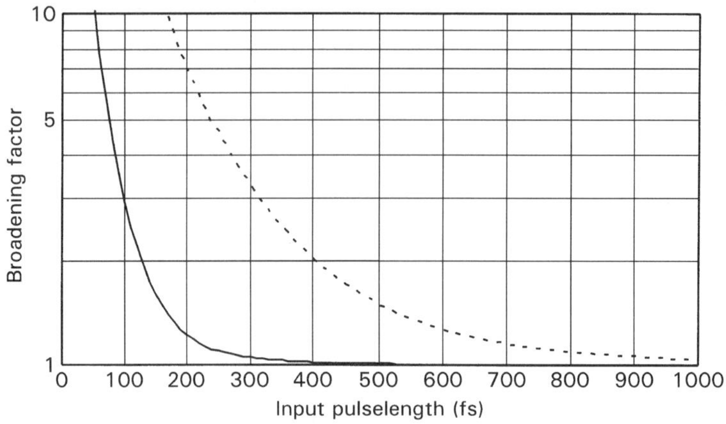

FIGURE 1 The diagram shows the pulse broadening due to group-velocity dispersion, if the laser is free-space- (solid line) or fiber-coupled (dashed line) to the LSM 510 NLO (see Section 7). The dispersions for the free-space- and the 3 m fiber-coupled system are $10^4\ \mathrm{fs}^2$ and $10^8\ \mathrm{fs}^2$, respectively.

In the case of a bandwidth-limited input pulse, this equation simplifies to

$$\tau = \tau_0 \sqrt{1 + \left[\frac{4\ln(2)\,{}_2 z}{\tau_0^2}\right]^2}, \qquad C = 0. \tag{3$'$}$$

From the second term in Eq. (3), it can be seen that the pulse-broadening effect due to the optical material gets larger if the pulselength at the input to the material gets smaller. This is depicted in Figure 1.

The pulse-broadening effect of a bandwidth-limited pulse is shown on a logarithmic scale as a function of pulselength at the input of the optical system for two different dispersion parameters. The smaller parameter corresponds to the dispersion which was measured using a free-space-coupled laser and the second using a 3 m fiber to couple the laser to the LSM 510 NLO (see Section 7). In both set-ups there is a strong dependence of the broadening effect if the initial pulselength is smaller than 200 fs and smaller than 600 fs using direct coupling and fiber coupling, respectively.

However, the pulse-broadening effect is undesirable, since it will lower the multiphoton fluorescence signal from the specimen for a given average power. As discussed above, this is because the peak power equals the pulse energy divided by the pulselength. To increase the peak power, which will increase the efficiency of excitation, there are two options. The first is to use higher average power, but this would also increase the heating effect of the sample and would eventually result in damage to the specimen. Another option is to use suitable pulse compression techniques to shorten the pulse length. This will be described in the following section in more detail.

4 COMPENSATION OF PULSE BROADENING

As stated above, compensation of the pulse-broadening effect is most useful if pulse-lengths smaller than 200 fs and 600 fs are desired in a direct-coupled set-up and a fiber-coupled set-up respectively. Although short laser pulses are broadened and gain a positive chirp in normal dispersive media, there are methods referred to as pre-chirping to overcome this effect. By looking at Eq. (3), it can be seen that the pulsewidth decreases with increasing z if the chirp multiplied by the dispersion parameter is negative. Since optical materials in the visible and near-infrared wavelength range have a positive dispersion parameter, the chirp of the initial pulse must be negative in order to compress the pulse to its minimum value.

A negative chirp can be achieved using optical systems containing dispersive components such as prisms or gratings. Depending on the amount of chirp to be imparted on the pulses, prism stretchers (low effective dispersion), prism–grating hybrid stretchers (medium effective dispersion), or grating stretchers (high effective dispersion) are used. Using a prechirp unit consisting of prisms is sufficient to compensate for the dispersion of the optics inside an LSM 510 NLO, including the objective lens. The distance required to compensate for the dispersion within the LSM 510 NLO needs to be approximately 0.9 m (assuming that the prisms are SF 10).

However, for compensation inside a 3 m long optical fiber, a prechirp unit consisting of a pair of gratings needs to be used [6]. This is because a grating compressor has a 100-fold higher dispersion parameter in comparison to a prism compressor. If a grating with 561 lines per mm is used, the grating separation needs to be roughly 80 mm in order to compensate for the dispersion of the fiber-coupled LSM 510 NLO. Using prisms to achieve the same effect would require the prism separation to be roughly 8 m which is not practical.

In all prechirp units, frequency components blueshifted with respect to the center frequency are traveling faster than redshifted frequency components. In a set-up where two prisms or gratings are used, the amount of negative dispersion can be adjusted by changing the distance between the two dispersive elements. If the negative dispersion of the prechirp unit equals the positive dispersion of the microscope, the pulselength at the location of the specimen is at its minimum value assuming the laser produces bandwidth-limited pulses. More details of the different dispersion compensation methods can be found in [7, 8].

5 NONLINEAR EFFECTS IN OPTICAL FIBERS

In confocal microscopes, lasers are connected to the scan head via optical fibers. This is a convenient and safe way to deliver the light to the scan module, and makes it possible to arrange the lasers at a distance from the microscope work area, preventing vibration near the microscope. A similar approach can be used to deliver short laser pulses to the microscope for MPLSM. However, there are some constraints on using fiber delivery for this purpose.

To propagate a short-pulse laser through a fiber, special single-mode fibers must be used. This is because in a multimode fiber, different modes are propagated which travel, with very different velocities. This would result in complete distortion of the pulses [2]. The core diameter of a single-mode fiber depends on the center wavelength of the laser pulses, and would be about 5 μm for a conventional titanium : sapphire laser. Since the laser light inside the fiber is confined to this small area, nonlinear effects can be generated inside the fiber. Most relevant is self phase modulation (SPM), which effects the spectral components of the laser pulses. GVD, in contrast, does not affect the spectrum, but delays the spectral components with respect to each other. If a prechirp unit is used before the fiber, the spectral bandwidth is lowered by the SPM. This results in an increased minimum achievable pulselength at the specimen. Hence, these nonlinear effects must be avoided.

As in every nonlinear process, the efficiency of the process depends strongly on the peak intensity. In order to prevent SPM within the fiber, one needs to limit the peak intensity. For a pulsed laser, the peak intensity can be calculated from

$$I = \frac{P_{\text{avg}}}{\tau f A}, \tag{4}$$

where P_{avg}, f, A, and τ are the average power, the repetition rate, the cross-section of the beam, and the pulselength, respectively. The peak intensity should be below a critical value of $5\,\text{GW}\,\text{cm}^{-2}$ for a fused-silica fiber. This corresponds to a maximum average power of less than 8 mW, which can be delivered without SPM if a 100 fs pulse at 800 nm with a repetition rate of 80 MHz is assumed in a 5 μm fiber core diameter. However, applications in nonlinear microscopy often require more power to be delivered to the specimen.

As discussed above, the cross section of the beam is determined by the center wavelength of the laser pulses, owing to the need for single-mode propagation in the fiber. Since the repetition rate of a short-pulse laser cannot be changed easily, and the average power should be as high as possible, only the pulselength parameter is left to control the peak intensity within the fiber.

The influence of the pulselength on the SPM and hence on the spectral bandwidth can be understood by looking at Figure 2. If a laser pulse is first sent through a prechirp unit and then coupled into the fiber, the pulses become shorter as they propagate inside the fiber. This behavior is shown in Figure 2*a*. The plot shows the pulselength as a function of position within the fiber. Figure 2*b* shows the spectral bandwidth of the laser pulse at each position inside the fiber. Towards the end of the fiber, the spectral bandwidth becomes smaller. This is because of the high peak intensities at the end of the fiber, which give rise to SPM.

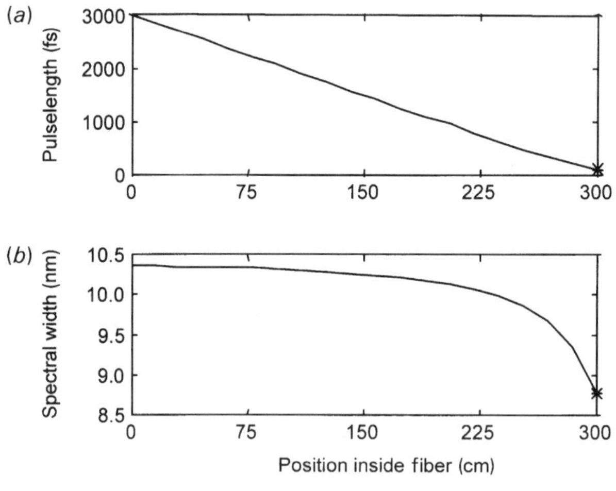

FIGURE 2 Propagation of prechirped short pulses within a single-mode fiber. As the pulses get shorter towards the end of the fiber, the spectral bandwidth of the pulses becomes narrower owing to the influence of self phase modulation. The average power and the dispersion parameter of the 3 m long silica fiber used to calculate the plots were 50 mW and $10^5\,\text{fs}^2$, respectively.

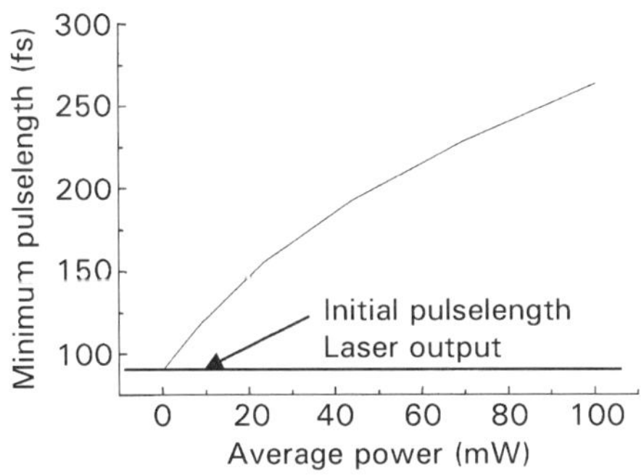

FIGURE 3 Minimum achievable pulselength in the presence of self phase modulation, showing the pulse broadening due to the narrowing of the spectral bandwidth as the average power launched into the fiber is increased.

Since the spectral bandwidth becomes smaller, the pulse can no longer be compressed to its initial pulselength measured at the output of the laser. This is shown in Figure 3 for a bandwidth-limited input pulse of 80 fs traveling through a 3 m long silica fiber. The graph shows the minimum achievable pulselength at the output of the fiber for different average powers launched into the fiber. It can be seen that in order to get a certain output pulselength, the average power that can be launched into the fiber is limited owing to the influence of SPM.

To avoid a strong influence of SPM, only comparatively long pulses should be propagated. This can be achieved by an increased prechirp and the use of additional dispersive elements behind the fiber. The pulses become compressed to the initial pulselength in the additional element in an area where the cross-section of the beam is much larger than the cross-section within the fiber. The higher the dispersion in the additional element, the longer is the pulse at the end of the fiber, if prechirping is compensating for the dispersion of the whole optical set-up and the more average power can be delivered.

6 PULSEFRONT DISTORTION IN LENSES

In addition to GVD and SPM, chromatic aberration of lenses leads to pulse distortions – specifically, a radius-dependent group delay is introduced [5, 10]. Therefore, different radial portions of the beam across the pupil of the objective lens arrive at different times at the focal region, and cause a temporal broadening of the pulse. This results in a lower peak intensity in the focal region (Figure 4). This effect is also referred to as propagation-time difference (PTD).

For lens systems the maximum group delay due to chromatic aberration, $df/d\lambda$, can be estimated using the following equation:

$$\Delta\tau = -\frac{\lambda_0}{2c_0 f_0} r^2 \left.\frac{df}{d\lambda}\right|_{\lambda_0}, \tag{5}$$

where f_0 and r are the focal length of the lens system and the radius of the pupil, respectively. In order to avoid PTD effects, objective lenses have to be corrected for small

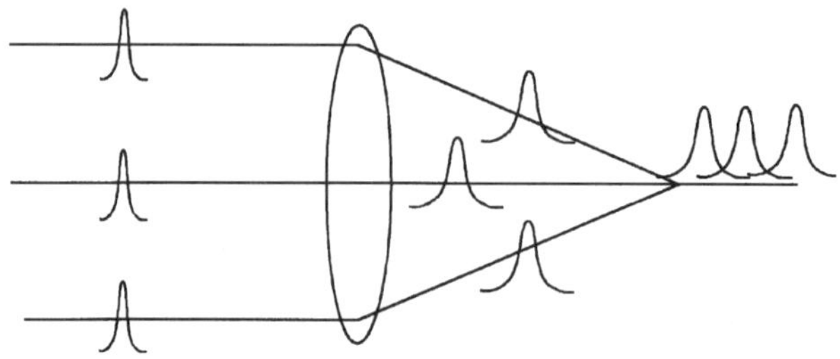

FIGURE 4 Schematic pulse transformation by a singlet lens, showing the influence of chromatic aberrations on the peak intensity due to propagation-time difference.

chromatic aberration. In [9], it was found that the group delay can be neglected if the pulselength is not much shorter than 100 fs for conventional objective lenses.

7 ZEISS LSM 510 NLO USING A FIBER-COUPLED COHERENT MIRA 900 F

The following is the optical set-up of the LSM 510 NLO (Carl Zeiss Jena GmbH) using a fiber-coupled Coherent Mira 900F (NLO here stands for nonlinear optics). The system used for the experiments consists mainly of four different modules, which are the Axioplan 2 microscope stand, the LSM 510 NLO scan module, and the laser modules for the visible and ultrashort lasers. The optical beam layout inside the four modules is depicted schematically in Figure 5. The laser module VIS produces all the excitation lines for conventional confocal single-photon fluorescence microscopy, while the laser module NLO consists of a titanium : sapphire laser that produces femtosecond pulses in the nearinfrared wavelength region and is used for nonlinear microscopy. Laser beams from the two laser modules are coupled to the scan module with two separate fibers via two movable collimation lenses and are directed towards the main dichroic beam splitter (MDB). The collimation lenses can be adjusted so that excitation using the VIS and the NLO laser can be accomplished within the same focal plane of the specimen. In front of the MDB splitter is a plate that reflects a small portion of the excitation light to a monitor diode, which enables measurement of the excitation power. The MDB splitter directs the light via two identical galvanometric scanners, the scan optics, the tube lens, and the microscope objective lens to the specimen. The signal from the specimen is collected by the objective lens, and is directed back through the tube lens, the scan optics, and the scanners to the MDB splitter. The MDB splitter passes the excited fluorescence signal, which is focused with the pinhole optics, to up to four independently controllable confocal apertures onto the photomultipliers. Up to four different fluorescence probes can be detected simultaneously. Owing to the inherent sectioning effect, the confocal aperture can be opened for the detection channels of two-photon excited fluorescence imaging, resulting in a higher collection efficiency.

In addition to the four standard descanned detection paths, other detectors can be used with the system. Non-descanned detectors can be used with two-photon excitation to collect the light directly after the objective lens to give better collection efficiencies for deep tissue imaging. Non-descanned transmitted excitation and fluorescence light can be further combined with descanned fluorescence signals. Also the transmitted light photo-multiplier (T-PMT) can be used to detect the second-harmonic signal from a specimen that is generated in the forward direction of the excitation light [12]. The fluorescence light can also be coupled into a multimode fiber after descanning and passing through a confocal aperture for easy access of the optical signal to external detectors for lifetime imaging and spectrally resolved detection.

In the following section, the fiber coupling of the short-pulse laser to the LSM 510 NLO is described in more detail. The schematic optical setup is shown in Figure 5. The Mira 900F laser is pumped by a Verdi 5 W laser (both from Coherent, Inc., Santa Clara) and produces laser pulses with a pulselength of about 100 fs in a wavelength range of 690–1064 nm. The maximum average power at 800 nm is about 700 mW. Before entering the fiber, the laser pulses are conditioned in a prechirp unit called a group dispersion compensator (GDC) (Coherent, Inc., Santa Clara), consisting of a single holographic grating element in a four-pass configuration. The optical layout of the prechirp unit is shown in Figure 6. Light from the laser is directed by the mirror M1 onto the dispersive grating, which spatially separates the different spectral components of the laser pulse.

The beam now bounces several times at different heights between the mirrors M2 and M3, so that the grating is passed four times. Owing to the angular dispersion of the grating, different frequency components travel different optical pathlengths and the laser pulses acquire a linear chirp, which can be used to cancel out the dispersion of glass materials. After the fourth bounce off the grating, the spectral components of the laser pulses are spatially overlaid again, and the beam passes over M1 and is coupled via an acousto-optical modulator (AOM) into a 3 m long single-mode and polarization-preserving fiber (Figure 5) with a core diameter of 5 μm. The amount of dispersion compensation can be controlled by changing the grating position. Furthermore the grating can be rotated to center the spectrum on the lens. By adjusting the tilt of the lens shown in Figure 6, third-order dispersion, which is mainly induced by the grating compressor itself, can be compensated. The AOM is used in its first diffraction order by the laser scanning microscope to block and to attenuate the beam at a pixel synchronized time base, which allows for region-of-interest (ROI) imaging and blanking of the beam during the flyback of the galvanometer scanner.

In order to prevent SPM within the fiber, the pulses travel seven times through a unit called a post-fiber compressor (PFC), which consists mainly of a TeO2 crystal. The PFC adds another dispersion of 10^5 fs^2 to the dispersion of the fiber (10^5 fs^2) and the LSM 510 NLO including the microscope stand and the objective lens (10^4 fs^2).

By adjusting the grating position, the amount of dispersion compensation of the prechirp unit can be adjusted, so that it balances the amount of dispersion induced by the fiber, the PFC, and the LSM 510 NLO including the microscope stand and the objective lens.

The pulselength at several characteristic points (see Figure 7) within the fiber delivery set-up will be investigated next for optimal compensation of the pulse broadening and hence for a condition where the pulselength at the location of the specimen is at its minimum. If a bandwidth-limited 100 fs pulse enters the GDC, it will be stretched by the grating compressor to about 6000 fs owing to negative chirp. As the pulse travels through the fiber experiencing positive GVD, the pulselength will decrease so that at the end of the fiber, it will be about 3000 fs. Thus, the minimum pulselength that the fiber sees is above 3 ps, resulting in a peak intensity that is below the threshold where SPM plays an important role for average powers of up to 50 mW. As the pulses travel through the PFC and the LSM 510 NLO, including the microscope stand and objective lens, the pulselength will decrease further, and at the location of the specimen will reach its minimum of approximately

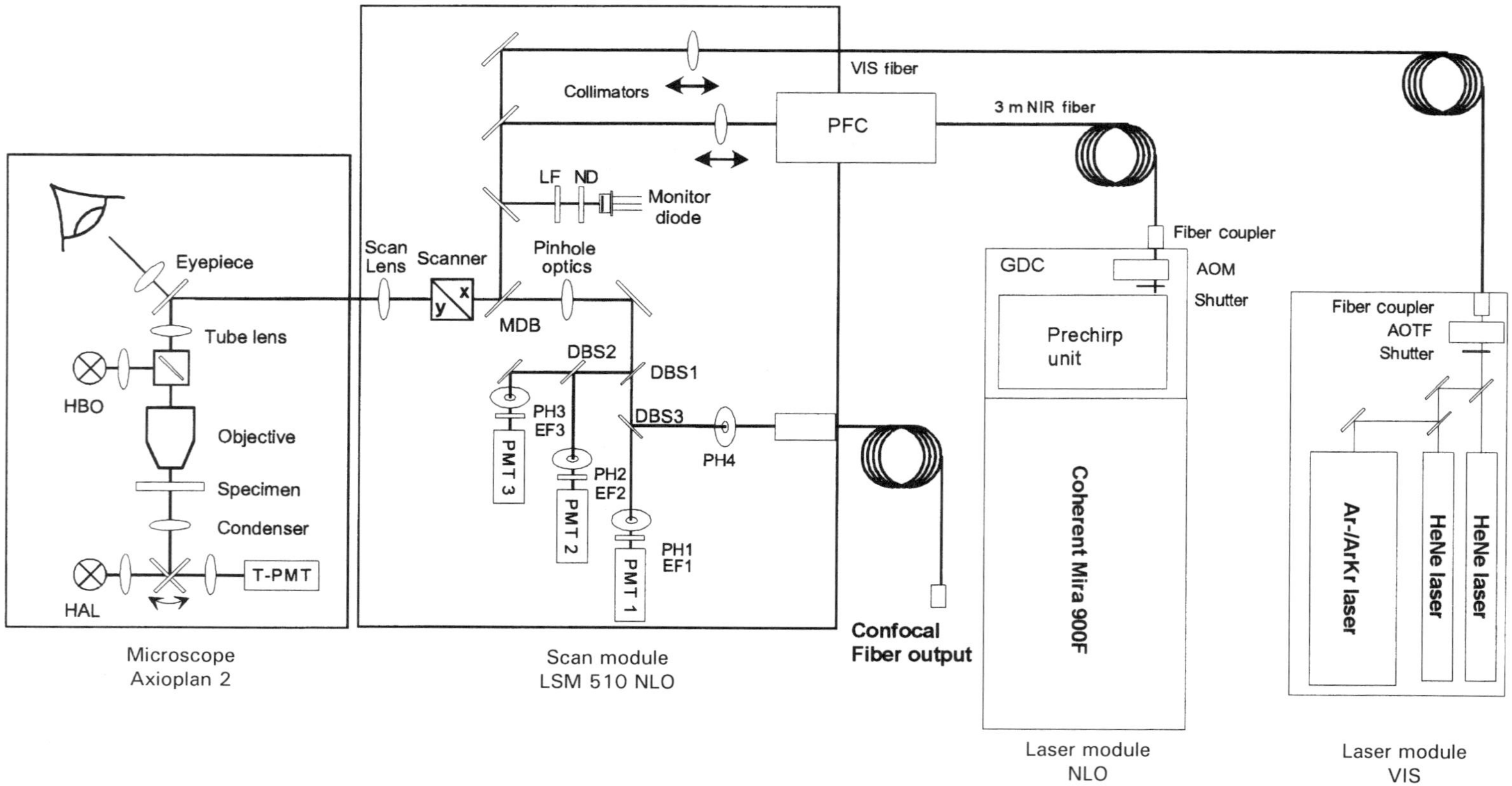

FIGURE 5 Schematic showing the optical beam path inside the Zeiss LSM 510 NLO with a Coherent Mira 900F fiber coupled to the system.

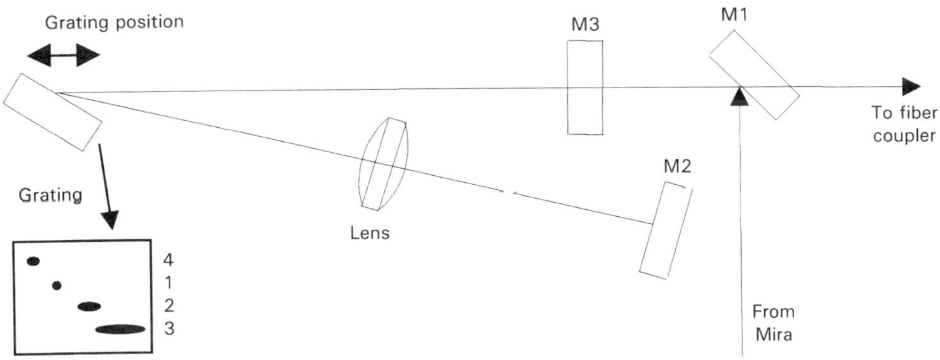

FIGURE 6 Schematic optical beam path inside the Coherent GDC for prechirping the laser pulses before entering the fiber and the LSM 510 NLO.

120 fs. Since the beam diameter within the PFC and the LSM 510 NLO is larger than 1 mm the peak intensity is low enough to avoid SPM.

Figure 8 shows the pulselength after the pulses have traveled through the fiber, the PFC, and the LSM 510 NLO including the microscope stand and the Zeiss Plan Neofluar 40 × /1.3 oil-immersion objective lens as a function of the grating position. The pulselength was measured in the object plane using the autocorrelation technique described in [6] in which the Michelson interferometer was set up between the Mira and the GDC, showing a minimum pulse length of 120 fs. The center wavelength was 800 nm and the bandwidth was 12.5 nm.

The reason for not reaching the input pulselength is loss of spectral bandwidth at the fiber coupling. The maximum power that can be delivered to the specimen in this condition is 50mW. If higher powers are used, the spectral bandwidth is decreased, and hence the minimum achievable pulselength increases owing to the influence of SPM. This can be seen by measuring the spectral bandwidth and the minimum achievable pulselength as a function of the average power coming out of the PFC. The two plots are shown in Figure 9. The strong influence of the SPM is visible at average powers higher than 50 mW. The point where SPM limits the pulselength depends strongly on the spectral bandwidth of the pulse

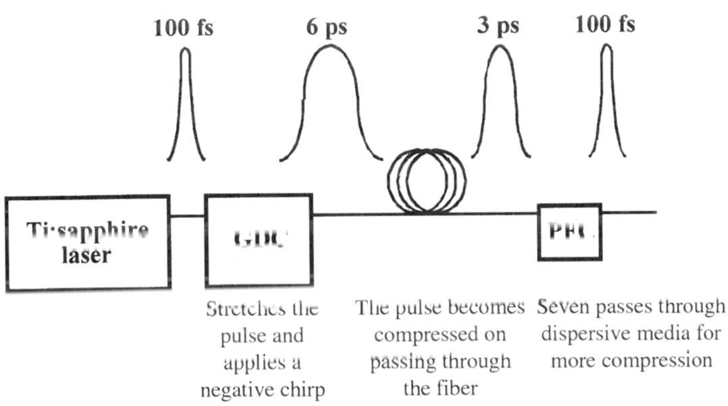

FIGURE 7 Block diagram showing characteristic pulselengths in a fiber delivery set-up for ultrashort-pulse delivery.

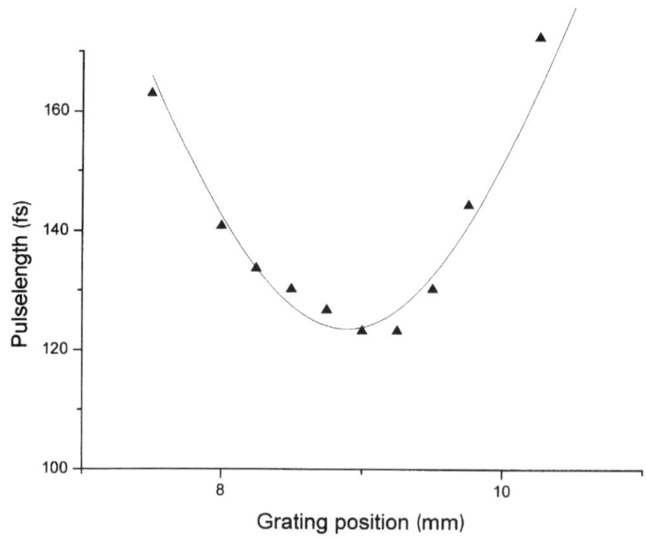

FIGURE 8 Pulselength measured after the Zeiss PN 40 × /1.3 oil-immersion objective lens as a function of the grating position at a wavelength of 800 nm and an average power of 50 mW. The spectral bandwidth of the input pulse was 12.5 nm.

going into the fiber. This is because, for the same dispersive material used after the fiber, a pulse with a broader spectral bandwidth (shorter pulselength) will be longer at the end of the fiber (see Figure 1); therefore, more average power can be delivered through the fiber without the influence of SPM. For that reason, the efficiency of beam coupling to the fiber is very crucial, since clipping of the spectrum of the pulses can occur. For the Coherent mid wave fiber, the wavelength range that can be delivered is limited to 750–950 nm, owing to the single-mode condition.

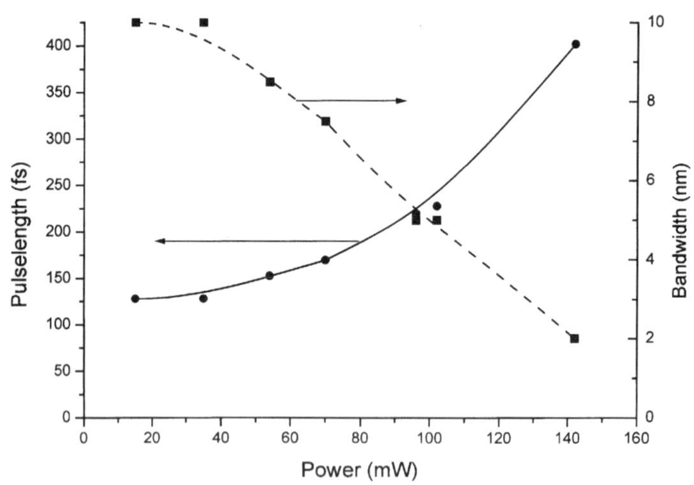

FIGURE 9 Pulselength (solid line) and spectral bandwidth (dashed line) as functions of the average power coming out of the PFC measured at 800 nm using a Zeiss PN 40 × /1.3 oil-immersion objective lens. The spectral bandwidth of the input pulse was 12.5 nm.

8 ZEISS OBJECTIVE LENSES RECOMMENDED FOR TWO-PHOTON EXCITATION

For nonlinear microscopy, objective lenses must be optimized for the following parameters: high transmission in the NIR and visible wavelength ranges, long working distance with a high numerical aperture, a uniform GVD across the pupil of the objective lens for the excitation wavelength range, and a small PTD.

The transmission of a objective lens depends strongly on the design of its antireflection coating. Typical transmission curves for two different objective lenses are shown in Figure 10. The Zeiss Achroplan series is coated with a UV/visible antireflection coating giving optimal performance from 350 nm to approximately 800 nm. In contrast, the Zeiss IR-Achroplan series has a visible/IR coating that allows high transmission from 420 nm to 1400 nm. The Achroplan objectives are best for two-photon applications where excitation is below 850 nm and the fluorescence from the specimen is in the ultraviolet or lower visible wavelength range. These criteria describe the imaging parameters for many ion indicators, nuclear dyes such as DAPI or Hoechst, and fluorescent proteins such as BFP. The IR-Achroplan lenses are best for two-photon applications requiring an excitation wavelength greater than 850 nm and where the fluorescence emission is above 420 nm. These latter criteria apply to many dyes, such as Fluorescein, DiA, Dil, and rhodamine and many of the fluorescent proteins, such as CFP, GFP, YFP, and dsRed. These examples are given to indicate ideal imaging criteria; however, for particularly bright specimens, nonoptimized parameters can often be used to produce high-quality images.

Table 1 summarizes the important characteristic parameters for objective lenses recommended for nonlinear microscopy and for the optics inside the LSM 510 NLO, including the microscope stand. The dispersion parameters were calculated using Eq. (1), the material data and the thickness of each optical element on axis. The data are consistent with the results achieved using the set-up described in [6]. For comparison, the dispersion parameter for the optics within the LSM 510 NLO, including the microscope stand and the AOM, is 7500 fs^2. The variation of the dispersion parameter for different beams across the

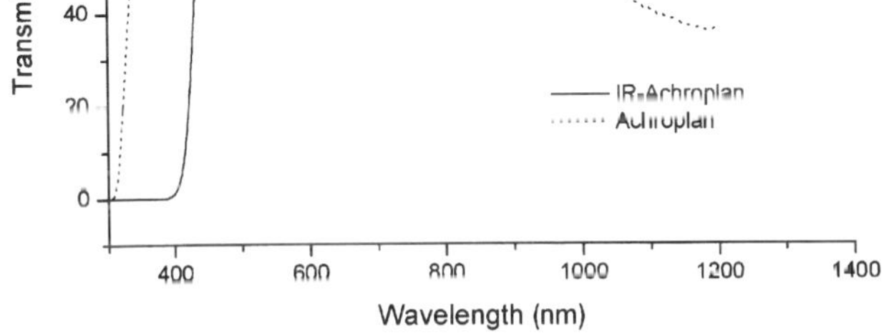

FIGURE 10 Typical transmission efficiency as a function of wavelength for UV/visible-corrected and visible/NIR-corrected Zeiss microscope objective lenses.

TABLE 1 Summary of specific parameters for Zeiss microscope objective lenses recommended for two-photon applications. All data were measured at 800 nm.

Objective lens	Working distance (mm)	Dispersion parameter (fs^2)		Pulse-broadening factor of 100 fs pulse, Eq. (3′)	Maximum PTD[a] (fs)
		On-axis	Variation		
IR-Achroplan 40 ×/0.8 water	3.61	1714	±20	1.14	−3
IR-Achroplan 63 ×/0.9 water	2.00	1494	±15	1.11	−9
Plan Neofluar 40 ×/1.3 oil	0.20	2328	±30	1.23	9
Plan Apochromat 20 ×/0.75	0.61	1531	±10	1.12	10

[a]The propagation-time difference (PTD) is calculated for the whole optical setup including the LSM 510 NLO and the microscope stand.

pupil of the objective lens is listed in the fourth column. It can be seen from both parameters that the pulse broadening is nearly independent of the particular objective lens used and the position of the beam across the pupil of the objective lens. The pulse broadening due to the GVD of the different objective lenses and the LSM 510 NLO including the microscope stand and the AOM was calculated using Eq. (3′) for a 100 fs pulse at a wavelength of 800 nm, and is listed in the fifth column. Since the pulse broadening is nearly independent of the objective lens, it can be compensated using a fixed amount of dispersion compensation on the prechirp unit.

The pulse broadening due to propagation-time difference (PTD) is listed in the last column. The PTD was calculated for the whole optical set-up, since it depends on the chromatic aberrations of the complete system, including the LSM 510 NLO and the microscope stand, for a wavelength of 800 nm. The PTD effect is negligible if the pulse length is not much shorter than 100 fs. It can be seen that the special IR-corrected objective lenses show a smaller PTD in comparison with the standard UV/visible-corrected objective lenses. Negative values indicate that pulses at the edge of the pupil are delayed with respect to pulses traveling on axis.

9 APPLICATIONS OF THE ZEISS LSM 510 NLO/COHERENT FIBER DELIVERY

The microscope design described above has been installed at the Biological Imaging Center (BIC) at Caltech, Pasadena, and is used routinely for a variety of imaging experiments. The BIC is a multi-user facility, with a strong emphasis on confocal and multiphoton laser scanning microscopy. Researchers within and outside the group are interested in studying tissue dynamics in a variety of animal systems, including the fruitfly, chick, mouse, rabbit, zebrafish, frog, sea urchin, and human. For these studies, MPLSM has several advantages. Since many of these studies involve characterizing cell dynamics in living embryos, it was important to utilize a system in which the peak power was optimized without having to use excessive average power. Thus, by using a system, such as the fiber delivery system, where the pulselength is kept optimum, less overall power can be used, which limits the potential for toxic effects during long-term imaging. In addition, MPLSM can be used to collect images with a high signal-to-noise ratio in order to characterize fine structures within specimens. This particular system has been used to study somite formation in the chick embryo, mouse heart and blood cell development, formation of axonal connections in the zebrafish brain and spinal cord, autofluorescence in the rabbit and human cornea, neuronal networks in the fruitfly, sea urchin embryogenesis, frog gastrulation, the potential of new fluorescent dye molecules, fluorescence resonance energy transfer, and fluorescent protein fusions, as well as a number of other applications by part-time users from outside our research group. Some examples of these studies are shown in this section, in order to provide some practical information about multiphoton imaging with this particular system.

9.1 Initial System Checks

For our initial characterization of the system, we measured the average power coming through the objective. We were able to measure average power levels of 42 mW at the sample depending on the wavelength using the Zeiss Plan-Neofluar 10 × /0.3 objective lens. Since these data only relate to the average power within the sample, we measured the pulselength at the objective lens using the approach detailed above and in [6]. Our measurements verified that the pulsewidth at the sample was approximately 120 fs.

Next, we were interested in characterizing two-photon excitation with respect to distance along the optical axis (Figure 11). For these studies, we employed a simple model. Using a 1.4 mm deep well constructed on a microscope slide, filled with a dilute solution of carboxyfluorescein in water (0.008 mg/ml), we collected optical sections in 10 μm increments through the solution. The data shown below were collected using the Zeiss IR-Achroplan 40 × /0.8 water-immersion objective lens which has a working distance of approximately 3.6 mm. These data indicate that there is a minimal decrease in two-photon excitation along the z axis under these conditions.

Our initial measurements showed that efficient two-photon excitation could be produced using this system. In order to confirm this and to illustrate the use of the AOM, we imaged pollen grains using 800 nm excitation. Pollen grains are a stable,

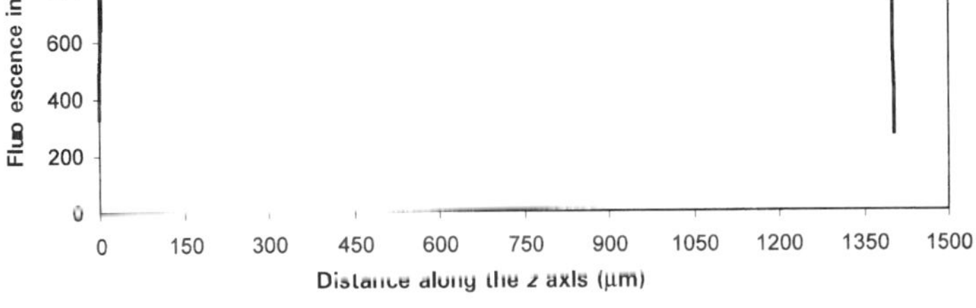

FIGURE 11 Two-photon excitation along the optical axis. The data above reflect the average intensity measurements from a z stack of images with optical sections recorded every 10 μm. Data were taken using the Zeiss IR Achroplan 40 × /0.8 water immersion objective lens.

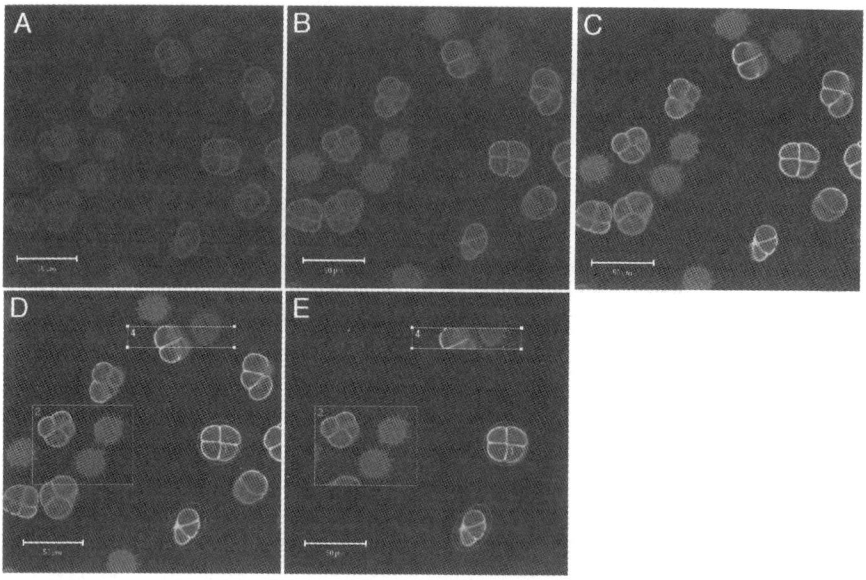

FIGURE 12 See ftp site for color figure. Using the AOM to modulate the percentage transmission of laser pulses and ROI scanning. (*A*), (*B*), and (*C*) show the same field imaged using 20%, 35% and 56% transmission of the laser pulses. (*D*) and (*E*) illustrate the use of ROI scanning to scan only the regions indicated by the colored rectangles. (M. Dickinson, unpublished observations.)

convenient standard used to optimize the imaging parameters in our confocal and MPLSM systems (mixed pollen grain slides are obtained from Carolina Biological, F6-B690). As can be seen in Figure 12, the AOM allows for precise control of the percent transmission of the laser pulses, and can be used to induce multiphoton excitation in defined areas, referred to as regions of interest (ROI). Although the BIC also has two other MPLSM systems, neither of these systems are equipped with an AOM and the titanium : sapphire laser intensity must be controlled crudely using neutral-density filters. Control of the AOM through the Zeiss LSM 510 software not only offers linear variation of titanium : sapphire transmission and ROI scanning, but is also used to control beam blanking on the flyback of the galvanometer scanner, to control localized bleaching, and to provide multidirectional scanning.

9.2 Applications in Developmental Embryology

Although our initial experiments showed that the system could be used to produce high-quality images of pollen grains, we were interested in showing that the system could be used to image commonly used dyes within turbid biological samples. To date, we have observed efficient excitation of many commonly used dyes such as, Hoechst 33342, DAPI, fluorescein, Oregon green, Alexa 488, DiO, DiI, DiA, CFP, GFP, YFP, rhodamine, and Texas red. Owing to the use of the fiber and the AOM, we are limited to a wavelength range between 780 nm and 900 nm, and we have not tried to image more redshifted dyes that optimally excite above 900 nm. Figure 13 shows images of the embryonic heart of a

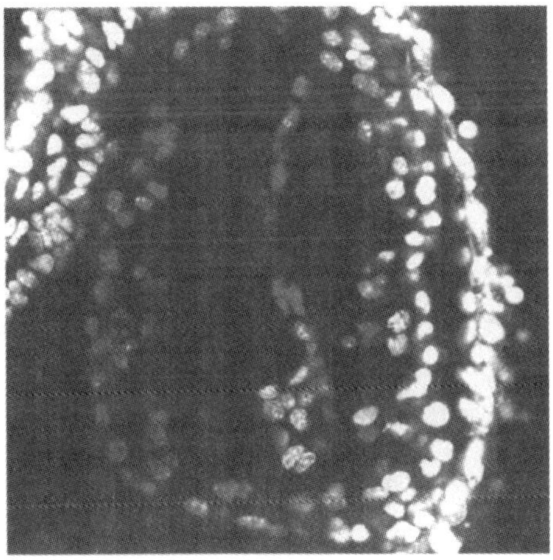

FIGURE 13 The embryonic heart of an 8.5-day postcoitum mouse embryo stained with Hoechst 33342. This image was collected using 800 nm two-photon excitation, a short-pass 680 nm emission filter, and a Zeiss C-Apochromat 40 × /1.2 water-immersion objective lens. This single optical section was taken from 210 μm down into the sample. More-interior regions of the heart are stained less strongly than outer cells owing to poor dye penetration. (M. Dickinson, unpublished observations.)

mouse embryo stained with Hoechst 33342. This single optical section was one of a stack of sections ranging from the surface of the sample to 230 μm in the z direction into the sample. Deeper imaging was prevented by the working distance of the lens. It was clear from this experiment, that deep imaging of embryonic tissues was indeed possible.

More recent experiments involve using this system to study fine filopodial projections between cells in the developing somite and the surrounding environment. During embryonic development, transient structures such as somites form on either side of the neural tube. These cells respond to cues from their environment, and are induced to form derivatives that contribute to dermal cells within the skin, the axial skeleton, and skeletal muscle. Using the fiber delivery system to image cells expressing GFP, we have been able to document many fine filopodial projections between somite cells and nearby cells within the neural tube and the overlying ectoderm (Figure 14) [14].

Our initial observations have indicated that this system can be used for many applications; however, only to the power limitations imposed by the use of the fiber, deep tissue imaging of weak or poorly excited fluorochromes in highly scattering samples may be limited using the system as described here. Many factors affect the ability to collect images from deep within a specimen. In particular, the concentration of dye within the cells being imaged, the two-photon cross-section of the dye, the scattering effects of the tissue, and the collection efficiency of both the objective lens and the detection system are key parameters. We have been able to successfully image up to 350 μm in chicken and mouse embryos, and up to 850 μm in more transparent samples such as the rabbit cornea (M. Dickinson, unpublished observation). Non-descanned detectors are commonly used in multiphoton microscopy to recover scattered emission photons, and the use of a non-descanned detector may improve the depth penetration of this system. In many cases, deep imaging is a trade-off between the amount of power one can safely apply to a specimen and

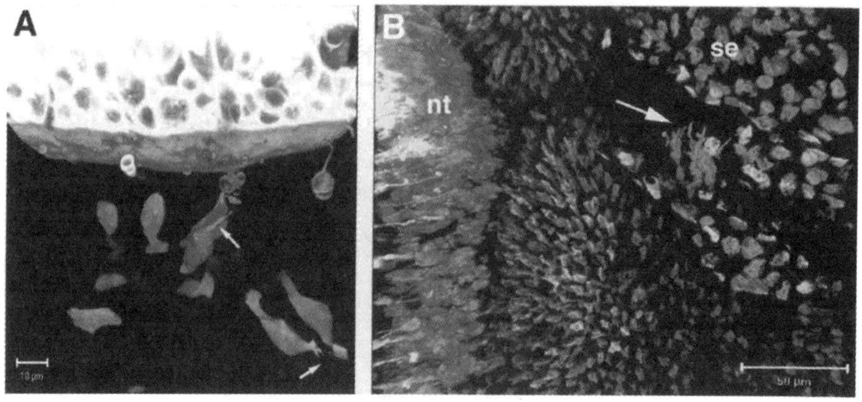

FIGURE 14 GFP-labeled somite cells in the chick embryo. (*A*) The solid mass of cells at the top of the image are the neural tube cells labeled with DiA, whereas individual somite cells in the lower half of the figure are labeled with GFP. This transparency rendering (with a threshold of 4) was made using the Zeiss LSM 510 software (R2.8), and consists of 166 images taken at 0.5 μm intervals. B is also a transparency rendering of 102 optical slices taken at 0.5 μm intervals (with a threshold of 6). GFP cells can be seen extending filopodia toward surface ectoderm (se) cells, which are labeled with Hoechst 33342. The neural tube (nt) is labeled with DiA. Many long filopodia (arrows) can be observed extending toward the neural tube or the surface ectoderm. Images were taken using 860 nm excitation, a short-pass 680 nm emission filter, and a Zeiss IR-Achroplan 40 × /0.8 water immersion objective lens.

the depth at which images can be recovered. Here, we have shown that efficient excitation can be achieved using minimal power levels.

10 CONCLUSIONS

We have described here how optical fibers can be used for the delivery of laser pulses for multiphoton microscopy. Compensating for the group-velocity dispersion (GVD) that occurs when ultrashort pulses encounter dispersive elements in optical fibers as well as the microscope has lead to the development of a multiphoton laser scanning microscope system (MPLSM) that can be used for multiphoton imaging with minimal average power levels. This system provides an example of how one important parameter in multiphoton imaging, the pulse length, can be controlled. MPLSM systems can also be configured using a direct-coupling approach, where the laser pulses are reflected into the scan module via mirrors placed on an optical table without using a prechirp unit. This approach is simpler and more common, because it provides for more power output and more freedom of the tuning range of the laser, and is easier for novice users to operate the light delivery of the short laser pulses. Currently, there are no commercial systems available for GVD compensation for a free-space-coupled laser. However, if care is taken not to introduce too many optical elements, it is possible to design a system where 200–300 fs pulses are delivered to the sample, providing short enough pulses for efficient imaging. These systems are optimized for optimal power delivery, whereas the system that we have described in this chapter is optimized for minimum pulselength delivery of about 100 fs. Thus, systems can be designed to maximize different parameters depending on the users needs.

With each generation of development, more and more efficient multiphoton systems are being built for microscopy. From the laser to the specimen, there is a large emphasis being placed on improving this technology. Alternate ultrafast laser sources are being developed, such as chromium : forsterite lasers [15]. More emphasis has been placed on developing the best objective lenses for nonlinear applications. New dyes are emerging that have large two-photon cross-sections, indicating that brighter biological probes are on the horizon [16]. In addition, MPLSM has been combined with other applications, such as fluorescence correlation spectroscopy, lifetime imaging and spectral emission detection, providing a combination of imaging and quantitation of fluorescence events (for a review, see [17]). As MPLSM has increased in popularity and has become more accessible to biologists, more and more applications are being realized, and there is a demand for more sensitive, light-efficient systems. Here, we have discussed one way of optimizing excitation efficiency as part of a global effort to extend this technology.

ACKNOWLEDGMENTS

We would like to thank Coherent, Inc., in particular Briggs Atherton and Bill Capinski, for their cooperation.

REFERENCES

1 König, K., Simon, U., and Halbhuber, K.-J., 3D resolved two-photon fluorescence microscopy of living cells using a modified confocal laser scanning microscope. *Cell. Mol. Biol.* **42**, 1181–1194 (1996).

2 Agrawal, G.P., *Nonlinear Fiber Optics*, 2nd edn. San Diego: Academic Press, 1995.

3 Feurer, T., and Sauerbrey, R., Characterization of short pulse lasers. *Experimental Methods in Physical Sciences* Vol. 29C (ed. F.B. Dunning and R.G. Hulet). San Diego: Academic Press, 1997.

4 Diels, J.C., Rudolph,W., Ultrashort Laser Pulse Phenomena. San Diego: Academic Press, 1996.

5 Diels, J., Fontaine, J., McMichael, I., and Simoni, F., Control and measurement of ultrashort pulse shapes (in amplitude and phase) with femtosecond accuracy. *Appl. Opt.* **24**, 1270–1282 (1985).

6 Wolleschensky, R., Feurer, T., Sauerbrey, R., and Simon, U., Characterization and optimization of a laser scanning microscope in the femtosecond regime. *Appl. Phys.* **B67**, 87–94 (1998).

7 Fork, R.L., Brito Cruz, C.H., Becker, P.C., and Shank, C.V., Compression of pulses to six femtoseconds by using cubic phase compensation. *Opt. Lett.* **12**, 483–485 (1987).

8 Treacy, E.B., *IEEE* **5**, 454 (1969).

9 Netz, R., Feurer, T., Wolleschensky, R., and Sauerbrey, R., Measurement of the pulse-front distortion in high numerical aperture lenses. *Appl. Phys.* **B70** (2000).

10 Kempe, M. et al., Impact of chromatic and spherical aberration on the focusing of ultrashort light pulses by lenses, *Opt. Lett.* **18** (1993).

11 Müller, M., Squier, J., Wolleschensky, R., Simon, U., and Brakenhoff, G.J., Dispersion precompensation of 15 femtosecond optical pulses for high-numerical aperture objectives. *J. Microsc.* **191**, 141–150 (1998).

12 Gauderon, R. and Sheppard, C.J.R., Effect of a finite-size pinhole on noise performance in single-, two-, and three-photon confocal fluorescence microscopy. *Opt. Lett.* **23** (1998).

13 Brakenhoff, G.J., Müller, M., and Squier, J., Femtosecond pulse width control in microscopy by two-photon absorption autocorrelation. *J. Microsc.* **179**, 253–260 (1995).

14 Dickinson, M.E., Longmuir, K.J., and Fraser, S.E., Analysis of somitogenesis using two-photon microscopy. *Proc. SPIE* **4262**, 18–26 (2001).

15 Robertson, A., Fuchs, H., Ernst, U., Wallenstein, R., Scheuer, V., and Tschudi, T., Prismless femtosecond Cr : forsterite laser. *J. Opt. Soc. Am.* **B17**, 668–671 (2000).

16 Albota, M., Beljonne, D., Bredas, J.L., Ehrlich, J.E., et al., Design of organic molecules with large two-photon absorption cross sections. *Science* **281**, 1653–1656 (1998).

17 So, P.T.C., Dong, C.Y., Masters, B.R., and Berland, K.M. Two-photon excitation fluorescence microscopy. *Annu. Rev. Biomed. Eng.* **2**, 399–429 (2000).

CHAPTER 9

Cellular and Subcellular Perturbations During Multiphoton Microscopy

KARSTEN KÖNIG and UDAY K. TIRLAPUR

Laser Microscopy Division, Institute of Anatomy II, Friedrich Schiller University Jena, Teichgraben 7, D-07740 Jena, Germany

1 INTRODUCTION

Nonlinear effects such as nonresonant multiphoton absorption require high light intensities in the range of megawatts to terawatts per square centimeter. In a variety of multiphoton applications such as multiphoton-mediated microprocessing of inorganic and organic materials, typically high-power laser sources (kilowatts or higher) in combination with focusing optics of low numerical aperture (NA) are used. However, when laser beams are focused to diffraction-limited spots by microscope objectives with NA > 1, moderate laser powers in the milliwatt range are sufficient to induce nonlinear effects, such as two-photon absorption. In fact, near-infrared (NIR) continuous-wave (CW) laser beams at a power of even less than 100 mW are able to induce visible fluorescence by nonresonant two-photon excitation (Hänninen et al., 1994; König et al., 1995, 1996a).

The major application of multiphoton laser scanning microscopes in the life sciences involves the excitation of routinely used intracellular fluorescent probes by the simultaneous absorption of two or three NIR photons (Denk et al., 1990). Most fluorophores possess no one-photon absorption band in the NIR. Their two-photon and three-photon absorption cross-sections of about 10^{-48}–$10^{-50}\,cm^4\,s/photon$ and 10^{-75}–$10^{-84}\,cm^6\,(s/photon)^2$ are relatively low (Xu et al., 1995; Maiti et al., 1997).

Almost all multiphoton laser scanning microscopes are based on the application of ultrashort laser pulses at high repetition rate. Because the multiphoton efficiency of an n-photon process depends on the peak power P according to a P^n relation, the use of pulsed laser beams enable efficient fluorophore excitation at a fast scanning rate with peak powers in the watt and kilowatt range. However, the mean power can be kept at a low level in the microwatt and milliwatt range.

When studying living systems, a laser wavelength in the so-called "optical window" of cells and tissues is desirable – more so because it constitutes the spectral range where the

Confocal and Two-Photon Microscopy: Foundations, Applications, and Advances, Edited by Alberto Diaspro. ISBN 0-471-40920-0

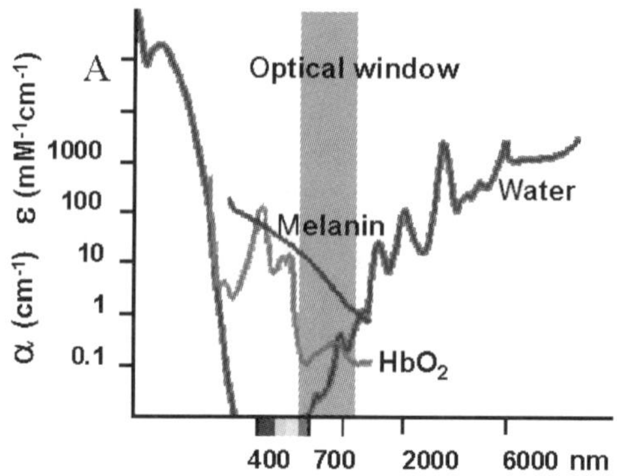

FIGURE 1 See ftp site for color figure. Absorption spectra of the cellular absorbers water, melanin, and hemoglobin. The range of low absorption and low scattering is from about 600 nm up to 1200 nm (optical window). At higher NIR wavelengths, linear absorption of water increases, and may induce destructive linear heating effects.

one-photon absorption and scattering coefficients of unstained cells and tissues are relatively low and where the light penetration depth is high (Fig. 1). Often, this window covers the spectral range from 600 nm to 1200 nm. In the out-of-focus regions where the light intensity is low, the object under investigation does not experience destructive effects owing to a lack of efficient absorption. For most cells, the out-of-focus structure appears transparent. Exceptions are pigmented cells, where linear absorption is likely to occur.

In the case of two-photon microscopy, the probability of two-photon absorption in a first approximation decreases with the distance d from the focal point according to a d^{-4} relation. When using high-NA objectives and appropriate power levels, two-photon effects can therefore be limited to a minute subfemtoliter volume. By contrast, microscopy with ultraoviolet (UV) or visible laser light implies possible destructive effects by one-photon absorption in out-of-focus illuminated regions. This problem is often encountered in conventional confocal laser scanning microscopy based on UV and visible CW lasers, as well as in conventional multiphoton-mediated intracellular surgery with pulsed UV and visible laser beams (Berns, 1998; Greulich, 1999). Furthermore, the use in fluorescence microscopes of noncoherent excitation radiation from high-pressure mercury lamps is known to induce significant cell damage, including DNA damage (König et al., 1996b).

Femtosecond laser microscopy on living specimens can be performed without phototoxic effects under certain conditions. This has been demonstrated by long-term scanning experiments on single cells with a peak intensity of about 200 GW cm^{-2}, where laser exposure for hours had no impact on cellular reproduction and vitality (König, 2000) or on long-term irradiation of living embryos. Hamster embryos were exposed for 24 hours to NIR femtosecond laser pulses of a multiphoton fluorescence microscope without impact on embryo development, in contrast to control studies performed with a conventional one-photon laser scanning microscope (Squirrel et al., 1999).

Nevertheless, NIR multiphoton microscopes can potentially induce undesirable destructive effects in the specimens under investigation due to to the high light intensity and subsequent multiphoton absorption within the focal volume (Diaspro and Robello, 1999; König, 2000). Indeed, during multiphoton microscopy, a further "optical window" exists for the safe, nondestructive observation of specimens of interest that is mainly defined by light-intensity thresholds. Outside this window, cells and tissues may undergo either (i) a slow destructive effect due to multiphoton excitation of cellular absorbers and subsequent cytotoxic photochemical reactions, (ii) photothermal damage in the case of pigmented cells, or (iii) an instantaneous destructive optomechanical effect by optical breakdown phenomena and plasma formation as a result of multiphoton-mediated ionization of molecules.

2 PHOTOCHEMICAL DAMAGE IN MULTIPHOTON MICROSCOPES

2.1 Absorbers and Targets in Biological Specimens

Photochemical damage can be induced by the nonlinear excitation of endogenous and exogenous absorbers followed by phototoxic reactions. These reactions include the formation of reactive oxygen species (ROS), which result in oxidative stress. A wide variety of endogenous cellular absorbers can be excited with multiphoton microscopes. Some of these are fluorescent and can be imaged by autofluorescence detection. Table 1 lists important naturally occurring endogenous absorbers in cells and tissues that can be excited with NIR radiation via two- or three-photon excitation processes.

For example, amino acids with an one-photon absorption maximum below 300 nm can be excited via a three-photon process with NIR photons. The reduced coenzymes NAD(P)H absorb intense NIR radiation in the 700–800 nm range by two-photon excitation. It is known from one-photon studies that excitation of amino acids as well as of NAD(P)H can induce irreversible DNA damage (Cunningham et al., 1985; Tyrell and Keyse, 1990; Bertling et al., 1996). Excitation of flavins, which can also be excited by a two-photon process, may also cause cell damage by the UV/blue-light-induced production of hydrogen peroxide and its metabolite, the hydroxyl radical (Hockberger et al., 1999).

Also, the excitation of cellular porphyrins leads to the formation of ROS. Singlet oxygen is produced by a type II photooxidation process (photodynamic reaction). Metal-free porphyrins such as coproporphyrin and protoporphyrin are efficient ROS producers, and are termed photosensitizers. As is known from photodynamic porphyrin therapy, ROS formation in mitochondria often leads to apoptosis, whereas ROS formation in lysosomes or other organelles may induce either necrosis or apoptosis (Doiron and Gomer, 1985; Moor, 2000). Photodynamic reactions are also the major source of phototoxic response when using fluorescent markers or other exogenous absorbers.

TABLE 1 Spectral characteristics (one-photon maxima) of important endogenous absorbers

Absorber	Absorption maximum (nm)	Fluorescence maximum (nm)
Water	1500, 1900, 2900	—
Hemoglobin	410, 540, 580, 760	—
Tryptophan	220, 280–290	320–350
NAD(P)H	340	450–470
Flavins	370, 450	530
Melanin	< 300	575
PP IX	405, 505, 540, 575, 630	635

2.2 Laser Exposure Parameters

In most experiments, we employed user-friendly compact solid state 80 MHz titanium : sapphire lasers (Vitesse from Coherent, U.S.A., and MaiTai from Spectra Physics, U.S.A.) as "turnkey" laser sources. In addition, a tunable argon-ion laser-pumped Mira system (Coherent, U.S.A.) was used. The laser was coupled to an adapter module (JenLab GmbH, Jena, Germany), which consisted of a 1 : 5 beam expander, a step-motor driven beam attenuator, a fast photodiode for power measurements, a trigger pulse supply, a control unit for laser operation, and a shutter. The expanded beam at tunable power was introduced in a modified laser scanning microscope (LSM 410, Carl Zeiss Jena GmbH, Jena, Germany) with baseport detector (König, 2000). The beam was focused to a diffraction-limited spot by conventional high-NA objectives onto the sample. The typical NIR transmission of the objective was found to be 50%, whereas the overall transmission was typically in the region of 30%. When studying living cells, a closed microchamber with 170 μm thick glass windows has been used (MiniCeM, JenLab GmbH, Jena, Germany).

We used the illumination modes of (i) scanning a 512 × 512 pixel area, (ii) scanning a defined region of interest (ROI), or (iii) single-point exposure, where the beam is parked at pixels of interest within the living cell.

The high sensitivity of the baseport photomultiplier enabled the detection of two-photon excited intracellular fluorophores with high fluorescence quantum yield (e.g., DAPI and Hoechst), with a mean power at the sample of 25 μW–1 mW at a frame rate of 1 Hz (512 × 512 pixels). More importantly, detection of endogenous fluorophores, such as fluorescent coenzymes, was possible with a mean excitation power of less than 2 mW at appropriate NIR wavelengths.

The laser power after transmission through the internal microscope optics was measured in air using a Fieldmaster power meter with the two detector heads LM-2 and LM-3. In order to correct for the true photon flux through oil and the chamber window into the medium as well as for the limited detection angle of the power detector heads, correction values of 1.1 and 2.0 had been determined using a sandwich system (König, et al., 1997a).

Optical dispersion results in pulse broadening during transmission through the microscope optics. Typically, the pulsewidth at the sample was about 150–200 fs (König, 2000). At 10 mW mean power, the peak power and the peak intensities reaches values of 0.8 kW and 1.2 TW cm^{-2}, respectively, when assuming a full-width at half-maximum (FWHM) beam size of $\lambda/2NA \approx 310$ nm. A typical beam dwell time (time of exposure per pixel) during one scan was about 30 μs, which resulted in a frame rate of 8 s per frame. At zoom setting 2, 512 × 512 pixels covered an area of 160 μm × 160 μm. For cell damage studies, typically one or more cells of interest were scanned 10 times in the same focal plane at 30 μs beam dwell time at zoom 2.

2.3 Evidence for NIR-Induced ROS Formation

With a view to determining whether multiphoton NIR laser scanning microscopy with 80 MHz laser radiation above certain intensity levels is able to induce ROS formation and DNA breaks in living cells, we have performed extensive studies on PtK2 cells.

Using 170 fs laser pulses at 800 nm, we found by standard in vivo analysis that a mean power of 7 mW or more was sufficient to produce ROS in PtK2 cells within a scanned

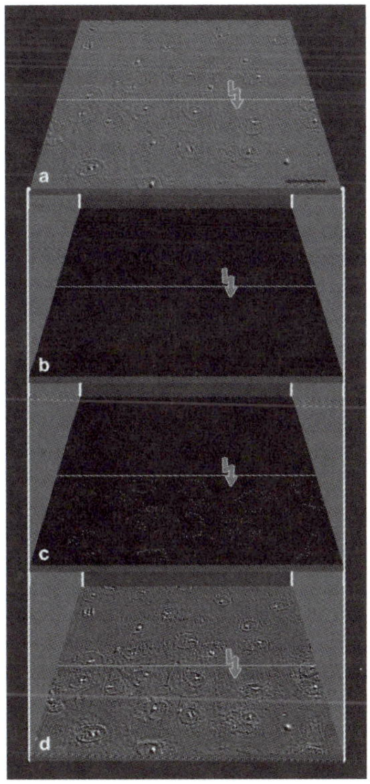

FIGURE 2 See ftp site for color figure. Detection of generation of ROS in one half of PtK2 cells irradiated with an 800 nm NIR laser at a mean power of 7 mW using a Jenchrom px blue. (*a*) Emboss-filtered transmission image of a monolayer of PtK2 cells before irradiation. (*b*) Autofluorescence image before NIR irradiation. (*c*) NIR femtosecond-laser-induced (at mean power of 25 μW) fluorescence image of the Jenchrom reaction product formed exclusively in one half of the cells exposed to a high laser power (7 mW). (*d*) Electronically merged transmission and fluorescence images, showing the discrete presence of punctate fluorescent products only in high-power (7 mW) NIR-irradiated cells.

region of interest. By contrast, no ROS formation was detected in the non-irradiated adjacent cells. In order to detect ROS, we used a recently developed fluorescent probe Jenchrom px blue (JenLab GmbH, Jena, Germany). As can be seen in Figure 2, it is obvious that the punctate fluorescence of the Jenchrom reaction product is exclusively confined to the irradiated cells.

In addition, we have also used the membrane-permeable dye dihydrofluorescein (DHF) for in situ detection of ROS (Hockberger et al., 1999) in PtK2 cells. We exposed one-half of the DHF loaded PtK2 cells with 170 fs pulses at 800 nm at mean powers of 2, 5, 7, 10, 12, and 15 mW. There was virtually no DHF fluorescence in cells exposed to a mean laser power of 5 mW or less; however, diffuse DHF fluorescence was present in cells exposed to higher mean laser powers, and was very intense in cells exposed to a mean laser power of 15 mW. Figure 3 demonstrates the effects after 2 mW and 12 mW laser exposure.

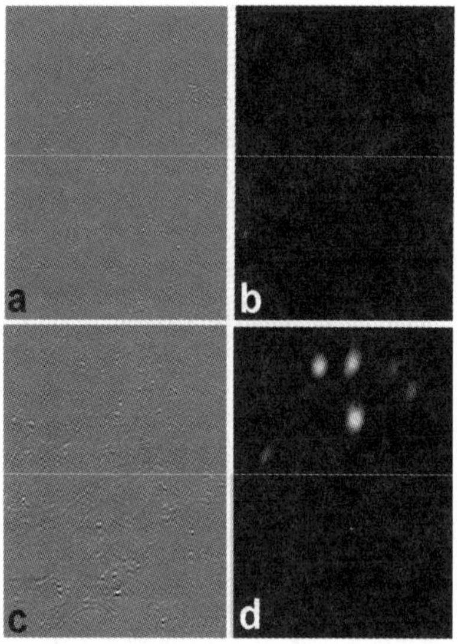

FIGURE 3 Detection of formation of ROS in one-half of PtK2 cells irradiated with an 800 nm NIR laser at a mean power of 2 mW (*a*, *b*) and at 12 mW (*c*, *d*) using the membrane-permeable dye dihydrofluorescein (DHF). Note the presence of DHF fluorescence only in cells (upper half) exposed to higher laser power.

2.4 Evidence for NIR-induced DNA Strand Breaks

Shafirovich et al. (1999) have demonstrated in vitro that excitation of pBR322 supercoiled plasmid DNA with intense 810 nm NIR femtosecond laser pulses at a pulsewidth of 140 fs, a 76 MHz pulse-repetition rate, and average intensities of 1.2 mW cm^{-2} (GW cm^{-2} peak intensities) results in single-strand breaks as a result of simultaneous multiphoton absorption.

However, it is not known at present whether such DNA strand breaks as identified in the in vitro system also occur in vivo when cells are exposed to intense NIR irradiation during multiphoton microscopy.

In order to understand NIR laser–target interactions, we have undertaken a more detailed analysis of the effect of 800 nm NIR femtosecond laser pulses at different mean laser powers on unlabelled PtK2 cells with a view to characterizing and ascertain DNA damage in situ using a well-established TdT-mediated dUTP-nick end-labeling (TUNEL) assay (Gavrieli et al., 1992; Negri et al., 1997). Cells were irradiated with an 800 nm (170 fs pulse duration) NIR scanning beam 10 times, with a beam dwell time of about 60 or 120 μs per scan, which corresponded to 16 or 32 s per frame of 512 × 512 pixels (at zoom setting 2) at mean powers of 2, 5, 10, and 20 mW.

Control PtK2 cells not exposed to an 800 nm laser beam and those exposed to 2–3 mW mean power did not show any TUNEL-positive green fluorescence in their nuclei (Fig. 4) indicating that lower mean laser powers do not induce any DNA strand breaks in situ, and hence are not deleterious.

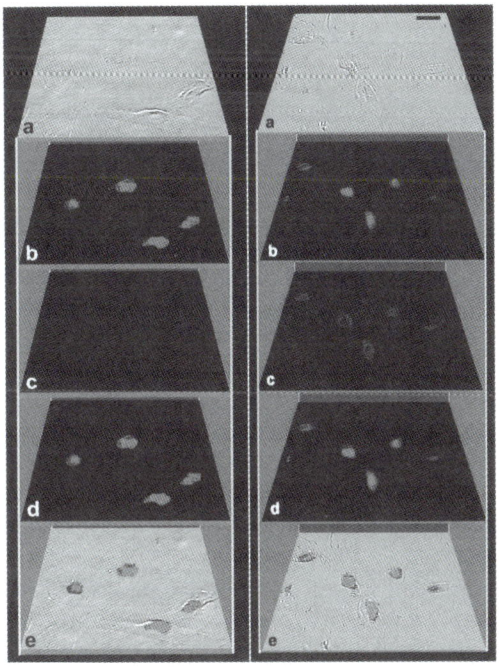

FIGURE 4 See ftp site for color figure. False-color-coded 3D representation of PtK2 cells exposed to NIR femtosecond laser pulses at a mean power of 2 mW. Two-photon images: (*a*) transmission; (*b*) PI fluorescence; (*c*) TUNEL-negative nuclei, without fluorescein fluorescence; (*d*) electronically merged PI and fluorescein fluorescence; (*e*) overlay of transmission and fluorescence images. The scale bar represents 15 μm.

However, it is quite evident that most of the PtK2 cells exposed to a mean power of 5 mW manifest TUNEL-positive green fluorescence, mostly in the nuclear area. With an increase in the mean laser power to 10–15 mW, almost all of the cells had TUNEL-positive nuclei (Fig. 5). The relative fluorescence of the green TUNEL-positive nuclei of cells exposed to 5 mW and 10 mW mean laser power was similar. Interestingly when cells were exposed to a further higher mean power of 20 mW, green fluorescence was seen in all nuclei with increased fluorescence intensity.

The pattern of this green fluorescence indicative of DNA strand breaks in high-power NIR-exposed cells is strikingly identical to those reported for other apoptotic mammalian cells (Negri, et al., 1997).

Our in vivo findings of NIR laser-induced DNA fragmentation are consistent, and corroborate those of Shafirovich et al. (1999), who demonstrated that NIR (810 nm) femtosecond laser pulses can induce DNA strand cleavage. Furthermore, nuclear events during programmed cell death (apoptosis) are known to begin with changes in the chromatin in the nuclear periphery (Tirlapur et al., 2001). Because the structural changes and DNA breaks in the nuclei in NIR-exposed cells are quite similar to those occurring during apoptosis in different cell types (Negri et al., 1997; Bedner et al., 2000), it is therefore conceivable that NIR-laser irradiation of PtK2 cells leads to apoptosis-like cell death.

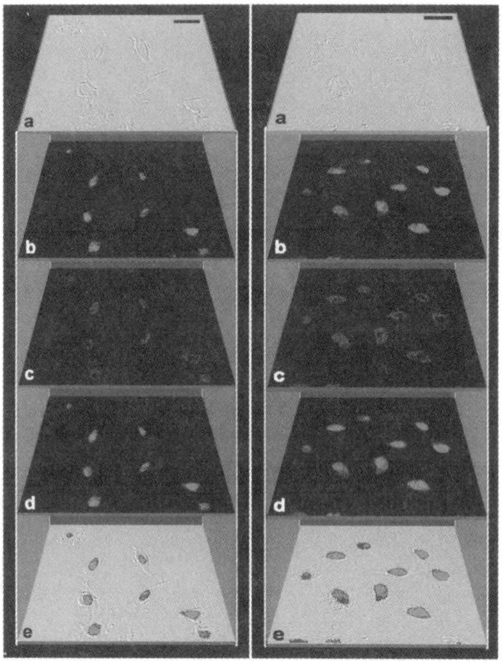

FIGURE 5 See ftp site for color figure. Pseudocolor-coded 3D representation of PtK2 cells exposed to NIR femtosecond laser pulses at a mean power of 5 mW. Two-photon images: (*a*) transmission; (*b*) PI fluorescence; (*c*) TUNEL-positive nuclei, showing fluorescein fluorescence; (*d*) electronically merged PI and fluorescein fluorescence; (*e*) overlay of transmission and fluorescence images. The scale bar represents 15 μm.

2.5 Photodynamically Induced Effects

We have studied the effect of two-photon excitation of porphyrins and subsequent interaction with molecular oxygen. Light-excited porphyrin photosensitizers react with oxygen by a type II photooxidation process where energy transfer occurs from the metastable triplet state to molecular oxygen. This results in the formation of reactive singlet oxygen which induces cytotoxic effects.

In particular, we have studied the effect of 780 nm-excited protoporphyrin IX, which was synthesized in the mitochondria of Chinese hamster ovarian (CHO) cells after external administration of the porphyrin precursor aminolevulinic acid (ALA, 1.5 $mg\,ml^{-1}$). Four hours after incubation, cells were exposed to 170 fs pulses at a mean power of 2 mW and a frame rate of 0.0625 s^{-1} (16 s per frame). This corresponds to a mean beam dwell time of about 400 ms per scan on a cell. Live/dead assays revealed lethal effects and fading of the red PP IX fluorescence in most of cells after 100 scans in the same focal plane. Control cells without ALA could be scanned more than 700 times permanently (3 hours) without impact on vitality (König et al., 1999a).

Nonlinear excitation of porphyrins can be used to induce desired photodynamic reactions in tumor tissue. The major advantages of NIR porphyrin photodynamic treatment are the excitation of higher-order singlet states (Soret-band excitation) and the high light-penetration depth. Two-photon excitation can also be used to induce type I photooxidation processes (charge transfer) (Fisher et al., 1997).

3 PHOTOTHERMAL DAMAGE

In the absence of precise knowledge of the intracellular one-photon and multiphoton endogenous absorbers of NIR radiation and their spatially resolved accumulation, water with a typical absorption coefficient of less than $0.1\ \text{cm}^{-1}$ is considered the most important absorber in the "optical window" of cells and tissues in nonpigmented cells.

Our studies on femtosecond laser- and CW laser-exposed CHO cells have shown that cell damage at 920 nm is less pronounced compared with that due to exposure with 780 nm radiation, where water absorbs less. It is therefore conceivable that photodamage by very intense femtosecond laser radiation is thus not exclusively based on linear water heating.

A typical low heating of 1 K/100 mW was determined for CHO cells during exposure to CW 1064 nm microbeams at megawatt per square centimeter intensity and gigajoule per square centimeter fluence (Liu et al., 1995). The temperature measurements were performed by microspectrofluorometry. CHO cells were labeled with the thermosensitive fluorescent membrane stain Laurdan, which exhibits a redshift of the fluorescence maximum with increasing temperature. Calibration enabled the determination of a mean intracellular temperature with an accuracy of about 0.1 K.

Also, temperature calculations for nonpigmented cells exposed to femtosecond and picosecond pulses reveal only a slight, nontoxic, increase in temperature (0.1 K/100 mW at 850 nm) (Schönle and Hell, 1998).

Nevertheless, pigmented cells such as cells with melanin and erythrocytes may undergo photothermal destruction. In the "optical window" of cells and tissues, the one-photon absorption coefficients of these pigments are higher than that of water. While the absorption of melanin decreases with increasing wavelength, the pigment hemoglobin has an absorption maximum in this spectral region. Reduced hemoglobin possess a 760 nm absorption band, with a one-photon molecular absorption coefficient of about $1\ \text{mM}^{-1}\ \text{cm}^{-1}$. Our studies have shown that human erythrocytes ($\alpha \approx 30\ \text{cm}^{-1}$) exposed to 780 nm CW microirradiation at room temperature undergo hemolysis within 120 s when the power exceeds 60 mW. A heating rate of about 60 K/100 mW can be estimated from these experimental data. The heating rate decreased with wavelength higher than 760 nm.

Interestingly, photoinduced hemolysis studies revealed that femtosecond laser pulses at high repetition rate cannot simply be considered as quasi-CW beams. Damage may occur at a much lower mean power of pulsed radiation at high repetition frequency compared with power values of CW radiation. Using 130 fs pulses (76 MHz) at the same wavelength, a mean power of less than 10 mW was found to be sufficient to induce hemolysis (König et al., 1996c).

4 DAMAGE BY OPTICAL BREAKDOWN

The high transient powers, particularly in the kilowatt range, combined with submicrometer exposure spots reveal in enormous light intensities in the terawatt per square centimeter range, which can cause optical breakdown phenomena resulting in plasma formation and destructive thermomechanical effects. Optical breakdown induces immediate cell damage.

The threshold for optical breakdown decreases with pulsewidth. In water, the threshold for 100 fs pulses was found to be two orders of magnitude lower than for 6 ns pulses (Vogel et al., 1999). The primary cause of optical breakdown in the case of femtosecond pulses is efficient multiphoton absorption. Inverse 'Bremsstrahlung' and collision processes are not dominant.

The onset of optical breakdown depends on the material. For some materials, femtosecond laser pulses at light intensities of about 0.1 TW cm^{-2} are sufficient to induce optical breakdown. Optical breakdown and plasma formation is often accompanied by a nearly white luminescence and complex subnanosecond emission decay (König et al., 1996d).

We have observed instantaneous white plasma luminescence and immediate damage of intracellular structures in CHO cells during scanning with 76 MHz, 170 fs pulses at mean powers higher than 10 mW. Interestingly, the optical breakdown occurred preferentially in the mitochondrial region, and did not start in the extracellular medium or in the nucleus. Evaluation with the fluorescent live/dead fluorophore kit (calcein and ethidium homodimer from Molecular Probes, Eugene, U.S.A.) confirmed immediate death in scanned cells with extended plasma formation. However, when the phenomenon of optical breakdown was strongly confined to a tiny perinuclear region, the cell apparently did tend to survive. In contrast, at higher laser powers, shock-wave-induced morphological damage was manifested by total cell fragmentation (König et al., 1996d).

5 MODIFICATIONS OF ULTRASTRUCTURE

Our ultrastructural studies on CHO cells reveal that mitochondria are the major targets of NIR irradiation. Femtosecond laser exposure induced swollen mitochondria accompanied by loss of christae and the formation of electron-dense bodies in the mitochondrial matrix (Oehring et al., 2000). In addition, the lumen of endoplasmatic reticulum was enlarged. For a mean laser power of more than 12 mW at 800 nm, complete destruction of mitochondria and their transformation to electron-dense structures was observed. At higher power levels, also the outer membranes ruptured. Figure 6 demonstrates NIR-induced damage of mitochondria in CHO cells.

6 INFLUENCE OF ULTRASHORT NIR PULSES ON REPRODUCTIVE BEHAVIOR

Squirrel et al. (1999) intermittently exposed Chinese hamster embryos for 24 hours to femtosecond laser pulses of gigawatt per square centimeter peak intensity without impact on embryo development, in contrast to embryos exposed to visible light, which did not survive. We have studied the influence of intense NIR exposure on the reproduction behavior of more than 3000 CHO cells. For this purpose, adherent CHO cells were maintained in a sterile cell chamber and exposed through a glass window. Following laser scanning, the cell division of the exposed cell as well as of its daughter cells, was monitored for up to 3 days. If the cell was not affected by the NIR light, division occurred every 12 hours and the cell produced a clone of up to 64 cells within 3 days.

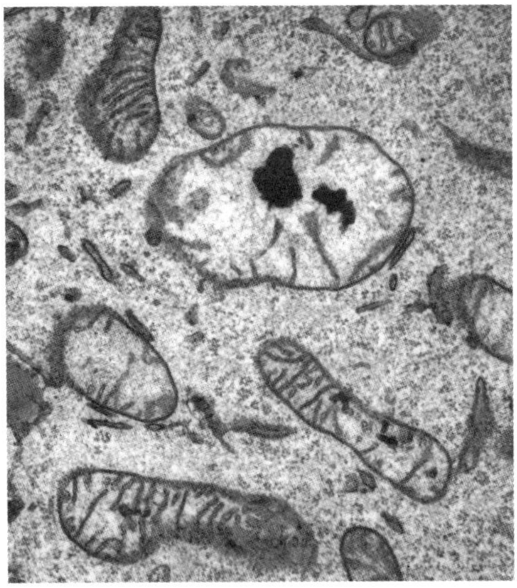

FIGURE 6 Electron microscopy reveals significant damage to mitochondria at high light intensities.

We found that CHO cells can be scanned with an 800 nm, 80 MHz femtosecond laser beam at 2 mW mean power for hours without damage to the exposed cells and their derivatives. In particular, we irradiated single CHO cells for a time period of 700 scans and 3 hours, respectively (16 s per frame). However, at higher mean powers and intensities, respectively, a significant decrease in cloning efficiency occurred. In order to study the dependence on power, wavelength, and pulsewidth, we used tunable laser beams that passed through pulse-stretching units based on prism pairs and grating pairs. Typically, single CHO cells were scanned at microsecond pixel dwell time 10 times in the same focal plane (König et al., 1997b, 1999b).

We found less damage at wavelengths in the 800–900 nm range compared with the 700–800 nm range. Interestingly, photodamage was found to be more pronounced in the case of femtosecond laser pulses compared with picosecond laser pulses (1–5 ps). At 780 nm, impaired cell division was found for 50% of the cells at 11 mW mean power (P_{50}) in the case of 2 ps pulses and at 3 mW for 240 fs pulses. CW scanning beams did not impair cellular reproduction up to 35 mW power. Using mean laser powers of less than 2 mW, CHO cells could be scanned with picosecond and femtosecond pulses for hours without impact on reproduction. This implies that below certain power (intensity) thresholds, "safe" multiphoton microscopy is possible.

When the P_{50} values are plotted versus the square root of the pulse broadening, a nearly linear graph is obtained (König et al., 1999b). This indicates that the photodamage process is likely based on a two-photon excitation process rather than a one-photon or three-photon event. This was independently confirmed by Koester et al. (1999), based on calcium measurements in neocortical neurons in rat brain slices.

In order to compare the photodamage in NIR multiphoton microscopes versus UV one-photon microscopes, the same cloning assay was performed on CHO cells using a 364 nm scanning beam of a conventional argon-ion laser scanning microscope. A 50% cloning efficiency was obtained at a low power of 4 μW. Therefore, UV-induced damage occurs at about three orders of magnitude lower powers (4 μW) compared with the mean values of NIR femtosecond laser pulses (2–8 mW, 750–800 nm).

7 NANOSURGERY

As pointed out in Section 5, the cell can survive if the damage induced by a very intense beam is confined to a minute subcellular volume. Exposure of a small region of interest, including single-spot exposure, may therefore be useful for intracellular processing of biological material.

We increased the mean power in femtosecond laser microscopes with high-NA objectives to values of 30–50 mW and light intensities of terawatts per square centimeter, to knock out intracellular material, to perform nanodissections (König et al., 1999c; König, 2000) and to "drill" holes.

By fine tuning of the laser power only the central part of the beam spot provided sufficient intensity for material removal. Therefore, we were able to drill holes and to cut structures with a precision below the diffraction limit – i.e., with a cut size below the size of the submicrometer beam spot.

For example, we have used the multiphoton-induced ablation process to cut human chromosomes. Measurements with a force scanning microscope reveal chromosome dissection with a FWHM cut size below 200 nm. In addition, partial removal (incisions) of chromosome material with a precision of 110 nm and an FWHM of 85 nm was observed, which is likely the smallest laser cut in biological material ever performed (König et al., 2001). Figure 7 shows examples of hole drilling and line cutting in human chromosomes.

Multiphoton laser microscopes can be therefore used as noncontact nanoscalpels for surgery inside the cell, inside the cell nucleus, or inside an organelle without affecting

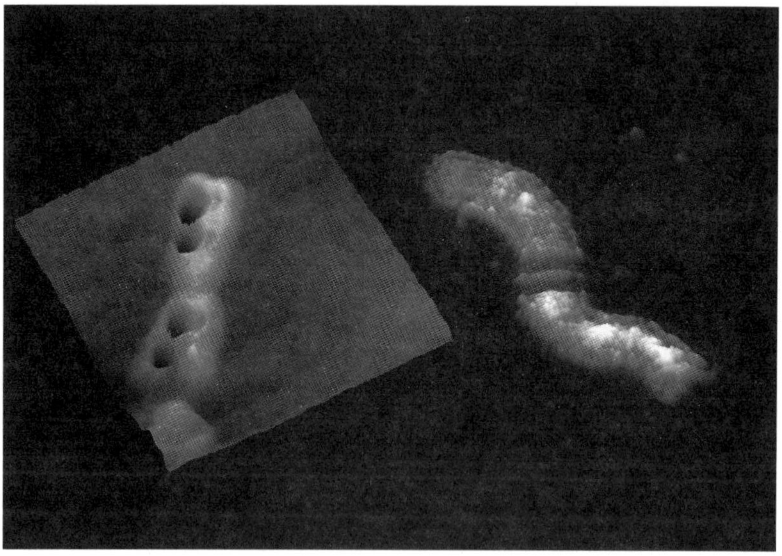

FIGURE 7 See ftp site for color figure. Nanoprocessing of human chromosomes with 80 MHz NIR femtosecond laser pulses at 800 nm.

other cellular compartments. Indeed, we were able to cut chromosomes within a living PtK cell without loss of vitality. The cells completed cell division after laser surgery (König, 2000).

8 CONCLUSIONS

Multiphoton microscopes can be used in live-cell fluorescence microscopy at moderate laser powers without damage to the target. Repeated scanning of the biological object of investigation over extended time periods is possible without impact on reproductive behavior and without lethal effects.

Above certain intensity thresholds, which depend on the laser wavelength and the pulsewidth, apoptosis-like gradual cell damage by two-photon photochemistry may occur. When studying pigmented cells and tissues such as erythrocytes or skin with a high amount of melanin, photothermal damage has to be considered. Immediate damage occurs at high light intensities (e.g., in the terawatt per square centimeter range) by optical breakdown phenomena. Plasma-induced ablation can be used for intracellular nanosurgery.

ACKNOWLEDGMENTS

We should like to thank I. Riemann (University Jena) for her help in various ways and H. Oehring (University Jena) for his assistance with electron microscopy. We are grateful to Peter So (MIT) and E. Gratton (University of Illinois), in whose laboratories the first studies of damage by femtosecond laser pulses were performed. This work was supported in part by grants from the German Science Foundation (DFG), the Federal Ministry of Education, Science, Research, and Technology (BMBF), the Ministry of Science, Research, and Culture of the State of Thuringia, Zeiss Jena, Coherent, and JenLab GmbH (Jena).

REFERENCES

Bedner, E., Smolewski, P., Amstad, P., and Darzynkiewicz, Z. (2000). Activation of caspases measured in situ by binding of fluorochrome-labeled inhibitors of caspases (FLICA): correlation with DNA fragmentation. *Exp. Cell Res*. **259**, 308–313.

Berns, M.W. (1998). Laser scissors and tweezers. *Scientific American* **278**, 62–67.

Bertling, C.J., Lin, F., and Girotti, A.W. (1996). Role of hydrogen peroxide in the cytotoxic effects of UVA/B radiation on mammalian cells. *Photochem. Photbiol*. **64**, 137–142.

Cunningham, M.L., Johnson, J.S., Giovanazzi, S.M., and Peak, M.J. (1985). Photosensitised production of superoxide anion by monochromatic (290–405 nm) ultraviolet irradiation of NADH and NADPH coenzymes. *Photochem. Photobiol*. **42**, 125–128.

Denk, W., Strickler, J.H., and Webb, W.W. (1990). Two-photon laser scanning microscope. *Science* **248**, 73–76.

Diaspro, A., and Robello, M. (1999). Two-photon excitation of fluorescence in three-dimensional microscopy. *Eur. J. Histochem*. **43**, 169–178.

Doiron, D.R., and Gomer, C.J. (1984). *Porphyrin Localisation and Treatment of Tumours*. New York: Alan R. Liss.

Fisher, W.G., Partridge, W.P., Dees, C., and Wachter, E.A. (1997). Simultaneous two-photon activation of type I photodynamic therapy agents. *Photochem. Photobiol.* **66**, 141–155.

Gavrieli, Y., Sherman, Y., and Ben-Sasson, S.A. (1992). Identification of programmed cell death in situ via specific labelling of nuclear DNA fragmentation. *J. Cell Biol.* **119**, 493–501.

Greulich, K.O. (1999). *Micromanipulation by Light in Biology and Medicine*. Basel: Birkhäuser.

Hänninen, P.E., Soini, E., and Hell, S.W. (1994). Continuous wave excitation two-photon fluorescence microscopy. *J. Microsc.* **176**, 222–225.

Hockberger, P.E., Skimina, T.A., Centonze, V.E., Lavin, C., Chu, S., Dadras, S., Reddy, J.K., and White, J.G. (1999). Activation of flavin-containing oxidases underlies light-induced production of H_2O_2 in mammalian cells. *Proc. Natl. Acad. Sci. USA* **96**, 6255–6260.

König, K. (2000). Invited review: Multiphoton microscopy in life sciences. *J. Microsc.* **200**, 83–104.

König, K., Liang, H., Berns, M.W., and Tromberg, B.J. (1995). Cell damage by near-IR microbeams. *Nature* **377**, 20–21.

König, K., Liang, H., Berns, M.W., and Tromberg, B.J. (1996a). Cell damage in near infrared multimode optical traps as a result of multiphoton absorption. *Opt. Lett.* **21**, 1090–1092.

König, K., Krasieva, T., Bauer, E., Fiedler, U., Berns, M.W., Tromberg, B.J., and Greulich, K.O. (1996b). Cell damage by UVA radiation of a mercury microscopy lamp probed by autofluorescence modifications, cloning assay, and comet assay. *J. Biomed. Opt.* **1**, 217–222.

König, K., Simon, U., and Halbhuber, K.J. (1996c). 3D resolved two-photon fluorescence microscopy of living cells using a modified confocal laser scanning microscope. *Cell. Mol. Biol.* **42**, 1181–1194.

König, K., So, P.T.C., Mantulin, W.W., Tromberg, B.J., and Gratton, E. (1996d). Two-photon excited lifetime imaging of autofluorescence in cells during UVA and NIR photostress. *J. Microsc.* **183**, 197–204.

König, K., Liang, H., Kimel, S., Svaasand, L.O., Tromberg, B.J., Krasieva, T., Berns, M.W., Halbhuber, K.J., So, P.T.C., Mantulin, W.W., and Gratton, E. (1997a). Cell damage in UVA and cw/femtosecond NIR microscopes. *Proc. SPIE.* **2983**, 37–44.

König, K., So, P., Mantulin, W.W., Tromberg, B.J., and Gratton, E. (1997b). Cellular response to near infrared femtosecond laser pulses in two-photon microscopes. *Opt. Lett.* **22**, 135.

König, K., Riemann, I., and Fischer, P. (1999a). Photodynamic therapy by non-resonant two-photon excitation. *Proc. SPIE.* **3592**, 43–49.

König, K., Becker, T.W., Fischer, P., Riemann, I., and Halbhuber, K.J. (1999b). Pulse-length dependence of cellular response to intense near-infrared laser pulses in multiphoton microscopes. *Opt. Lett.* **24**, 113–115.

König, K., Riemann, I., Fischer, P., and Halbhuber, K.J. (1999c). Intracellular nanosurgery with near infrared femtosecond laser pulses. *Cell. Mol. Biol.* **45**, 195–201.

König, K. Riemann, I., and Fritzsche, W. (2001). Nanodissection of human chromosomes with near-infrared femtosecond laser pulses. *Opt. Lett.* **26**, 819–821.

Koester, H.J., Baur, D., Uhl, R., and Hell, S.W. (1999). Ca^{2+} fluorescence imaging with pico- and femtosecond two-photon excitation: signal and photodamage. *Biophys. J.* **77**, 2226–2236.

Liu, Y., Cheng, D., Sonek, G.J., Berns, M.W., Chapman, C.F., and Tromberg, B.J. (1995). Evidence for localised cell heating induced by near infrared optical tweezers. *Biophys. J.* **68**, 2137–2144.

Maiti, S., Shear, J.B., Williams, R.M., Zipfel, W.R., and Webb, W.W. (1997). Measuring serotonin distribution in live cells with three-photon excitation. *Science* **275**, 530–532.

Moor, A.C.E. (2000). Signaling pathways in cell death and survival after photodynamic therapy. *J. Photochem. Photobiol.* **57**, 1–13.

Negri, C., Donzelli, M., Bernardi, R., Rossi, L., Bürkle, A., and Scovassi, I.A. (1997). Multiparametric staining to identify apoptotic human cells. *Exp. Cell. Res.* **234**, 174–177.

Oehring, H., Riemann, I., Fischer, P., Halbhuber, K.J., and König, K. (2000). Ultrastructure and reproduction behaviour of single CHO-K1 cells exposed to near infrared femtosecond laser pulses. *Scanning* **22**, 263–270.

Schönle, A., and Hell, S. (1998). Heating by absorption in the focus of an objective lens. *Opt. Lett.* **23**, 325–327.

Shafirovich, V., Dourandin, A., Luneva, N.P., Singh, C., Kirigin, F., and Geacintov, N.E. (1999). Multiphoton near-infrared femtosecond laser pulse-induce DNA damage with and without the photosensitizer proflavine. *Photochem. Photobiol.* **68**, 265–274.

Squirrel, J.M., Wokosin, D.L., White, J.G., and Bavister, B.D. (1999). Long-term two-photon fluorescence imaging of mammalian embryos without compromising viability. *Nature Biotechnol.* **17**, 763–767.

Tirlapur, U.K., König, K., Peuckert, C., Krieg, R., and Halbhuber, K.J. (2001). Femtosecond near-infrared laser pulses elicit generation of reactive oxygen species in mammalian cells leading to apoptosis-like death. *Exp. Cell. Res.* **263**, 88–97.

Tyrell, R.M., and Keyse, S.M. (1990): The interaction of UVA radiation with cultured cells. *J. Photochem. Photobiol.* **4**, 349–361.

Vogel, A., Nahen, K., Theisen, D., Birngruber, R., Thomas, R.J., and Rockwell, B.A. (1999). *Appl. Phys.* **B68**, 271.

Xu, C., Guild, J., Webb, W.W., and Denk, W. (1995). Determination of absolute two-photon excitation cross sections by in situ second-order autocorrelation. *Opt. Lett.* **20**, 2372–2374.

CHAPTER 10

Practical Multiphoton Microscopy

JOHN M. GIRKIN[1] and DAVID L. WOKOSIN[2]

[1]Institute of Photonics, University of Strathclyde, Wolfson Centre, 106 Rottenrow, Glasgow G4 0NW, U.K.

[2]Centre for Biophotonics, Strathclyde Institute for Biomedical Sciences, University of Strathclyde, 27 Taylor Street, Glasgow G4 0NR, U.K.

1 INTRODUCTION

Since the first use of multiphoton excitation for fluorescence microscopy (Denk et al., 1990) the technique has gained acceptance as the preferred method of imaging biological samples at depth and with minimal disturbance of the sample. In the intervening years since this first report of two-photon imaging, significant theoretical and practical work has been undertaken to maximize these benefits, primarily directed towards minimally invasive imaging of intact and living tissue (Denk, 1996; Denk and Svoboda, 1997). In this chapter, we aim to present the most important considerations that have emerged for successful multiphoton imaging of biological preparations. The key requirement to ensure long-term viability of the sample is to reduce the excitation power onto the sample, thus reducing unwanted phototoxic and thermal damage and minimizing photobleaching effects within the sample.

Ever since the invention of the original microscope over 400 years ago, biological researchers have found microscopes a very valuable asset to their work. This has lead to the desire to obtain ever-higher-quality images, and more recently to develop three-dimensional imaging without having to physically dissect the sample (Brakenhoff et al., 1985; Carlson et al., 1985; White et al., 1987). The confocal microscope was developed for exactly this purpose, and if one wishes to image relatively thin fixed samples, then it is still the instrument of choice. Confocal systems have developed rapidly over the past 15 years, and the development of the Yokogawa spinning-disc instrument means that high-speed confocal imaging is now possible. This confocal system uses a multiplicity of microlenses to scan the sample with the excitation beam and then image the fluorescence back through an array of confocal apertures. More conventional confocal systems utilize a single-point-scanning system to obtain the image. This produces a more flexible optical arrangement,

Confocal and Two-Photon Microscopy: Foundations, Applications, and Advances, Edited by Alberto Diaspro.
ISBN 0-471-40920-0

but with limitations on speed. The point-scanned microscope does, however, have the potential for higher signal-to-background discrimination, giving the instrument the ability to image deeper into the sample than the multiple-point arrangement. Multiphoton microscopy then enhances this signal-to-background ratio further, since the excitation volume is inherently localized in space. This means that the excitation and emission optical systems can actually be decoupled, giving the system greater flexibility and the ability to image even deeper.

For these reasons, the majority of multiphoton microscopes utilize a point-scanning system in which the light source forms a planar image inside the sample. This plane is defined by the nonlinear excitation of the fluorescence. In all biologically practical systems, an ultrashort-pulse laser is used, where femtosecond pulses are most commonly applied for dynamic fluorescence imaging via two-photon and/or three-photon excitation. The pulsed laser is critical to providing the localized excitation volume and hence the inherent three-dimensional resolution (Stelzer et al., 1994; Williams et al., 1994). Frequently, however, the new laser is the focus of far too much attention. If an existing laser scanning confocal system is converted for multiphoton applications by only switching the laser source, then the resulting system is likely to be worse than the original confocal microscope. The longer wavelength used for the excitation may lead to a reduction in lateral resolution, which will not be compensated for by the normal advantages of multiphoton microscopy – namely, in-depth imaging with good contrast and minimal sample disturbance. In addition, axial longitudinal chromatic aberration may further degrade the image by reducing the fluorescent light collection.

It is our view that the emission-collection optimization should be the paramount consideration. The primary mandate for all fluorescence microscopy is to collect every excited photon, and then to actually detect as many of the collected photons as possible. In multiphoton excitation, compared with confocal imaging, nearly every fluorescent photon is a "useful" photon (can be used to build up the image), since it can only have originated from the optical plane under investigation, making the total light collection very important. (In practice, there is sometimes some long-wavelength, one-photon-generated fluorescence, but this can easily be removed using the correct blocking filters.) As a demonstration that this emission collection is important, the majority of the key niche applications of two-photon excitation microscopy (applications not possible with laser scanning confocal imaging) have required enhanced emission collection to enable the visualization (Yuste and Denk, 1995; Bennett et al., 1996; Svoboda et al., 1997; Squirrell et al., 1999). Multiphoton microscopy is no different from more conventional microscopy in that the objective lens chosen is critical to the overall performance of the system. This lens should be very carefully matched to the biological problem, taking into account the resolution actually required, the depth of imaging, and the exact status of the biological preparation.

In the following, we shall prioritize the practical aspects of obtaining the most sensitive multiphoton systems for both deep and live-cell imaging, and discuss how each factor can be optimized. The overall aim is to collect and detect the fluorescent light as efficiently as possible, which, along with correctly selected optical elements, reduces the requirements on the laser source and decreases the risk of significant damage to the sample. We shall also discuss the current limitations to obtaining the maximal multiphoton excitation signals from longer-wavelength (>850 nm) near-infrared excitation.

2 EMISSION COLLECTION: POTENTIAL SIGNAL

The single most important aspect of fluorescence microscopy is the collection of every excited photon possible (Pawley, 1995). This axiom is true for fixed cells (where photobleaching is the limited factor), as well as live cells, where ultimately phototoxicity imposes the limits on light microscopy. Point-scanning laser excitation systems often have short pixel dwell times (microseconds), which necessitates very high source intensities for sufficient signal-to-noise imaging. These high intensities have a correspondingly high risk of fluorophore bleaching and saturation potential. Every emission photon possible should be included in the final image in order to maximize the signal-to-noise and signal-to-toxicity balance. This is a two-stage process involving the collection of the light and then the conversion of the light into electronic signals that can be stored and built into an image. Once the best-quality image possible has been obtained, sophisticated mathematical algorithms can be applied to enhance the features of interest to the biological researcher.

2.1 The Role of the Numerical Aperture

The maximum emission-light collection depends critically on the cone of light accepted into the objective lens. For point-scanning (condensing) systems, this light cone is related to the numerical aperture (NA) of the lens and the refractive index n of the immersion medium (oil, air, or water). The numerical aperture is given by the formula

$$\mathrm{NA} = n \sin\theta,$$

and

$$\sin\theta = \frac{y}{l},$$

where y is the radius of the collection area and l is the distance from sample to lens. Therefore, $y = l\sin\theta = l\,\mathrm{NA}/n$, and since the collection area is given by πy^2, the light collected varies as $(\mathrm{NA}/n)^2$.

Figure 1 shows the light-collection performance of a range of lenses compared with a 1.4 NA 60× oil lens. The 1.4 NA 60× oil and 1.2 NA 60× water lenses have the best collection potential, the 1.3 NA oil lenses can collect 85% compared to the best two lenses, while the 0.75 NA air lenses and 1.0 NA water-dipping lens only collect 66% as much light.

As is usual in science, a compromise must be reached, since the higher-NA lenses gain light-collection power at expense of a loss of working distance, as shown on the right-hand axis of Figure 1. Typical working distances for oil-immersion lenses with NA = 1.4 are around 200 μm. This makes them unsuitable for really deep imaging of intact tissue, but excellent for maintaining live cells within thin or single cell preparations. For long working distances, one needs to move to either the water-dipping lenses or to high-NA (around 0.75) air lenses, depending on sample preparation. As a general rule, one selects a lens with as large a numerical aperture as possible giving the working distance required for the preparation under investigation. The 1.4 NA lens should be the first choice for confocal

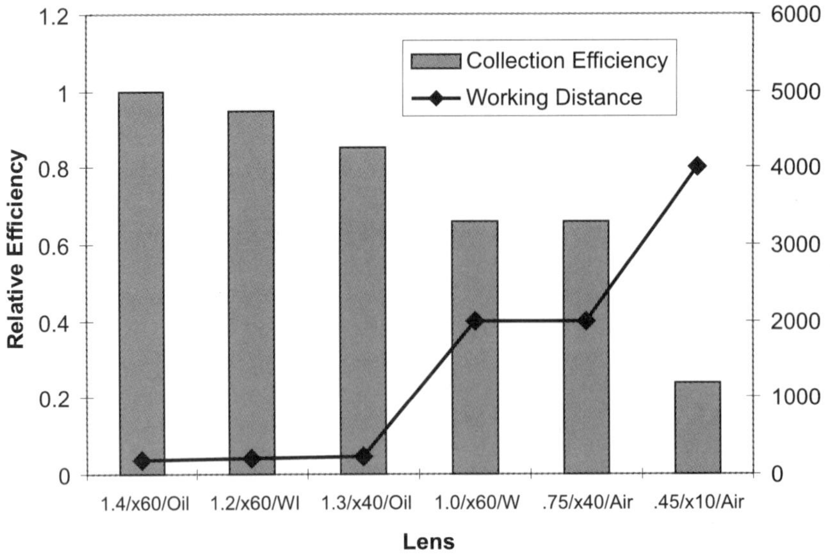

FIGURE 1 Objective-lens collection efficiency and working distance.

imaging, and provide a sensible starting point for multiphoton lens selection. However, the overriding reality is to match the lens to the biological sample under investigation.

The numerical aperture of a lens also determines the ultimate resolution of the optical system. As will be discussed later, in actual biological preparations, the limitation on actually seeing features within a sample is normally compromised by other factors, such as scattering, leading to a loss of contrast in the image. The resolution of the objective can be approximated by

$$\text{lateral resolution} = \frac{0.61\lambda}{\text{NA}}, \qquad \text{axial resolution} \propto \frac{n\lambda}{(\text{NA})^2}.$$

It should be noted at this point that the wavelength λ also appears in these formulas; it is the excitation wavelength, since this determines the size of the excitation volume within the sample (full details may be found in Wilson and Shepherd, 1984). Owing to the use of longer wavelengths for multiphoton excitation, the result is that the theoretical resolution of a multiphoton imaging system can be lower than that of a comparable confocal microscope. However, for deep sectioning, light scattering within a sample also compromises the confocal performance. These factors all influence whether a particular sample should be imaged with a confocal, multiphoton, or even wide-field camera imaging system. The highest priced option, the multiphoton microscope, should not automatically be assumed to be the best system for every biological imaging challenge.

In addition, the numerical aperture of the lens also has a significant contribution in determining the signal-to-background and signal-to-noise ratios. These factors play a large role in determining whether a particular sample can be successfully imaged, and in particular on the maximum depth that can be visualized (Sandison et al., 1995).

2.2 Magnification and Scanning Performance

The real aim in fluorescence microscopy is simply to visualize fine structures (often 1 μm or less) within samples via light-intensity contrast. This visualization process has three components: a probe, a sample interaction, and detection. In this chapter, we are interested in laser light probes exciting fluorochromes, which in turn emit light to be collected then detected. In epi-configuration confocal systems, the objective lens both forms the diffraction-limited spot inside the sample and also collects and re-images the fluorescence onto a parfocal aperture. In epi-configuration multiphoton systems, the emission does not need to be re-imaged, since the localization of the excitation volume is sufficient to determine the three-dimensional (3D) resolution. The challenge with 3D datasets in a point-scanned system is therefore to match the computer-controlled pixel volume (resel) to the excitation volume (voxel) in the sample (Pawley, 1995). When the computer acquisition is properly matched to the diffraction-limited volume, image resolution is optimized, and the maximum spatial information can be obtained from the excited fluorescence.

In the excitation plane, the acquisition match is a two-step process: first, the diffraction-limited spot (the excitation area in focal plane) should be half of the size of the features of interest in order to be able to resolve these features. Then, the displayed image pixels should again be half the size of the diffraction-limited spot in order to accurately describe the lens spot. In a 3D dataset, the focus will then be changed to acquire additional planes; the separation of these planes should also be half of the axial resolution to avoid missing information from the sectioned sample. The important parameters for this matching process are as follows:

$$\text{lateral (in-plane) resolution} = \frac{0.61\lambda}{\text{NA}},$$

$$\text{axial (focal-plane) resolution} \propto \frac{\lambda n}{(\text{NA})^2},$$

$$\text{maximum view through system} = \frac{\text{eyepiece field of view} \times 0.9}{\sqrt{2}}$$

$$= 15.9\ \text{mm for a 25 mm eyepiece}$$

$$\text{lens field of view (FOV)} = \frac{\text{maximum view}}{\text{lens magnification}},$$

$$\text{pixel size on screen} = \frac{\text{FOV}}{\text{Zoom} \times \text{number of pixels on screen}}.$$

The final goal is therefore to set the electronic zoom to produce (square) pixels with sides half the size of the lateral resolution and to set the focus-motor step size to half of the axial resolution. Note that the axial resolution is often three times worse than the lateral resolution – this can be a significant problem when trying to match small spherical features inside samples. Subsequent image restoration or deconvolution algorithms can greatly enhance the morphology of 3D rendered samples but these algorithms require accurate acquisitions to ensure that no sample data is missed.

2.3 Summary of Lens Selection for High Collection Efficiency

In order to maximize the collection efficiency, one should select a lens with the highest numerical aperture possible to allow for imaging into the required depth of the sample. A guideline on lens selection is given in Table 1 which provides a starting point for the user to optimize the system for a particular biological problem. It will be shown later that this is

TABLE 1 Data for Nikon CFI-60 objective-lens series

Samples on coverslips (#1.5 : 160–180 μm)

Working distance	Lens[a]	NA	Coll. (%)	BA dia. (mm)	Lat. res[b] (μm)	Nyq. zoom[b]	Cam. pixel[b] (μm)	Relative cost	Notes
<10 μm	60× PA, oil	1.4	46	9.3	0.25	3.97	7.39	$$$$	(1)
<220 μm[c]	60× PA, water	1.2	45	8	0.29	3.41	8.63	$$$$$$$$$	(2)
<220 μm	**40× SF, oil**	**1.3**	**43**	**13**	**0.27**	**5.54**	**5.31**	**$$$**	**(3)**
<170 μm	100× plan, oil[d]	1.25	41	5	0.28	2.13	13.8	$	—
<1 mm[e]	**20× PA[d]**	**0.75**	**34.5**	**15**	**0.46**	**6.4**	**4.6**	**$$**	**(4)**
<2 mm[e]	20 PF[d]	0.5	23	10	0.69	4.26	6.9	$	—
<4 mm[e]	10× PA[d]	0.45	20.5	18	0.77	7.68	3.83	$$	—
<16 mm[e]	10× PF[d]	0.3	14	12	1.15	5.12	5.75	$	—

Samples without coverslips (upright microscope only)

Working distance	Lens	NA	Coll. (%)	BA dia. (mm)	Lat. res (μm)	Nyq. zoom	Cam. pixel (μm)	Relative cost	Notes
<2 mm[f]	**60 × fluor**	**1.0**	**37.5**	**6.7**	**0.35**	**2.84**	**10.35**	**$$$**	**(5)**
<0.48 mm[e]	40 × plan NCG	0.65	30	6.5	0.53	2.77	10.62	$	—
<10 mm[e]	10 × plan[d]	0.25	11.5	10	1.38	4.26	6.9	$	—

Abbreviations: NA, numerical aperture of objective; Coll., lens light-collection efficiency, assuming isotropic emission; BA dia., diameter of back aperture, or entrance pupil of lens; Lat. res., lateral resolution; Nyq. zoom, zoom setting on Bio-Rad 1024 MP for accurate Nyquist sampling; Cam. pixel, required camera pixel size for accurate Nyquist sampling.

[a]Plan, field flat to 25 mm, parfocal for 2 wavelengths in the visible; PF, plan fluor, as plan, plus designed for high visible-light transmission; PA plan apochromat, as plan, plus parfocal for 3 wavelengths in the entire visible spectrum; SF, super fluor, designed for maximum UV transmission down to 340 nm and visible; Fluor, as SF, designed for UV and IR transmission.

[b]Derived with 800 nm illumination via two-photon excitation.

[c]Water-based sample.

[d]These lenses satisfy the paraxial approximation.

[e]Noncontact.

[f]Water-based sample; water-dipping lens.

Notes: Every objective lens listed can be used on upright microscopes. The lenses indicated in **bold** are those that we use often in each category: oil-, water-, and air-immersion. The practical sample depth limit for multiphoton imaging is approximately 500 μm.

(1) The best confocal lens; all PA lenses are designed for confocal imaging.
(2) Pre-sort cover slips to select those with 170 μm.
(3) Best near-infrared transmission spectrum; best general multiphoton lens.
(4) Very high pixel brightness; best working distance for good light collection with air lens.
(5) Best NA for water-dipping lenses; good near-infrared transmission spectrum.

an oversimplification, since it ignores the infrared light transmission through the lenses and aberrations present, but the approximation provides a very practical starting point for lens selection in a multiphoton microscope.

3 FOCAL-PLANE MISMATCH: CHROMATIC ABERRATION

Having selected the most suitable lens for the task in hand, one now needs to consider the limitations that the selection may bring. Multiphoton microscopy is now forcing microscope objectives to be used with a very wide range of wavelengths passing through their optical elements. One of the results of this is that one of the perennial problems for objective-lens designers, namely chromatic aberration, is reappearing. Modern objectives from all of the major microscope manufactures are excellently corrected for chromatic aberration in the visible portion of the spectrum (400–700 nm); however, until recently, there was no requirement to extend this further into the infrared. Conventional infrared lenses are corrected, but here the wavelength range is 700–1100 nm. Multiphoton imaging utilizes light from 700 nm to 1050 nm for excitation and from 350 nm to 700 nm for fluorescence detection, and Figure 2 illustrates the effect produced by chromatic aberration in a multiphoton microscope.

The illumination light (infrared) is focused into a plane, from which the fluorescence (visible) is generated. The longitudinal chromatic aberration, however, means that the peak collection point for the optical system is in a different focal plane. This has two effects on the performance of the system. Initially, the lens is not operating at its maximum numerical aperture for light collection, leading to a loss of signal. Secondly, the light that is collected is not correctly imaged back through the optical system. As an illustration, instead of being collimated from the back of the objective, the beam will diverge, which may lead to vignetting of the beam by the optical elements back through the scanning system.

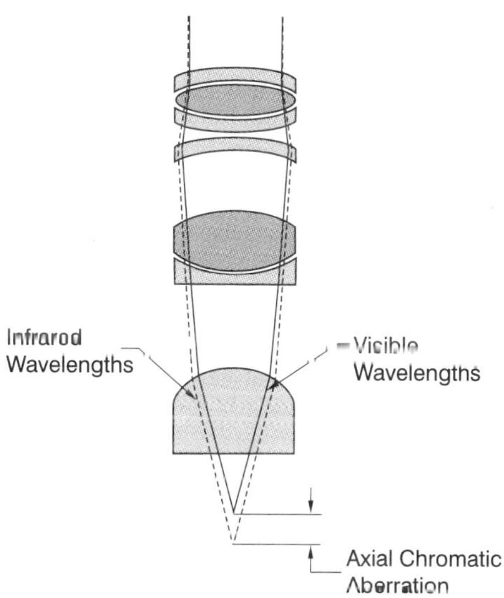

FIGURE 2 Axial chromatic aberration.

This situation is further deteriorated in a converted confocal microscope. Confocal detection (a parfocal pinhole) restricts the axial depth of field of the detector to discriminate or remove background fluorescence. Even when the pinholes are fully open, such systems are still quasi-confocal, or have a limited detection depth of field (Sandison et al., 1995). There are, however several approaches that can be used to overcome these problems without having to force the lens designers to produce ever more expensive and esoteric optical designs.

The aim of all of the approaches to solving this problem is to remove the optical effect of the pinhole as a limiting aperture. The first method is to detect the light as soon as possible after the objective lens. Using the so-called non-descanned detector scheme with multiphoton excitation can significantly reduce the effects of this axial longitudinal chromatic aberration (aLCA) by increasing the detection depth of field (Wokosin et al., 1998a,b). Such detector schemes are discussed in detail later in this chapter.

In a new development by Bio-Rad, the problem is approached in an alternative manner. The Radiance MP system includes a lens designed to increase the effective collecting aperture of the detectors to partially compensate for the chromatic errors and also collect scattered emission light. This "demagnifying" lens is placed after the descanning optics but before the light enters the detectors (or, in this instrument, the fiber coupling to the detector). Such a system provides a lower-cost solution to the problem, since one does not need to purchase extra detectors, and it is not as sensitive to ambient light sources as non-descanned detector configurations.

An alternative approach, using the currently available objectives, is to alter the incoming excitation light. The divergence of the source can be adjusted such that the near-infrared excitation focal plane overlaps with the visible emission. When this divergence matching is used in conjunction with an improved detector scheme, the effect of aLCA can be minimized. The divergence-matching option can, however, be hard to achieve for a tunable Ti : sapphire laser. With this tunable source, the effective focal plane can be different for each objective and each wavelength range. Optimizing the system for one objective and excitation wavelength (with 10–24 nm bandwidth) is one solution, but this is not attractive for a shared instrument, since it may make the situation significantly worse for another user at a different wavelength. Multiphoton systems are now becoming available in which a motorized lens is used to overlap the excitation and emission planes, but to date the performance of these systems in a range of biological preparations has not been evaluated. In general, it is better to improve the efficiency of the detection system before adding the complexity of a lens for divergence matching unless the commercial system already includes this feature.

3.1 Summary of Focal-Plane Mismatch

The focal-plane mismatch between the near-infrared excitation wavelength(s) and the visible emission wavelength(s) is a major concern for converted confocal systems where the emission light is re-imaged (descanned) back through the scan head. The use of correctly placed non-descanned detectors can significantly improve the situation without overly complicating the multiphoton microscope. Unfortunately, such systems are prone to detecting light from throughout the room (with monitors often causing diagonal banding on the image), but the added signal does help to compensate for this drawback. The use of excitation and emission plane matching optics is more complex, and the full benefit has yet to be determined.

4 LIGHT DETECTION: ACTUAL SIGNAL

4.1 Emission Path Efficiency

Having collected as much of the light as possible with the best objective, the aim is now to direct as many of the photons as possible into the light detectors. In a conventional point scanning confocal imaging system, the light has to travel back through the optical path to be descanned before passing through the confocal aperture and reaching the detector. In a typical scanned system (Fig. 3), this can require around six mirrored surfaces and two dichroic filters. Even with 95% reflectivity on all surfaces, one could only detect at best 75% of the emitted light, assuming no other losses through transmissive optics, and in practice this figure is often nearer 50% (Sandison et al., 1995; Wokosin and White, 1997). The critical feature of multiphoton excitation – of the emission only coming from a point – means that the pinhole, and hence descanning, is not required to produce optically sectioned images. In addition, the near-infrared excitation wavelengths are spectrally far removed from the emission in the visible portion of the spectrum, so little compromise has to be made with the dichroic separation filters to prevent the source light from reaching the detectors. A significant reduction in fixed optical losses can be realized through the use of more efficient non-descanned detection (Williams et al., 1994; Denk et al., 1994). A typical system is shown in Figure 4. The emission path for such a detector system requires of the only a dichroic mirror and an excitation-source-blocking filter, which can result in 85% of the light collected by the objective reaching the detector. In addition, one gains in collection efficiency owing to the focal-plane mismatch described above. This higher efficiency may not, however, be realized if the non-descanned detection systems are designed to be far away from the optical microscope and therefore require large and/or numerous optical elements (fibers) and filters. As John White has expressed it, "The best optics are no optics!."

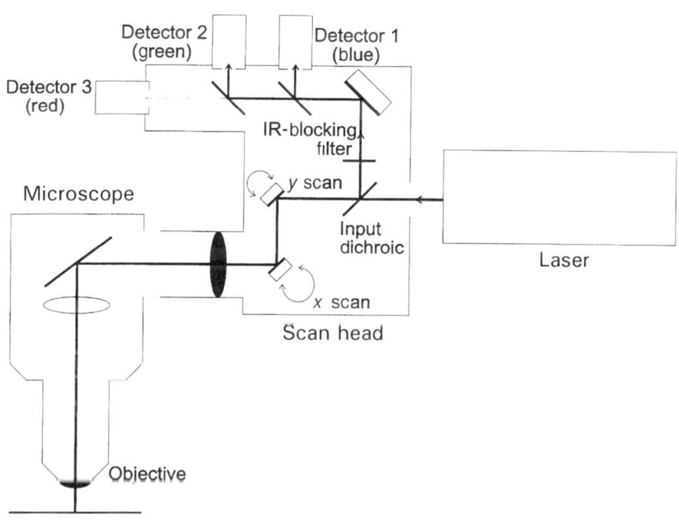

FIGURE 3 Descanned detection.

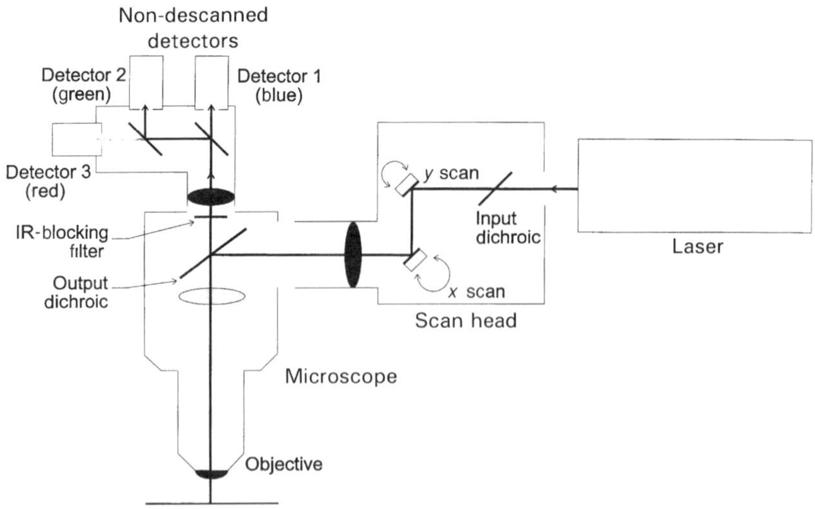

FIGURE 4 Non-descanned detection.

4.2 Multiple Labeling

Multiphoton microscopy of multiply labeled samples produces the same fluorophore emission overlap compromises as in any form of fluorescence microscopy. It has been demonstrated that the fluorophore emission spectrum is independent of the excitation mode (Xu et al., 1996). In fact, multiple labeling with overlapping fluorophores may be easier with conventional laser scanning confocal systems – specifically those using AOTF excitation control. In such systems, sequential imaging, with a range of source wavelengths, can be undertaken easily without changing filters in order to maximize emission collection whilst minimizing fluorophore crosstalk. With their broader spectral output, femtosecond lasers, however, can sometimes excite all the fluorophores present simultaneously. In addition, rapidly changing the wavelength of excitation is not easy to achieve using the current range of femtosecond laser sources. However, good blue/red and green/red fluorophore combinations are available for high-efficiency multiphoton excitation imaging, though the red fluorophores are not well matched to the normally used Ti : sapphire wavelengths (750–920 nm). The combination of Ti : sapphire and Nd : YLF (1047 nm) lasers is, however, well suited to such combinations. An alternative, though as yet not widely used, approach is to select fluorophores where one is excited via a two-photon process and the other via a three-photon route. Figure 5 demonstrates the use of this technique using the Nd : YLF laser as the excitation source and Alexa 568 (two-photon) and DAPI (three-photon) as the fluorophores.

Improvements in optical-filter design and production techniques have, however, led to filters that can separate closely emitting fluorophores (Stanley, 2001). Instrumentation is also under development in which the emitted light is collected and spectroscopically resolved with a resolution of around 10 nm (Zeiss Inc., Personal presentation, 2001). This may enable closely spaced fluorophores to be resolved, but there is a loss of overall detection efficiency, and so this is unlikely to become the technique of choice for most multiply labeled samples.

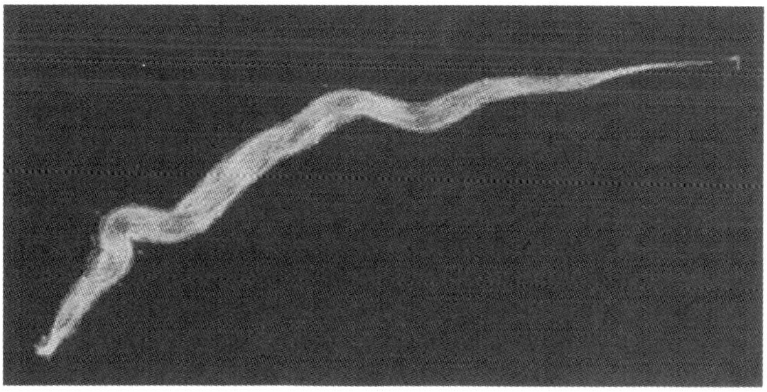

FIGURE 5 See ftp site for color figure. Dual-labelled smooth muscle cell. DAPI three-photon excitation; Alexa 568 two-photon excitation.

4.3 Detector Conversion Efficiency

With the maximum possible quantity of light now entering the detectors, the next problem to overcome in a practical system is to maximize the conversion of photons into an electrical signal. A range of options are available for detection, and full details of detectors are discussed elsewhere (Pawley, 1995; Inoue and Spring, 1997), but a short outline of the considerations for practical systems will be presented here.

In principle, all low-light-level detectors work on the same basis. The incoming photon is converted as efficiently as possible into an electron. The highest possible conversion rate here would be 100%, but this is never achieved, and typical figures range from 1% to 40% for photomultiplier devices and up to 80% for silicon photo detectors. However, these values are strongly affected by the wavelength being detected. This figure is known as the quantum efficiency (QE) of the device (40% conversion being a QE of 0.4). The electrons produced by the photocathode are then directed into a section of the device in which they are multiplied to give the overall gain of the device. For a typical photomultiplier, the gain is around 10^6. This signal is them amplified by conventional electronics before being converted to a digital signal in the system computer.

For very low light levels, the signal from the photomultiplier can be analyzed using the detectors and electronics running in a photon-counting mode. Here, the output from the photomultiplier is amplified, and only signal voltages above a threshold figure are detected by the computer. At each pixel within the scan, the number of counts above this threshold is used for the image brightness. Such a method has advantages with very low signal levels, but repeated scans are required to build up the image. Photon counting is therefore not applicable to rapid imaging, but is for use at low light levels, with the imaging being built up from almost zero background. In order to obtain a high-quality image using this technique, one often requires between 10 and 20 scans. Currently only Bio-Rad offer photon counting on commercial confocal or multiphoton systems.

The detector cathode material and its quantum efficiency are very important in optimizing the actual signal generated from the incoming, collected light. The spectral responses of a range of detectors are given in Figure 6, and it can be seen that bi-alkali cathodes are much more effective for blue and green (DAPI/FITC) imaging than the broad-spectrum S20 (multi-alkali) cathode photomultiplier tubes. In addition, bi-alkali cathode have significantly lower sensitivity beyond 700 nm, and hence require orders of

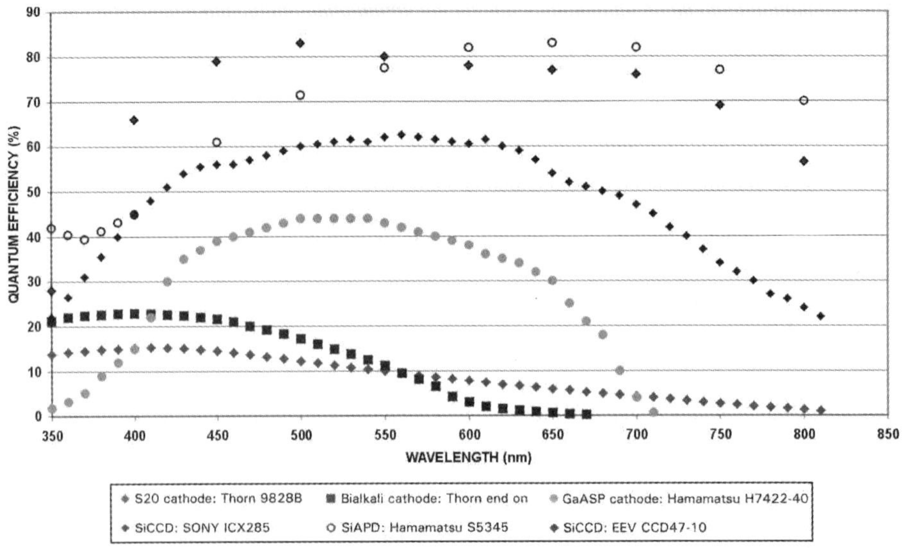

FIGURE 6 Spectral responses of selected photodetectors.

magnitude less source blocking than S20 cathodes. Most commercial scan heads are fitted with S20 cathodes, since visible confocal laser systems are set up for the imaging of triply labeled (green/red/far-red) samples. The current multiphoton excitation wavelengths are best suited for blue/green/red imaging – hence the requirement for a different photocathode. With the advent of new fluorescent compounds (in particular, the yellow and red fluorescent proteins), there may be good reason for equipping at least one channel with a longer-wavelength S20 photocathode, even for multiphoton imaging.

An alternative approach to photomultiplier detector is through the use of an avalanche photodiode. In principle, these devices operate in similar manner, with the conversion of photons to electrons, but the electrons are then accelerated very rapidly, causing a rapid multiplication as they rush towards the collector. This avalanche means that the detector is very sensitive, but does not normally have the dynamic range of a more conventional photomultiplier. The spectral responses of such devices can be similar to those of photomultipliers.

A new hybrid device is currently being developed by Hamamatsu. This consists of a GaAsP cathode photomultiplier with a hybrid avalanche photodiode detector attached behind the photomultiplier section (alternatively, the GaAsP cathode photomultiplier signal is available without the avalanche photodiode). These devices are very promising with 40% QE figures through the visible and even less source blocking required than for the bi-alkali cathodes (see the Hamamatsu 1999 datasheet for GaAsP cathode PMT and hybrid detectors). However, the cathodes on these devices are very sensitive to current (signal) overload, which is caused by high light intensities on the device. Large signals can permanently damage the detector, although now a limiting circuit is available for these devices. Such detectors cannot, therefore, be effectively used where large signal intensity changes are expected. Currently, these devices are being used for very low-light-level work, in particular two-photon fluctuation correlation spectroscopy (FCS). The hybrid device has also been used for multiphoton imaging, where it is switched into the optical path as the emission detector after focusing deeply into the samples (W. Zipfel, Personal communication January 2001; see also the Cornell website http://www.englib.cornell.edu/drbio/).

In addition to the detectors listed above, newer semiconductor-based detectors (CCDs) are now improving to the point where they can be considered for quantified imaging of bright samples. These devices are normally built into electronic cameras, and hence are actually imaging detectors rather than point detectors such as photomultipliers; therefore, a different set of noise considerations is required. Such devices do not yet have the same sensitivity and low noise as the detectors described above. In multiphoton systems where the beam is not point-scanned, CCD-based detectors have been used for rapid multiphoton imaging (Straub and Hell, 1997).

4.4 Minimizing Noise

In any system, the final limitation is whether the signal can be detected above the noise present in the system. This ratio can be tipped in favor of the user by taking care with the detectors. They should not be exposed to bright sources of light with a voltage applied, and they should be run with the output saturated for as short a time as possible. Photomultiplier detectors also take a little time to reach thermal equilibrium, and hence it is often worthwhile waiting while the detectors "warm up" for around 30 minutes before starting to take images. As most multiphoton systems are point-scanned, the imaging time at each pixel is short and the background can be assumed to be constant over this period. This means that the dark background and thermal noise of the detectors can be removed using a black offset. For many practical applications, this is not critical, but it can be important when the signal is very low.

4.5 Summary on Light Detection

The aim should always be to direct the light to the detector as soon as possible, with the minimum number of optical components – hence the preference for correctly implemented non-descanned detectors. Multiple labeling is routinely possible with multiphoton excitation, but care should be exercised for the best combinations of fluorophores. The detector should be selected to match the emission wavelength of the fluorophore and dichroic filter to increase the overall collection and detection efficiency. Care should also be taken with detectors to ensure that they are not being affected by stray room lights or the image being displayed on the monitor.

In a multiple-channel detection system, at least one detector should have a short-wavelength bi-alkali detector, and this should be the one to which all light is sent if the system is operating with no wavelength-selective optics (dichroic filters). In the more standard three-channel system, the best combination would seem to be two bi-alkali detectors and one S20 detector.

Whenever possible, the full dynamic range of the detector system should be used, but care must be taken to ensure that the brightest pixels are not too close to saturation.

5 LASER SOURCES

If the system has been optimized as described above, the requirements on the laser source can be significantly reduced for most standard multiphoton imaging applications. In general, there are three important parameters to be considered in the selection of the laser for any imaging problem: wavelength, pulsewidth, and average power. In general, the peak power on the sample should be kept to a minimum in order to reduce potential damage to the sample. Each of these features will be examined in turn, with some indication being given of the starting point for practical imaging.

5.1 Practical Sources

Since the original multiphoton experiments in microscopy, significant advances have been made in the field of ultrashort-pulse lasers. In general, the aim has been to lower the technical skill required by the operator and to make the systems simpler to maintain. The argon-pumped dye lasers that were used in the original work were soon replaced with argon-pumped Ti : sapphire lasers (Curley et al., 1992), and more recently they have been superseded by all-solid-state sources requiring minimum cooling and standard mains electrical supply. A more detailed description of these sources is contained in a review article by Wokosin et al. (1996). Although multiphoton microscopy has been undertaken using a continuous-wave source (Hannienen et al., 1994, 1995), for practical biological imaging, ultrashort pulses are preferred. The aim of the continuous-wave image was an attempt at reducing the source cost, but to date suitable biological applications are limited to DAPI imaging. The output from the ultrashort-pulse laser should then be delivered by directly coupled optics rather than through an optical fiber, in order to minimize pulse-stretching problems.

Suitable sources based around Ti : sapphire or Nd : YLF laser materials can now be described as "turnkey" systems. In order to make the sources easier to operate, there has been some loss of flexibility. The tuning range either has been restricted, in the case of the Ti : sapphire systems, or is not there for the Nd : YLF- based sources. In addition, in the newer Ti : sapphire lasers, the dispersion compensation within the laser cannot be adjusted. These few disadvantages are probably outweighed by the benefit of having a system that just requires switching on rather than hours of tender loving care to obtain high-quality two-photon images.

5.2 Wavelength

The first prerequisite for the center wavelength at which to operate the laser has to be that which will excite the fluorophore being used in the biological preparation. However, it may be preferable not to always tune the laser to the peak of the absorption of the fluorophore. Many fluorophores have a wide absorption band, some with a full-width half-maximum absorption greater than 100 nm, and this can be used to the benefit of the biology being undertaken to reduce the unwanted effects of the imaging process. However, this is not always the case; for example, eGFP has a much smaller absorption profile, and the flexibility in the exciting wavelength is thus reduced. The actual wavelength used for the imaging probably has the largest single effect on the long-term viability of a live sample. Each wavelength and each sample has a certain toxicity threshold that can only be determined by experimental assay, but a general consensus is growing that the longer the wavelength that can be used, the better for the sample. Once the toxic threshold has been exceeded in terms of peak power at a given wavelength, the sample viability will deteriorate rapidly. It has been shown that the toxicity is related to the actual two-photon excitation events in the UVA to blue (350–500 nm) portion of the optical spectrum (König, 2000). The number of excitation events is a complex combination of the sample, wavelength, pulsewidth, pulse shape, peak power, and repetition rate.

Tables of fluorophore cross-sections are now being published, but these should only be used as a guide to the wavelength for the laser; see Xu et al. (1996) and Chapter 4 of this book (Xu, 2002). Carefully controlled experiments should be undertaken to determine the correct operating parameters if the sample viability is important to the experimental result. Table 2 lists some of the commonly used fluorophores and a suitable starting wavelength for excitation. In selecting the wavelength, it should also be borne in mind that, between 850 nm and 950 nm, many microscope objectives have a dip in their optical transmission. A consequence of this is that as the laser wavelength is tuned, the fluorescent image may dim – but this could be due to a fall in power on the sample rather than moving away from the peak of the absorption curve. The excitation wavelength may also be selected such that two or more fluorophores are excited simultaneously, although each is away from its peak excitation.

It should also be remembered that, in effect, multiphoton excitation takes place with a "comb" of wavelengths. For a source at 1050 nm, there are potential effects due to the fundamental wavelength (predominately thermal), two-photon events at an equivalent wavelength of 525 nm (normally fluorescence excitation), and three-photon events at 350 nm (potential for phototoxic UV transitions). For a source operating at 700 nm, the potential for damage through this comb of wavelengths may be worse, since one now has to consider possible effects at 700 nm, 350 nm, and 233 nm (with the potential to significantly damage DNA).

5.3 Pulsewidth

The original concept in multiphoton imaging was to use the shortest pulses possible in order to reduce the average power on the sample. Whilst this is a sensible approach in some situations (where the heating effect of the average power is the predominate abuse to

TABLE 2 Selected fluorophore and excitation wavelengths[a]

Fluorophore	One-photon excitation (nm)	Two-photon excitation (nm)	Peak emission (nm)
DAPI	350	800 (2P) 1047 (3p)	470
EGFP	470	900–920	510
Alexa 488	488	830–850	525
FITC	490	830–850	520
YFP	525	>950, 1047	540
Propidium iodide	536	>900, 1047	623
FM4-64	543	1047	640
Cy3	554	1047	568
Texas red	596	1047	620

[a]These are suggested wavelengths, and may not be the most suitable for every system. Some consideration has been given to the lens transmission and the laser performance, in addition to the absorption profile of the fluorophore.

the sample), it can overcomplicate the optical configuration. The femtosecond pulses normally used for multiphoton imaging have bandwidths in excess of 15 nm, and when these pulses are passed through optical elements (predominately the objective lens in a multiphoton microscope), dispersion takes place, effectively stretching the pulse in time and reducing its peak power and hence potential fluorescence signal; (see Guild et al. (1997) and Chapter 8 of this book (Wolleschensky et al., 2002). Once the pulse has started to be stretched, the effect will continue, since it is due to the spectral bandwidth and not the actual pulse length at that position in the optical train.

The use of optical systems to compensate for pulse lengthening has been tried by several groups (see, e.g., Soeller and Cannell 1996) who wished to have very short pulses on the sample. All of these compensation systems used to date do yield laser-limited pulsewidths after the objective, but with a greatly increased complexity to the imaging system. The dispersion compensation is wavelength-dependent – and so, as the laser is tuned, the optical train has to be adjusted to again minimize the pulse length on the sample. For most biological imaging laboratories, this is adding an unnecessary complication to the system. Such compensation is also required if an optical fiber is used to deliver the light to the scan head. Here, the situation is further complicated, since the fiber also limits the power that can be delivered owing to nonlinear effects within the fiber with such short pulses. The one possible exception to this is with a version of the Nd : YLF laser where the optical fiber is required as part of the process to produce the femtosecond pulses. In this situation, fiber coupling can be an option.

In order to minimize dispersion problems, the trend has been to work with pulses of around 150–200 fs (König 2000). At these pulselengths and spectral bandwidths, the dispersion in the system will stretch a 200 fs pulse to around 220 fs at around 850 nm, only lowering the peak power by 10%. Picosecond pulses have been used for imaging (Bewersdorf and Hell, 1998), which then makes fiber coupling of the laser source to the scan head a viable option. In general, the use of long pulses tends to mean that one reaches the thermal-damage barrier three times sooner than with the more conventional 200 fs pulses. In addition, one currently pays the same price for femtosecond or picosecond laser systems, so one might as well select the shorter pulses. Figure 7 demonstrates the increase in laser power required as the pulsewidth increases. As an approximation, imaging with picosecond pulses requires 30% more average power for the same signal when compared with femtosecond imaging. This extra average power normally limits the depth to which the sample can be imaged, without compromising the viability (Wokosin et al., 1998 a,b).

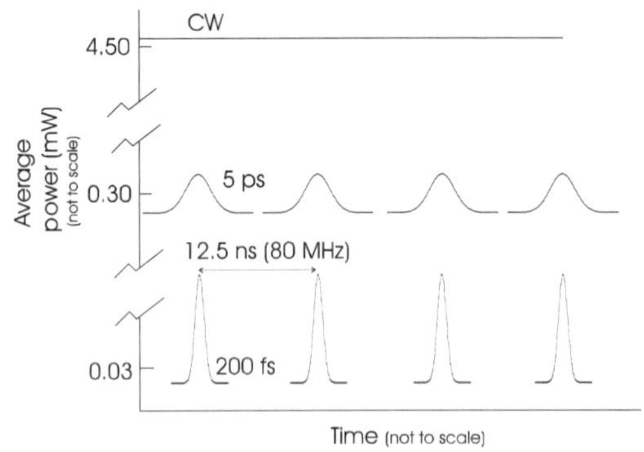

FIGURE 7 Effect of changing pulselength on peak power.

5.4 Average Power

Multiphoton imaging typically requires about 500 times more average power on the sample than conventional confocal laser scanning microscopy in order to achieve a similar number of excitation events within the focal volume. The effect of the average power on the sample (apart from affecting the peak power) is to cause heating within the sample due to linear absorption. This was first noticed in optical-trapping experiments, where the choice of the trapping wavelength can have a significant effect on the sample viability (König et al., 1996). Below 850 nm, heating is not normally a significant problem unless the sample is heavily pigmented. As the wavelength rises beyond 900 nm, single-photon water absorption can become an issue, and may be the predominating factor in limiting the use of multiphoton imaging to live samples.

The spectral width of the absorption curves and the tunability of Ti : sapphire lasers can, however, be used to an advantage in certain situations to aid the problems caused by the imaging on the biological function of the sample. In one experiment (D.W. Piston, personal communication, 2000), the excitation was moved off the absorption peak because of thermal complications. If one just considered the absorption profile, then eGFP would be excited at 950 nm, but the performance of Ti : sapphire lasers is beginning to fall off here. Typically, therefore, eGFP is excited at around 920 nm; however, for this deep-imaging application, when the power has to be increased, linear absorption and heating in water is a problem. By tuning the laser to around 880 nm, the eGFP is still excited (although not as efficiently), but the water heating is reduced by a factor of three, leading to an overall signal increase in this case of 20%.

Typically for femtosecond sources, the power on the sample is under a few milliwatts for most normal samples. This figure is, however, significantly affected by the samples, the excitation wavelength, and the numerical aperture of the objective. As one images deeper into the preparation, it may be necessary to increase the power as scattering events start to degrade the focus of the incoming beam and scatter the emitted light away from the collection optics. This is one of the major features of deep imaging by multiphoton excitation, where, without the linear absorption of the excitation light, the illumination can be increased to maintain the image quality.

5.5 Summary on Laser Parameters

As a general rule, the aim should be to keep the peak power low and the wavelength long. For the majority of practical imaging situations, one need not worry about pulselength compensation, and can select direct beam delivery rather than using optical fibers. Pulselengths of around 150–200 fs provide the best compromise between a high peak power and minimal dispersion loss in the optics.

6 PRACTICAL LIVE CELL SAMPLE IMAGING

With all of the above parameters adjusted, one has established a multiphoton imaging system in which the biologist has the best chance of imaging samples with minimal disturbance to biological function. In all situations, it should, however, be remembered that no imaging process is noninvasive. When shining light onto the sample, there is a probability of affecting the function or process of the sample. Multiphoton imaging in no way removes the need to undertake carefully controlled biological experiments in order to determine which effects are desired and which are induced by the imaging process. In particular, most samples have an exogenous fluorophore added, which may be affecting the systems within the sample. Through the use of optimized light collection and wavelength selection, one has the best chance to minimize the harmful effects of the imaging process.

In live cells, there are several routes by which the imaging process can cause damage to the sample. At the fundamental laser wavelength, there is the potential of one-photon absorption and hence thermal damage. At the two-photon wavelength, there is the risk of indirect phototoxic damage, probably due to fluorophore breakdown, though this will be limited to the focal volume within the sample. Three-photon absorption is also possible, and this has an effective wavelength in the ultraviolet portion of the spectrum, leading to the possibility of direct RNA or DNA damage. This possibility has been demonstrated with the recent development of second- and third-harmonic microscopy, where light is produced at the doubled or tripled fundamental wavelength without the presence of a fluorophore.

One of the early investigations, which examined the effects of multiphoton imaging in thin samples, was undertaken on Chinese Hamster ovarian cells (König et al., 1997). The results here demonstrated that the two photon imaging process can affect cell function and that the multiphoton technique should not, therefore, be viewed as a panacea for all live cell imaging. For thinner cells, the excitation volume in confocal and two-photon imaging may be very much the same (see the Cornell website http://www.englib.cornell.edu/drbio). Therefore, when changing only the laser, the two techniques will require roughly the same number of excitation events for the same detected signal within thin cells. If chromatic-aberration effects are present and included, then the required excitation events at the sample will need to be increased to maintain the same detected signal. As an added concern, with pulsed illumination excitation, the bleaching in the excited volume can also be nonlinear in nature (Patterson and Piston, 2000). In short, the "new" two-photon excitation system may be less sensitive, and more toxic, than the original confocal microscope.

For live-cell imaging, establishing, and then staying below, the magic toxicity or damage threshold is critically important. The aim should always be to maximize the signal-to-toxicity ratio through the use of the guidelines discussed above. For example, the addition of non-descanned emission detection can increase the system sensitivity and therefore reduce the required excitation events for the same detected signal. This has the effect of increasing the buffer between the threshold for high-quality imaging and the damage threshold. By careful optimization of all of the system parameters, quite spectacular long-term imaging is indeed possible (Squirrell et al., 1999). In this work, the use of longer-wavelength sources and appropriate emission-collection enhancements lead to the long-term imaging of developing hamster embryos and the successful re-implantation and subsequent birth of live young. In turn this young was reproductively viable, producing a litter of 13.

7 DEEPER IMAGING

The other major benefit, along with live-cell work, linked to multiphoton imaging is that of being able to image deeper within samples. Without doubt this has been demonstrated in a wide range of samples, but system optimization is also required to maximize this benefit, in particular the use of non-descanned detection schemes. In addition to the enhanced collection options and the use of longer-wavelength sources previously listed, there are several additional factors that should be considered.

7.1 Spherical Aberration

This is discussed in some detail in another chapter of this book, but a short consideration of the problems that it causes for practical imaging is worth considering here. The main goal for excitation in point-scanning fluorescence microscopy is to achieve a diffraction-limited excitation volume (at depth) inside the sample. This goal is compromised when the optical properties of the bulk sample differ from the objective-lens design parameters. Any mismatch between the expected and actual index of refraction or optical properties of the sample can cause serious problems when deeper imaging is attempted. This can be demonstrated simply by using an air objective (with an NA above 0.25) designed to work with a coverslip and comparing the image resolution of a sample with and without a coverslip. Even the 170 μm of glass can significantly affect the quality of the image. The effect is even more noticeable with high-NA lenses, and indeed it has been shown that oil-immersion lenses used with a water medium produce spots that are no longer diffraction-limited (Jacobsen, 1994). This reduces the system resolution, and can significantly decrease system sensitivity (emission detection from the diffraction-limited spot). It has been demonstrated (Centonze and White, 1998) that as one progresses into the sample, the effect of spherical aberration can be seen in the optical sections. Figure 8 shows the spherical intensity variations of vertical sections through a fixed tissue sample imaged with an oil-immersion lens. In order to minimize the effects of spherical aberration in live, and hence water-containing, samples, one should probably use a water-matched lens for imaging beyond 10 μm. In essence, this is for imaging beyond tissue-cultured cell applications.

7.2 Source Scattering

In confocal imaging, the excitation light can be significantly reduced before it reaches the focal volume within the sample owing to absorption outside the focal region. In multi-photon imaging, this is not normally a problem, but the local changes or heterogeneity in the refractive index can actually remove photons from the excitation spot by changing their direction (Fig. 9). This scattering effect is path length-dependent, so deeper imaging is more sensitive to these effects. The effect of this scattering is to remove potential exciting photons from the excitation volume within the sample, and hence lower the energy density. The consequence of this is then observed as an apparent loss of signal from within the sample. The normal solution is to increase the laser power, since the detectors should already be at their optimal setting. This has potential consequences for sample damage, and can be a critical limiting factor in the imaging of deep, intact tissue samples.

The mechanism or mode of the scattering is very sample-dependent, and often a mixture of several different types and is the subject of many books and papers. All of the scattering effects are, however, a function of the excitation wavelength and optical path (focus depth) as the tissue type changes. Generally, the shorter wavelengths are scattered significantly more than the longer wavelengths. This wavelength dependence is often quoted as being exponential and close to a power of 2.5 for biological tissue in the wavelength regions of interest for multiphoton microscopy.

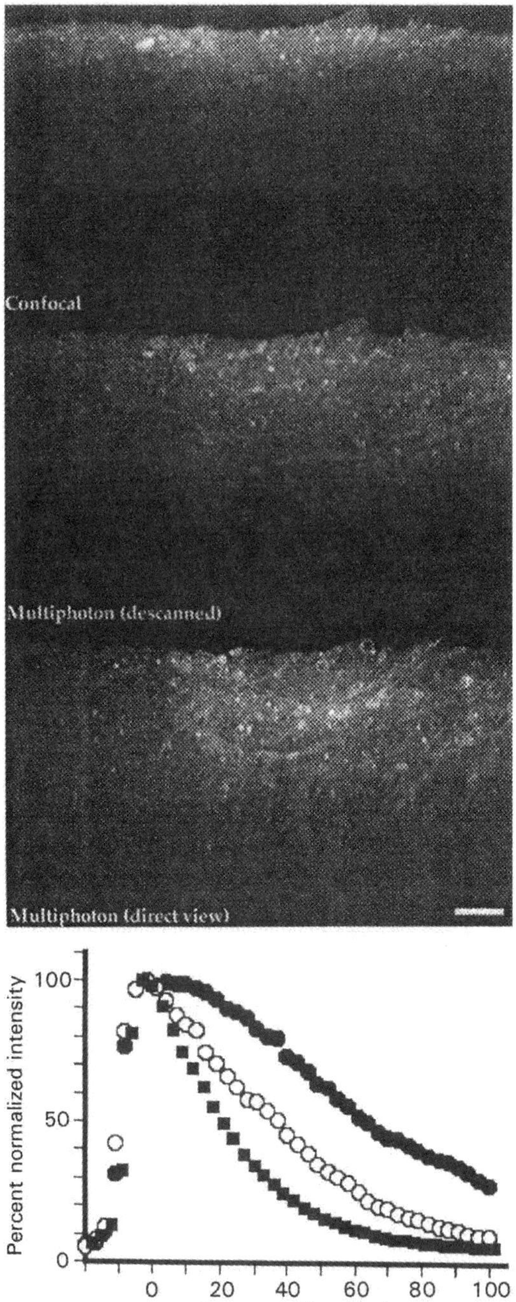

FIGURE 8 Spherical aberration.

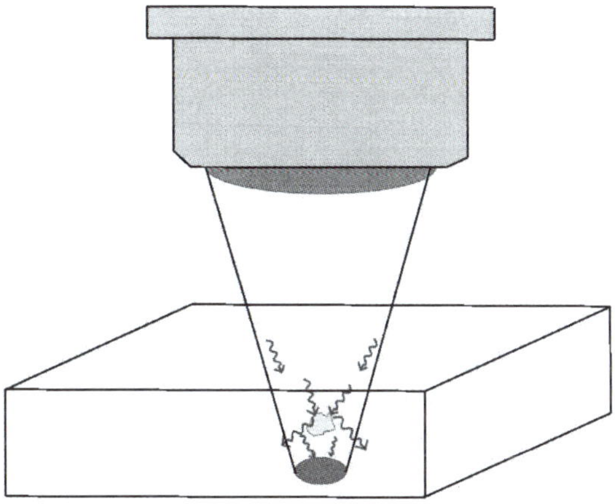

FIGURE 9 Source photons removed by scattering.

For UV excitation systems, source scattering is a real concern and is predominated by Rayleigh scattering with a λ^4 dependence. The scattered UV or blue excitation light will still excite fluorophores after a scattering event, leading to the potential for phototoxic and photobleaching effects away from the focal volume. Added to this, the fluorescence generated by these scattered photons has the potential to add background to the detected signal, which reduces the clarity or contrast of the image.

With multiphoton excitation systems, the source scattering is effectively not an issue, since the isolated photons will not cause bulk excitation, and, as described above, the source average power can be increased to compensate for lost photons. In this situation, emission scattering tends to be the limiting factor.

7.3 Emission Scattering

Once the incoming excitation light has generated the fluorescence, what is critical to the overall performance of the system is the ability to detect this light. As has been described above, this should be the main focus of attention in any multiphoton system. Owing to the localized nature of the excitation volume (and therefore emission volume), even scattered emitted light is useful, since it can only have emanated from the focal plane. In addition to the scattering, the fluorescent light is often emitted in all directions equally, so if the detectors are only looking for light coming back through the sample then one is ignoring in excess of 50% of the possible light. Several schemes have been implemented to try to improve this situation. In the growing technique of second-harmonic imaging, this problem is even worse, since the generated light is colinear with the excitation source.

In one configuration, an oiled microscope illumination condenser is used to collect the fluorescent light traveling out of the sample (Fig. 10). The resulting collimated light is then reflected back through the sample to be collected by the objective lens and presented to the standard detectors on the other side of the microscope objective (Wokosin et al., 1998). This has been shown to increase the signal by up to 70%, depending on the lens. In an alternative configuration, the mirror is replaced with a second detector, and the signal from this is added to that of the normal detector in order to enhance the signal (Svoboda et al., 1997)

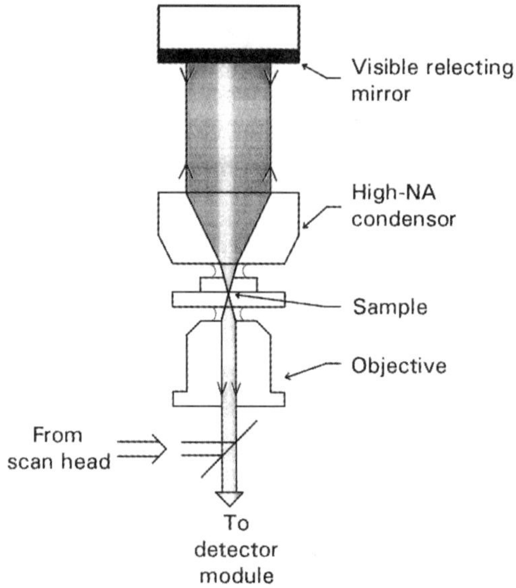

FIGURE 10 Enhanced emission collection using the condenser lens and mirror.

Since 1994, the ability to image deeper into scattering samples has been broadcast as one of the major benefits of multiphoton imaging. Centonze and White (1998) have directly compared confocal and two-photon excitation imaging to determine a main limitation for deeper imaging. The contrast of the detected signal is a key variable, and emission scattering is the main mechanism degrading the image. For confocal imaging, as the focus depth increases, the photons from brighter pixels are scattered away from the pinhole, while background photons from out-of-focus excitation are scattered into the pinhole. Thus, the image contrast is shrinking from both ends as the focus depth increases. In multiphoton excitation systems (since there is no background excitation and the scattered emission photons can be used as signal), the average excitation power can be increased to combat source photon losses and aberrations without affecting the image contrast. Figure 11 illustrates the comparison between confocal and multiphoton imaging deeply within a sample. As a guide, if one wishes to image deeper than 30–40 μm into a scattering sample, then the multiphoton microscope should be the instrument of choice over the confocal system.

7.4 Lens Working Distance

Pragmatically, with multiphoton imaging, we are now faced with the situation where the objective-lens working distance is often the limitation. The best high-NA (1.3–1.4) oil lenses seem to have a reasonable working distance (200–220 μm). Large aberrations become visible if the glass thickness is reduced. The good news is that the working-distance figures (at least for Nikon lenses) are often conservative. This can be seen in the bottom of Figure 3*C* from Centonze and White (1998). Here the Nikon 60× (1.4) PA lens has a quoted working distance of 210 μm (with $1\frac{1}{2}$ coverslip), but the available imaging depth was closer to 100 μm. The next jump up, from the 100 μm to the 300 μm regime, is a quantum leap to 1 mm with air lenses (0.75 NA) or 2 mm with water-dipping lenses

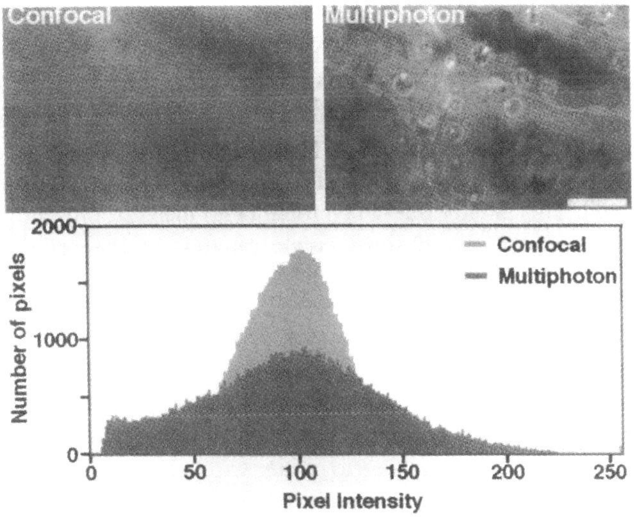

FIGURE 11 Confocal verses multiphoton imaging with depth.

(0.8 NA to 1.0 NA). As has been described above, these lenses have dramatically less emission collection capacity, and one is back to making a compromise between imaging depth and collection efficiency.

8 PRACTICAL LIMITATIONS

8.1 Overfilling Objective Lens

The primary focus of this chapter so far has been on emission collection as it enables live imaging and deep imaging. With the emission collection optimized and the objective lens matched to the sample, the desired image may still not be within easy reach. The next goal is to produce a diffraction-limited spot, at depth, inside the sample with as little loss of excitation photons as possible. The role of scattering and objective aberration in this regard has already been discussed, but it is also possible that the optical system is not using the laser power as effectively as possible. The most important variable in nonlinear excitation microscopy is the average power, the number of photons available to be compressed in space and time. The photons from the pulsed laser (temporal confinement) must actually enter the lens and then traverse through the lens to contribute to the diffraction-limited spot (spatial confinement). In the limit with no power on the sample, you have no image!

Maximizing the system excitation efficiency is particularly important for longer-wavelength imaging (875–1050 nm) where the Ti : sapphire systems have considerably less average power available. In addition, at these longer wavelengths, the diffraction-limited spot size rises as λ; hence, the excited volume rises as λ^3. The result of this is that one has less light spread over a larger volume, leading to fewer excitation events per unit volume. As one images deeper into the sample, therefore, the requirement for some "spare" average power becomes increasingly important to keep the signal levels and contrast high.

The most commonly encountered signal limitation is simply overfilling the objective lens back aperture (entrance pupil). The size of the back aperture can be easily measured or calculated: twice the tube length multiplied by the NA divided by the magnification (Inoue and Spring, 1997). The normal range of objective lenses, even within a single lens type (e.g., plan fluor), can result in a wide variety of aperture sizes. A 100× (1.3 NA) lens will have a dramatically smaller back aperture than the 40× (1.3 NA) oil lens for the same emission collection efficiency. For a laser-beam size matched to the larger 40× oil lens aperture, the difference in area, and hence laser throughput, between the lenses is squared to determine the potential two-photon excitation. With the example quoted above, the potential two-photon signal with the 100× lens is only 6% of the signal with the 40× lens unless some action is undertaken to utilize all the laser power, which is otherwise simply heating up the back of the objective.

Using simple optics (often a single lens for each different objective), the incoming spot size can be changed so that its area can be made to better match the back aperture of each objective. In a multiple-user facility, where a wide range of objective lenses are likely to be used, this is an important and simple step that can be implemented to make much more efficient use of the laser power. As each objective is used, a single lens in the incoming beam path can be changed to match the objective. Taken to the extreme, this may even mean that the green source pumping the Ti : sapphire laser can be of a significantly lower power and hence cost. In summary, some mechanism should be in place to match the laser-beam spot size (and shape) to the circular aperture of the objective lens(es) being used.

8.2 Objective-Lens Near-Infrared Transmission

The situation with the objective-lens transmission is probably not as serious as has been thought over the last few years. Once into the lens, the photons still need to travel through the multiple layers of lens glass and coatings. Most objectives (especially fluor lenses) are designed and optimized for UV/visible wavelengths. These coatings have been shown to have very reasonable extended tails into the near infrared, so the lens transmission values can be above 70% out to 1050 nm. For the Nikon CFi-60 series, studied in detail by the authors, and according to data from Olympus Inc., the major lens transmission concerns occur at or near 900 nm and are due to a single type of glass and a compromise made during the design of the lenses. Unfortunately, the region between 900 nm and 950 nm is ideal for exciting eGFP with two-photon excitation and is also where Ti : sapphire lasers are falling off in their performance. Below 850 nm and near 1064 nm, most of the lens transmission values are quite reasonable. There currently seems to be diminishing returns in investing in better coatings for the new Nikon Eclipse series objective lenses. The likely 10% improvement in longer-wavelength NIR transmission is not currently justified with the available high-peak-power reserve Ti : sapphire and Nd : YLF mode locked laser systems. The investigation into collection of the scattered light by the lens, and the removal of the interior baffles inside the lens, has more merit for lens improvements (Cornell website: http://www.englib.cornell.edu/drbio).

For users who wish to evaluate their lens transmission, a simple experimental arrangement is possible to measure the average power through the lens. The laser should be set to the required wavelength and where possible the scanning beam parked in the center of the imaging field. A laser-power meter should then be placed after the lens such that it is not at the focus but where all of the light is entering the meter. (Oil or water should not be placed on the power meter, since the effect of this can be compensated for and damage can occur to the meter.) The laser should be blocked, following which, the lens should be removed and the back aperture of the objective measured or calculated. A mechanical iris aperture should be adjusted to have the same diameter as the back aperture of the objective and placed over the power meter; the average power coming through the system without the lens should then be measured. The average power, the meter and aperture should be as close to the normal position and value of the objective as possible.

Using the figures measured above, a simple calculation will yield the air-lens transmission. For oil and water lenses, an increase of around 4% should be allowed for the loss due to the non use of fluid. In addition, a correction factor of $(NA/0.95)^2$ should be used to allow for the change in NA through the non use of index-match fluid. Using this technique, it can quickly be demonstrated that for most lenses, the infrared transmission is not as bad as has been previously thought for many multiphoton imaging applications. This work is currently subject to a detailed investigation by the authors.

9 SIMPLE ADDITIONAL INSTRUMENTATION

9.1 Two-Photon Dose Quantification

One of the major requirements in undertaking any form of quantitative biological or other scientific investigation is to ensure that the conditions when experiments are repeated are the same. When this requirement is placed on multiphoton imaging, there are a very large number of variables present that need to be measured. With respect to imaging, the optical system can be assumed to be the same, but the laser parameters can change from day to day, and even during a single day. The critical requirements on the laser source are:

1. Wavelength
2. Average Power
3. Pulse length

In the newer computerized laser systems, the wavelength is selected via the front panel, and so this is being measured by the system. On older laser systems, most users have a small scanning spectrometer that is used to set the wavelength and to ensure good mode-locking of the source. The average power can be measured using a standard power meter normally supplied with the laser. The pulselength is, however, significantly harder to measure.

For an accurate measurement, an autocorrelator should be used, since this gives a precise measurement of the actual pulselength. Such instruments are not easy to use, and they are not suitable for measuring the pulselength at the sample (after the objective). A simple system is used by the authors to monitor the two-photon dose on the sample, based upon nonlinear absorption in a semiconductor device (Ranka et al., 1997).

A GaAsP infrared-blind photodiode (e.g., a Hamamatsu G1116) can be mounted in a holder behind a focusing lens (12 mm diameter and focal length between 10 mm and 30 mm). A small portion of the incoming laser beam is sent into the detector before the beam enters the scan head but after any intensity-controlling optics (neutral density filters, AOTFs or Pockels' cell); see Figure 12. A 100 kΩ resistor is mounted across the diode and

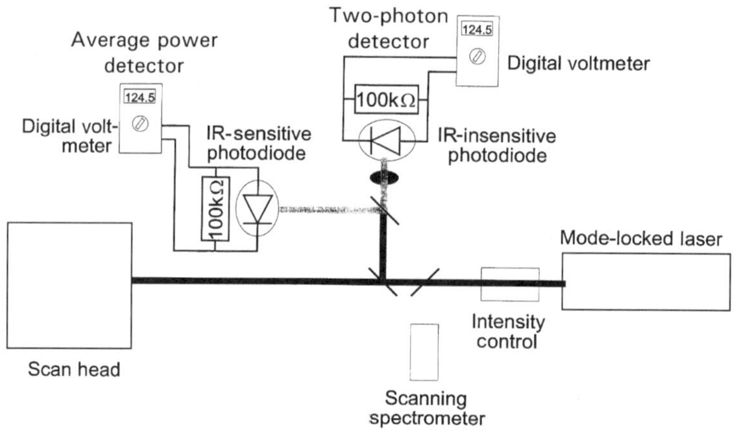

FIGURE 12 Average and peak laser-power monitors.

a 200 mV digital voltmeter connected across this output. When the Ti : sapphire pulses are incident on the detector then the fundamental gives no signal on the meter, but when the focal volume is matched to the depletion region of the diode then two-photon excitation of electrons into the conduction band takes place and a signal can be detected. This voltage is directly related to the peak power in the same way as the two-photon excitation fluorescent signal in the sample. As the pulse lengthens, so the voltage will fall. Through the correct choice of resistor, or focusing lens, up to four decades of useful range can be found. This gives a quantified measure of the output from the laser related to the pulselength, and with a measurement of the average (one-photon) power, the laser output is characterized.

A second diode can then have its outer "can" removed and be mounted, with a similar resistor and meter, after the objective; this will then give a measurement of the two-photon "dose" on the sample. By undertaking a simple calibration process, changing the power and wavelength, the dose on the sample stage can be linked to that measured by the first detector. The stage detector can then be removed, and the user can make a measurement of the incoming two-photon dose and relate this to the actual sample dose.

In the authors' multiphoton system, such a device is mounted for every laser source, along with a small infrared-sensitive photodiode, which has been calibrated to measure the average power. Conditions for each image can now be recorded and care taken that for live-cell studies the threshold damage power for the biological system is not exceeded. Such a device is of low cost, but enables quantifiable imaging to be undertaken. Detailed investigations of these devices is currently under investigation by the authors.

10 SUMMARY

- Select the highest numerical aperture lens suitable for the biological experiments being undertaken.
- Optimize the light-detection system, preferable through the use of non-descanned detectors.
- Where possible, correct for the chromatic aberration in the objective, due to the long-wavelength excitation, through the use of a mode-matching lens.

- Select the correct photomultiplier for the emission wavelength of the fluorophore being excited.
- Minimize the number of optical components before the lens and the detector.
- Select a laser wavelength to optimize the biology; this may not be at the peak of the fluorophore absorption.
- Keep the laser peak power on the sample to a minimum.
- Monitor for biological changes caused by the imaging process.
- Consider water lenses for imaging living cells beyond 10 μm.
- Use coverslips of the correct thickness.
- For in depth imaging beyond 40 μm, multiphoton microscopy is likely the method of choice.

ACKNOWLEDGMENTS

We acknowledge funding support from the Wellcome Trust, SHEFC, JIF, EPSRC, and the University of Strathclyde.

REFERENCES

Bennett, B. D., Jetton, T. L., Ying, G., Magnuson, M. A., and Piston, D. W. (1996). Quantitative subcellular imaging of glucose metabolism within intact pancreatic islets. *J. Biol. Chem.* **271**, 3647–3651.

Brakenhoff, G. J., van der Voort, H. T. M., van Sprosen, E. A., Linnemans, W. A. M., and Nanninga, N. (1985). Three-dimensional chromatin distribution in neuroblastoma nuclei shown by confocal scanning laser microscopy. *Nature* **317**, 748–749.

Bewersdorf, J. and Hell, S. W. (1998). Picosecond pulsed two-photon imaging with repetition rates of 200 and 400 MHz. *J. Microsc.* **191**, 28–38.

Carlsson, K., Danielsson, P. E., Lenz, R., Liljeborg, A., Majlof, L., and Aslund, N. (1985). Three-dimensional microscopy using a confocal laser scanning microscope. *Opt. Lett.* **10**, 53–55.

Centonze, V. E. and White, J. G. (1998). Multiphoton excitation provides optical sections from deeper within scattering specimens than confocal imaging. *Biophys. J.* **75**, 2015–2024.

Curley, P. F., Ferguson, A. I., White, J. G., and Amos, W. B. (1992). Application of a femtosecond, self-sustaining, mode-locked, Ti : sapphire laser to the field of laser scanning confocal microscopy. *Opt. Quantum Electron.* **24**, 851–859.

Denk, W. (1996). Two-photon excitation in functional biological imaging. *J. Biomed. Opt.* **1**, 296–304

Denk, W. and Svoboda, K. (1997). Photon upmanship: why multiphoton imaging is more than a gimmick. *Neuron* **18**, 351–357.

Denk, W., Strickler, J. H., and Webb, W. W. (1990). Two-photon laser scanning fluorescence microscopy. *Science* **248**, 73–76.

Denk, W., Delaney, K. R., Gelperin, A., Kleinfield, D., Stowbridge, B. W., Tank, D. W., and Yuste, R. (1994). Anatomical and functional imaging of neurons using 2-photon laser scanning microscopy. *J. Neurosc. Meth.* **54**, 151–162.

Guild, J. B., Xu, C., and Webb, W. W. (1997). Measurement of group delay dispersion of high numerical aperture objective lenses using two-photon excited fluorescence. *Appl. Opt.* **36**, 397–401.

Hannienen, P. E., Soini, E., and Hell, S. W., (1994). Continuous wave excitation two-photon fluorescence microscopy. *J. Microsc.*. **176**, 222–225.

Hannienen, P. E., Schrader, M., Soni, E., and Hell, S. W. (1995). Two-photon excitation fluorescence microscopy using a semiconductor laser. *Bioimaging* **3**, 70–75.

Inoue, S. and Spring, K. (1997). *Video Microscopy: The Fundamentals*, 2nd edn. New York: Plenum Press.

Jacobsen, H., Hanninen, P., Soini, E., and Hell, S. W. (1994). Refractive-index-induced aberrations in two-photon confocal fluorescence microscopy. *J. Microsc.* **176**, 226–230.

König, K. (2000). Multiphoton microscopy in the life sciences, *J. Microsc.*. **200**, 83–104.

König, K., Tadir, Y., Patrizio, P., Berns, M. W., and Tromberg, B. J. (1996). Effects of ultraviolet exposure and near infrared laser tweezers on human spermatozoa. *Hum. Reprod.* **11**, 2162–2164.

Mohler, W. A., Simske, J. S., Williams-Masson, E. M., Hardin, J. D., and White, J. G. (1998). Dynamics and ultrastructure of developmental cell fusions in the *Caenorhabditis elegans* hypodermis. *Curr. Biol.* **8**, 1087–1090.

Patterson, G. H., and Piston, D. W., (2000). Photobleaching in two-photon excitation microscopy. *Biophys. J.* **78**, 2159–2162.

Pawley, J. (Ed.) (1995). *The Handbook of Biological Confocal Microscopy*, 2nd edn. New York: Plenum Press.

Ranka, J. K., Gaeta, A. L., Baltuska, A., Pschenichnikov, M. S., and Wiseman, D. A. (1997). Autocorrelation measurement of 6 fs pulses based on the two-photon-induced photocurrent in a GaAsP photodiode. *Opt. Lett.* **22**, 1344–1346.

Sandison, D. R., Piston, D. W., Williams, R. M., and Webb, W. W. (1995). Quantitative comparison of background rejection, signal-to-noise ratio, and resolution in confocal and full-field laser scanning microscopes. *Appl. Opt.* **34**, 3576–3588.

Soeller, C. and Cannell, M. B. (1996). Construction of a two-photon microscope and optimisation of illumination pulse duration. *Pfluegers Arch. Eur. J. Physiol.* **432**, 555–561.

Squirrell, J. M., Wokosin, D. L., White, J. G., and Bavister, B. D. (1999). Long-term two-photon fluorescence imaging of mammalian embryos without compromising viability. *Nature Biotechnol.* **17**, 763–767.

Stanley, M. (2001). Mirrors and filters for multiphoton microscopy. *Proc. SPIE* **4262**, 52–61.

Steltzer, E. H. K., Hell, S., Lindek, S., Stricker, R., Pick, R., Storz, C., Ritter, G., and Salmon, N. (1994). Nonlinear absorption extends confocal fluorescence microscopy into the ultra-violet regime and confines the illumination volume. *Opt. Commun.* **104**, 223–228.

Straub, M. and Hell, S. W. (1997). Fluorescence lifetime three-dimensional microscopy with picosecond precision using a multifocal multiphoton microscope. *Appl. Phys. Lett.* **73**, 1769–1771.

Svoboda, K., Denk, W., Kleinfeld, D., and Tank, D. W. (1997). In vivo dendritic calcium dynamics in neocortical pyramidal neurons. *Nature* **385**, 161–165.

White, J. G., Amos, W. B., and Fordham, M. (1987). An evaluation of confocal versus conventional imaging of biological structures by fluorescence light microscopy. *J. Cell Biol.* **105**, 41–48.

Williams, R. M., Piston, D. W., and Webb, W. W. (1994). Two-photon molecular excitation provides intrinsic 3-dimensional resolution for laser-based microscopy and microphotochemistry. *FASEB J*, **8**, 804–813.

Wilson, T. and Sheppard, C. J. R. (1984). *Theory and Practice of Scanning Optical Microscopy.* New York: Academic Press.

Wokosin, D. L. and White, J. G. (1997). Optimization of the design of a multiple-photon excitation laser scanning fluorescence imaging system. *Proc. SPIE* **2984**, 24–29.

Wokosin, D. L., Centonze, V. E., White, J. G., Armstrong, D., Robertson, G., and Ferguson, A. I. (1996). All-solid-state ultra-fast lasers facilitate multiphoton excitation fluorescence imaging. *IEEE* **2**, 1051–1065.

Wokosin, D. L., Dymott, M., and White, J. G., (1998a). Pulse width considerations for multiphoton excitation laser scanning fluorescence imaging. *Proc. SPIE* **3260A**, 115–122.

Wokosin, D. L., Amos, W. B., and White, J. G., (1998b). Detection sensitivity enhancements for fluorescence imaging with multiphoton excitation microscopy. *Proc. IEEE Eng. Biol. Med. Soc.* **20**, 1707–1714.

Wolleschensky, R., Dickinson, M. E., and Fraser, S.E. (2002). Group-velocity dispersion and fiber delivery in multifocal laser scanning microscopy. In: *Confocal and Two-Photon Microscopy: Foundations, Applications, and Advances* (ed. A. Diaspro). New York: Wiley-Liss, Chapter 8.

Xu, C. (2002). Cross-sections of fluorescent molecules in multiphoton spectroscopy. In: *Confocal and Two-Photon Microscopy: Foundations Applications, and Advances* (ed. A. Diaspro). New York: Wiley-Liss, Chapter 4.

Xu, C., Zipfel, W., Shear, J. B., Williams, R. M., and Webb, W. W. (1996). Multiphoton fluorescence excitation: new spectral windows for biological nonlinear microscopy. *Proc. Natl. Acad. Sci. U.S.A.* **93**, 10763–10768.

Yuste, R. and Denk, W. (1995). Dendritic spines as basic functional units of neuronal integration. *Nature* **375**, 682–684.

CHAPTER 11

Sampling, Resolution, and Digital Image Processing in the Spatial and Fourier Domains

KENNETH R. CASTLEMAN

Advanced Digital Imaging Research, LLC, 2525 South Shore Blvd., League City, TX 77573, U.S.A.

1 INTRODUCTION

Images occur in the physical universe in analog (continuous) form, but data can be processed by computer only in digital (numerical) form. Thus, it is necessary to create a digital image to represent the physical image if we are to take advantage of computer power. The processing that we implement may parallel other processes that can be accomplished by physical means, or it may be impossible except when done by a computer.

In this chapter, we concern ourselves with guidelines that can help avoid the pitfalls that can result from digital processing. For our purposes here, the sampling and resolution questions have been handled satisfactorily if an image can be converted into digital form and back again to continuous form without the introduction of disturbing degradations. Notice that our goal is not to avoid degradation altogether, since this may well be impossible. We are satisfied if the digital processing does not degrade the usefulness of the image for its intended purpose.

2 DIGITIZING AN IMAGE

2.1 Continuous and Discrete Functions

A function that exists at all values of its arguments is said to be continuous. A discrete function is only defined for integer values of its arguments. An optical image is continuous, and a digital image is discrete.

A continuous image is converted to digital form by the rather brutal processes of sampling followed by quantization. Then, as an array of integers, it can be processed by

Confocal and Two-Photon Microscopy: Foundations, Applications, and Advances, Edited by Alberto Diaspro.
ISBN 0 471 40920-0

computer techniques that are specified mathematically and implemented numerically. Some of these techniques model physical processes, such as those that can be implemented photographically, while other techniques do not correspond to any physically realizable process. The image must be converted back to continuous form (e.g., by a CRT display) in order to be seen. Properly done, digital processing introduces little evidence that the image has been treated so harshly.

2.2 Sampling

"Sampling" means to measure the amplitude (brightness or color) of an image at a collection of individual points of small extent. The arrangement of the sample points is called the "sampling pattern," or "sampling grid." Proper sampling allows us to reconstruct a continuous image for viewing, without introducing distracting evidence that the image has spent time in digital form. This is one test of proper sampling. Another is whether accurate quantitative information can be extracted from the image. In this section, we address how sampling can be properly done.

Pixel Size The sampling spot is generally circular, and its shape is the point spread function (PSF) of the imaging system (defined below). In a later section, we examine the PSF of a microscope in more detail.

Pixel Spacing The spacing between adjacent pixels is generally set approximately equal to the nominal pixel diameter. Larger spacings are dangerous, because they tend to discard information and can give rise to the artifact of "aliasing" (Moiré effects). Smaller spacings merely waste storage media since the amount of information in the image does not increase with the larger number of pixels. Exceptions to this rule are noted below.

Pixel spacing at the sensor (image plane) scales down to the specimen plane by the magnification factor (see below). Ideally one would measure, rather than calculate, pixel spacing in a digital imaging microscope since manufacturers' specifications are nominal and approximate.

2.3 Quantization

"Quantization" means to reduce a measured value to an integer (e.g., between 0 and 255 for 8-bit images). It is the process of generating an integer that is proportional to the amplitude (brightness) of the image at the corresponding pixel location. It is usually accomplished with an electronic circuit called an "analog-to-digital converter."

3 MICROSCOPY

A diagram of a simple imaging system is shown in Figure 1. The diameter of the lens is a. While a typical microscope is considerably more complex than this, it serves the purposes of this discussion. We consider the simple situation where a microscope objective forms an image on a CCD array.

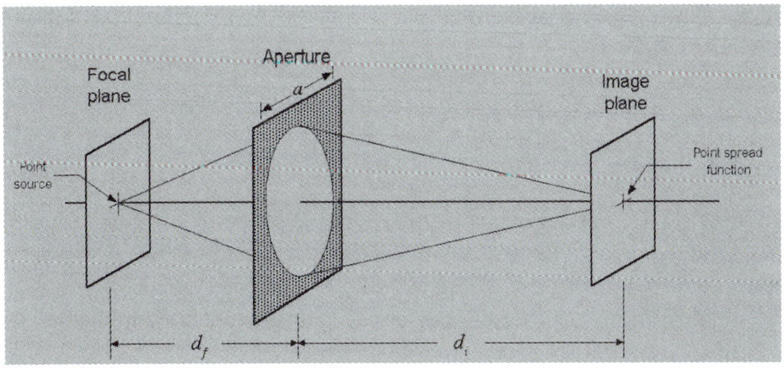

FIGURE 1 A simple imaging system. The lens forms a bright spot in the image plane due to the point source in the focal plane.

3.1 Imaging Geometry

In Figure 1, a point source in the focal plane produces a small spot of light in the image plane. This spot is called the "point spread function" (PSF) of the lens. The two distances are related by the focus equation

$$\frac{1}{d_i} + \frac{1}{d_f} = \frac{1}{f}, \tag{1}$$

where f is the focal length of the lens. Since, in a microscope, d_i is fixed, the focal plane is located at

$$d_f = \frac{fd_i}{d_i - f}. \tag{2}$$

The magnification factor is

$$M = \frac{d_i}{d_f}. \tag{3}$$

When the point source moves off the origin by an amount d, its image moves by the amount Md. In many cases, additional magnification is introduced by intermediate lenses located between the objective and the camera. We can still use the simple model in Figure 1, provided that the actual magnification is accounted for. The microscope eyepieces, which figure into conventional computations of "magnification," have no effect on the digitized image.

It is customary to specify a microscope objective not by its focal length and aperture diameter, but by its "power" (magnification, Eq. (3)) and "numerical aperture." The latter is given by

$$\text{NA} = n \sin \alpha \approx \frac{a}{2d_f} \approx \frac{a}{2f}, \tag{4}$$

where n is the refractive index of the medium (air, water or immersion oil) located between the specimen and the lens, and

$$\alpha = \arctan\left(\frac{a}{2d_f}\right) \tag{5}$$

is the angle between the optical axis and a marginal ray from the origin of the focal plane to the edge of the aperture. The first and second approximations in Eq. (4) assume small aperture and high magnification, respectively. Microscope manufacturers commonly engrave the magnification power and numerical aperture on their objective lenses, and the actual focal length and aperture diameter are little used.

3.2 Working in the Specimen Plane

With microscopy, it is convenient to perform size calculations in the focal plane rather than the image plane, since that is where the objects of interest reside. This involves a 180° rotation and a scaling by the factor M, Eq. (3).

Since $d_f \approx f$, the imaging parameters become simpler if we scale them into the focal plane. Then the pixel spacing and resolution can be specified in units of micrometers at the specimen, and spatial frequencies can be specified in cycles per micrometer in the focal plane. In this chapter, we refer imaging parameters to the focal plane.

3.3 The Point Spread Function

Assuming an ideal (diffraction-limited) lens with a circular aperture, operating in narrowband incoherent light, it can be shown (O'Neill, 1963; Harris, 1964; Goodman, 1968; Castleman, 1996; Born and Wolf, 1999) that the PSF is given by

$$h(r) = \left[2\frac{J_1\left(\pi \frac{r}{r_0}\right)}{\pi \frac{r}{r_0}} \right], \tag{6}$$

where $J_1(x)$ is the first-order Bessel function of the first kind. Since we are working in the focal plane, the constant r_0, which serves as a dimensional scaling factor, is given by

$$r_0 = \frac{\lambda d_f}{a}, \tag{7}$$

and r is the radial distance measured from the origin of the focal plane and is given by

$$r = \sqrt{x^2 + y^2}, \tag{8}$$

where x and y are coordinates in the focal plane.

The two-dimensional PSF is circularly symmetric, and its shape is plotted in Figure 2. This is the familiar "Airy disc" pattern, which has a main central peak and a series of small amplitude side lobes of decreasing intensity.

Ideally (for best resolution), the PSF would be extremely narrow, and without side lobes. We see from Eq. (6) that the scale factor r_0 is a specification of the PSF width. It

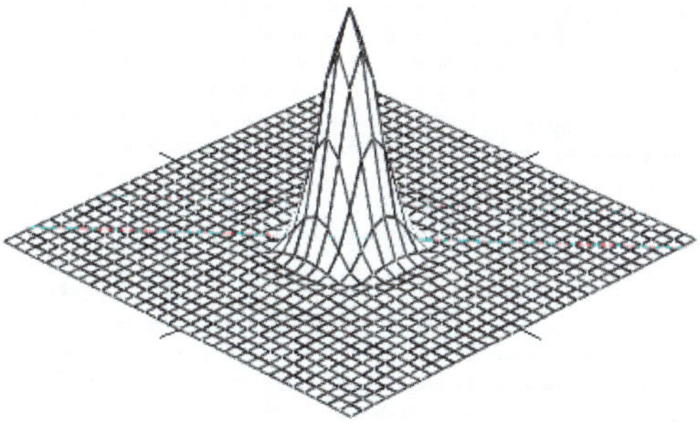

FIGURE 2 The point spread function (PSF). The brightness distribution of the spot image has this shape.

becomes smaller with larger numerical aperture, and it takes on smaller values at shorter illumination wavelengths, Eq. (7). The side lobes cannot be eliminated.

4 THE SPATIAL FREQUENCY DOMAIN

4.1 Linear Systems

A linear system is one whose output is linearly proportional to its input. In other words, doubling the amplitude of the input will double the amplitude of the output. An optical imaging system, such as a microscope, is, to a good approximation, a linear system. A linear system can change the amplitude and phase of a sinusoidal input, but not its characteristic shape. A two-dimensional linear system is completely characterized by its PSF.

4.2 The Fourier Transform

The Fourier transform $F(s)$, of a one-dimensional function $f(t)$ is defined as

$$F(s) = \mathscr{F}\{f(t)\} = \int_{-\infty}^{\infty} f(t)e^{-j2\pi st}\,dt, \tag{9}$$

where $j^2 = -1$. The Fourier transform is a linear integral transformation that, in the general case, takes a complex function of n real variables into another complex function of n real variables. The inverse Fourier transform of $F(s)$ is defined as

$$f(t) = \mathscr{F}^{-1}\{F(s)\} = \int_{-\infty}^{\infty} F(s)e^{j2\pi st}\,ds. \tag{10}$$

The only difference between the direct and inverse Fourier transformations is the sign of the exponent.

Fourier's integral theorem states that

$$f(t) = \int_{-\infty}^{\infty} \left[\int_{-\infty}^{\infty} f(t)e^{-j2\pi st}\,dt\right] e^{j2\pi st}\,ds. \tag{11}$$

This means that the transformation is reciprocal, and

$$\mathcal{F}\{f(t)\} = F(s) \Rightarrow \mathcal{F}^{-1}\{F(s)\} = f(t). \tag{12}$$

The functions $f(t)$ and $F(s)$ are called a "Fourier transform pair". For any $f(t)$, the Fourier transform $F(s)$ is unique, and vice versa. $F(s)$ is referred to as the "spectrum" of $f(t)$.

The Two-Dimensional Fourier Transform

For functions of two dimensions, the direct and inverse Fourier transforms are defined by

$$F(u, v) = \int_{-\infty}^{\infty}\int_{-\infty}^{\infty} f(x, y)e^{-j2\pi(ux+vy)}\, dx\, dy \tag{13}$$

and

$$f(x, y) = \int_{-\infty}^{\infty}\int_{-\infty}^{\infty} F(u, v)e^{j2\pi(ux+vy)}\, du\, dv, \tag{14}$$

respectively, where $f(x, y)$ is an image and $F(u, v)$ is its spectrum. $F(u, v)$ is, in general, a complex-valued function of two real frequency variables, u and v. The variable u corresponds to frequency along the x-axis, and the same is true for v and the y-axis.

4.4 Some Properties of the Fourier Transform

Linearity The spectrum of the sum of two functions is the sum of their individual spectra. Multiplying a function by a constant multiplies its spectrum by the same value.

Symmetry In the general case, a complex-valued function of a single real variable has a Fourier transform that is also a complex-valued function of a real variable. However, if a function is even, then its spectrum is even. If a function is odd, then its spectrum is odd. If a function is real, then its spectrum has an even real part and an odd imaginary part. If a function is real and even, then its spectrum is real and even. If a function is real and odd, then its spectrum is imaginary and odd.

Periodicity If a function is periodic, then its spectrum is discrete (sampled). Likewise, if a function is discrete, then its spectrum is periodic. If a function is both periodic and discrete, then its spectrum is both periodic and discrete.

Since a digital signal or image is of finite extent, it can be considered to be one cycle of a periodic signal or image. Since it is discrete, it must have a periodic spectrum. Thus, we can compute its spectrum as one cycle of a periodic discrete function. Thus, both a digital image and its spectrum are single cycles of periodic discrete functions.

4.5 The Discrete Fourier Transform

If f_i, $0 \leqslant i \leqslant N-1$, is a sampled (discrete) function, then its spectrum can be computed by the discrete Fourier transform (DFT)

$$F_k = \frac{1}{\sqrt{N}} \sum_{i=0}^{N-1} f_i e^{-j2\pi ki/N}, \tag{15}$$

and the inverse DFT is (Cizek, 1986)

$$f_i = \frac{1}{\sqrt{N}} \sum_{k=0}^{N-1} F_k e^{-j2\pi ik/N}. \tag{16}$$

If we define the $N \times N$ matrix W with elements

$$w_{i,k} = \frac{1}{\sqrt{N}} e^{-j2\pi ki/N} \tag{17}$$

then the transforms can be expressed as simple matrix operations

$$\mathbf{F} = W\mathbf{f} \qquad \text{and} \qquad \mathbf{f} = W^{-1}\mathbf{F} \tag{18}$$

The two-dimensional DFT is given by

$$\mathbf{G} = W\mathbf{g}W \qquad \text{and} \qquad \mathbf{g} = W^{-1}\mathbf{G}W^{-1} \tag{19}$$

where $\mathbf{g}$ is an image and $\mathbf{G}$ is its spectrum. Since W is a unitary matrix, its inverse is simply its conjugate transpose. Equation (19) implies that the two-dimensional DFT can be implemented as a concatenation of two one-dimensional DFTs. One first transforms the rows of the image, then the columns of the resulting array (Castleman, 1996).

4.6 The Fast Fourier Transform

Implementing Eq. (18) directly requires N^2 multiplication operations for an N-point one-dimensional DFT. The matrix W defined in Eq. (17), however, possesses considerable symmetry. As a result it is possible to implement the DFT in as few as $N \log_2 N$ multiplications, provided that $N = 2^p$ is a power of 2. Such an efficient implementation of the DFT is called a fast Fourier transform, or FFT (Brigham, 1988).

4.7 The Centered Fourier Transform

The DFT, whether implemented by Eq. (15) or by an FFT algorithm, produces a spectrum that is arranged in a rather inconvenient form. In one dimension, the left and right halves of the spectrum are interchanged in the output vector. In two dimensions, the four quadrants of the spectrum become interchanged in the output matrix, with the zero-frequency component located in the upper left corner, rather than at the center of the matrix. This usually means that an "unpacking" step is required before the spectrum can be viewed or used.

If we redefine the transform matrix, Eq. (17), as

$$w_{i,k} = \frac{1}{\sqrt{N}} e^{-j2\pi(k-a)(i-a)/N}, \tag{20}$$

where $a = \frac{1}{2}(N-1)$ and N is odd, then the origin of frequency space occurs at the center of the output vector or matrix. We do, however, lose the speed benefits of the FFT with this implementation.

4.8 The Optical Transfer Function

The optical transfer function (OTF) is the two-dimensional Fourier transform of the PSF (Bracewell, 1999), and the inverse Fourier transform of the OTF is the PSF. Thus, having either, one can obtain the other. Since the OTF is unique for any particular PSF, it likewise serves as a complete characterization of a linear imaging system.

The OTF is a complex-valued function of two orthogonal spatial frequency variables. It specifies how well the lens can reproduce, in the image plane, sinusoidal features that occur in the focal plane. The value of the OTF, at a particular frequency, is the factor by which the contrast of sinusoidal features in the specimen, at that frequency, is multiplied in the imaging process. The modulus (magnitude) of the OTF is called the modulation transfer function (MTF) of the lens.

Symmetry It is a property of the two-dimensional Fourier transform that, if the PSF is an even function (symmetrical about the x and y axes), then the OTF will be real-valued and, likewise, an even function (Castleman, 1996; Bracewell, 1999). Since the lens is circularly symmetric, the image it forms of a point source is likewise symmetrical. Thus, we see that a symmetrical PSF and a real-valued OTF go hand-in-hand. This also means that the OTF reduces to the (real-valued) MTF. For our purposes then, we can use the terms OTF and MTF interchangeably.

The OTF of a diffraction-limited lens is given by (O'Neill, 1963; Harris, 1964; Goodman, 1968; Castleman, 1996; Born and Wolf, 1999).

$$H(q) = \frac{2}{\pi - 2}\left\{\arccos\left(\frac{q}{f_c}\right) - \sin\left[\arccos\left(\frac{q}{f_c}\right)\right]\right\}, \tag{21}$$

where q is the spatial frequency variable, measured radially in two-dimensional frequency space. It is given by

$$q = \sqrt{u^2 + v^2}, \tag{22}$$

where u and v are the spatial frequencies in the x and y directions of the focal plane respectively. The parameter f_c is called the "cutoff frequency," and since we are working in the focal plane, it is given by

$$f_c = \frac{1}{r_0} = \frac{a}{\lambda d_f} \approx \frac{2\mathrm{NA}}{\lambda}. \tag{23}$$

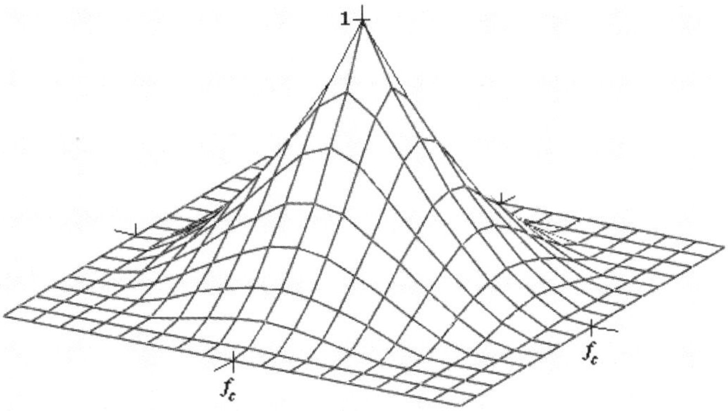

FIGURE 3 The Optical Transfer Function (OTF). The circularly symmetric OTF takes the value unity at zero frequency, and decreases to zero at the cutoff frequency.

The OTF for a lens with a circular aperture is plotted in Figure 3. It starts at unity at zero frequency (large objects), and decreases steadily to zero at the cutoff frequency. Sinusoidal structures having frequencies above f_c do not appear in the image. This indicates that any lens has a limit on how small are the structures that it can resolve.

It is evident in Eq. (23) that the cutoff frequency can be improved (increased) by going to a higher numerical aperture or a shorter wavelength. For a 100×, 1.2 NA objective in green light, $f_c = 4.36$ cycles μm^{-1} in the focal plane.

There is a property of the Fourier transform known as the "similarity theorem." This states, loosely, that the broader a function is, the narrower will be its Fourier transform, and vice versa (Castleman, 1996; Bracewell, 1999). Since the PSF and OTF are related by the Fourier transform, we see that a narrow PSF and a broad OTF go hand-in-hand. Clearly, the better imaging system is the one able to form a smaller spot image from a point source. Likewise, the better imaging system has a broader OTF and thus allows higher frequencies to pass through.

5 RESOLUTION

The concept of resolution sometimes gives rise to considerable confusion. The principal concept of "resolution" is to specify how small an object can be reproduced by a particular imaging system. There are several ways in which one can do this, and we touch on a few of them here.

5.1 Definition

In general, the resolution of an imaging system is a specification of how well it can reproduce objects of different sizes. Often this is reduced to a single number that purports to specify the smallest object size that can be reproduced. This is not always totally satisfying, however, since the system gradually loses its imaging ability as the objects become smaller and smaller.

5.2 Specification

Here, we discuss three approaches to specifying the resolution of an imaging system by a single number, building on the ideas presented earlier in this chapter.

Abbé Resolution To a good approximation, the half-amplitude diameter of the central peak of the image-plane PSF is given by the "Abbé distance" (Abbé, 1873)

$$D_i = r_0 = \frac{\lambda d_i}{a}. \tag{24}$$

For a microscope objective, the Abbé distance in the focal plane is

$$D_f = \frac{1}{M}\lambda\frac{d_i}{a} = \lambda\frac{d_f}{a} \approx \frac{\lambda}{2\mathrm{NA}}. \tag{25}$$

This value can be taken as an indicator of the resolution of the system. Notice that resolution improves (D_f becomes smaller) at shorter wavelength and higher numerical aperture.

Rayleigh Resolution For a lens with a circular aperture, the first zero of the image-plane PSF occurs at a radius of

$$\delta_i = 1.22r_0 = 1.22\frac{\lambda d_i}{a}, \tag{26}$$

which is called the radius of the Airy disk. According to the "Rayleigh criterion of resolution" (Rayleigh, 1896), two point sources can just be resolved if they are separated, in the image, by the distance $1.22r_0$ (Castleman, 1987, 1993, 1996; Bradbury, 1984; Inoue, 1986).

In optics terminology, the Rayleigh distance defines circular "resolution cells" in the image, since two point sources can be resolved if they do not fall within the same resolution cell.

For a microscope objective, the Rayleigh distance (resolution cell diameter) in the focal plane is

$$\delta_f = 1.22\frac{\lambda d_f}{a} \approx 0.61\frac{\lambda}{\mathrm{NA}}. \tag{27}$$

Again the resolution improves (δ_f becomes smaller) at shorter wavelength and higher numerical aperture. For a $100\times$, 1.2 NA objective in green light, the Rayleigh resolution is approximately 0.28 μm.

Cutoff Frequency For a microscope objective, the incoherent optical cutoff frequency, in the focal-plane coordinate system, is given by Eq. (23). This can be used as a basis for comparison of imaging systems.

The approximations in Eqs. (23) and (27) begin to break down at low power and high numerical aperture, but these normally do not occur together. One can quantify the degree of approximation by comparing f and a, or by comparing the angles $\arctan(a/2d_f)$ and $\arcsin(\mathrm{NA}/n)$.

The OTF A more complete specification of resolution is the OTF, Eq. (21). Since the PSF is symmetrical, the OTF is real-valued.

The actual OTF of a real microscope imaging system will fall short of the ideal form of Eq. (21) and Figure 3. The OTF can be measured in several ways (Castleman, 1996). One can use a fluorescent microsphere as an approximation to a point source, and the resulting image is the (magnified) PSF. A Fourier transform then yields the OTF. One can use a sine-wave target to measure the frequency response (OTF) directly. Since such calibration targets are difficult to manufacture, a bar target can be substituted. One can also differentiate the image of a sharp edge. The Fourier transform of that is a one-dimensional profile of the OTF (Castleman, 1996).

6 THREE-DIMENSIONAL IMAGING

In Figure 1, it was assumed that the objects of interest are located in the focal plane. Microscopic specimens are typically three-dimensional. Thus, any single image may well contain contributions from objects located outside the focal plane. Since optical imaging is a linear operation, the images of all objects sum together in the image plane.

If the image is not in focus, that is, if Eq. (1) is not satisfied, then the PSF becomes broader, the OTF becomes narrower, and the image becomes blurrier. For a plane located Δz beyond the focal plane (Δz is positive moving away from the lens), the OTF is given, to a good approximation, by (Castleman, 1996)

$$H(w, q) = \frac{2}{\pi - 2}(\beta - \sin\beta)\,\mathrm{jinc}\left[4kw\left(1 - \frac{|q|}{f_c}\right)\frac{q}{f_c}\right], \tag{28}$$

where

$$\mathrm{jinc}\,x = 2\frac{J_1(x)}{x}, \qquad \beta = \cos^{-1}\left(\frac{q}{f_c}\right), \qquad q = \sqrt{u^2 + v^2}, \qquad f_c = \frac{a}{\lambda d_f} \approx \frac{2\mathrm{NA}}{\lambda}, \tag{29}$$

and, for reasonably high magnification and small defocus, the defocus path length error is $w = \Delta z \mathrm{NA}^2/2$. Figure 4 shows what happens to the PSF and the OTF under defocus conditions. The amount of defocus is specified in terms of w, that is, wavelengths of pathlength difference between central and marginal rays. Figure 4 shows the effect of $w = \frac{1}{2}\lambda$ and $w = \frac{3}{4}\lambda$.

6.1 Optical Sectioning

Since only objects in the focal plane are imaged in focus, one can effect the sectioning of a thick specimen by taking a series of images at different z-axis positions. This technique, of course, is plagued by the shadows of out-of-focus structures in the section images. The technique of three-dimensional deconvolution (Castleman, 1996) can be used to reduce this problem and clean up the individual optical section images. Deconvolution methods,

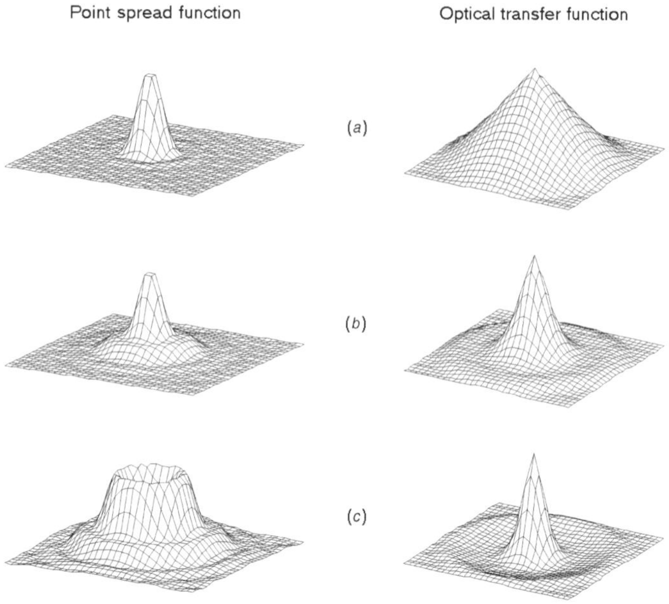

FIGURE 4 The effect of defocus. This graph shows the PSF and OTF of a diffraction-limited lens (*a*) with zero defocus, and with (*b*) $\frac{1}{2}$ wave, and (*c*) $\frac{3}{4}$ wave of defocus. (After Castleman (1996.))

ranging from simple subtraction of blurred adjacent plane images to the more complex constrained iterative deconvolution, are relatively effective, but somewhat computationally complex. The confocal microscope, by contrast, does not require deconvolution, because it images only those structures lying within, or quite near, the focal plane.

7 DIGITAL PROCESSING

We now examine some of the factors that are brought into play by digital processing of images.

7.1 Quantization Noise

Quantization introduces noise. The quantized image, $\boldsymbol{q}(x, y)$, differs from the original, $\boldsymbol{f}(x, y)$ by a maximum amplitude of one-half gray level. This difference is conveniently quantified by the mean square error of quantization

$$N_q = E\{[f(x, y) - q(x, y)]^2\}, \tag{30}$$

where $E\{\cdot\}$ is the expectation operator. We can define the signal-to-noise ratio (in decibels) due to quantization as

$$\text{SNR} = 10 \log_{10}\left(\frac{\sigma^2}{N_q}\right), \tag{31}$$

where σ^2 is the variance of the image. For uncorrelated (random) images, this is a linear relationship with a slope of 6 dB per bit. That is,

$$\mathrm{SNR} = 6B, \tag{32}$$

and each additional bit of gray level adds 6 dB to the SNR. For the more common case of images wherein adjacent pixels are correlated, the resulting SNR is slightly higher, namely

$$\mathrm{SNR} = 6B + 11. \tag{33}$$

So the actual quantization-noise level, as a fraction of the signal amplitude, is

$$\frac{N_q}{\sigma^2} = 10^{(6B+11)/10}. \tag{34}$$

7.2 The Sampling Theorem

According to the Shannon sampling theorem, a continuous signal can be reconstructed, without error, from its sample points provided that (i) the sample spacing is no greater than one-half the wavelength of the highest-frequency component present in the spectrum of the signal, and (ii) the interpolation function has the proper form ($\sin ax/ax$). This result generalizes easily to two dimensions and the case of digitized images. If the sample spacing is larger than that stated above, then the reconstruction will not be accurate, and Moiré effects (aliasing) may be visible in the reconstructed image. If an interpolation function of another form is used, such as the Gaussian-shaped CRT display spot commonly employed, the reconstruction accuracy may suffer as well.

Since the OTF of a microscope is zero for all frequencies above the optical cutoff frequency f_c, we can be assured that there will be no energy in the image at frequencies greater than f_c. Thus, we can avoid aliasing effects by selecting the sampling frequency to be at least $2f_c$. Stated differently, the sample spacing at the specimen must be no greater than $1/2f_c = \lambda/4\mathrm{NA}$.

7.3 Quantitative Image Analysis

It is frequently required to locate and isolate the objects in an image and to extract quantitative information (size, shape, color, etc.) from each of them. Here the question of pixel spacing is somewhat different from that associated with resolution and aliasing as discussed above.

When a digital image is segmented, each object in the image is a connected set of pixels. The boundary of the object is defined by the subset of pixels that are boundary pixels, namely, those that have a neighbor outside the object. Thus, the boundary is a polygonal approximation to the actual (continuous, smooth) boundary of the original object. How well that polygon approximates the true boundary depends on the pixel spacing.

Even if the pixel spacing is set so as to avoid aliasing, and the Rayleigh resolution is adequate for viewing the image, the boundary polygon may still be too rough to support accurate measurement of area, perimeter, and similar features. For example, consider a small cell that occupies only ten pixels in the image. If even one pixel is erroneously included in, or dropped from the object, the result is a 10% change in measured area. Thus, a separate analysis is required to ensure that extracted measurements will be sufficiently accurate. One can generate geometric shapes of known size but different position and orientation, and measure them to determine the accuracy of a particular configuration.

There are three ways to deal with the situation of pixel spacing that is too large for accurate measurement extraction. These are oversampling, resampling, and curve fitting.

Oversampling The simplest way to resolve the problem is to reduce the pixel spacing. This can be implemented simply by increasing the magnification of the system. Often it is necessary to sample the image at one-half, or even less, of the maximum pixel spacing that avoids aliasing.

Resampling Another way is to resample the digital image to increase its size. For example, one can use interpolation with the sinc function to double the size of an image by inserting new pixels between the existing ones, vertically and horizontally (Castleman, 1996). This does not add any new information to the image, and it may not look any better when displayed. Measurement accuracy, however, may well be improved, simply because the boundary polygon becomes a better approximation to object shape.

Curve Fitting The third way to handle the problem is to build curve fitting into the measurement algorithms. For example, instead of taking object perimeter to be just that of the boundary polygon, one can fit arcs of smooth curves (e.g., quadratics) through sets of sequential boundary points and sum the resulting arclengths. While this considerably complicates the software, it can produce more accurate measurement data.

8 SUMMARY

In this chapter, we have touched on some theory that sets forth basic guidelines for digital processing of microscope images. Quantization noise can be reduced to tolerable levels by using enough bits of gray scale. The Rayleigh resolution criterion, Eq. (27), is a convenient way to specify resolution with a single number. The OTF, Eq. (21), provides a more complete specification of resolution. The sampling theorem specifies how one can set the pixel spacing to avoid the artifact of aliasing. Oversampling, resampling, and curve fitting can pave the way for the extraction of accurate measurements.

A well-designed digital imaging system can reproduce images well enough to accomplish its intended purposes without wasting resources. Digital processing, properly done, need not degrade an image in any significant way.

REFERENCES

Abbé, E. (1873). "Beiträge zur Theorie des Mikroskops und der mikroskopischen Wahrnehmung," *M. Schultze's Arch. Microsc. Anat.* **9**, 413–468.

Born, M. and Wolf, E. (1999). *Principles of Optics*, 7th (revised) edn. Cambridge: Cambridge University Press.

Bracewell, R. (1999). *The Fourier Transform and Its Applications*. New York: McGraw-Hill.

Bradbury, S. (1984). *An Introduction to the Optical Microscope*. New York: Oxford University Press.

Brigham, E.O. (1988). *The Fast Fourier Transform and Its Applications*. Englewood Cliffs, NJ: Prentice-Hall.

Castleman. K.R. (1987). Spatial and photometric resolution and calibration requirements for cell image analysis instruments. *Appl. Opt.* **26**, 3338–3342.

Castleman, K.R. (1993). Resolution and sampling requirements for digital image processing, analysis and display. In: *Electronic Light Microscopy* (ed. D. Shotton). New York: Wiley-Liss, 71–93.

Castleman, K.R. (1996). *Digital Image Processing*. Englewood Cliffs, NJ: Prentice Hall, Chapters 10, 15, 16, and 22.

Cizek V. (1986). *Discrete Fourier Transforms and Their Applications*. Bristol: Adam Hilger.

Goodman, J.W. (1968). *Introduction to Fourier Optics*. New York: McGraw-Hill.

Harris, J.L. (1964). Diffraction and resolving power. *J. Opt. Soc. Am.* **54**, 931–936.

Inoue, S. (1986). *Video Microscopy*. New York: Plenum Press.

O'Neill, E.L. (1963). *Introduction to Statistical Optics*. Reading MA: Addison-Wesley.

Rayleigh, Lord (1896). On the theory of optical images, with special reference to the microscope. *Phil. Mag.* **54**(5), 167–195.

CHAPTER 12

Image-Restoration Methods: Basics and Algorithms

PATRIZIA BOCCACCI AND MARIO BERTERO

INFM, DISI, University of Genoa, Via Dodecaneso, 35, 16146, Genoa, Italy

1 INTRODUCTION

Several methods can be used for investigating fluorescent three-dimensional specimens: wide-field, confocal, and two-photon microscopes have been and still are powerful tools for the noninvasive imaging of biological samples [1].

All of these optical systems can be modeled in the same way. A fluorescent sample, illuminated with a light of given wavelength, emits light of longer wavelength. The detection of the emitted light through the optical system of the microscope provides a fluorescence image of the target.

In the case of isoplanatism, the incoherent nature of fluorescence light allows one to model the microscope as a linear space-invariant optical system whose behavior is fully characterized by its *point spread function* (*PSF*). Therefore, the fluorescence image can be viewed as the convolution of the sample with the PSF of the microscope. The effect is a blurring of the sample, and this effect can be annoying if the PSF has a large *full-width at half-maximum* (*FWHM*), as in the case of a wide-field microscope. In addition, two other factors degrade the image: the presence of the background, due to autofluorescence, photon scattering, and detector offsets, and the noise, due to photon counting (Poisson noise) and light intensity conversion (Gaussian readout noise).

The goal of image restoration is to reduce the effect of the blurring, given an estimate of the PSF and, possibly, of the background.

The PSF of the microscope can be obtained from a theoretical description of the process of image formation or can be measured by detecting the image of a pointlike object (i.e., an object with extension smaller than the resolution distance of the instrument). While the PSF is the image of a pointlike object, its Fourier transform, the *optical transfer function* (*OTF*), characterizes the response of the optical system to the Fourier components of the sample.

Confocal and Two-Photon Microscopy: Foundations, Applications, and Advances, Edited by Alberto Diaspro.
ISBN 0-471-40920-0

The OTF is a complex-valued function. However, since high-quality lenses, designed to introduce a minimum phase shift, are in general used in microscopy, the modulus of the transfer function (MTF) can be conveniently used to discuss imaging properties of the system.

The purpose of this chapter is to discuss the basic features of the problem of image restoration and to introduce the standard methods that have been proposed for its solution.

ILL-CONDITIONING AND NOISE

The development of a mathematical model describing the processes of image formation and sampling is the starting point of any restoration method, and the purpose of this section is to formulate such a model and discuss those features that have important consequences for the restoration problem.

We assume that the volume to be imaged is partitioned into $J_1 \times J_2 \times J_3$ voxels, characterized by the triple of indices $\{j_1, j_2, j_3\}$, with $0 \leqslant j_1, j_2, j_3 \leqslant J_1 - 1, J_2 - 1, J_3 - 1$. We denote by $g = g(j_1, j_2, j_3)$ the value of the image at the voxel $\{j_1, j_2, j_3\}$, and we denote its discrete Fourier transform (DFT) by the corresponding capital letter, that is, $G = G(l_1, l_2, l_3)$, where $\{l_1, l_2, l_3\}$ is the triple of indices characterizing the voxels in Fourier space. Then, if the isoplanatism condition is satisfied over the field of the microscope, the image g is given by

$$g = k * f + b + n, \tag{1}$$

where

- f is the physical quantity to be imaged (e.g., fluorochrome density);
- k is the PSF of the microscope; the DFT of k, namely K, is the OTF;
- b is the background;
- n is the noise contamination from any number of sources (counting, readout noise, etc.), modeled as a zero-mean and additive, though not necessarily signal-independent, contribution;
- $k * f$ denotes the discrete convolution product defined by

$$(k * f)(j_1, j_2, j_3) = \sum_{j'_1, j'_2 j'_3} k(j_1 - j'_1, j_2 - j'_2, j_3 - j'_3) f(j'_1, j'_2, j'_3). \tag{2}$$

All of these assumptions imply that g is the result of a random process with mean $k * f + b$.

As is well known, computations are greatly simplified if the PSF is replaced by a periodic one, because, in such a case, the convolution product can be computed by means of the FFT algorithm. Since the PSF is, in general, significantly different from zero only in a small region around the central voxel, the periodic approximation may generate artifacts restricted to the boundaries of the image. These effects, which may be important, especially in the axial direction, can be eventually reduced by a suitable zero-padding of the image. In conclusion, if we take the DFT of both sides of Eq. (1), we get

$$G = KF + B + N. \tag{3}$$

Finally, in order to complete our notation, we define the *subtracted image* g_s and its DFT G_s as follows:

$$g_s = g - b, \qquad G_s = G - B. \tag{4}$$

The image-restoration problem can now be formulated in the following way: *given* g, k, and b, *determine an estimate* f_{res} *(the restored image) of the unknown physical quantity* f. A pictorial representation of the process of image formation followed by that of image restoration is given in Figure 1.

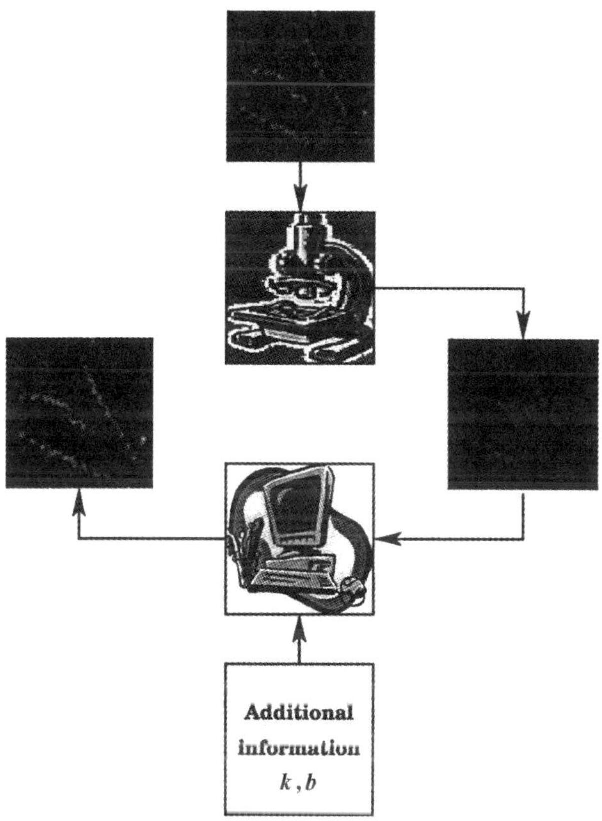

FIGURE 1 Pictorial representation of the process of image formation followed by image restoration implemented on a computer. Additional inputs to the computer are the PSF and the background.

At first glance, the most natural approach to the solution of this problem seems to consist in neglecting the noise term n in Eq. (1) and define $f_{\rm res}$ as the solution of the linear equation

$$k * f_{\rm res} = g_s. \tag{5}$$

However, this equation does not have a solution if the MTF is zero somewhere. In such a case, one gets a solution by considering a least-squares approach: $f_{\rm res}$ is defined as the object that minimizes the *discrepancy functional*

$$\delta^2(f) = \|k * f - g_s\|^2 = \frac{1}{J} \sum_{j_1, j_2, j_3} |KF - G_s|^2, \tag{6}$$

where $J = J_1 J_2 J_3$, $\|.\|$ denotes the usual Euclidean norm, and, in addition, Parseval's equality is used. We find that, in the voxels where $K(l_1, l_2, l_3) \neq 0$, $F_{\rm res}$ is given by

$$F_{\rm res} = \frac{G_s}{K}, \tag{7}$$

while in the voxels where $K(l_1, l_2, l_3) = 0$, $F_{\rm res}$ is not defined. This is the well-known *nonuniqueness* of the solution of the problem due to the band-limiting of the imaging system (the existence of invisible objects – i.e., objects producing a zero image) [2]. Anyway, a solution can be obtained by putting $F_{\rm res} = 0$ at the voxels where the MTF is zero. In this way, $F_{\rm res}$ is defined everywhere, and, by means of an inverse DFT, one obtains a solution $f_{\rm res}$ of the least-squares problem.

However, the restored image $f_{\rm res}$ obtained in such a way does not make sense from a physical point of view as a consequence of the combined effect of the noise and of the so-called *ill-conditioning* (or ill-posedness) of the restoration problem [2]. Ill-conditioning is due to the fact that the MTF becomes small at the boundary of the band of the imaging system (in the continuous model, the MTF tends to zero at the boundary). If we derive G_s from Eq. (3) and substitute the result into Eq. (7), we obtain

$$F_{\rm res} = F + \frac{N}{K}; \tag{8}$$

therefore $F_{\rm res}$ coincides with F except for a term representing the propagation of the noise from the image g to the restored image $f_{\rm res}$. This noise propagation can be negligible at the voxels, in Fourier space, where the MTF is large (in general, at low spatial frequencies), but it can be dominant at the voxels where the MTF is small (in general, at high spatial frequencies). As a consequence, the contribution of f to $f_{\rm res}$ is contaminated by wild noise fluctuations.

3 LINEAR FILTERING

The previous discussion indicates that the main difficulty is due to the high-frequency components of the noise, therefore, these noise artifacts could be reduced by a suitable filtering of $f_{\rm res}$. A first attempt in this direction can be the following. If we denote by $\langle |F|^2 \rangle$

the power spectrum of the object f and by $\langle |N|^2 \rangle$ that of the noise n, and if we assume that noise and object are uncorrelated (this assumption is certainly false in the case of Poisson noise, but it simplifies the discussion), then the power spectrum of the subtracted image is given by

$$\langle |G_s|^2 \rangle = |K|^2 \langle |F|^2 \rangle + \langle |N|^2 \rangle. \tag{9}$$

If we further assume that the power spectra of object and noise are approximately constant over the band of the imaging system, that is, $\langle |F|^2 \rangle \simeq S_f^2$ and $\langle |N|^2 \rangle \simeq S_n^2$, we get

$$\langle |G_s|^2 \rangle = |K|^2 S_f^2 + S_n^2. \tag{10}$$

This equation clearly shows that the Fourier components of the image corresponding to very small values of the MTF contain only noise. Since the information about the object is coming from the Fourier components corresponding to sufficiently large values of the MTF, it appears reasonable to keep Eq. (7) only for those frequencies such that the signal term is larger than the noise term, namely such that

$$|K|^2 > \frac{S_n^2}{S_f^2}, \tag{11}$$

and to set $F = 0$ elsewhere.

This procedure is clearly a linear filtering of the image, with a filter that depends both on the MTF and on a *signal-to-noise ratio* (*SNR*) defined as the ratio between the power spectrum of the object and that of the noise: $\mathrm{SNR} = (S_f/S_n)^2$. The restored image obtained in such a way is free of strongly oscillating artifacts due to high-frequency noise, but it may contain other artifacts due to the discontinuity introduced in the DFT of the restored object by the filtering condition (11).

A much more efficient and flexible filtering is provided by the so-called *regularization methods*, introduced by the Russian mathematician A.N. Tikhonov [3] for providing a unified approach to the solution of ill-conditioned/ill-posed problems [4–6]. The simplest of these methods consists in minimizing the following functional

$$\phi_\mu(f) = \|k * f - g_s\|^2 + \mu \|f\|^2, \tag{12}$$

where μ is a positive number, called the *regularization parameter*. The first term measures the "discrepancy" between the computed image $k * f$, corresponding to a given object f, and the detected image g_s, while the second measures the "energy" of the object f. Therefore, if f_μ is an object that minimizes (12), then, for small values of μ, its discrepancy is small while its energy is large. The situation is reversed when μ is large. This property is the origin of the term used for μ, because it determines the "regularity" of the solution.

By means of Parseval's equality for the DFT, we can rewrite Eq. (12) as

$$\phi_\mu(f) = \frac{1}{J} \sum_{j_1, j_2, j_3} (|KF - G_s|^2 + \mu |F|^2); \tag{13}$$

then, by means of elementary algebra, it is easy to show that there exists a unique minimum, denoted by F_μ and given by

$$F_\mu = \frac{K^*}{|K|^2 + \mu} G_s, \tag{14}$$

where K^* is the complex conjugate of K. It follows that F_μ is zero where the MTF is zero. It also follows that, when $\mu \to 0$, F_μ tends to the restored object discussed in Section 2, and therefore the value $\mu = 0$ must be discarded.

The inverse DFT of F_μ, f_μ, is called a *regularized solution* of the restoration problem. The procedure outlined above provides a family of solutions, one for each value of μ, and therefore, in order to get a restored image $f_{\rm res}$, we must understand how we can choose μ.

If we rewrite Eq. (14)

$$F_\mu = \frac{|K|^2}{|K|^2 + \mu} \frac{G_s}{K}, \tag{15}$$

then we find that F_μ is obtained by filtering the solution of Section 2, with a filter given by

$$K_\mu = \frac{|K|^2}{|K|^2 + \mu}. \tag{16}$$

This filter takes values close to 1 at the voxels where $|K|^2$ is large with respect to μ, it is exactly equal to $\frac{1}{2}$ at the voxels where $|K|^2 = \mu$, and it becomes small at the voxels where $|K|^2$ is much smaller than μ. Therefore, μ controls the filtering of the solution discussed in Section 2: small values of μ imply weak filtering, while large values of μ imply strong filtering.

The observations at the beginning of this section imply that a suitable choice could be to take μ of the order of $(S_n/S_f)^2$; this quantity, however, is not known in most cases, and therefore the problem arises of estimating an optimal value of μ.

It is important to note that such an optimal value exists in the following sense: if we define a *relative restoration error* as

$$\varrho(\mu) = \frac{\|f_\mu - f\|}{\|f\|}, \tag{17}$$

where f is the unknown object, then it is possible to prove that there exists a unique value of μ, $\mu_{\rm opt}$, that minimizes $\varrho(\mu)$ [2]. More precisely, as μ increases from 0 to ∞, $\varrho(\mu)$ first decreases, reaches the minimum, and then increases monotonically.

It is obvious that $\mu_{\rm opt}$ can be computed only in the case of numerical simulations. In the case of real images, one can only try to obtain an estimate of $\mu_{\rm opt}$. Several methods have been proposed for this purpose [2]; here we only mention the so-called *discrepancy principle*: if σ_n^2 is the average variance of the noise per pixel, then $\mu_{\rm dis}$ is the unique value of μ such that

$$\|k * f_\mu - g_s\| = \sqrt{J}\sigma_n. \tag{18}$$

It is known from numerical experiments that this method provides acceptable results in many cases even if μ_{dis} is, in general, a bit larger than μ_{opt}. Since, for a given value of μ, the computation of the regularized solution is simple and fast, an alternative to the automatic methods for the choice of μ can be an interactive method based on the visualization of the regularized solutions for different values of μ and on the decision of the user about the best restoration.

We conclude this section by observing that the implementation of the method is as follows:

```
TIK algorithm {
    preprocessing
    {
        compute and store K and G_s;
        choose a value of μ;
    }
    for each μ
    {
        compute F_μ as given in Eq. (14);
        compute f_μ by inverse FFT.
    }
}
```

Therefore, for a given value of μ, the computation of f_μ requires, basically, the computation of one inverse DFT. If the size of the 3D image is not excessive, this computation can be performed in a few seconds or less, allowing for a visual choice of μ.

4 CONSTRAINED LEAST-SQUARES PROBLEMS

Linear filtering provides satisfactory results in many circumstances (see, e.g., [7] for an application to confocal microscopy), in particular when the object does not contain bright spots or sharp intensity variations, otherwise the quality of the restored image can be degraded by annoying ringing effects.

If the sharp variations of intensity are with respect to a black background, then it is known that the ringing effects can be greatly reduced by methods searching for non-negative solutions of the restoration problem. Since these solutions cannot fit exactly the detected image (the best fit is provided by the unphysical solution of Section 2), they can be defined in a least-squares sense, that is, as the non-negative object $f \geqslant 0$ that minimize the discrepancy functional of Eq. (6).

In the case of a band-limited system, this problem may have several solutions, but it can be proved that there exists a unique non-negative solution with minimal Euclidean norm [8]. This solution, however, can be strongly contaminated by the noise corrupting the detected image in a way similar to that in Section 2. It is possible to overcome this difficulty by searching for non-negative minimizers of the functional of Eq. (12). It is easy to prove that there exists a unique non-negative minimizer for any value of μ, and several iterative methods for its computation have been proposed.

We mention that developed by Carrington [9,10] which is also described in Chapter 16 of this book [11] and others which can be classified as projected-gradient methods [12,13].

A comparison of some of these iterative methods in the case of confocal microscopy is given by van Kempen et al. [14].

Some of these methods can also be applied directly to the computation of non-negative minimizers of the discrepancy functional, and, in such a case, they have a very interesting property: when the number of iterations increases, the restoration error first decreases, reaches a minimum, and then grows. An example of this behavior is given in Figure 2. This property, which is called *semiconvergence*, implies that the number of iterations acts as a regularization parameter, so that, by a suitable stopping criterion, it is possible to obtain regularized versions of the non-negative least-squares solutions.

For this reason, we believe that it may be convenient to apply these iterative methods directly to the functional (6) rather than to the functional (12). In such a way, one has to control only one parameter, namely the number of iterations, and is not obliged to use the method several times for different values of μ as would happen in the case of the functional (12) in order to optimize μ.

In this section, we describe two of these methods: the *projected Landweber method* (PL) [8] and the *iterative space reconstruction algorithm* (ISRA) [15].

PL is a gradient method with projection, at each iteration, on the cone of the non-negative objects. The algorithm is as follows:

$$f^{(m+1)} = [f^{(m)} + \tau k^T * (g_s - k * f^{(m)}]_+ \tag{19}$$

where m is the number of iterations, $[\cdot]_+$ denotes the non-negative part of the image (obtained by keeping the non-negative values and by setting to zero the negative ones), k^T denotes the transposed PSF ($k^T(j_1, j_2, j_3) = k(J_1 - j_1, J_2 - j_2, J_3 - j_3)$), and τ is a *relaxation parameter* satisfying the conditions

$$0 < \tau < \frac{2}{|K|^2_{\max}}, \tag{20}$$

$|K|^2_{\max}$ being the maximum value of $|K|^2$.

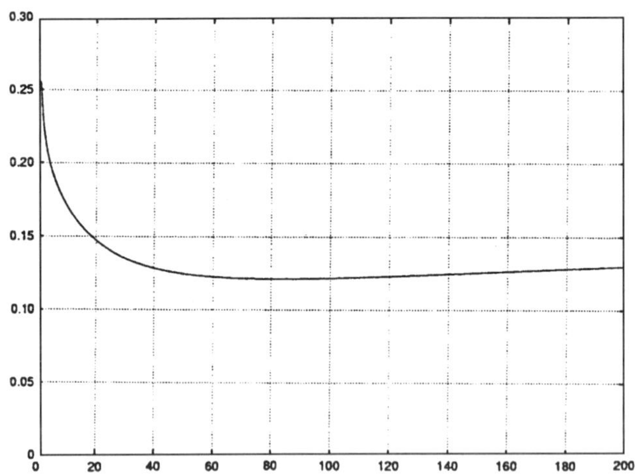

FIGURE 2 A picture of the behavior of the (relative) restoration error as a function of the number of iterations, as shown by iterative methods with the semiconvergence property.

An important point is the initialization of the iterative procedure, namely the choice of $f^{(0)}$. It has been proved that, for any given $f^{(0)}$, the process converges to a non-negative least-squares solution but it is not known what choice of $f^{(0)}$ provides that with minimal Euclidean norm [8]. However, it is reasonable to assume that it can be obtained by taking the initial guess as small as possible, namely $f^{(0)} = 0$. This is the usual choice in the practical applications of the method.

Since the method has the semiconvergence property, the relative restoration error, defined by

$$\varrho^{(m)} = \frac{\|f^{(m)} - f\|}{\|f\|}, \tag{21}$$

has a unique minimum corresponding to a number of iterations that can be called the optimum one, m_{opt}. Therefore, as in the case of the regularization parameter, one should find criteria, which are usually called *stopping rules*, for estimating such a value. This problem is essentially open, even if one can easily extend to PL the discrepancy principle that was formulated in the case of the regularized solutions. Numerical practice indicates that this stopping rule in general stops the iterations a bit too early.

The implementation of PL is as follows:

PL algorithm {
  preprocessing
  {
    choose the value of τ (as a rule of thumb one can take $\tau = 1.8/|K|^2_{\max}$);
    compute and store $\tau K^* G_s$ and $(1 - \tau|K|^2)$;
    initialize with $f^{(0)} = 0$;
  }
  for each iteration m:
  {
    given $f^{(m)}$ compute $F^{(m)}$ by FFT;
    compute $H^{(m)} = \tau K^* G_s + (1 - \tau|K|^2)F^{(m)}$;
    compute $h^{(m)}$ by inverse FFT;
    compute $f^{(m+1)} = [h^{(m)}]_+$.
  }
}

Therefore, each iteration requires the computation of two FFTs. Since the optimal number of iteration is, in general, rather large (even if an acceleration of the method by means of preconditioning has been considered [16]), the computational cost of this method is much higher than that of the regularization method of Section 3. The advantage is that it can provides much better results in the case of sharp intensity variations with respect to a black background. It must be observed, however, that the number of iterations increases with the SNR, so that it can be small in the case of images with small SNR. We point out that this is a rather general property of all the iterative methods considered in this chapter.

We conclude this short presentation of the PL method by observing that it can be modified in order to introduce also an upper bound to the values taken by the restored image [2] – at least when such an upper limit is available.

ISRA is another iterative method for determining non-negative least-squares solutions. Indeed, it has been proved [17] that, for any initial guess, the iterations converge to a minimum non-negative point of the functional (6). The iterative scheme is defined as follows:

$$f^{(m+1)} = f^{(m)} \frac{k^T * g_s}{k^T * k * f^{(m)}}, \tag{22}$$

where product and division of images are intended voxel by voxel.

The algorithm applies to the case where the PSF is non-negative and so is the image. In the applications to microscopy where the subtracted image g_s is used, it may happen that g_s is negative in some voxels owing to fluctuations of the noise affecting the background. These negative values may generate negative values of $k^T * g_s$, so that, in order to avoid instabilities of the algorithm, these negative values must be set to zero. In other words, $k^{T'} * g_s$ must be replaced by $[k^T * g_s]_+$ in Eq. (22). We point out that if $[k^T * g_s]_+$ is zero in a certain voxel, then all the iterates $f^{(m)}$ are zero in that voxel.

Another point is the choice of the initial guess $f^{(0)}$. It is convenient to choose a strictly positive $f^{(0)}$, and, for this reason, one usually takes a uniform image (e.g., equal to 1 everywhere). Obviously, in the absence of uniqueness, the result may depend on the initial guess.

Numerical experience suggests that ISRA also has the semiconvergence property, and therefore the problem of searching for stopping rules also arises in this case. In addition, one finds that PL, initialized with $f^{(0)} = 0$, and ISRA, initialized with a uniform image, provide restorations with essentially the same quality.

The implementation of ISRA is as follows:

```
ISRA algorithm {
    preprocessing
    {
      compute and store |K|^2 and [k^T * g_s]_+ = g_b;
      initialize with f^(0) = constant;
    }
    for each iteration m:
    {
      given f^(m) compute F^(m) by FFT;
      compute H^(m) = |K|^2 F^(m);
      compute h^(m) by inverse FFT;
      compute f^(m+1) = f^(m) g_b / h^(m).
    }
}
```

Again, each iteration requires the computation of two FFTs, and, in general, the number of iteration required by ISRA is a bit larger than that required by PL.

5 MAXIMUM-LIKELIHOOD

As is well known, the least-squares approach to the restoration problem can be derived from a *maximum-likelihood* approach in the case of additive, zero-mean, and Gaussian white noise. Indeed, if the variance of the noise at each pixel is σ_n^2, then the log-likelihood function is given by

$$l(f) = -\frac{1}{2\sigma_n^2} \|k * f - g_s\|^2 + \text{constant}. \tag{23}$$

The properties above are basically true for the readout noise but not for the photon noise. This is related to Poisson statistics, and an alternative to the previous methods consists in taking into account only this kind of noise.

The basic assumption is that the values of the image g in distinct voxels are the realizations of independent Poisson processes and that, in each voxel, the mean value of g is given by $k * f + b$, where f is the unknown object and b is the background. Then, by neglecting a term independent of f, the log-likelihood function is given by

$$l(f) = \sum_{j_1, j_2, j_3} [g \ln(k * f + b) - (k * f + b)], \tag{24}$$

where ln $(k * f + b)$ denotes the image obtained by taking the logarithm of $k * f + b$ voxel by voxel and the product of images $g \ \ln(k * f + b)$ is also a product voxel by voxel. As in the case of ISRA, we must assume that both the image and the PSF are non-negative.

The restoration problem can now be formulated as the problem of maximizing the function (24). It has been proved [1] that $l(f)$ is bounded and concave, so that it certainly has global maxima (in general, many). The problems that must be considered in such a case are as follows:

- Does a global maximum provide a sensible solution of the image-restoration problem?
- How can global maxima be computed?

As far as the first point is concerned, a precise mathematical answer has not yet been provided. There is evidence that the log-likelihood function may have several maxima; in addition, there is numerical evidence that these maxima may be affected by noise propagation. In other words, the problem of maximizing the log-likelihood function is ill-posed. This is not surprising, because it is true in the case of white Gaussian noise and there is no argument indicating that Poisson noise is more favorable than the Gaussian noise from the point of view of the restoration problem. In conclusion, some form of regularization should be introduced also for the maximization of $l(f)$.

A possible solution to this problem is provided by the method that has been proposed for computing the maxima of $l(f)$. This is a particular case of a general method, known in statistics as *expectation-maximization* (*EM*) [18]. It is also known as the *Lucy–Richardson* (*LR*) *method* [19,20] in the literature on image restoration problems. Therefore we shall call it the LR/EM method.

The LR/EM method is defined by the following iterative scheme:

$$f^{(m+1)} = \frac{1}{\eta} f^{(m)} k^T * \frac{g}{k * f^{(m)} + b}, \tag{25}$$

where the product and quotient of images are again defined voxel by voxel and the parameter η is given by

$$\eta = K(0, 0, 0) = \sum_{j_1, j_2, j_3} k(j_2, j_2, j_3). \tag{26}$$

If we assume that the PSF has been normalized in such a way that the sum of the voxel values is one, then we have $\eta = 1$.

It has been proved that the LR/EM iterates always converge to a maximum of $l(f)$ [18], independently of the initial guess $f^{(0)}$ (the unique condition on $f^{(0)}$ is that it must be strictly positive). Of course, if $l(f)$ has several maxima, the limit will depend on the choice of $f^{(0)}$, but, in practical applications, this point is ignored and the most popular choice is to take a constant array (e.g., 1 in all voxels). We discuss in a moment what has been learned from numerical practice. First, we point out some interesting features of the methods.

The first is that, if we start with a positive $f^{(0)}$, then all the $f^{(m)}$ will be non-negative. In addition, if it happens that, for a certain m, $f^{(m)}$ becomes zero in the voxel $\{j_1, j_2, j_3\}$, then all the subsequent iterates will be zero at that voxel.

The second property applies to the case of background zero ($b = 0$), and is obtained by summing the voxel values of both parts of Eq. (25). If $\eta = 1$, it is easy to obtain that, for any m,

$$\sum_{j_1, j_2, j_3} f^{(m)}(j_1, j_2, j_3) = \sum_{j_1, j_2, j_3} g(j_1, j_2, j_3), \tag{27}$$

and therefore each iterate is characterized by a total number of photons equal to that of the detected image.

A few comments should be made about the behavior of the iterates for increasing m. At the beginning, one obtains rather smooth restorations, which become sharper and sharper as m increases; but if the number of iterations is pushed too far, then artifacts related to noise appear. Such behavior is similar to that of the iterative methods discussed in the previous section, described as the semiconvergence property, and suggests that LR/EM has a regularizing effect on the maxima of $l(f)$. More precisely, if one computes the restoration error, as defined in Eq. (21), one finds that $\varrho^{(m)}$ first decreases, takes a minimum value, and then grows. In some cases, the minimum is so flat that the limit of the iterates seems to be reached. In any case, the number of iterations required for reaching the minimum is rather high, and increases with the SNR value. It is obvious that, also in this case, we have the problem of introducing stopping rules, but, up to now, no general solution to this problem has been found.

Finally, the implementation of LR/EM is as follows:

LR/EM algorithm {
  preprocessing
  {
    set to zero the negative values of k and g;
    evaluate the background of g;
    compute and store K normalized to $K(0, 0, 0) = 1$;
    initialize with $f^{(0)}$ = constant;
  }
  for each iteration m:

{
given $f^{(m)}$ compute $F^{(m)}$ by FFT;
compute $G^{(m)} = KF^{(m)}$;
compute $g^{(m)}$ by inverse FFT;
compute $h^{(m)} = g/(g^{(m)} + b)$;
compute $H^{(m)}$ by FFT;
compute $Q^{(m)} = K * H^{(m)}$;
compute $q^{(m)}$ by inverse FFT;
compute $f^{(m+1)} = f^{(m)}q^{(m)}$.
}
}

It follows that each iteration requires the computation of four FFTs, and therefore the computational cost of one LR/EM iteration is approximately twice the cost of one PL or ISRA iteration.

6 BAYESIAN METHODS

Regularization or iterative methods allow control of noise propagation from the data to the solution, since this propagation could be excessive as a consequence of the ill-conditioning of the problem. If one looks carefully at these methods, one understands that they are based on the use of some additional information on the solution, so-called *a priori information*: the object must be not too large, sufficiently smooth, non-negative, etc. This information is expressed in terms of rigid constraints on the solution, but it is also possible to introduce a sort of *probabilistic* a priori information by assuming that the components of the unknown object f are the values of random variables whose density functions are given. This is an old and fruitful idea [21], which, when combined with the ML approach, leads to the so-called *Bayesian methods*. Indeed, the likelihood function is interpreted as the conditional density function of the image g given a value f of the object, while the given density function of f is interpreted as the marginal density function of the object. Then, from Bayes' theorem [22], one can derive the joint probability distribution of the variables g and f as well as the conditional density function of the object f given a value g of the image. If it is possible to compute the global maximum of this conditional density function then one obtains the so-called *maximum a posteriori* (*MAP*) estimate of the unknown object f.

The simplest case is obtained by assuming that noise and object are independent Gaussian processes. The procedure outlined above then leads to the so-called *Wiener filter*, which has been shown to provide results similar to those obtained with the regularization method of Section 3. Indeed, a strict relationship between regularization methods and Wiener filter has been rigorously established [23].

Bayesian methods are very flexible, and allow the introduction of sophisticated constraints. The difficulty is that they are, in general, very expensive from a computational point of view. As far as we know, a rigorous analysis of the improvements that could be provided by these methods is not yet available.

7 SUPERRESOLUTION

Superresolution is a keyword used in several contexts and with quite different meanings. In most cases, it means an enhancement in resolution beyond the limit described for instance by the FWHM of the PSF of the optical instrument. Such an enhancement can be obtained, in some cases, by mean of a postprocessing of the image, using one of the restoration methods described in the previous sections. However, a more restrictive definition of superresolution is often used.

It is well known that the FWHM of the PSF is related to the band of the optical instrument. Therefore, an improvement in resolution could be obtained by extrapolation of the Fourier transform of the image outside this band. For this reason, the term superresolution is usually applied to attempts at recovering the object spectrum outside the band of the instrument [24].

Out-of-band extrapolation can be performed, at least in principle, under the assumption that the object has a finite spatial extent, because, in such a case, the Fourier transform of the object is an entire analytic function and hence is determined uniquely by the part of the Fourier transform transmitted by the optical system [25]. This argument, however, does not take into account the limitations due to the instability of analytic continuation in the presence of noise, which, in general, does not allow an improvement of resolution [26].

More recently, it was found [27] that a considerable amount of out-of-band extrapolation is possible, even in the presence of noise, if the linear dimensions of the object are not much greater than the resolution distance of the instrument (given, e.g., by the FWHM of the PSF). The potential applications of these results to confocal microscopy were soon recognized; the results of this research are summarized in Chapter 24 of this book [28].

As far as superresolution by data inversion is concerned, this can be obtained if the restoration method can implement the constraint on the support of the object. This property is possessed by some of the algorithms described in the previous sections, namely PL, ISRA, and LR/EM.

Let us assume that $D_1, D_2 \ldots, D_p$ are the domains (sets of voxels) where f is nonzero; the diameter of each domain must be not much larger than the FWHM of the PSF, while the distance between any pair of domains must be larger than the FWHM. Let D be the union of these domains, $D = D_1 \cup d_2 \cup \cdots \cup D_p$, that is, the support of the object. Then, as far as PL is concerned, it can be modified by zeroing the result of each iteration at the voxels outside D. For ISRA and LR/EM, the support constraint can be implemented by initializing the iterations with an image that is zero outside D. Indeed, as we have remarked, if the result of a certain iteration is zero at a certain voxel, then the results of all subsequent iterations will be zero at that voxel.

It is interesting to note that all of these methods also implement the constraint of positivity. Indeed, it is known from numerical practice that the combined use of positivity and support constraints can provide a considerable improvement of resolution [2].

8 BLIND DECONVOLUTION

All restoration methods, assume that the PSF of the optical system is known. As already remarked in Section 1, the latter can be computed or eventually measured by imaging a sphere with a radius smaller than the FWHM. However, the result of such a measurement can be poor because the SNR is too low. In such a case, it may be convenient to use methods designed to extract both the PSF and the underlying object f from the recorded image g, which are usually called *blind deconvolution* (*BD*) methods. These methods have been developed mainly in the case of astronomical images, but their application to microscopy could also be considered.

BD methods can be divided into two classes. In the first class, BD is attempted by minimizing an error function designed to optimize both the restored object and the PSF and to implement various kinds of constraints (positivity, support, etc.) [29,30]. The underlying object and the PSF are restored simultaneously by these methods.

The second class consists of methods that restore the object and the PSF separately in an iterative form: within each cycle, either the object or the PSF is held fixed while the other is updated, using one of the usual deconvolution techniques (remember that the convolution product is symmetric with respect to exchange of f and k). The cycles are repeated by alternately updating f and k, and therefore these methods are referred to as *iterative blind deconvolution* (IBD).

IBD was first proposed with the Wiener filter as deconvolution method [31]. However most of the IBD methods are based on the LR/EM deconvolution algorithm [32–35], so that inner LR/EM iterations are performed within each IBD cycle. In the simplest case, only one LR/EM is considered in each cycle, so that, if m is the index characterizing the cycles, we have (assume $K(0, 0, 0) = 1$ and put $G_0 = G(0, 0, 0)$)

$$k^{(m+1)} = \frac{1}{G_0} k^{(m)} [f^{(m)}]^T * \frac{g}{k^{(m)} * f^{(m)}}, \tag{28a}$$

$$f^{m+1} = f^{(m)} [k^{(m+1)}]^T * \frac{g}{k^{(m+1)} * f^{(m)}}. \tag{28b}$$

Here, it has been assumed that the background is zero. The initialization of the iterations is discussed in the papers mentioned above. It is obvious, however, that within each cycle, one can perform multiple LR/EM iterations, using eventually a different number of iterations for the object and the PSF. It is also obvious that one can use different deconvolution methods in the two cases. For instance, an IBD method based on projected Landweber methods with different kinds of constraints has been proposed [36].

We conclude by remarking that, if the form of the PSF is known in the sense that it depends on a (small) number of unknown parameters, then the problem is to estimate these parameters as well as the underlying object. For such a case, a method of *semiblind* deconvolution has been proposed [34].

9 CONCLUDING REMARKS

In this chapter, we have introduced four methods (linear filtering, the projected Landweber method, ISRA, and LR/EM) that can be considered as standard for approaching the image-restoration problem. In addition, we have briefly discussed their relationship to some non-standard topics such as Bayesian methods, superresolution, and blind deconvolution. Except in some particular cases, we believe that at least one of these methods can be conveniently used for practical restorations. However, we should like to conclude by stressing one important point. In general, three-dimensional images in microscopy are very large (e.g., $512 \times 512 \times 64$), so that the computational cost of their processing is an important issue. As shown in this chapter, when an estimate of the regularization parameter is available, linear filtering is the least-demanding method, because it requires only one inverse FFT. In addition, linear filtering works well in cases where the object to be restored is sufficiently smooth without too sharp intensity variations in the field. Therefore, linear filtering can be conveniently used for processing sets of images with this property and characterized by approximately the same signal-to-noise ratio, since the regularization parameter does not depend strongly on this quantity.

Iterative methods, which are much more demanding from the computational point of view, should be used when additional constraints are important for improving the quality of the restoration. From this point of view, PL is the most flexible one, because it allows the implementation of relevant constraints such as lower and upper bounds, support of the object, etc. In addition, the computational cost of one PL iteration is approximately one half the computational cost of one LR/EM iteration. The latter method is especially convenient in the case of dominant Poisson noise with low SNR.

REFERENCES

1. A. Periasamy, (Ed.) (2001). *Methods in Cellular Imaging*. Oxford University Press. New York.
2. Bertero, M. and Boccacci, P. *Introduction to Inverse Problems in Imaging. Bristol*: IOP Publishing, 1998.
3. Tikhonov, A.N. Solution of incorrectly formulated problems and the regularization. *Soviet Math Dokl.* **4**, 1035 (1963).
4. Tikhonov, A.N. and Arsenin, V.Y. *Solutions of Ill-Posed Problems*. New York: Winston/Wiley, 1977.
5. Groetsch, C.W. *Inverse Problems in the Mathematical Science*. Braunschweig: Vieweg, 1993.
6. Engl, H.W., Hanke M., and Neubauer, A. *Regularization of Inverse Problems*. Dordrecht: Kluwer, 1996.
7. Bertero, M., Boccacci, P., Brakenhoff, G. J, Malfanti, F., and van der Voort, H. T. M. Three-dimensional image restoration and super-resolution in fluorescence confocal microscopy. *J. Microsc.* **157**, 3 (1990).
8. Eicke, B., Iteration methods for convexly-constrained ill-posed problems in Hilbert space. *Num. Funct. Anal. Optim.* **13**, 413 (1992).
9. Carrington, W.A. Image restoration in 3D microscopy with limited data. *Proc. SPIE* **1205**, 72 (1990).
10. Carrington, W.A., Lynch, R.M., Moore, E.M., Isenberg, G., Fogartiy, K.E., and Fay, F.S. Superresolution three-dimensional images of fluorescence in cells with minimal light exposure. *Science*, **268**, 1483 (1995).
11. Carrington, W., Imaging live cells in 3D using wide-field microscopy with image restoration. In: *Confocal and Two-Photon Microscopy: Foundations, Applications and Advances* (ed. A. Diaspro). New York: Wiley-Liss, 2002, Chap. 16.
12. Lagendijk, R.L., Biemond, J., and Boekee, D.E. Regularized iterative image restoration with ringing reduction. *IEEE Trans. Acoust. Speech, Sig. Process.* **ASSP-26**, 1874 (1988).
13. Biemond, J, Lagendijk, R.L., and Marsereau, M. Iterative methods for image deblurring. *Proc. IEEE* **78**, 856 (1990).
14. van Kempen, G.M.P., van Vliet, L.J., Verveer, P.J., and van der Voort, H.T.M. A quantitative comparison of image restoration methods for confocal microscopy. *J. Microsc.* **185**, 354 (1997).
15. Daube-Whiterspoon, M.E. and Muehllerhner, G. An iterative image space reconstruction of algorithm suitable for volume ECT. *IEEE Trans. Med. Imag.* **MI-5**, 61 (1986).

16. Piana, M. and Bertero, M. Projected Landweber and preconditioning. *Inverse Problems* **13**, 441 (1997).
17. De Pierro, A.R. On the convergence of the iterative image space reconstruction algorithm for volume ECT. *IEEE Trans. Med. Imag.* **MI-6**, 174 (1987).
18. Shepp, L.A. and Vardi, Y. Maximum likelihood reconstruction for emission tomography. *IEEE Trans. Med. Imag.* **MI-1**, 113 (1982).
19. Richardson, W.H. Bayeisn-based iterative method of image restoration. *J. Opt. Soc. Am.* **62**, 55 (1972).
20. Lucy, L. An iterative technique for the rectification of observed distributions. *Astron. J.* **79**, 745 (1974).
21. Turchin, V.F., Kozlov, V.P., and Malkevich, M.S. The use of mathematical statistics methods in the solution of incorrectly posed problem. *Soviet Phys. Usp.*, **13**, 681 (1971).
22. Papoulis, A. *Probability, Random Variables and Stochastic Processes*. New York: McGraw-Hill, 1965.
23. Bertero, M., De Mol, C., and Viano, G. The stability of inverse problems. In: *Inverse Scattering Problems in Optics* (ed. H.P. Baltes). Springer-Verlag, 1980, p. 161.
24. Bertero, M., and De Mol. C. Super-resolution by data inversion. In: *Progress in Optics*, Vol. XXXVI (ed. E. Wolf). Amsterdam: Elsevier, 1996, p. 129.
25. Wolter, H. On basis analogies and principal differences between optical an electronic information. In: *Progress in Optics*, Vol. I, (ed. E. Wolf). Amsterdam: North-Holland, 1961, p. 155.
26. Toraldo di Francia, G. Degrees of freedom of an image. *J. Opt. Soc. Am.* **59**, 799 (1969).
27. Bertero, M. and Pike, E.R. Resolution in diffraction-limited imaging, a singular value analysis-I. The case of coherent illumination. *Optica Acta* **29**, 727 (1982).
28. Pike, R. Superresolution in fluorescence confocal microscopy and in DVD optical storage. In: *Confocal and Two-Photon Microscopy: Foundations, Applications, and Advances* (ed. A. Diaspro). New York: Wiley-Liss, 2002, Chap. 24.
29. Lane, R.G. Blind deconvolution of speckle images. *J. Opt. Soc. Am.* **A-9**, 1508 (1992).
30. Jefferies, S.M. and Christou, J.C. Restoration of astronomical images by iterative blind deconvolution. *Astrophys. J.* **415**, 862 (1993).
31. Ayers, G.R. and Dainty, J.C. Iterative blind deconvolution method and its applications. *Opt. Lett.* **13**, 547 (1988).
32. Holmes, T.J. Blind deconvolution of quantum-limited incoherent imagery: maximum-likelihood approach. *J. Opt. Soc. Am.* **A-9**, 1052 (1992).
33. Tsumuraya, F., Miura, N., and Baba, N. Iterative blind deconvolution method using Lucy's algorithm. *Astron. Astrophys.* **282**, 699 (1994).
34. Fish, D.A., Brinicombe, A.M., and Pike, E.R. Blind deconvolution by means of the Richardson–Lucy algorithm. *J. Opt. Sci. Am.* **A-12**, 58 (1995).
35. Biggs, D.S.C. and Andrews, M. Asymmetric iterative blind deconvolution of multiframe images. *Proc. SPIE* **3461**, 33 (1998).
36. Bertero, M., Bindi, D., Boccacci, P., Cattaneo, M., Eva, C., and Lanza, V. A novel blind-deconvolution method with an application to seismology. *Inverse Problems* **14**, 815 (1998).

CHAPTER 13

Two-Photon Excitation Fluorescence Microscopy Imaging in *Xenopus* and Transgenic Mouse Embryos

AMMASI PERIASAMY,[1,2] COLTEN NOAKES,[1,2] PAUL SKOGLUND,[2]
RAYMOND KELLER,[2] and ANN E. SUTHERLAND[3]

[1]W.M. Keck Center for Cellular Imaging and [2]Department of Biology, Gilmer Hall, University of Virginia, Charlottesville, VA 22903, U.S.A.
[3]Department of Cell Biology, University of Virginia Health Sciences Center, Charlottesville, VA 22908, U.S.A.

1 INTRODUCTION

The optical microscope has been an essential tool for generations of biologists, revealing details of individual cells and organisms invisible to the naked eye. Advances in optics, digital computers, digitizers and image-processing software, and low-light-level photodetectors have improved light-microscope technology (Bright and Taylor, 1986; Periasamy and Herman, 1994; Periasamy, 2001, Pawley, 1995, Inoué and Spring, 1997; Herman, 1998). The development of a large number of fluorescent dyes that reveal the chemical and molecular dynamics of intact cells has significantly broadened the usefulness of optical microscopy, allowing real-time observation and measurement of the activities of dye-marked constituents in cells and tissues (Tsien, 1989; Chalfie et al., 1994; Cubitt et al., 1995; Xu and Webb, 1996). A major problem in visualizing the structures of thick tissue has been the contribution of signal from above and below the plane of focus, which produces a background glow and results in degradation of the image. To remedy these problems and produce more reliable three-dimensional data, the most widely used techniques are digital deconvolution and confocal microscopy (White et al., 1987; Shaw, 1993; Chen et al., 1995; Diaspro et al., 1996). Laser scanning confocal microscopy (LSCM) uses a pinhole aperture to restrict the "out of focus" flare reaching a single detector, the photomultiplier tube (PMT). Because of the pinhole aperture, the entire fluorescence signal collected by the objective lens is not used (Wilson and Sheppard, 1984; Shotton, 1989; Pawley, 1995). In order to increase the signal, increased excitation laser intensity is used, but this will often introduce photobleaching and photodamage. Unlike the

Confocal and Two-Photon Microscopy: Foundations, Applications, and Advances, Edited by Alberto Diaspro.
ISBN 0-471-40920-0

confocal configuration, digital deconvolution microscopy (DDM) uses the entire fluorescence signal collected by the objective lens, without the use of pinhole or ultraviolet optics, to deliver the emitted light to two-dimensional high-sensitivity and linear CCD detectors. The "out-of-focus" flare introduced into the image at different optical sections is reversed by computer deconvolution through the use of a point spread function (PSF) of the imaging system (Agard et al., 1989).

In contrast, with two-photon excitation microscopy (TPEM), only a diffraction-limited spot is illuminated in the specimen, thereby reducing photobleaching and photodamage outside this limited area (Denk et al., 1990; Piston et al., 1994; Potter et al., 1996; So et al 1998; Wokosin et al., 1996; Hell et al., 1998). This two-photon absorption was theoretically predicted by Göppert-Mayer (1931) , and it was experimentally observed for the first time in 1961 by using a ruby laser as the light source (Kaiser and Garrett, 1961). Denk et al. (1990) demonstrated laser scanning two-photon fluorescence microscopy for biological applications for the first time using a mode-locked dye laser producing a 100 fs stream of pulses at 80 MHz and a BioRad MRC500 confocal scan head.

The two-photon system uses infrared pulsed laser light that illuminates only the focal plane, thus reducing autofluorescence and photobleaching. Although this technique is superior to others in fluorescence imaging, the two-photon images can still have a considerable amount of background glow deep down in the tissue. We propose that digital deconvolution software can improve these signals by eliminating the background noise in the two-photon images (Periasamy, 1998; Periasamy et al., 1999). We describe in this chapter how the deconvolution of two-photon images improved the visualization of optic vesicles of EGFP transgenic mouse embryos at about 90 μm deep. Also, we describe the comparison of three microscopy techniques (TPEM, LSCM, and DDM) for GFP–GAP-43 fusion protein images in developing *Xenopus* embryos.

2 MATERIALS AND METHODS

2.1 Two-Photon Excitation Microscopy

When an infrared laser beam is focused on a specimen, it illuminates a single point and the fluorescence emission is localized to the vicinity of the focal volume. The fluorescent intensity then falls off rapidly in the lateral and axial directions. In the confocal microscope, one-photon illumination occurs throughout the excitation-beam path. This results in absorption along this path, giving rise to substantial fluorescence emission both above and below the focal plane (Fig. 1). Such out-of-focal-plane excitation contributes to photobleaching and photodamage in the specimen planes that are not involved in two-photon imaging. Finally, the infrared illumination in two-photon imaging introduces considerably less autofluorescence and penetrates deeper into the living specimen than visible-light excitation.

For example, a fluorophore might absorb one photon at 365 nm and fluoresce at a blue wavelength around 460 nm. In contrast, two-photon (2p) excitation occurs when two photons of energies $h\omega$ and $h\omega'$ are absorbed simultaneously, and a molecule is excited to the state of energy $E_e = h\omega + h\omega'$. In other words, the fluorophore exhibits 2p absorption at approximately twice (730 nm) the one-photon absorption wavelength, while the wavelength of 2p emission is the same as that of one-photon emission (460 nm). The probability of two-photon absorption depends on the co-localization of two photons within

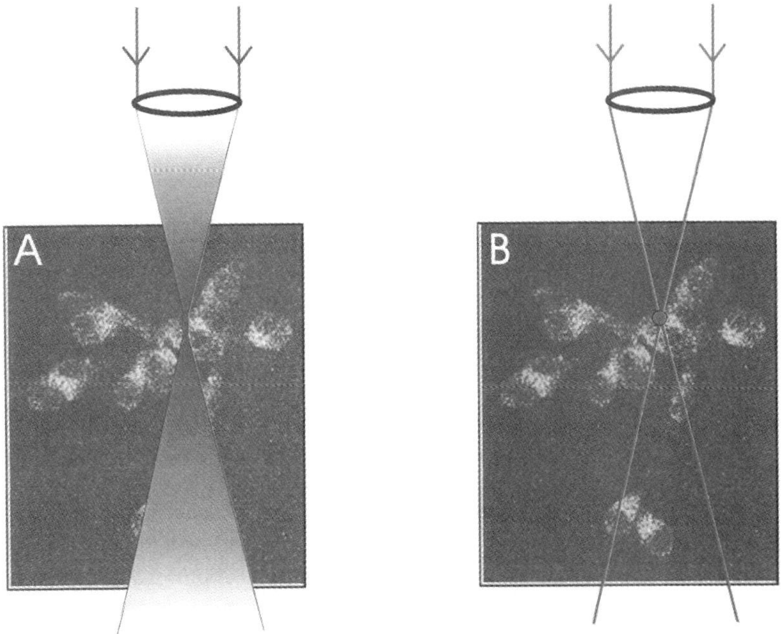

FIGURE 1 See ftp site for color figure. Illustration of excitation-light path in wide-field epifluorescence or confocal (*A*) microscopy and two-photon excitation microscopy (*B*). In (*A*), the illumination occurs throughout the field of view, whereas in two-photon-excitation fluorescence microscopy, it is limited to a spot at the focal plane of the infrared laser beam.

the absorption cross-section of the fluorophore. In addition, the rate of excitation is proportional to the square of the instantaneous intensity. Two-photon excitation is made possible by the extremely high local instantaneous intensity. This provided by the combination of diffraction-limited focusing of a single laser beam in the specimen plane and the temporal concentration of a femtosecond mode-locked laser – typically of the order of 10^{-50}–10^{-49} cm^4 s/photon/molecule (Denk et al., 1995). As we know, infrared-light excitation can penetrate deep inside tissue, and the emission signal is in the visible spectrum. The visible light is more scattered deep inside the tissue, so the light is decreased when it reaches the objective lens. Usually, the signals are noisy when they are acquired deep inside the tissue, but by deconvolving these signals, one can obtain a better signal-to-noise (S/N) ratio.

2.2 Instrumentation

The instrumentation of two-photon excitation microscopy (TPEM), laser scanning confocal microscopy (LSCM), and digital deconvolution microscopy (DDM) has been described in the literature (Periasamy et al., 1999). A brief introduction to the instrumentation used for the experiments is given below.

The TPEM system consists of a Nikon TE300 epifluorescent microscope with a 100 W mercury arc lamp for transillumination. A PlanFluor 20×, NA 0.75 multi-immersion (water and oil) objective lens was used for the image acquisition. A built-in lens and the dichroic in the filter cube were used to couple the TE300 externally (with no physical

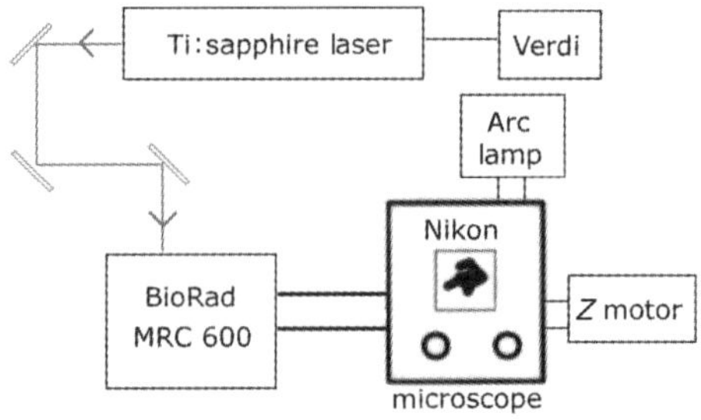

FIGURE 2 Schematic illustration of two-photon excitation fluorescence microscopy.

connection between the microscope and the confocal scan head) to the confocal laser head. A Coherent 5 W Verdi-pumped, tunable (model 900 Mira) mode-locked ultrafast (76 MHz) pulsed laser (Coherent Inc., Santa Clara, CA) was coupled to the laser port of a BioRad MRC600 (BioRad, Hercules, CA) laser scanning confocal scan head through beam-steering optics (Fig. 2). This laser was equipped with *x*-wave optics for an easily tunable range of the entire wavelength (720–920 nm). The silver-coated mirrors for maximum reflections in the infrared and visible spectrum (Chroma Technology Corp.) replaced the MRC600 conventional mirrors M1, *x*- and *y*-galvo, and concave (1 and 2) mirrors. The controller card of the MRC600 and the DOS-based software COMOS 7.0a were installed in a 450 MHz Pentium operating Windows '98. Confocal Imaging Assistant was used to convert the BioRad PIC file into separate TIF files for analysis and digital deconvolution.

The Nikon PCM2000 LSCM and DDM microscope were installed in one platform Nikon TE200 epifluorescence microscope (Nikon Inc., Melville, NY) to localize the GFP–GAP-43 fusion proteins in developing *Xenopus* embryos as shown in Figure 5 below. The hardware and software of the LSCM and DDM were interfaced to a personal computer through controllers and driven by the C-Imaging (Compix, Inc., Cranberry Township, PA) software (Periasamy et al., 1999). The DDM and LSCM use an arc lamp and lasers (argon-ion 488 nm (for GFP or EGFP) and 457 nm (for CFP), HeNe green, 543 nm, and HeNe yellow 598 nm) as light sources, respectively. An Orca-1 CCD camera (Hamamatsu Photonics System, Middlesex, NY) was installed in another port of the TE200 microscope for DDM. The quantum efficiency in the blue/green spectral region of the Orca-1 is about 50%. The data from the same region of the specimen for LSCM and DDM was acquired on the same optical platform. All images were acquired using filters from Chroma Technology (Brattleboro, VT).

2.3 Specimen Preparation

2.3.1 Xenopus Embryos To construct cpg–GFP fluorescent dye, double-stranded cDNAs of champignon (CDS) were cloned into pCS2mt–UGP. mRNA synthesis was achieved when cpg–GFP was linearized with *Not*I and used as a DNA template for capped mRNA synthesis using SP6 mMESSAGE mMACHINETM kit (Ambion). Eggs were obtained by injecting *Xenopus laevis* females with 400 units of human chorionic gonadotropin and were fertilized in vitro. Embryos were then dejellied with 2% L-cysteine hydrochloride monohydrate (Sigma) (pH 8.0) and cultured in 0.3X MBS. Developmental stages were determined according to Nieuwkoop and Faber (1967). At the appropriate stage, a total of 1 ng of the synthetic mRNA was microinjected into the marginal zone of four blastomeres of a 4-cell embryo in 0.3X MBS (including 4% Ficoll) and cultured until gastrula stage (stage 10). Open-faced explants of the dorsal marginal zone (DMZ) of early gastrula-stage embryo (stage 10) were then made by standard methods using a knife of eyelash hair and a hairloop (Wilson and Keller, 1991). Both sides of the explants were faced on coverglass, and the explants were cultured in DFA.

The preparation of the specimen for localizing the GFP–GAP-43 fusion protein in developing *Xenopus* using TPEM, LSCM, and DDM has been explained in the literature (Periasamy et al., 1999). In brief, capped in vitro transcribed RNA was prepared using the mMessage mMachine kit (Ambion). Plasmid encoding GFP–GAP-43 was linearized with *Not*I and transcribed with SP6 (Kim et al., 1998). This fusion protein, produced in the lab of Paul Garrity, combines a palymitoylation signal from GAP-43 with GFP (Chalfie et al., 1994), localizing GFP to the cell membrane. Embryos were injected with 5–20 pg RNA in three or four separate injections in the animal half of the embryo for animal caps and in various locations for intact embryo experiments, including vegetal injections. Injections were by standard methods (Kintner, 1988). Briefly, needles pulled (Pul-1, WP Instruments Inc.) from 0.75 mm internal-diameter glass capillaries (Harvard Apparatus) are mounted on a pressure-injection system (Medical Systems Corp.) and in vitro synthesized RNA is front-loaded into the needle. Drop sizes are calibrated by means of a normalized reticule, and embryos are injected in 1X MBS, 5% Ficoll, and transferred to 1/3X MBS after a minimum of 2 hours. MBS is 88 mM NaCl, 0.82 mM $MgSO_4$, 2.4 mM $NaHCO_3$, 1.0 mM KCl, 0.33 mM $Ca(N_3)_2$, 0.41 mM $CaCl_2$, and 10 mM Hepes pH 7.4.

2.3.2 Mouse Embryos Female C57BL/6 mice were mated to male mice of the strain C57BL/6 TgN(ACTbEGFP)1Osb (Jackson Laboratory), which are heterozygous for the EGFP transgene under the control of the chicken β-actin promoter. This transgene causes constitutive expression of EGFP in all cells of the embryo (Cubitt et al., 1995). The embryos resulting from this cross were harvested at day 9.5 of gestation, and those carrying the EGFP transgene were identified using a stereoscope equipped with fluorescence optics. Embryos expressing EGFP were placed into Eagle Plus medium (Stephens et al., 1995) in a coverslip chamber for examination.

3 DATA ACQUISITION

3.1 Point Spread Function (PSF)

The PSF was acquired in TPEM and LSCM to remove the optical noise in deep-tissue imaging of the biological living specimen. The PSF was also acquired for the conventional or wide-field microscopy to remove the out-of-focus information as explained by Periasamy et al. (1999). Digital deconvolution of all the images were performed on a Silicon Graphics Indigo2 (R10000 processor) running Deltavision 2.10 (Applied Precision, Issaquah, WA). The Deltavision used the algorithm called the constrained-iteration method (based on work by D.A. Agard and J.W. Sedat, University of California, San Francisco, CA; Agard et al., 1989). It uses an iterative process to predict the correct image and then convolution with the OTF to match the observed data.

The PSF images were acquired with a PlanFluor 20×, NA 0.75 multi-immersion objective lens (Fig. 3). The Ti : sapphire laser was tuned to an excitation wavelength of 780 nm and the two-photon image was acquired using a 650 nm short-pass (SP) filter

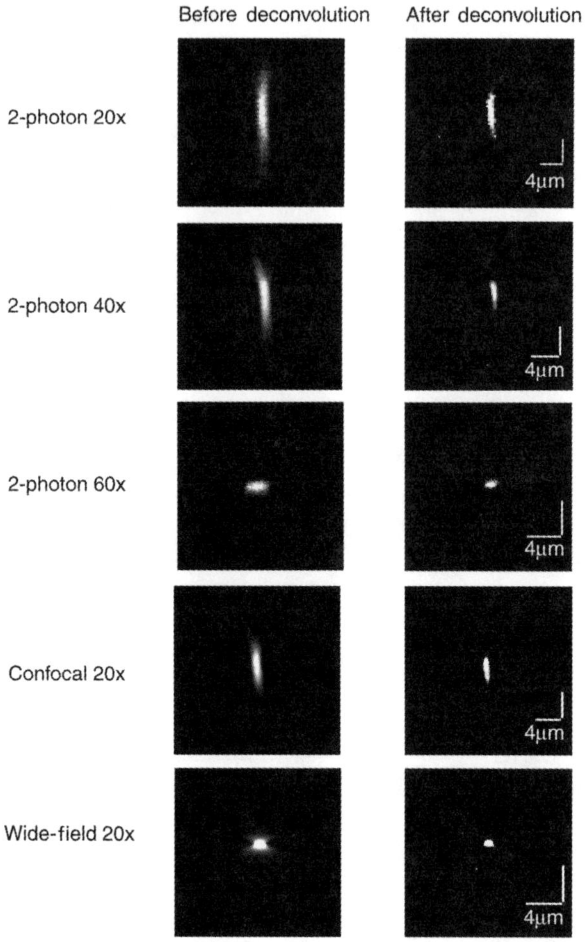

FIGURE 3 Point spread functions (PSFs) for various objective lenses and microscope systems (TPEM, LSCM, and DDM), with and without digital deconvolution. The PSFs were volume-rendered (360°) using Deltavision software (see Section 3.1 for details). The volume-rendered image clearly indicates approximately a single point (or spot) for DDM after removing the out-of-focus information. The volume-rendered images after deconvolution processing for 20×, 40×, and 60× objective lenses clearly demonstrate the depth of focus in TPEM, LSCM, and DDM microscope systems.

(Chroma Technology Corp., Brattleboro, VT). Using a slide of 0.3 μm to 1 μm beads, we selected a field of view that had a number of clearly fluorescing 1 μm beads in the same focal plane. After finding a suitable selection of beads, we exposed the beads to two photon absorption and measure the range of focus to do a z-series collection of all the focal planes at 0.5 μm slices. It was important to find a well-formed, even bead for optical sectioning, since the uniform blurring of out of focus information from this bead would be important when we deconvolved the living specimen. After collection and separation of the BioRad PIC file into separate TIF images, we transferred the files to a processing machine and isolated one bead to use as the PSF. A single bead was selected in a 16 × 16 pixel area from the acquired two-photon optical sections. The stack of 116 TIF files was inserted into the Deltavision PSF file converter to make an OTF algorithm. This algorithm was then saved in order to apply to the postcollection analysis of our living specimen data. Figure 3 demonstrates the volume-rendered (360°) PSF images for three objective lenses and for TPEM, LSCM, and DDM, with and without deconvolution. The image restoration of these kinds of PSF images has been described in the literature (Diaspro et al., 2000).

3.2 Image Acquisition – Biological Specimens

Two separate biological specimens were examined: explants of *Xenopus laevis* embryos, and E9.5 mouse embryos. The explants were fixed to coverslips adhered to the bottom of a Petri dish with holes cut for the coverslips to be viewed by the microscope. The Petri dish was filled with a solution of DFA. This set-up allowed us a number of different explants to image without having to remove and readjust the focus of the microscope. The mouse embryos were floating freely in Eagle Plus culture medium in a coverslip chamber, and we selected the region of the optic vesicle in the developing head for examination. We located the explant or embryo using an arc-lamp-illuminated GFP filter set. The excitation wavelength of 870 nm was tuned from the Ti : sapphire laser to excite the *Xenopus* explant or transgenic mouse embryo. We used the two-photon excitation to image the explant or the embryo at different optical sections using 650 nm SP as an emission filter. As the laser scanned, we focused down into the sample slowly at 4× zoom in order to see the depth measurement. When one focal extreme was realized, we focused back up through the sample to find the other focal extreme. Setting these two points in the COMOS software and adjusting the z section to 1 µm we collected 80 optically sectioned images at a size of 768 × 512 pixels. The pixel size at 4× zoom on the PlanFluor 20×, NA 0.75 multi-immersion lens is 0.214 µm/pixel, which made our field of view 164 µm × 110 µm. Using the CAS Confocal Assistant software, each BioRad PIC file was divided into separate TIFF stacks for digital deconvolution. After transferring the TIF files from the TPEM computer, we used the Deltavision software (Agard et al., 1989) to process them into a proprietary file format for the Unix-based software package. Figure 4 demonstrates a comparison between the processed and unprocessed optical sections of the living specimen at varying depths in the sample of *Xenopus* embryos.

The images were collected in developing *Xenopus* embryo to localize the GFP–GAP-43 fusion proteins using TPEM, LSCM, and DDM as explained in Periasamy et al. (1999). The objective lens (PlanFluor 20×, NA 0.75 multi-immersion lens) used for the data collection is the same as what we used for the TPEM images. The deconvolution processing also implemented as explained above for the GFP–GAP-43 fusion proteins (see Fig. 5).

4 RESULTS AND DISCUSSION

Applying digital deconvolution processing to images collected under two-photon excitation microscopy has three major advantages over unprocessed two-photon image acquisition: (1) it can allow us to see sections of the living specimen with higher spatial resolution than we could see before processing; (2) it decreases background noise globally and uniformly based on calibrated algorithms of the PSF, (3) it reverses any degradation of the image due to convolution, and thus improves the S/N ratio. In order for digital deconvolution processing to provide these advantages, however, several important criteria

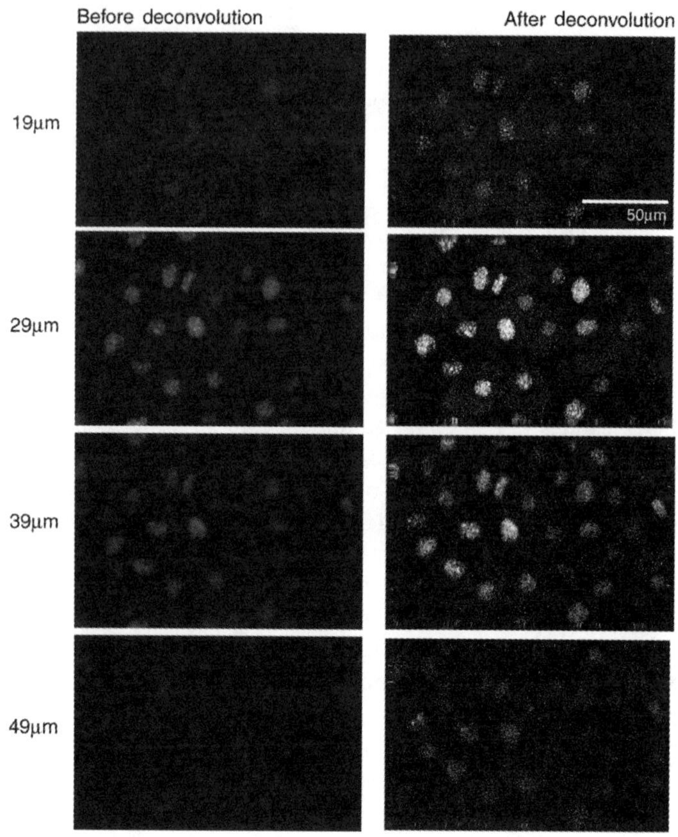

FIGURE 4 Comparison of unprocessed and processed images of the nuclei and cytoplasm outlines acquired from *Xenopus laevis* explants at different depths within the living specimen.

must be met. First, great care must be taken to ensure that the PSF acquired is an accurate representation of the flare of the lens and microscope. This must be practiced and tested on a number of beads to account for irregularities in the beads and/or light source. A good PSF should be taken for each lens in the microscope system for future work, to use as an image-processing standard. For some work, however, it may be necessary to customize the collection of the PSF and to oversee the acquisition of images under the same conditions as the living specimen. It is important to acquire the PSF and living specimen with the same lens, zoom, and pixel size. We used the acquired PSF data and processed using digital deconvolution software to visualize the true depth of focus of the respective objective lenses used for the image acquisition in TPEM, LSCM, and DDM systems as shown in Figure 3.

The digital deconvolution of images acquired from the living specimen at a depth of greater than 30 μm resulted in superior spatial resolution and background reduction than the unprocessed images. Even at the focal plane, where spatial resolution was fairly high to begin with, we reduced the background and increased signal to the point where we could view the nuclei in addition to the cytoplasm and suggestions of cell processes that could not be distinguished clearly before digital deconvolution. At the focal extremes in the

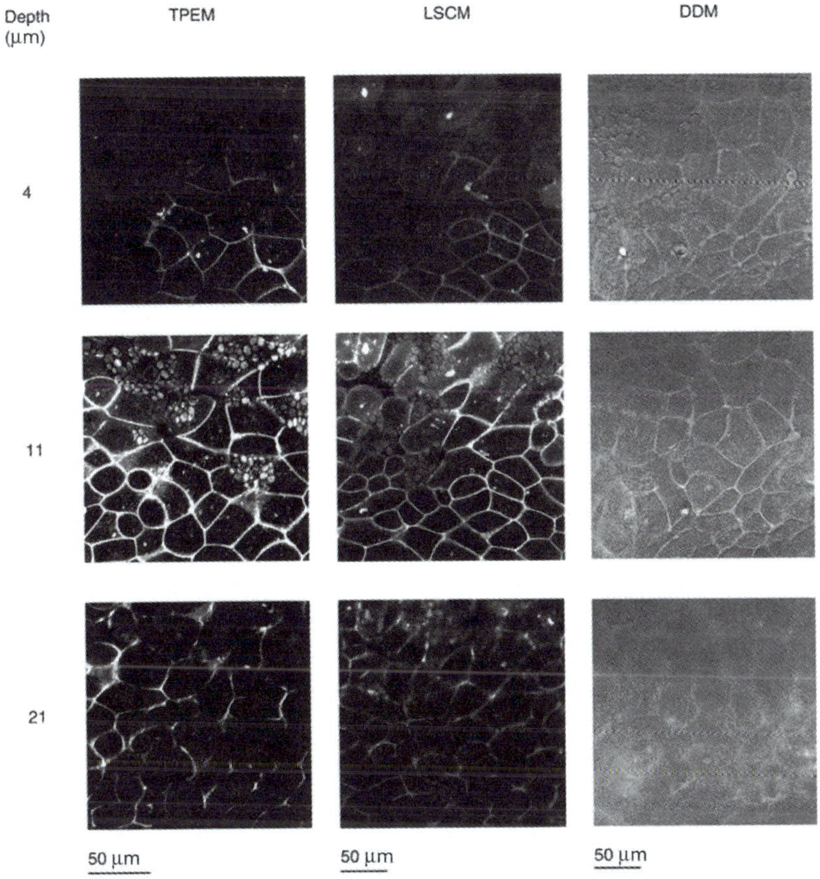

FIGURE 5 Localization of GFP–GAP-43 fusion protein images in developing *Xenopus* embryos was acquired with same objective lens using Nikon epifluorescence-based TPEM, LSCM, and DDM microscope systems. The details of the images are superior in TPEM mode. Generally, deterioration and less detail of information of GFP–GAP-43 proteins were observed in all sections of the tissue in LSCM mode. The DDM system is not good for tissue imaging, as demonstrated in the figure (images were deconvolved (15 iterations) using Deltavision software). (Adapted from Periasamy et al., 1999.)

Xenopus embryo explant, 19 μm and 49 μm, the upper and lower sides of the nuclei can be clearly distinguished from the background signal in the deconvolved images (Fig. 4). In the unprocessed images, the nuclei in these focal planes cannot clearly be distinguished from background signal, and thus provide no data.

Figure 5 shows the comparison of TPEM, LSCM, and DDM at different depths for localization of GAP-43 fusion proteins in developing *Xenopus* embryo. In our study, the deterioration of GFP–GAP-43 protein signal is clearly seen at 11 and 21 μm depths in LSCM. In DDM, the localization of GFP–GAP-43 protein image was clear at 11 μm but not at 21 μm. In all sections, two-photon signals are far superior to LSCM or DDM. Moreover, we did observe photobleaching of the specimen, which was very low in the TPEM compared with LSCM and DDM. It is important to adjust the excitation intensity and the gain of the PMT to obtain a better S/N ratio in LSCM. The DDM will be the ideal imaging system in cases where it is difficult to improve the S/N ratio of the image. The DDM is not good for real-time calcium or deep-tissue imaging, but it is good for three-dimensional double or triple labeling of cellular imaging. The TPEM is an ideal system for deep-tissue imaging and for UV fluorophore excitation compared with LSCM. The comparison of these three systems listed in the literature (Periasamy et al., 1999; Diaspro, 2001) clearly demonstrates the advantages of TPEM over LSCM and DDM (except in price). Currently, the price of the digital deconvolution system including the software for image acquisition and processing is almost equivalent to or sometimes more than LSCM.

In the living mouse embryo, we were able to resolve individual cells within the epithelium of the optic vesicle, at depths of up to 90 μm. In this case, the epithelial layer was only poorly distinguished from the background at 70–90 μm in the unprocessed images, but is well resolved in the deconvolved images, as shown by the arrows in Figure 6. The digital deconvolution process clearly provides greater imaging depth penetration within the living specimen and at a higher spatial resolution that the unprocessed two-photon images.

The enhanced imaging potential will make a number of new investigations possible, especially in the mouse. Owing to inaccessibility and complicated geometry, cell behavior

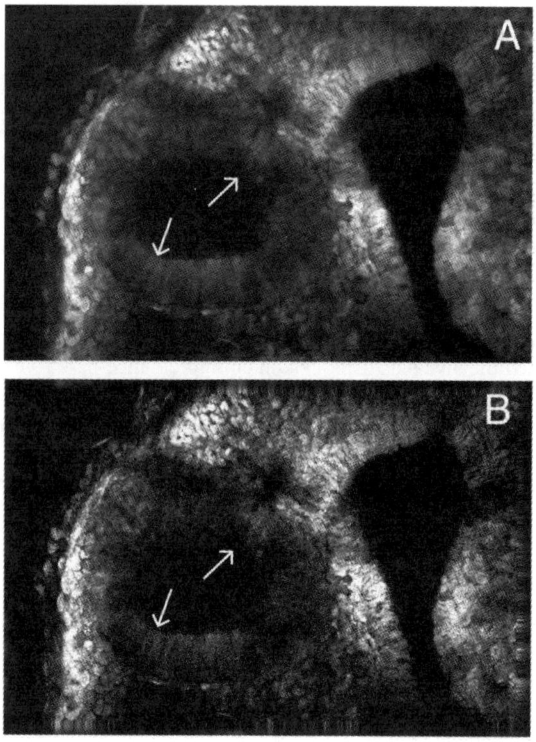

Depth = 90μm

FIGURE 6 Comparison of unprocessed and processed images of a region of the head (the optic vesicle is slightly left of center) acquired from E9.5 EGFP transgenic mouse embryos at 90 μm depths within the living specimen. (A) Before deconvolution; (B) after deconvolution. The digital deconvolution algorithm software clearly helps to resolve the optical vesicles (indicated by arrows) in a mouse embryo at 90 μm.

during morphogenesis in the mouse embryo has not been examined. For example, little is known about the behavior of mesoderm cells during gastrulation in the mouse embryo, owing to the difficulty of imaging at the midpoint of the cylindrically shaped embryo. At this stage, the embryo is approximately 150–200 μm in total thickness, which means that imaging the primitive streak at the midpoint should be well within the imaging capacity of the deconvolved two-photon system. Similarly, direct observation of the behavior of trophoblast cells during implantation into the uterus has not been possible, owing to the thickness and opacity of the decidual mass. Application of two-photon microscopy and subsequent deconvolution of time-lapse images will allow us to make some novel observations in both of these areas.

5 CONCLUSIONS

The contrast between true fluorescence signal and background absorption is increased to the maximum, which optimizes the two-photon data to see into the living specimen as deeply and as clearly as possible. This increase in contrast and resolution due to deconvolution processing software will allow three-dimensional details of the placement of cells in the tissue, intercellular processes and growth within the tissue, and even time-lapse imaging of deep-tissue events.

ACKNOWLEDGMENTS

This work was supported by a grant from the W.M. Keck Foundation and a grant RO1HD034807 to A.E.S.

REFERENCES

Agard, D. A., Hiraoka, Y., Shaw, P., and Sedat, J. W. (1989). Fluorescence microscopy in three dimensions. *Meth. Cell Biol.* **30**, 353–377.

Bright, G. R. and Taylor, D. L. (1986). Imaging at low light level in fluorescence microscopy. In: *Application of Fluorescence in the Biomedical Sciences* (ed. D. L. Taylor, F. Lanni, A. S. Waggoner, R. F. Murphy, and R. R. Birge). New York: Alan R. Liss, pp. 257–288.

Chalfie, M., Tu, Y., Euskirchen, G., Ward, W. W., and Prasher, D. C. (1994). Green fluorescent protein as a marker for gene expression. *Science* **263**, 802–805.

Chen, H., Swedlow, J. R., Grote, M., Sedat, J. W., and Agard, D. A. (1995) The collection, processing, and display of digital three-dimensional images of biological specimens. In: *Handbook of Biological Confocal Microscopy*. 2nd edn, (ed. J. M. Pawley). New York: Plenum Press, pp. 197–210.

Cubitt, A. B., Heim, R., Adams, S. R., Boyd, A. E., Gross, L. A., and Tsien, R. Y. (1995). Understanding, improving and using green fluorescent proteins. *Trends Biochem Sci.* **20**, 448–455.

Denk, W., Strickler, J. H., and Webb, W. W. (1990) Two-photon laser scanning fluorescence microscopy, *Science* **248**, 73–76.

Denk, W., Piston, D. W., and Webb, W. W. (1995). Two-photon excitation in laser scanning microscopy. In: *Handbook of Biological Confocal Microscopy* 2nd edn (ed. J. B. Pawley), New York: Plenum Press: pp. 445–458.

Diaspro, A. (2000). Building a two-photon microscope using a laser scanning confocal architecture. In: *Methods in Cellular Imaging* (ed. A. Periasamy), New York: Oxford University Press, pp. 162–179.

Diaspro, A., Beltrame, F., Fato, M., and Ramoino, P. (1996). Characterizing biostructure and cellular events in 2D/3D. *IEEE Eng. Med. Biol.* **15**, 92–100.

Diaspro, A., Annunziata, S., and Robello, M. (2000). Single-pinhole confocal imaging of sub-resolution sparse objects using experimental point spread function and image restoration. *Microsc. Res. Tech.* **51**, 464–468.

Göppert-Mayer, M. (1931). Über Elementarakte mit Quantenspreungen. *Annln. Phys.* **9**, 273–295.

Hell, S. W., Booth, M., and Wilms, S. (1998). Two-photon near- and far-field fluorescence microscopy with continuous-wave excitation. *Opt. Lett.* **23**, 1238–1240.

Herman, B. (1998). *Fluorescence Microscopy*, 2nd edn. New York: Springer-Verlag.

Inoué, S. and Spring, K. (1997). *Video Microscopy*, 2nd edn. New York: Plenum Press.

Kaiser, W. and Garrett, C. G. B. (1961). Two-photon excitation in $CaF_2 : Eu^{2+}$. *Phys. Rev. Lett.* **7**, 229–231.

Kim, S. H., Yamamoto, A., Bouwmeester, T., Agius, E., and De Robertis, E. M. (1998). The role of paraxial protocadherin in selective adhesion and cell movements of the mesoderm during *Xenopus* gastrulation. *Development* **125**, 4681–4691.

Kintner, C. (1988). Effects of altered expression of the neural cell adhesion molecule, N-CAM, on early neural development in *Xenopus* embryos. *Neuron* **1**, 545–555.

Nieuwkoop, P. D. and Faber, J. (1967). *Normal Table of* Xenopus Laevis. Amsterdam: North-Holland.

Pawley, J. B. (Ed.) (1995). *Handbook of Biological Confocal Microscopy*, 2nd edn. New York: Plenum Press.

Periasamy, A. (1998). Digital deconvolution FRET microscopy: 3D visualization of protein–protein interactions in a single living cell. *Proc. SPIE* **3260**, 106–114.

Periasamy, A., (Ed.) (2001). *Methods in Cellular Imaging*. New York: Oxford University Press.

Periasamy, A., Skoglund, P., Noakes, C., and Keller, R. (1999). An evaluation of two-photon excitation versus confocal and digital deconvolution fluorescence microscopy imaging in *Xenopus* morphogenesis. *Microsc. Res. Techn.* **47**, 172–181.

Periasamy, A. and Herman, B. (1994). Computerized fluorescence microscopic vision in the biomedical sciences. *J. Comput. Assist. Microsc.* **6**, 1–26.

Piston, D. W., Kirby, M. S., Cheng, H., Lederer, W. J., and Webb, W. W. (1994). Two-photon excitation fluorescence imaging of three-dimensional calcium-ion activity. *Appl. Opt.* **33**, 662–669.

Potter, S.M., Wang, C. M., Garrity, P. A., and Fraser, S. E. (1996). Intravital Imaging of green fluorescent protein using two-photon laser-scanning microscopy. *Gene* **173**, 25–31.

Shaw, P. J. (1993). Computer reconstruction in three-dimensional fluorescence microscopy. In: *Electronic Light Microscopy* (ed. D. Shotton). New York: Wiley-Liss, pp. 211–230.

Shotton, D. M. (1989). Confocal scanning optical microscopy and its applications for biological specimens. *J. Cell. Sci.* **94**, 175–206.

So, P. T. C., Kim, H., and Kochevar, I. E. (1998). Two-photon deep tissue ex vivo imaging of mouse dermal and subcutaneous structures. *Opt. Exp.* **3**, 339–350.

Stephens, L. E., Sutherland, A. E., Klimanskaya, I. V., Andrieux, A., Meneses, J., Pedersen, R. A. and Damsky, C. H. (1995). Deletion of beta 1 integrins in mice results in inner cell mass failure and peri-implantation lethality. *Genes Dev.* **9**, 1883–1895.

Tsien, R.Y. (1989). Fluorescent indicators of ion concentrations. *Meth. Cell Biol.* **30**, 127–156.

van der Voort, H. T. M. and Strasters, K. C. (1995). Restoration of confocal images for quantitative image analysis. *J. Microsc.* **178**, 165–181.

White, J. G., Amos, W. B., and Fordham, M. (1987). An evaluation of confocal versus conventional imaging of biological structures by fluorescence light microscopy. *J. Cell Biol.* **105**, 41–48.

Wilson, T. and Sheppard, J. R. (1984). *Theory and Practice of Scanning Optical Microscopy.* New York: Academic Press.

Wilson, P. and Keller, R. (1991). Cell rearrangement during gastrulation of *Xenopus*: direct observation of cultured explants. *Development* **112**, 289–300.

Wokosin, D. L., Centonze, V. E., White, J., Armstrong, D., Robertson, G., and Ferguson, A. I. (1996). All-solid-state ultrafast lasers facilitate multiphoton excitation fluorescence imaging. *IEEE J. Select. Top. Quantum Electron.* **2**, 1051–1065.

Xu, C. and Webb, W. W. (1996). Measurement of two-photon excitation cross-sections of molecular fluorophores with data from 690 nm to 1050 nm. *J. Opt. Soc. Am.* **B13**, 481–491.

CHAPTER 14

Multiple Color Fluorescence Imaging in Microscopy

K. R. CASTLEMAN

Advanced Digital Imaging Research, 2525 South Shore Blvd., League City, TX 77573, U.S.A.

1 INTRODUCTION

One of the most exciting recent developments in cytogenetics is multicolor fluorescence in situ hybridization (M-FISH). This technique employs labeling with multiple DNA probes, each using a different fluorophore. As many as seven (Ried et al., 1992) or eight (Morrison, 1998) different probes have been used, labeling each of the human chromosomes with a different combination of fluors (Nederlof et al., 1989), adding an Alu-banding pattern (Baldini and Ward, 1991; Matera and Ward, 1992), and a DNA counterstain.

This chapter covers the basics of fluorescence microscope imaging and digital processing of the resulting multispectral images.

2 FLUORESCENCE MICROSCOPE IMAGING

2.1 Fluorescence Microscope Optics

Figure 1 shows the basic elements of a digitizing epi-illuminated fluorescence microscope. The polychroic mirror reflects light at the excitation wavelengths down through the objective while passing emitted light up to the image sensor. The excitation filter blocks illuminating light at emission wavelengths, while the emission filter prevents light at excitation wavelengths from reaching the sensor.

Confocal and Two-Photon Microscopy: Foundations, Applications, and Advances, Edited by Alberto Diaspro.
ISBN 0-471-40920-0

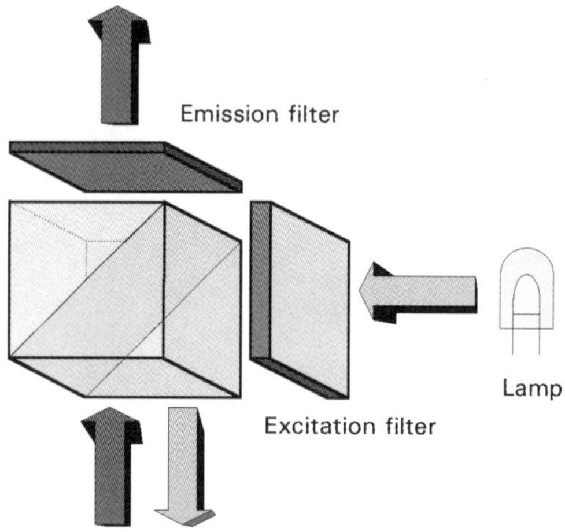

FIGURE 1 Fluorescence microscope optics. The polychroic mirror reflects excitation light down through the objective and passes emitted light up to the image sensor. The excitation filter blocks light at emission wavelengths, while the emission filter blocks light at excitation wavelengths.

2.2 Beer–Lambert Law

Fluorescence microscope imaging obeys, to a good approximation, the Beer–Lambert law. The intensity (gray level) recorded at a single pixel is (Garini et al., 1999; Castleman, et al., 2000)

$$I = tC \int_{\lambda} L(\lambda) F_{\mathrm{ab}}(\lambda) T_{\mathrm{ex}}(\lambda)\, d\lambda \int_{\lambda} R(\lambda) F_{\mathrm{em}}(\lambda) T_{\mathrm{em}}(\lambda)\, d\lambda, \tag{1}$$

where t is the integration time, $R(\lambda)$ is the proportion of photons emitted from the pixel volume of the specimen that are collected by the CCD pixel well and converted into photoelectrons, $F_{\mathrm{em}}(\lambda)$ is the emission spectrum of the fluor, normalized to unit area under the curve, $T_{\mathrm{em}}(\lambda)$ is the transmission spectrum of the emission filter, $L(\lambda)$ is the normalized spectrum of the excitation, $F_{\mathrm{ab}}(\lambda)$ is the normalized absorption spectrum of the fluor, and $T_{\mathrm{ex}}(\lambda)$ is the transmission spectrum of the excitation filter. The first integral quantifies the degree to which the fluor is excited, while the second accounts for emission and collection of the light.

The quantity

$$C = lS\rho\varepsilon\Phi L_0 A_D \tag{2}$$

collects the wavelength-independent constants. In this expression, l is the optical depth of the specimen, S is the area of a single pixel at the specimen, and ρ is the molar concentration of the fluor. Thus, the product lS is the volume of the pixel, and ρlS represents the quantity of fluor present at that pixel location. This is usually the quantity of interest, to be deduced from the observed brightness. The factor Φ is the quantum yield of the fluor, and A_D is the conversion factor between the number of photoelectrons collected by a pixel well on the CCD chip and the gray level of the corresponding pixel in the digital image. The extinction coefficient (probability of photon absorption) for the fluor at wavelength λ is $\varepsilon F_{ab}(\lambda)$. The excitation source flux is $L_0 L(\lambda)$.

2.3 Multicolor Fluorescence

In modern fluorescence microscopy, it is often useful to label different molecular components with different fluorophores. This permits visualization of several cellular components in spatial relation to each other inside the cell. One usually records several images, each taken with a different set of optical filters, where each filter set has been designed to match one of the fluors in use. This yields a multispectral image with one color channel per fluor.

2.4 Beer–Lambert Law for Multicolor Fluorescence Microscopy

For multicolor fluorescence microscopy, the Beer–Lambert law can be written as follows. The intensity (gray level), recorded at a single pixel in color channel i, that is due to the brightness of fluor j is

$$I_{i,j} = t_i C_j \int_\lambda L(\lambda) F_{ab,j}(\lambda) T_{ex,i}(\lambda)\, d\lambda \times \int_\lambda R(\lambda) F_{em,j}(\lambda) T_{em,i}(\lambda)\, d\lambda, \tag{3}$$

where t_i is the integration time in channel i, $F_{em,j}(\lambda)$ is the emission spectrum of fluor j, normalized to unit area under the curve, $T_{em,i}(\lambda)$ is the transmission spectrum of the emission filter for color channel i, $F_{ab,j}(\lambda)$ is the normalized absorption spectrum of fluor j, and $T_{ex,i}(\lambda)$ is the transmission spectrum of the excitation filter for color channel i. $R(\lambda)$ and $L(\lambda)$ are defined as before.

In the expression

$$C_j = lS\rho_j \varepsilon_j \Phi_j L_0 A_D, \tag{4}$$

ρ_j is now the molar concentration of the jth fluor and Φ_j is the quantum yield of the jth fluor. The extinction coefficient for the jth fluor at wavelength λ is $\varepsilon_j F_{ABS,j}(\lambda)$. The factors l, S, and A_D are as defined before.

The total intensity (gray level) recorded in color channel i is then

$$I_i = \sum_{j=1}^{N} I_{i,j}, \tag{5}$$

where N is the number of color channels (and of fluorophores) in use. These formulas hold for each pixel in the image separately.

3 IMAGE QUALITY

3.1 Spectral Overlap

It is clear from Eq. (5) that each fluor contributes to the recorded brightness in each of the color channels. If the fluor spectra do not overlap, and the excitation and transmission filter passbands are well designed, this overlap will be minimal. It is difficult, however, to achieve this ideal situation in practice. Many of the available fluorophores have relatively broad spectra. Fluor spectra commonly overlap, and ideally suited filters are seldom available off the shelf. Given the overlap of absorption and emission spectra, it is difficult to design a system so that each fluorophore appears completely isolated to a single color channel. Thus one is forced to work with multispectral images in which each color channel is contaminated with brightness from adjacent channels.

3.2 Color Compensation

If the degree of cross-channel coupling can be calculated (e.g., from Eqs. (3)–(5)) or measured (e.g., using images of single fluors), then it is possible to reduce or remove the effects of spectral overlap by image processing. We call this technique "color compensation" (Castleman, 1993, 1994, 1996). Digitally implemented color compensation can reduce or eliminate color spreading, making it practical to identify a larger number of cellular components in a single specimen than is otherwise possible.

3.2.1 Color Spreading Even with a monochrome camera and carefully selected fluorophores and filters, spectral overlap still spreads each fluorophore across the color channels to some extent. With broad-spectrum fluorophores, and especially when using a color CCD camera (where the sensitivity spectra are set by the chip manufacturer to match those of the human eye), the situation becomes even more severe, and can result in significant color spreading. Color spreading makes it more difficult to process the different fluorophore images independently, and it reduces the color saturation of FISH images when they are displayed.

3.2.2 Color compensation The linear preprocessing technique of "color compensation" (Castleman, 1993, 1994) can remove the spreading of multiple fluorophores among color channels and isolate each fluorophore to a single channel. It reduces or eliminates the aforementioned mixing of colors in the digitized image, yielding separate, quantitative maps of the distribution of each fluorophore. This simplifies subsequent image analysis, since each labeled molecular component of the specimen can then be addressed independently in a separate monochrome digital image that is in register with the other images of the set.

Color compensation is an image-processing step that operates on a color digital image immediately after acquisition. For an N-color system, it requires that each $N \times 1$ pixel vector be premultiplied by an $N \times N$ color compensation matrix. The elements of this matrix must be computed or determined experimentally for a particular fluorescent microscope imaging system and set of fluors.

The concept employed is that the light emitted from each fluorophore is distributed, by the imaging process, among the different color channels. We model the color spreading phenomenon as a linear transformation.

3.2.3 *The Color Spread Matrix* Let the $N \times N$ matrix $\mathbf{C}$ specify how the fluorophore brightnesses are spread among the color channels. Each element c_{ij} is the proportion of the brightness from fluorophore i that appears in color channel j of the digitized image. Ideally, $\mathbf{C}$ would be the identity matrix, but in practice this is not the case. Values for the color spread matrix elements can be calculated from Eqs. (3)–(5) or determined experimentally from digitized images of specimens stained with single fluorophores.

If $\mathbf{x}$ is the $N \times 1$ vector of actual fluorophore brightness values at a particular pixel, then the vector of gray levels recorded by the digitizer at that pixel is

$$\mathbf{y} = \mathbf{C}\mathbf{x} + \mathbf{b}, \tag{6}$$

where the vector $\mathbf{b}$ accounts for the black-level offset of the digitizer. That is, b_i is the gray level that corresponds to zero brightness in channel i.

3.2.4 *The Color Compensation Matrix* From Eq. (6), the matrix of true brightness values is

$$\mathbf{x} = \mathbf{C}^{-1}(\mathbf{y} - \mathbf{b}). \tag{7}$$

This equation can be used to correct the recorded image for spectral overlap. We call the inverse of the color spread matrix the "color compensation matrix", since it corrects the gray-level vector for the effects of color spreading. This development assumes that gray levels are linear with brightness. If necessary, a point operation (Castleman, 1996) can establish linearity prior to color compensation.

3.2.5 *Unequal Exposure Times* The foregoing analysis assumes that the exposure times are the same for each color channel – or at least the same as those used in the calibration study that determined the color spread matrix.

Sometimes, it is necessary to use different exposure times to accommodate large brightness variations among the fluors used. We can account for this in the following way. Let the diagonal matrix $\mathbf{E}$ specify the relative exposure time used in each channel in a particular digitization. That is, e_{ii} is the ratio of the current exposure time for color channel i to the exposure time used to calculate the color spread matrix. Equation (6) then becomes

$$\mathbf{y} = \mathbf{E}\mathbf{C}\mathbf{x} + \mathbf{b}, \tag{8}$$

which can be solved for

$$\mathbf{x} = \mathbf{C}^{-1}\mathbf{E}^{-1}(\mathbf{y} - \mathbf{b}). \tag{9}$$

Since $\mathbf{E}$ is a diagonal matrix, its inverse is also diagonal, having diagonal elements that are simply the reciprocals of the corresponding elements of $\mathbf{E}$. Further, $\mathbf{C}^{-1}\mathbf{E}^{-1}$ can be thought of as a modified color compensation matrix. It is merely $\mathbf{C}^{-1}$ after each ith column has been divided by e_{ii}. Thus, there is a simple way to modify the color compensation matrix to account for variation in exposure time. One simply divides the elements in column i by the factor by which the exposure has been changed in channel i.

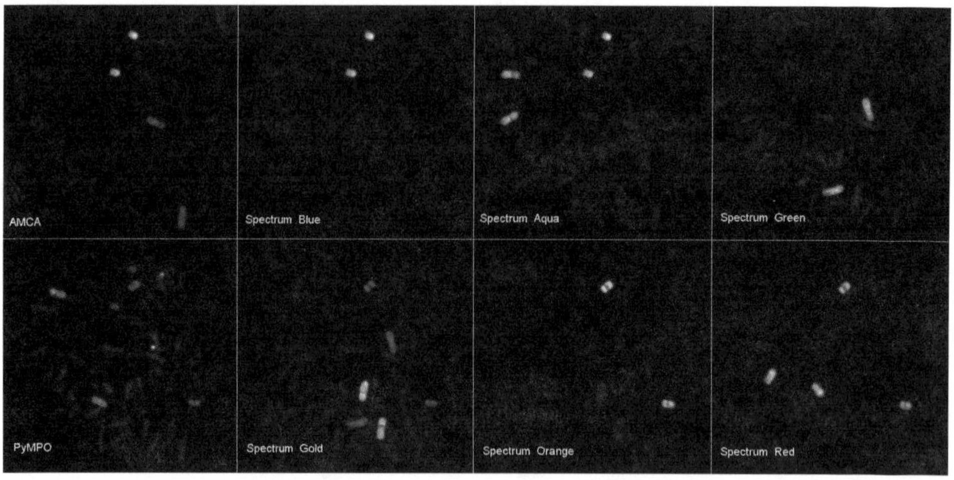

FIGURE 2 An eight-channel image of a human metaphase chromosome spread (Morrison 1998). The eight fluors used are AMCA, spectrum blue, spectrum aqua, spectrum green, PyMPO (yellow), spectrum gold, spectrum orange, and spectrum red.

3.3 An Example

Figures 2 and 3 show an example of color compensation in an eight-channel fluorescence microscope image. Notice, for example, that the two chromosomes in channel 2 also show up brightly in channels 1 and 3 prior to compensation, but they are effectively removed in Figure 3. Likewise, the two orange chromosomes are removed from the red channel.

3.4 Noise

There are four categories of noise that affect fluorescent microscope images. These are photon noise, readout noise, dark-current noise, and biochemical noise. Photon noise

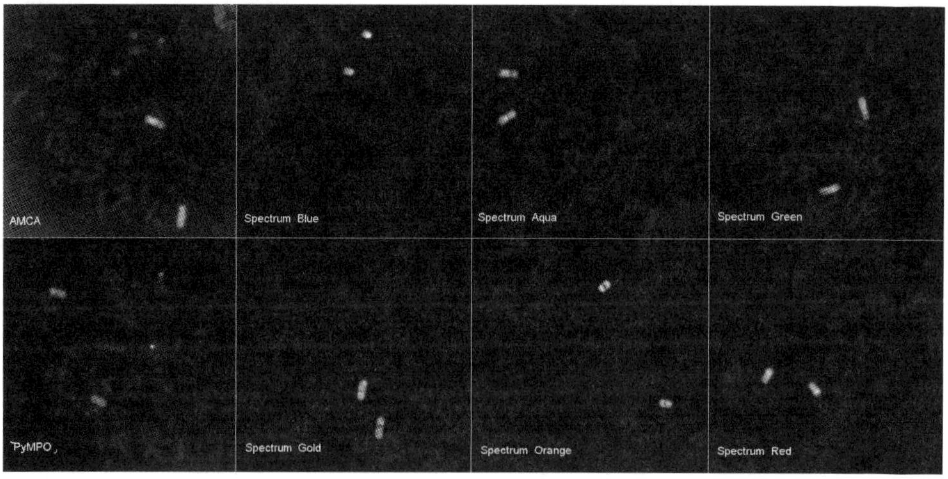

FIGURE 3 Color-compensated output of the eight-channel image of Figure 2 (Morrison 1998).

stems from the statistical nature of light. Each camera pixel captures a finite number of photons, and this number has a statistical distribution. This type of noise is random with a Poisson distribution that can usually be assumed to be Gaussian (normal). Its effect decreases at longer exposures.

Readout noise is also random, and originates in the circuitry on the CCD chip. It is at its worst at short exposures and rapid image readout rates.

Dark-current noise results from the leakage, on the CCD chip, of electrons that are not due to photon absorption. It manifests itself as a fixed "starfield" pattern superimposed on the image. Since it is fixed, it can be recorded, stored, and subtracted out. It is worse at long exposures.

One wishes to set the exposure time long enough that photon and readout noise are swamped by the signal, but short enough that dark-current noise is not a problem.

Biochemical noise results from nonuniform hybridization of the specimen, and can only be controlled at the specimen preparation bench.

4 IMAGE SEGMENTATION

It is often necessary to isolate the individual objects in a multispectral fluorescence image. We shall consider two approaches to segmenting such an image into discrete objects.

4.1 Thresholding

One straightforward approach to segmenting a multispectral fluorescence image is simply to perform a thresholding operation (Castleman, 1996) in each color channel. Then, depending on which fluors are present or absent, the different colored objects can be identified. Thresholding is subject to errors caused by nonuniform excitation and uneven hybridization. Preprocessing the images with normalization steps can significantly improve the segmentation results.

4.2 Color Histograms

In a two-color image (e.g., blue cells with red dots on a black background), it is often useful to display a color histogram. This is a scatterplot showing how many pixels have each possible combination of brightness in the two channels (see Fig. 4). The different colored objects in the image will produce corresponding peaks in the histogram. The locations of these peaks can guide the selection of thresholds for segmentation. One can segment the image by establishing nonlinear boundary lines in the histogram to define regions in color space that correspond to each type of object.

One can measure the area of certain colored objects by integrating the volume under the corresponding peak. One can even "weigh" the objects by integrating the product of the gray level with the histogram value (Castleman, 1996).

As an example of measuring objects from a color histogram, consider the case of a cell labeled with red and green fluors (Fig. 4). It is desired to determine what proportion of the red fluor overlies the green fluor. One would expect to find four peaks in the red-versus-green color histogram, one corresponding to isolated red fluor, one corresponding to isolated green fluor, one corresponding to overlapping red and green fluors (yellow), and a

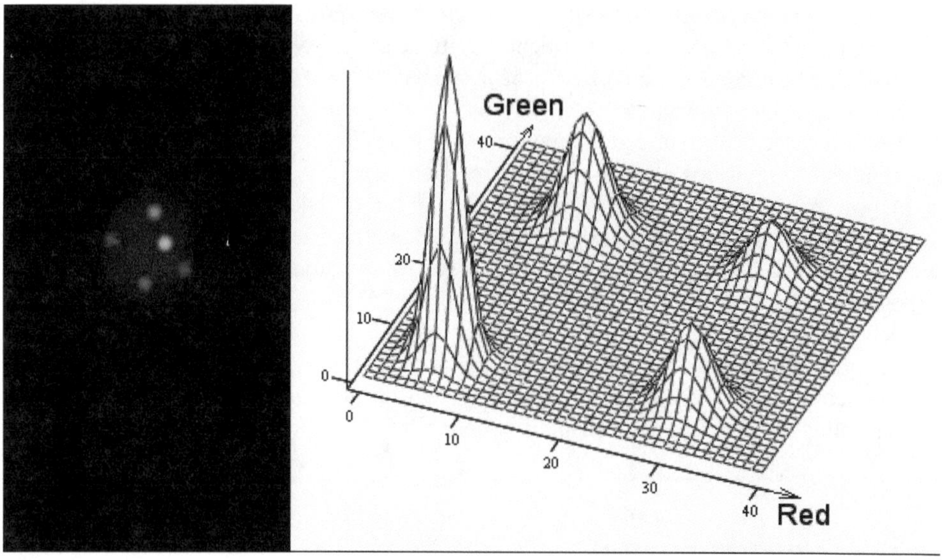

FIGURE 4 See ftp site for color figure. Use of the two-color histogram (a simulation). The cell at the left has two red and two green dots, and one yellow dot where the red and green fluors overlap. The red-versus-green histogram shows peaks corresponding to background, red dots, green dots, and the yellow dot. The volume under each peak is the total area of the corresponding dots.

background peak. The volume under each peak indicates the area of the corresponding region in the image.

4.3 Classification in Color Space

Each pixel can be considered to represent a point in N-dimensional color space, where N is the number of color channels digitized. Thus, each of the different colored objects in the image will correspond to a "cloud" of points in color space, and segmentation becomes the task of isolating these clusters.

4.3.1 Bayes Classifier One straightforward and quite powerful approach is the use of the Bayes maximum-likelihood classifier. It generates second-order surfaces that partition the color space into disjoint regions, one for each object type (i.e., for each color). Assuming Gaussian distributions for the clusters of points in color space, the Bayes classifier maximizes the probability that each pixel will be assigned correctly.

We illustrate the use of the Bayes classifier with a simple example. We assume that a three-color system (red, green, blue) is used to digitize a fluorescent microscope image. The vector of gray levels at a single pixel location is

$$\mathbf{x} = \begin{bmatrix} x_1 \\ x_2 \\ x_3 \end{bmatrix}, \qquad \text{where} \qquad j = \begin{cases} 1 \Rightarrow \text{red}, \\ 2 \Rightarrow \text{green}, \\ 3 \Rightarrow \text{blue}. \end{cases} \tag{10}$$

We further assume that the images contain two types of objects: normal and abnormal cells. Both types of cells bind the blue fluor, the normals also bind the green fluor, but the abnormals pick up the red fluor instead. One would expect a three-dimensional histogram (scatterplot) of the color space to show three clusters of points: one each for background, normal cells and abnormal cells. Since the background is dark, its cluster will fall near the origin of color space. The normal cells will give rise to a cloud of points near the blue–green corner, while the abnormals will fall near the red–blue corner.

4.3.2 Prior Probabilities Let us assume that the area of a typical image is 90% occupied by background and 10% by cells. Further assume that, overall, only 10% of the cells are abnormal. Thus the vector of prior probabilities, where $P_i = P\{\text{pixel belongs to class } i\}$, is

$$\mathbf{P} = \begin{bmatrix} 0.90 \\ 0.09 \\ 0.01 \end{bmatrix}, \qquad \text{where} \qquad I = \begin{cases} 1 \Rightarrow \text{background.} \\ 2 \Rightarrow \text{normal,} \\ 3 \Rightarrow \text{abnormal.} \end{cases} \tag{11}$$

4.3.3 Classifier Training The first step is to train the classifer to recognize the three types of pixels. For this, we require a training set containing pixels that fall in the background, in normal, and in abnormal cells. It is the statistics of these training-set pixels that constitute the knowledge that the classifer has about the problem.

4.3.4 The Mean Vector Using the training set, we calculate, for each class i, the mean brightness, in each color j, of a pixel. That is,

$$\mu_{ij} = \frac{1}{N_i} \sum_{k=1}^{N_i} x_{ijk}, \tag{12}$$

and x_{ijk} is the value, in color j, of pixel k in class i of the training set. The mean vector for class i is

$$\boldsymbol{\mu}_i = \begin{bmatrix} \mu_{i1} \\ \mu_{i2} \\ \mu_{i3} \end{bmatrix}. \tag{13}$$

This vector, the mean vector of the training set, is an estimate of the mean of the entire population of class i pixels. Unless the training set is both adequately large and representative of the population, it will be a poor estimate, and classifier accuracy will suffer.

4.3.5 Covariance We calculate the covariance matrix for each class as

$$\mathbf{S}_{ij_1j_2} = \frac{1}{N_i - 1} \left(\sum_{k=1}^{N_i} x_{ij_1k} x_{ij_2k} - N_i \mu_{ij_1} \mu_{ij_2} \right), \tag{14}$$

4.3.6 Variance and Standard Deviation The diagonal elements of the covariance matrix are the variances of the features, for that class. The variance is the square of the standard deviation. That is,

$$\sigma_{ij}^2 = S_{ijj}, \qquad \sigma_{ij} = \sqrt{S_{ijj}}. \tag{15}$$

4.3.7 Correlation From the covariance matrix we can compute the correlation matrix for each class. For class i, this is

$$\mathbf{C}_{ij_1j_2} = \frac{\mathbf{S}_{ij_1j_2}}{\sigma_{ij_1}\sigma_{ij_2}}. \tag{16}$$

The elements of the correlation matrix are bounded by ± 1. A correlation of 1.0 means that the corresponding two features are proportional to each other. A correlation of -1 means that each is proportional to the negative of the other. A correlation of zero means that the two features are uncorrelated (statistically independent).

Not only is the use of highly correlated features redundant, but also it can degrade classifier accuracy. One can either combine correlated features (e.g., by averaging), or simply discard all but one of them.

4.3.8 The pdf The probability density function for class i is

$$p_i(\mathbf{x}) = \frac{1}{\sqrt{(2\pi)^3|\mathbf{S}_i|}}\exp[-\tfrac{1}{2}|(\mathbf{x}-\boldsymbol{\mu}_i)^T\mathbf{S}_i^{-1}(\mathbf{x}-\boldsymbol{\mu}_i)|], \tag{17}$$

where $\mathbf{x}$ is the vector of RGB values for a pixel, as in Eq. (10). The superscript T indicates the matrix transpose, and $\mathbf{S}_i^{-1}$ is the inverse of the covariance matrix for class i.

4.3.9 Classification Each class of pixels is now characterized by its prior probability, mean vector, and covariance matrix. The classifier has been trained. We now have enough statistical information about the problem to begin classifying pixels. By Bayes' rule, the likelihood that an unknown pixel having color vector $\mathbf{x}$ belongs to class i is

$$L_i = P_i\frac{\mathrm{pdf}_i(\mathbf{x})}{p(\mathbf{x})}, \qquad \text{where} \qquad p(\mathbf{x}) = \mathrm{pdf}_1(\mathbf{x}) + \mathrm{pdf}_2(\mathbf{x}) + \mathrm{pdf}_3(\mathbf{x}), \tag{18}$$

or

$$L_i = \frac{P_i}{p(\mathbf{x})\sqrt{(2\pi)^3|\mathbf{S}_i|}}\exp\left[-\tfrac{1}{2}|(\mathbf{x}-\boldsymbol{\mu}_i)^T\mathbf{S}_i^{-1}(\mathbf{x}-\boldsymbol{\mu}_i)|\right]. \tag{19}$$

Thus, we can compute the three likelihoods (one for each class) and assign the pixel to the most likely (largest likelihood) class. If the pdfs are, as we have assumed, Gaussian (normal) density functions, then no other partitioning of color space will result in lower overall error rates (Castleman, 1996).

4.3.10 *Log-Likelihoods* Since the logarithm is a monotonic function, we can take the logarithm of both sides of Eq. (19) and use the resulting value for classification purposes. Equation (19) becomes

$$\ln L_i = \ln P_i - \tfrac{1}{2}|(\mathbf{x} - \boldsymbol{\mu}_i)^T \mathbf{S}_i^{-1} (\mathbf{x} - \boldsymbol{\mu}_i)| - \tfrac{3}{2}\ln 2\pi - \tfrac{1}{2}\ln|\mathbf{S}_i| - \ln p(\mathbf{x}). \tag{20}$$

The third and fifth terms are constants, and, for classification purposes, can be dropped, leaving

$$\mathrm{LL}_i = \ln P_i - \tfrac{1}{2}|(\mathbf{x} - \boldsymbol{\mu}_i)^T \mathbf{S}_i^{-1} (\mathbf{x} - \boldsymbol{\mu}_i)| - \tfrac{1}{2}\ln|\mathbf{S}_i|. \tag{21}$$

The first term accounts for the prior probabilities, while the third term accounts for the within-class variation of the features. The larger this variation is, the less confidently one can assign the pixel to that class. The second term is the square of the "Mahalanobis distance". It represents the variance-normalized distance, in feature space, from the unknown color to the class mean.

4.3.11 *Distance Classifier* We can simplify the Bayes classifier further by computing only the Mahalanobis distance from an unknown pixel to each class mean

$$D_i = \sqrt{\tfrac{1}{2}|(\mathbf{x} - \boldsymbol{\mu}_i)^T \mathbf{S}_i^{-1} (\mathbf{x} - \boldsymbol{\mu}_i)|}, \tag{22}$$

and assigning each pixel to the class having the nearest mean. This results in what is called a "distance classifier". It corresponds to the special case where the prior probabilities are equal among the classes, and likewise for the within-class variations. A distance classifier is sometimes used when the prior probabilities are unknown and the covariance matrix cannot be estimated accurately.

4.3.12 *Uncorrelated Features* While 30 or so pixels per class in the training set might be sufficient to estimate the means and variances, considerably more might be required to estimate the off-diagonal elements of the covariance matrix. One solution, if the training set is small, is to set the off-diagonal elements to zero. This is equivalent to assuming that the features are uncorrelated. Under pressure of inadequate training-set size, this can yield a more stable and better performing classifier than one designed around inadequately estimated covariances. A distance classifier, which automatically assumes uncorrelated features, results in an even simpler classifier. Since there are so many pixels in an image, however, it is seldom difficult to accumulate training sets that are quite large, and covariances can usually be estimated accurately.

4.4 A Numerical Example

To illustrate the operation of the Bayes classifier we include a numerical example having six pixels from each class in the training set. While this is hopelessly inadequate for any real case, it serves to illustrate the calculations. This example will permit readers who choose to implement a Bayes classifier to check their implementation for numerical accuracy.

Assume that the training set is

$$\begin{bmatrix} 30 & 34 & 47 & 57 & 18 & 6 \\ 45 & 8 & 21 & 38 & 58 & 16 \\ 61 & 50 & 34 & 22 & 33 & 4 \end{bmatrix}, \quad \begin{bmatrix} 6 & 28 & 47 & 20 & 59 & 38 \\ 96 & 115 & 127 & 84 & 73 & 99 \\ 69 & 82 & 97 & 117 & 92 & 121 \end{bmatrix},$$
$$\begin{bmatrix} 76 & 115 & 125 & 74 & 136 & 56 \\ 13 & 51 & 34 & 69 & 24 & 19 \\ 81 & 101 & 147 & 78 & 70 & 111 \end{bmatrix} \tag{23}$$

for the RGB color values of six pixels each of background, normal, and abnormal, respectively. The mean vectors for the three classes are then

$$\boldsymbol{\mu}_1 = \begin{bmatrix} 32 \\ 31 \\ 34 \end{bmatrix}, \quad \boldsymbol{\mu}_2 = \begin{bmatrix} 33 \\ 99 \\ 96 \end{bmatrix}, \quad \boldsymbol{\mu}_3 = \begin{bmatrix} 97 \\ 35 \\ 98 \end{bmatrix}. \tag{24}$$

The covariance matrices are

$$\mathbf{S}_1 = \begin{bmatrix} 346 & -7.4 & 94.4 \\ -7.4 & 365.6 & 69.8 \\ 94.4 & 69.8 & 406 \end{bmatrix} \quad \mathbf{S}_2 = \begin{bmatrix} 364 & -17.6 & 101.8 \\ -17.6 & 390 & -54.8 \\ 101.8 & -54.8 & 405.6 \end{bmatrix}$$
$$\mathbf{S}_3 = \begin{bmatrix} 1056 & 33.4 & 123.6 \\ 33.4 & 454.8 & -41.4 \\ 123.6 & -41.4 & 810.4 \end{bmatrix}, \tag{25}$$

from which the standard deviations are

$$\boldsymbol{\sigma}_1 = \begin{bmatrix} 18.6 \\ 19.1 \\ 20.1 \end{bmatrix}, \quad \boldsymbol{\sigma}_2 = \begin{bmatrix} 19.1 \\ 19.7 \\ 20.1 \end{bmatrix}, \quad \boldsymbol{\sigma}_3 = \begin{bmatrix} 32.5 \\ 21.3 \\ 28.5 \end{bmatrix}, \tag{26}$$

and the correlation matrices are

$$\mathbf{C}_1 = \begin{bmatrix} 1 & -0.021 & -0.252 \\ -0.021 & 1 & 0.181 \\ 0.252 & 0.181 & 1 \end{bmatrix}, \quad \mathbf{C}_2 = \begin{bmatrix} 1 & -0.047 & 0.265 \\ -0.047 & 1 & -0.138 \\ 0.265 & -0.138 & 1 \end{bmatrix},$$
$$\mathbf{C}_3 = \begin{bmatrix} 1 & 0.048 & 0.134 \\ 0.048 & 1 & -0.068 \\ 0.134 & -0.068 & 1 \end{bmatrix}, \tag{27}$$

and the determinants of the covariance matrices are

$$|\mathbf{S}_1| = 4.629 \times 10^7, \quad |\mathbf{S}_2| = 5.251 \times 10^7, \quad |\mathbf{S}_3| = 3.792 \times 10^8. \tag{28}$$

Suppose that we have an unknown pixel having color vector

$$\mathbf{x} = \begin{bmatrix} 33 \\ 99 \\ 95 \end{bmatrix}. \tag{29}$$

The three likelihoods Eq. (19), are $L_1 = 7.976 \times 10^{-5}$, $L_2 = 0.09$, and $L_3 = 4.064 \times 10^{-6}$, and the pixel would be assigned to class 2. The log-likelihoods, with constant terms dropped, Eq. (21), are $\mathrm{LL}_1 = -18.3$, $\mathrm{LL}_2 = -11.3$, and $\mathrm{LL}_3 = -21.3$. Again, the pixel would be assigned to class 2. The Mahalanobis distances to the class means, Eq. (22), are $D_1 = 3.065$, $D_2 = 0.037$, and $D_3 = 2.612$. This pixel would be assigned to class 2 by a distance classifier as well.

A well-trained Bayes classifier can be quite effective at multispectral pixel classification, even in the presence of considerable noise. One should pay particular attention to numerical-precision issues since some of the parameters in the probability calculations can become quite small or quite large.

REFERENCES

Baldini, A. and Ward, D. C., (1991). In situ hybridization banding of human chromosomes with Alu-PCR products: a simultaneous karyotype for gene mapping studies. *Genomics* **9**, 770–774.

Castleman, K. R. (1993). Color compensation for digitized FISH images. *Bioimaging* **1**, 159–165.

Castleman, K. R. (1994). Digital image color compensation with unequal integration periods. *Bioimaging* **2**, 160–162

Castleman, K. R. (1996). *Digital Image Processing*. Englewood Cliffs, NJ: Prentice-Hall.

Castleman, K. R., Eils, R., Morrison, L., Piper, J., Saracoglu, K., Mark A. Schulze, M. A., and Speicher, M. R. (2000). Classification Accuracy in Multiple Color Fluorescence Imaging Microscopy. *Cytometry* **41**, 139–147

Garini, Y., Gil, A., Bar-Am, I., Cabib, D., and Katzir, N. (1999). Signal to noise analysis of multiple color fluorescence imaging microscopy. *Cytometry* **35**, 214–226.

Matera, A. G. and Ward, D. C. (1992). Oligonucleotide probes for the analysis of specific repetitive DNA sequences by fluorescence in situ hybridization. *Hum. Mol. Genet.* **1**, 535–539.

Morrison, L. (1998). Spectral overlap corrections in fluorescence in situ hybridizations employing high fluorophore densities. Presented at the XIX Congress of the International Society for Analytical Cytology, Colorado Springs, CO. *Cytometry*, Suppl. 9, Abstr. CT 108, p. 140.

Nederlof, P. M., Robinson, D., Abuknesha, R., Wiegant, J., Hopman, A. H. N., Tanke, H. J., and Raap, A. K. (1989). Three-color fluorescence in situ hybridization for the simultaneous detection of multiple nucleic acid sequences. *Cytometry* **10**, 20–27.

Ried, T., Baldini, A., Rand, T. C., and Ward D. C. (1992). Simultaneous visualization of seven different DNA probes by in situ hybridization using combinatorial fluorescence and digital imaging microscopy. *Proc. Nat. Acad. Sci. USA* **89**, 1388–1392.

CHAPTER 15

Stereological Methods for Estimating Geometrical Parameters of Microscopic Structure by Three-Dimensional Imaging

LUCIE KUBÍNOVÁ, JIRÍ JANÁČEK, and IVAN KREKULE

Institute of Physiology, Academy of Sciences of the Czech Republic, Vídeňská 1083, 14220, Prague 4, Czech Republic

1 INTRODUCTION

Measurements of geometrical parameters of structural components of organs, tissues, or cells are the main prerequisite for quantitative analysis in a number of studies in biological research, especially when the relationships between function and structure are analyzed.

Digital images of perfectly registered stacks of serial optical sections from thick specimens recorded by confocal and two-photon microscopes represent suitable data for quantitative measurements as well as for computer three-dimensional (3D) reconstructions that can be made without the necessity of solving the tedious problem of alignment of images of successive sections (Russ, 1990; Pawley, 1995). While the role of confocal microscopy in computer-aided 3D imaging is well known, its advantages for measuring geometrical characteristics of microscopic structure have not yet been fully recognized (Rigaut et al., 1992). However, especially in comparative studies, it is often more desirable to obtain quantitative information on the geometry of the structure of interest (e.g., the volume and/or surface area of cells or tissue components) than its visualization. Digitized images of registered serial optical sections recorded by a confocal microscope represent 3D data especially suitable for such measurements – not only because of the increased axial resolution of confocal images but also because of direct computer support during the measurements. In such cases it is not necessary to perform a 3D reconstruction, but is sufficient just to evaluate the optical sections captured by the confocal microscope.

There is a wide variety of methods for estimating the geometrical characteristics of microscopic structure based on the evaluation of stacks of perfectly registered sections (Kubínová et al., 1999). Each of these techniques has specific drawbacks and advantages. We shall summarize the most promising of these methods and discuss which are the most suitable for specific applications. Interactive stereological methods and automatic digital

Confocal and Two-Photon Microscopy: Foundations, Applications, and Advances, Edited by Alberto Diaspro.
ISBN 0-471-40920-0

methods will be introduced and compared. In general, automatic methods are faster, but require automatic segmentation (i.e., recognition) of analyzed objects in the images. We shall present two groups of automatic digital methods – the first based on image processing and the second based on surface triangulation.

1.1 Stereological Methods

Stereological methods are precise tools for the quantitative evaluation of the structure of 3D objects. They were originally based mainly on observations made on 2D sections (Weibel, 1979), applying 0D (i.e., point), 1D (i.e., linear), or 2D (i.e., planar) test probes and counting, for example, the number of test points falling into the given structure or the number of intersection points of test lines with the structure surface. In contemporary design-based stereology, there is a new trend towards spatial estimators evaluating small 3D samples of structure under study. A 3D sample of examined tissue can be analyzed if a rectangle within a field of view of a microscope is focused through. The digitized images of serial optical sections from thick specimens recorded by confocal and two-photon microscopes are especially suitable for the application of stereological spatial estimators. By using special software, it is possible to generate different virtual test probes with arbitrary predefined (e.g., random) position and orientation within the stack of sections, and apply them directly to this image data.

The idea of estimating stereological parameters from optical sections within a thick slice was first used for counting particles according to the optical disector principle (Gundersen, 1986; Sterio, 1984) and the unbiased sampling brick rule (Howard et al., 1985), and was then applied to many other stereological methods – for example, to the nucleator (Gundersen, 1988) and planar rotator (Jensen and Gundersen, 1993) applied to a stack of optical sections for estimating the mean particle volume; to the optical rotator (Kiêu and Jensen, 1993; Tandrup et al., 1997), spatial grid (Sandau, 1987), vertical spatial grid (Cruz-Orive and Howard, 1995), and fakir (Kubínová and Janáček, 1998) methods for estimating the surface area; and to the vertical slice (Gokhale, 1990), total vertical projection (Cruz-Orive and Howard, 1991), and global spatial sampling (Larsen et al., 1998) methods for length estimation.

Some of these methods still require randomization of the direction of the thick slice, while others can be applied to slices cut in an arbitrary direction. Nevertheless, in practice, it may be essential to take sections in a particular fixed direction. For example, it is not easy to orientate cells in a suspension or a cell culture at random, and it can be easier to determine the mutual localization of different tissues in sections with a specific direction. Therefore, we have focused on methods without assumptions on thick-slice orientation, enabling us to estimate such important parameters as volume, number, surface area, and length. The main idea here is to use virtual, randomly oriented spatial grids of points, lines, or planes that can be generated by a computer and visualized together with the structure under study. Thus, both test grid and microscopic structure are "focused through" by browsing through the stack of optical sections, and so interactions between the grid and structure can be evaluated.

1.2 Methods Based on Image Processing

Automatic measurements of geometric characteristics of 3D objects can be applied directly to their binary images obtained by automatic segmentation of the grayscale images captured by a confocal microscope: Automatic segmentation is a procedure for processing the source digital grayscale image (defined as a data structure of numerical values in the spatial grid of image elements called pixels in 2D or voxels in 3D), resulting in a binary image in which the foreground elements belong to the objects under study (Serra, 1982). The validity of the results of such automatic measurement (i.e., their unbiasedness and precision) depends on how precisely the model describes the object under study. Appropriate spatial resolution and high image quality are necessary for geometrical measurements. High-quality contrast images enable segmentation by simply thresholding the image values, while image inhomogeneities due to uneven staining or heterogeneous acquisition conditions require more advanced techniques of segmentation based on region- and edge-detection techniques. The images distorted by noise must be preprocessed by filtration. Heavy signal distortion may complicate or even preclude the exact segmentation. The exact spatial calibration of the image, enabling one to calculate the position of individual elements, is also a prerequisite for the measurement of any dimensional characteristic. If a usual regular orthogonal grid is used, the calibration is defined by the lengths of intervals between the image elements in all orthogonal directions.

A 3D image formed by a dense sequence of exactly registered 2D images may contain more complete spatial information on the object under study; for example, separate 2D sections do not contain information on gradients in the directions perpendicular to the image plane. Further, topological properties, as the continuity of objects, cannot be judged from separate sections, and the counting of objects requires 3D information in general. Finally, 3D image processing, using more spatial information, can be more effective and robust than 2D processing of individual slices. Basic algorithms for 3D image processing can be derived from those used in 2D image processing in a straightforward way (Meyer, 1992).

1.3 Methods Based on Surface Triangulation

Triangulated surfaces are currently used to model 3D objects in computer graphics. Such surfaces can be obtained from 3D grayscale digital images by isosurface detection using the marching-cube algorithm (Lorensen and Cline, 1987) or by detecting the object contours in subsequent horizontal slices first and then connecting the contours by triangulated stripes (Oliva et al., 1996). The contours can be detected interactively or by (semi-) automatic tracking of the object boundaries in the slices (Baba, 1992).

Geometric characteristics of the models delimited by the triangulated surfaces can be used as estimates of the characteristics of the objects under study (Guilak, 1994).

2 METHODS FOR COUNTING PARTICLES

2.1 Stereological Methods

An unbiased counting or sampling of 3D particles (e.g., cells) can be achieved by using stereological methods: the disector (Sterio, 1984; Gundersen, 1986), the fractionator (Gundersen, 1986), or the unbiased sampling brick rule (Howard et al., 1985). In all of these methods, 3D probes are used. In a physical disector or fractionator, only two parallel physical sections representing the top and bottom of the probe are examined, which means that it is necessary to find corresponding locations in two separate sections. The optical versions of the methods are based simply on focusing through the 3D probe within a thick section, and so they are usually more efficient. If segmentation of particles is feasible, automatic methods can also be used for their number estimation.

It should be stressed that estimating the cell number by counting cell profiles in a single tissue section is not appropriate, because higher cells are more likely to be hit by the section than smaller ones (Fig. 1); that is, the number of cell profiles depends on the height of cells normal to the plane of sectioning.

2.1.1 Optical-Disector Principle, Unbiased Sampling Brick Rule, and Optical Fractionator

The disector and unbiased sampling brick are 3D probes that sample particles with a uniform probability in 3D space, irrespective of particle size and shape (Sterio, 1984; Howard et al., 1985; Gundersen, 1986).

In the optical disector (Gundersen, 1986; West and Gundersen, 1990), the particle is sampled if it fulfills both of the following conditions: (i) its top profile lies inside the thick section within the physical slice; (ii) this top profile is sampled by the unbiased sampling frame placed into the reference plane moving through the thick section. In practice, the thick section is simply focused through, and particle profiles that come into sharp focus are observed whether they are sampled by the sampling frame or not (Fig. 2). The optical disector is especially suitable for counting human or animal cells by their nuclei using a conventional transmission microscope (see, e.g., West and Gundersen, 1990), because such nuclei are usually convex and are nearly spherical in shape, and only their central optical sections come to a sharp focus. For counting objects of more complicated, even nonconvex shapes, or objects with boundaries seen in sharp focus within their entire depth (e.g., plant cells), it is recommended that one uses instead the unbiased sampling brick rule and a confocal microscope.

With the unbiased sampling brick rule, the particle is sampled if it lies at least partly in the sampling brick and at the same time is not intersected by the exclusion planes (Fig. 2). In practice, one is focusing through the particles lying within the thick slice and checking if any of the particle profiles is intersected by the exclusion line of the sampling frame displayed on the optical sections (Fig. 2). When applying the unbiased sampling brick rule in a transmission microscope, there is often a problem seeing the caps of cells sharply enough. In confocal microscopy, clear and sharp images of the cell profiles can usually be observed, and so the method is applicable.

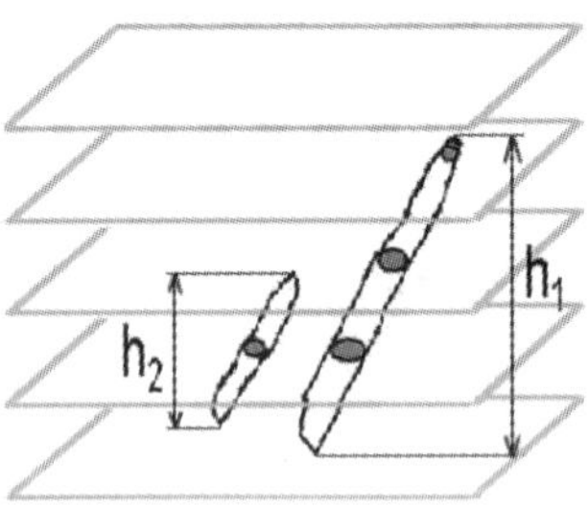

FIGURE 1 Schema showing that the number of cell profiles depends on the height of cells normal to the plane of sectioning. The longer cell with height h_1 is hit by three planes, while the smaller cell with height h_2 is hit by just one plane; therefore its profile is less likely to be present in a single tissue section. This means that estimating the cell number by counting cell profiles in a single tissue section is biased.

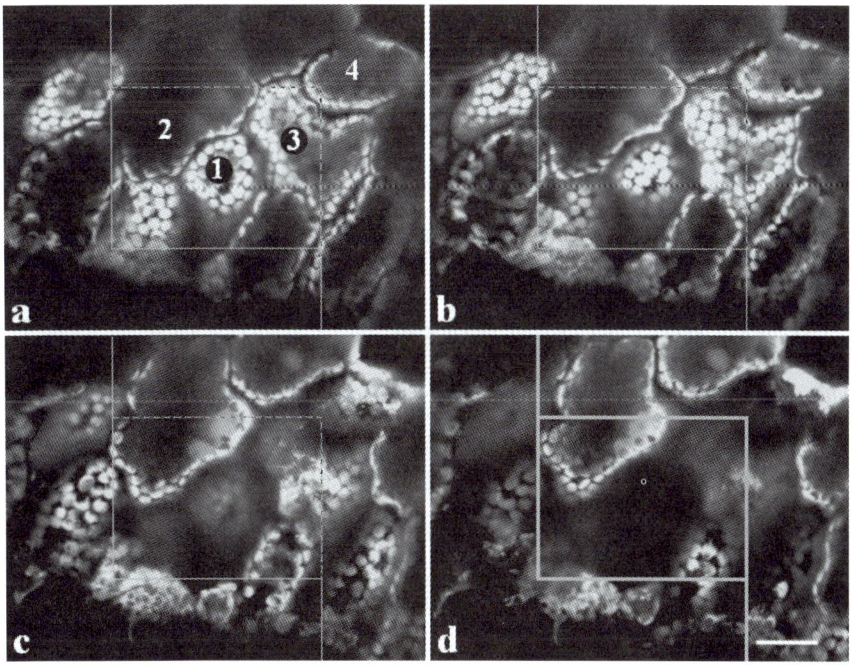

FIGURE 2 Counting mesophyll cells of a Norway spruce needle using the optical-disector principle and the unbiased sampling brick rule. Four serial sections of mesophyll cells, 3 μm apart, are shown in (*a*)–(*d*). In (*a*), four cell profiles are sampled by the unbiased sampling frame (the cell profiles must not be intersected by the filled exclusion line). During focusing through, cells 1 and 4 disappear; that is, they do not intersect the look-up, exclusion plane in (*d*), nor are they intersected by exclusion lines. Thus, these two cells are sampled by the disector as well as the unbiased sampling brick. On the other hand, cells 2 and 3 can still be seen in the exclusion plane (*d*), and so they are not counted. The height of the disector (unbiased sampling brick) is 9 μm here. The scale bar represents 20 μm.

In both methods, if the positions of the sampling frames are uniform over all possible positions in the reference space and a point grid of p points is placed in the frame, the number of particles N_{par} can be estimated by a practically unbiased estimator:

$$N_{\text{par,est}} = \frac{\sum_{i=1}^{n} Q^{-}_{i,\text{par}}}{\sum_{i=1}^{n} P_{i,\text{ref}}} \frac{p}{a'h} V_{\text{ref}}, \tag{1}$$

where n is the number of disectors (sampling bricks), $Q^{-}_{i,\text{par}}(i = 1, \ldots, n)$ is the number of particles sampled by the ith disector (ith sampling brick), $P_{i,\text{ref}}(i = 1, \ldots, n)$ is the number of points of the p-point grid in the ith frame hitting the reference space, a' is the actual area of the frame, h is the height of the disector (sampling brick), and V_{ref} is the volume of the reference space.

The fractionator method (Gundersen, 1986) is based on direct counting of particles in a known predetermined fraction of the containing space (e.g., an animal organ). In an optical fractionator, the containing space is usually first divided arbitrarily into blocks, then a given fraction (bsf) of blocks is sampled and the sampled blocks are cut exhaustively into thick sections. Further, a fraction (ssf) of the thick sections is sampled, and a fraction (asf) of the area of the sections is sampled by unbiased counting frames of the 3D probes (disector or unbiased sampling brick) having their height equal to the fraction (hsf) of the mean section thickness $\bar{t}$. Finally, the particles are counted by the disector or unbiased sampling brick rule (Fig. 2). The number of particles can be estimated by

$$N_{\text{est}} = \frac{1}{\text{bsf}} \frac{1}{\text{ssf}} \frac{1}{\text{asf}} \frac{1}{\text{hsf}} \sum Q^-, \tag{2}$$

where $\sum Q^-$ is the total number of particles counted.

2.1.2 Example: Counting Mesophyll Cells from Norway Spruce Needles

The systematic transverse sections of the needle, about 50 μm thick and 2 mm apart, were cut by a hand microtome and put into a droplet of distilled water between two coverglasses. They were examined using a Bio-Rad MRC600 confocal laser scanning microscope with a Nikon water-immersion objective (60×, NA 1.2) so that the refractive indices of the tissue and the immersion liquid matched. Series of 40 optical sections (with the pixel size of 0.275 μm × 0.275 μm), 1 μm apart, were recorded, as well as an image at 4× magnification showing the entire needle section. The total number of mesophyll cells in the needle N_{mcell} can be estimated by the optical fractionator, Eq. (2), or by the unbiased sampling brick rule (Fig. 2) using Eq. (1), if the needle volume V_{leaf} is estimated by the Cavalieri principle (see Section 5.1 and Gundersen and Jensen, 1987) applied to low-magnification images of sections. The cells were counted using our DISECTOR module (Tomori et al., 2001) running in the IMAGE TOOL (UTHSCA, U.S.A.) freeware environment. The size and density of sampling frames were chosen so that on average, five sampling bricks per needle section were evaluated. One measurement took about five minutes; that is, one section was evaluated in half an hour, while the analysis of the entire needle took two or three hours since five or six serial sections per needle were taken.

2.2 Automatic Counting of Particles

The number N and number density N_V of 3D objects (particles) can be estimated by counting the connected components in a 3D binary image, which means that automatic segmentation of analyzed objects from the original grayscale 3D image is required.

Unbiased methods for counting particles in an image should sample all particles with the same probability. This can be ensured by associating a unique point with each object and by subsequent counting of the associated points within the region. Direct counting of all segmented components in the image would be biased, because usually some of them are not included completely within the image (i.e., the analyzed region), and counting them would yield an overestimate of the true numerical density N_V, while excluding them would result in underestimation of N_V.

The unbiased counting by image processing is done in a volume of interest (VOI) inside the 3D image. The VOI (usually a parallelepiped) should be surrounded by a guard area inside the image that is large enough to contain entire particles intersecting the VOI. The point associated with the object can be the first point of the object visited during systematic scanning of all image voxels. The algorithm for unbiased counting of connected components in VOI consists in scanning the image and labeling the components connected to unlabeled foreground voxels by unused labels. The unlabeled foreground voxels contained within VOI are counted. The connected components are labeled by recursive labeling of neighboring voxels. To obtain an estimate of N_V, the resulting number of associated points is divided by the volume of the VOI.

Under the assumption that the connected objects are simple (i.e., without holes or tunnels), their number can be estimated by the 3D Euler–Poincaré characteristic of the image foreground (Serra, 1982). The Euler–Poincaré characteristic of one simple object is equal to unity, and the correct result of counting is achieved owing to the additivity of the Euler–Poincaré characteristic with respect to the disjoint union of sets. The advantage of this approach is that the Euler–Poincaré characteristic can be calculated by counting specific local configurations by a simple algorithm (Nikolaidis and Pitas, 2001).

3 METHODS FOR LENGTH ESTIMATION

3.1 Stereological Methods

The unbiased estimation of the length of curves in 3D (e.g., a capillary network) is based on counting intersection points between the curves and planar probes. The orientation of the planar probes can be arbitrary only if the curves have the same geometrical properties in all directions of space – that is, they are isotropic (Weibel, 1979). Otherwise, in general, isotropic random orientation of the planar probes is required. This means that every orientation must have the same probability of being chosen for the measurement. Apart from stereological methods for length estimation using isotropic thin physical sections (Mattfeldt, 1985, 1990), methods applied to thick sections with fewer or no requirements on physical section orientation have recently been developed. Let us mention the methods of vertical slices (Gokhale, 1990) and total vertical projections (Cruz-Orive and Howard, 1991), which require vertical orientation (vertical sections are sections longitudinal to a fixed but arbitrary axial direction and randomized around this vertical axis), and the "slicer" method (called "global spatial sampling" by Larsen et al., 1998), which uses arbitrarily oriented thick sections. If the series of digital images of optical sections obtained by a confocal or two-photon microscope is recorded, the slicer method can be easily implemented (see below). In some special cases, even the method of total vertical projections is applicable to arbitrarily oriented thick sections. The vertical projections of the series are made by special software, for example, the freeware Confocal Assistant (ftp://ftp.genetics.bio-rad.com/Public/confocal/cas/SETUP402.EXE), which enables one to generate projections under a predefined angle of rotation. Then the cycloid-shaped test lines are superimposed on the image of the vertical projection (using special software), with the cycloid minor axis perpendicular to the vertical axis. The numbers of intersection points of projected curves and test lines are counted and the curve length estimated as shown in Cruz-Orive and Howard (1991). An example of such a measurement is shown in Figure 3, where the cycloid arcs were generated by our STESYS software system (Karen et al., 1998; Tomori et al., 2000). However, this measurement is applicable only if the curves do not overlap too much in the projection (Fig. 3). The slicer method thus appears to be more universal.

3.1.1 *Slicer (Global Spatial Sampling) Method*

The length $L(X_1)$ of a linear structure (fiber) X_1 in 3D space $\mathbb{R}^3$ can be expressed by the Crofton formula (Santaló, 1976)

$$L(X_1) = 2E_{T_1 \in G_{3,1}}\left[\int_{y \in T_1} Q(X_1 \cap (T_1^{\perp} \oplus y))\, dy\right], \tag{3}$$

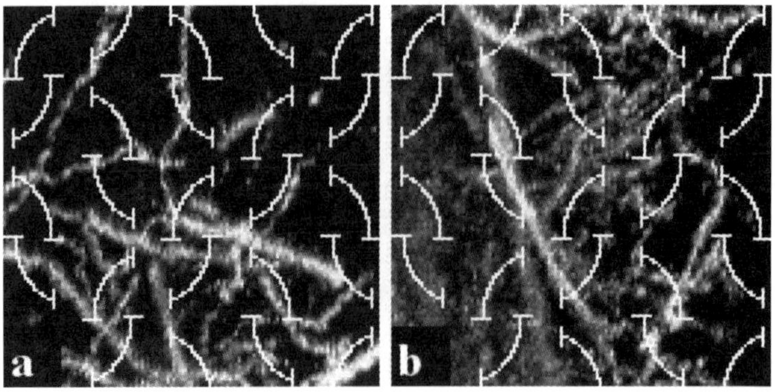

FIGURE 3 See ftp site for color figure. The method of total vertical projections applied to Golgi-stained nerve cells. (*a*) In this total projection, the cell extensions do not overlap too much, so that it is possible to count intersections between cycloid arcs and cell extensions. (*b*) The cell extensions overlap too much in this projection.

where $Q(X_1 \cap (T_1^\perp \oplus y))$ is the number of intersections between the fiber X_1 and the plane $T_1^\perp \oplus y$ perpendicular to the line T_1 and containing the point $y \in T_1$; the mean $E[\cdot]$ is calculated over the lines of all directions in 3D space, that is, over the Grassmann manifold $G_{3,1}$ (Santaló, 1976).

In practice, the length of a linear structure (fiber) X_1 can be estimated by Eq. (3) if a finite set of directions is selected and, for each direction, a system of parallel test planes is chosen (see Fig. 4). For simplicity, we shall call this system of parallel test planes the "slicer probe" (called the "systematic set of isotropic virtual planes" by Larsen et al., 1998). A spatial grid of planes can be then defined as the n-tuple $(T_2^1, T_2^2, \ldots, T_2^n)$ of slicer probes T_2^i in R_3, each of which can be expressed parametrically as follows:

$$T_2^i = \{\mathbf{o}^i + f\mathbf{d}^i + \alpha\mathbf{h}^i + \beta\mathbf{v}^i : f \text{ integer}; \alpha, \beta \text{ real numbers}\}, \qquad i = 1, \ldots, n; \tag{4}$$

that is, the probe consists of planes parallel to the directions $\mathbf{h}^i$ and $\mathbf{v}^i$, mutually congruent by translations $f\mathbf{d}^i$; the probe position is given by $\mathbf{o}^i$. The length of X_1 is then estimated by the formula

$$L_{\text{est}}(X_1) = \frac{2}{n}\sum_{i=1}^{n} d_i Q_i(X_1 \cap T_2^i), \tag{5}$$

FIGURE 4 Slicer probe used for length estimation.

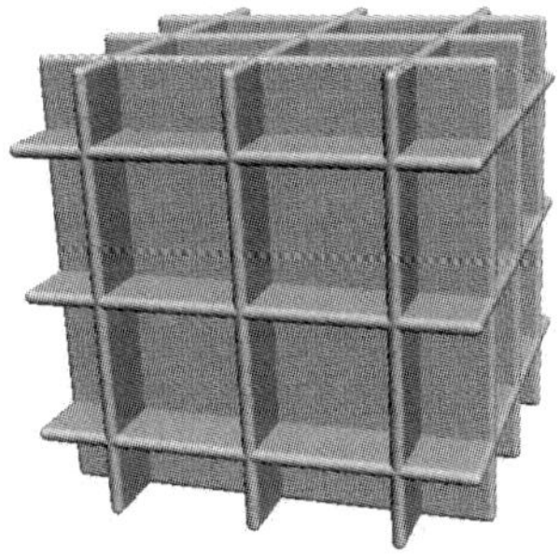

FIGURE 5 Cubic spatial grid consisting of three mutually perpendicular slicer probes.

where d_i is the distance between the planes of the ith slicer probe and $Q_i(X_1 \cap T_2^i)$ is the number of intersections between the fiber and the ith slicer probe ($i = 1, \ldots, n$). In practice, it is efficient to use a cubic spatial grid consisting of three mutually perpendicular slicer probes (Fig. 5) given by $T_{2,0}^1$, $T_{2,0}^2$, $T_{2,0}^3$) defined by vectors $\mathbf{o}_0^i$, $\mathbf{d}_0^i$, $\mathbf{h}_0^i$ and $\mathbf{v}_0^i$ ($i = 1, 2, 3$), specified in Table 1, by random shifting of its origin, proper scaling, and IUR (Isotropic Uniformly Random) rotation of the entire spatial grid.

When estimating, for example, the capillary length, the intersections between the isotropic virtual test planes and capillaries in the stack of optical sections within a thick tissue slice are counted. If a cubic spatial grid consisting of three mutually perpendicular slicer probes (see above and Fig. 5) is used, the length of capillaries L_{cap} in the given stack of sections can be estimated by the following formula:

$$L_{\text{cap,est}} = \tfrac{2}{3} d(Q_1 + Q_2 + Q_3), \tag{6}$$

where d is the distance between neighboring parallel planes, and $Q_j (j = 1, 2, 3)$ is the number of intersections between the jth slicer probe and the capillaries within the given stack. This length estimator is unbiased if the orientation of the spatial grid of planes is isotropic random, its position is uniform random, and the intersections are counted properly (see Larsen et al., 1998). The total length of capillaries in a muscle can be estimated if the slicer method is combined with the fractionator method (Gundersen, 1986).

The above measurement requires one to use specific software generating an isotropic set of virtual slicer probes (e.g., C.A.S.T.-GRID by Olympus, Denmark, or SLICER software described below) applied to stacks of transmission, or preferably, confocal images.

TABLE 1 List of vectors defining the grid ($T_{2,0}^1$, $T_{2,0}^2$, $T_{2,0}^3$) that is used for generating the orthogonal triplet of planar probes; the vectors $\mathbf{e}_1$, $\mathbf{e}_2$, and $\mathbf{e}_3$ are unit coordinate vectors

i	$\mathbf{o}_0^i$	$\mathbf{d}_0^i$	$\mathbf{h}_0^i$	$\mathbf{v}_0^i$
1	0	$\mathbf{e}_1$	$\mathbf{e}_2$	$\mathbf{e}_3$
2	0	$\mathbf{e}_2$	$\mathbf{e}_1$	$\mathbf{e}_3$
3	0	$\mathbf{e}_3$	$\mathbf{e}_1$	$\mathbf{e}_2$

3.1.2 Example: Length of Capillaries in a Rat Skeletal Muscle A muscle sample of an extensor digitorum longus (EDL) rat muscle was processed as follows (Kubínová et al., 2001). After being fixed in 2% paraformaldehyde with addition of sodium periodate, L-lysine, and sucrose, and rinsing in phosphate-buffered saline, the muscle sample was cut into slices (< 17 mm thick). Tissue samples were pretreated with 5% normal swine serum (Dako, Denmark) to reduce background staining. Further, they were incubated with rabbit anti-mouse immunoglobulin (Ig) raised against fibronectin (1 : 1300) (Dako, Denmark), immersed in FITC-conjugated swine anti-rabbit IgG (1 : 20) (Dako, Denmark), and mounted in SlowFade Antifade (Eržen and Maravić, 1993). All antibodies were diluted in phosphate-buffered saline. Serial optical sections (1 μm apart) inside a thick transverse slice of muscle were captured by a Bio-Rad MRC600 confocal laser scanning microscope using an oil-immersion objective (40×, NA 1.3). Then the stacks of digitized images (with a pixel size of 0.412 μm × 0.412 μm) of 40 optical serial sections were sampled by systematic uniform random sampling. The captured stacks were used for estimating the length of capillaries around a muscle fiber in the stack by the slicer method using our SLICER software written in TurboPascal 6 for an IBM PC under MS DOS operating system (Fig. 6) (Kubínová et al., 2001). A cubic spatial grid consisting of three mutually perpendicular slicer probes (Fig. 5) was used and Eq. (6) was applied. The distance between slicer planes was chosen to be 10 μm, ensuring about 20–50 capillary intersections per fiber. One measurement took 10–20 minutes, depending on the complexity of the course of capillaries.

3.2 Automatic Length Measurements Based on Image Processing (Digital Crofton Method, DC)

The Crofton formula (3) can also be used in automatic length measurements based on image processing (Meyer, 1992). The number of intersections of the object and the test plane can be calculated using the Euler–Poincaré characteristic of the intersection of the

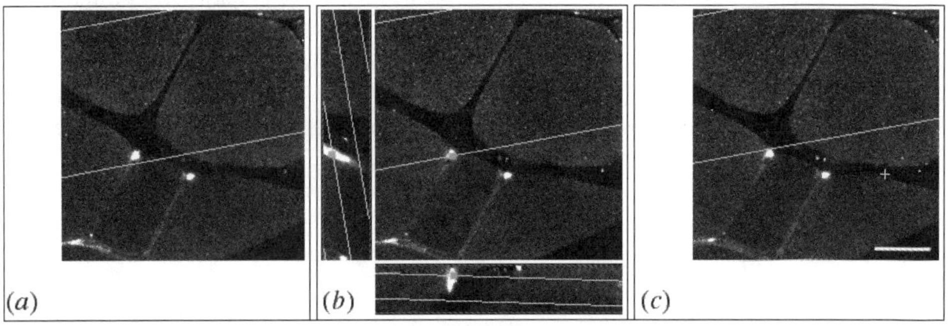

FIGURE 6 See ftp site for color figure. Estimation of muscle capillary length by the slicer method using the SLICER program. Three successive serial sections of the muscle, 1 μm apart, are shown (*a*), (*b*), and (*c*). Gray lines represent the intercepts of the slicer planes with the current section. In the first section (*a*), the slicer plane is going near the capillary profile (bright spot). In the second section (*b*), the slicer plane is intersecting the capillary – the intersection is marked by a gray point. The orthogonal views of the muscle slice containing the intersection point are shown on the left and bottom side (*b*). In the third section (*c*), the slicer plane is already leaving the capillary profile. The scale bar represents 20 μm. Reproduced by permission from Kubínová and Janáček (2001). Copyright 2001, John Wiley & Sons, Inc.

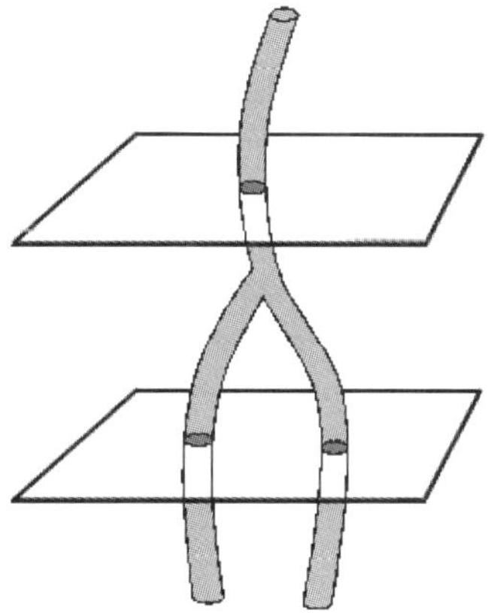

FIGURE 7 The length of fibers can be estimated from the Euler–Poincaré characteristic of the intersection of the fibers and the test plane, which is equal to the number of fiber profiles in the plane. Here, the Euler–Poincaré characteristic is equal to 1 in the upper test plane and to 2 in the lower test plane.

object and the test plane (Fig. 7), if the intersection consists of simple profiles (e.g., in the case of fibers). The definition of length, using the Crofton formula with the Euler–Poincaré characteristic, enables easy implementation of the automated method. However, for hollow objects, such a definition can lead to counterintuitive results – that is, it is not in accordance with the intuitive concept of length: the Euler–Poincaré characteristic (equal to the number of components minus the number of holes) of planar sets with holes is less than the number of sets. The length measured as described above is equal to the integral of the mean curvature of the surface divided by π (Santaló, 1976).

The length of a feature set in a 3D binary image can be estimated from its intersections with the principal planes of the orthogonal grid. The principal planes of the grid are the planes intersecting the triplets of mutually neighboring vertices; the densities of the principal planes of the grid, S_V (i.e., the ratio of the test-plane areas to the volume; the reciprocal $1/S_V$ corresponds to the distance between successive planes), are listed in Table 2.

The intersection of an orthogonal grid with planes intersecting any orthogonal line forms a planar rectangular grid. The intersection of the orthogonal grid with the planes that

TABLE 2 The densities S_V of principal planes of the orthogonal grid of points with coordinates (iu_1, ju_2, ku_3) for integers i, j, k and calibration constants $u_1 > 0$, $u_2 > 0$, and $u_3 > 0$. The densities are listed for all three types of plane (i.e., for those intersecting only rows, rows and surface diagonals, or only surface diagonals of the cubic lattice)

Plane type	S_V	
Rows	u_i^{-1}	$1 \leqslant i \leqslant 3$
Rows and diagonals	$(u_i^{-2} + u_j^{-2})^{1/2}$	$1 \leqslant i \leqslant j \leqslant 3$
Diagonals	$(u_1^{-2} + u_2^{-2} + u_3^{-2})^{1/2}$	

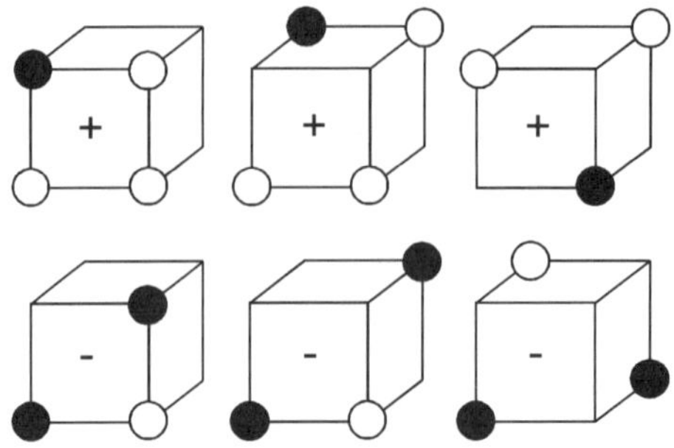

FIGURE 8 Examples of configurations of neighbors in the cubic lattice used for length measurements. All relevant local configurations of vertices in the general orthogonal lattice can be obtained by orthogonal rotations and deformations of the examples. The configurations should be counted with the respective signs shown.

do not intersect any orthogonal line forms a hexagonal planar lattice. The Euler–Poincaré characteristics of binary objects in the rectangular lattice and the hexagonal lattice can be calculated by counting local configurations

$$N\begin{pmatrix}1 & 0\\ 0 & 0\end{pmatrix} - N\begin{pmatrix}\cdot & 1\\ 1 & 0\end{pmatrix} \quad \text{and} \quad N\begin{pmatrix}0 & & 0\\ & 1 & \end{pmatrix} - N\begin{pmatrix} & 0 & \\ 1 & & 1\end{pmatrix},$$

respectively (Serra, 1982). The resulting 3D configurations, which should be counted with respective signs during length measurement using the digitized projection formula, are shown in Figure 8; the remaining configurations can be obtained by orthogonal rotations and proper deformations of these configurations. The length of the segmented object is then estimated as

$$L_{\text{est}} = 2\frac{1}{m}\sum_{i=1}^{m}\frac{1}{S_{V_i}}n_i, \tag{7}$$

where m is the number of directions, n_i is the configuration count, and S_{Vi} is the density of principal planes (Table 2) for the ith direction.

3.3 Automatic Length Measurements Based on Surface Triangulation (Triangulation Method, T)

After delimitation of the object in a 3D image by triangulated surfaces, the object length can be estimated by summing the lengths of the edges, multiplied by the angles between the outer normals to neighboring triangles and divided by 2π. The definition of length used in this measurement is based on the integral of mean curvature of the object surface (Santaló, 1976), which can lead to counterintuitive results in the case of hollow objects.

3.4 Comparison of Stereological and Automatic Digital Methods

3.4.1 Test Images: Neuronal Extensions A series of 8-bit grayscale images of Golgi-stained nerve cells (courtesy of Bio-Rad Microscience, Ltd.) with resolution

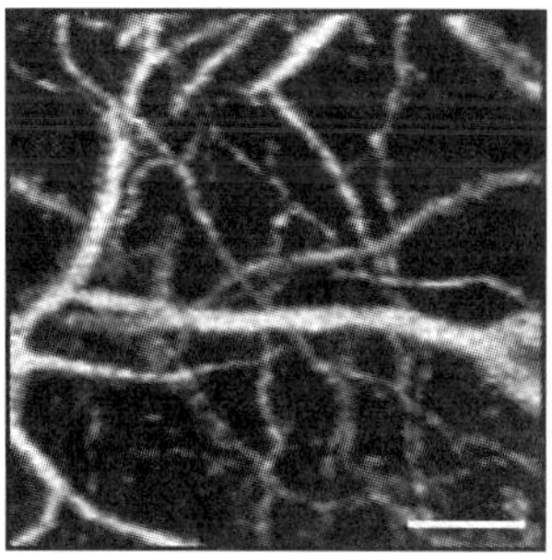

FIGURE 9 Maximum-intensity projection of one of the six original grayscale images ($128 \times 128 \times 66$ voxels) of Golgi-stained nerve cells (courtesy of Bio-Rad Microscience, Ltd.) used for comparison of different methods of length estimation. The scale bar represents 20 m.

0.741 μm × 0.741 μm and $384 \times 256 \times 66$ voxels was used for comparison of different methods of length estimation (Fig. 9). The image was divided into six parts, $128 \times 128 \times 66$ voxels each; thus, six series of equal size were available for the measurements. The total length of neuronal extensions within each series was estimated. The neurons were enhanced by the top-hat transform (Meyer, 1992), using the residuum of grayscale opening by reconstruction. The binary image was obtained by thresholding and removing objects with surface area less than 50 μm^2.

3.4.2 Methods Compared The slicer method (S) was applied manually to both binary (S01) and original grayscale (SG) data (Table 3) using our SLICER software. Each measurement was performed independently five times and the corresponding coefficients of error were calculated. On average, more than 200 intersections between the slicer probes and the neuronal extensions in one series were counted, which took about 20 minutes.

Both the image-processing method and the method based on surface triangulation were applied to four sets of image data differing in their degrees of smoothing and resolution. For this purpose, the above binary image data were processed in the following way. First, the binary image was converted to grayscale (foreground 255, background 0), and then it was smoothed by a Gaussian filter with standard deviation 1.5 μm operating on a 3^3

TABLE 3 The total length of neuronal extensions (in μm) within six tested samples of nerve cells measured by the slicer method applied manually to both original grayscale (SG) and binary (S01) data, and by the digital Crofton (DC) and triangulation (T) methods applied to data with no smoothing and no resampling (DC01, T01), with smoothing made once and no resampling (DC11, T11), with no smoothing and resolution decreased twice (DC02, T02), and to the once-smoothed data with the resolution decreased twice (DC12, T12)

Sample No.	SG	S01	DC01	DC11	DC02	DC12	T01	T11	T02	T12
1	1302	1082	1468	975	1059	937	1825	816	1284	851
2	2290	1956	3055	1843	2051	1772	3725	1289	2301	1370
3	2552	1652	2481	1387	1738	1385	3047	882	1972	937
4	1438	1184	1630	1027	1159	977	2076	811	1326	812
5	2232	1618	1893	1285	1482	1269	2269	946	1700	974
6	2450	1656	2264	1443	1654	1431	2712	1031	1848	1092

neighborhood applied m times ($m = 0, 1$). Further, the resolution was decreased n times ($n = 1, 2$), taking the maximum grayscale value of the original voxels as the new voxel value.

The digital Crofton (DC) method based on image processing was then applied to the four sets of binary data obtained after thresholding (at 80) the smoothed and resampled images (Fig. 10). The results of length measurements made on images with smoothing and resolution parameters m and n (denoted by DCmn) are listed in Table 3 for DC01 (i.e., measurement made on the data used also for slicer measurement, with no smoothing and no resampling applied), DC11 (smoothing made once and no resampling), DC02 (no smoothing, resolution decreased twice), and DC12 (smoothed once, resolution decreased twice).

The method based on surface triangulation (T) was also applied to four sets of grayscale image data, with smoothing applied m times and resolution decreased n times (Fig. 11). The surfaces of fibers were triangulated by isosurface detection with threshold 128. The results of length measurements made on images with smoothing and resolution parameters m and n (denoted by Tmn, with $m = 0$, and $n = 1, 2$) are listed in Table 3.

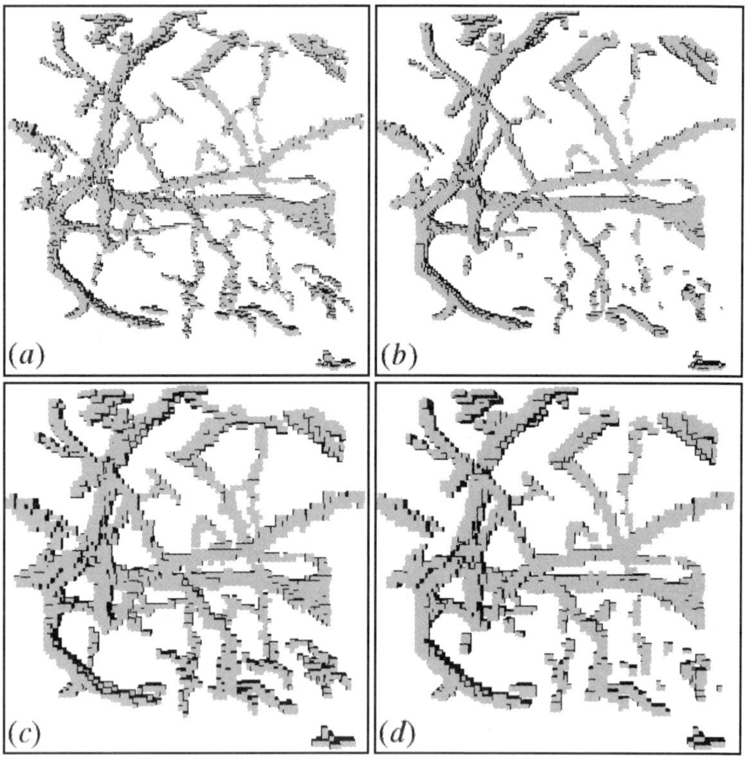

FIGURE 10 The 3D reconstructions of image data used for length measurements by the digital Crofton (DC) method. (*a*) Data used in DC01 with no smoothing and no resampling applied. (*b*) Data used in DC11 (smoothing made once and no resampling). (*c*) Data used in DC02 (no smoothing, resolution decreased twice). (*d*) Data used in DC12 (smoothed once, resolution decreased twice). For the corresponding original grayscale image of Golgi-stained nerve cells, see Figure 9.

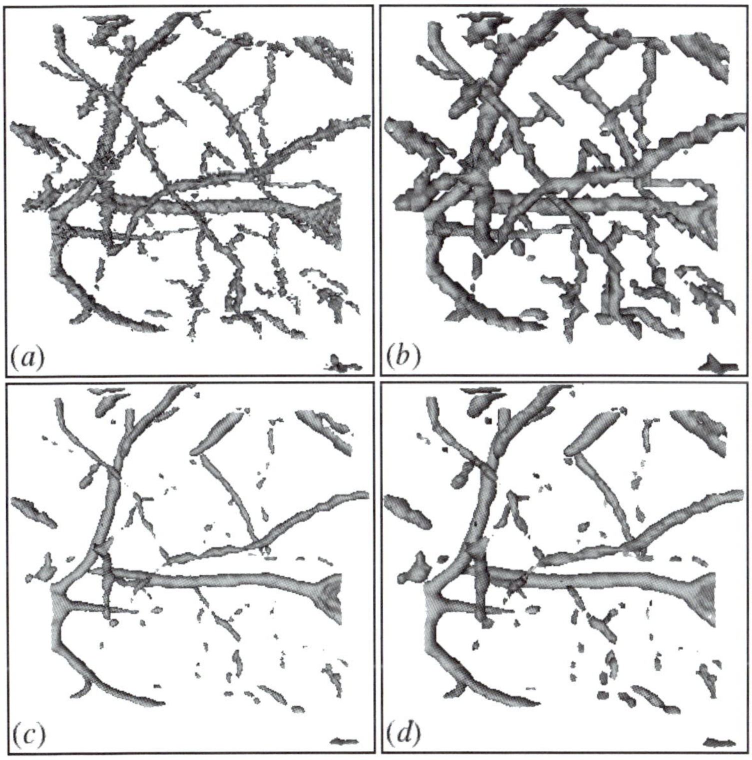

FIGURE 11 The 3D reconstructions of image data used for length measurements by the method based on surface triangulation (T). (*a*) Data used in T01 with no smoothing and no resampling applied. (*b*) Data used in T11 (smoothing made once and no resampling). (*c*) Data used in T02 (no smoothing, resolution decreased twice). (*d*) Data used in T12 (smoothed once, resolution decreased twice). For the corresponding original grayscale image of Golgi-stained nerve cells, see Figure 9.

All of the above image processing procedures and implementations of DC and T methods were performed on an SGI INDY workstation using our special custom-made modules developed in the IRIS Explorer visualization environment.

3.4.3 Results of Comparison The results show that the length measurement was very sensitive to segmentation: the results obtained by the slicer method applied to segmented images were about 30% lower than those from the original grayscale data (Table 3). The precision of segmentation is, of course, dependent on the quality of the original grayscale image – so better results could be expected for images with less noise and higher contrast. However, the image data chosen for the test (Fig. 9) are of a standard quality encountered in confocal microscopy imaging. The coefficients of error were on average 0.029 for SG and 0.028 for S01.

The triangulation-based measurement was not suitable for length measurements in our dataset. The differences in results for images with different degrees of smoothing and resampling are very large (Table 3).

On the other hand, the digital Crofton method appeared to be more suitable for automatic length measurement. When applied to binary data with no smoothing or resampling (i.e., DC01), the results were not significantly different (at a 0.05 level) from those obtained by the slicer method applied to the original grayscale data (Table 3). The results of the slicer method used for the length estimation of neuronal extensions from the binary data with no smoothing or resampling (i.e., S01) correspond to the measurements obtained by the digital Crofton method (DC02) applied to the binary data with half resolution.

In general, the results of the comparison presented here show that automatic methods for length measurement are very sensitive to resolution and to the image processing, usually smoothing, applied to decrease noise. Before applying any automatic method, it should be very carefully tested for the given type of image data to determine whether it gives reasonably precise results.

4 METHODS FOR SURFACE-AREA ESTIMATION

4.1 Stereological Methods

The stereological estimation of the surface area of a 3D object (e.g., a cell-surface area) is based on counting intersection points between the object surface and linear probes or on measurement of the length of the intersection between the surface and test planes going through the object. The orientation of test lines, planes, or projections can be arbitrary only if the structure has the same geometrical properties in all directions of space (i.e., if it is isotropic). Otherwise, isotropic random orientation of test probes is required. This means that every orientation must have the same probability to be chosen for the measurement. The most frequently used stereological methods for the unbiased estimation of surface area suitable for practical application on 2D sections are the method of vertical sections (Baddeley et al., 1986) ensuring the isotropic orientation of test lines, and the orientator method (Mattfeldt et al., 1990), ensuring the isotropic orientation of test planes.

Recently, highly efficient methods based on using spatial grids of test lines (Sandau, 1987; Cruz-Orive and Howard, 1995; Kubínová and Janáček, 1998) have emerged. We shall focus on the fakir method (Kubínová and Janáček, 1998), placing no requirements on the orientation of the thick sections.

4.1.1 Fakir Method The fakir method is based on the Crofton formula for the surface area of objects (Santaló, 1976), in which the area $S(X_2)$ of the surface $X_2 \subset \mathbb{R}^3$ is given by

$$S(X_2) = 2E_{T_1 \in G_{3,1}}\left[\int_{y \in T_1^{\perp}} I(X_2 \cap (T_1 \oplus y))\,dy\right], \tag{8}$$

where $I(X_2 \cap (T_1 \oplus y))$ is the number of intersections between the surface X_2 and the lines $(T_1 \oplus y)(T_1 \subset \mathbb{R}^3)$ perpendicular to the plane of projection $T_1^{\perp}$ and containing point the $y \in T_1^{\perp}$; the mean $E[\cdot]$ is calculated over the lines of all directions in 3D space, that is, over the Grassmann manifold $G_{3,1}$ (Santaló, 1976).

In practice, the surface area can be estimated by Eq. (8) if a finite set of directions is selected and, for each direction, a system of parallel test lines (called the fakir probe – see Cruz-Orive (1993) and Fig. 12) is chosen. A spatial grid of lines is defined as the n-tuple

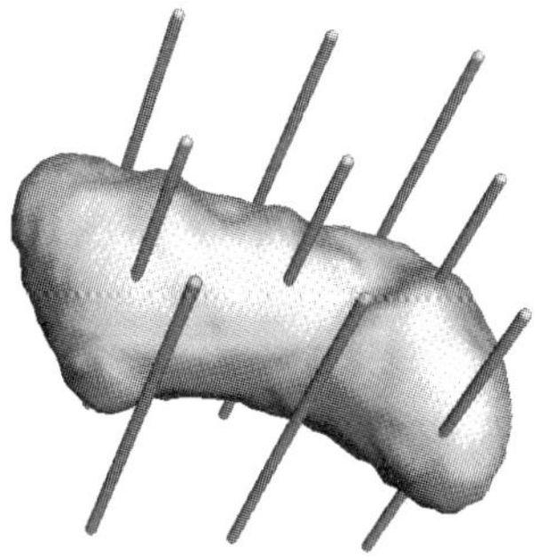

FIGURE 12 Fakir probe piercing a 3D object. Reproduced by permission from Kubínová and Janáček (2001). Copyright 2001, John Wiley & Sons, Inc.

$(T_1^1, T_1^2, \ldots, T_1^n)$ of fakir probes T_1^i in $\mathbb{R}^3$, each of which can be expressed parametrically as follows:

$$T_1^i = \{\mathbf{o}^i + f\mathbf{v}^i + g\mathbf{h}^i + \alpha\mathbf{d}^i : f, g \text{ intergers; } \alpha \text{ a real number}\}, \qquad i = 1, \ldots, n; \tag{9}$$

that is, the probe consists of parallel lines with direction $\mathbf{d}^i$, mutually congruent by translations $f\mathbf{v}^i + g\mathbf{h}^i$; the probe position is given by $\mathbf{o}^i$. The area of the surface X_2 is then estimated by the formula

$$S_{\text{est}}(X_2) = 2\frac{1}{n}\sum_{i=1}^{n}\frac{v}{l_i}I_i(X_2 \cap T_1^i), \tag{10}$$

where l_i/v is the length of the ith fakir probe per unit volume and $I_i(X_2 \cap T_1^i)$ is the number of intersections between the surface and the ith probe $(i = 1, \ldots, n)$.

In principle, Eq. (10) holds for a simple random as well as a systematic random sampling of directions of fakir probes. Systematic sampling appears to be more efficient, since it usually results in a decreased variance when estimating the area of most of anisotropic surfaces from total projections (by Eq. (8)), due to the negative covariance of projections (Mattfeldt et al., 1985). It can be shown that a highly efficient grid is a spatial grid consisting of three mutually perpendicular fakir probes (with IUR direction of the first) that are halfway-shifted with respect to each other in order to fill the space "with minimum voids" (see Fig. 13) (Kubínová and Janáček, 1998). This orthogonal triplet of shifted fakir probes (OTSFP) is generated from the grid $(T_{1,0}^1, T_{1,0}^2, T_{1,0}^3)$ defined by the vectors $\mathbf{o}_0^i$, $\mathbf{v}_0^i$, $\mathbf{h}_0^i$, and $\mathbf{d}_0^i$, $(i = 1, 2, 3)$ specified in Table 4, by random shifting of its origin, proper scaling, and IUR rotation of the entire spatial grid. If the grid constant u (i.e., the shortest distance between lines of the fakir probe) is the same for all three fakir probes (which means that $l/u = u/u^3 = u^{-2}$), then the surface area can be estimated by the formula

$$S_{\text{est}}(X_2) = \tfrac{2}{3}u^2(I_1 + I_2 + I_3), \tag{11}$$

where I_i is the number of intersections between the surface and the ith probe $(i = 1, 2, 3)$. The above estimator is unbiased if the orientation of the spatial grid is isotropic random

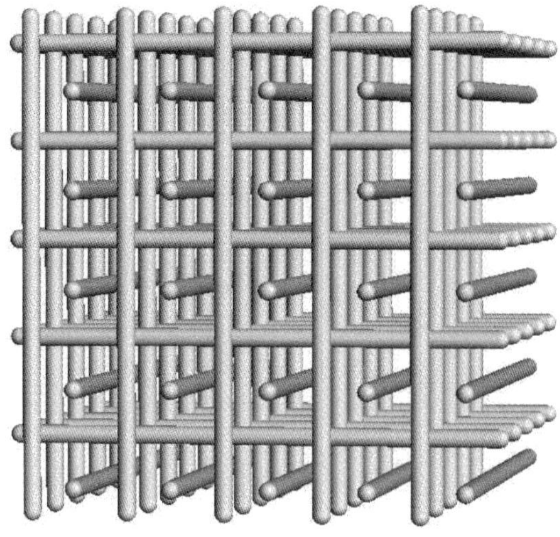

FIGURE 13 Orthogonal triplet of shifted fakir probes (OTSFP). Reproduced by permission from Kubínová et al. (1999). Copyright; 1999, Wiley-Liss, Inc., a subsidiary of John Wiley & Sons, Inc.

(i.e., each orientation in 3D has the same probability of being chosen as the orientation of the test grid – see, e.g., Weibel, 1979) and its position is uniform random.

The measurement can be easily performed using our interactive FAKIR and GEN3GRID programs (written in TurboPascal 6 for an IBM PC under MS DOS), which generate an isotropic set of virtual fakir probes, and so it is not necessary to randomize the direction of the stack of sections.

4.1.2 Example: Total Area of Exposed Surface of Mesophyll Cells in Barley Leaf The practical implementation of the fakir method using FAKIR software, if a stack of optical sections captured by a confocal microscope is available, can be demonstrated by estimating the total area of exposed surface of mesophyll cells in a barley (*Hordeum vulgare* L.) leaf. The specimens of thick transverse sections of the barley leaf used in a former study (Kubínová, 1993) were used. The images of serial optical sections within these specimens, 1 μm apart, were recorded by a confocal laser scanning microscope (CLSM): a Bio-Rad MRC 600 (using an excitation wavelength of 488 nm) equipped with an oil-immersion Nikon CF N Plan Fluor objective (40×, NA, 1.3). The pixel size was 0.412 μm × 0.412 μm.

TABLE 4 List of vectors defining the grid ($T^1_{1,0}$, $T^2_{1,0}$, $T^3_{1,0}$) that is used for generating the orthogonal triplet of shifted fakir probes; the vectors $\mathbf{e}_1$, $\mathbf{e}_2$, and $\mathbf{e}_3$ are unit coordinate vectors

i	$\mathbf{o}^i_0$	$\mathbf{v}^i_0$	$\mathbf{h}^i_0$	$\mathbf{d}^i_0$
1	0	$\mathbf{e}_3$	$\mathbf{e}_2$	$\mathbf{e}_1$
2	$\frac{1}{2}\mathbf{e}_3$	$\mathbf{e}_1$	$\mathbf{e}_3$	$\mathbf{e}_2$
3	$\frac{1}{2}(\mathbf{e}_1 + \mathbf{e}_2)$	$\mathbf{e}_1$	$\mathbf{e}_2$	$\mathbf{e}_3$

When estimating the total area of a surface, that is too large to lie entirely within a single thick slice, a combination with the fractionator principle (Gundersen, 1986) can be useful. In our case of a barley leaf, this means that thick transverse slices are taken in a systematic uniform random manner along the leaf (see Kubínová, 1993) and a system of frames is placed also uniformly at random on these slices. Thus, only small blocks "cut out" of the tissue are evaluated. In each of them, the surface area of mesophyll cell walls exposed to intercellular spaces is estimated independently by the fakir method. If the blocks are sampled with a probability $1/p$, then the total area of exposed surface of mesophyll cells in the leaf S_{int} (the so-called internal leaf surface area) can be estimated by

$$S_{\text{int,est}} = p \sum_{j=1}^{m} S_{\text{int},j,\text{est}} \tag{12}$$

where m is the number of tissue blocks and $S_{\text{int},j,\text{est}}$ is the estimate of exposed surface area in the jth slice ($j = 1, \ldots, m$).

The FAKIR software can be directly applied to a series of digitized images (Fig. 14). If the intersections between the exposed surfaces of mesophyll cells and fakir probes are counted in each of the sample tissue blocks and if the cubic OTSFP is used, it is possible to estimate $S_{\text{int},j}$ ($j = 1, \ldots, m$) by Eq. (11) and then the total internal leaf surface area S_{int} by Eq. (12).

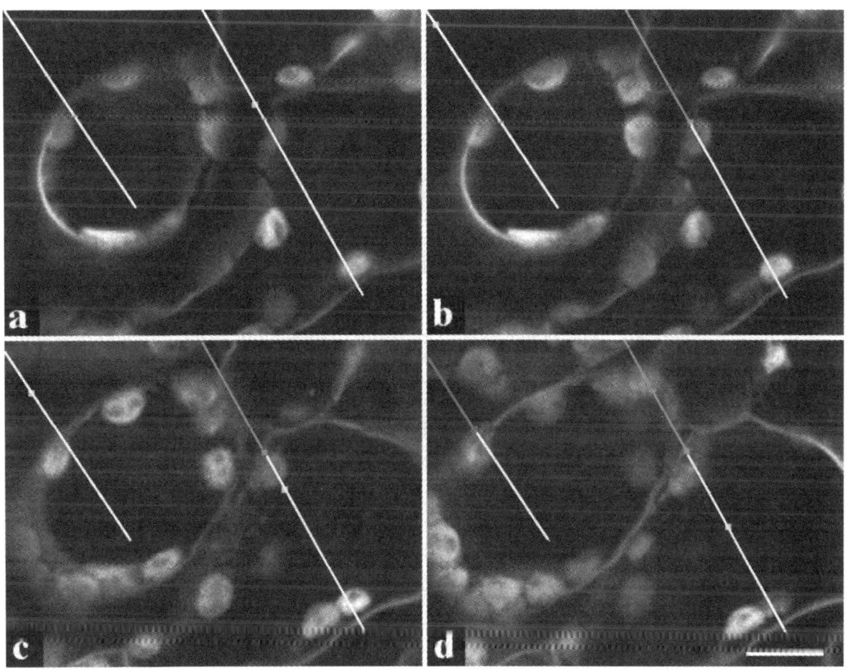

FIGURE 14 See ftp site for color figure. The fakir method applied to the estimation of the area of exposed surface of mesophyll cells in barley leaf using the FAKIR program. Four serial optical sections 2 μm apart are shown. The test lines of the fakir probe piercing the mesophyll cells are white at the beginning of the measurement (a). The white points denote the intersection points between the test lines and the current section. As soon as such a point comes into contact with the cell wall during focusing-through, it is marked by a mouse as a gray point and the part of the test line above this point becomes gray (b, d). The scale bar represents 10 μm.

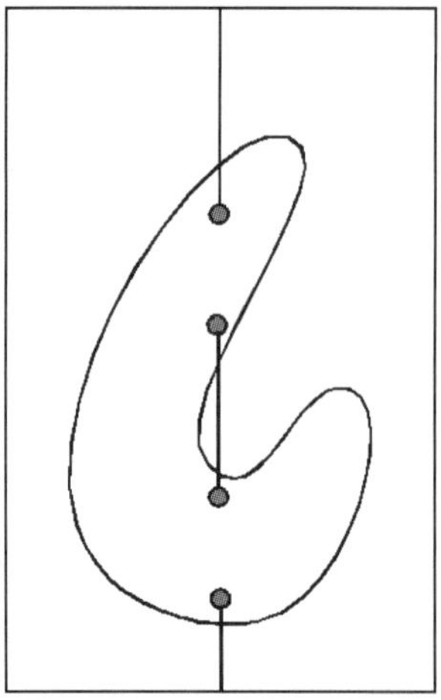

FIGURE 15 The surface area of an object can be estimated by counting the intersections of the object surface with linear probes. There are four intersection points between the test line and the object surface here.

4.2 Automatic Surface-Area Measurements Based on Image Processing (Digital Crofton Method, DC)

In automatic surface-area measurements based on image processing, it is also advantageous to use the Crofton formula (8) (Meyer, 1992), exploited in interactive stereological methods for estimating the surface area using spatial grids. The method based on this formula (the digital Crofton method, DC), estimating the surface area of the objects by counting intersections of the object surface with linear probes (Fig. 15), should be used for measurements of 3D digital images rather than measuring the area of the surface of a digitized object directly: in digitized images the smoothness of the boundary is not preserved, and thus the direct measurement overestimates the actual surface area (Fig. 16).

The intersections of the object surface with the rows and the diagonals of an orthogonal grid, called the principal lines, are used for digital measurements of surface area employing the Crofton formula. Three directions of principal lines connecting neighbors differing in

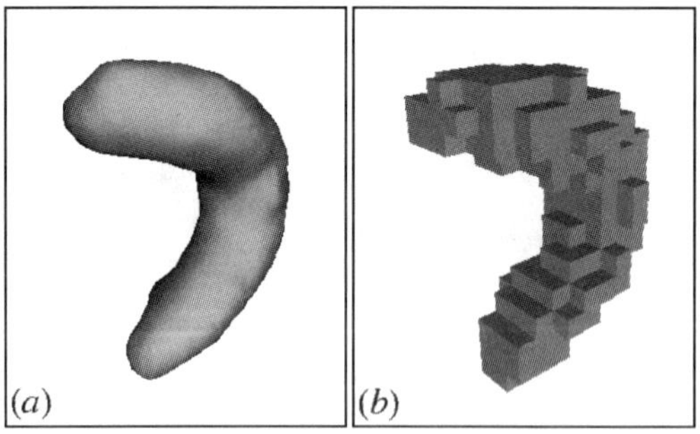

FIGURE 16 Tobacco cell chain surface (*a*) and its digitized image with resolution 8 μm^3, where the voxels are represented as cubes (*b*). The 3D reconstruction is based on image data captured by a confocal microscope.

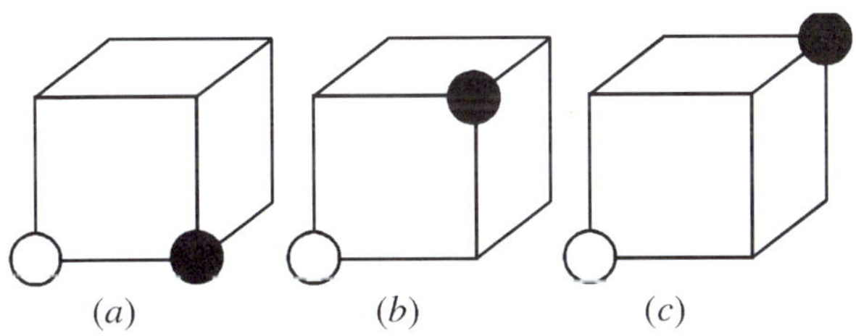

FIGURE 17 Examples of configurations of neighbors in the orthogonal lattice used for surface-area measurements by systematic total projections into the principal directions of the cubic lattice. The configurations correspond to the principal lines in the direction of rows (*a*), voxel surface diagonals (*b*), and voxel solid diagonals (*c*). All relevant local configurations of neighboring points with alternating values in a general orthogonal lattice can be obtained by orthogonal rotations and deformations of these examples.

one coordinate are the directions of the orthogonal voxel axes (rows), six directions of principal lines connecting neighbors differing in two coordinates are the directions of the voxel surface diagonals, and four directions of principal lines connecting neighbors differing in three coordinates are the directions of the voxel solid diagonals (Fig. 17).

According to the digital Crofton method, the surface area can be estimated by summing pairs of neighbours with alternating values:

$$S_{\text{est}} = 2\frac{1}{m}\sum_{i=1}^{m}\frac{1}{L_{Vi}}n_i, \tag{13}$$

where m is the number of directions, n_i is the number of configurations of neighboring points with alternating values (Fig. 17), and L_{Vi} is the density of principal lines in the ith direction (i.e., the length of principal lines per unit volume) (Table 5).

4.3 Automatic Surface-Area Measurements Based on Surface Triangulation

The triangulated surfaces are obtained from 3D grayscale digital images by isosurface detection or by connecting the object contours by triangulated strips. The surface area is then obtained by summing the areas of all the triangles.

4.4 Comparison of Stereological and Automatic Digital Methods

4.4.1 Test-Images: Tobacco Cell Chains The tobacco cell line BY-2 (*Nicotiana tabacum* L., cv. Bright Yellow) was cultured in vitro in a liquid culture medium containing

TABLE 5 The densities L_V of principal lines of the orthogonal grid of points with coordinates (iu_1, ju_2, ku_3) for integers i, j, k and calibration constants $u_1 > 0$, $u_2 > 0$, and $u_3 > 0$; the densities are listed for all three types of line types (i.e., for those going in the direction of rows, surface diagonals, or solid diagonals of the cubic lattice)

Line type	L_V	
Rows	$u_i/u_1u_2u_3$	$1 \leqslant i \leqslant 3$
Surface diagonals	$(u_i^2 + u_j^2)^{1/2}/u_1u_2u_3$	$1 \leqslant i < j \leqslant 3$
Solid diagonals	$(u_1^2 + u_2^2 + u_3^2)^{1/2}/u_1u_2u_3$	

2,4-dichlorophenoxyacetic acid (2,4-D) as the only growth regulator (10^{-5} M). The cell samples were taken from an early exponential-growth phase and stained without prefixation by adding an aliquot part of a 0.01% solution of fluorescein diacetate (FAD) to the culture medium. Fourteen individual cells or short cell chains were selected, with the aim of obtaining a set of test objects having as large variability of shapes as possible (Fig. 18). The images of serial optical sections of each cell chain, 1 μm apart, were recorded by a Bio-Rad MRC 600 confocal laser scanning microscope (using an excitation wavelength of 488 nm) equipped with an oil-immersion Nikon CF N Plan Fluor objective (40 ×, NA 1.3). The pixel size was 0.412 μm × 0.412 μm. In all cases, the stack of sections encompassed the entire cell chain.

The volume datasets originating from the capture series of grayscale images were resampled to a spatial resolution of 1 μm^3, closed by 2D morphological closing (Serra, 1982), and smoothed by a 3D Gaussian filter using a 3^3 neighborhood applied twice with standard deviation $\sigma = 1.5$ μm. The binary volume data sets were obtained by thresholding, morphological closure with 2^3 structural element, filling of cavities, and opening by reconstruction (Fig. 19). The algorithms for the above morphological operations were designed and their implementation developed as custom-made modules in an IRIS Explorer visualization environment run on an SGI INDY workstation.

4.4.2 Methods Compared The cell-chain surface area was estimated by the fakir (F) method (Fig. 20) using the OTSFP probe (Fig. 13) and our interactive FAKIR and GEN3GRID programs. The fakir method was applied to grayscale (FG) as well as binary (F01) volume data, while each measurement was performed independently five times and the corresponding coefficients of error were calculated. On average, 200 intersections between the fakir probes and the chain surface were counted. In this way, the surface area of one chain was estimated in about 30 minutes.

Both the image-processing method and the method based on surface triangulation were applied to four sets of image data differing in their degrees of smoothing and resolution. For this purpose, the above binary image data were processed in the following way. First, the binary image was converted to grayscale (foreground 255, background 0), and then it was smoothed by a Gaussian filter with standard deviation 1.5 μm operating on a 3^3 neighborhood applied m times ($m = 0, 1$). Further, the resolution was decreased n times ($n = 1, 2$), taking the average grayscale value of the original voxels as the new voxel value.

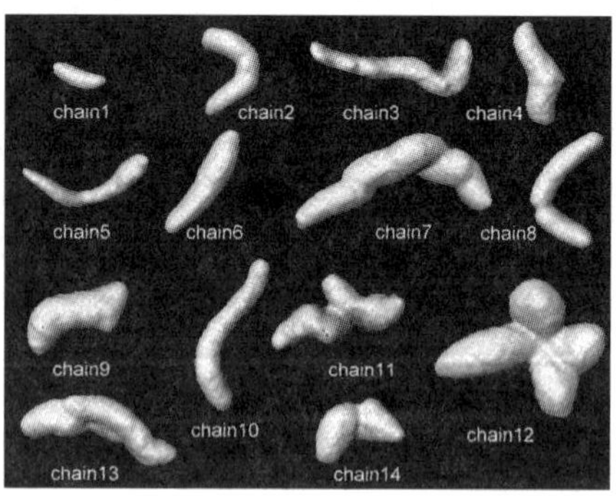

FIGURE 18 The 3D reconstructions of 14 tobacco cell chains used for the measurements. Reproduced by permission from Kubínová et al. (1999). Copyright 1999, John Wiley & Sons, Inc.

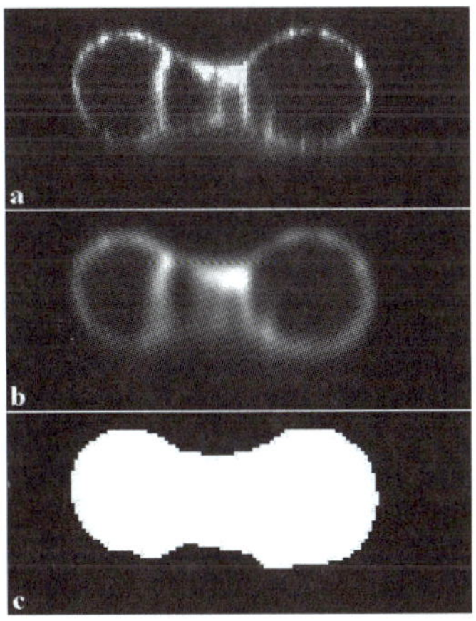

FIGURE 19 Image processing of tobacco cell-chain image shown on x–z section of chain2. (a) Original grayscale image after resampling to a spatial resolution of 1 μm^3. (b) Grayscale image after 2D morphological closing and smoothing by a 3D Gaussian filter using a 3^3 neighborhood applied twice with standard deviation $\sigma = 1.5\,\mu m$. (c) Binary image obtained from (b) by thresholding, morphological closure with 2^3 structural element, filling of cavities, and opening by reconstruction. Reproduced by permission from Kubínová et al. (1999). Copyright 1999, John Wiley & Sons, Inc.

The digital Crofton (DC) method, based on automatic counting of intersections between the surface and lines in 13 principal directions of the cubic lattice, was then applied to the four sets of binary data obtained after thresholding the smoothed and resampled images. The results of surface-area measurements made on images with smoothing and resolution parameters m and n (denoted by DCmn) are listed in Table 6 for DC01 (i.e., measurement made on the data used also for slicer measurement, with no smoothing and no resampling applied), DC11 (smoothing made once and no resampling), DC02 (no smoothing, resolution decreased twice), and DC12 (smoothed once, resolution decreased twice).

The method based on surface triangulation (T) was also applied to four sets of grayscale image data, with smoothing applied m times and resolution decreased n times. The surfaces of cells were triangulated by isosurface detection with threshold 128. The results of length measurements made on images with smoothing and resolution parameters m and n (denoted by Tmn, with $m = 0, 1$ and $n = 1, 2$) are listed in Table 6.

All of above image processing procedures and implementations of DC and T methods were performed on an SGI INDY workstation using our special custom-made modules developed in the IRIS Explorer visualization environment.

4.4.3 Results of Comparison

It was found that the triangulation method with no smoothing or resampling (T01) gave larger results in comparison with the triangulation method (T02) with no smoothing and resampling to $2 \times 2 \times 2$ (Table 6). The results of the T01 and T02 methods were larger than those of the rest of tested methods. The interactive fakir method applied to grayscale images (FG) did not yield results different from those obtained with its binary version (F01), and its results also did not differ significantly from those of the DC01, DC11, and T11 methods. The digital Crofton method without

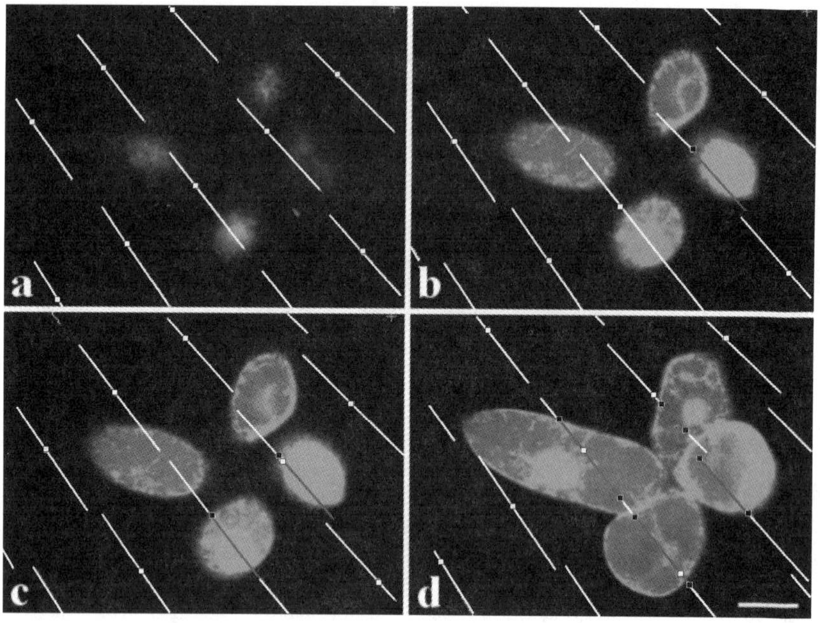

FIGURE 20 See ftp site for color figure. The fakir method applied to surface-area and volume estimation of a tobacco cell chain using the FAKIR program. The process of marking intersection points between the cell surface and the fakir probe is shown during focusing through the cell chain. Three serial optical sections of the cell chain are shown in (*a*), (*b*), and (*c*). The test lines of the fakir probe piercing the cell chain are white at the beginning of the measurement (*a*). The white points denote the intersection points between the test lines and the current section. As soon as such a point comes into contact with the cell profile during focusing-through, it is marked by a mouse as a black point and the part of the test line above this point becomes gray (*b,c*). After marking all intersection points, the parts of the fakir probe lying inside the cell chain are gray (*d*). The length of these gray intercepts is used for cell chain volume estimation, while the number of black intersection points is proportional to the cell chain surface area. The scale bar represents 20 μm. Reproduced by permission from Kubínová and Janáček (2001). Copyright 1999, John Wiley & Sons, Inc.

smoothing (DC01) gave larger results than that with smoothing (DC11) and the triangulation method with smoothing and no resampling (T11). However, the results of the F01, DC01, DC11, T11, and T02 methods were very close to each other, on average differing by about 1%. The coefficients of error of the fakir method were low (0.020 for FG and 0.012 for F01). The results of DC02 were lower than those of DC01 (by less than 1%), but higher than those of DC12 (by 0.5%). Similarly, the T12 method yielded a lower value of cell-chain surface area than T11 by less than 1%. The T01 method overestimated the chain surface area by about 8%.

The results show that the surface-area measurement was not very sensitive to segmentation – at least in the case of tested image data with relatively smooth surfaces. The triangulation-based measurement as well as the digital Crofton method were quite suitable for automatic surface-area measurements, and the differences in results for images with different degrees of smoothing and resampling were not large (Table 6). However,

TABLE 6 The surface area of 14 tobacco cell chains (in μm^2) measured by the fakir method applied manually to both grayscale (FG) and binary (F01) data, and by the digital Crofton (DC) and triangulation (T) methods applied to data with no smoothing and no resampling (DC01, T01), with smoothing made once and no resampling (DC11, T11), with no smoothing and resolution decreased twice (DC02, T02), and to the once-smoothed data with the resolution decreased twice (DC12, T12)

Chain No.	FG	F01	DC01	DC11	DC02	DC12	T01	T11	T02	T12
1	6 240	6 030	6 180	6 130	6 170	6 100	6 710	6 140	6 160	6 070
2	18 700	18 200	18 700	18 500	18 400	18 300	20 200	18 500	18 600	18 400
3	31 600	30 000	31 100	31 000	30 700	30 700	33 800	31 100	31 400	30 900
4	20 100	20 100	20 200	20 000	19 900	19 800	22 000	20 100	20 200	19 900
5	17 500	18 300	17 300	17 200	17 200	17 000	18 800	17 200	17 300	17 100
6	20 500	20 600	20 600	20 600	20 600	20 500	22 300	20 600	20 700	20 500
7	49 600	49 300	50 600	50 400	50 400	50 200	55 000	50 500	50 700	50 200
8	25 300	25 300	24 200	24 100	24 100	24 000	26 200	24 200	24 300	24 000
9	28 300	28 300	28 600	28 500	28 400	28 300	31 000	28 500	28 600	28 300
10	25 800	25 900	26 300	26 100	26 100	26 000	28 500	26 200	26 300	26 000
11	37 500	36 800	38 000	37 700	37 700	37 400	41 200	37 800	37 800	37 400
12	66 800	66 200	67 400	67 100	67 000	66 700	73 100	67 300	67 500	66 900
13	38 500	37 700	37 800	37 600	37 500	37 400	41 000	37 700	37 800	37 400
14	24 500	23 600	24 400	24 200	24 100	24 000	26 400	24 300	24 300	24 100

before applying any of these automatic methods, it should be carefully tested for the given type of image data to determine whether it gives reasonably precise results.

5 METHODS FOR VOLUME ESTIMATION

5.1 Stereological Methods

The volume of a 3D object can be expressed by the integral

$$V = \int_0^h A(x)\,dx, \tag{14}$$

where $A(x)$ is the area of the section of the object passing through the point $x \in [0, h]$ and h is the caliper diameter of the object perpendicular to section planes.

If systematic uniform random points are selected from the interval $[0, h]$, it is possible to estimate the object volume by the Cavalieri estimator (Gundersen and Jensen, 1987):

$$V_{\text{est}} = T \sum_{j=1}^{n} A_j, \tag{15}$$

where T denotes the distance between sections, A_j is the area of the jth section ($j = 1, \ldots, n$), and n is the number of sections. In practice, the area A_j is usually estimated by the point-counting method, where a point test system is superimposed on the

jth section and the number of test points (P_j) falling into the section is counted. The following relation holds:

$$A_{j,\text{est}} = aP_j, \tag{16}$$

where a is the area corresponding to a single test point. Thus, it is possible to estimate the object volume by the formula

$$V_{\text{est}} = Ta \sum_{j=1}^{n} P_j. \tag{17}$$

The Cavalieri principle is usually used for volume estimation from series of physical sections. If digitized images of perfectly registered serial optical sections encompassing the object are available, efficient stereological methods using spatial grids (spatial grid of points, fakir method) can be used also for volume estimations. Such methods can be applied to estimate, for example, the volume of cells from optical sections, provided that the cell profiles are visible and sharp, as they are under confocal and two-photon microscopes.

5.1.1 Spatial Grid of Points The method of volume estimation using a spatial grid of points (Cruz-Orive, 1997) is a modification of the Cavalieri principle. If a cubic spatial grid of points (Fig. 21) is applied, the object volume can be estimated by the formula

$$V_{\text{est}} = u^3 P, \tag{18}$$

where u is the grid constant (the distance between two neighboring points of the grid) and P is the number of grid points falling into the object. The position of the point grid should be uniform random.

5.1.2 Fakir Method The fakir probe (introduced in Section 4.1.1) can be used also for estimating the object volume, when the total length of intercepts between the object and the probe is measured (Cruz-Orive, 1993, 1997). Different combination of fakir probes

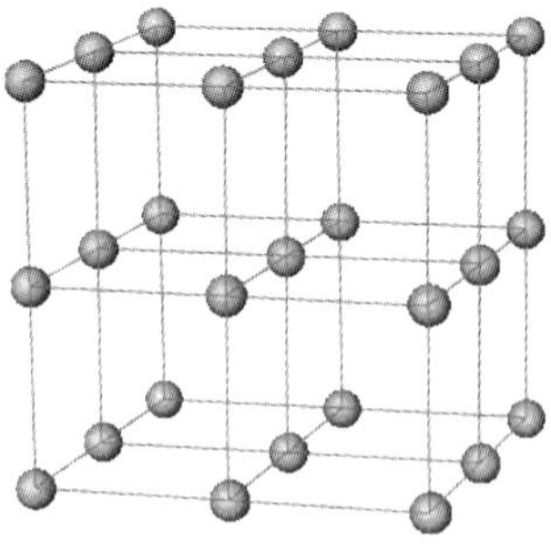

FIGURE 21 Cubic spatial grid of points used for volume estimation. Reproduced by permission from Kubínová and Janáček (2001). Copyright 1999, John Wiley & Sons, Inc.

can be used to increase the efficiency (i.e., to lower the variance of the method). If the spatial grid consists of n fakir probes, the object volume can be estimated by the formula

$$V_{\text{est}} = \frac{1}{n} \sum_{i=1}^{n} \frac{v}{l_i} L_i, \tag{19}$$

where l_i/v is the length of the ith fakir probe per unit volume and L_i is the length of intercepts between the object and the ith probe ($i = 1, \ldots, n$). This means that if an orthogonal triplet of shifted fakir probes (OTSFP; Fig. 13) is used, the object volume V can be estimated by the following formula:

$$V_{\text{est}} = \tfrac{1}{3} u^2 (L_1 + L_2 + L_3), \tag{20}$$

where L_i ($i = 1, 2, 3$) is the length of the intercept between the object and the ith probe, and u is the grid constant (i.e., the distance between neighboring parallel lines of the triple grid). The position of the spatial grid should be uniform random.

5.1.3 *Example: Volume of a Muscle Fiber Within a Thick Muscle Slice*

Thick transverse muscle slices were processed and captured by a confocal microscope as described above (see Section 3.1.2).

The volume of the ith fiber in the muscle slice ($V_{i,\text{fib}}$) was estimated using the cubic spatial grid of points. The grid constant u was chosen to be 10 µm, ensuring that about 50–100 points hit the fiber. The method was implemented using the SANDAU program (written in TurboPascal 6 for an IBM PC under MS DOS), generating a square point grid and superimposing it on sections. The test points (given by the knot points of the test grid) falling into the fiber profile in the parallel systematic optical sections were then counted (see Fig. 22).

$V_{i,\text{fib}}$ was also estimated by the fakir method using OTSFP (Fig. 13). Here, the length of the intercepts between the fiber slice and the ith probe was counted by our FAKIR software (written in TurboPascal 6 for an IBM PC under MS DOS). Here the grid constant u was chosen to be 10 µm ensuring that about 10–20 intercepts went through the fiber.

5.2 Automatic Methods for Volume Measurements Based on Image Processing

The volume of the object is proportional to the number n of foreground image elements:

$$V_{\text{est}} = \frac{1}{N_V} n = u_1 u_2 u_3 n, \tag{21}$$

where N_V is the density of grid elements and u_i ($i = 1, 2, 3$) are the lengths of intervals between grid elements in the directions of the respective orthogonal axes. This method can be called a voxel-counting method (VC), since the volume of the binary volume dataset is estimated simply by summing the volumes of all the relevant voxels.

The quality of results depends again on the quality of segmentation (Hermann, 1991).

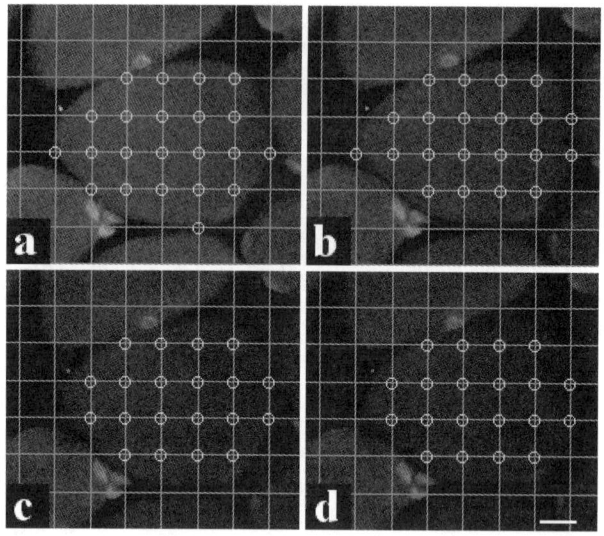

FIGURE 22 See ftp site for color figure. The method using a cubic spatial grid of points applied to muscle-fiber volume estimation using the SANDAU program. The test points (given by the knot points of the test grid) falling into the fiber within the 40 μm thick muscle slice are counted. The grid constant is 10 μm here; the estimated volume of the muscle fiber slice is thus equal to $(22 + 21 + 20 + 20) \times 1000\,\mu m^3 = 83\,000\,\mu m^3$.

5.3 Automatic Methods for Volume Measurements Based on Surface Triangulation

The volume of objects delimited by the triangulated surfaces can be calculated by summing properly signed volumes of all tetrahedra with bases in the surface triangles and one common vertex (Santaló, 1976).

5.4 Comparison of Stereological and Automatic Digital Methods

5.4.1 Test Images: Tobacco Cell Chains Tobacco cell chains were processed and captured by a confocal microscope as described above (see Section 4.4).

5.4.2 Methods Compared The cell-chain volume was estimated by the spatial grid of points (GP) using the SANDAU program, generating a square point grid and superimposing it on sections. The test points (given by the knot points of the test grid) falling into the cell chain were counted. The method was applied to grayscale (GPG) as well as binary volume (GP01) data and repeated independently five times. The corresponding coefficients of error were calculated. The grid constant was chosen to be 10 μm, ensuring on average that 300 test points fell inside the cell chain. Measurement of one chain took 5–10 minutes. For results, see Table 7.

Further, the cell-chain volume was estimated by the fakir method (F) using OTSFP (Fig. 13) and our FAKIR software (Fig. 20). The method was applied to grayscale (FG) as well as binary (F01) volume data, while each measurement was performed independently five times and the corresponding coefficients of error were calculated (Table 7). The grid

TABLE 7 The volume of 14 tobacco cell chains (in μm^2) measured by the spatial grid of points (GP) and by the fakir (F) methods applied manually to both grayscale (GPG, FG) as well as binary (GP01, F01) volume data, and by the voxel-counting (VC) and triangulation (T) methods applied to data with no smoothing and no resampling (VC01, T01), with smoothing made once and no resampling (VC11, T11), with no smoothing and resolution decreased twice (VC02, T02), and to the once smoothed data with the resolution decreased twice (VC12, T12)

Chain No.	GPG	GP01	FG	F01	VC01	VC11	VC02	VC12	T01	T11	T02	T12
1	33.6	34.2	34.6	33.7	34.9	34.8	35.0	34.6	34.9	34.6	34.4	34.1
2	135.8	143.8	145.3	143.6	141.4	140.9	140.1	140.2	141.2	140.6	140.4	139.7
3	256.2	251.6	254.3	246.5	248.4	248.0	242.5	245.6	248.2	247.3	247.2	246.2
4	185.0	183.0	198.6	187.9	186.1	185.6	184.6	184.7	186.0	185.3	185.0	184.4
5	105.0	109.0	104.2	111.3	108.4	108.0	108.0	107.2	108.2	107.6	107.2	106.5
6	201.8	194.2	190.8	196.7	194.8	194.6	194.9	193.6	194.6	194.0	193.5	192.9
7	570.6	580.2	565.1	582.0	580.1	579.6	578.6	577.7	579.8	578.6	577.8	576.5
8	190.4	195.6	195.4	206.3	196.4	196.1	196.4	195.3	196.2	195.4	195.0	194.2
9	329.8	329.2	321.7	332.1	331.6	331.2	331.5	330.7	331.4	330.7	330.2	329.4
10	214.4	213.4	212.6	218.3	216.0	215.6	215.6	214.4	215.8	215.0	214.4	213.6
11	374.4	372.2	369.4	371.7	370.8	370.3	371.4	369.3	370.6	369.6	368.7	367.7
12	898.4	901.8	894.5	928.7	911.4	910.8	910.3	908.9	911.1	909.6	908.9	907.2
13	392.8	401.2	403.5	406.0	399.5	399.1	399.4	398.1	399.3	398.3	397.4	396.5
14	258.2	263.6	268.4	260.9	264.1	263.7	263.3	262.9	263.9	263.3	262.9	262.2

constant was chosen to be 20 μm, ensuring on average 30 intercepts per measurement. The evaluation of one cell chain took about 10 minutes.

The voxel-counting (VC) method was applied to the same four sets of binary data as were used for surface-area measurements by the DC method (Section 4.4.1). The procedure was included in the program module for the implementation of the digital Crofton method developed in the IRIS Explorer visualization environment and performed on an SGI INDY workstation. The results for VC01, VC11, VC02, and VC12 are listed in Table 7.

Finally, the algorithm for volume measurement based on surface triangulation (T) using the method of marching cubes (Lorensen and Cline, 1987) was applied to the same data as used for surface-area measurements by the T method (Section 4.4.1). The method was again run on an SGI INDY workstation in the IRIS Explorer visualization environment. The results for T01, T11, T02, and T12 are listed in Table 7.

5.4.3 Results of Comparison All of the methods tested, namely the spatial grid of points (GP), fakir (F), voxel-counting (VC), and triangulation (T) methods, did not differ significantly. On average, their results differed in the range of 1%. The interactive fakir method applied to grayscale images (FG) yielded results that were not different from those obtained by its binary version (F01). The same holds for the spatial grid of points (GPG, GP01) method. The coefficients of error of GP and F methods were very low (0.012 for GPG, 0.008 for GP01, 0.017 for FG, and 0.006 for F01).

The results show that volume measurement was not sensitive to segmentation – at least in the case of tested image data of relatively simple shapes. All tested methods were quite suitable for volume measurements. Although volume measurements are relatively robust, they should be tested before applying to the given type of image data to determine whether reasonably precise results are obtained.

6 DISCUSSION

It has been demonstrated how useful confocal microscopy can be in estimating different geometrical parameters of microscopic structure. In comparison with conventional optical microscopy, confocal and especially two-photon microscopy offer not only higher resolution and examination of thicker specimens but also a better possibility to analyze living, fresh, or more easily prepared specimens.

Confocal microscopy, like any other technique, has some drawbacks. Sometimes, it can be difficult to find a suitable fluorescence staining of structures to be examined. Here, it should be noted that in histological specimens, a nonspecific staining such as eosin can often be used (Jirkovská et al., 1998), while the cell surface can be visualized by immunofluorescence techniques when antibodies are bound to integral membrane proteins of cellular plasma membranes and labeled by a common fluorescent dye. Another drawback consists in the fact that the axial resolution (although higher than in a conventional optical microscope) is lower than the lateral resolution. The shape of the point spread function of a confocal microscope is elongated in the direction of the z axis (Shaw, 1994), which causes defocusing that can possibly result in overestimation of the surface area and volume of the examined objects. This can be eliminated by applying special deconvolution algorithms (Holmes et al., 1995) to perform 3D deblurring of the images before measurements are made. In two-photon excitation fluorescence microscopy, the axial resolution is higher than in confocal microscopy (Denk et al., 1990; Diaspro, 1999; Nakamura, 1999), and it is also possible to penetrate more deeply into the specimen with decreased bleaching of fluorescence dyes. However, possible deblurring of 3D images should be considered even here. Aberrations, especially the axial displacement due to unmatched refractive indices (Sheppard and Török, 1997), should also be taken into account in measurements of microscopic structures. Further, deformations and possible tissue shrinkage due to the technical processing of specimens should be carefully tested (Andersen and Gundersen, 1999).

In practical application of the presented methods, the question of efficiency is important, since it is always desirable to get sufficiently precise results with the least effort. Therefore, more and more attention is being paid to the study of variances of stereological estimators. The variances of the estimators based on the Cavalieri principle, the spatial grid of points, and the disector principle have been studied using Matheron's theory of regionalized variables (Matheron, 1965); see, for example, Gundersen and Jensen (1987), Cruz-Orive (1989, 1993, 1999), Kiêu et al. (1998), and Gundersen et al. (1999). For the estimators based on measuring intersections of the object with isotropic uniform random grids (i.e., the fakir and slicer methods), the variance can be split into a component due to the grid orientation and a residual component due to the grid position (Hahn and Sandau, 1989). The first component depends on the mutual orientation of the fakir probes applied and on the shape of the object under study, namely the anisotropy of the surface, which can be expressed by the rose of directions of normals to the surface. The second, residual component of variance is dependent mainly on the grid density and arrangement, (e.g., on the mutual shift of the fakir probes). The first component of variance is clearly the highest for a totally anisotropic object, which rose of directions is represented by a single vector (i.e., for a flat surface in 3D). It can be proved that in this special, "worst" case, the coefficient of variance of the surface-area estimate obtained by applying a single fakir probe is 57.74%, while for the estimate obtained with three orthogonal fakir probes (Fig. 13) it is only 10.16% (Mattfeldt, 1985; Janáček, 1999). For less-anisotropic surfaces, the first component of variance is decreasing; for example, for the triple grid estimator of the surface area of an ellipsoid with diameter ratios of 1 : 4 : 4, the coefficient of variance was calculated to be 4.90%, while an ellipsoid with diameter ratios of 1 : 1.6 : 1.6, it is already very close to zero (Hahn and Sandau, 1989). The second, residual component of variance reflects the arrangement of the spatial grid applied. It can be shown that the shifted OTSFP (Fig. 13) is much more efficient than a nonshifted orthogonal triplet of fakir probes (Janáček, 1999).

A number of interactive stereological methods and automatic digital methods for estimating geometrical characteristics of microscopic structure have been introduced and compared from the point of view of their precision and efficiency. It should be stressed that no absolutely universal method optimal for any type of structure exists. In general, the automatic methods are faster than interactive stereological methods, but they require automatic segmentation (i.e., recognition) of analyzed objects from the images, or at least a high contrast between the object and the background. For surface-area estimation, the fakir method appears to be the most universal of all the methods presented, and so it might be recommended for testing the applicability of other, less time-consuming methods. The method using a spatial grid of points is very good for interactive volume estimation. The triangulation method applied to grayscale images appears to be suitable for measuring the volume and surface area of isotropic as well as anisotropic objects, provided that a high contrast between the object and the background is achieved. If object segmentation is feasible, the voxel-counting method is suitable for volume measurements and the digital Crofton method for surface-area measurements. Unlike volume and surface-area measurements, it appears that length measurement is very sensitive to segmentation, resolution, and degree of smoothing. The triangulation-based measurement was not suitable for length measurements in our model dataset, while the digital Crofton method gave better results. It should be stressed that before applying any automatic method, it should be very carefully tested whether it gives reasonably precise results for the given type of image data.

All the methods presented were implemented using our programs or software modules. Programs implementing stereological methods described above are included in the 3DTOOLS software package (for details, see Kubínová and Janáček, 2001). All programs of this package were written in TurboPascal 6 for an IBM PC under the MS DOS operating system. The input data are assumed to be in PIC image format used with BioRad confocal microscopes; however, utilities for conversions between TIFF and PIC formats are offered by the PICCONV program, included in the 3DTOOLS package. The SANDAU, FAKIR, FAKIRD, and SLICER programs enable one to browse rapidly through a series of optical sections and apply various virtual test probes to this stack of images. Then, the interactions of the structure under study (e.g., intersection points between a test line and the object surface) are sampled interactively. The sampling is facilitated by the possibility of browsing through the stack of sections (with test probe applied) not only in the original lateral, x–y direction but also in the orthogonal (i.e., axial) x–z and y–z directions. The images of sections in the three main orthogonal directions can be generated very quickly, which means that browsing through the image volume resembles focusing directly under a microscope. The programs of the 3DTOOLS software package are available as a freeware via the Internet at the website http://www.biomed.cas.cz/fgu/fakir/3dtools.htm. Counting particles by the optical disector or unbiased sampling brick rule is supported in our newly developed DISECTOR module (Tomori et al., 2001) running in the IMAGE TOOL (UTHSCA, U.S.A.) freeware environment together with our older module STESYS (Tomori et al., 2000; http://www.saske.sk/~tomori/) for generating stereological test systems. The image processing of 3D image data and implementation of automatic digital methods were performed on an SGI INDY workstation using our special custom-made modules developed in the IRIS Explorer visualization environment.

Some of the stereological methods presented are also included in commercial software systems for stereology, e.g. C.A.S.T.-GRID (Olympus, Denmark), Stereo Investigator (MicroBrightfield, U.S.A.), and Digital Stereology (Kinetic Imaging, U.K.).

ACKNOWLEDGMENTS

We wish to thank Professor Ida Eržen (Institute of Anatomy, Medical Faculty, University of Ljubljana) for providing us with muscle specimens, Dr. Jana Albrechtová (Department of Plant Physiology, Faculty of Science, Charles University, Prague) for preparing Norway spruce needle sections, Dr. Z. Opatrný (Department of Plant Physiology, Faculty of Science, Charles University, Prague) for providing us with tobacco cell chains, and Anna Smallcombe (Bio-Rad Microscopy Division, Bio-Rad Microscience Ltd.) for images of Golgi-stained nerve cells.

The study presented here was supported by the Grant Agency of the Academy of Sciences of the Czech Republic (Grant A5011810) and by the Grant Agency of the Czech Republic (Grants 304/01/0257 and 304/99/0680) and by the Ministry of Education, Youth and Sports of the Czech Republic (KONTAKT grant No. 184).

REFERENCES

Andersen, B. B. and Gundersen, H. J. G. (1999). Pronounced loss of cell nuclei and anisotropic deformation of thick sections. *J. Microsc.* **196**, 69–73.

Baba, N. (1992). Computer aided three-dimensional reconstruction from serial section images. In: *Image Analysis in Biology* (ed. D.-P. Häder). Boca Raton, FL: CRC Press, pp. 251–270.

Baddeley, A. J., Gundersen, H. J. G., and Cruz-Orive, L. M. (1986). Estimation of surface area from vertical sections. *J. Microsc.* **142**, 259–276.

Cruz-Orive, L. M. (1989). On the precision of systematic sampling: a review of Matheron's transitive methods. *J. Microsc.* **153**, 315–333.

Cruz-Orive, L. M. (1993). Systematic sampling in stereology. *Bull. Int. Statist. Inst.* **55**, 451–468.

Cruz-Orive, L. M. (1997). Stereology of single objects. *J. Microsc.* **186**, 93–107.

Cruz-Orive, L. M. (1999). Precision of Cavalieri sections and slices with local errors. *J. Microsc.* **193**, 182–198.

Cruz-Orive, L. M. and Howard, C. V. (1991). Estimating the length of a bounded curve in three dimensions using total vertical projections. *J. Microsc.* **163**, 101–113.

Cruz-Orive, L. M. and Howard, C. V. (1995). Estimation of individual feature surface area with the vertical spatial grid. *J. Microsc.* **178**, 146–151.

Denk, W., Strickler, J. H., and Webb, W. W. (1990). Two-photon laser scanning fluorescence microscopy. *Science* **248**, 73–76.

Diaspro, A. (Ed.) (1999). Two-photon excitation microscopy. *Microsc. Res. Tech.* **47**, 163–214.

Eržen, I., Maravić, V. (1993). Simultaneous histochemical demonstration of capillaries and muscle fiber types. *Histochemistry* **99**, 57–60.

Gokhale, A. M. (1990). Unbiased estimation of curve length in 3D using vertical slices. *J. Microsc.* **159**, 133–141.

Guilak, F. (1994). Volume and surface area measurement of viable chondrocytes in situ using geometric modelling of serial confocal sections. *J. Microsc.* **173**, 245–256.

Gundersen, H. J. G. (1986). Stereology of arbitrary particles. A review of unbiased number and size estimators and the presentation of some new ones, in memory of William R. Thompson. *J. Microsc.* **143**, 3–45.

Gundersen, H. J. G. (1988). The nucleator. *J. Microsc.* **151**, 3–21.

Gundersen, H. J. G. and Jensen, E. B. (1987). The efficiency of systematic sampling in stereology and its prediction. *J. Microsc.* **147**, 229–263.

Gundersen, H. J. G., Jensen, E. B. V., Kiêu, K., and Nielsen, J. (1999). The efficiency of systematic sampling in stereology – reconsidered. *J. Microsc.* **193**, 199–211.

Hahn, U. and Sandau, K. (1989). Precision of surface area estimation using spatial grids. *Acta Stereol.* **8**, 425–430.

Hermann, G. T. (1991) Quantitation using 3D images. In: *3-D Imaging in Medicine* (ed. J. K. Udupa and G. T. Hermann). Boca Raton, FL: CRC Press, pp. 145–161.

Holmes, T. J., Bhattacharyya, S., Cooper, J. A., Hanzel, D., Krishnamurthi, V., Lin, W., Roysam, B., Szarowski, D. H., and Turner, J. N. (1995). Light microscopic images reconstructed by maximum likelihood deconvolution. In: *Handbook of Biological Confocal Microscopy*, 2nd edn. (Ed. J. B. Pawley) New York: Plenum Press, pp. 389–402.

Howard, C. V., Reid, S., Baddeley, A., and Boyde, A. (1985). Unbiased estimation of particle density in the tandem scanning reflected light microscope. *J. Microsc.* **138**, 203–212.

Janáček, J. (1999). Errors of spatial grids estimators of volume and surface area. *Acta Stereol* **18**, 389–396.

Jensen, E. B. V. and Gundersen, H. J. G. (1993). The rotator. *J. Microsc.* **170**, 35–44.

Jirkovská, M., Kubínová, L., Krekule, I., and Hach, P. (1998). Spatial arrangement of fetal placental capillaries in terminal villi: a study using confocal microscopy. *Anat. Embryol.* **197**, 263–272.

Karen, P., Kubínová, L., and Krekule, I. (1998). STESYS software for computer-assisted stereology. *Physiol. Res.* **47**, 271–278.

Kiêu, K., and Jensen, E. B. V. (1993). Stereological estimation based on isotropic slices through fixed points. *J. Microsc.* **170**, 45–51.

Kiêu, K., Xiong, W., and Trubuil, A. (1998). Precision of systematic counts. Rapport Technique 1998-1, Unité de Biométrie, INRA-Versailles.

Kubínová, L. (1993). Recent stereological methods for the measurement of leaf anatomical characteristics: estimation of volume density, volume and surface area. *J. Exp. Bot.* **44**, 165–173.

Kubínová, L. and Janáček, J. (1998). Estimating surface area by isotropic fakir method from thick slices cut in arbitrary direction. *J. Microsc.* **191**, 201–211.

Kubínová, L. and Janáček, J. (2001). Confocal microscopy and stereology: estimating volume, number, surface area and length by virtual test probes applied to three-dimensional images. *Microsc. Res. Tech.* **53**, 425–435.

Kubínová, L., Janáček, J., Guilak, F., and Opatrný, Z. (1999). Comparison of several digital and stereological methods for estimating surface area and volume of cells studied by confocal microscopy. *Cytometry* **36**, 85–95.

Kubínová, L., Janáček, J., Ribarič, S., Čebašek, V., and Eržen, I. (2001). Three-dimensional study of the capillary supply of skeletal muscle fibres using confocal microscopy. *J. Musc. Res. Cell Motil* (in press).

Larsen, J. O., Gundersen, H. J. G., and Nielsen, J. (1998). Global spatial sampling with isotropic virtual planes: estimators of length density and total length in thick, arbitrarily orientated sections. *J. Microsc.* **191**, 238–248.

Lorensen, W. E. and Cline, H. E. (1987). Marching cubes: a high resolution 3D surface construction algorithm. *Comput. Graph.* **21**, 163–169.

Matheron, G. (1965). *Les Variables Régionalisées et leur Estimation*. Paris: Masson.

Mattfeldt, T., Möbius, H.-J., and Mall, G. (1985). Orthogonal triplet probes; an efficient method for unbiased estimation of length and surface of objects with unknown orientation in space. *J. Microsc.* **139**, 279–289.

Mattfeldt, T., Mall, G., Gharehbaghi, H., and Moller, P. (1990). Estimation of surface area and length with the orientator. *J. Microsc.* **159**, 301–317.

Meyer, F. (1992). Mathematical morphology: from two dimensions to three dimensions. *J. Microsc.* **165**, 5–28.

Nakamura, O. (1999). Fundamental of two-photon microscopy. *Microsc. Res. Tech.* **47**, 165–171.

Nikololaidis, N. and Pitas, I. (2001). *3-D Image Processing Algorithms*. New York: Wiley.

Oliva, J.-M., Perrin, M., and Coquillart, S. (1996). 3D reconstruction of complex polyhedral shapes from contours using a simplified generalized Voronoi diagram. In: *Eurographics 96* (ed. J. Rossignac and F. Sillion). Oxford: Blackwell, pp. 397–408.

Pawley, J. B. (Ed.) (1995). *Handbook of Biological Confocal Microscopy*, 2nd edn. New York: Plenum Press.

Rigaut, J. P., Carvajal-Gonzales, S., and Vassy, J. (1992). 3-D image cytometry. In: *Visualization in Biomedical Microscopies* (ed. A. Kriete). Weinheim: VCH, pp. 205–248.

Russ, J. C. (1990). *Computer Microscopy: The Measurement and Analysis of Images*. New York: Plenum Press.

Sandau, K. (1987). How to estimate the area of a surface using a spatial grid. *Acta Stereol.* **6**, 31–36.

Santaló, L. A. (1976). *Integral Geometry and Geometric Probability*. Reading, MA: Addison-Wesley.

Serra, J. (1982). *Image Analysis and Mathematical Morphology*. London: Academic Press.

Shaw, P. (1994). Deconvolution in 3-D optical microscopy. *Histochem, J.* **26**, 687–694.

Sheppard, C. J. R. and Török, P. (1997). Effects of specimen refractive index on confocal imaging. *J. Microsc.* **185**, 366–374.

Sterio, D. C. (1984). The unbiased estimation of number and sizes of arbitrary particles using the disector. *J. Microsc.* **134**, 127–136.

Tandrup, T., Gundersen, H. J. G., and Jensen, E. B. V. (1997). The optical rotator. *J. Microsc.* **186**, 108–120.

Tomori, Z., Krekule, I., Kubínová, L. (2001). DISECTOR program for unbiased estimation of particle number, numerical density and mean volume. *Image Anal. Stereol.* **20**, 119–130.

Tomori, Z., Matis, L., Karen, P., Kubínová, L., and Krekule, I. (2000). STESYS2: extended STESYS software for MS Windows. *Physiol. Res.* **49**, 695–701.

Weibel, E. R. (1979). *Stereological Methods*, Vol. 1: *Practical Methods for Biological Morphometry.* London: Academic Press.

West, M. J. and Gundersen, H. J. G. (1990). Unbiased stereological estimation of the number of neurons in the human hippocampus. *J. Comput. Neurol.* **296**, 1–22.

CHAPTER 16

Imaging Live Cells in 3D Using Wide-Field Microscopy with Image Restoration

WALTER CARRINGTON

University of Massachusetts Medical Center, Shrewsbury Campus, 222 Maple Ave., Shrewsbury, MA 01545, U.S.A.

1 INTRODUCTION

Green fluorescent protein (GFP) has presented great opportunities for the use of fluorescence to study dynamic cellular processes in live cells. The production of fluorescently labeled proteins in live cells has now become a routine tool of molecular biology. It has also presented great challenges for imaging technologies. In this chapter, we discuss the requirements for imaging dynamic processes involving small numbers of molecules, with an emphasis on achieving the highest possible spatial and temporal resolution and sensitivity. Our approach to high speed three-dimensional (3D) fluorescence microscopy is to use wide-field microscopy with advanced high-speed, low-noise CCD cameras, fast focus change, and sophisticated computational techniques for image restoration (deconvolution). The instrument described was designed originally for very high-speed imaging of calcium ions (Ca^{2+}) in live cells. The speed it achieves, 543 frames s^{-1}, is faster than needed for GFP imaging, but allows us to explore the requirements for successful GFP imaging.

A cell is three-dimensional, and the knowledge of the 3D distribution of molecules is often critical to the understanding of the relationship of different molecular species. The motion of an organelle or molecular cluster is not always constrained to a single plane and cannot always be followed with 2D imaging. Thus, high-resolution 3D imaging over time is frequently required.

A live cell is in constant internal motion. Even in cases where the cellular process being studied is slow, resolution will be degraded by random or directed motion. Thus, 3D imaging speed must be fast enough to minimize motion blur. Our current image-restoration methods provide 100 nm transverse resolution. When an organelle or cluster of molecules moves, the resolution of the restored image degrades. Imaging an organelle moving at 1 μm s^{-1} at 30 frames s^{-1} doubles the resolvable distance for that organelle! Clearly, for

Confocal and Two-Photon Microscopy: Foundations, Applications, and Advances, Edited by Alberto Diaspro.
ISBN 0-471-40920-0

high-resolution imaging, faster frame rates are desirable. For following organelles or molecules moved by molecular motors at the highest possible resolution, frame rates up to 100 frames s^{-1} are desirable. Even if the dynamics of the process being studied are slow, internal motion of the cell may require significant speed to obtain high resolution across the whole cell.

This requirement of high speed implies extremely high pixel rates. For a typical 640×480 field of view, 100 frames s^{-1} requires more than 3×10^7 pixels s^{-1}. At these speeds, fluorescence saturation severely limits the signal levels that can be achieved by scanning microscopes – for a conventional laser scanning confocal microscope, at this frame rate, the pixel dwell time is only 0.03 μs. *A typical fluorophore will emit fewer than 10 photons in this dwell time*. Scanning confocal microscopes will therefore be severely photon-starved when imaging small numbers of fluorophores at these speeds. The pixel dwell time at this 100 frames s^{-1} rate is 10 ms with wide-field illumination. Fluorescence saturation is not a limitation for a microscope using conventional wide-field illumination.

Similarly, in order to image clusters of small numbers of molecules, high sensitivity and low noise are necessary. The high-speed cameras that we describe have quantum efficiencies of 65–70% and readout noises of 6–8.5 electrons. This is comparable to the best slow-scan CCD cameras that have readout rates 30–100 times slower. The collection efficiency of a conventional epifluorescence microscope (including our high-speed microscope) is roughly 0.07; the collection efficiency of a typical confocal microscope is roughly 0.002 (Sandison et al., 1995). The comparison with various confocal technologies will be discussed further later in this chapter.

2 HIGH-SPEED 3D FLUORESCENCE MICROSCOPE

A high-speed 3D fluorescence microscope has been developed by our group at the University of Massachusetts Medical School. This instrument was originally designed for 3D imaging of high-speed calcium events in isolated smooth muscle cells. Because of its speed and sensitivity, it has proven very useful for imaging dynamic cellular processes in 3D with GFP in live cells. The instrument is based on a sensitive high-speed CCD camera, high-speed focus change, wide-field laser illumination, and 3D image restoration.

2.1 High-Speed CCD Camera

In collaboration with the MIT Lincoln Laboratories, we developed a 128×128 high-speed CCD camera. This camera operates at rates up to 543 frames s^{-1}, with a readout noise of only 6 electrons and 70% quantum efficiency. The camera has four readout ports. This camera is a frame-transfer device; each image is electronically transferred to a nonimaging part of the CCD chip, where it is read out while the imaging portion of the chip is acquiring the next image. This frame-transfer capability enables the use of continuous imaging with essentially 100% duty cycle until the readout buffer is filled (200 images). This camera is close to ideal for high-speed calcium imaging.

The area of this camera is too small for high-resolution imaging of a whole cell, and the frame rate of this camera is greater than is needed for GFP imaging. Therefore, with the Lincoln Laboratories, we have developed a 640×480 CCD camera that operates at 94 frames s^{-1} (full frame), with readout noise no more than 8.5 electrons and quantum efficiency of at least 65% at 500 nm. This camera is also a frame-transfer CCD with four readout ports. This camera is now being integrated into a microscope optimized for imaging GFP. The software developed in our laboratory for this camera allows streaming images at full speed to a RAID disc array until disc space is exhausted.

These cameras both have sensitivity and noise comparable to those of good slow cameras. There appear to be no commercially available low-noise large-format cameras with full frame rates greater than about 30 frames s^{-1}. There are a number of CCD cameras that are excellent when speed is not necessary. Quantum efficiencies of more than 80% with readout noise of 5 or 6 electrons or less are readily available, with readout rates of one or a few frames per second. These are excellent for faint fixed samples. However, even for studies of slow events in living cells, there is a danger of losing resolution because of internal motion in the cell. For studies of many transport phenomenon, this temporal resolution is not adequate.

There are currently no commercially available CCD cameras that equal the high-speed performance of these two state-of-the-art cameras. However, the characteristics of commercially available cameras are evolving rapidly. As of March 2001, there are several cameras available that are worth consideration on the basis of the manufacturers' claimed performance; we have not verified these claims. The Hamamatsu Orca ER and the Photometrics Quicksnap HQ are fairly similar in their specifications. Both use progressive scan interline transfer CCD chips that allow the same electronic shuttering as frame-transfer cameras. They have 12-bit A/D, quantum efficiencies of 60–70%, readout noise of 8 electrons, and full-frame rates of 8.3 frames s^{-1} (Orca HQ, 1280×1040) and 10 frames s^{-1} (Quicksnap HQ, 1392×1040). Both allow binning and subarray readout, and have frame rates of about 16 frames s^{-1} at 640×512 or about 30 frames s^{-1} at 256×256. Both of these smaller sizes are quite usable. An older camera that is capable of higher speeds when binned or subarrayed is the Hamamatsu C4880-80. This camera has higher noise at high speed and a lower quantum efficiency of 40%. Not all light falling on the camera reaches a pixel; it has a pixel fill ratio of about 75%, which gives it an effective quantum efficiency of about 30%. The pixel fill ratio of the other cameras is not provided, but has likely been improved over this older camera. Since manufacturers continue to improve their cameras, they should be contacted for information on new cameras.

2.2 High-Speed Focus Change

For our high-speed small-format calcium imaging system, we designed and built a high-speed piezo focus changer that changes focus at a rate of 1 $\mu m\, ms^{-1}$. When the lens is coupled to the coverslip by immersion oil, substantial flexing of the coverslip was measured, which made the actual focus change uncertain and inaccurate. A stiffer chamber was constructed to avoid coverslip flexing, and allowed focus change at the full rate, but it provided too limited a field of view. With conventional chambers and the protocol of change focus, wait for the coverslip to recover from the impulse, take an image, and repeat, it was necessary to extend the time for an optical section to 10–20 ms rather than the 1 or 2 ms rate of the rest of the system. Therefore, we developed a protocol of continuous change of focus while imaging; the objective moves at a constant velocity and accelerates only at the top and bottom of the 3D image stack (Tuft et al., 1997; Kirber et al., 2001). This substantially reduces the effect of the motion on the coverslip. With a water-immersion lens, the effect is further reduced, and measurements of coverslip position with an eddy current sensor show that 1 μm/2 ms results in negligible motion of the coverslip. However, it imposes an axial motion blur on the image, which we have characterized and incorporated into the image-restoration process. For high resolution where the focus change is 200–500 nm per image, the resulting loss in z resolution after image restoration is negligible. Of course, use of this continuous-motion focus-change protocol requires the use of interline or frame-transfer cameras for electronic shuttering.

There is a commercially available device for changing focus – the PIFOC piezo device from P.I., Inc. Commercial vendors appear to use this device in the step, pause, step protocol. With this stepping protocol, this device seems to be limited to about 15 frames s^{-1}; this is the rate claimed by several vendors (e.g., Noran's Ultrafast 3D option). At higher speeds, this protocol with this device can result in discrepancies of several micrometers or more in focus position because of coverslip flex and ringing. We have tested the continuous-focus-change protocol with an older version of the PIFOC with a water-immersion lens. Our measurements on an older, smaller version of the PIFOC device show that there is negligible ringing when used with a ramp voltage at 1.0 μm/20 ms. Our protocols for high-resolution imaging use 0.5 μm or smaller focus intervals, and the newer, larger PIFOC device has nearly double the resonant frequency of the device we tested. Therefore, with water-immersion objectives and with the continuous-motion focus-change protocol, this higher-frequency device should reach rates of 0.5 μm/10 ms accurately and without ringing. Thus, it will likely perform adequately for GFP imaging. For higher-speed imaging as used for calcium experiments, accurate measurements of the newer device will be needed before any conclusion can be drawn on its capability.

2.3 Wide-Field Laser Illumination

With wide-field illumination, all pixels are illuminated simultaneously and fluorescent saturation is not a limitation, even at high frame rates. To provide sufficient illumination power for very high frame rates and low fluorophore counts, laser illumination is necessary. This system uses a separate laser for ultraviolet (UV) illumination for fura-2 or photolytic uncaging of caged compounds. It uses an argon–krypton laser from Coherent for visible illumination. In this microscope's usual configuration, the maximum excitation power at visible wavelengths delivered to the sample is about 1 mW. Usually the power used is well below 1 mW. For special purposes, 20 mW power at visible wavelengths has been delivered to the sample, and higher power is available, but is not usually needed.

3 HIGH-RESOLUTION IMAGE RESTORATION

Conventional wide-field microscopes with cooled CCD cameras have very efficient light-collection paths, and can acquire images with substantially less light exposure to the sample than conventional scanning microscopes. However light from out-of-focus planes obscures details of 3D objects and their spatial relationships. Quantitative fluorescence measurements are often substantially in error because of this out-of-focus light. Image-restoration methods (Agard and Sedat, 1983; Carrington and Fogarty, 1987; Carrington, 1990; Agard et al., 1989; Carrington et al., 1990a,b, 1995, 1999; Fay et al., 1986, 1988, 1989, 1990) combine quantitative measurements of the microscope 3D blurring characteristics with optically sectioned data to computationally reverse the blurring of the microscope, thus reducing out-of-focus haze and improving resolution. By combining this knowledge with 3D data, an improved image can be calculated that is consistent with this knowledge and consistent with any additional a priori information that may be available.

Several methods have been applied to image restoration of 3D fluorescence images of biological samples. Linear methods such as linear filters or nearest-neighbor have been applied (Weinstein and Castleman, 1971; Agard et al., 1989; Diaspro et al., 1990; Castleman, 1995). These have the advantage of requiring little computer power, but are less accurate than iterative methods that incorporate a priori information such as the non-negativity of a fluorescent dye density. Several iterative methods that incorporate the non-negativity are routinely used with success by biologists (Agard and Sedat, 1983; Agard et al., 1989; Conchello, 1994; Carrington et al., 1995; Holmes et al., 1995; Verveer and Jovin, 1997).

Image restoration or deconvolution is a computational method that uses optical sectioning and quantitative knowledge of the microscope blurring process in the form of a point source function to compute a more accurate estimate of the true 3D fluorescence distribution. The most straightforward way to obtain this information is to measure the point spread function (PSF) of the instrument – that is, to measure the pattern of 3D blurring of a fluorescent point source. Usually this is done by obtaining the 3D image (optical sections) of a small fluorescent bead – typically 100–200 nm in diameter. There is a tradeoff between the size of the bead (ideally a diameter of 0) and the signal level needed (a larger diameter means more fluorescence). Some investigators use an ideal PSF calculated from optical theory (Gibson and Lanni, 1991). However, the PSF changes with minor variations in the optical elements of the microscope; this, the measured PSF of an instrument often differs from that predicted by theory, because of minor misalignment or differences between examples of filters, objectives, or other optical elements that should be identical but are not. An interesting alternative is "blind deconvolution," which calculates the PSF from the cell data itself in a joint process with the deconvolution of that cell data (Krishnamurti et al., 1995). In order to operate properly, this approach places reasonable a priori constraints on the PSF. Although this has been used with success, its developers find a small improvement in results of the deconvolution of the cell data when a measured PSF is used. In our laboratory, 3D bead images are normally used as the PSF. However, for applications where the highest possible resolution is required, we calculate a PSF on a subpixel grid from a 3D bead image (Carrington et al., 1995).

The microscope's detector or camera collects light from the whole sample. Each pixel of the camera is focused on some point in the sample, say (x_i, y_i, z_i); for a scanning microscope, at each scan position, the detector is focused on some point in the sample, (x_i, y_i, z_i). The fraction of the light originating near the focus position (x_i, y_i, z_i) is greater than the fraction of light detected from a point far out of focus and far away, transversely, from the point on which the pixel is focused. In terms of the PSF, the center is where the pixel is focused, so this is equivalent to saying that the PSF is brightest in the center and dimmer farther away from the center. The pixel (or detector) collects a weighted sum of the light emitted by the whole sample. Each pixel has a separate weighting function. If the imaging system is translation-invariant, then it is described by a single PSF, and the weighting function at pixel number i focused at (x_i, y_i, z_i) is then given by $k_i(x, y, z) = h(x_i - x, y_i - y, z_i - z)$, where $h(x, y, z)$ is the PSF. However, the assumption of translation invariance is not necessary in our formulation. Then, the model of the imaging process is given by the following equation:

$$g_i = \iiint_R k_i(x, y, z) f(x, y, z)\, dx\, dy\, dz, \qquad i = 1, 2, 3 \ldots, N. \tag{1}$$

Here, $f(x, y, z)$ is the fluorescence dye density, R is the volume where fluorescent light originates, g_i is the light detected at pixel i (or scan position i), and N is the total number of pixels of data in the 3D image. The integration is truly a continuous integral. Cameras have pixels, but cells do not. A cell is a continuum object, and should be so modeled. The detection or measuring process is finite and discrete. We make no assumptions on the geometry of the data sampling. With a CCD camera, the sampling in x and y is set by the magnification and camera geometry. The sampling in z is usually in uniform steps extending from the top of the cell to the bottom, but this is not assumed in our formulation. Indeed, we have made use of a nonuniform z spacing and small number of sections in z to speed up image acquisition when using slow cameras (Carrington et al., 1995; Lynch et al., 1996). Also, for the highest resolution, we make use of this formulation to compute the image restoration on a finer, subpixel, grid than the original data.

In order to estimate the dye density, we could attempt to solve Eq. (1). However, we do not know the noise-free data, but only a noisy version, $\tilde{g}_i = g_i + \text{noise}$. If Eq. (1) were solved directly, the presence of noise in the data would be amplified in the solution. If a least-squares solution were attempted, the error in the estimate of dye density would still be large. Some sort of smoothing must be placed upon the estimate of dye density. In our approach, this is done by minimizing the following expression, subject to the constraint that a dye density must be non-negative.

$$O(f) = \sum_{i=1}^{N} |\tilde{g}_i - \iiint_R k_i(x, y, z) f(x, y, z)\, dx\, dy\, dz|^2 + \alpha \iiint_R |f(x, y, z)|^p\, dx\, dy\, dz. \quad (2)$$

The first term in Eq. (2) measures how well the dye density f fits the measured data. This term by itself would lead to a least-squares solution and would not be accurate. The second term is a measure of how smooth the solution is. A large value of the parameter α would force the estimate to be very smooth with no detail. The parameter, p, is usually chosen as $p = 2$, though values $1 < p < 2$ are sometimes used. A very small value of α would make the estimate too noisy. Other investigators use different measures of smoothness or different measures of how well the estimate fits the data. A numerical algorithm for this constrained minimization is given by Carrington (1990) and Carrington et al. (1995).

Figure 1 shows an image restoration computed on the same grid as the data. Figures 2, 3 and 4 show image restorations computed on a finer grid than the data sampling (Carrington et al., 1995; Rizzuto et al., 1998a,b; Young et al., 1998; Young and Doxsey, 2000). This algorithm is referred to below as a subpixel or superresolution method. Improvements in this approach have led to the ability to resolve structures separated transversely by 100 nm or less in live or fixed cells. In particular, we developed a method for testing resolution that uses an accurate piezo driver to translate a 40 nm fluorescent bead through a known distance. The bead is imaged twice with a 100 nm displacement to produce an image equivalent to that of two beads separated by 100 nm. Image restoration of this two-bead image results in full resolution of the two beads (i.e., both beads are surrounded by a black region). Image restoration of both fixed and living cells show details within the cell that are consistent with a resolution of 100 nm.

This image restoration method was applied to both fixed and living cells. In fixed cells, for example, it showed a previously unseen cagelike structure of pericentrin and gamma tubulin in the centrosome (Dictenberg et al., 1998); see Figure 2. Figures 3 and 4 illustrate this approach applied to live-cell imaging of proteins labeled with GFP. 3D time series images are shown in Figure 3 (top) that show centrosomes with GFP–pericentrin in live

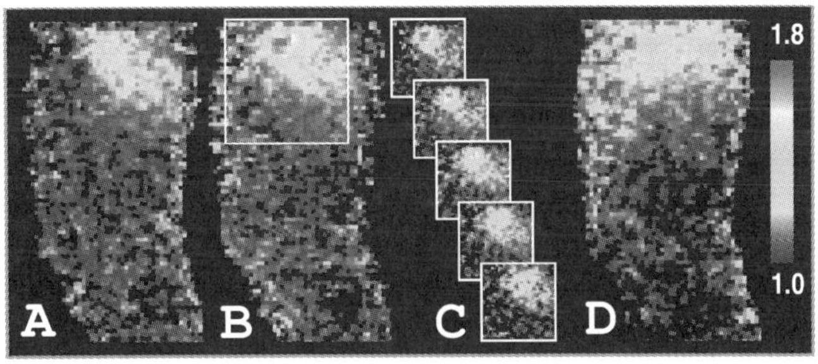

FIGURE 1 See ftp site for color figure. Three selected pseudocolored 3D images of [Ca^{2+}] in a smooth muscle cell during onset and decay of a calcium spark, reflecting a brief release of [Ca^{2+}] from internal stores of [Ca^{2+}] (Kirber et al., 2001). These are taken from a time series of 32 three-dimensional images. The pseudocoloring gives the ratio of each image to the 3D image at time 0. (*a*) is at time 265 ms; (*b*) is at time 285.5 ms; (*d*) is at time 503.5 ms. In (*a*), (*b*), and (*d*), each image is a projection of five focus planes taken while continuously moving the objective with 1 μm of motion during each 2D image. Each 2D image is a 5 ms exposure; each 3D image took 26.5 ms, including focus turnaround time at the top or bottom of each image. Each 3D image is restored, incorporating the motion blur of the continuous motion focus change into the point spread function used in the image restoration. (*c*) is a view of the five separate planes of the image restoration of the spark in (*b*). While the spark appears in the volume projection to be in the center of the cell, the view of the individual planes of the restored spark image in (*b*) shows that it is actually near the bottom of the cell.

cells; each time point is a 3D projection of a 3D image restoration; 24 three-dimensional images are obtained, with only 23% bleaching. Figure 3 (lower right) shows a high-resolution view of one of these time points. In Figure 4, we show a 3D projection of a live cell with GFP with a mitochondrial targeting sequence. Substructure in the distribution of the GFP in the cylindrical mitochondria is clearly visible (Rizzuto et al., 1998a,b). These

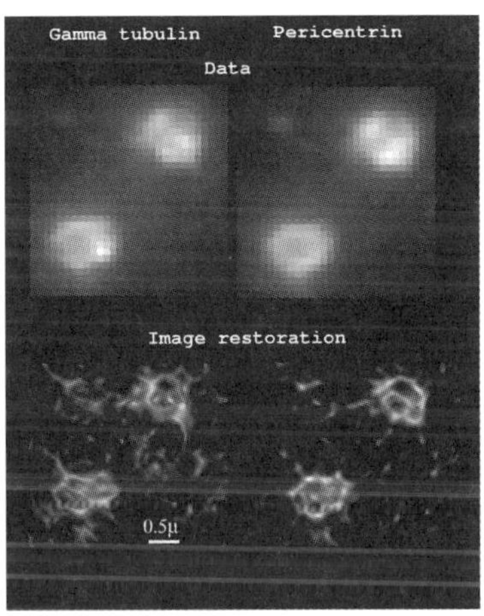

FIGURE 2 Image restoration of prophase centrosome in a dual-labeled fixed CHO cell showing the centrosomal proteins gamma-tubulin (left) and pericentrin (right). The top images are volume renderings of the data. The bottom images are volume renderings of the corresponding image restorations. The image restorations are calculated on a grid that is sampled twice as finely as the original data. The sampling in the original data was on a uniform grid with voxel size, 94 nm × 94 nm × 100 nm; in the restored image, the voxel size was 47 nm × 47 nm × 50 nm.

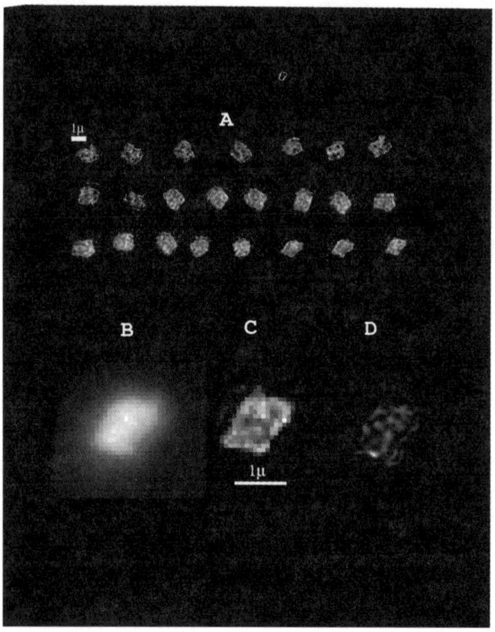

FIGURE 3 A 3D movie of GFP–pericentrin. The scale bars are 1 μm. In (*A*), each small image is a 3D volume rendering of an image restoration. Each image restoration is calculated on a two-times-finer grid; the data is 80 nm × 80 nm × 100 nm voxels and the restoration is 40 nm × 40 nm × 50 nm voxels. These are images of GFP–pericentrin over time. The 3D images are taken every 750 ms; each 2D image had a 20 ms exposure, with an additional 5 ms for coverslip and piezo oscillations to settle (the stepping protocol was used, not continuous-motion focus change). Each 3D image consisted of 26 planes of data; there was a 100 ms pause between each 3D image. Other images in Young et al. (1998) and Young and Doxsey (2000) show faster motion as pericentrin particles move to join onto the centrosome at speeds of 1 $\mu m\, s^{-1}$. (*B*) The data of the last image of the time series in (*A*) shown in a volume rendering; the voxel size is 80 nm × 80 nm × 100 nm. (*C*) Image restoration of (*B*) on the same grid as the data; the voxel size is 80 nm × 80 nm × 100 nm. (*D*) Image restoration of (*B*) on a three-times-finer grid than the data; the voxel size is 26.7 nm × 26.7 nm × 50 nm.

high-resolution images would not be possible without the use of our high-speed microscope with its high-speed, high-quantum-efficiency, low-noise CCD camera and fast focus changer. In each of these GFP images, the apparent transverse resolution is about 100 nm.

4 SPEED REQUIREMENTS FOR HIGH-RESOLUTION GFP IMAGING IN 3D

In experiments with GFP targeted to mitochondria (Rizzuto et al., 1998a,b) (see Fig. 4) and in experiments with GFP–pericentrin, a centrosomal protein (Young et al., 1998; Young and Doxsey, 2000) (see Fig. 3), the highest possible resolution was required. In both experiments, the resolvable distance after image restoration was about 100 nm in the transverse direction. However, in both cases there is considerable motion. In 3D movies of GFP–pericentrin, we are now able to watch the assembly process as clusters of pericentrin move to the centrosome and join on. This motion can occur at rates as fast as molecular motors move – sometimes more than 1 $\mu m\, s^{-1}$ (Young et al., 1998; Young and Doxsey,

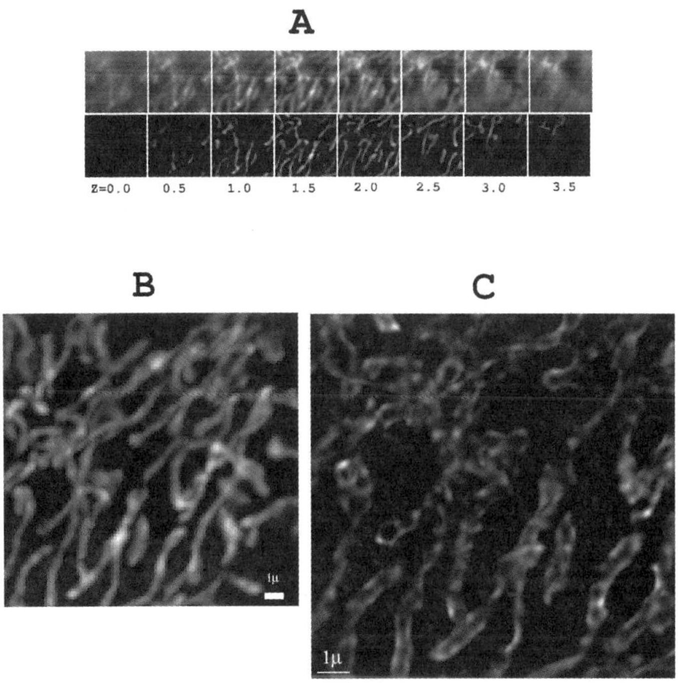

FIGURE 4 GFP with mitochondrial targeting sequence in live HeLa cells. (*A*) Set of optical sections before and after normal image restoration (top is the original data, bottom is the restored data). Only part of the data is shown. (*B*) Volume rendering of the image restoration of the data in (*A*). (*C*) Volume rendering of an image restoration of the data in (*A*) and (*B*) calculated on a grid three times more finely sampled than the original data in x and y, and two times more finely sampled in z. The resolution is sufficient to begin to show the hollow cylindrical structure of the mitochondria.

2000). In addition, Theurkauf has observed transport of particles containing a GFP–exuperantia fusion protein at speeds up to 2 μm s^{-1} (Theurkauf and Hazelrigg, 1998). While these motions may seem modest, it is significant when compared with the resolution of about 100 nm with image restoration. *Simulations show that the resolution deteriorates by roughly the amount the object moves during the time it takes to obtain optical sections through about 1 μm of axial depth.* For high resolution, we use z steps of 500 nm or less (100–200 nm is preferred), or at least three images in 1 μm. Thus, imaging a motion of 1 μm s^{-1} at 10 ms/frame results in the deterioration of resolution from 100 nm to 130 nm; at the preferred z sampling for a high resolution of 0.2 μm, the resolution deteriorates to 160 nm. Clearly there is a trade-off here for optimizing resolution. Of course, most of the cell is not moving this fast, so it could be imaged more slowly; if imaged more slowly, the image of the cell deteriorates only in the vicinity of the fast motion. However, in pericentrin imaging for centrosome assembly, the time and place of most interest is when pericentrin particles join to the centrosome. So, in the place where the most resolution is needed, something is moving rapidly. In the mitochondrial experiments, motion is slower: 250 nm s^{-1} can occur, with 100 nm s^{-1} being common (Rizzuto et al., 1998b). Even this slower motion requires imaging at rates of 20–60 ms/image to avoid loss of resolution due to motion blur. The images in Figures 3 and 4 were acquired somewhat more slowly than the ideal speed, because this analysis had not yet been done when the experiments were performed (or was not yet known to the experimenter). In general, speeds of 20–200 frames s^{-1} are necessary to achieve the highest resolution across a whole sample.

5 ALTERNATIVE APPROACHES

In this section, we discuss alternative approaches. These include conventional laser scanning microscopes that scan a single diffraction-limited spot, recent advances in tandem-scanning confocal microscopes, line-scan confocal microscopes, two-photon excitation versions of these confocal microscopes, and recent hybrid microscopes.

Laser scanning confocal microscopes provide biologists with a convenient way to improve images of 3D objects by optically rejecting out-of-focus light (Sheppard and Wilson, 1978; Wilson and Sheppard, 1984). Confocal microscopes can provide clear images in 2D or 3D of thick fluorescent specimens unobscured by this out-of-focus haze; the elimination of the haze substantially improves the accuracy of quantitative fluorescence measurements. It also allows confocal microscopes to image thick specimens whose image would be dominated by out-of-focus haze in a wide-field microscope. Potentially, confocal microscopes also provide improved resolution; however, to collect enough light for practical fluorescence imaging, the confocal pinhole is usually opened up to an extent that both transverse and axial resolution are essentially the same as a wide-field microscope. Confocal microscopes suffer from limitations for 3D fluorescence imaging because of photobleaching and photodamage, especially for single living cells (Sandison et al., 1995). This is caused by the relatively low quantum efficiency of the photomultipliers used in laser scanning confocal microscopes, loss of light to the confocal pinhole, losses due to chromatic aberration, and losses of additional optical elements in the light-collection path. Fluorescence saturation limits the number of photons emitted per frame by the sample in point-scanning confocals where pixel dwell time is measured in microseconds. For high-speed imaging, where the pixel dwell time may be 0.03 μs, the number of photons emitted by a fluorophore will be less than 10, and such a single-point scanning confocal microscope will detect only a tiny fraction of a photon per fluorophore. For example, fluoroscein emits about 200 photons μs^{-1} at saturation (Tsien and Waggoner, 1995). At 640×480 pixels at 100 frames s^{-1} there is a dwell time of 0.03 μs. When operated at one-third of saturation power, on the order of 2 photons will be emitted per fluorophore. Using 0.002 for a detection efficiency – see Table 1 (Sandison et al., 1995) – roughly 0.004 photons will be detected from each fluorophore (Table 2). (While there have certainly been improvements in confocal microscopes since this study – for example, BioRad claims improved PMT efficiency – the qualitative conclusions that we make are unchanged by factors of 2.) Thus, a molecular cluster must contain thousands of fluorophores before it is detectable in a laser scanning confocal microscope at these speeds. This is clearly too few photons to be useful. Using the measurements from Table 1 (Sandison et al., 1995), the same microscope optics with the confocal add-ons omitted and with a CCD camera with 70% quantum efficiency, gives a collection efficiency of 0.074. This is about 37 times greater than in confocal mode. With the same average power delivered to the sample (about 0.3 mW for 10 ms), the wide-field microscope detects about 0.16 photons. At this power level, a molecular cluster must contain hundreds of fluorophores to be detectable. However, the wide-field microscope, unlike the confocal, permits substantial increase in excitation power delivered to the sample without fluorescence saturation. With 100 mW delivered to the field of view (i.e., $300\times$ the 0.3 mW), a single fluorophore will produce about 48 photons detected; this is a usable signal. The laser used in our microscope will produce this power level.

TABLE 1 Detection efficiency of a laser scanning confocal microscope (LSCM) versus a wide field epifluorescence microscope with CCD[a]

	LSCM	Wide field
Solid angle of collection, NA= 1.25	0.2	0.2
Objective transmission	0.75	0.75
Nikon eyepiece	0.96	0.96
Dichroic mirror	0.8	0.8
Barrier filter	0.92	0.92
MRC-600 mirrors	0.55	1
Scanning duty cycle	1	1
Quantum efficiency (600 nm)	0.06	0.7
Detector aperture	0.6	1
Overall detection efficiency	0.002	0.074

[a] Taken from Sandison et al. (1995), based on a BioRad MRC600 with modifications.

Table 2 data illustrate a potential problem in many imaging situations. If there are clusters consisting of low numbers of the molecule of interest, conventional imaging methods will not detect them, particularly when motion is present in the sample. This statement also applies to our high-speed microscope when it is operated at its normal power levels (a maximum of about 1 mW to the sample) and 10 ms exposures. By increasing the power level in our high-speed microscope, we can potentially detect a single fluorophore – although a single fluorophore is likely to be obscured by cell autofluorescence. Preliminary measurements with this high-speed system indicate that several fluorophores in a cluster may be detectable above autofluorescence levels in live cells at imaging rates of 100 frames s^{-1}.

Recent tandem scanning microscopes with multiple lenslets efficiently illuminate multiple spots in the sample. Each spot must still be below the saturation power, but the total power delivered to the sample is multiplied by the number of illumination spots in the field of view (Ichihara et al., 1996). The detection efficiency of a single-photon tandem-scanning confocal will be somewhat lower than that of the epifluorescence microscope with CCD; with two-photon excitation, it is not necessary to descan, and the unmodified epifluorescence light path may be used (Brewersdorf et al., 1998).

TABLE 2 Number of photons detected as a function of the average power delivered to the sample versus collection efficiency[a]

Collection Efficiency	Frames s^{-1}	Photons detected at power $S/3$	Photons detected at $25 \times S/3$	Photons detected at $60 \times S/3$	Photons detected at $300 \times S/3$
0.002	15	0.029	0.723	1.736	8.68
0.002	100	0.004	0.109	0.260	1.30
0.074	15	1.071	26.7	64.2	321.2
0.074	100	0.161	4.01	9.6	48.2

[a] Here S is the saturation power at a diffraction-limited spot, or approximately 1 mW for fluorescein. It is assumed that for a confocal microscope a power of $S/3$ is the maximum practical to avoid nonlinear effects as saturation is approached. The detection efficiencies used are for confocal (0.002) and epifluorescence microscopes (0.074), and are taken from Table 1. Pixel dwell times (not shown) are based on a 640×480 pixel field of view.

However, the pulse rate of the laser (typically less than 100 MHz) used in multiple-photon systems imposes a limit of one photon per fluorophore per pulse, or somewhat less because of saturation limits. So in both cases the detection efficiency will be one-third to one-half of the epifluorescence microscope. Applying this to Table 2, since Brewersdorf et al. (1998) use about 25 illumination spots in the field of view, we find about 8–13 photons detected per fluorophore at 15 frames s^{-1} and 1 or 2 photons detected per fluorophore at 100 frames s^{-1}. At 15 frames s^{-1}, it is marginal for detecting clusters of a couple fluorophores, but adequate for 5–10 fluorophores. At 100 frames s^{-1}, it would require tens of fluorophores in cluster to be detectable. Adding more illumination spots helps with saturation limits, but at 300 illumination spots, the distance between spots (with 100 nm pixels) is less than 3 μm. There is then substantial crosstalk between the illumination spots; they blur into each other at about 1.5 μm out of focus. The rejection of out-of-focus light is then only marginally better than wide-field imaging.

A similar argument can be made for line-scan confocal imaging with a single- or multiple-photon illumination (Amos and White, 1995; Brakenhoff and Visscher, 1995; Brakenhoff et al., 1996). However, the conclusion is different. Several hundred times as much power can be placed in an illumination line as in a single spot without saturation problems. So there is the potential of detecting small numbers of fluorophores in motion, if sufficient laser power is available. However, for the thin cultured cells typical of GFP experiments, it is not clear that the benefits outweigh the cost and complexity. In particular, although two-photon line scan is attractive in its high detection efficiency, it excites a volume 3–6 μm in depth (Brakenhoff et al., 1996). The thickness of the excited region is a large enough fraction of the thickness of a typical cultured cell used in GFP imaging that not very much out-of-focus light is rejected, and nearly the whole cell is bleached in each image. Thus, while line-scan multiphoton microscopes offer advantages over wide-field imaging for thick samples, they offer little advantage for imaging single cultured cells. Unfortunately, the line-scan confocal microscopes once manufactured by BioRad and Meridien are no longer manufactured.

There are also hybrid microscopes that have recently been developed that provide some of the benefits of confocal microscopes while not being limited by fluorescence saturation. Wilson has developed a tandem-scanning microscope based on illumination by an uncorrelated sequence of dots on a scanning disk; Verveer has used a spatial light modulator to implement the same idea (Verveer et al., 1998). This method produces an image that is the sum of a wide-field image and a confocal image. Half of the area of the cell is illuminated at once, so fluorescence saturation does not limit signals. In addition, Wilson has also developed a grating illumination system to provide optical sectioning; this also illuminates a large enough area at once that saturation is not a limit (Neil et al., 1998). This method also can use a CCD camera as a detector.

ACKNOWLEDGMENT

This work was supported by Grant Number GM61981 from N.I.H. Its contents are solely the responsibility of the author and do not represent the official views of the N.I.H.

REFERENCES

Agard, D.A. and Sedat, J.W. (1983). *Nature* **302**, 676.

Agard, D.A., Hiraoka, Y., Sedat, J.W., and Shaw, P. (1989). Fluorescence microscopy in three dimensions. *Meth. Cell Biol.* **30**, 353–377.

Amos, W.B. and White, J.G. (1995). Direct view confocal imaging systems using a slit aperture. In: *Handbook of Biological Confocal Microscopy*, 2nd edn (ed. J.B. Pawley). New York: Plenum Press, pp. 403–414.

Brewersdorf, J., Pick, R., and Hell, S. (1998). Multifocal multiphoton microscopy. *Opt. Lett.* **23**.

Brakenhoff, G.J. and Visscher, K. (1995). Real-time confocal microscopy. In: *Handbook of Biological Confocal Microscopy*, 2nd edn. (ed. J.B. Pawley). New York: Plenum Press, pp. 355–361.

Brakenhoff, G.J., Squier, J., Norris, T., Bliton, A.C., Wade, M.H., and Athey, B. (1996). *J. Microsc.* **181**, 253.

Carrington, W.A. (1990). Image restoration in 3D microscopy with limited data. In: *Proceedings on Bioimaging and Two Dimensional Spectroscopy* (ed. L.C. Smith). SPIE, pp. 72–83.

Carrington, W.A. and Fogarty, K.E. (1987). 3D molecular distribution in living cells by deconvolution of optical sections using light microscopy. In: *Proceedings of the 13th Annual Northeast Bioengineering Conference* (ed. K. Foster). IEEE, pp. 108–111.

Carrington, W.A., Fogarty, K.E., Lifshitz, L.M., and Fay, F.S. (1990a). Three-dimensional imaging on confocal and wide-field microscopes. In: *Handbook of Biological Confocal Microscopy*, 1st edn (ed. J. Pawley). New York: Plenum Press, pp. 151–161.

Carrington, W.A., Fogarty, K.E., and Fay, F.S. (1990b). 3D fluorescence imaging of single cells using image restoration. In: *Non-Invasive Techniques in Cell Biology*. New York: Wiley-Liss, pp. 53–72.

Carrington, W.A., Lynch, R.M., Moore, E.D.W., Isenberg, G., Fogarty, K.E., and Fay, F.S. (1995). Superresolution three-dimensional images of fluorescence in cells with minimal light exposure. *Science* **268**, 1483–1487.

Carrington, W.A., Fogarty, K.E., Lifshitz, L.M., and Tuft, R.A. (1999). High resolution 3-D imaging of living cells by image restoration. In: *Imaging Living Cells* (ed. R. Rizzuto and C. Fasolato). Heidelberg: Springer-Verlag.

Castleman, K.R. (1995). *Digital Image Processing*, Prentice-Hall, Englewood Cliffs, New Jersey.

Conchello, J.-A. (1994) *Proc. SPIE* **2302**, 369.

Denk, W., Piston, D.W., and Webb, W.W. (1995). Two-photon molecular excitation in laser-scanning microscopy. In: *Handbook of Biological Confocal Microscopy*, 2nd edn (ed. J.B. Pawley). New York: Plenum Press, pp. 445–457.

Diaspro, A., Sartore, M., and Nicolini, C. (1990). Three-dimensional representation of biostructures imaged with an optical microscope: I. Digital Optical Sectioning. *Image and Vision Computing* **8**, 130–141.

Dictenberg, J.B., Zimmerman, W., Sparks, C.A., Young, A., Vidair, C., Zheng, Y., Carrington, W.A., Fay, F.S., and Doxsey, S.J. (1998). Pericentrin and γ-tubulin form a protein complex and are organized into a novel lattice at the centrosome. *Cell Biol.* **141**, 163–174.

Elliott, D.J., Bowman, D.S., Abovich, N., Fay, F.S., and Rosbash, M. (1992). A yeast splicing factor is localized in discrete subnuclear domains. *EMBO J.* **11**, 3731–3736.

Fay, F.S., Fogarty, K.E., and Coggins, J.M. (1986). Analysis of molecular distribution in single cells using a digital imaging microscope. In: *Optical Methods in Cell Physiology* (ed. P. DeWeer and B. Salzberg). New York: Wiley, pp. 51–62.

Fay, F.S., Carrington, W.A., and Fogarty, K.E. (1988). Three-dimensional molecular distribution in single cells analyzed using the digital imaging microscope. *J. Microsc.* **153**, 133–149.

Fay, F.S., Carrington, W.A., Lifshitz, L.M., and Fogarty, K.E. (1989). Three-dimensional analysis of molecular distribution in single cells using the digital imaging microscope. *Proc. SPIE* **1161**, 12–23.

Fay, F.S., Fogarty, K.E., Carrington, W.A., and Lifshitz, L.M. (1990). Extraction of information from biological systems using the digital imaging microscope: analysis of white blood cell chemotaxis. In: *Optical Microscopy for Biology* (ed. Herman and Jacobson). New York: Wiley-Liss, pp. 419–435.

Gibson, F.S. and Lanni, F. (1991). Experimental test of an analytic model of aberration in an oil-immersion objective lens used in three-dimensional microscopy. *J. Opt. Soc. Am.* **A8**, 1601–1613.

Holmes, T.J., Bhattacharyya, S., Cooper, J.A., Hanzel, D., Krishnamurthi, V., Lin, W.-c., Roysam, B., Szarowski, D.H., and Turner, J.N. (1995). Light microscopic images reconstructed by maximum likelihood deconvolution. In: *Handbook of Biological Confocal Microscopy*, 2nd edn (ed. J.B. Pawley). New York: Plenum Press, pp. 389–400.

Ichihara, A., Tanaami, T., Isozake, K., Sugiyama, Y., Kosugi, Y., Mikuriya, K., Abe, M., and Uemura, I. (1996). *Bioimaging* **4**, 52.

Kirber, M., Etter, E.F., Bellve, K.A., Lifshitz, L.M., Tuft, R.A., Fay, F.S., Walsh, J.V., and Fogarty, K.E. (2001). Relationship of Ca^{2+} sparks to STOCS studied with 2D and 3D imaging in feline oesophageal smooth muscle cells. *J. Physiol.* **531**, 315–327.

Krishnamurti, V., Liu, Y.-H., Holmes, T.J., Roysam, B., and Turner, J.N. (1992). *Proc. SPIE* **1660**, 95.

Lynch, R.M., Carrington, W.A., Fogarty, K.E., and Fay, F.S. (1996). Metabolic modulation of hexokinase association with mitochondria in living smooth muscle cells. *Am. J. Physiol.* **270**, C488–C499.

Neil, M.A.A., Wilson, T., and Juskaitis, R. (1997). *Opt. Lett.* **22**, 1905.

Neil, M.A.A., Wilson, T., and Juskaitis, R. (1998). A light efficient optically sectioning microscope. *J. Microsc.* **189**, 114–117.

Rizzuto, R., Carrington, W., and Tuft, R.A. (1998a). Digital imaging microscopy of living cells. *Trends Cell Biol.* **8**, 288–292.

Rizzuto, R., Pinton, P., Carrington, W., Fay, F.S., Fogarty, K.E., Lifshitz, L.M., Tuft, R.A., and Pozzan, T. (1998b). Close contacts with the endoplasmic reticulum as determinants of mitochondrial Ca^{2+} responses. *Science* **280**, 1763–1766.

Sandison, D.R., Williams, R.M., Wells, K.S., Strickler, J., and Webb, W.W. (1995). Quantitative fluorescence confocal laser scanning microscopy (CLSM). In: *Handbook of Biological Confocal Microscopy*, 2nd edn (ed. J.B. Pawley). New York: Plenum Press, pp. 39–51.

Sheppard, C.J.R. and Wilson, T. (1978). Depth of field in the scanning microscope. *Optics Lett.* **3**, 115–117.

Theurkauf, W.E. and Hazelrigg, T.I. (1998). In vivo analysis of cytoplasmic transport during *Drosophila* oogenesis: characterization of a multi-step anterior localization pathway. *Development* **125**, 3655–3666.

Tsien, R.Y. and Waggoner, A. (1995). Fluorophores for confocal microscopy. In: *Handbook of Biological Confocal Microscopy*, 2nd edn (ed. J.B. Pawley). New York: Plenum Press, pp. 267–273.

Tuft, R.A., Bowman, D.S., Carrington, W.A., Fogarty, K.E., and Fay, F.S. (1997) High-speed digital-imaging microscopy with continuous focus scanning. *Biophys. J.* **72**, A295.

Verveer, P.J. and Jovin, T.M. (1997). Efficient superresolution restoration algorithms using maximum a posteriori estimations with application to fluorescence microscopy. *J. Opt. Soc. Am.* **A14**, 1696–1706.

Verveer, P.J., Hanley, Q.S., Verbeek, P.W., Van Vliet, L.J., and Jovin, T.M. (1998). Theory of confocal fluorescence imaging in the programmable array microscope (PAM). *J. Microsc.* **189**, 192–198.

Weinstein, M. and Castleman, K.R. (1971). Reconstructing 3-D specimens from 2-D section images. *SPIE Proc.* **26**, 131–138.

Wilson, T. and Sheppard, C.J.R. (1984). *Theory and Practice of Scanning Optical Microscopy.* Academic Press, London.

Young, A., Tuft, R.A., Carrington, W.A., and Doxsey, S. (1998). Centrosome dynamics in living cells. *Meth. Cell Biol.* **58**, 223–238.

Young, A. and Doxsey, S.J. (2000). Centrosome assembly of pericentrin and γ tubulin is dynein mediated and required for microtubule nucleation and spindle function in somatic cells. *Mol. Biol. Cell* **11**, 2047–2056.

CHAPTER 17

Confocal Microscopy in the Study of the Cell Nucleus

LUCA M. NERI[1,2], SILVANO CAPITANI,[1] and ALBERTO M. MARTELLI[2,3]

[1] Dipartimento di Morfologia ed Embriologia, Sezione di Anatomia Umana Normale, Università di Ferrara, via Fossato di Mortara 66, 44100 Ferrara, Italy
[2] Istituto di Citomorfologia Normale e Patologica del C.N.R., c/o I.O.R., via di Barbiano 1/10, 40137 Bologna, Italy
[3] Dipartimento di Scienze Anatomiche Umane e Fisiopatologia dell'Apparato Locomotore, Facoltà di Farmacia, Università di Bologna, Via Irnerio, 48, 40126 Bologna, Italy

1 INTRODUCTION: THE CELL NUCLEUS

The cell nucleus is one of the best known but least understood and most complicated of cellular organelles. Even 170 years after its description (Qumsiyeh, 1999), the structural and functional organization of the nucleus remains the subject of heated debate. On one side, the nucleus has been proposed to possess its own skeleton and distinct suborganelles; on the other, it has been considered as a largely disordered bag of chromatin, immersed in a homogeneous nucleoplasm, in which all "structures," except for the nucleoli and the double-layered nuclear envelope, are no more than transient complexes that form and dissolve as a result of DNA transcription and replication, and RNA processing activities (Lamond and Earnshaw, 1998). An obvious reason for such a view is that there are no membrane structures inside the nucleus.

Nevertheless, apart from very prominent domains such as the nuclear envelope with its pore complexes (Panté and Aebi, 1995), the nucleolus with its subcompartments (Hernandez-Verdun, 1991), and heterochromatin, the nucleus also contains a heterogeneous group of inclusions, such as interchromatin granules (Clevenger and Epstein, 1984; Thiry 1993, 1995; Ferreira et al., 1997), nuclear bodies (Brasch and Ochs, 1992), and coiled bodies (Thiry, 1994), which have been known for a long time.

Thanks to the outstanding advances made in molecular biological techniques, our knowledge about the genetic machinery and its regulation and how DNA is replicated is continuously growing. Now genes can be identified, purified, sequenced, changed at our will, reintroduced into cells, and expressed there as proteins. However, an overall view of how functions are structurally integrated within the nucleus is still lacking.

Confocal and Two-Photon Microscopy: Foundations, Applications, and Advances, Edited by Alberto Diaspro.
ISBN 0-471-40920-0

For a long time, investigations of the nuclear compartment have relied exclusively on the information given by the transmission electron microscope, which provided fundamental data for clarifying some aspects of chromatin organization. However, over the past decade, a new burst of interest in the cell nucleus took place in parallel with the development and application of confocal microscopy techniques (Wilson, 2002; Diaspro and Sheppard, 2002). Furthermore, confocal microscope was not alone in this exploration of the nucleus, but was accompanied by an explosion in the number of new probes for the detection of specific nuclear constituents. In this chapter, we shall focus on the complex interactions between structural and functional aspects/molecules in the nucleus. In fact, we believe that for a complete comprehension of the nucleus, it is necessary to know both its architectural and functional organization. Regarding nuclear architecture, we shall consider nuclear domains and nuclear matrix.

As far as functional aspects are concerned, we shall not examine DNA- or RNA-related processes, but rather the existence and roles of nuclear signaling pathways based on inositol lipids. In both fields, confocal microscopy has been an excellent tool, and has contributed to the demonstration that the nucleus is not an amorphous bag containing a mass of DNA, RNA, and protein, but rather is a highly dynamic and organized organelle that carries out a myriad of essential functions with speed, efficiency, and unsurpassed fidelity, in response to the metabolic requirements of the cell. It may be worth remembering here that, thanks to the recent development of the methodologies for the in situ detection of nucleic acids, there is now accumulating evidence that interphase chromosomes occupy spatially distinct regions of the nucleus, referred to as chromosome territories, separated by interchromosomal channels. A very useful technique for demonstrating individual chromosomal territories during interphase is fluorescence in situ hybridization (FISH) (Visser and Aten, 1999), using both gene- and chromosome-specific probes. Individual chromosomes appear to occupy defined territories, but have irregular shapes, and the homologues are not associated throughout interphase (Ferreira et al., 1997).

2 NUCLEAR ARCHITECTURE DURING INTERPHASE

2.1 Nuclear Domains

The list of suborganelles and functional domains of the nucleus is steadily growing. We have no doubt that in the near future, new and unexpected nuclear domains will be discovered and, similarly, new nuclear functions will emerge. Each of these domains, which are not necessarily stable but dynamically variable, has been shown to contain specific proteins that play the role of "markers" for identifying these nuclear substructures both at the light- and the electron-microscope levels. However, the exact functions of most of these domains are only marginally known.

Coiled Bodies Although its name derives from studies in the electron microscope, the coiled body (CB) was originally identified in the light microscope around the start of the 20th century by Santiago Ramón y Cajal, who termed them nucleolar accessory bodies (Ramón y Cajal, 1903). CBs are highly enriched in several classes of small nuclear ribonucleoproteins (snRNPs: U1, U2, U4, U5, U6, and U7), nucleolar (fibrillarin, Nopp 140, NAP 57, and ribosomal protein S6) and cell-cycle control (WT1, SPH-1, cyclin E-

CDK2) proteins, as well as basal transcription factors (TFIIF and TFIIH) (Matera, 1999). Nevertheless, the only epitope reported to date that can be used as an unambiguous marker for CBs is a protein called p80 coilin (Alliegro and Alliegro, 1998). CBs are involved in snRNP trafficking, play a major role in directing intranuclear snRNA sorting, and might even be an obligate waystation in the biogenetic pathway of snRNPs (Matera, 1999). p80 coilin seems to be essential for the presence of CBs, because microinjection of anti-p80 antibodies in living human cells led to the disappearance of these organelles within 6 hours. These elegant experiments were performed using a confocal microscope (Almeida et al., 1998). Recently, through the use of overexpressed p80 coilin fused to green fluorescent protein (GFP), it was possible to demonstrate that all Cajal bodies move within the nucleoplasm. The movements included translocations through the nucleoplasm, joining of bodies to form larger structures, and separation of smaller bodies from larger Cajal bodies. It was also observed, by time-lapse fluorescence microscopy, that there was movement of bodies to and from nucleoli. These data suggested that there may be at least two classes of Cajal bodies, differing in their size and dynamic behavior (Platani et al., 2000).

PML Bodies These inclusions comprise a dense ring that contains PML protein, known as the signature polypeptide of this nuclear body and first detected in promyelocytic leukemia cells. PML bodies, approximately 0.5 μm in diameter, contain a number of interesting components, the most prominent of which are the retinoblastoma protein (Rb), SP100, and PIC1/SUMO-1 (Matera, 1999). In acute promyelocytic leukemia, PML protein, resulting from the t(15;17) chromosomal translocation combines with the retinoic acid receptor α (RARα), generating a fusion product (de Thé et al., 1991). Confocal microscopy analysis showed that the PML–RARα hybrid displays an altered nuclear localization when compared with that of its normal PML and RARα counterparts (Weis et al., 1994). In leukemic cells, the major proportion of PML nuclear bodies are disorganized into several and aberrant microstructures containing both PML and PML–RARα. Importantly, treatment of the neoplastic cells with retinoic acid restores both the normal morphology and the number of bodies (Koken et al., 1994).

PML protein appears to function as a negative growth regulator and a tumor suppressor, and it is also involved in transcription regulation (Zhong et al., 2000). Moreover PML bodies are the target for a variety of virus-derived proteins, and subsequent loss of organization of these domains is an important phase of the viral infection cycle (Cao et al., 1998). The SP100 protein was originally identified as an autoantigen in primary biliary cirrhosis (Sternsdorf et al., 1997).

GATA Bodies These contain the hematopoietic GATA transcription factors 1, 2, and 3, and were originally described by confocal microscopy in erythroblasts and megakaryocytes. They are not involved in RNA transcription and do not contain splicing factors. Their function still remains elusive (Elefanty et al., 1996).

The Polycomb Group (PcG) Domain This contains the heterochromatin protein known as HP1, a structural adapter whose role is to assemble macromolecular complexes in chromatin. HP1 interacts with multiple nuclear proteins, including the lamin B receptor, the nuclear receptor cofactor TIF1-α, βLAP, and SP100. The HP1 protein is also involved in gene silencing. Therefore, PcG domains may represent silencing compartments for the inactivation of specific genes in euchromatin (Lamond and Earnshaw, 1998; Satyn et al.,

1997). Confocal analysis with an antibody to RING1 (another component of PcG bodies) revealed evidence of a microgranular signal comprising numerous small nuclear foci. For example in fetal lung fibroblast MRC-5 cells, the nucleus is consistently stained with about 50–100 small foci that are highly concentrated into two or three areas of the nucleus (Saurin et al., 1998).

The Gems Bodies These structures, which resemble coiled bodies, have been called gemini of coiled bodies (gems). They are frequently paired with coiled bodies, and sometimes the two types of nuclear inclusions overlap (Liu and Dreyfuss, 1996). Gems contain the SMN (survival of motor neurons) protein, encoded by the gene responsible for a severe inherited form of human muscular wasting disease, spinal muscular atrophy. Double-labeling experiments analyzed by confocal microscopy revealed the SMN protein to colocalize with the Sm class of snRNP proteins and with a polypeptide called SIP1, which is involved in snRNP biogenesis. The SMN protein is required for pre-mRNA splicing, and likely serves to assembly and maintain the splicing machinery in an active form. Therefore, defects in snRNP assembly may be involved in the pathogenesis of spinal muscular atrophy (Liu et al., 1997; Fischer et al., 1997).

Sam68 Nuclear Bodies (SNBs) These peculiar structures are predominantly found in transformed cells. They contain Sam68 protein, which is an RNA-binding polypeptide whose binding activity is negatively regulated by tyrosine phosphorylation. SNBs also contain two Sam68-like proteins called SLM-1 and SLM-2, and are also enriched in nucleic acids, but are not the site of mRNA transcription. Confocal microscopy revealed dynamic changes of SNBs in response to transcriptional inhibitors and during mitosis. In particular, actinomycin D, at concentrations known to inhibit RNA polymerases I and II, caused SNBs to disassemble. Similar results were obtained with another transcription inhibitor, α-amanitin (Chen et al., 1999).

Heat Shock Factor 1 (HSF1) Granules These structures are stress-dependent and are enriched in HSF1. This is the ubiquitous stress-responsive transcriptional activator that is essential for the inducible transcription of genes encoding heat-shock proteins and molecular chaperones (Jolly et al., 1997). Confocal microscopy demonstrated that HSF1 granules are large (1.5–25 μm) clustered or ring-like granular structures that appear in the nuclei of cells exposed to heat shock, heavy metals, and amino acid analogues. It is found, by immunostaining and confocal analysis, that they do not contain other heat-shock transcription factors, such as HSF2, HSF3 and HSF4 (Cotto et al., 1997). These bodies also contain HAP, a novel hnRNP protein that is identical both to SAF-B, a component of the nuclear scaffold, and to HET, a transcriptional regulator of the gene for heat shock protein 27 (Chiodi et al., 2000).

Nuclear Speckles These structures correspond to interchromatin granule clusters revealed by electron microscopy. They are highly enriched in the conserved splicing factor SC-35. Staining of nuclei with an antibody to SC-35 showed in combination with confocal microscopy that mammalian cells commonly contain 20–50 speckles (Zeng et al., 1997). Nuclear speckles also contain a hyperphosphorylated form of the largest subunit of RNA polymerase II, the snRNPs, and elevated concentrations of proteins involved in transcription and cellular transformation (Mortillaro et al., 1996). The role of speckles in transcription is still debated. Indeed, some authors have reported that over 90% of speckles have moderate to high levels of associated transcription activity. The transcription sites were found along the periphery as well as inside the speckles (Wei et al., 1999). Others have ruled out that speckles are involved in pre-mRNA transcription (Cmarko et al., 1999). Speckles may be involved in mRNA maturation or export. An alternative scenario is that speckles function either as depots supplying splicing factors to active gene loci, or way-stations accumulating snRNPs bound either to partially spliced pre-mRNA or to excised introns after release of mRNA from the spliceosome (Lamond and Earnshaw, 1998).

Perinucleolar Compartment (PNC) This is a unique nuclear structure localized at the periphery of the nucleolus. The PNC has an irregular shape with a large variability in size (0.25–1.0 μm in diameter). This nuclear structure is predominantly found in transformed cells, and is rarely observed in normal primary cells. Time-lapse studies by confocal microscopy have revealed that the size and shape of the PCN is dynamic over time. The PCN dissociates at the beginning of mitosis and reforms at late telophase in the daughter nuclei (Huang et al., 1997). It was also shown, by employing a confocal laser scanning microscope, that PNC contains several small RNAs transcribed by RNA polymerase III and heterogeneous ribonucleoproteins, and it is the site of RNA synthesis. The structure of the PNC was found to be altered upon inhibition of RNA synthesis, suggesting that the integrity of the PNC reflects its transcriptional activity (Huang et al., 1998).

2.2 The Nuclear Matrix

The possible existence of an internal, nonchromatin nuclear proteinaceous structural framework that facilitates gene functions as a set of spatially concerted reactions has become a popular but controversial field of investigation (Pederson, 1998). Countless studies have demonstrated that when the nucleus is stripped of most DNA, RNA, and soluble proteins (mainly histones), a structure remains that has been variously referred to as the nuclear matrix or scaffold, or the nucleoskeleton (Berezney et al., 1995; Penman, 1995; Martelli et al., 1992; Pederson, 2000). Such a structure, acting as a framework, would provide attachment sites for the structural domains (or DNA loops) in which the eukaryotic genome is subdivided during interphase. Indeed, it is bewildering to realize that, in a typical diploid mammalian cell, about 2 m of double-stranded DNA are packed within a nucleus with a diameter of approximately 10 μm. To achieve this, the DNA double helix is first coiled into nucleosomes (11 nm diameter), then the nucleosomes into solenoids (30 nm diameter), which are thought to be folded into loops of tens of kilobasepairs (Cook, 1995). It is clear that only a highly structured organelle could handle the topological problems arising, for example, during duplication of DNA.

However, there has always been uncertainty as to whether or not the nuclear matrix, isolated in vitro, could really represent a framework of the nucleus in vivo. In fact, the only nuclear framework whose existence is universally accepted is the nuclear lamina, a continuous thin fibrous layer that underlies the inner nuclear membrane and is mainly composed of three related proteins named lamin A, B, and C (Moir and Goldman, 1993). Evidence mostly deriving from in vitro experiments has indicated that the insoluble nuclear matrix may be involved in genome duplication, RNA synthesis and processing, anchoring of DNA loops, regulation of gene expression, protein phosphorylation, and a variety of other functions (for a review, see Berezney, 1991). The nuclear scaffold is composed of nonhistone proteins, plus variable amounts of DNA and RNA. Ultrastructural investigations have shown that three different domains can be identified in the scaffold: an outer lamina, an inner fibrogranular network, and residual nucleoli (Berezney, 1991).

The nuclear matrix should also be considered in light of the sometimes-overlooked evidence that each step in its isolation may cause nuclear components to bind to it by new interactions that do not exist in vivo or to change their spatial distribution. Our laboratory has repeatedly demonstrated that, during the isolation procedure that lead to the final nuclear matrix, several nuclear polypeptides change their localization (Neri et al., 1994a, 1995, 1997a–c; 1999a). However, we have also provided evidence that some nuclear-matrix proteins, such as NuMA, topoisomerase IIα, and lamins, are more resistant to isolation-induced spatial redistribution (Neri et al., 1997a, 1999b,c). NuMA (for an updated review, see Zeng, 2000), apart from binding scaffold/matrix-associated regions (Ludérus et al., 1994), has also been proposed to act as a bridge between the nucleoskeleton and the splicing apparatus (Zeng et al., 1994a). However, the predicted structural features of NuMA, such as a globular head, tail domains, and a central-helical rod domain, are similar to those of the members of the family of intermediate filament proteins (Yang et al., 1992). Therefore, NuMA could oligomerize through coiled-coil interactions to create a filamentous network, possibly corresponding to the core filaments of the nuclear matrix, of which NuMA is a component (Zeng et al., 1994b). In this connection, it should be emphasized that recent data from Saredi et al. (1996) indicate that when NuMA is overexpressed in the cytoplasm, it assembles into an extensive filamentous network, composed of interconnected 5 nm filaments. Therefore, NuMA protein might be a true structural component of a nuclear framework (Zeng, 2000; Zhang et al., 1994) and not only a linkage between the sites of RNA splicing metabolism and the nuclear matrix. If this hypothesis is confirmed, NuMA will be the first structural constituent of the matrix identified so far. Other investigations, in which a mutated form of NuMA was overexpressed, have suggested that it could indeed play a major structural role in the architecture of the normal interphase nucleus (Gueth-Hallonet et al., 1998), even if more recent results have questioned such a conclusion (Taimen et al., 2000), because NuMA can be lost by nonproliferating cells without dramatic effects on nuclear structure and shape. Concerning DNA topoisomerase IIα, its role in acting as a structural linkage anchoring the basis of chromatin loops to the nuclear matrix or chromosome scaffold has been repeatedly proposed in mammalian cells (see, e.g., Gasser and Laemmli, 1986; Adachi et al., 1989), even though more recent evidence collected from *Xenopus* oocytes has challenged this hypothesis (Hirano and Mitchison 1993; Fischer et al., 1993).

On the other hand, lamin proteins are the major constituents of the nuclear lamina, a 10–50 nm thick fibrous protein network lining the nucleoplasmic face of the inner nuclear membrane (for a review, see Moir et al., 1995). In most mammalian cell types, the lamina is composed of two major types of polypeptides: the acidic B-type lamins (B1 and B2) and the neutral A-type lamins (lamins A and C in mammals) (Moir et al., 1995). Even though lamins have been localized to the nuclear periphery, they have also been found within the nucleoplasm during certain phases of the cell cycle (G_1 and S) or in some pathological conditions (see, e.g., Moir et al., 1994; Machiels et al., 1995; Hozak et al., 1995). In particular, Hozak et al. (1995) showed that in HeLa cells, lamin A forms part of a diffuse skeleton that ramifies throughout the nuclear interior. Internal lamin A is undetectable in whole cells or in isolated nuclei, because it is buried in dense chromatin and is inaccessible to antibodies. However, lamin A becomes accessible when chromatin is removed by nuclease digestion and electroelution. We have extended these observations, using K562 human erythroleukemia cells, in which we demonstrated that both lamin A and B1 are constituents of the internal nuclear matrix, and that lamin A, but not B1, is associated with some types of nuclear inclusions such as coiled and GATA bodies (Neri et al., 1999b), see Figure 1.

Therefore, in our opinion, the fact that these polypeptides are resistant to various isolation protocols, as demonstrated by morphological investigations and confocal analysis, reinforce the contention that they serve as structural components of an internal nucleoskeleton.

3 NUCLEAR SIGNALING DURING INTERPHASE

3.1 The Nucleus as a Separate Signaling Compartment

Extracellular signals regulate gene expression by triggering signal-transduction cascades that result in the modulation of transcription-factor activity. Evidence collected in the past decade has highlighted that the cell nucleus can be regarded as a separate signaling compartment. This evidence comes from the existence of phospholipids in the nucleus that resist washing of nuclei with detergents. It was first proposed that some of these serve a structural function, for example as components of chromatin or of the laminar layer that surrounds the nucleus.

At present, the physicochemical form of intranuclear lipids are poorly understood. They are presumably not in a bilayer membrane, but instead form proteolipid complexes similar to the association of phosphatidylinositol 4,5-bisphosphate (PtdIns(4,5)P_2) with cytoskeletal proteins such as ezrin (Barret et al., 2000). Furthermore, it has been demonstrated that lipid kinases are located inside the nucleus, rather than in the nuclear envelope, and that their products change upon stimulation of the cells. Thus, the overall conclusion is that there is nuclear inositide synthesis within the nucleus that is distinct from that in the classic lipid inositol cycle in the plasma membrane, and that can be regulated in some way separately from the latter cycle (Neri et al., 1999d; Martelli et al., 1999a). We have many new techniques and tools, including confocal microscopy, to understand the pathways and processes that occur in this compartmentalized organelle, resembling a microcosm of lipid signaling and exhibiting a level of complexity and refinement similar to that found in the cytoplasm.

3.2 Inositol Lipids

PtdIns(4,5)P_2 has been immunocytochemically located by means of a specific monoclonal antibody (KT10) either in the nuclear membrane (Tran et al., 1993) or in the internal domains of the nucleus (Mazzotti et al., 1995). The results reported by Mazzotti et al. (1995) identified intranuclear phospholipids by confocal microscopy of isolated rat hepatocyte nuclei. In membrane-deprived nuclei or in nuclei whose membranes have been permeabilized with detergents or by freezing and thawing, this technique indicated that PtdIns(4,5)P_2 antigenic sites exist distributed diffusely throughout the nucleoplasm and not only at the nuclear membrane. Other authors (Boronenkov et al., 1998) identified nuclear internal domains, which likely correspond to "nuclear speckles," containing PtdIns(4,5)P_2 (as detected by means of the well-established monoclonal antibody AM212) that gave a fluorescent signal that co-localized with SC-35. The discrepancies are likely to depend on the different fixation and/or permeabilization techniques employed

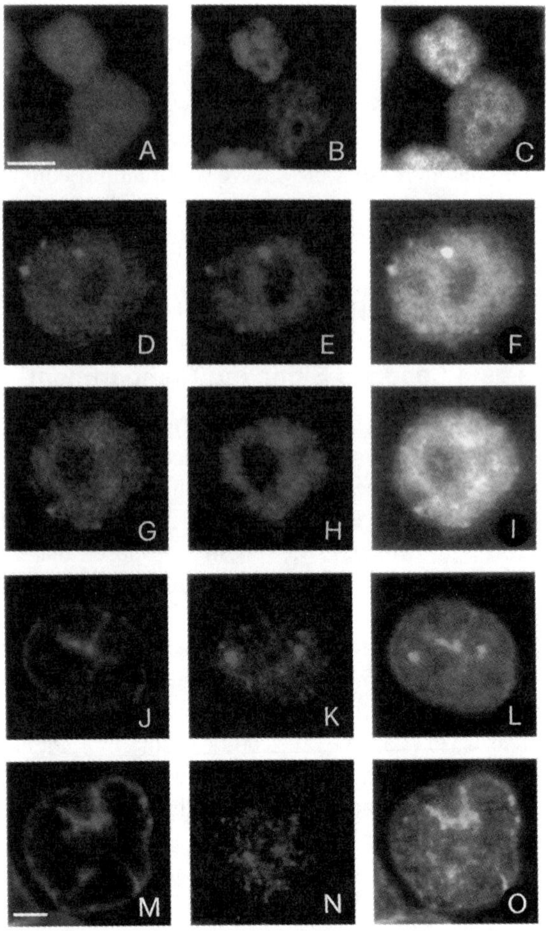

FIGURE 1 See color insert. Double labeling immunofluorescence imaged by confocal microscopy to verify whether intranuclear lamin A colocalized with a ribonucleoprotein particle (RNP) antigen known to be a constituent of the nuclear matrix. As shown in (*A*)–(*C*), we observed that in 37°C-stabilized, $(NH_4)_2SO_4$-extracted nuclear matrices, lamin A shared an almost-complete colocalization with RNP, as detectable from the superimposed yellow fluorescence. Next, we wanted to analyze in isolated matrices the distribution of both lamin A and B1 relative to either p80 coilin, a typical component of coiled bodies, or the transcription factor GATA-1, which is particularly expressed in erythroid cells and is a constituent of a class of nuclear inclusions.

In addition to coiled bodies, antiserum to p80 coilin stained quite diffusely the nuclear matrices except for nucleolar remnant and the periphery (*E*, *K*). Double staining with anti-lamin A antibody revealed a strong co-localization of this protein with p80 coilin, at the level of both the coiled bodies and the internal meshwork (*F*).

Immunostaining with anti-GATA-1 (*H*, *N*) antibody revealed that some foci of intense fluorescence were present, in addition to a diffuse granular nucleoplasmic fluorescence. Double-labeling experiments showed that lamin A co-localized with the bodies and the diffuse staining (*I*). On the other hand, as shown in (*L*), there was no co-localization between lamin B and p80 coilin. Similarly, we did not detect significant co-localization between lamin B and GATA-1 protein (O).

For this set of experiments, we used nuclear matrices extracted with 0.25 M $(NH_4)_2SO_4$. (*A*, *D*, *G*) lamin A; (*J*, *M*) lamin B1; (*B*) RNP; (*E*, *K*) coilin; (*N*) GATA-1. Co-localization of lamin A with RNP (*C*), coilin (*F*), or GATA-1 (*I*). Co-localization of lamin B1 with coilin (*L*) or GATA-1 (*O*). In (*A*–*C*), the scale bar represents 5 μm. For all the other pictures, the scale bar (as indicated in *M*) represents 2 μm.

in these two studies rather than on the type of monoclonal antibody to PtdIns(4,5)P_2, because Mazzotti et al. (1995) also found that KT10 stained the nuclear interior. Interestingly, Boronenkov et al. (1998) were unable to find an association of either phosphatidylinositol 4-phosphate 5-kinase (PIPK) or PtdIns(4,5)P_2 with invaginations of the nuclear envelope or nuclear membrane structures, as those described by Fricker et al. (1997). However, the functions played by inositol lipids present in the speckles remain to be clarified.

3.3 The PI 3-Kinase Family and Related Enzymes

Using an antibody to p85 regulatory subunit and immunocytochemical staining, Neri et al. (1994b, 1999e) demonstrated a rapid translocation of phosphatidylinositol 3-kinase (PI3-K) to the nucleus after incubation of PC-12 rat pheochromocytoma cells in the presence of nerve growth factor (NGF). Also, in HL-60 cells in situ, morphological analysis by confocal microscopy showed the translocation to the nucleus of PI3-K when cells were treated with differentiating doses of all-*trans*-retinoic acid (ATRA) (Marchisio et al., 1998; Bertagnolo et al., 1999) or vitamin D_3 (Neri et al., 1999f). Interleukin-1 (IL-1) caused a rapid and transient intranuclear translocation of GFP-labeled PI3-K in Saos-2 cells (Bavelloni et al., 1999). In MC-3T3 E1 osteoblast-like cells, the translocation of PI3-K to the nucleus and the nuclear matrix was caused by growth factors such as insulin-like growth factor I (IGF-I) and platelet-derived growth factor (PDGF) (Martelli et al., 2000).

Metjian et al. (1999) have shown that in HepG2, cells, the γ isoform of the p110 catalytic subunit of PI3-K was present in the cytoplasm under resting conditions. Upon stimulation with serum, the enzyme translocated to the nucleus in a manner that was sensitive to pertussis toxin.

Recently, the nuclear occurrence of D-3 phosphorylated inositol lipids has also been demonstrated by immunological techniques employing monoclonal antibodies specific for the lipids themselves (Yokogawa et al., 2000). A very recent report has highlighted the presence of PI-3K within nuclear and nucleolar transcription sites of higher plants (Bunney et al., 2000).

Another enzyme that is attracted to the nuclear compartment by the nuclear production of PtdIns(3,4,5)P_3 is PKB/Akt, as demonstrated in MC-3T3 E1 osteoblast-like cells stimulated with IGF-I and PDGF (Borgatti et al., 2000). It should be noted that there is a time lapse of PKB/Akt, when its nuclear translocation is compared with that of PI-3K in the same cell system (Martelli et al., 2000; Borgatti et al., 2000).

3.4 The PLC Family

Immunocytochemical analysis showed the presence of phospholipase C $\beta 1$ (PLC-$\beta 1$) in the nucleus of quiescent 3T3 fibroblasts (Martelli et al., 1992; Zini et al., 1993). This localization has been confirmed in PC12 cells (Zini et al., 1994), MEL cells (Martelli et al., 1994), rat hepatocytes (Bertagnolo et al., 1995), COS-7 cells (Kim et al., 1996), and C6Bu-1 glioma cells (Bahk et al., 1998). Also, other members of the PLC-β family ($\beta 2$ and $\beta 3$) have been shown to be present in isolated nuclei of HL-60 (Bertagnolo et al., 1997). Also, PLC-$\gamma 1$ and $\gamma 2$ have been described to localize within the nuclear compartment of PC-12 cells (Zini et al., 1994), rat hepatocytes (Neri et al., 1997d), HL-60 cells (Bertagnolo et al., 1997), A431 cells, HeLa cells, mouse hepatoma MH 22A cells, and rat Zajdela ascitic hepatoma cells (Diakonova et al., 1997).

3.5 Other Elements of Inositol Lipid Metabolism

Additional information on the intranuclear localization of polyphosphoinositide metabolism has come from a very comprehensive study performed by Boronenkov et al. (1998). These authors detected by means of immunofluorescence staining two forms of PIPK, Iα and IIα, in the nuclei of mammalian cells. The same authors demonstrated that these kinases were associated with a distinct subnuclear domain, identified as "nuclear speckles," which also contain pre-mRNA processing factors, such as SC-35, and are associated with the nuclear matrix.

Nuclear localization of diacylglycerol (DAG) kinase IV was shown by immunofluorescent staining and immunoblotting analysis in COS-7 cells overexpressing a FLAG epitope-tagged cDNA. DAG kinase IV is the rat homologue of the recently cloned human DAG kinase-ζ (Bunting et al., 1996).

Topham et al. (1998) demonstrated that 15–30% of total cellular DAG kinase-ζ was located in the nucleus of both COS-7 cells overexpressing the DAG kinase-ζ cDNA and of A172 cells, a glioblastoma-derived line that constitutively expresses DAG kinase-ζ.

3.6 The PKC Family

Several protein kinase C (PKC) isoforms can be found in the nucleus or translocate there. Most PKC isoforms can be found at one time or another in the nucleus, either in resting or stimulating cells. Correlations of nuclear localization with proliferation and differentiation are highly indicative of a role of nuclear PKC in growth regulation (Martelli et al., 1999b,c). Since nuclear PKC levels were found to be higher in breast carcinoma cells, in diethylnitrosamine-induced hepatocarcinoma cells, and in hyperproliferative guinea pig epidermis cells, several authors have associated nuclear PKC with carcinogenesis (see, e.g. La Porta et al., 1997). These studies were facilitated by widespread use of immunofluorescent staining coupled with confocal laser scanning microscopy, which allowed for an accurate dissection of the involved subnuclear compartments, and supported the data coming from biochemical fractionation experiments. In Swiss 3T3 cells, IGF-I, PDGF, or epidermal growth factor (EGF) caused an increase in PKC-α within the nucleus, except for nucleoli. Fractionation experiments and immunoblotting analysis were in agreement with morphological findings (Neri et al., 1994c). As shown in Figure 2, in quiescent cells PKC-α was mainly detected in the cytoplasm and in the perinuclear region. Nuclei were immunolabeled to a much lesser extent (Fig. 2*A*). Incubation of cells with IGF-I for 30 minutes produced a striking increase in PKC-α labeling in the nuclear interior (Fig. 2*B*).

A very detailed analysis of the subnuclear distribution of PKC isoforms was performed in NG-108-15 nerve cells by Beckmann et al. (1994). It was observed that the α isozyme was located in the nucleus except for nucleoli, whereas the δ isotype was found only in the nucleoli. Finally, PKC-ε was localized in the pore complexes at the nuclear envelope. Although present in the cytoplasm, the β_{II} isoform was not detectable in nuclei.

Alternative techniques were employed to study by immunofluorescence the distribution of PKC isozymes. Bastiaens and Jovin (1996) covalently labeled purified recombinant PKC-β_I with the fluorescent sulfoindocyanine succinimide ester, Cy3. After labeling, the lipid/Ca^{2+}-dependent activity was fully retained, and it was also possible to determine the loci of the Cy3 labels on the protein. The regulatory domain of the kinase exhibited a 3.5-fold higher specific activity than the catalytic domain. Subsequently, the labeled PKC-β_I was microinjected into Balb/c 3T3 cells to follow its time-dependent redistribution in

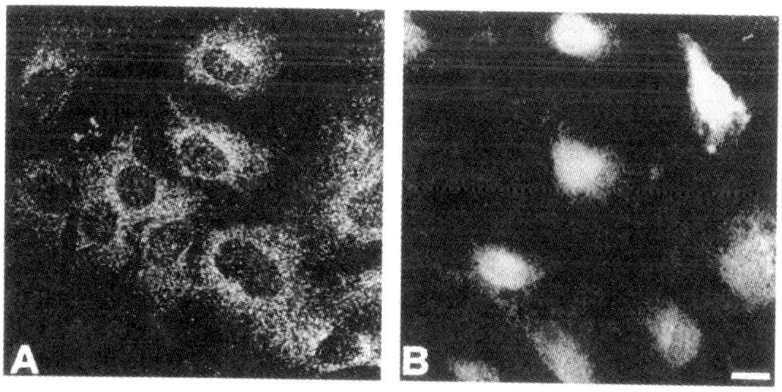

FIGURE 2 Single confocal laser scanning microscope optical sections showing the subcellular distribution of PKC-α in Swiss 3T3 cells: (A) quiescent cells; (B) 30 minutes of IGF-I stimulation. A polyclonal antibody to PKC-α was employed, as described by Neri et al. (1998). The scale bar represents 10 μm.

response to phorbol esters. The combined use of conformal laser scanning microscopy, fluorescence lifetime imaging microscopy, and novel determinations of fluorescence resonance energy transfer based on photobleaching digital imaging microscopy allowed assessment of the localization to the nucleus of the kinase. Simultaneous labeling with a polyclonal antibody recognizing the C-terminal catalytic domain of the kinase indicated that, after phorbol-ester treatment, PKC was fragmented and that the regulatory, but not the catalytic domain, was retained predominantly in the nucleus.

It may also be worth recalling that intracellular localization of PKC could be analyzed by using fluorescently tagged 4-phorbol-12-myristate 13-acetate (PMA) in living cells (Godson et al., 1996) or fluorescent derivatives of the bisindolylmaleimide inhibitors (Chen and Poenie, 1993).

All the PKC isoforms can now be stably overexpressed as hybrids tagged with GFP, and this will undoubtedly allow for more accurate investigations of their intracellular localization and redistribution following diverse stimuli (Almholt et al., 1999).

Finally, PKC-δ has been recognized by confocal microscopy to be an apoptotic lamin kinase that co-localizes with lamin B during apoptosis, phosphorylating lamin B (Cross et al., 2000).

4 CONCLUSIONS

Developments in visualization approaches have stimulated renewed interest in understanding the functional organization of the cell nucleus. The use of increasingly sophisticated molecular techniques and the availability of a large number of antibodies and also nucleic acid probes have advanced our understanding of the temporal and spatial organization of nuclear functions, as well as revealing the strikingly complex nature of the mammalian cell nucleus. Many successes have arisen from fresh initiatives to study nuclear organization. Confocal and other imaging techniques have been trained on this most enigmatic organelle in a renewed attempt to understand its organization. The reward is that we can now appreciate some of the unique stratagems devised by the nucleus to coordinate its work, such as its tight compartmentalization.

The consequences of breaches in nuclear structure–function interrelationships are observed in an expanding series of diseases, providing options for high-resolution diagnosis and targeted therapy (Stein et al., 1999). Therefore, the challenge that we are presently facing is to define in detail the mechanisms that mediate each component of gene regulation in relation to nuclear structure–function relationships. For example, cancer is associated with gross alterations to the organization of the nuclear matrix, and is therefore likely to affect both DNA metabolism and subnuclear organization (Hughes and Cohen, 1999; Koteny and Getzenberg, 1999; Stein et al., 2000). In fact, many cancers are typified by complex recombinations, often involving three or more chromosomes. Since chromosomes reside in distinct territories of the interphase nucleus, we might expect recombination events to be more likely between adjacent chromosomes. The loss of $p21^{WAF/CIP}$ alone is sufficient to reorganize these domains (Linares-Cruz et al., 1998), so it is possible that a cell lacking a critical checkpoint gene might have shuffled the position of its chromosome territories to favor such events. This would have important consequences, given that there is growing recognition that placement of regulatory components of gene expression must be temporally and spatially coordinated to optimally support biological control. These and others are all crucial questions that need to be answered for a better comprehension of the cell nucleus.

ACKNOWLEDGMENTS

This work was supported by AIRC, Italian CNR PF Biotecnologie, and Italian MURST Cofin-1999 grants to S. Capitani, by an Italian MURST 60% grant and a grant for biomedical research Azienda Ospedaliera "Arcispedale S. Anna" to the University of Ferrara, and by Funds from the Italian Ministry for Health "Ricerca Finalizzata" to A.M. Martelli.

REFERENCES

Adachi, Y., Kas, E., and Laemmli, U. K. (1989). Preferential, cooperative binding of DNA topoisomerase II to scaffold-associated regions. *EMBO J.* **13**, 3997–4006.

Alliegro, M. C. and Alliegro, M. A. (1998). Protein heterogeneity in the coiled body compartment. *Exp. Cell. Res.* **239**, 60–68.

Almeida, F., Saffrich, R., Ansorge, W., and Carmo-Fonseca, M. (1998). Microinjection of anticoilin antibodies affects the structure of coiled bodies. *J. Cell Biol.* **142**, 899–912.

Almholt, K., Arkhammer, P. O. G., Thastrup, O., and Tullin, S. (1999). Simultaneous visualization of the translocation of protein kinase Cα-green fluorescent protein hybrids and intracellular calcium concentrations. *Biochem. J.* **337**, 211–218.

Bahk, Y. Y., Song, H., Baek, S. H., Park, B. Y., Kim, H., Ryu, S. H., and Suh, P.-G. (1998). Localization of two forms of phospholipase C-β1, a and b, in C6Bu-1 cells. *Biochim. Biophys. Acta* **1389**, 76–80.

Barret, C., Roy, C., Montcourrier, P., Mangeat, P., and Niggli, V. (2000). Mutagenesis of the phosphatidylinositol 4,5-bisphosphate (PIP(2)) binding site in the NH_2-terminal domain of ezrin correlates with its altered cellular distribution. *J. Cell Biol.* **151**, 1067–1080.

Bastiaens, P. I. H. and Jovin, T. M. (1996). Microspectroscopic imaging tracks the intracellular processing of a signal transduction protein: fluorescent labeled protein kinase C β_1. *Proc. Natl. Acad. Sci. USA* **93**, 8407–8412.

Bavelloni, A., Santi, S., Sirri, A., Riccio, M., Faenza, I., Zini, N., Cecchi, S., Ferri, A., Auron, P., Maraldi, N. M., and Marmiroli, S. (1999). Phosphatidylinositol 3-kinase translocation to the nucleus is induced by interleukin 1 and prevented by mutation of interleukin 1 receptor in human osteosarcoma Saos-2 cells. *J. Cell Sci.* **112**, 631–640.

Beckmann, R., Lindschau, C., Haller, H., Hucho, F., and Buchner, K. (1994). Differential nuclear localization of protein kinase C isoforms in neuroblastoma x glioma hybrid cells. *Eur. J. Biol. Chem.* **222**, 335–343.

Berezney, R. (1991). The nuclear matrix: a heuristic model for investigating genomic organization and function in the cell nucleus. *J. Cell Biochem.* **47**, 109–123.

Berezney, R., Mortillaro, M. J., Ma, H., Wei, X., and Samarabandu, J. (1995). The nuclear matrix: a structural milieu for genomic function. *Int. Rev. Cytol.* **162A**, 1–65.

Bertagnolo, V., Mazzoni, M., Ricci, D., Carini, C., Neri, L. M., Previati, M., and Capitani, S. (1995). Identification of PI-PLC $\beta 1$, $\gamma 1$, and $\delta 1$ in rat liver: subcellular distribution and relationship to inositol lipid nuclear signalling. *Cell. Signal.* **7**, 669–678.

Bertagnolo, V., Marchisio, M., Capitani, S., and Neri, L. M. (1997). Intranuclear translocation of phospholipase C $\beta 2$ during HL-60 myeloid differentiation. *Biochem. Biophys. Res. Commun.* **235**, 831–837.

Bertagnolo, V., Neri, L. M., Marchisio, M., Mischiati, C., and Capitani, S. (1999). Phosphoinositide 3-kinase activity is essential for all-*trans*-retinoic acid-induced granulocytic differentiation of HL-60 cells. *Cancer Res.* **59**, 542–546.

Borgatti, P., Martelli, A. M., Bellacosa, A., Casto, R., Massari, L., Capitani, S., and Neri, L. M. (2000). Translocation of Akt/PKB to the nucleus of osteoblast-like MC3T3-E1 cells exposed to proliferative growth factors. *FEBS Lett.* **477**, 27–32.

Boronenkov, I. V., Loijens, J. C., Umeda, M., and Anderson, R. A. (1998). Phosphoinositide signaling pathways in nuclei are associated with nuclear speckles containing pre-mRNA processing factors. *Mol. Biol. Cell* **9**, 3547–3560.

Bunney, T. D., Watkins, P. A. C., Beven, A. F., Shaw, P. J., Hernandez, L. E., Lomonossoff, G. P., Shanks, M., Peart, J., and Drobak, B. K. (2000). Association of phosphatidylinositol 3-kinase with nuclear transcription sites in higher plants. *Plant Cell* **12**, 1679–1687.

Bunting, M., Tang, W., Zimmerman, G. A., McIntyre, T. M., and Prescott, S. M. (1996). Molecular cloning and characterization of a novel human diacylglycerol kinase zeta. *J. Biol. Chem.* **271**, 10230–10236.

Cao, T., Duprez, E., Borden, K. L. B., Freemont, P. S., and Etkin, L. D. (1998). Ret finger protein is a normal component of PML nuclear bodies and interacts directly with PML. *J. Cell Sci.* **111**, 1319–1329.

Chen, C.-S. and Poenie, M. (1993). New fluorescent probes for protein kinase C. *J. Biol. Chem.* **268**, 15812–15822.

Chen, T., Boisvert, F.-M., Bazett-Jones, D. P., and Richard, S. (1999). A role for the GSG domain in localizing Sam68 to novel nuclear structures in cancer cell lines. *Mol. Biol. Cell* **10**, 3015–3033.

Chiodi, I., Biggiogera, M., Denegri, M., Corioni, M., Weighardt, F., Cobianchi, F., Riva, S., and Biamonti, G. (2000). Structure and dynamics of hnRNP-labelled nuclear bodies induced by stress treatments. *J. Cell Sci.* **113**, 4043–4053.

Cmarko, D., Verschure, P. J., Martin, T. E., Dahmus, M. E., Krause, S., Fu, X.-D., van Driel, R., and Fakan, S. (1999). Ultrastructural analysis of transcription and splicing in the cell nucleus after bromo-UTP microinjection. *Mol. Biol. Cell* **10**, 211–223.

Cook, P. R. (1995). A chromomeric model for nuclear and chromosome structure. *J. Cell Sci.* **108**, 2927–2935.

Cotto, J. J., Fox, S. G., and Marimoto, R. I. (1997). HSF1 granules: a novel stress-induced nuclear compartment of human cells. *J. Cell Sci.* **110**, 2925–2934.

Cross, T., Griffiths, G., Deacon, E., Sallis, R., Gough, M., Watters, D., and Lord, J. M. (2000). PKC-delta is an apoptotic lamin kinase. *Oncogene* **19**, 2331–2337.

de Thé, H., Lavau, C., Marchio, A., Chomienne, C., Degos, L., and Dejean, A. (1991). The PML–RARα fusion mRNA generated by the t(15;17) translocation of acute promyelocytic leukemia encodes a functionally altered RAR. *Cell* **76**, 333–343.

Diakonova, M., Chilov, D., Arnaoutov, A., Alexeyev, V., Nikolski, N., and Medvedeva, N. (1997). Intracellular distribution of phospholipase C γ1 in cell lines with different levels of transformation. *Eur. J. Cell Biol.* **73**, 360–367.

Diaspro, A. and Sheppard, C. S. R. (2002). Chapter 3 *this book.*

Elefanty, A. G., Antoniou, M., Custodio, N., Carmo-Fonseca, M., and Grosveld, F. G. (1996). GATA transcription factors associate with a novel class of nuclear bodies in erythroblasts and megakaryocytes. *EMBO J.* **15**, 319–333.

Ferreira, J., Paolella, G., Ramos, C., and Lamond, A. I. (1997). Spatial organization of large-scale chromatin domains in the nucleus: a magnified view of single chromosome territories. *J. Cell Biol.* **139**, 1597–1610.

Fischer, D., Hock, R., and Scheer, U. (1993). DNA topoisomerase II is not detectable on lampbrush chromosomes but enriched in the amplified nucleoli of *Xenopus* oocytes. *Exp. Cell Res.* **209**, 255–260.

Fischer, U., Liu, Q., and Dreyfuss, G. (1997). The SMN–SIP1 complex has an essential role in the spliceosomal snRNP biogenesis. *Cell* **90**, 1023–1029.

Fricker, M., Hollinshead, M., White, N., and Vaux, D. (1997). Interphase nuclei of many mammalian cell types contain deep, dynamic, tubular membrane-bound invaginations of the nuclear envelope. *J. Cell Biol.* **136**, 531–544.

Gasser, S. M. and Laemmli, U. K. (1986). The organisation of chromatin loops: characterization of a scaffold attachment site. *EMBO J.* **5**, 511–518.

Godson, C., Masliah, E., Balboa, M. A., Ellisman, M. H., and Insel, P. A. (1996). Isoform-specific redistribution of protein kinase C in living cells. *Biochim. Biophys. Acta* **1313**, 63–71.

Gueth-Hallonet, C., Wang, J., Harborth, J., Weber, K., and Osborn, M. (1998). Induction of a regular lattice by overexpression of NuMA. *Exp. Cell Res.* **243**, 434–452.

Hirano, T. and Mitchison, T. J. (1993). Topoisomerase II does not play a scaffolding role in the organization of mitotic chromosomes assembled in *Xenopus* egg extracts. *J. Cell Biol.* **120**, 601–612.

Hozak, P., Sasseville, A. M.-J., Raymond, Y., and Cook, P. R. (1995). Lamin proteins form an internal nucleoskeleton as well as a peripheral lamina in human cells. *J. Cell Sci.* **108**, 635–644.

Huang, S., Deerinck, T. J., Ellisman, M. H., and Spector, D. L. (1997). The dynamic organization of the perinucleolar compartment in the cell nucleus. *J. Cell Biol.* **137**, 965–974.

Huang, S., Deerinck, T. J., Ellisman, M. H., and Spector, D. L. (1998). The perinucleolar compartment and transcription. *J. Cell Biol.* **143**, 35–47.

Hughes, J. H. and Cohen, M. B. (1999). Nuclear matrix proteins and their potential applications to diagnostic pathology. *Am. J. Clin. Pathol.* **111**, 267–274.

Jolly, C., Marimoto, R. I., Robert-Nicoud, M., and Vourc'h, C. (1997). HSF1 transcription factor concentrates in nuclear foci during heat shock: relationship with transcription sites. *J. Cell Sci.* **110**, 2935–2941.

Kim, C. G., Park, D., and Rhee, S. G. (1996). The role of carboxyl-terminal basic amino acids in the $G_q\alpha$-dependent activation, particulate association, and nuclear localization of phospholipase C-β1. *J. Biol. Chem.* **271**, 21187–21192.

Koken, M. H. M., Puvion-Dutilleul, F., Guillemin, M. C., Viron, A., Linares-Cruz, G., Stuurman, N., de Jong, L., Szostecki, C., Calco, C., Chomienne, C., Degos, L., Puvion, E., and de Thé, H. (1994). The t(15;17) translocation alters a nuclear body in a retinoic acid-reversible fashion. *EMBO J.* **13**, 1073–1083.

Koteny, B. R. and Getzenberg, R. H. (1999). Nuclear structural proteins as biomarkers of cancer. *J. Cell Biochem. Suppl.* **32/33**, 183–191.

Lamond, A. I. and Earnshaw, W. C. (1998). Structure and function in the nucleus. *Science* **280**, 547–553.

La Porta, C. A., Tessitore, L., and Comolli, R. (1997). Changes in PKC α, δ and in nuclear β isoform expression in tumor and lung metastatic nodules induced by diethylnitrosamine in the rat. *Carcinogenesis* **18**, 715–719.

Linares-Cruz, G., Bruzzoni-Giovanelli, H., Alvaro, V., Roperch, J. P., Tuynder, M., Schoevert, D., Nemani, M., Prieur, S., Lethrosne, F., Piouffre, L., Reclar, V., Faille, A., Chassoux, D., Dausset, J., Amson, R. B., Calvo, B., and Telerman, A. (1998). pW21 WAF-1 reorganizes the nucleus in tumor suppression. *Proc. Natl. Acad. Sci. USA* **95**, 1131–1135.

Liu, Q. and Dreyfuss, G. (1996). A novel nuclear structure containing the survival of motorneuron protein. *EMBO J.* **15**, 3555–3565.

Liu, Q., Fischer, U., Wang, F., and Dreyfuss, G. (1997). The spinal muscular atrophy disease gene product, SMN, and its associated protein SIP1 are in a complex with spliceosomal snRNP proteins. *Cell* **90**, 1013–1021.

Ludérus, M. E., den Blauween, J. L., de Smit, O. J., Compton, D. A., and van Driel, R. (1994). Binding of matrix attachment regions to lamin polymers involves single-stranded regions and the minor groove. *Mol. Cell. Biol.* **14**, 6297–6305.

Machiels, B. M., Broers, J. L. V., Raymond, Y., de Ley, L., Kuijpers, H. J. H., Caberg, N. E. H., and Ramaekers, F. C. S. (1995). Abnormal A-type lamin organization in a human lung carcinoma cell line. *Eur. J. Cell Biol.* **67**, 328–335.

Marchisio, M., Bertagnolo, V., Colamussi, M. L., Capitani, S., and Neri, L. M. (1998). Phosphatidylinositol 3-kinase in HL-60 nuclei is bound to the nuclear matrix and increases during granulocytic differentiation. *Biochem. Biophys. Res. Commun.* **253**, 346–351.

Martelli, A. M., Gilmour, R. S., Bertagnolo, V., Neri, L. M., Manzoli, L., and Cocco, L. (1992). Nuclear localisation and signalling activity of phosphoinositidase C β1 in Swiss 3T3 cells. *Nature* **358**, 242–245.

Martelli, A. M., Billi, A. M., Gilmour, R. S., Neri, L. M., Manzoli, L., Ognibene, A., and Cocco, L. (1994). Phosphoinositide signaling in nuclei of Friend cells: phospholipase C β down-regulation is related to cell differentiation. *Cancer Res.* **54**, 2536–2540.

Martelli, A. M., Capitani, S., and Neri, L. M. (1999a). The generation of lipid signaling molecules in the nucleus. *Prog. Lipid Res.* **38**, 273–308.

Martelli, A. M., Sang, N., Borgatti, P., Capitani, S., and Neri, L. M. (1999b). Multiple biological responses activated by nuclear protein kinase C. *J. Cell. Biochem.* **74**, 499–521.

Martelli, A. M., Capitani, S., and Neri, L. M. (1999c). Lipid second messengers as regulators of nuclear protein kinase C isoforms. *Curr. Top. Biochem. Res.* **1**, 235–242.

Martelli, A., Borgatti, P., Manfredini, M., Massari, L., Capitani, S., and Neri, L. M. (2000). Phosphatidylinositol 3-kinase translocates to the nucleus of osteoblast-like MC3T3 cells in response to proliferative and not apoptotic stimuli. *J. Bone Miner. Res.* **15**, 1716–1730.

Matera, A. G. (1999). Nuclear bodies: multifaceted subdomains of the interchromatin space. *Trends Cell Biol.* **9**, 302–309.

Mazzotti, G., Zini, N., Rizzi, F., Rizzoli, R., Galanzi, A., Ognibene, A., Santi, S., Matteucci, A., Martelli, A. M., and Maraldi, N. M. (1995). Immunocytochemical detection of phosphatidylinositol 4,5-bisphosphate localization sites within the nucleus. *J. Histochem. Cytochem.* **43**, 181–191.

Metjian, A., Roll, R. L., Ma, A. D., and Abrams, C. S. (1999). Agonists cause nuclear translocation of phosphatidylinositol 3-kinase gamma. A $G\beta\gamma$-dependent pathway that requires the p110gamma amino terminus. *J. Biol. Chem.* **274**, 27943–27947.

Moir, R. D. and Goldman, R. D. (1993). Lamin dynamics. *Curr. Opin. Cell Biol.* **5**, 408–411.

Moir, R. D., Montag-Lowy, M., and Goldman, R. D. (1994). Dynamic properties of nuclear lamins: lamin B is associated with sites of DNA replication. *J. Cell Biol.* **125**, 1201–1212.

Moir, R. D., Spann, T. P., and Goldman, R. D. (1995). The dynamic properties and possible functions of nuclear lamins. *Int. Rev. Cytol.* **162B**, 141–182.

Mortillaro, M. J., Blencowe, B. J., Wei, X., Nakaysu, H., Du, L., Warren, S. L., Sharp, P. A., and Berezney, R. (1996). A hyperphosphorylated form of the large subunit of RNA polymerase II is associated with splicing complexes and the nuclear matrix. *Proc. Natl. Acad. Sci. USA* **93**, 8253–8257.

Neri, L. M., Santi, S., Marugg, R. A., Riederer, B. M., Capitani, S., Cataldi, A., and Martelli, A. M. (1994a). In vitro heat exposure induces a redistribution of nuclear matrix proteins in K562 erythroleukemia cells. *Exp. Cell Res.* **213**, 275–285.

Neri, L. M., Milani, D., Bertolaso, L., Stroscio, M., Bertagnolo, V., and Capitani, S. (1994b). Nuclear translocation of phosphatidylinositol 3-kinase in rat pheochromocytoma PC12 cells after treatment with nerve growth factor. *Cell. Mol. Biol.* **40**, 619–626.

Neri, L. M., Billi, A. M., Manzoli, L., Rubbini, S., Gilmour, R. S., Cocco, L., and Martelli, A. M. (1994c). Selective nuclear translocation of PKC α in Swiss 3T3 cells treated with IGF-I, PDGF and EGF. *FEBS Lett.* **347**, 63–68.

Neri, L. M., Riederer, B. M., Marugg, R. A., Capitani, S., and Martelli, A. M. (1995). The effect of sodium tetrathionate stabilization on the distribution of three nuclear matrix proteins in human K562 erythroleukemia cells. *Histochem. Cell Biol.* **104**, 29–36.

Neri, L. M., Fackelmayer, F. O., Zweyer, M., Kohwi-Shigematsu, T., and Martelli, A. M. (1997a). Subnuclear localization of S/MAR-binding proteins is differently affected by in vitro stabilization with heat or Cu^{++}. *Chromosoma* **106**, 81–93.

Neri, L. M., Riederer, B. M., Marugg, R. A., Capitani, S., and Martelli, A. M. (1997b). Nuclear scaffold proteins are differently sensitive to stabilizing treatments by heat or Cu^{++}. *J. Histochem. Cytochem.* **45**, 295–305.

Neri, L. M., Zweyer, M., Falcieri, E., Bortul, R., and Martelli, A. M. (1997c). Changes in the subnuclear distribution of two RNA metabolism-related proteins can be detected in nuclear scaffold or matrix prepared by different techniques. *Histochem. Cell Biol.* **108**, 525–536.

Neri, L. M., Ricci, D., Carini, C., Marchisio, M., Capitani, S., and Bertagnolo, V. (1997d). Changes of nuclear PI-PLC $\gamma 1$ during rat liver regeneration. *Cell. Signal.* **9**, 353–362.

Neri, L. M., Borgatti, P., Capitani, S., and Martelli, A. M. (1998). Nuclear diacylglycerol produced by phosphoinositide-specific phospholipase C is responsible for nuclear translocation of protein kinase C-α. *J. Biol. Chem.* **273**, 29738–29744.

Neri, L. M., Bortul, R., Zweyer, M., Tabellini, G., Borgatti, P., Marchisio, M., Bareggi, R., Capitani, S., and Martelli, A. M. (1999a). Influence of different metal ions on the ultrastructure, biochemical properties and protein localization of the K562 cell nuclear matrix. *J. Cell. Biochem.* **73**, 342–354.

Neri, L. M., Raymond, Y., Giordano, A., Capitani, S., and Martelli, A. M. (1999b). Lamin A is part of the internal nucleoskeleton of human erythroleukemia cells. *J. Cell. Physiol.* **178**, 284–295.

Neri, L. M., Raymond, Y., Giordano, A., Borgatti, P., Marchisio, M., Capitani, S., and Martelli, A. M. (1999c). Spatial distribution of lamin A and B1 in the K562 cell nuclear matrix stabilized with metal ions. *J. Cell. Biochem.* **75**, 36–45.

Neri, L. M., Capitani, S., Borgatti, P., and Martelli, A. M. (1999d). Lipid signalling and cell responses at the nuclear level. *Histol. Histopathol.* **14**, 321–335.

Neri, L. M., Martelli, A. M., Borgatti, P., Colamussi, L., Marchisio, M., and Capitani, S. (1999e). Increase in nuclear phosphatidylinositol 3-kinase activity and phosphatidylinositol (3,4,5) trisphosphate synthesis precede PKC-zeta translocation to the nucleus of NGF-treated PC12 cells. *Faseb J.* **13**, 2299–2310.

Neri, L. M., Marchisio, M., Colamussi, M. L., and Bertagnolo, V. (1999f). Monocytic differentiation of HL-60 cells is characterized by the nuclear translocation of phosphatidylinositol 3-kinase and of definite phosphatidylinositol-specific phospholipase C isoforms. *Biochem. Biophys. Res. Commun.* **259**, 314–320.

Pederson, T. (1998). Half a century of nuclear matrix. *Mol. Biol. Cell.* **11**, 799–805.

Pederson, T. (2000). Thinking about a nuclear matrix. *J. Mol. Biol.* **277**, 147–159.

Penman, S. (1995). Rethinking cell structure. *Proc. Natl. Acad. Sci. USA* **92**, 5251–5257.

Platani, M., Goldberg, I., Swedlow, J. R., and Lamond, A. I. (2000). In vivo analysis of Cajal body movement separation and joining in live human cells. *J. Cell Biol.* **151**, 1561–1574.

Ramón y Cajal, S. (1903). Un sencillo método de coloración selectiva del retÍculo protoplásmico y sus efectos en los diversos órganons nerviosos de vertebrados e invertebrados. *Trab. Lab. Invest. Biol. (Madrid)* **2**, 129–221.

Saredi, A., Howard, L., and Compton, D. A. (1996). NuMA assembles into an extensive filamentous structure when expressed in the cell cytoplasm. *J. Cell Sci.* **109**, 619–630.

Satyn, D. P. E., Gunster, M. J., van der Vlag, J., Hamer, K. M., Schul, W., Alkema, M. J., Saurin, A. J., Freemont, P. S., van Driel, R., and Otte, A. P. (1997). RING1 is associated with the polycomb-group protein complex and acts as a transcriptional repressor. *Mol. Cell Biol.* **17**, 4105–4113.

Saurin, A. J., Shields, C., Williamson, J., Satijn, D. P. E., Otte, A.P, Sheer, D., and Freemont, P. S. (1998). The human polycomb group complex associates with pericentromeric heterochromatin to form a novel nuclear domain. *J. Cell Biol.* **142**, 887–898.

Stein, G. S., Montecino, M., van Wijnen, A. J., Montecino, M., Stein, J. L., and Lian, J. B. (1999). Nuclear structure/gene expression interrelationships. *J. Cell. Physiol.* **181**, 240–250.

Stein, G. S., van Wijnen, A. J., Stein, J. L., and Lian, J. B. (2000). Nuclear structure-gene expression interrelationships: implications for aberrant gene expression in cancer. *Cancer Res.* **60**, 2067–2076.

Sternsdorf, T., Jensen, K., and Will, H. (1997). Evidence for covalent modification of the nuclear dot-associated proteins PML and SP100 by PIC1/SUMO-1. *J. Cell Biol.* **139**, 1621–1634.

Taimen, P., Viljamaa, M., and Kallajoki, M. (2000). Preferential expression of NuMA in the nuclei of proliferating cells. *Exp. Cell Res.* **256**, 140–149.

Topham, M. K., Bunting, M., Zimmerman, G. A., McIntyre, T. M., Blackshear, P. J., and Prescott, S. M. (1998). Protein kinase C regulates the nuclear localization of diacylglycerol kinase-ζ. *Nature* **394**, 697–700.

Tran, D., Gassard, P., Berthon, B., Fukami, K., Takenawa, T., Giraud, F., and Claret, M. (1993). Cellular distribution of polyphosphoinositides in rat hepatocytes. *Cell. Signal.* **5**, 565–576.

Visser, A. E. and Aten, J. A. (1999). Chromosomes as well as chromosomal subdomains constitute distinct units in interphase nuclei. *J. Cell Sci.* **112**, 3353–3360.

Wei, X., Somanathan, S., Samarabandu, J., and Berezney, R. (1999). Three-dimensional visualization of transcription sites and their association with splicing factor-rich nuclear speckles. *J. Cell Biol.* **146**, 543–558.

Weis, K. S., Rambaud, S., Lavau, C., Jansen, J., Carvalho, M., Carmo-Fonseca, M., Lamond, A. I., and Dejean, A. (1994). Retinoic acid regulates aberrant nuclear localization of PML RARα in acute promyelocytic leukemia cells. *Cell* **76**, 345–356.

Wilson, T. (2002). Chapter 2 *this book*.

Yang, C. H., Lambie, E. J., and Snyder, M. (1992). NuMA: an unusually long coiled-coil related protein in the mammalian cell nucleus. *J. Cell Biol.* **116**, 1303–1317.

Yokogawa, T., Nagata, S., Nishio, Y., Tsutsumi, T., Ihara, S., Shirai, R., Morita, K., Umeda, M., Shirai, Y., Saitoh, N., and Fukui, Y. (2000). Evidence that 3′-phosphorylated polyphosphoinositides are generated at the nuclear surface: use of immunostaining technique with monoclonal antibodies specific for PI3,4-P(2). *FEBS Lett.* **473**, 222–226.

Zeng, C. (2000). NuMA: a nuclear protein involved in mitotic centrosome function. *Microsc. Res. Tech.* **49**, 467–477.

Zeng, C., He, D., Berget, S. M., and Brinkley, B. R. (1994a). Nuclear-mitotic apparatus protein: a structural protein interface between the nucleoskeleton and RNA splicing. *Proc. Natl. Acad. Sci. USA* **91**, 1505–1509.

Zeng, C., He, D., and Brinkley, B. R. (1994b). Localization of NuMA protein isoforms in the nuclear matrix of mammalian cells. *Cell Motil. Cytoskel.* **29**, 167–176.

Zeng, C., Kim, E., Warren, S. L., and Berget, S. M. (1997). Dynamic relocation of transcription and splicing factors dependent upon transcriptional activity. *EMBO J.* **16**, 1401–1412.

Zhang, G., Taneja, K. L., Singer, R. H., and Green, M. R. (1994). Localization of pre-mRNA splicing in mammalian nuclei. *Nature* **372**, 809–812.

Zhong, S., Salomoni, P., and Pandolfi, P. P. (2000). The transcriptional role of PML and the nuclear body. *Nature Cell Biol.* **2**, E85–E90.

Zini, N., Martelli, A. M., Cocco, L., Manzoli, F. A., and Maraldi, N. M. (1993). Phosphoinositidase C isoforms are specifically localized in the nuclear matrix and cytoskeleton of Swiss 3T3 cells. *Exp. Cell Res.* **208**, 257–269.

Zini, N., Mazzoni, M., Neri, L. M., Bavelloni, A., Marmiroli, S., Capitani, S., and Maraldi, N. M. (1994). Immunocytochemical detection of the specific association of different PIC isoforms with cytoskeletal and nuclear matrix compartments in PC12 cells. *Eur. J. Cell Biol.* **65**, 206–213.

CHAPTER 18

Confocal Imaging of Neuronal Growth and Morphology in Brain Slices

IRINI SKALIORA[1] and STAMATIS N. PAGAKIS[2]

[1] University Laboratory of Physiology, University of Oxford, Parks Road, Oxford OX1 3PT, U.K.
[2] Confocal Microscopy and Image Analysis Laboratory, National Institute for Medical Research, The Ridgeway, Mill Hill, London NW7 1AA, U.K.

1 INTRODUCTION

The fundamental advantage of confocal microscopes is their ability to reject out-of-focus light. Consequently, it is possible to obtain clearly separated optical (as opposed to physical) sections of labeled specimens, largely uncontaminated by information coming from above or below. This in turn leads to (i) increased spatial resolution and (ii) the capability of three-dimensional (3D) imaging. Both of these properties are especially important when studying complex multicellular tissues such as the central nervous system.

For our studies in neural development, it was particularly important to be able to obtain good resolution 3D images deep within brain slices. In one project, we wanted to monitor the behavior of thalamic axons as they grew towards their cortical targets (Skaliora et al., 2000). It is known that axon guidance is a highly dynamic process, which is controlled by active and continual interactions of growth cones with the various signals they encounter on their way. It is also increasingly apparent that the effect of guidance cues is not invariant, but is highly dependent on the cellular and molecular context. Furthermore, there is considerable evidence that growth cone morphology, especially when examined together with the accompanying behavior, reflects the nature of the immediate local environment. We needed, therefore, to be able to document the changes in morphology and growth patterns of thalamic axons at each portion of their trajectory as they grew through a normal physiological environment. We did not want to have to rely on successive static images, but wanted to observe the axons in real time and monitor the dynamic behavior of growth cones as they encountered the cues presented to them by the extracellular matrix and intermediate targets. We therefore used a living slice preparation that contained the entire uninterrupted pathway from the thalamus to the cerebral cortex; we labeled thalamic

Confocal and Two-Photon Microscopy: Foundations, Applications, and Advances, Edited by Alberto Diaspro.
ISBN 0-471-40920-0

axons with fluorescent tracers and acquired 2D and 3D images of growth cones at regular time intervals.

In another project our aim was to investigate the factors that control the fine dendritic morphology of layer V pyramidal cells in the cortex. We needed to have accurate measurements of large portions of the dendritic tree and also to be able to resolve the exact size and morphology of dendritic spines. In this case, it was more important to have high-resolution 3D images and not so important to observe this process in real time. Therefore, we decided to use fixed slices in which the specific neurons were labelled with fluorescent tracers. Using fixed tissue improves the spatial resolution considerably, because (i) the tissue can be embedded in an optically correct medium, (ii) there are no motion artifacts, and (iii) one can use oil-immersion objectives, which have the highest numerical aperture (NA).

In this chapter, we shall describe methods for labeling and imaging neurons for these two applications and discuss the various problems often encountered.

2 EXPERIMENTAL DESIGN

The particulars of specimen preparation will obviously depend on the experimental question. For both our projects, we used brain slices because they combined accessibility to imaging (and pharmacological manipulations, not discussed here) with a relatively unperturbed cellular and molecular environment.

2.1 Preparation of Specimen – How Thick Should the Slices Be?

For experiments with living tissues, a compromise has to be reached: smaller and thinner slices will be the most viable; larger and thicker slices will contain more of the relevant structures, but (owing to the limits of perfusion) will tend to become necrotic in the middle. For the time-lapse imaging of axon outgrowth, we used embryonic forebrain slices cut at a thickness of 400 µm. This thickness seemed to be optimal in terms of mechanical stability, viability, and preservation of the axonal trajectories.

In the experiments with fixed tissue, the requirements were different. Neither stability nor, obviously, viability were issues of concern, and, therefore, we were free to choose *optically* optimal slices. For the studies of dendritic development, we wanted a large field of view (to include as much of the dendritic field as possible and thus minimize the time-consuming task of montages) and maximal spatial resolution (in order to resolve and make quantitative measurements on individual dendritic spines). Slices were cut at 200 µm, which was the best compromise between preserving as much of the dendritic tree as possible in one slice (and thus minimizing the need to montage the images in depth) and maintaining the area of interest within the working distance of the objective (see also Section 3.2).

Surgical Procedures For the time-lapse experiments, embryos were delivered by caesarean section from pregnant rats deeply anesthetised with an intraperitoneal injection of pantobarbital (Nembutal, 100 mg/kg). The brains were dissected in cold saline (Hanks' balanced salt solution, Sigma) supplemented with extra glucose (to a final concentration of 6.0 g/l). Whole forebrain slices including the ventrolateral thalamic nucleus and somatosensory cortex were obtained using a procedure modified from the literature (Bernardo et al., 1986; Agmon and Connors, 1991). Brains were embedded in 3% low-melting agarose in order to improve stability and handling. They were then placed with the ventral face down and a vertical cut through the tissue was made at a 35°–45° angle to the midline (depending on the age of the animal) through the most anterior part of the brain. The tissue rostral to the cut was discarded, and the remaining tissue was glued on to the vibratome stage. Slices were cut at a thickness of 400 μm and left to equilibrate at room temperature for 30–60 minutes.

For the experiments with fixed tissue, animals were perfused with ice-cold saline until clear of blood, followed by 4% paraformaldehyde (see Section 2.5 for choice of fixative).

2.2 Where Should Slices Be Placed?

For the axon outgrowth time-lapse studies, the slices were placed in 35 mm petriPERM culture dishes (Sigma, Z37,6744) coated with polylysine (Skaliora et al., 2000). These dishes have a gas-permeable membranous base, which has two main advantages over the standard plastic one: (i) it is very thin, optically clear, and with low autofluorescence, and therefore it is possible to acquire images through it; (ii) it allows for more efficient exchange of gases. (As will be noted later, using anything other than a coverglass between the specimen and the objective will compromise the quality of the images, but petriPERM dishes significantly improved the specimen longevity, so we chose to accept the optical aberrations as a trade-off.) During the imaging sessions, the slices were not perfused, because we found that the movement associated with the flow of the solution was detrimental to the quality of the pictures. Instead, the slices were kept in a sealed environment: the lid of the dishes was removed and a thin film of silicone oil was poured gently over the culture medium. This arrangement was more efficient at reducing evaporation compared with the plastic cover, and also optically clearer for bright field microscopy. In long experiments, lasting up to 15 hours, the film of silicone was removed and the culture medium was replaced every 4 hours.

The fixed slices were mounted in saline (since neither glycerol nor oil are compatible with the lipophilic dyes used to label the cells), between two coverglasses (No. 1.5) separated by a 220 μm thick rectangular spacer (Fig. 1). In our experiments, such spacers were improvised from plastic sheets; however, it is now possible to order similar devices with precise specifications (e.g., Secure Seal Imaging Chambers, Grace Biolabs). The reason for this mounting arrangement was to allow access for both surfaces of the slice to be imaged, since it was not possible to know a priori how deep within the slice the area of interest would be. Furthermore, collecting images through a standard glass slide (about 1 mm thick) would make the use of high-NA objectives impossible, thereby seriously deteriorating the image quality (see also Section 3.2).

2.3 Which Dye Should Be Used?

Fluorescent dyes are used to increase the contrast between the structure of interest and the surrounding tissue. Ideally, they should be bright, target-specific, and neutral to the physiological functions of cells. Such dyes come in a variety of specifications in terms of brightness, excitation/emission (Ex/Em) wavelengths, and chemical properties (ion-specific tracers, lipophilic dyes, degree of membrane permeability, etc.). Therefore, the fluorophore of choice will depend on the type of cells used and the experimental questions.

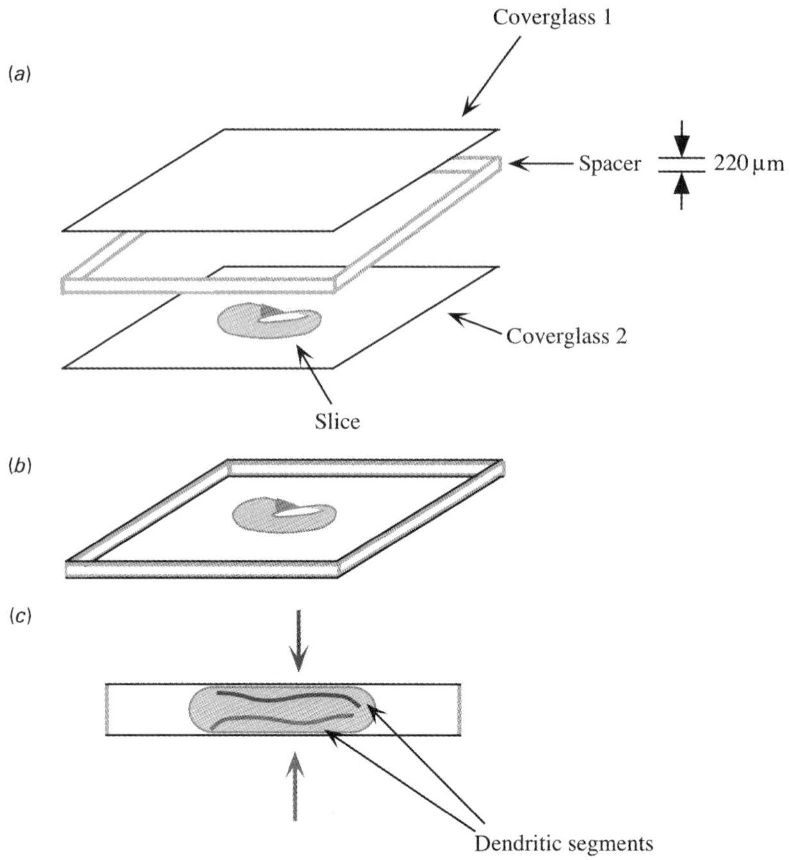

FIGURE 1 See ftp site for color figure. Fixed slices were embedded in saline, between two coverglasses, in order to allow optical access from both sides. (*a*) The slice was placed at the center of one coverglass, directly on the surface (with as little saline as possible between the slice and the glass). The rectangular spacer was glued directly on the coverglass, and the second coverglass was placed on top. (*b*) Diagrammatic view of the slice embedded between the two coverglasses. (*c*) Side view of the slice, indicating the position of two dendritic segments, at different depths within the slice. The dendrite drawn in red would not be visible from the top side due to the limits of the working distance of the objective and other optical problems (Section 4.1.2). For clarity of the illustration, the thicknesses of the spacer and the coverglass (which in fact are quite similar) are not shown to scale.

In principle, brighter (high-quantum-efficiency) dyes will require less intense illumination during image acquisition. Furthermore, using dyes that have longer excitation wavelengths will be (i) less damaging to the cells (because of the lower energy of the excitation laser), and (ii) more easily detectable deeper inside the tissue (owing to less scattering of longer wavelengths). However, one also has to keep in mind that confocal microscopes are usually equipped with either an argon-ion laser (457, 488, and 514 nm lines) or an argon/krypton mixed-gas laser (488, 568, and 647 nm lines), and therefore are limited to these excitation wavelengths. Consequently, a dye with lower quantum efficiency but with an excitation maximum closer to the wavelength of the laser might require less illumination than a brighter dye with a more disparate Ex. Finally, one has to consider the issue of autofluorescence – in other words, the property that all tissues have, to a greater or lesser degree, to emit an intrinsic fluorescence. For reasons that we are not able to expand on presently, this problem is worse with shorter wavelengths. Thus, when there is a choice, it may be better to choose a dye as close to the red part of the spectrum as possible. The trade-off in using long wavelengths is a relative reduction in the theoretical resolution (see Section 4.1.3).

For our studies, where the goal was to monitor axonal and dendritic form in space and time, we used lipophilic dyes that migrate fast within the lipid bilayers, are not toxic, and can easily diffuse within the cell membranes and reveal the shape of cells and processes. Furthermore, they are not affected by fixation. Popular lipophilic tracers are the carbocyanine dyes, such as DiIC18(3) (1,1′-dioctadecyl-3,3,3,3′-tetramethylindocarbocyanine perchlorate, Molecular Probes, Molecular Weight (MW) 934, Ex 549 nm, Em 565 nm) and DiOC18(5) (3,3′-dioctadecyloxcarbocyanine perchlorate, Molecular Probes, MW 882, Ex 484 nm, Em 501 nm). Other members of the family exist that are less lipophilic (C12DiI, C16DiO), or faster diffusing (Fast DiI and Fast DiO), or redshifted (DiIC18(5): Ex 644 nm, Em 665 nm or DiIC18(7) : Ex 750 nm, Em 780 nm). We tried several of these, and found DiI to work best overall: it labeled both axons and dendrites brightly, had low autofluorescence, and has been reported not to spread from one neuron to the next (Mills et al., 1994). An additional advantage of DiI, which was not explored in our studies, is that it can be photoconverted to a stable reaction product for permanent staining.

2.4 How Should the Cells be Labeled?

There are several methods for labeling neurons with DiI, including iontophoretic or pressure injection of the dye dissolved in ethanol, DMSO, or formamide. The method that worked best in our hands was insertion of DiI crystals directly into the tissue. For the time-lapse experiments, thalamic axons and their growth cones were labeled by inserting minute crystals of the dye into the thalamus with a tungsten needle under a dissecting microscope. The slices were returned to the incubator (37°C) and left overnight to allow for dye transport. This incubation period (12–24 hours) was more than adequate for the dye to be actively transported to the growing tips of the axons (up to 1–2 mm away) and brightly label the entire growth cone, including filopodia and lamellipodia. With this method, we were able to obtain small numbers of labeled thalamic axons in each slice, which were clearly identifiable and could be traced for long distances back towards the thalamus. For the purposes of our experiment, a relatively low yield of labeled neurons was advantageous, since we wanted to be able to follow axons as close to their source of origin as possible and also to have only one or few growth cones in the field of view during imaging (Skaliora et al., 2000). In cases where either single or, conversely, large numbers of neurons need to be labeled, different methods might be more suitable (for a somewhat more detailed discussion of the various techniques, see Cline et at., 2000).

For the fixed tissue experiments, the procedure was quite similar, except that larger and multiple crystals were used and they were inserted into the corpus callosum of hemisected fixed brains. This method required an incubation of several weeks at room temperature (22–25°C) for the passive diffusion of the dye back along the axon, so that the entire dendritic tree, including spines, was brightly labeled. It is possible to accelerate the passive diffusion by incubating the tissue at higher temperatures (e.g., 35°C), but we found that signal-to-background ratio was significantly compromised.

With older animals, this method of DiI labeling was not successful. This is likely to be due to (i) the longer axonal distances and (ii) the increased myelination of the adult brain, both of which make passive diffusion slower and incomplete. To get around this problem, callosally projecting neurons were first retrogradely labeled in vivo, by injection of rhodamine latex microspheres into the contralateral hemisphere. Subsequently, brains were perfused, and sectioned coronally at 200 μm, and individual backfilled cells were injected intracellularly with lucifer yellow (LY) to reveal the full dendritic morphology. LY (Ex 428 nm, Em 545 nm) is a hydrophilic probe which has been a popular neuroanatomical tracer. It can be excited (although marginally) by the 488 nm line of the argon laser, it is very bright, and, like DiI, it is not affected by fixation. In all cases, slices were counterstained with the nuclear stain Hoechst 33342 (Molecular Probes, U.S.A.; Ex 350 nm, Em 460 nm) to reveal the cortical layers. There is no optical interference (i.e., crosstalk) between LY and Hoechst 33342.

2.5 How Should the Slices Be Fixed?

For experiments that are not designed to monitor dynamic processes, fixation has several advantages: (i) it allows the mounting medium of the specimen to be matched to the immersion medium of the objectives; (ii) it allows the use of antifading agents without concern that they might affect the health of the tissue; (iii) it allows better control of specimen preparation, such as hydration levels; and (iv) it facilitates antibody penetration due to permeabilization and selective removal of unwanted materials.

However, not all fixation protocols (historically designed for light microscopy) are suitable for fluorescence confocal microscopy. The problems have to do with autofluorescence and the preservation of 3D cellular structure. Many commonly used fixatives (gluteraldehyde, formaldehyde, ethanol, methanol, and acetone) produce high levels of autofluorescence, due to the denaturation or chemical modification of macromolecules, and therefore are not compatible with fluorescence studies (Bacallao et al., 1995). The degree of autofluorescence varies, and it is only a problem when it is at levels comparable to the signal from the fluorescent dye. Therefore, the choice of fixative will depend indirectly on the chosen fluorophore.

Ideally, the fixation process should work very fast and preserve the cellular 3D structure before the cell can react to the chemicals and produce structural artifacts. In our experiments, animals were perfused with cold saline solution followed by 4% freshly made paraformaldehyde. This method produced relatively low levels of autofluorescence, which was easily suppressed by the gain and offset controls of the confocal microscope.

3 EQUIPMENT

3.1 Microscopes Although upright microscopes have a simpler light path and therefore marginally better optical performance, inverted microscopes are preferable for experiments with living tissue. This is because (i) the top of the dish is free for manipulations and (ii) it is possible to use high-resolution water-immersion objective lenses. An exception is when slice imaging needs to be combined with electrophysiological recordings that require mechanical (as well as optical) access to the top of the specimen. In this case, one can use "dipping" objectives, which are designed to work without coverglass and have extra-long working distance. The trade-off is that they have low to medium numerical apertures and thus lower spatial resolution and brightness.

For the experiments described in this chapter, we used two confocal systems on inverted microscopes: a Leica CLSM with an argon-ion laser (lines at 488 and 514 nm) and a Leica TCS 4D with an argon/krypton laser (lines at 488, 568, and 647 nm).

3.2 Objective Lenses

Ideally, an objective lens should have a high NA (for high light-collection efficiency and spatial resolution), long working distance (to be able to focus on structures deep in the slice), good color correction (to be able to focus at the same focal plane at both excitation and emission wavelengths), and, finally, high light transmission. Unfortunately, these requirements are often contradictory. For example, good color correction tends to reduce light transmission, and, in order to achieve a high NA, the working distance needs to be reduced.

Irrespective of the type of lens chosen, it is important that it be used according to the manufacturer's specifications. Failure to do so will compromise the quality of the images. For example (unless the lens has a coverglass thickness correction collar), the correct type of coverglass should always be used (usually with a refractive index (RI) of 1.52 at a wavelength of 589 nm, and a thickness of 170 μm – No. 1.5). Also, in the case of oil-immersion objectives, the specified immersion oil with the correct RI at the temperature of the experiment should be used. The sensitivity of high-NA objectives to changes in their specifications is so high that even a 3–5°C difference in the temperature of the immersion medium, or a 10% difference in coverglass thickness, will produce optical errors that will reduce both the usable NA and the usable working distance (and thus the ability to image deep into the slice – see Section 4.1.2).

Lenses work optimally when the RI of their immersion medium matches as closely as possible that of the mounting medium of the specimen. For this reason, water-immersion objectives are strongly recommended for experiments with living tissues. With fixed samples, either oil- or water-immersion objectives can be used, depending on the mounting medium of the specimen.

For our studies, which required detection of relatively faint signals deep inside the tissue in watery medium, we had to use water-immersion, long-working-distance lenses, with as high an NA as the first two requirements would allow. They were Leitz Fluoreszenz 25 × /0.75 (working distance 300 μm) and 50 × /1 (working distance: 680 μm) and a Leica (HC PL APO) 63 × /1.2 (working distance 220 μm, and with a correction ring for a range of coverglass thickness between 140 and 180 μm).

Leitz 50× lens, which was mainly used for the time-lapse, temperature-controlled experiments, was fitted with a home-built ring heater. We found this very useful in maintaining the correct specimen temperature, since immersion lenses come into contact with the specimen. Owing to their high thermal capacity, unless heated, they act as heat sinks, cooling down the specimen just beyond the immersion medium.

Finally, the chromatic performance of the objective lens is a parameter that needs to be considered. In short, since the refractive properties of the lens depend on the wavelength of light, unless corrected, a lens will not be able to focus light of different colors to the same focal point. In wide-field microscopy, where the specimen is illuminated uniformly, the chromatic performance of the lens along the *z* axis is not critical. However, in confocal microscopy, chromatic aberrations are detrimental, since the objective fails to focus the excitation and emission wavelengths (i.e., light of two different colors) at the same point in the axial (z) dimension. In practice, this will appear as a significant reduction in image brightness and, in experiments with multiple colors, as a misalignment of the image from the different channels, so that fluorescence from one point within the tissue will appear to be coming from multiple points (Fig. 2).

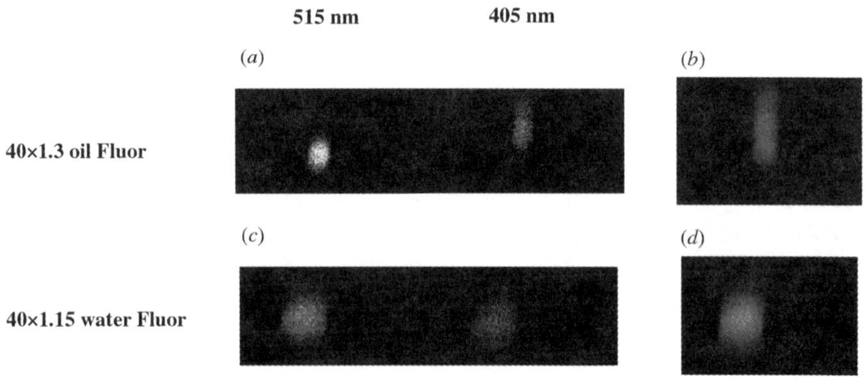

FIGURE 2 See ftp site for color figure. Illustration of the differences in chromatic correction of the objective lenses indicated. We imaged doubly labeled fluorescent beads, 4 μm in diameter, mounted in water. (*a*) and (*c*) show simultaneously acquired cross-sectional images (viewed along the x–z plane) of a single bead, at the emission wavelengths indicated. It can be seen in (*a*) that, because of inadequate color correction, the two fluorescence images, coming from the same bead, do not align along the z axis. In contrast, when the same bead is imaged with the other objective, the two images are perfectly aligned (*c*). Such a considerable difference in chromatic performance was unexpected, since both objectives had the same chromatic specification (Fluor). This result indicates that significant differences can exist even between objectives of similar specifications, and therefore individual testing is essential when maximum performance is required. Both objectives also have reduced ultraviolet transmission, since the image obtained at 405 nm is significantly dimmer than that obtained at 515 nm, with the oil objective being slightly brighter. For ease of comparison, the images from (*a*) and (*c*) are shown as green–red overlays in (*b*) and (*d*), respectively. Finally, as can be seen in (*a*), the oil-immersion objective lens has significant spherical aberration, because the beads were mounted in water. Thus, the spherical bead appears elongated along the z axis. It is possible to compensate for such distortions, by choosing an immersion oil with a different refractive index (Section 4.1.2). Data for the 40 × /1.15 objective lens were acquired with higher zoom, and hence the bead appears bigger. All images appear in pseudocolor.

3.3 Computer Programs Used for Data Analysis

Confocal microscopes are heavily used instruments where most of the available time is spent on data acquisition. In addition, it is our experience that no single software package is powerful in all aspects of image analysis. This is especially true for confocal software, which is primarily designed for image acquisition. Therefore, we routinely do the image analysis on separate workstations, using a suite of data-analysis software.

The time-lapse analysis, the measurements on dendrites and spines, as well as the basic 3D visualization (maximum-projection images) were performed on a Macintosh computer using the public domain NIH Image program (developed at the U.S. National Institutes of Health and available on the Internet at http://rsb.info.nih.gov/nih-image/). For advanced

3D visualization, we used the software Imaris (Bitplane, Switzerland) for the SGI computers running the Unix operating system.

For final image presentation, we used Adobe Photoshop (Adobe Systems Inc.) and Canvas (Deneba Systems Inc.). Photoshop has certain image-processing functions, mostly focusing on image presentation. Canvas is particularly good when images and graphics (e.g., annotations, drawings, or graphs) need to be combined for publication figures. The major limitation of both these packages is that they cannot process 3D data sets.

Recently, two Adobe Photoshop-compatible collections of programs, "Image Processing Tool Kit" and "Fovea Pro Tool Kit" (for 8- and 16-bit gray and 24- and 48-bit RGB color images, respectively) have become available (for information, look in http://members.AOL.com/FoveaPro/ and http://members.AOL.com/ImagProcTK). These programs (called plug-ins) add scientific image processing and analysis functionality to Photoshop, including filters to read in Biorad confocal images. Both tool kits also work with NIH Image (version 1.62 or later) on the Mac, but only with 8-bit grayscale images. NIH Image has recently been rewritten for the Java environment, and is compatible with all operating systems. It is available at http://rsb.info.nih.gov.

4 IMAGING METHODS

In this section, we shall first discuss the principles of some commonly encountered problems and examine their possible solutions. We shall then give a narrative account of the methods that we used for the requirements of our particular applications.

4.1 Commonly Encountered Problems

Although certain image defects can be at least partially corrected during data analysis, it is always preferable to try to solve or minimize their causes before and during acquisition. To do so, one needs to understand the problems and have an appreciation of the possible solutions.

4.1.1 Image Blurring Caused by Movement Acute slices situated in a liquid, often perfused, medium will inevitably tend to move slightly. Since the scanning process takes a finite amount of time, even small movements will result in blurring. At best, this will reduce the resolution of an image (slight nonvisible blur) and, in some cases, will make the image unusable (total blur). If movement proves to be a problem, one can either (i) increase the confocal scanning speed and/or (ii) take measures to limit the movement.

(i) Increasing scanning speed is not always possible (the confocal hardware comes with a maximum scanning speed) and, even if it is, may not be desirable. With faster scanning, proportionally smaller number of photons will be recorded per scan, leading to a lower signal-to-noise ratio and thus reduced resolution (Stelzer, 1998). To maintain resolution, one has to increase either the laser power (assuming the fluorophore is not already saturated) or the amount of fluorophore. Neither of these two options is safe for live specimens, and they should be used with care.

(ii) To limit specimen movement, several measures can be taken. In our experiments, we did the following:

(a) The dish membrane was coated with polylysine, which facilitates the adherence of the slice.

(b) The slices were not perfused, to eliminate movement associated with the flow (as stated previously, in long experiments, the medium was replaced regularly).

(c) The culture medium was covered with a thin film of silicone oil, to minimize evaporation and the resulting movement. This arrangement worked well with the embryonic tissue used in our experiments, which are known to have low oxygen requirements, but may not be recommended with more mature tissue. In cases where perfusion is essential, it is possible to reduce movement by placing a weight on top of the slice (e.g., a wire ring, which is thinner than the slice, and on which a membrane is attached; Potter, 2000). Nevertheless, none of these precautions will completely eliminate the need to frequently check and readjust the focus during long experiments (see also Section 4.2.1).

4.1.2 Aberrations Caused by the Need to Image Deep Inside the Slice The absolute limit for the depth at which images can be successfully acquired (depth penetration) is the *nominal* working distance of the objective. In practice, however, attenuation of the light as it travels towards and through the specimen and then back to the objective lens will dramatically reduce the image contrast, causing the *effective* working distance to be much shorter.

The degree of light attenuation with depth depends on three parameters: refraction, absorption, and scattering. In experiments where the *amount* of light is used for measurements, attenuation has to be quantified, modeled, and then corrected. Good analyses can be found in Rigaut and Vassy (1991) for light-intensity measurements and in Errington and White (1999) for volume measurements. In studies of neuronal structure and form such as ours, it was not necessary to correct for light attenuation, since the intensity of the fluorescence was not being measured.

The refraction of light along the optical path (caused by the difference in the RI at the interfaces between lens, lens immersion medium, coverglass, mounting medium, and specimen) results in an optical error termed spherical aberration. Spherical aberration is minimal if the RIs of all segments of the optical path are the same. Although we can select the RI of the other segments, the RI of the specimen is unknown and also spatially variable. Therefore, the specimen is the primary cause of spherical aberrations in an otherwise-optimized system. As the focal point is moved deeper inside the specimen, gradually increasing refraction causes a progressive reduction in the recorded brightness, and eventually the image becomes fuzzy and dim.

The first step to optimize a system for spherical aberration is to use the objective lenses according to their specifications, that is, with the correct coverglass and immersion medium. In this section, we shall assume that these parameters are selected correctly (see Section 3.2 for more information), and we shall discuss some practical advice to reduce the aberrations caused by the specimen and the mounting medium.

There are three basic steps:

(i) Improve the optical properties of the specimen in order to make it as homogeneous as possible. Naturally, this will be feasible only with fixed samples.

(ii) Place the specimen in a mounting medium with the correct RI for the objective used.

(iii) Position the specimen as close to the coverglass as possible. This is particularly significant in cases when (ii) is difficult to achieve. In such a situation, any layers of mounting medium between the coverglass and the biological specimen will increase light refraction and reduce the quality of the acquired image.

In fixed preparations, we have fairly good control of all of these parameters. The specimen can be "cleared up" (by dehydrating it and/or selectively removing unwanted material) in order to improve its optical properties and reduce light absorption and scattering. It can then be embedded in mounting medium with an appropriate RI (i.e., water 1.33; glycerol 1.47; or oils, with a range of RI around 1.52). Finally, it is easier to position the specimen exactly on the coverglass. With such carefully prepared samples, high-resolution images can be obtained from deeper inside a thick preparation.

In experiments with living tissues, a significant source of optical errors comes from the difficulty to optimize all the above parameters. The optical properties of the specimen cannot be improved, there is no choice of mounting medium, and the tissue cannot be placed exactly on the coverglass (frequently, the slice is not perfectly flat and a layer of saline exists between them). To put this in perspective, an oil objective with an NA of 1.3 would collect only 55% of the light intensity it would normally collect from a specimen that sits right onto the coverglass, if the specimen were located 10 μm away from it; and at a distance of 20 μm, the recorded intensity would drop to 40% (Keller, 1995). Therefore, the ideal objectives for live preparations are water-immersion ones, since the layer of saline between the coverglass and the specimen will not affect their performance. Unfortunately, good-quality water-immersion objectives are very expensive and might not always be available. In this case, it might still be possible to obtain decent images using oil-immersion objectives, provided one tries to compensate for aberrations as much as possible. For instance, a good procedure would be to image a fluorescent bead under similar conditions as those of the specimen (e.g., beads of known diameter inserted inside the tissue) and try immersion oils with different RIs until the image is optimized. If a 3D dataset can be acquired, the best conditions would be those under which a side view of the dataset shows an image of the bead that is symmetric around the focal point.

If the image quality allows for deeper penetration, but there is no extra working distance because the lens touches the coverglass, a thinner coverglass could be used. As the thickness deviates from the specifications, increased aberrations will occur. Replacing the lens immersion oil with one of different RI could reduce these aberrations. Obviously, this will not be possible with water-immersion objectives, but they frequently come with an aberration correction ring. We used such an objective (HC PL APO 63 × /1.2), which was adjusted by focusing on a labeled structure inside the slice and turning the correction ring until the brightness and the clarity of the structure were maximized.

If even longer working distance is needed, noncoverglass dipping objectives can be used. They usually require an upright microscope, although an organ bath for inverted microscopes that can accommodate such lenses has been reported (Cody and Williams, 1997). These objectives compromise resolution (and possibly color and flat-field correction) in return for compact size and long working distance.

4.1.3 Photo-bleaching Caused by the Need to Maximize Temporal and/or Spatial Resolution The NA of the objective lens sets the limits on the spatial resolution of a microscope – and it obviously has to accommodate the smallest structure that needs to be detected. To match this resolution when creating the digital image, sampling theory states that the emitted light should be sampled twice as fine (Nyquist criterion). For instance, an image from a confocal microscope using a lens with an NA of 1.0 (which provides a resolution of about 0.2 μm at 500 nm, $0.61\lambda/1.4$ NA), needs to be sampled at 0.1 μm per pixel. The pixel size is selected by using the zoom control of the confocal microscope.

Similarly, the time-lapse frequency should be twice as fast as the fastest event one wants to record. This is adjusted by selecting the appropriate waiting period on the time-lapse function of the confocal software.

Undersampling, in either the spatial or the temporal domain, will reduce the respective resolution. Oversampling will not offer any additional information. Instead, it will increase photodamage, delay the acquisition process, and produce huge data sets that could become unmanageable.

Having said that, sampling at the ideal frequencies is often not possible, and suboptimal sampling parameters need to be chosen empirically. To do this, it is important to understand that achieving both high-speed *and* high-resolution imaging might not be possible, because of the limited number of photons that can be produced from the fluorophore per unit time. With increasing scanning speed, the images become dimmer. To restore image brightness (detect more photons), without increasing the intensity of illumination that might be harmful for the specimen, one is forced to increase the pixel size in all three dimensions (voxel). This will reduce the spatial resolution substantially, but it will achieve increased brightness, since each bigger voxel accumulates the signal from several (depending on the zoom factor and the size of the confocal pinhole) smaller ones.

Speed requirements are less severe while imaging slower events, such as the axon growth in our experiments, or cell migration. Therefore, our sampling regime concentrated on maximizing the spatial resolution, while maintaining the well-being of the tissue; to increase the image brightness, we first increased the photomultiplier gain (rather than the laser power or the fluorophore concentration). If the signal was still not high enough, we increased the pinhole size or decreased the zoom. This would reduce both axial and lateral resolutions, but it improved the image brightness without additional photodamage to the specimen.

It is common practice in confocal microscopy to average several frames for each image produced. This procedure reduces the image noise and helps maintain resolution in cases of very dim samples. An equivalent method is to slow down the scanning speed of the confocal system by a similar factor to the frames averaged. Although there have been lengthy discussions, there is no clear consensus on which method is better for delivering the light to the sample and the fluorophore (Dailey, 2000). We chose to use a fast scanning speed and use an average of 4–8 frames per image (see Section 4.2.1).

For fixed specimens the sampling can be optimized, with only one limitation: the exhaustion of the fluorophore. The gain of the system, the laser power and the confocal pinhole can all be set to the optimal values, and lengthy scans are acceptable since there is no specimen movement involved. Fluorophore antifading agents (e.g., DABCO or *n*-propyl gallate) can also be added in the mounting medium to reduce fading and increase imaging time (Florijn et al., 1995). Commercial mounting media, already containing antifading agents, are also available (Citifluor in PBS or glycerol, Agar Scientific; Vectashield and Vectamount from Vector Labs, U.S.A.; Prolong, or Slowfade Molecular Probes, U.S.A.).

4.2 Imaging Protocols

4.2.1 Axon Outgrowth (Time-Lapse) Studies For experiments with living tissue, the main concern is how often one can acquire an image without compromising the tissue's health. In our experiments on axon growth and growth cone behavior, it was important to image frequently in order to obtain an accurate view of the dynamic process of navigation.

Based on the conflicting needs as outlined in the previous section, we chose the following compromise (Fig. 3):

(i) *Frequent acquisition of 2D images of the growth cone*: high power images of the distal segment of the axon were acquired every 60–120 s, in one plane, with the pinhole maximally open in order to maximize the amount of collected light, and thus minimize the laser intensity.

(ii) *Regular acquisition of 3D images of the distal segment of the axon*: lower-power series of images of a larger portion of the axon, collected every 30–60 minutes with a smaller pinhole to provide depth discrimination. This allowed us to view an axon in the context of other axons, clarified the relative positions of neuronal elements in case of uncertainty (Fig. 4), and provided stable measuring points (see also Section 5).

Briefly, the procedure was as follows:

1. Scan quickly in low power (10×) and with the pinhole maximally open until an appropriate preparation and area of interest (AOI) is identified (Fig. 5). This was usually done through the eyepieces, because it was more convenient and faster (thus reducing the likelihood of photodamage), but could equally be done using the confocal.

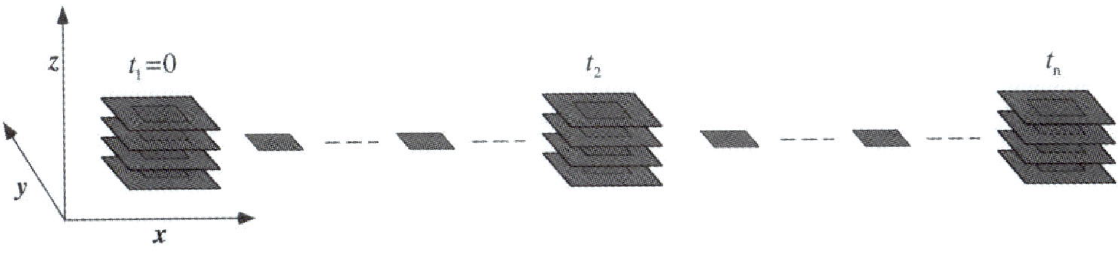

FIGURE 3 See ftp site for color figure. Paradigm of image acquisition during the time-lapse experiments. A series of optical sections in depth (shown in blue) was acquired at the beginning of the experiment ($t_1 = 0$) to provide the baseline location and morphology of one or a few axons within the region of interest. A maximum-projection image of such a data set is shown in Figure 6. The procedure was repeated at regular intervals ranging between 30 and 60 minutes ($t_2, \ldots, t_n$). In between, high-power (shown in red) single images were acquired at short intervals (60–120 s) to document the changes in morphology and behavior of one or two growth cones within the region of interest. Such data are shown in Figure 7. The difference in size of the blue and red images is meant to reflect the different magnifications (selected by the zoom factor of the confocal).

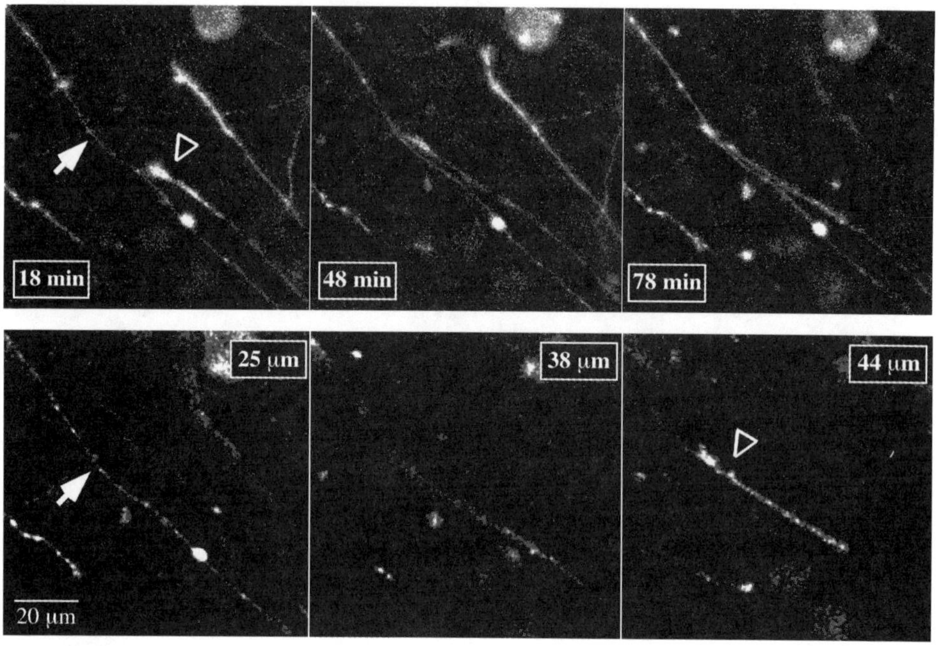

FIGURE 4 See ftp site for color figure. The panels in the top row are maximum-projection images, generated from a series of confocal optical sections spanning 30 μm in depth and separated by time intervals of 30 minutes. It can be seen that the axon towards the centre (open arrowhead) is extending forward and, in the third top panel, appears to be growing on top of another axon (closed arrow). However, analysis of the single optical sections (bottom row) reveals that the two axons are in fact separated by about 20 μm. Maximum-projection images were created by a "brightest-point algorithm" that performs a pixel-by-pixel comparison of a series of optical sections and uses the brightest point from each one in order to create the projection image. Projection algorithms (using either the brightest point or the average) are an integral part of all 3D image-analysis software packages presently available, as well as some of the image-acquisition software packages that accompany confocal systems.

2. Once the AOI was identified, the objective was switched to the 50×, the pinhole was closed down, and a series of optical sections in depth were acquired. Each image was the average of several frames (usually 4, sometimes 8, depending on the quality of label and location within the tissue). Such images, spanning 20–40 μm in depth and up to hundreds of micrometers in height and width, were acquired at regular intervals (usually hourly), and allowed us to have a snapshot of the axons in space and time (Fig. 6).
3. The zoom was increased to allow for better resolution of the tip of the axon with the advancing growth cone and side branches. The pinhole was set to be maximally open, and single (i.e. 2D) images were acquired at short intervals (Fig. 7). Although such images provide good estimates of growth cone morphology and size, some quantitative aspects of the 3D structure (especially in the z dimension) would inevitably be difficult to extract. This was a necessary compromise in order to allow for longer experiments and more frequent sampling. For our studies, intending to compare growth cone behavior in different regions of the axon pathway, highly

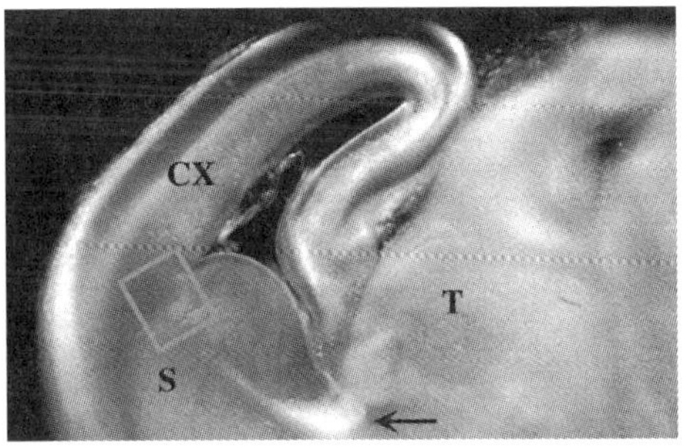

FIGURE 5 See ftp site for color figure. Photomicrograph of a thalamocortical slice obtained under transillumination. The red box indicates identification of the area of interest to be subsequently imaged. CX, cortex; T, thalamus; S, striatum. The arrow indicates the trajectory of thalamic axons as they exit the thalamus, traverse the striatum, and head towards their cortical targets.

quantitative measurements of growth cone morphology were less important than the dynamic structure and rate of instantaneous growth. Contrary to this, in studies where the fine structure of neurons is being examined, 3D imaging would be required (see Section 4.2.2).

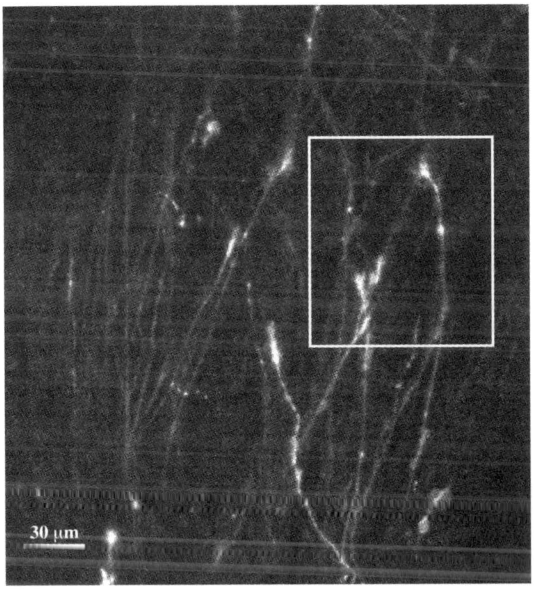

FIGURE 6 Maximum-projection image of a group of six axons within the area of interest shown in Figure 5. Each of these growth cones is located at a slightly different plane of focus. In this particular example, we used 22 sections spanning 40 μm in depth. The image has not been further processed in any way, except to adjust the contrast for optimal visibility. The box indicates the region that was subsequently imaged at a single plane as indicated in Figure 7.

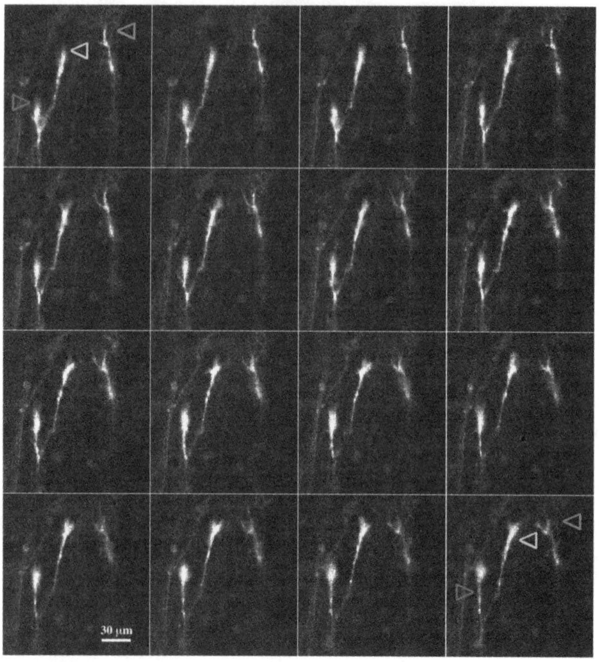

FIGURE 7 See ftp site for color figure. Time-lapse sequence of three growth cones located at the same focal plane within the region of interest (their relative positions compared with Figure 3 have changed in the 90 minutes since the previous image was acquired). Each panel is a single confocal image collected at 90 s intervals. Arrowheads indicate the position of the leading tip of the axons at the onset of the imaging session. It can be seen that for the duration of the experiment illustrated here, the growth cone at the right-hand corner of each panel does not advance at all (blue arrowhead), whereas the other two move forward by 18 μm (green arrowhead) and 30 μm (red arrowhead), respectively. Interestingly, the first axon, which had been "static" for about 6 hours, later began a vigorous forward extension (Skaliora et al., 2000). An animated version of the static growth cone can be found at http://nucleus.med.upatras.gr/hsn/. Such comparisons are useful because they indicate that the reduction or even the arrest of extension is part of the normal growth process and does not in itself indicate a damaged or unhealthy axon.

A problem frequently encountered with 2D time-lapse imaging is that the AOI tends to move out of the plane of focus. This may be due to physiological reasons (e.g., axons growing at an angle to the surface of the dish) or to technical ones (as discussed in Section 4.1.1). Although every effort must be made to minimize the latter, it is never possible to completely eliminate loss of focus, and therefore it is important to continually monitor and readjust the focal plane throughout the experiment. In unattended imaging (i.e., experiments running overnight) we found it essential to acquire *series* of images at every time point, in order to maximize the likelihood of maintaining the AOI within the data set. In this case, images were acquired less frequently (e.g., every 5–10 minutes). The precise ratio of sampling frequency to planes of imaging has to be determined experimentally.

4.2.2 *Dendritic Morphology Studies* For experiments with fixed tissue, it is obviously easier to acquire optimal images, since one does not have to worry about the health of the tissue. Nevertheless, there is still the issue of fluorophore bleaching with time, which makes it essential to store the fixed slices in the dark and take the images as quickly as possible after preparation of the slices.

Again the slide was scanned in low power in order to identify well-labeled and completely preserved cells. The objective was then switched to a 63×/1.2 water-immersion objective that (when the zoom was set to 1.0) gave a field of view of 160 μm × 160 μm. The pinhole was adjusted depending on the brightness of the label, the depth of the dendrite within the slice and the required axial resolution. Stacks of images (each image was an average of 8 or 16 frames) were collected along the axial (z) dimension, by gradually moving the focus in 1 μm steps. This separation was adequate for accurate spine measurements. However, for complete 3D reconstruction, finer sampling in depth would be required. Depending on the position of the dendritic branches in relation to the cutting plane, these stacks could span between 10 and 50 μm. The stage was then moved to a neighboring portion of the dendrite, and the procedure was repeated until the entire dendritic tree was covered. In acquiring these successive stacks of images, it was important to allow for enough overlap to be able to subsequently perform an accurate montage of the entire dendritic segment (Fig. 8). At the end of the imaging session, the

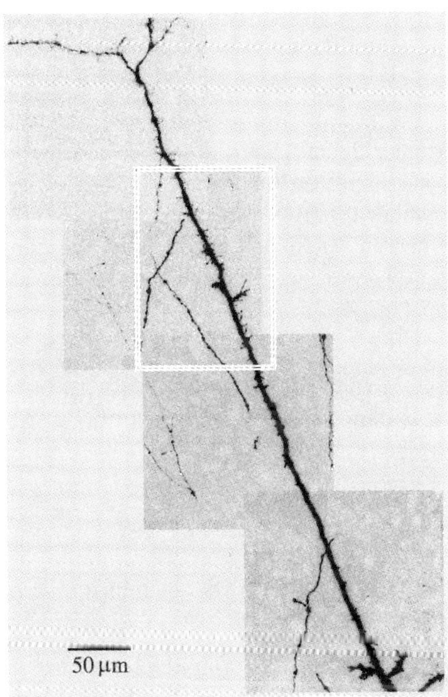

FIGURE 8 Montage of four maximum-projection images spanning the apical dendrite of a layer V pyramidal cell in the cortex. Each image has been generated by the projection of 25 single confocal optical sections acquired at 1 μm z steps. The individual maximum-projection images have been subsequently combined to create a montage of the entire dendritic segment. The area within the box appears in the right panel of Figure 11 as a 3D reconstruction. These images have been inverted (fluorescence appears black and background white) to improve visibility.

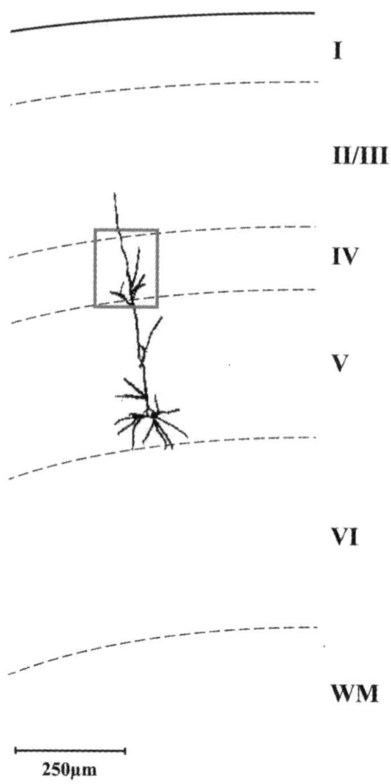

FIGURE 9 See ftp site for color figure. Digitized version of camera-lucida drawing of a layer 5 pyramidal cell, after it had been scanned on the confocal. Such drawings were made in order to have a complete view of the cells whose dendrites were included in the quantitative analysis. To obtain confocal records of the entire neuron would have been too time-consuming and, at least in this case, irrelevant to the aim of the study. The area within the red box appears in the left panel of Figure 11 as a 3D reconstruction.

(somewhat faded) labeled cells were drawn using a camera lucida to keep a record of the position of the cell body and dendrite with respect to the cortical layers (Fig. 9).

5 DATA ANALYSIS

In both these sets of experiments, our goal was to measure different aspects of neuronal structure and morphology.

For the time-lapse experiments, we wanted to measure axon elongation, size and shape of growth cone, numbers and size of side branches, and angle of turning. However, since a brain slice is an artificial and unnatural condition, first it is important to take all precautions possible to ensure that the axons and growth cones analyzed were healthy and representative of fibers growing in the intact brain. Axons were excluded from further study if any of the following applied: (i) beaded appearance, which might indicate photodamage induced by exposure to the fluorescent light; (ii) growth cone escaping to the surface of the slice: in order to avoid artifacts induced by the cutting procedure, all imaged growth cones included in the analysis were located at least 30 μm deep into the slice; (iii) steady and gradual changes in rates of growth, since this could indicate a gradual deterioration of the axon due to either the culture conditions or damage induced by the laser. Furthermore, all imaging sessions began at least 12 hours after slice preparation and labeling, since it has been shown that axons can still grow for at least 3 hours after they have been severed from the parent cell body (Harris et al., 1987).

For measurements, we used the freely available software NIH Image, which has inbuilt tools for measuring between selected points or alongside an irregular shape. The acquired images were first calibrated to the final magnification used (objective × confocal zoom factor), thereby assigning a correspondence between pixel size and actual distance. The length of the distal segment of the axon was measured from a stable and distinctive point along the axon, such as a branch point or distinct inflection (but not a varicosity, since these often were seen to move within the axon shaft), to the distal tip of the growth cone. For these measurements, the outline of the axon was followed with the mouse cursor on the screen in order to obtain the *real* length of the axon, beyond the reference point (Fig. 10). We chose this procedure, instead of marking only the final position of the growth cone, in order to avoid errors due to movements of the entire slice in the dish. These length values were divided by the total observation time to give a value for the net overall *rate* of advance. As an independent measurement, rates of growth were also calculated separately

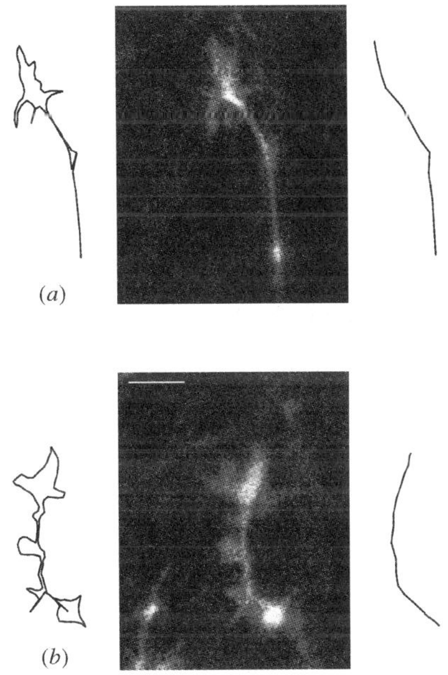

FIGURE 10 See ftp site for color figure. Examples of types of measurements made for the analysis of axon outgrowth. The pseudocolored images in the middle are single confocal pictures of the distal segment of an axon and its growth cone at different points in time. The black lines on the right of each image illustrate the kind of measurement used to calculate axonal extension: the outline of an axon is followed from a stable measuring point to the most distal tip of the growth cone. The lines on the left of each image illustrate the measurements used to calculate growth cone activity: in this case, all visible side branches and filopodia were incorporated in the measurements. The scale bar represents 10 μm.

for shorter, defined periods for comparison between different growth patterns. These "instantaneous" rates were determined over periods of at least 10 minutes, in order to avoid miscalculations due to "imaging noise" (i.e., the tip of the growth cone moving in and out of the focal plane). To determine growth cone *activity*, even if not accompanied by forward extension, the total length of all visible side branches and filopodia was measured (Fig. 10; see also http://nucleus.med.upatras.gr/hsn/ for an animated time-lapse sequence of a growth cone displaying particularly conspicuous "stationary motility"). Finally, to assess the percentage of time spent in advance, pause, or retraction, the normalized axon length from each frame was plotted as a function of time, and the duration of time spent in each mode was expressed as a function of the total time of observation (Skaliora et al., 2000).

For the fixed-tissue experiments, we wanted to measure the shape, size, and density of dendritic spines. As these images often spanned several tens of micrometers in depth and therefore treating them as two-dimensional would compromise the accuracy of the measurements, we implemented an algorithm that took into account the z dimension as well. In this, the entire stack of images was used to calculate the length of the dendritic segment by allowing measurements to span different frames within the stack. Briefly, the images were first calibrated in all x, y, and z axes. The x and y calibrations were defined by the pixel size, whereas the z calibration was defined by the separation between images in depth (usually 1 μm). Subsequently, a macro was created in NIH Image software that

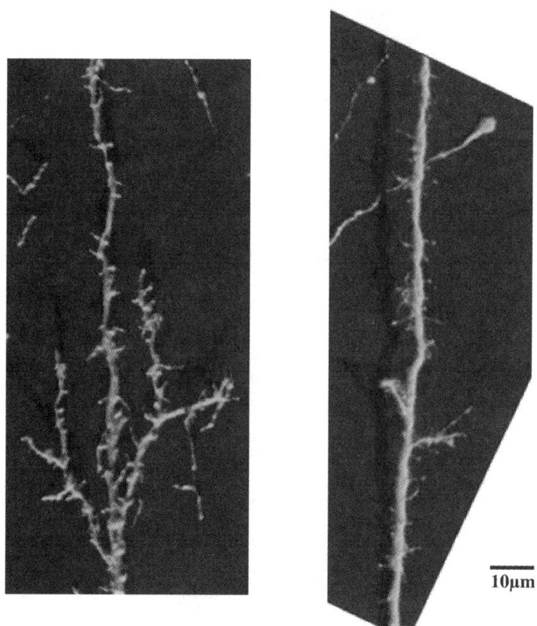

FIGURE 11 See ftp site for color figure. Three-dimensional visualization of dendritic segments is obtained by combining a series of single confocal images in depth. To create these reconstructions, the visualization method "Simulated Fluorescent Process" of the Imaris software (see Section 3.3) was used. In short, each voxel of the 3D dataset is assigned an absorption and an emission factor based on its intensity in the original dataset. This information, together with the depth information, is used to assign the transparency and the shadow-casting properties of the final image (Messerli et al., 1993).

allowed the measuring tool to be used on different frames within the stack, so that the selected points carried with them not only x and y but also z coordinates. For 3D reconstruction and visualization, confocal stacks of images were imported into Imaris software, and the depth information was used to create the shadows as illustrated in Figure 11.

6 FUTURE DIRECTIONS

There are some major system-related limitations that are often encountered in confocal imaging of living tissues: photodamage caused by the intense laser illumination, limited depth penetration of the laser beam, and slow image-acquisition speed. Current developments seem to provide solutions to some of these issues.

6.1 Multiphoton Confocal Microscopy

Multiphoton laser excitation is based on the idea of achieving excitation of a fluorophore with laser beams of longer wavelengths, and thus lower energy. This is accomplished by using a pulsed, high-power, long-wavelength (800–1000 nm) laser beam that excites the fluorophore only at the focal point, because only there the photon density would be high enough to cause the simultaneous absorption of two (or more) photons by the same molecule of fluorophore. In other words, the fluorescent molecule will absorb energy equivalent to that of one photon of shorter wavelength. Thus, fluorophores that require blue or green excitation wavelengths can be excited with lower-energy red lasers.

There are several advantages to this technique. Because the fluorophore is excited only at the focal point, photodamage of the overlaying and underlying structures is minimal and no out-of-focus light is generated. As a result, no confocal pinhole is required and the chromatic correction of the objective lens is not as critical. Furthermore, since long excitation wavelengths are used, deeper sample penetration is achieved and less photodamage is caused. All of these properties are very advantageous to the imaging of thick living specimens, and they outweigh the major disadvantage compared with single-photon confocal microscopes, which is reduced spatial resolution.

6.2 Fast Confocal Scanners

Point-scanning confocal systems provide the highest possible resolution, since they couple the ideal excitation (a point source) with confocal emission optics. However, their limitation is slow image-acquisition speed (see also Section 4.1.3). Faster-scanning confocal systems do exist, but they tend to compromise either the excitation or the emission accuracy, in return for speed.

Currently, a system build by Yokogawa, using a spinning (Nipkow) disc with multiple pinholes coupled to a second spinning disc with microlenses, appears attractive for live biological imaging (see http://www.mcbinst.com/yokogawa.htm). The microlenses focus the laser excitation light on to the pinholes on the Nipkow disc, thus improving the light throughput, which was the main limitation of the spinning-disc technique when it first appeared. We believe that such a system could prove very useful for dynamic experiments similar to those presented in this chapter. It provides fast scanning speeds (up to 320 frames per second) with near-confocal optics and a cooled CCD camera as sensor. This arrangement allows imaging with lower excitation intensities, because of the image-integration capability of the CCD camera chip. Such a technique could help reduce the phototoxicity and fluorophore saturation caused by the high laser power used in point scanning systems. A disadvantage of such a system is that it has fixed pinhole size, optimized only for one objective lens.

REFERENCES

Agmon, A. and Connors, B. W. (1991). Thalamocortical responses of mouse somatosensory (barrel) cortex in vitro. *Neuroscience* **41**, 365–379.

Bacallao, R., Kiai, K., and Jesaitis, L. (1995). Guiding principles of specime preservation for confocal fluorescence microscopy. In: *Handbook of Biological Confocal Microscopy* (ed. J. B. Pawley). New York: Plenum Press, pp. 311–323.

Bernardo, K. L., Ma, P. M., and Woolsey, T. A. (1986). In vitro labeling of axonal projections in the mammalian central nervous system. *J. Neurosci. Meth.* **16**, 89–101.

Cline, H. T., Edwards, J. E., Rajan, I., Wu, G., and Zou, D. (2000). In vivo imaging of CNS neuron development. In: *Imaging Neurons: A Laboratory Manual.* Cold Spring Harbor, NY: Cold Spring Harbor Laboratory Press, pp. 13.1–13.12.

Cody, S. and Williams, D. (1997). A novel organ bath design enables the use of saline-immersible lenses on inverted microscopes. *J. Microsc.* **185**, 94–97.

Dailey, M. (2000). Maintaining live cells and tissue slices in the imaging setup. In: *Imaging Neurons: A Laboratory Manual.* Cold Spring Harbor, NY: Cold Spring Harbor Laboratory Press, pp. 10.1–10.17.

Errington, R. J. and White, S. S. (1999). Measuring dynamic cell volume in situ by confocal microscopy. In: *Confocal Microscopy Methods and Protocols.* Totowa, NJ: Humana Press, pp. 315–340.

Florijn, R. J., Slats, J., Tanke, H. J., and Raap, A. K. (1995). Analysis of antifading reagents for fluorescence microscopy. *Cytometry* **19**, 177–182.

Harris, W. A., Holt, C. E., and Bonhoeffer, F. (1987). Retinal axons with and without their somata, growing to and arborizing in the tectum of *Xenopus* embryos: a time-lapse video study of single fibres in vivo. *Development* **101**, 123–133.

Keller, H. E. (1995). Objective lenses for confocal microscopy. In: *Handbook of Biological Confocal Microscopy* (ed. J. B. Pawley). New York: Plenum Press, pp. 111–125.

Messerli, J. M., van der Voort, H. T., Rungger-Brandle, E., and Perriard, J. C. (1993). Three-dimensional visualization of multi-channel volume data: the SFP algorithm. *Cytometry* **14**, 725–735.

Mills, L. R., Niesen, C. E., and Kerr, R. (1994). Confocal imaging of living neurons and organelles. In: *Three-Dimensional Confocal Microscopy: Volume Investigation of Biological Systems.* New York: Academic Press, p. 507.

Potter, S. (2000). Two-photon microscopy for 4D imaging of living neurons. In: *Imaging Neurons: A Laboratory Manual.* Cold Spring Harbor, NY: Cold Spring Harbor Laboratory Press, pp. 20.1–20.16.

Rigaut, J. P. and Vassy, J. (1991). High-resolution three-dimensional images from confocal scanning laser microscopy. Quantitative study and mathematical correction of the effects from bleaching and fluorescence attenuation in depth. *Anal. Quantit. Cytol. Histol.* **13**, 223–232.

Skaliora, I., Adams, R., and Blakemore, C. (2000). Morphology and growth patterns of developing thalamocortical axons. *J. Neurosci.* **20**, 3650–3662.

Stelzer, E. H. K. (1998). Contrast, resolution, pixelation, dynamic range and signal-to-noise ratio. *J. Microsc.* **189**, 15–24.

CHAPTER 19

FISH Imaging

MICHAL KOZUBEK

Laboratory of Optical Microscopy, Faculty of Informatics, Masaryk University, Botanická 68a, 602 00 Brno, Czech Republic

1 INTRODUCTION

The in situ hybridization technique enables the detection of specific DNA or RNA sequences in the cytoplasm, organelles, chromosomes, or cell nuclei. This ability is of great importance in the analysis of chromosome and genome structure and aberrations, the construction of physical maps of chromosomes, the determination of the spatial and temporal expression of genes, and the localization of transformation sequences, oncogenes, etc.

The DNA or RNA sequences are visualized by means of DNA or RNA probes. *Probes* are DNA or RNA sequences that are complementary to the target sequence. Both the probe and the target sequence must be single-stranded, and that is why the first step of the FISH technique is denaturation of both the probe and the target sequence in cells, which are usually fixed to a microscope slide. Then, the probe is let to anneal to the target sequences by dropping the probe on the slide and letting the mixture hybridize for some time in a suitable environment. The probe that did not anneal to any target is then washed away in post-hybridization washing steps. Thus, the remaining probe is annealed exactly to the target sequences.

In order to visualize the probes, they must be fluorescently labeled (the so-called *direct method*) or chemically modified (the so-called *indirect method*) before hybridization. The indirect method means that the probe is modified with molecules of biotin or digoxigenin, and then (after hybridization) labeled anti-biotin or anti-digoxigenin molecules are applied (this is the same approach as in immunocytochemistry). The most frequently used labels are fluorescent labels, although some nonfluorescent (e.g., light-absorbing) labels also exist. If fluorescent labels are used, the technique is called *fluorescence in situ hybridization (FISH)*.

The ideal length of the probe is 100–300 bp. Shorter probes result in lower nucleic acid hybrid stability and larger probes have difficulties in penetrating into cell structures.

Confocal and Two-Photon Microscopy: Foundations, Applications, and Advances, Edited by Alberto Diaspro.
ISBN 0-471-40920-0

However, by using many probes simultaneously (i.e., by visualization of many targets simultaneously), it is possible to visualize large objects such as whole chromosomes. The final targets (which we want to see in the microscope) can be *unique or low-copy sequences* such as protein-coding genes or transformation sequences, *repetitive sequences* such as satellite sequences in centromeric regions, *specific chromosomes* or *genomes, mRNA sequences*, or *viral sequences*.

It is possible to use several types of probes simultaneously. Different types have different labels that are observed at different excitation/emission wavelengths (i.e., different colors) in a fluorescence microscope. This approach is called *multicolor FISH*. It is also convenient to paint total DNA (besides the probes) with a certain dye in order to see the nuclear boundary. This stain is called a *counterstain*, and mostly propidium iodide (PI) or 4′,6-diamidine-2-phenylindole (DAPI) are used.

Two other modifications of the FISH method are *comparative genomic hybridization (CGH)* and *primer-mediated DNA synthesis (PRINS)*. CGH provides a comprehensive analysis of imbalanced chromosomal material of the entire genome. It results in a ratio profile of two fluorescent signals. The relative strengths of the signals indicate chromosomal subregions that may be amplified or deleted. In research, CGH is used to examine chromosomal aberrations in several types of tumors to identify chromosomal regions containing genes that may contribute to the disease. The procedure includes the extraction of DNA from the tissue of interest, DNA labeling with fluorochrome–dUTP, and hybridization of metaphase chromosomes with the first fluorochrome-labeled specimen DNA and the second fluorochrome-labeled normal reference DNA (Wilkens et al., 1998; Weiss et al., 1999).

The PRINS method is based on primer-mediated DNA synthesis. It starts with the annealing of an oligonucleotide DNA primer adjacent to the DNA region of interest. This annealed oligonucleotide serves as a primer for Taq DNA polymerase. The polymerase incorporates four nucleotides – dATP, dGTP, dCTP, and dUTP – of which the dUTP is labeled with a fluorochrome. The applications of the PRINS methodology are focused on the detection and localization of different genes in the cell nucleus (Koch et al., 1989; Koch, 1992).

Different probes are suitable for different purposes. Probes for repetitive sequences produce the brightest signal, and are therefore suitable for automatic detection and scoring. Probes for specific chromosomes are important in structural aberration scoring. Unique sequence probes are invaluable for the detection of specific translocations occurring in different malignant diseases.

Probes are mostly produced by *cloning*: the DNA is inserted into a *vector*, and both the vector and the insert are amplified inside appropriate host cells. The amplified DNA is then extracted. The detection of genetic loci by means of FISH requires techniques for the efficient cloning of large DNA fragments. In the beginning, the most frequently used vectors were cosmids (derived from λ phage) and yeast artificial chromosomes (YACs). Both these clones have serious disadvantages. Cosmids accept comparatively small inserts (35–45 kb) and the clones are relatively unstable (Collins and Hohn, 1978; Craig et al., 1990). YAC systems enable the cloning of inserts up to 1 Mb (Burke et al., 1987; Burke and Olson, 1991); however, the inserts are occasionally unstable and manipulation with the long DNA is difficult. Efforts to overcome the limitations of cosmids and YACs have resulted in alternative large-insert cloning approach using a bacteriophage P and the *Escherichia coli* fertility plasmid (F-factor). The bacteriophage P1 cloning system

(PAC) can accept inserts in the 70–100 kb range (Pierce et al., 1992; Ioannou et al., 1994). The F-factor-based BAC vectors can accept inserts up to 300 kb (Hosoda et al., 1990).

DNA probes can be also prepared and amplified using the long-template polymerase chain reaction (PCR) system. The method uses a mixture of DNA Taq polymerase and a thermostable DNA polymerase with proofreading activity to overcome limitations in the length of amplified fragments (Cheng et al., 1994). The system can amplify fragments of 20 kb and more from isolated genomic DNA using specific primers consisting of 20–30 nucleotides suitable for the region of interest. The procedure includes denaturation of template, annealing of specific primers, and primer extension by DNA polymerases using dNTP, where dUTP is labeled with a fluorochrome. The yield of labeled DNA depends on the number of PCR cycles.

Synthetic oligonucleotides can also be used as probes. They can be tailor-made to hybridize to specific sequences, but their synthesis is quite complicated and only 10–50 bp long oligonucleotides are feasible to produce. Because of the short length, they are easily washed away during the post-hybridization washing steps, and must therefore be handled carefully.

The FISH technique is the most frequently used technique in the structural cytology of the cell nucleus. It provides a reliable means for studying the genetic composition of cells in mitosis as well as in interphase (Cremer et al., 1988). Presently, FISH has reached a high detection sensitivity (individual genes can be detected) and a high multiplicity (several different probes can be applied to the same nucleus) (Tanke et al., 1995). It is used both in research and in clinical applications for the identification and quantification of numerical and structural chromosomal aberrations. Interphase cytogenetics is of great value for the detection of disease-related translocations such as the t(9,22) translocation observed in chronic myeloid leukemia (Tkatchuk et al., 1990), in prenatal diagnosis (Pinkel et al., 1988), or for studying various solid tumors (Hopman et al., 1991). In research, it can be used to study the spatial organization of the genetic material inside cell nuclei, the relative positions of different genetic loci, organization of chromosomes, etc. (van den Engh et al., 1992; Cremer et al., 1993; Eils et al., 1995; Ferreira et al., 1997; Kozubek S. et al., 1997; Croft et al., 1999; Munkel et al., 1999; Sadoni et al., 1999; Skalnikova et al., 2000). The influence of different carcinogenic or differentiation agents on nuclear topography (Bartova et al., 2000a,b; Jirsova et al., 2001) and the correlation between chromatin structure and malignant diseases (Kozubek S. et al., 1999a; Nikiforova et al., 2000) can also be studied. Not only isolated cells, but also cells in tissues can be hybridized (Koutna et al., 2000), as well as living cells (Zink and Cremer 1998). In order to increase the number of probes, the same slide can be re-hybridized several times with different sets of probes each time, and the same objects can be re-acquired and superimposed in computer memory (Lukasova et al., 1999; Cafourkova et al., 2001).

The latest trend is to perform many parallel hybridizations in a grid on one slide (Pinkel et al., 1998; Bubendorf et al., 1999). The slide with the grid is called a *microarray* or "*DNA chip*." This technology enables the analysis of human single-nucleotide polymorphisms and the detection of disease-causing mutations in large population studies.

Because this book is not aimed at a description of the methods of molecular biology, the FISH protocols will not be presented here. Instead, this chapter will concentrate on the imaging aspects of FISH studies: the available optical systems, and image-acquisition and image-analysis techniques.

2 OPTICAL SYSTEMS FOR FISH VISUALIZATION

Optical systems for FISH visualization can be divided into two categories: flow systems and slide-based systems. In *flow systems*, FISH-stained objects (cells and/or their compounds) are prepared in suspension and the suspension flows in a narrow stream across a laser beam. The objects flow individually (i.e., one after another), and a detector records their characteristics (fluorescence intensities) while they pass the laser beam. In *slide-based systems*, FISH-stained objects are fixed to a microscope slide and observed as a static image. Both laser and non-laser light sources can be used in this case. Instruments of both these types will be described and compared. Special attention will be paid to their calibration and performance optimization.

2.1 Flow Cytometry

Typical representatives of the flow systems are flow cytometers of different types. Flow cytometry (FCM) is so far the fastest approach to FISH-stained cell studies. It provides less information than the slide-based systems do, but it is sufficient for certain tasks. The power of flow cytometry lies in the ability to measure several parameters on tens of thousands of individual cells within a few minutes. Moreover, flow cytometers can be equipped with a sorting unit that can divide the cells online into different groups according to the measured parameters.

Two FCM instruments will be described in the following text: a conventional flow cytometer including a sorting unit and a slit-scan flow cytometer that enables measurements of individual chromosome profiles and online chromosome sorting according to the centromeric index or other attributes.

2.1.1 Conventional Flow Cytometry Flow cytometry was first used to study isolated cells, and in 1975 it was adapted also to the observation of isolated chromosomes (Gray et al., 1975). By the end of the 20th century, the applications of flow cytometry had spread through all branches of biological science.

In order to observe cells or chromosomes using flow cytometry, they must be prepared using special protocols and stained using FISH or an immunofluorescence technique. As the stained objects pass the detection point, their fluorescence and light scattered from them are measured.

A basic flow cytometer consists of the following parts:

- a *light source* (usually a laser) used to illuminate the flowing cells or chromosomes;
- a *flow chamber* (*or flow cell*) (the heart of the instrument), where the objects flow individually after each other and are illuminated by a focused beam of light;
- *optical components* used to focus the light beam to the flowing objects, collect the light coming from the objects, and split different color components to the appropriate detectors;
- *detectors* (photomultipliers) used to measure the light of different colors;
- a *computer interface* that passes the information from the detectors to the computer;
- a *computer* equipped with appropriate software to process the data;
- a *sorting unit* (optional) that divides the flowing objects into different reservoirs.

The instrument works as follows: the flow chamber hydrodynamically focuses the stream of cells or chromosomes, delivering them singly to the point of detection with an accuracy of ±1 μm or better. The light from the light source is focused to the flowing objects (the diameter of the beam is typically 50 μm), and fluorescence and scattered light are measured. The scattered light is measured either in the illumination-path direction (*forward scattering*) or in the orthogonal direction (*side scattering*). Individual color components of fluorescence are separated using appropriate beam splitters and barrier filters. The photomultipliers measure the intensity of the light components, and the analog signal is digitized and passed to the computer. The histograms of interest can be displayed on computer monitor and refreshed online.

Conventional flow cytometers thus enable the measurement of integrated values over whole objects (cells or chromosomes), such as total fluorescence of a specific dye. This is used mainly to measure the cell DNA content, determining the cell-cycle stage distribution or the chromosomal DNA content (or A–T versus G–C content) creating karyotypes that are different for normal cells and cells with various chromosomal anomalies. Flow cytometry is thus very suitable for clinical cytogenetics, and is used in hospitals all over the world.

If required, flowing cells or chromosomes can be sorted into different reservoirs according to one of the measured parameters. For example, cells at different stages of the cell cycle can be divided into separate reservoirs. The most common method of cell sorting is by electrostatic deflection of charged droplets that are produced by the vibration of the flow chamber. Sorting is a very useful feature of the FCM instrument, because it enables a choice of objects of interest for further studies (e.g., slide-based microscopy).

Further material about flow cytometry can be found, for example, in Ormerod (1994).

2.1.2 Slit-Scan Flow Cytometry

In conventional flow cytometry, the laser beam generally illuminates the whole chromosome or the whole cell. Thus, only the total amount of bound dye can be measured. In slit-scan flow cytometry, fine focusing of the laser beam is used (about 1 μm instead of 50 μm), and only a narrow slice of the cell or chromosome is illuminated. This allows measurement of the fluorescence profile of each cell or chromosome as it passes through the illuminated area and determination of, for example, the centromeric index of each chromosome (Lucas and Gray, 1987).

As a result of progress in computing technology, each chromosome profile can be determined in real time at a speed sufficient to allow online sorting based on centromeric index or other parameters. As each chromosome has its specific profile, online karyotyping is feasible. Furthermore, chromosomes with a translocation or other special cases of interest can be screened online (Hausmann et al., 1992).

2.2 Conventional Wide Field Microscopy

The history of optical microscopy dates back to the end of the 16th century. The first optical microscope was built in about 1590 in Holland. In the 17th century, the Dutch biologist Anthony Leeuwenhoek improved the optical microscope and observed biological samples for the first time. The most important improvement was made, however, only at the end of the 19th century by the German physicist Ernst Karl Abbé, who constructed the type of optical microscope that is still used nowadays. This type of microscope is called *conventional* or *wide-field*. Abbé also created a proper mathematical theory of light propagation in the optical microscope. Nowadays, conventional microscopy provides a whole range of high-quality tools for industry as well as for research.

The drawback of the conventional approach is the difficulty with three-dimensional (3D) observations, and therefore usually only 2D studies are performed. Reconstruction of 3D information from 2D slices acquired at different axial positions is possible using deconvolution techniques, but either the result is of poor quality or the time consumption of the deconvolution algorithm is unacceptable. However, in special cases, such as 3D hybridization dot studies in FISH imaging, it is possible to determine spatial positions quickly and with a high precision (Kozubek M. et al., 1999).

2.3 Scanning Microscopy

In the middle of the 20th century, the idea arose of scanning the specimen point by point, one line after another, instead of viewing the whole image at the same time (Young et al., 1951). Thus, the scanning light microscope (SLM) or scanning optical microscope (SOM) came into being. A mathematical theory of image formation in such a microscope was developed. The scanning method of image acquisition and display is widely used also, for example, in TV cameras, TV screens, and computer monitors, where the image is refreshed in exactly the same way.

Scanning microscopes were, however, not used until the invention of the laser and its employment as a light source in microscopy. A laser light source is necessary in SLM because it is not possible to concentrate light from an ordinary source to a sufficient intensity at one point. Moreover, the main advantage of the scanning method is that it is possible to achieve a confocal effect by placing a pinhole aperture in front of the detector. Unfortunately, this further reduces the light intensity. Therefore, nonlaser light sources are not used nowadays in SLM.

Because the image is formed point by point and only one intensity value (i.e. 0D information) has to be digitized at one time, photomultiplier tubes (PMTs) are used to detect the 0D light-intensity value. The image is formed progressively in computer memory, and cannot be seen instantly in an eyepiece as in the case of conventional microscopy.

2.4 Confocal Tandem Scanning Microscopy

In 1966, a so-called tandem-scanning microscope (TSM) or direct-view microscope was constructed (Petran et al., 1968). The idea was to scan not just one point but thousands or tens of thousands of points at the same time. Because the scanning is done in parallel, the image is formed faster. The scanning in TSM is performed using a rotating disc with pinholes arranged in such a way that the specimen is scanned completely during one rotation of the disc. Moreover, if the pinholes are far from each other, a confocal image is formed. This can be realized, for example, using a Nipkow disc (Nipkow, 1884).

The advantage of TSMs is the production of a confocal effect at low price (no laser source is needed). The disadvantage is a very low light efficiency (usually less than 1%). Thus, only bright specimens can be observed, and most TSM microscopes are not suitable for fluorescence observations.

TSMs have not been widely used, because confocal laser scanning microscopes have usually provided a better alternative. However, they have been used for industrial inspection, and are still commercially available. Recently, Nikon has even developed a brand new type of TSM (Optiphot 200C) for industrial inspection.

The U.S. company Atto Instruments Inc. has begun to produce a TSM that is also suitable for biological observations in fluorescence. The system is called CARV, and is available as a module for Zeiss microscopes (see http://www.atto.com). The light efficiency of the CARV system is slightly better (about 5%) than that of other TSMs. The confocality (i.e., optical-sectioning property) is comparable to that of confocal laser scanning microscopy (see Fig. 1). Nevertheless, acquisition of images in fluorescence (e.g., FISH images) requires long exposure times.

However, if the idea of using TSMs equipped with special rotating discs and state-of-the-art CCD cameras turns out to be applicable to biological research and FISH imaging, it is possible that TSMs will even come into fashion 30–40 years after their invention (just

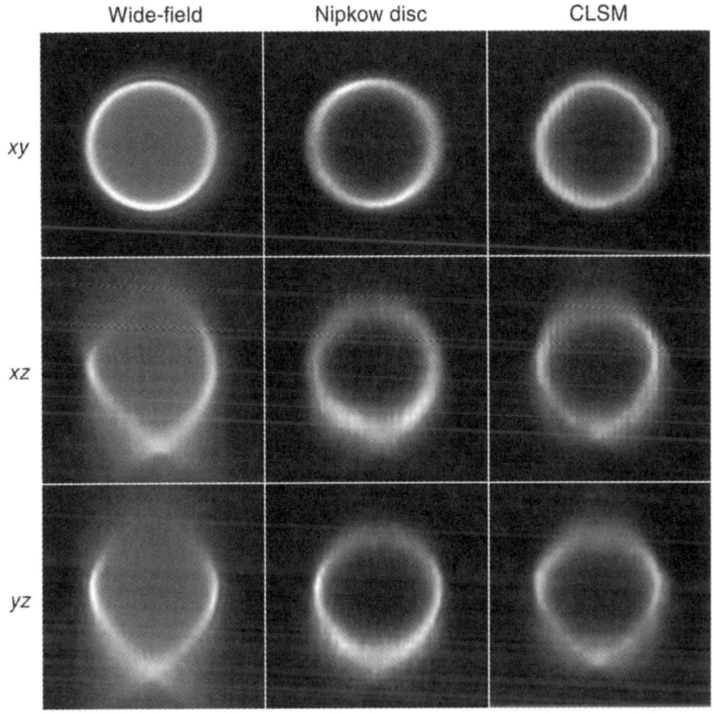

FIGURE 1 Comparison of the optical-sectioning capabilities of three different microscopy techniques: wide-field microscopy (left column), confocal microscopy based on a Nipkow disc (center column), and confocal laser scanning microscopy (right column). The instruments used in this study were the Axiovert 100S (Zeiss), the Axiovert 100S with CARV module (Zeiss), and the Sarastro 2000 CLSM (Molecular Dynamics), respectively. The confocality of each system is demonstrated using a polystyrene bead with fluorescence-labeled shell. The diameter of the beads is 6.0 μm, the peak emission wavelength is 515 nm. The beads were dried onto a slide surface, poured with embedding medium (VectaShield), and covered with a coverglass. For each microscopy type, three orthogonal central planes are shown: *xy* (top row), *xz* (center row) and *yz* (bottom row). The out-of-focus blur in wide-field microscopy is clearly visible. The bottom of the bead is always visible better, owing to reflections from the embedding-medium–slide interface, where the refractive index changes. Reprinted from Kozubek M. et al. (2001).

like SLMs came into fashion equipped with laser light sources 30–40 years after their invention).

2.5 Confocal Laser Scanning Microscopy

The confocal effect can be easily obtained using a scanning optical microscope with a small pinhole in front of the detector (see Chapter 2 of this book (Wilson, 2002)). The smaller the pinhole, the stronger is the confocal effect. However, small pinholes imply very low light efficiency. Therefore, confocal scanning light microscope (CSLM) needs a lot of light and bright specimens. Only after laser light sources became available, did confocal microscopes begin to be of use to biologists observing faint fluorescent objects. Nowadays, only laser light sources are used, and the technique is therefore called confocal laser scanning microscopy (CLSM).

The main advantage of CLSM is the possibility of obtaining confocal images even for faint fluorescent objects and, consequently, 3D images of the objects. The disadvantage of CLSM is the high price of the instruments and a relatively long acquisition time (as compared with wide-field microscopy). Also, laser illumination can be harmful to biological specimens observed in vivo. This drawback can be improved using much less harmful *two-photon excitation* (see Chapter 3 (Diaspro and Sheppard, 2002)).

2.6 Laser Scanning Cytometry

A recently constructed laser scanning cytometer (Kamentsky et al., 1997a,b) provides an interesting compromise between flow cytometry and conventional wide-field microscopy. *Laser scanning cytometry (LSCM)* is a tool for the fast evaluation of FISH-stained interphase nuclei fixed on a microscope slide and for hybridization dot counting. It is similar to confocal scanning light microscopy because it also uses a laser light source, point-by-point scanning, and two to four photomultipliers (plus appropriate beam splitters and filters) to detect signals from different fluorescent dyes. However, the laser-beam spot is somewhat larger (2.5–3.0 μm in diameter for a 40× objective), and the system uses nonconfocal imaging in order to increase the depth of field. This is necessary, because only 2D images are recorded. Furthermore, laser-beam scanning is performed in only one lateral direction, with the scanning in the other lateral direction being performed using stage movement. The stage moves from one end of the slide to the other, and consequently a long scan strip is acquired. Then, a neighboring strip is acquired, and then the next one, and so forth until the whole slide is scanned. Typically, a 40× objective, 2.5 μm beam-spot size, and 170 μm strip width are used. Appropriate software performs online image analysis and plots histograms or scattergrams similar to flow cytometry. Up to 1000 nuclei per minute can be processed using this technique.

The LSCM approach is fast, but deals only with low-resolution images (because of the large beam-spot size). Low resolution is acceptable for nuclei segmentation (nuclei are large and we are not interested in fine details), but problems arise when hybridization dots have to be detected. Images of two dots that are close to each other merge, and the dots are detected as one dot. Thus, in LSCM, only large targets are detectable, such as pericentromeric regions, while unique sequences are not detectable.

Nevertheless, LSCM is a good compromise when large quantities of cells are to be analyzed at high speed and information provided by flow cytometry is not sufficient. An example of an application is numerical aberration scoring, where pericentromeric probes can be used and only the number of chromosomes within each nucleus is to be computed. LSCM is now a commercially available technique, and can be a good supplement to flow cytometers in clinical cytogenetics.

2.7 High-Resolution Cytometry

High-resolution cytometry (HRCM) was introduced in 1999 (Kozubek M. et al., 1999), and is a technique that enables one to perform automatic acquisition and analysis of a large number of cells with high resolution using a computer-driven microscope. In HRCM, a special set of automated microscope components is used, with special driving software. All components and all software modules are optimized for speed and quality. High resolution is obtained by means of high-numerical-aperture (NA) Plan-Apochromat objectives typically combined with a cooled high-resolution digital CCD camera.

So far, software for the automated acquisition and analysis of FISH-stained genetic loci inside interphase nuclei has been developed. This software works in conventional wide-field mode (Kozubek M. et al., 1999) as well as in confocal mode (Kozubek M. et al., 2001), and the two modes can be combined. The speed of the acquisition and analysis depends on the mode used, and ranges from 50–100 nuclei per minute in 2D (wide-field mode) to 0.5–2 nuclei per minute in 3D (confocal mode). The system can automatically find cell nuclei in 2D or 3D images and FISH-stained genetic loci (genes, centromeres, and chromosomes) inside cell nuclei. Not only dot counting but also complex topological analysis can be performed, including the analysis of gene position inside the corresponding chromosome domain. The same nuclei can be re-hybridized with a different set of probes, re-allocated, and re-acquired. The analysis results of different hybridizations can be combined in computer memory.

2.8 Optimization of Optical System Performance

The performance of each optical system depends on many factors, such as the type of microscopy, the NA of the objective, the type of objective correction, the refractive index of the embedding medium (refractive-index mismatch), the coverglass thickness, the depth of the observed objects below the coverglass, the emission wavelength, the detector pinhole size in confocal microscopy, and the ratio between emission and excitation wavelengths in fluorescence microscopy. These factors influence the basic features of the optical system, such as the point spread function (PSF), the resolution, optical sectioning (confocality), the light throughput, and the degree of monochromatic and chromatic aberrations.

In FISH imaging, all of the above-mentioned features are of great importance. We need good resolution (and consequently a "narrow" PSF) to be able to resolve hybridization dots (e.g., FISH-stained genes) located close to each other. We need good optical sectioning when we observe FISH-stained tissue samples in 3D. We need a lot of light, because FISH signals are often very faint. And we also need to eliminate chromatic aberrations, because we want to overlay individual channels as precisely as possible.

First, let us summarize how we can optimize the optical system so that the above-mentioned features become better. Then (in the next section), we shall discuss the possibilities of compensation for imperfect optical system performance.

2.8.1 Resolution The word "resolution" is used with many different meanings. We can speak of camera resolution, image resolution, microscope-stage resolution, and optical resolution. *Camera resolution* is determined by the number of pixels of the CCD chip. It determines the lateral dimensions of the acquired image, and it can be maximized only at the time of purchase of the camera. *Image resolution* is the number of micrometers (or nanometers) per pixel. It is the inverse of the sampling frequency, which determines how fine are the smallest details that we can record; see Chapter 11 (Castleman, 2002). In a CCD-based system, image resolution can be improved either by increasing the magnification or by buying a CCD chip with smaller pixel size. In a PMT-based system, image resolution can be usually set in the software. *Microscope-stage resolution* is the smallest distance that the stage can travel in the x, y, or z direction. The axial stage resolution is especially important because of the small steps between neighboring planes in 3D imaging.

The most important feature of an optical system is the *optical resolution*. It is mostly defined as the shortest distance between two theoretical or real objects such that the two objects are distinguishable in the image formed by the optical system. Thus, optical resolution determines the shortest distance between objects that can be measured using a particular optical system. It is important to distinguish between theoretical resolution defined for point light sources and experimental resolution obtained when real objects are imaged. The latter resolution is often substantially worse, because real objects are far from being points. The difference is demonstrated in Figure 2. More details can be found in Kozubek M. (2001). This work shows that optical resolution for real objects can be improved by using the correct coverglass thickness (No. 1.5 coverglasses if possible), by using a suitable embedding medium (with refractive index close to that of the immersion oil), and, most importantly, by minimizing the depth of the observed objects below the coverglass (in order to minimize spherical aberration).

It should be noted that the previous paragraph describes resolution for two objects of the same color. If we speak about resolution of two objects of different spectral signature (stained with different fluorescent dyes), then resolution is infinitely small – we can resolve the two objects even if they coincide (zero distance).

2.8.2 Optical Sectioning (Confocality) An example of the optical sectioning properties of three different microscope types has already been shown in Figure 1. In general, confocality is primarily determined by the pinhole size in front of the detector (see Chapter 6: Török and Sheppard, 2002) in CLSM and by the pinhole size in the Nipkow disc in TSM. The smaller the pinhole, the better the confocality of the optical system. However, small pinholes decrease light throughput. Therefore, one must always find a compromise between confocality and light throughput.

2.8.3 Light Throughput There are several ways in which to increase the light throughput: one can buy a brighter lamp or laser (unfortunately, there is an upper limit), reduce confocality (see above), or increase the NA of the objective. If the signal is still too faint, another fluorochrome should be considered (a table of fluorochromes can be found, for example, at http://fluorescence.biorad.com).

2.8.4 Monochromatic and Chromatic Aberrations Modern objectives are quite well corrected for different types of monochromatic as well as chromatic aberrations. In general, the more expensive the objective, the better the corrections. Nevertheless, even the

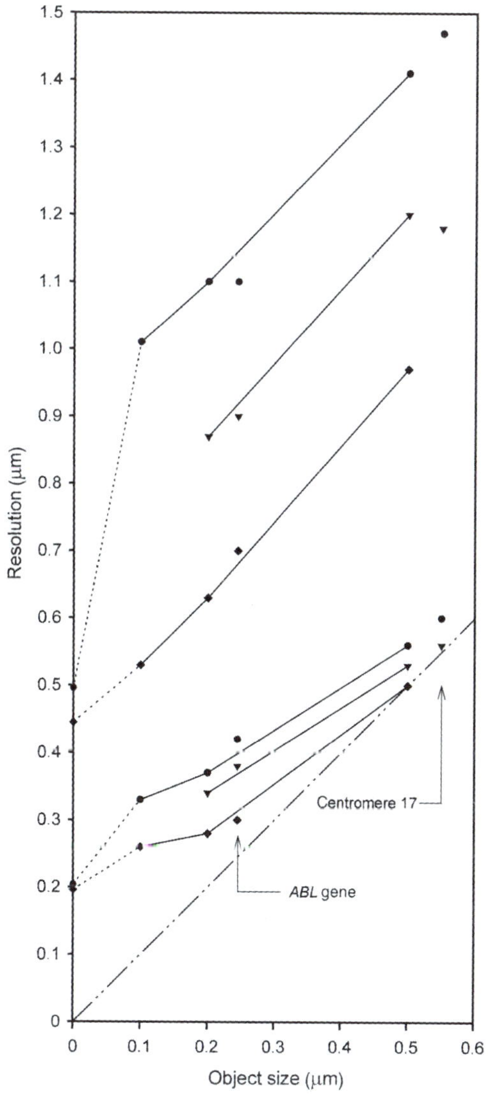

FIGURE 2 Resolution as a function of object size for a Plan-Apochromat 63×/1.4 oil objective. Resolution was computed as the full-width at half-maximum (FWHM) of the intensity profile. Three types of microscopy are compared: conventional wide-field (circles), confocal using a 5% Nipkow disk (triangles), and confocal laser scanning (CLSM) (diamonds). The detector aperture diameter of the CLSM was 66 μm. The upper halves of the curves represent axial resolution, and the lower halves represent lateral resolution. The dash-dotted line shows the theoretical limit. FWHMs for theoretical PSFs (generated by Huygens system, SVI, Netherlands) are shown as object size 0 μm. FWHMs for fluorescent beads were measured for 0.1 μm, 0.2 μm, and 0.5 μm bead diameter (nominal bead size). FWHMs for the *ABL* (FITC) and *BCR* (rhodamine) genes are shown as having approximately 0.25 μm size, and FWHMs for centromeres 17 (FITC) and 6 (rhodamine) are shown as having approximately 0.55 μm and 0.57 μm sizes, respectively. Straight lines connect only values for fluorescent beads and theoretical PSFs. The genes and centromeres are not connected with lines, because their size was not known – it was assessed based on their resolution and fluorescent-bead measurements. Note that the theoretical PSFs are highly underestimated for wide-field microscopy (real PSFs should be obviously worse than the computed ones) – this is most probably due to the fact that Huygens system generates PSFs valid near the coverslip – medium interface and does not take into account spherical aberrations and other effects in deeper layers. Reprinted from Kozubek M. (2001).

best objectives available on the market exhibit some residual aberrations. It is important to realize that most aberrations are specimen-dependent. For example, they depend on the coverglass thickness, the refractive index of the embedding medium (refractive-index mismatch), and the depth of the objects below the coverglass. Therefore, in order to minimize aberrations, the same rules apply as to resolution improvement (see above). The most important rule is to keep the objects as close to the coverglass as possible. In FISH imaging, this means applying only a small amount of the embedding medium and pushing all excess fluid between slide and coverglass away. On the other hand, care must be taken in order not to distort the cells while pushing the coverglass towards the slide.

2.9 Calibration of the Optical System

Certain types of artifacts caused by imperfect optics can be compensated using special calibration techniques. Mostly the calibration is performed using fluorescent beads. Some types of artifacts and their correction will now be presented.

2.9.1 Intensity Variation In some FISH applications, we need to monitor the incident light intensity and make sure that several images are recorded under the same conditions (i.e., each image is illuminated with approximately the same number of photons). The conditions are never the same, because the light-source intensity varies with time (as well as with the amount of dust in optical parts). When we perform repeated scanning of images of the same object (e.g., time series) or sequential exposures at different axial positions (3D acquisition), variations in intensity level can complicate further image processing.

A good solution to this problem is the use of fluorescent beads manufactured so that the emitted intensities are always equal (such as InSpeck beads from Molecular Probes). Using such beads, the actual intensity level can be measured and the exposure time can be calibrated accordingly. This approach will, however, not eliminate fading fluorescence of some FISH signals. The fading should be reduced by means of one of the available antifade reagents or by decreasing the light intensity.

2.9.2 Crosstalk Sometimes it is not possible to select filters for FISH imaging so that their spectra do not overlap. And even if they do not overlap, the spectra of individual fluorescent dyes usually do. The problem is that the excitation and emission spectra of most dyes are very broad, and the dyes are slightly excited also when other dyes are imaged. And because their emission spectra are also usually broad, the dyes are also recorded in other channels (where they do not belong). Moreover, the emission spectrum of the currently observed dye may overlap with the excitation spectrum of another dye. The problem of observing and recording an "unwanted" dye together with the "wanted" dye is called *crosstalk* between these two dyes.

In order to compensate for the crosstalk, it is necessary to estimate the fraction of the "unwanted" dye in the image. This can be done by recording some reference images and calculating the appropriate coefficient (fraction). The "unwanted" signal can be then subtracted. This technique is often called *color compensation*. However, it is always better to eliminate the crosstalk by choosing other filters or dyes, if possible, in order to obtain better signal quality.

2.9.3 Color Shifts In older microscopes, there used to be a problem with a slight shift of the field of view when switching to a different color channel. This problem was typically caused by an imperfect alignment of filter cubes in filter cube revolver. After acquisition, a *color alignment* used to be performed; that is, a lateral shift of individual channels was performed (according to some reference objects visible in all channels) so that all fields of view matched. In modern microscopes, this problem is usually not present any more.

2.9.4 Chromatic Shifts Besides an unstable field of view, there may be another reason for the mismatch between positions of the same object acquired at different colors. According to a basic principle of optics, the focal length f of a lens varies with wavelength λ. A straightforward consequence of this phenomenon is an axial offset between focal planes of different wavelengths. Hence, the same feature in the observed specimen changes its axial in-focus position when the wavelength is changed. This mismatch is called *axial (or longitudinal) chromatic aberration*. According to another principle of optics, the magnification M is proportional to $1/f$. If we combine the two principles together, we find that M varies with λ. Hence, the same feature in the observed specimen also changes its lateral off-axis position when imaged at different wavelengths. This magnification error is called *lateral chromatic aberration*. Chromatic aberrations are also called *chromatic shifts*. They are caused by imperfect chromatic correction of the optical system (usually by imperfect correction of the objective). Although manufacturers try to compensate for chromatic aberrations in their optical components (mostly objectives), even the best objectives (and microscopes) exhibit chromatic aberrations to some extent.

Chromatic aberrations are not negligible in some types of FISH studies. A typical example of such a study is a short-distance measurement between two targets visualized using two different fluorescent stains (such as a pair of heterologous genes involved in a translocation). Such measurements are often performed in the range 0–0.5 μm (nowadays mostly in 3D) and a precision of better than 0.1 μm (in both lateral and axial direction) is highly desirable. Obviously, such a precision cannot be obtained using a fluorescence microscope with chromatic aberrations of typically 0.1–0.3 μm in the lateral and 0.2–0.5 μm in the axial direction.

A method for measuring and software correction of chromatic aberrations has recently been described by Kozubek M. and Matula (2000). The method uses fluorescent beads stained with several fluorescent dyes simultaneously (such as TetraSpeck from Molecular Probes) for the determination of the aberration function (which is dependent on the lateral position in the field of view) and subsequent software correction. An example of chromatic-aberration correction is shown in Figure 3.

3 ACQUISITION OF FISH IMAGES

3.1 Detectors for FISH Image Capture

There exist 0D, 1D, and 2D light detectors. 3D detectors are unfortunately not available. The most common 0D detectors are photomultipliers (PMTs), the most common 1D detectors are line-scan CCD cameras, and the most common 2D detectors are CCD cameras. In FISH imaging, either PMTs or CCD cameras are used. The line-scan cameras are very rare, and are used in special industrial applications such as material inspection.

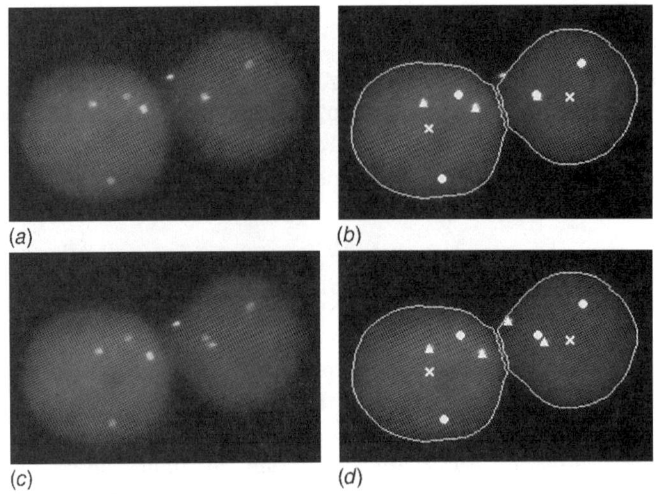

FIGURE 3 See ftp site for color figure. Example of chromatic-aberration correction. Study of the topology of the genes *ABL* (green) and *BCR* (red) in human cell nuclei (blue). The genes are visualized using the FISH technique. Image data (*a*, *c*) and analysis results overlaid over the image data (*b*, *d*) are shown. Uncorrected images (*a*, *b*) are compared with the images corrected for chromatic aberration (*c*, *d*). Only part of the whole image is shown. The nuclei are about 10 μm in diameter. The analysis results are visualized in yellow: the *ABL* genes are shown as triangles, the *BCR* genes as circles, and the nuclear boundaries as closed curves. Note that one of the *ABL* genes is out of the nucleus before correction, whereas after correction it gets into the correct place. Consequently, this gene was rejected from the analysis without correction, whereas after correction it was safely detected. Note also that the *ABL* and *BCR* genes coincide in the right nucleus before correction, but are well separated after correction. This means that, without correction, the right nucleus is wrongly classified as Philadelphia (Ph)-positive (carrying the t(9;22) translocation), whereas after correction it is correctly classified as Ph-negative (no t(9;22) translocation present). Reprinted from Kozubek and Matula (2000).

PMTs are usually used with laser light sources (in CLSM or flow cytometry), CCD cameras are used with mercury or other lamps (in wide-field microscopy or TSM).

While PMTs cannot be chosen (they are supplied by the manufacturer of the CLSM or flow-cytometry system), CCD cameras are usually purchased separately from the microscope, and the user has to choose from a large number of models produced by a variety of manufacturers. Here are some hints on how to select a suitable camera for FISH imaging.

1. It is best to choose a *monochrome (black-and-white) camera*, because of sensitivity (FISH-stained objects are often faint, so high sensitivity is advantageous). Sequential scanning of individual fluorescent dyes using a monochrome camera is slower than parallel acquisition of all dyes using a color camera, but the quality of the result is much better.
2. Further, the camera should be capable of integration of the image for up to several tens of seconds in order to be able to increase the exposure time if necessary. Cameras with integration possibility are called *slow-scan CCDs*. They have slower readout time than normal CCDs (typically 3 frames s^{-1} instead of 30 frames s^{-1}) but are optimized for low-light imaging and produce much better results.

3. It is also advantageous if the camera is cooled. For exposure times of up to several seconds, *air cooling by Peltier element* is sufficient. Water cooling or even liquid nitrogen cooling are not necessary.
4. If high-resolution studies are performed (such as precise gene position determination), then the CCD should have as *small a pixel size* as possible (in order to increase the sampling frequency). The pixel size varies from approximately 7 to 24 μm. The best CCDs have a pixel size of 6.7 or 6.8 μm.
5. The *dynamic range* of the camera should be 12-bit. Lower values (8-bit or 10-bit) are usually insufficient for FISH imaging. Higher values do not improve the image quality any more, because of low light levels (for low light levels, photon shot noise does not allow accurate determination of fluorescence level).
6. The *camera resolution* (dimensions of the CCD chip in pixels) is application-dependent. Usually, 1300 × 1000 pixels are adequate. For some FISH applications, larger chips have to be purchased (e.g., to capture a large part of a tissue with high resolution).
7. If precise fluorescence-intensity measurements have to be performed, it is also advantageous to buy a CCD chip of a *better grade*. Grade indicates how many defective pixels the chip contains. The lower the number of defective pixels, the more expensive the chip. Chips with no defects (usually marked as grade 0) are extremely expensive. A good compromise is to buy chips with no cluster defects (usually marked as grade 1), which means that the chip contains defective pixels but they do not occur in clusters.
8. Last, but not least, the *programmability* of the camera is important if one is to write one's own software. Especially useful is the *binning* feature that allows grouping of pixels into superpixels. This is extremely important for software autofocusing.

3.2 Sources of Noise in FISH Images

Noise is a phenomenon that is very unpleasant, but unfortunately occurs in each imaging system. Everybody who acquires images has to continually struggle with noise and try to minimize it. There are different types of noise. The three main types that influence imaging in optical microscopy are readout noise, dark-charge noise, and photon shot noise.

3.2.1 Readout Noise This is the noise caused by uncertainty in quantifying the electronic signal. It depends mainly on preamplifier design and postamplifier electronics, detector temperature (the lower the temperature, the lower the noise), and readout rate (the faster the readout rate, the higher the noise). The readout noise is specified by the producer.

3.2.2 Dark-Charge Noise This is the noise caused by statistical fluctuations of dark-charge signal, that is, signal generated internally in the detector even if no light is falling on it (e.g., in a dark room). Dark-charge signal is proportional to the exposure time. The producer specifies dark charge generated per unit time. If this constant is multiplied by the exposure (acquisition) time, the total dark-charge signal is obtained. Because dark charge obeys Poisson statistics, the dark-charge noise is equal to the square root of the dark-charge signal. Dark-charge noise is strongly dependent on the detector temperature: the lower the temperature, the lower the noise. Thus, dark-charge noise can be considerably reduced by cooling.

3.2.3 Photon Shot Noise This is the noise caused by statistical fluctuations of the acquired signal level. The signal level is a product of the photon flux (light intensity), the quantum efficiency of the detector (the average number of electrons generated by one photon), and the exposure time. Because the signal also obeys Poisson statistics, the photon shot noise is equal to the square root of the signal level. Thus, increasing the signal level can decrease the percentage of the photon shot noise. The signal level can be increased by increasing the fluorescence light intensity (this is not always easy in FISH imaging), or by increasing the quantum efficiency of the detector (this is possible only within a certain range at the time of purchase), or by increasing the exposure time (longer integration is usually possible, but is limited by fading of the fluorescent dyes).

3.2.4 Total Noise This is computed as the square root of the sum of squares of individual noise types. The total noise together with the acquired signal level determines the image quality, which is measured as the ratio of the signal level and the total noise. This is called the *signal-to-noise ratio* (SNR). In FISH imaging, we deal with low light levels, and therefore the dominant noise is photon shot noise. The image quality (SNR) in FISH imaging can be consequently improved only by choosing brighter fluorescent dyes or by increasing the exposure (integration) time for each image.

3.3 Choice of Acquisition Parameters

For each acquisition software, the user has to select many acquisition parameters that determine how the image should be captured. Let us have a look at the most important parameters that must be set before FISH image acquisition.

3.3.1 Exposure (Integration) Time As discussed above, the longer the exposure time, the better the image quality (SNR). However, the exposure time is limited by fading of fluorescent dyes. So a reasonable compromise must be found for each particular application. In older CLSMs and older CCD-based systems, the exposure time cannot be varied. Therefore, another approach is used – namely *multiple exposures*. This means that several images of the same field of view are acquired (with the same acquisition parameters) and averaged. The effect is comparable with longer integration, but the image has to be read several times, which means that the time taken increases and additional readout noise is added.

3.3.2 Sampling Frequency The sampling frequency (i.e., the number of pixels per unit length in the x, y, or z direction) should be set so that the *Nyquist criterion* is met; that is, the sampling frequency should be at least twice the largest frequency that we want to record. In practice, this means that the finer are the details that we want to resolve in the image, the higher should the sampling frequency be set. When fine details are not important (e.g., in the case of counterstain image of cell nuclei), the sampling frequency can be lower. But when small features have to be resolved (e.g., hybridization dots) a higher sampling frequency is desirable.

In PMT-based systems, the sampling frequency is changed by setting a different step for the rotation of mirrors that tilt the laser beam in the x and y directions. In CCD-based systems, the sampling frequency is simply equal to the total magnification divided by the camera pixel size (or the superpixel size if binning is used). As the spacing of the pixels in the CCD chip cannot be varied, the only way to change the sampling frequency is to change the overall magnification. This can be done either by changing the objective or by changing the intermediate zoom, if the microscope contains a zoom unit. The axial (z) sampling frequency is usually easily changed by setting a different step of the z-stepper motor or piezoelectric device.

3.3.3 Pixel (Voxel) Size The pixel (voxel) size in the x, y (x, y, z) direction is the inverse of the x, y, (x, y, z) sampling frequency. In PMT-based systems, the software provides the user with the possibility to set sampling frequency (pixels μm^{-1}) or pixel (voxel) size (μm pixel^{-1}). The effect of both approaches is identical. In CCD-based systems, such controls are usually not available, because of the constant sampling frequency for a given magnification (see above). The only exception is the use of the binning feature of a CCD chip. This enables to group squares of $N \times N$ camera pixels into larger superpixels (i.e., to increase the pixel size N times).

3.3.4 Image Size Usually the user wants to acquire as large a field of view as possible. Each system has some limitation imposed on the maximum number of pixels in each direction. This maximum number of pixels divided by the sampling frequency determines the maximum field of view. So mostly the maximum number of pixels is set. However, sometimes one may want to acquire only a small subregion (e.g., one cell) of the whole field of view. The majority of modern systems allow this. The only constraint posed on the subregion in such systems is that it be rectangular. This is because of the easy storage and handling of rectangular matrices in computer memory.

3.3.5 Bit Depth The bit depth says how many bits are used for storing the light-intensity value of one pixel in computer memory. This number is usually set so that it corresponds to the dynamic range of the detector. However, if the dynamic range of the detector (in bits) is not a multiple of 8 (e.g., if we have a 10- or 12-bit camera), then the bit depth is set to the closest higher multiple of 8 (e.g. 16 for 10- or 12-bit cameras). This means that some zero bits are appended to each pixel. It is also possible to omit 2 or 4 least significant bits and transform a 10- or 12-bit image into an 8-bit image. Some information is, however, lost using this approach. The multiple of 8 is required because computers work with bytes. Most software packages, however, do not provide the user with the possibility to change the bit depth.

3.3.6 Light Intensity For most light sources, the light intensity can be varied. The user changes either the power of a laser or the voltage of a lamp source. For mercury lamps, the intensity cannot be changed, but it is possible to regulate the intensity using different neutral-density filters. On the one hand, the intensity should be as high as possible in order to obtain good image quality (SNR). On the other hand, the intensity of fluorescence fades rapidly for high intensity levels. Using low light levels (at which the fluorescence does not fade) and long exposure times, one can obtain images of higher quality than using a short snap at a high intensity level. Therefore, a compromise must be found experimentally.

3.3.7 Gain Some systems allow the user to select the analog gain of the detector (PMT or CCD). The gain determines the degree of signal amplification. High gain settings improve image brightness, but at the expense of image quality. The reason is that the higher the gain, the larger the readout noise. So a good idea is to keep the gain settings at some medium values (usually recommended by the producer) and amplify the signal digitally.

3.4 Multidimensional Image Acquisition

The image information to be recorded can be 0D (one light-intensity value), 1D (illuminated light track), 2D (illuminated area), or 3D (illuminated volume). Some authors speak of 4D or even 5D imaging, denoting light wavelength and time as the other dimensions. As discussed earlier, light detectors can be 0D (photomultipliers), 1D (arrays of light-sensitive pixels), or 2D (different kinds of cameras).

In order to record nD light information, an mD light detector must be used, where $m \leqslant n$. If $m = n$, then one record is sufficient. If $m = n - 1$, then a series of records must be acquired. If $m = n - 2$, then scanning in two dimensions is necessary. Thus, for example, 2D information is usually recorded using a 2D light detector, but can be recorded using a 0D light detector combined with scanning (e.g., in scanning light microscopy). 3D information in optical microscopy can be recorded using axial scanning with a 2D detector (e.g., in tandem scanning microscopy) or both lateral and axial scanning using a 0D detector (e.g., in confocal laser scanning microscopy). A summary of the possible types of multidimensional image acquisition can be found in Table 1.

Thus, the main decision in image acquisition is which imaging device (light detector) to choose, whether 0D, 1D, or 2D. And also, if the dimension of the detector is less than the dimension of the image to be acquired, a method of scanning in the other dimensions must

TABLE 1 Multidimensional image acquisition: a summary of the possibilities of recording different types of image data; for each type, available detectors and corresponding number of dimensions to be scanned are presented

Information type	Variables	Image type	Detector/Scanning	Typical application
0D		Single value	0D/0D	Single measurement
1D	x	1D gray-scale	0D/1D	Flow cytometry
			1D/0D	Material inspection
2D	x, y	2D gray-scale	0D/2D	Scanning microscopy
			1D/1D	Not used
			2D/0D	All cameras
3D	x, y, z	3D gray-scale	0D/3D	Scanning microscopy
			1D/2D	Not used
			2D/1D	Wide-field microscopy
4D	x, y, z, λ	3D color	0D/4D	Scanning microscopy
			1D/3D	Not used
			2D/2D	Wide-field microscopy
4D	x, y, z, t	Time series of 3D gray-scale	0D/4D	Scanning microscopy
			1D/3D	Not used
			2D/2D	Wide-field microscopy
5D	x, y, z, λ, t	Time series of 3D color	0D/5D	Scanning microscopy
			1D/4D	Not used
			2D/3D	Wide-field microscopy

be chosen. Because 1D devices are not commonly used, the decision is reduced to the choice between PMT-based acquisition and CCD-based acquisition. There is no general consensus regarding which choice is better. Both systems have their advantages and disadvantages. The advantages of using a CCD detector are quicker and simpler acquisition and precise spacing of the pixels. The disadvantage is fixed sampling frequency (given by the pixel size of the CCD chip). In commercial systems, the manufacturer makes the choice: usually in laser-based systems, PMTs are used, and in non-laser-based systems, CCDs are used. However, there are exceptions to this rule: for example, some current commercially available TSM systems use laser light sources and CCD detectors.

4 PREPROCESSING OF FISH IMAGES

State-of-the-art microscope systems and image detectors usually provide images of sufficient quality, and therefore no or only a little preprocessing is required. If we do not take into account brightness and contrast adjustments for visualization purposes, there are just two types of preprocessing operations performed routinely on FISH images: noise suppression and deconvolution.

4.1 Noise Suppression

As explained earlier, the dominant type of noise in FISH imaging is photon shot noise. Because this type of noise produces spikes in the image, the best way to eliminate it is a median filter (a 3×3 matrix is sufficient). This filter also eliminates defective image pixels produced by defective CCD chip pixels (also spikes). The median filter is also known for its useful property of preserving edges in the image (Pratt, 1991). This feature is important for further image processing.

Another possibility is the use of a Gaussian filter as the best averaging filter. This also does the job – but only if photon shot noise is not too high, since otherwise the intensity extremes (spikes) in the image are spread into their surroundings. This is the basic difference between median and Gaussian filters: while a median filter completely removes spikes, the Gaussian filter just blurs them. So a Gaussian filter is used only if the noise level is low or as a preprocessing step for certain image-analysis algorithms (such as watershed) that require a smoothed image.

4.2 Deconvolution

Deconvolution is a process of obtaining a deblurred image from a blurred one. There are a variety of different algorithms for image deconvolution (see Chapter 12). Theoretically, it is possible to compute the original (unblurred) image data even from a wide-field axial series of optical slices, provided the blurring function is known. Unfortunately, the blurring function (PSF) is never precisely known, and moreover it changes dramatically with position in the image (especially along the axial direction). Also, the better the algorithm, the slower it is. Good deconvolution algorithms take hours per image. Therefore, only a small amount of image data can be processed, and the results are not always satisfactory (the image quality often does not change significantly). Nevertheless, in some types of FISH image data, the deconvolution process may be helpful.

5 ANALYSIS OF FISH IMAGES

The main task of the image-analysis process is to find objects in the image (so-called image segmentation) and to perform measurements on them. There are three main groups of objects that are analyzed using the FISH technique: chromosomes in metaphase spreads, isolated interphase nuclei, and tissue samples. Because segmentation algorithms differ for each group, segmentation will be discussed separately for the three cases. Finally, measurements on FISH images will be discussed together for all three cases.

5.1 Metaphase Spreads

In *conventional cytogenetics*, chromosomes in metaphase spreads are stained with a common dye (e.g., Giemsa, Hoechst, or DAPI). Different staining protocols produce different banding types (G-banding, Q-banding, C-banding, R-banding, and T-banding). The chromosomes are then classified (assigned the corresponding numbers or signs such as 1–22, X, and Y in the human genome) according to the banding pattern. This classification process is called *karyotyping*. Although several software systems exist for chromosome karyotyping in conventional cytogenetics, the reliability of the results is often not satisfactory, and a lot of human intervention is required to solve problems with touching or overlapping chromosomes.

In FISH imaging, the situation is much better, because individual chromosomes are stained with different dyes (each chromosome pair has its own color). There are two possibilities: either only selected pairs of chromosomes are stained and classified (leaving the remaining chromosomes unanalyzed) or all pairs of chromosomes are identified. In the latter case, the number of dyes required to stain all chromosome pairs is usually reduced using a combinatorial method: individual chromosomes are not stained with one dye, but each chromosome pair is stained with a combination of several dyes. This method is called *M-FISH*. The dyes are chosen so that each chromosome pair has its own unique combination of dyes. If we have N dyes, then we are able to obtain 2^N different combinations. Thus, for the human genome, for example, 5 dyes should be sufficient (yielding 32 different combinations). However, it is not easy to find dyes (or, better to say, stained DNA sequences) so that each chromosome pair has its own unique combination. Therefore, in practice, the number of dyes is slightly higher than the theoretical value (e.g., 6–8 for the human genome). The classification of the chromosomes is then performed according to the color signature, which is much more precise than the classification according to bands performed in conventional cytogenetics.

Using FISH, one can also stain individual genes on mitotic chromosomes. In this case, first the corresponding chromosome (which contains the given gene) is usually segmented, and then the gene is found within the chromosome area. It is also possible to locate only genes without locating the corresponding chromosomes (e.g., if we are interested only in the number of occurrences of the given gene). In both cases, the gene segmentation is usually performed using simple thresholding.

5.2 Isolated Interphase Nuclei

The segmentation of isolated interphase nuclei is a rather simple image-analysis task because the nuclei are well separated from each other. Problems occur only if the nuclei touch each other or overlap (e.g., if the density of the nuclei on the slide is high or if the nuclei tend to form clusters).

Usually the nuclei are segmented using a simple thresholding algorithm (i.e., equal intensity threshold for all objects in the image) applied to the counterstain image. The optimal threshold can be computed automatically using bilevel histogram analysis (Pratt, 1991). This approach assumes that there are two peaks in the intensity histogram: one corresponding to the background and one to the objects (in our case nuclei). For most FISH studies, this condition is satisfied. First, the two maxima are determined (mean intensity and standard deviation for each), and then the optimal threshold is computed as the global minimum between the two maxima. This approach, however, is appropriate only for images containing one type of nuclei at the same stage of the cell cycle. Otherwise, the nuclei differ in intensity, and simple thresholding does not produce satisfactory results (see Fig. 4*a–c*). The solution is adaptive (local) thresholding (Fig. 4*d*). Local thresholding allows the segmentation of each nucleus with its own intensity threshold level. As a result, correct boundaries can be determined for all nuclei independent of their intensity. More details about local thresholding can be found in Kozubek M. et al. (2001).

After thresholding of interphase nuclei, it is possible to improve the segmentation result further. First of all, small objects (dust) or incomplete nuclei should be excluded from the segmentation result. Then, for example, one can apply smoothing of nuclear boundaries using a modified Jarvis march in order to transform the "ragged" boundary curve (caused by thresholding) into a smooth one. Also, it is possible to separate slightly touching nuclei (if the overlap is not too serious) using a watershed or break-pairs method. These methods of advanced segmentation of interphase nuclei can be found in Kozubek M. et al. (1999).

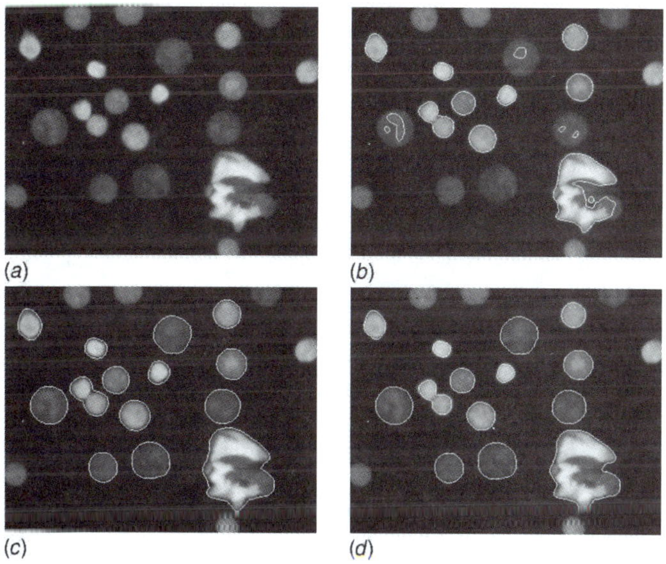

FIGURE 4 An example of global and local intensity thresholding of cell nuclei stained with DAPI. Cells are at different stages of the cell cycle, and therefore have different DNA contents and consequently different fluorescence intensities. (*a*) Original image. (*b*) Boundaries obtained by global thresholding with high threshold appropriate for bright cells. (*c*) Boundaries obtained by global thresholding with low threshold appropriate for faint cells. (*d*) Boundaries obtained by local thresholding with individual threshold for each cell.

After the segmentation of nuclei has been done, segmentation of FISH-stained signals (genes, chromosomes) within individual nuclei can commence. The algorithms (or, better to say, the principles of the algorithms) do not usually differ for 2D and 3D imaging. The easiest method of the segmentation of FISH-stained signals is thresholding. The threshold can be set for each dye and each nucleus separately. However, the segmentation results are usually not satisfactory. Therefore, more sophisticated approaches should be used. The approaches differ for chromosomes (large objects) and genes (small pointlike objects), but one thing is common to both: all FISH-stained signals are searched within the area (or volume) enclosed by the nuclear boundary. This means that if the boundary is underestimated, some signals may be omitted; if the boundary is overestimated, signals from neighboring nuclei may be wrongly included. Therefore, precise nuclear-boundary determination is crucial for the correct FISH-signal segmentation.

5.2.1 Chromosomes If probes specific to a certain pair of chromosomes are used, the resulting image contains painted volumes (ideally two) within the cell nucleus, where each volume consists of a large number of small spots corresponding to the places where the probe annealed. Owing to the heterogeneous inner structure of the chromosome domains, the painted volume is not continuous, and it is necessary to merge the intensively painted regions into one large domain corresponding to the whole chromosome. In order to find the intensively painted regions, a thresholding algorithm can be used where the threshold can be determined by intensity-histogram analysis: the mean background intensity $I_{\text{background}}$ is determined, as is its standard deviation $\text{Dev}_{\text{background}}$, and the threshold is computed as $T = I_{\text{background}} + k\text{Dev}_{\text{background}}$ where k should be set to 1.5–2.5 but can vary for different types of images. In order to merge the neighboring spots together into one larger domain, mathematical morphology (Serra, 1982) should be used – for example, the closing transformation (i.e., dilation followed by erosion). The strength of the transformation should be adjusted appropriately. After the closing operation, objects smaller than a predefined threshold size should be deleted, and the remaining larger volumes then correspond to chromosome domains.

5.2.2 Genes In the case of genes, centromeres and other pointlike objects (hybridization dots), the thresholding technique is of no use, because dot intensities vary even within the same nucleus and a universal threshold does not exist. If the threshold is high, some faint dots are not detected. If the threshold is lower (so that the faint dots are detected), the bright dots are oversegmented and often merge with each other and are detected as a single dot. That is why a more sophisticated segmentation technique must be used. In order to detect all hybridization dots that form at least a slight intensity maximum in a certain local neighborhood, it is advantageous to use a certain modification of the watershed algorithm. The *watershed algorithm* (Serra, 1982; Vincent and Soille, 1991) finds boundaries between individual objects in the image by flood simulation. This process has recently been adapted to hybridization dot segmentation (Netten et al., 1996; Kozubek M. et al., 1999). The advantage of the adapted algorithms over the standard watershed techniques is that they do not require so-called *markers*, that is, an initial assessment of local maxima (dot) positions. Another advantage of the watershed-based approaches is that the results are very accurate, and all local intensity maxima of a certain predefined height are detected.

5.3 Tissue Samples

While the segmentation of FISH-stained genetic loci inside cell nuclei is identical for tissues and isolated cells, the task of nuclear segmentation is much more difficult in tissues. Several semiautomatic approaches have been suggested (Ortiz de Solórzano, 1998a; Adiga and Chaudhuri, 1999) but no satisfactory solution has yet been found. It is also possible to combine manual segmentation used for the determination of nuclear outlines followed by the automatic analysis of FISH-stained genetic loci inside manually segmented nuclei (Kozubek M. et al., 2001). Work on better semiautomatic algorithms for the segmentation of cells in tissues is in progress in several laboratories.

5.4 Repeatedly Hybridized Cells

The FISH technique enables repeated staining of the same cells (the same microscope slide) several times with different sets of probes each time (Lukasova et al., 1999). The old probes are washed away before the next staining, and the new set of probes is added and hybridized with the target DNA. This technique of *repeated FISH* can be applied to metaphase spreads as well as to interphase nuclei. After the first hybridization, the user (or computer) selects specific fields of view that contain cells of interest. These fields of view are acquired, and the corresponding stage positions are saved to a hard disk. After the second, third, etc. hybridization, the stored stage positions are re-allocated, and exactly the same cells are acquired each time. This approach allows the visualization of a large number of probes within one cell simultaneously.

The merging of the analysis results from individual hybridizations is performed according to certain *reference points* in the image, such as nuclear weight centers. An example of repeated FISH and the merging process is shown in Figure 5. More details can be found in Kozubek M. et al. (2001).

5.5 Measurements on FISH Images

Measurements on segmented objects are performed in order to obtain object descriptions needed for consequent object classification or statistical analysis. For each nucleus, coordinates of the center of mass, size (area or volume), perimeter, and circularity are usually computed. In addition, the nuclear boundaries (chains of pixels) form part of the description of nuclei. The boundary chains are usually transformed into a *Freeman code* (Freeman, 1961), which is actually a vector representation of the boundary. The Freeman code reduces memory requirements for storing boundary chains. For each hybridization dot, certain attributes are computed: maximum intensity, dot height, dot size (or volume), and position relative to the nuclear center (the relative position is computed as a vector corresponding to the dot displacement from the nuclear center). A so-called *relative radius* of each dot can also be computed as the displacement vector length divided by the local radius, where the local radius is equal to the distance from the nuclear center to the membrane in the direction of the given dot. The relative radius ranges from 0% to 100%. This value is important because it is independent of nuclear size (volume) and shape, and enables the comparison of topological results for nuclei of different sizes. In the case of painted chromosomes, their size, the position of their center of mass, and maximum and average intensities are usually computed. The whole dataset can then be imported into a spreadsheet program or a statistical package for further evaluation.

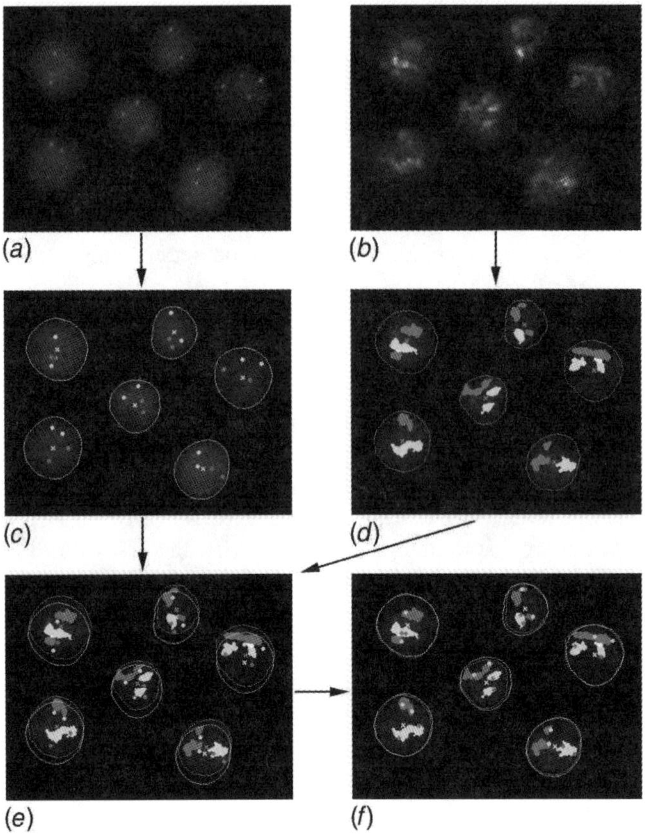

FIGURE 5 See ftp site for color figure. Merging of results from two different hybridizations of the same microscope slide. In the first hybridization (*a*), the *ABL* gene was stained with FITC (green), the *BCR* gene was stained with rhodamine (red), the counterstain was DAPI (blue), and the image was acquired in wide-field mode. In the second hybridization (*b*), the territory of chromosome 9 was stained with rhodamine (red), the territory of chromosome 22 was stained with FITC (green), the counterstain was DAPI (blue), and the image was acquired in confocal mode. Only a part of the whole field of view is shown (about 60 μm×45 μm) for both hybridizations in so-called autofocus (maximum-projection) mode. Each hybridization was analyzed separately. The results of the analysis are visualized using small circles for genes (*c*) and territory highlighting for chromosomes (*d*). The visualization color was set identical to the original emission color of the genetic loci. The nuclear boundaries (closed curves) and weight centers (crosses) are highlighted in cyan (first hybridization) and magenta (second hybridization). The two independent analysis results were subsequently overlaid (*e*), and corresponding pairs of nuclei were determined. Because of microscope-stage imprecision, the positions of nuclear weight centers do not match – they are shifted by a certain vector. After the determination of this vector, the nuclei can be correctly overlaid (*f*), enabling visualization of the gene positions within chromosome territories. Reprinted from Kozubek M. et al. (2001).

6 AUTOMATION OF FISH IMAGING

Manual evaluation of FISH images as well as their manual acquisition (i.e., acquisition requiring manual location of the objects, manual focusing, manual setting of acquisition parameters for each image, etc.) is often tedious, fatiguing, biased, and time-consuming.

Therefore, at least partial automation of these routine tasks is highly desirable. State-of-the-art hardware (microscopes, cameras, and computers) provides means of automation of the whole process. However, there is a lack of suitable control software. Not all FISH applications can be fully automated, but the vast majority can be at least semiautomated. If commercial software for a given task is not available, the only feasible solution is to write one's own tailor-made software for a given specific purpose. The assumption is that the biologist who needs the software finds a skilled programmer to do the job, and they both find a common language. This means that the biologist has to learn something about computers and the programmer has to learn something about the biology – otherwise they usually don't understand each other.

6.1 Automation of Microscope Components

The majority of commercially available microscopes can be automated or are even sold with automated parts. If a specific part of the microscope is not automated, it can be usually replaced by an automated one supplied either by the microscope manufacturer or by a third-party company specializing in automated microscope accessories, such as Ludl (http://www.ludl.com), Märzhäuser (http://www.marzhauser.com), Prior (http://www.prior.com) or Physik Instrumente (http://www.physikinstrumente.com). Let us have a look how individual microscope parts can be automated (the data are summarized in Table 2).

Microscope stages are usually motorized using stepper motors (one stepper motor for each direction). *x–y* motorized stages (two stepper motors) and *x–y–z* motorized stages (three stepper motors) are available. Important features of the motorized stages are travel range (maximum travel dimensions), resolution (the shortest step), and repeatability (how precisely one can re-allocate a specific position after moving away from it). For *focus control*, either a stepper motor can be used (e.g., the *x–y–z* motorized stage can be purchased) or a piezoelectric nanopositioning device can be bought (best of all a closed-loop one). Piezoelectric devices usually move objectives up and down rather than the stage. They are quicker and much more precise than stepper motors, but have only a

TABLE 2 Automation of microscope components: a summary of the available hardware possibilities and the corresponding response times

Microscope component	Usage	Response time[a]
Motorized objective revolvers	Slow but precise exchange of objectives	≈1–2 s
Motorized filter-cube revolvers	Slow but precise exchange of filter cubes	≈1–2 s
Motorized filter wheels	Fast exchange of excitation/emission filters	~100 ms
Motorized focus drives	Stepper motor (50–100 nm step) control or piezoelectric (10 nm step) control	≈200 ms/step ≈10 ms/step
Motorized stages (*x–y* or *x–y–z*)	Stepper motor (50–100 nm step) control	≈500 ms/100 μm
Electronic shutters	Slow or fast shutter possibility	≈1–100 ms
Lamp-intensity controllers	Voltage control via a D/A converter	≈10 ms

[a] The times include the so-called settling time (time to settle down at the new position); lateral and axial stage movements are not linear, and practical values are shown

limited travel range (typically 100–200 μm). *Filters* can be automated in two ways: either two motorized filter wheels are purchased (one for excitation and one for emission) or the whole filter cubes are moved using a motorized revolver (simultaneous exchange of dichroic mirror and excitation and emission filters). In the first case, a dual- or triple-band-pass dichroic mirror is needed (because one dichroic mirror must serve for several different dyes). This may slightly reduce image quality, but, on the other hand, filter wheels are much quicker than filter-cube revolvers. Also, *objectives* can be changed using a motorized revolver. If a computer-controlled *shutter* has to be placed into the optical path, either an integrated filter wheel with shutter or even a separate electronic shutter can be purchased. The control of *light intensity* is usually provided by the microscope manufacturer (the laser power is controlled – or the lamp voltage in the case of ordinary lamps). The intensity can be also controlled using a filter wheel carrying different neutral-density filters instead of filters selecting a certain wavelength range.

6.2 Automation of Image Acquisition

Image acquisition can be automated much more easily than image analysis. If suitable hardware components and a skilled programmer are available, it is not a problem to write a macro into an existing software system or new software for automated or semiautomated image acquisition. The driving computer can perform one, two, or more of the following tasks automatically: lateral scanning of the microscope slide, software focusing of selected (or each) field of view, evaluation of the given field of view (image quality at the given slide position, presence of objects of interest), axial scanning at the selected (or each) lateral position, scanning in wavelength dimension (sequential acquisition of several dyes through appropriate filters), selection of acquisition parameters (exposure time, region of interest, etc.), capture and storage of the image into computer memory, and compression of the acquired image data.

6.2.1 Lateral Scanning Each optical set-up has a limited *field of view*; that is, only a fraction of the objects to be acquired can be recorded during one image capture. For example, a 2D CCD camera is often used for the acquisition of 2D information. But the camera (CCD chip) sees only a fraction of the specimen, and scanning is necessary in order to capture the whole specimen. The final image then consists of many fields of view acquired sequentially. The scanning here does not increase the number of dimensions of the acquired image (if compared with axial scanning, for example), but effectively enhances the field of view. In optical microscopy, the scanning cycle is usually implemented by moving the stage with the specimen according to a certain pattern. The most commonly used pattern is a *meander pattern*, which means that the stage moves horizontally forward from one end to the other, then backward along the neighboring line, then again forward one line further, and so forth. In this way, the whole slide can be scanned (e.g., overnight). It is a good idea to provide a small *overlap* of the neighboring fields of view. The overlap should be roughly equal to the dimensions of the observed objects (cells). This enables one to analyze also cells that are incomplete (and therefore excluded from the analysis) at a given stage position. On the other hand, the overlap should not exceed the smallest object dimension – otherwise one object can be found in two neighboring images (and counted twice into the resulting statistical summary of the whole slide).

6.2.2 *Software Focusing* Because the depth of focus is usually small (typically about 0.3 μm for high-NA objectives), it is advantageous to focus the specimen before acquisition and to refocus the image each time the stage moves (e.g., in an automated lateral scanning cycle), because even the best microscope slides are not ideally flat. Automated focusing is possible both in direct (nonfluorescence) light microscopy (where even hardware autofocusing is available) and in fluorescence microscopy using appropriate software. The focusing algorithms take several image samples at different axial stage positions and compute a so-called *focus function* for each image. The focus function has its absolute maximum in the focal position, and decreases in both directions. The stage is finally moved to the axial position corresponding to this maximum. Several functions can be chosen for the focusing; comparisons can be found in Groen et al. (1985), Firestone et al. (1991), and Boddeke et al. (1994).

The time for focusing one field of view can be reduced to 2–3 s using state-of-the-art hardware such as CCD cameras with binning feature (Kozubek M. et al., 1999). If focusing is time-consuming, *interpolation or extrapolation techniques* can be used in the scanning cycle, so that focusing is performed only on each third field of view or even more rarely. Both techniques assume that the objects (mostly cells) lie within one plane that is tilted relative to the optical axis of the microscope. In the interpolation case, several points (usually corner points plus some inner points) are focused, and the algorithm interpolates among them. In the extrapolation case, the algorithm computes the correct axial position according to several previous neighboring measurements. In both cases, proper focusing is applied from time to time (usually at each third image) in order to correct the prediction obtained by interpolation or extrapolation.

6.2.3 *Evaluation of the Current Field of View* Before a 3D series is acquired and before different wavelengths are recorded, it is useful to assess whether or not the acquisition result will be worth the effort. This check is important especially when unsupervised lateral scanning of the slide is performed. For example, if the current field of view does not contain objects or contains only incomplete objects (objects that fall only partially into the current field of view) or distorted objects, it does not make sense to spend time acquiring a 3D series. The best way of checking the image quality and contents is to acquire one 2D image and perform some easy online image analysis (such as object segmentation using a thresholding approach). Because we are not interested in details at this stage, the 2D image can be of low resolution (e.g., a low-resolution picture of counterstain). Low-resolution pictures can be acquired much faster (e.g., using binning of a CCD camera) than proper high-resolution ones. Therefore, an image check can be done in a fraction of a second. In this way, empty or dusty parts of the microscope slide can be skipped quickly during the lateral scanning. This substantially reduces the total acquisition time for the whole slide.

6.2.4 *Axial Scanning* In order to record 3D images, axial scanning is required. The majority of microscope suppliers usually automate axial scanning so that the whole 3D image can be recorded by pushing one button. There are two parameters that the user must supply: the axial step and the number of slices. The *axial step* determines the voxel size in the z direction. Its value usually corresponds to the size of the smallest object that one wants to record (so that no object is missed). For high-resolution imaging, the axial step should be roughly equal to the depth of focus (i.e., typically about 0.3 μm for high-NA objectives). The *number of slices* determines the z dimension (the number of voxels in the z

direction) of the acquired 3D image. It is usually set so that the axial scanning range (the number of slices multiplied by the axial step) covers the axial dimensions of the largest objects on the slide. Both the axial step and the number of slices are usually constant for the whole slide.

6.2.5 Scanning in Wavelength Dimension Wavelength is usually considered as a special dimension (similarly to time), and many authors add this dimension to the final number of dimensions and speak of 4D or 5D imaging. In FISH imaging, the wavelength dimension is always present, and several different dyes (stains) have to be acquired and combined into one image. Either all dyes are acquired simultaneously (using a color CCD or several PMTs) or individual dyes are acquired sequentially through the appropriate highly specific filters (using a monochrome CCD). The second approach is slower, but gives better results because of the high specificity of the filters. It requires sequential scanning in wavelength dimension and combining (merging) of the individual monochrome images in computer memory into one pseudocolored image for visualization purposes. This can be easily implemented, provided that a 2D acquisition procedure is available.

In the case of 3D imaging, there are two possibilities for the acquisition of a multicolor FISH specimen: either to acquire each stain at the given z position and then move the stage axially, or to acquire all z positions for the first stain, then all z positions for the second stain, etc. The second approach is better as a rule, because the time to move the stage (or objective) one step axially is usually shorter than the time to exchange filters and consequently the total acquisition time is shorter. In order to maximize the reproducibility of the axial positions and keep the acquisition conditions identical for all stains, the stage (or objective) should return to the same initial position after each axial scan, so that the next series of optical slices (for the next stain) is acquired from the same starting point and according to the same pattern.

6.2.6 Selection of Acquisition Parameters The acquisition parameters are usually set once by the user for a given slide and kept constant. In some cases, however, it may be helpful to change some of these parameters during the lateral scanning process. For example, it may be useful to change the *exposure (integration) time*. Some parts of the slide may be dimmer than other parts, and the intensity of fluorescence may vary for different fields of view. Therefore, the initial exposure time may be appropriate only for a certain part of the whole slide. It may happen that the images from other parts of the slide are overexposed if the exposure time is constant. If the computer detects an overexposed (underexposed) image (easily detectable using intensity-histogram analysis), then it should reduce (increase) the exposure time appropriately. Another parameter that could be changed is the *region of interest (ROI)*. For speed purposes, it may be useful to determine ROIs containing objects of interest in the field of view during the image quality check (see Section 6.2.3). Then only subregions of the whole field of view can be acquired in 2D, 3D, or 4D (3D + wavelength). This substantially reduces readout time and also acquisition time in CLSM. Unfortunately, in the case of CCD-based systems, the acquisition time is not reduced if a subregion of the whole CCD chip is used for the acquisition.

6.2.7 Capture and Storage of the Image Usually the capture of a 2D or 3D image is initiated by pushing one button and the storage of the captured image to the hard disk is initiated by pushing another button. If the whole slide is scanned automatically (lateral scanning), no buttons are pushed. Therefore, the image capture automatically starts after the image-quality check, and the image storage automatically starts after the image capture. The user must select only the file format and the location (directory) for image storage before launching the lateral scanning. The only problem is how to choose names for the images to be stored. The most common way is to use numbers as file names. For example, the first image is named 0001.tif, the second one 0002.tif, etc. The number actually represents the stage position. It is always a good idea to store also stage coordinates for each stage position, so that the given stage position can be re-allocated according to the number. If 3D image data are acquired, then a pair of numbers is usually used: the first number represents the stage position and the second number represents the number of slice. For example, if the number of slices is 40, then the 3D image at the seventh stage position can be stored as a sequence of 40 images named 0007_01.tif, 0007_02.tif, . . . , 0007_40.tif. Some software producers have their own 3D file formats and store all slices into one file. Unfortunately, there exists no standard for such 3D image files, so the export of 3D images into a different software package is problematic.

6.2.8 Compression of the Acquired Image Data Data compression is a good tool for reducing memory requirements for image storage. Especially in the case of 3D imaging, the image data occupy a large amount of space on the hard disk. A typical 8-bit monochrome 3D image consisting of 40 slices where each slice is a 2D image consisting of 1300 × 1000 pixels occupies about 50 MB of disk space. If we acquire three dyes, the color 3D image will occupy 150 MB. And if we now launch lateral scanning and acquire several hundreds of such color 3D images, then we need several tens of gigabytes of disk space (i.e., an entire large hard disk) per slide. This is obviously unacceptable. Therefore, data compression is highly desirable. There are two types of compression methods available: *lossless* (which do not change image quality) and *lossy* (which reduce image quality). Lossy methods naturally achieve better compression ratios (depending on how much the quality is changed), but at the expense of blurring the image. They are used when the details in the image are not important. In FISH imaging, we usually need to keep the original data, because we are interested in details and blurring is undesirable. Moreover, blurring is mostly unacceptable if consequent image analysis has to be performed. Hence, lossless methods should be used. Unfortunately, even the best lossless compression methods do not achieve good compression ratios on real data. The best algorithms reduce the amount of data to 50–70%. This does not help much.

The best way to solve this problem is to store only those *subregions (subvolumes)* of the whole image that contain objects of interest. For example, if isolated cells are imaged, then it is possible to store only the cells and their immediate surroundings. If metaphase spreads are imaged, then only the area containing the chromosomes can be extracted. The selection of the regions of interest can be done either automatically using some online image segmentation if the image is simple (e.g., isolated cells) or by the user using a mouse if the image is complicated (e.g., tissue). The extraction of objects of interest can reduce the memory requirements by one or even two orders of magnitude.

6.3 Automation of Image Analysis

Similarly to image acquisition, there are several image-analysis tasks that can be considered for automation: selection of image-analysis algorithms and their sequence, selection of the parameters of a specific algorithm, and processing of a large number of images using the same approach. Let us have a look what can be done about automation of these tasks.

6.3.1 Selection of Algorithms and Their Sequence Automatic selection of suitable algorithms and their proper sequence for the acquired data is a very difficult problem (not only in fluorescence microscopy), and it has not yet been solved. Therefore, this task will still be carried out by humans in the future (at least for a few years). To facilitate this task, some systems enable recording of the sequence of algorithms that the biologist has applied to a particular image, and provide some sequences recorded by the developer of the system. The user's or developer's sequence can be used for the analysis of similar images (images of the same type) in the future. The sequence is usually written into a text file using a *macro language*. The user can further edit the text file – a so-called *macro* – in a text editor. Some algorithms can be deleted and others added to the sequence. Macro programming is an indispensable part of most software packages.

6.3.2 Selection of Algorithm Parameters Each algorithm has several parameters that influence its behavior. The parameters can be selected automatically (i.e., they are computed by a previous algorithm in the sequence) or they must be selected by a user. The user usually alters the default parameters specified by the developer of the algorithms so that the algorithm works well for his or her own image. It is sometimes difficult to understand how some parameters influence the algorithm's behavior. Therefore, some systems enable the user to run the same algorithm with several different parameter settings generated automatically. Then the user selects the best result instead of selecting individual parameters. The parameters that led to the best result can be stored to the hard disk and used for the analysis of other images.

6.3.3 Processing of a Large Number of Images Often we need to apply the same analysis approach to several (or even many) images. For example, we may want to analyze all images in a certain directory using a specified macro (as discussed above, a macro is the best way of describing the whole analysis approach). This task is called *batch processing*, and a number of software packages include this feature. However, it is always a good idea to check the analysis results for individual images visually after the batch processing. The visual inspection can usually be performed quite quickly by overlaying the analysis results (e.g., object boundaries or various signs representing object locations) over the original image data (as in Fig. 5). Most software packages also allow correction of the analysis results during the visual inspection. At least the possibility of deleting unwanted objects should be included in each package. Some packages also allow modification of analysis results (e.g., nuclear boundary or gene position) or addition of a new object (e.g., a new nucleus). It is, however, not always advantageous to modify the results or add new objects, because of the bias introduced by the human factor.

7 SOFTWARE FOR FISH IMAGING

There are several types of software that are needed for FISH imaging. First of all, there is the operating system (OS). The choice of OS determines which software packages can be purchased. Secondly, one needs some software for driving individual hardware components (so-called *drivers*), such as microscope and camera. Drivers use functions provided by the OS, and therefore they are OS-specific. Then one needs image acquisition software. This software uses drivers for communication with the microscope hardware. Finally, one needs software for image analysis that can analyze images stored on the hard disk. There also exist combined acquisition and analysis software packages that allow the user to analyze freshly acquired images without storing them to the hard disk.

7.1 Software for Driving Microscopes

The basic software for communication with a specific hardware component (a driver) must be written and supplied by the manufacturer of this component. Unfortunately, most microscope hardware producers as well as camera producers provide drivers only for Microsoft Windows operating systems. In this way, they, rather than the user, make the choice of the operating system. Without an appropriate driver, it is not possible to communicate with the given component, even if one's own software is being created. The manufacturer can, however, provide a low-level specification of the communication protocol (e.g., codes or words sent via a serial port if the given component is connected to a serial port). In this case, it is possible to write one's own driver for the operating system that one wants to use. Therefore, it is important to make sure that correct drivers or low-level specification are available before the purchase of new hardware.

7.2 Software for Image Acquisition

Similarly to drivers, software for image acquisition is usually supplied by the microscope or camera manufacturer (and included in the price). However, often only basic functions are implemented, and better software packages must be paid for. The number of good software packages that support the given hardware component varies from one to several tens. In the case of cameras, for example, many packages can be found that support a given camera (software producers try to include support for all commonly available cameras). Motorized stages also are quite well supported by different software producers. On the other hand, in the case of CLSM microscopes, the user mostly has no choice – there is just one software package delivered. In general, when one wants to buy new hardware, it is important to find out how this hardware is supported in different software packages before the purchase.

7.3 Software for Image Analysis

Many software packages for image analysis are available on the market. Because image analysis does not depend on microscope and acquisition hardware, packages for different operating systems can be obtained. Some packages are even sold for different operating systems. The price of the software depends on the number of different algorithms that are included. Basic image-processing packages can be obtained for free, whereas good packages can cost up to several tens of thousands of dollars. Some packages provide complex solutions (i.e., algorithms of all kinds are included), while others are specialized for one type of algorithm (e.g., deconvolution). Therefore, the selection of a particular software package depends on what one actually wants to do with the images. Because each research group needs different algorithms, tailor-made solutions are never found among the commercially available packages. Whatever software package one buys, the vast majority of its features are not needed, and only a few percent are really used. A real tailor-made solution can be made only as a result of cooperation between biologists and programmers. This takes time and money, but in some cases it is the only solution.

7.4 Software for Automated Imaging

Different software packages provide different means of automation of the acquisition and analysis process. Most packages provide the user with the possibility of macro programming (as discussed earlier), which is very useful for recording the acquisition and/or analysis process. Packages specialized for 3D imaging provide means of automated axial scanning and capturing a whole 3D image by pushing one button. There exist also packages that allow lateral scanning of the microscope slide using meander or other user-defined pattern. So, different degrees of automation can be achieved in different packages.

Software for fully automated acquisition and analysis is still very rare. Either it is not commercially available or it is very expensive. It is provided with some specialized instruments such as laser scanning cytometers or flow cytometers. Metaphase spread acquisition and analysis can also be automated using special packages (including FISH-stained metaphase spreads). Software systems for the automated acquisition and analysis of FISH-stained interphase nuclei have also begun to appear. The following companies offer automated acquisition and analysis solutions for FISH imaging (metaphase spreads, interphase nuclei or both):

- Applied Imaging, UK (http://www.aii.co.uk)
- Imstar, France (http://www.imstar.fr)
- Loats Associates, USA (http://www.loats.com)
- MetaSystems, Germany (http://www.metasystems.de)

An alternative approach is to write one's own software package for a specific task. For example, several packages for the automated acquisition and analysis of FISH-stained interphase nuclei had appeared before this type of software became commercially available (Netten et al., 1997; Ortiz de Solórzano et al., 1998b; Kozubek M. et al., 1999). The advantage of this approach is that one can easily modify the existing package or add a new feature to it. The disadvantage is that it costs time and money during the development stage, as discussed above. Both commercial and custom software systems for automated imaging have the disadvantage that they are made only for a particular hardware set-up and cannot be used with other hardware components. Moreover, the above-mentioned companies usually supply complex solutions – that is, software and hardware together, including the microscope. So installation of the software for automated imaging into a current hardware set-up may be impossible. Therefore, care must be taken when purchasing the microscope hardware components if automated imaging is planned in the future.

8 SELECTED APPLICATIONS

Because the range of FISH applications is very large, it is difficult to summarize all research activities in this field. Therefore, only several examples will be described. Typical FISH images which are processed in these examples are shown in Figure 5 (isolated cells) and Figure 6 (cells in tissue).

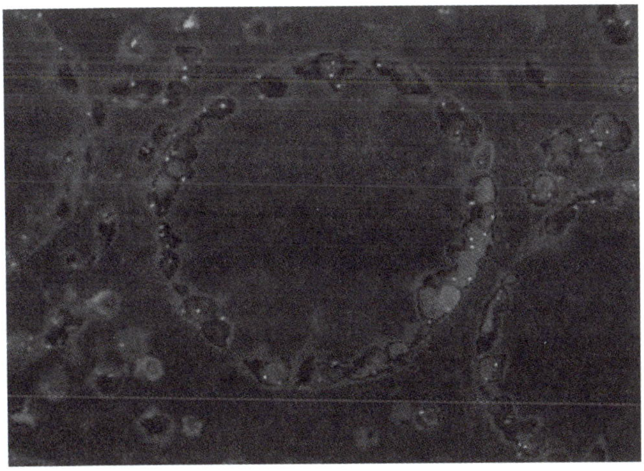

FIGURE 6 See ftp site for color figure. Typical example of a maximum-projection image of 3D fixed 4 μm thick colon tissue section with FISH signals of the chromosome 7 centromere. The nuclei were counterstained with DAPI. Reprinted from Koutna et al. (2000).

8.1 Investigation of Higher-Order Chromatin Structure

It has been established that chromosomes occupy distinct territories (Cremer et al., 1982, 1993), which can be clearly seen using the FISH technique. Recent results suggest that there are two compartments in the cell nucleus: a central one consisting of early replicating and transcriptionally active euchromatin (active genes) and a peripheral one consisting of later replicating and inactive heterochromatin (Ferreira et al., 1997; Sadoni et al., 1999). The inner structure of the cell nucleus, the arrangement of chromosomes within its volume, the spatial positions of individual genes, interactions between them, and cell-cycle changes of nuclear structure are, however, still poorly understood. Discovery of further principles of chromatin organization would have a strong impact on all biological disciplines, and would also be of great importance for clinical practice.

Centromeric regions can be visualized using antibodies; however, experiments with FISH staining have allowed selective detection of particular pair of centromeres and investigations of their positioning inside cell nuclei (Popp et al., 1990; Höfers et al., 1993; Hulspas et al., 1994; Skalnikova et al., 2000). These experiments showed that the positioning of centromeres of individual chromosomes in cell nuclei is nonrandom, can be markedly variable, and can differ for various chromosomes. Centromeres were found mostly near the nuclear periphery. The visualization of genes enabled direct detection of coding sequences, and showed that they are located on the periphery of chromosome territories (Kurz et al., 1996) but in central parts of the cell nucleus (Kozubek S. et al., 1997; Parreira et al., 1997). Genetic structures such as genes, centromeres, and chromosomes were found mostly in specific subregions of cell nuclei that form distinct layers or shells at definite distances from the center of the nucleus to the locus (Kozubek S. et al., 1999a; Skalnikova et al., 2000). Inside these layers, the genetic loci are distributed randomly. The compartmentalization of cell nuclei as well as the positioning of genetic loci in the nuclei is basically conserved in different cell types and at various stages of the cell cycle (Sadoni et al., 1999; Skalnikova et al., 2000). Some features of nuclear structure are conserved even in solid tissues (Koutna et al., 2000), differentiated cells (Bartova et al., 2000a), and during cellular repair after irradiation (Jirsova et al., 2001), although shifted positioning of genetic loci was systematically observed during these processes (Bartova et al., 2000b; Jirsova et al., 2001).

The cells growing attached to the slides can be regularly observed, and their movement can be traced. In this way, the sister cells can be determined and the location of their chromosomes in interphase nuclei can be investigated after hybridization. The comparison shows how regular the movement of individual genes in the cell cycle is and to what extent the positions of genes are influenced by random motion. It was found that there is a good correlation between the distances of homologous centromeres in sister cells and that the positions of chromosomes are also mostly symmetrical about a certain point of symmetry (Kozubek S. et al., 1999b; Sun and Yokota, 1999). These results show that the movement of chromatin in cell nuclei is substantially restricted. It seems that the kinetics of chromatin motion during the cell cycle is determined at the moment when the cell is born. The "program" of chromatin movement is the same in both sister cells.

Nuclear higher-order structure is related to gene expression. Transcriptionally active regions are located in the inner parts of cell nuclei (Sadoni et al., 1999), and the transition of some genetic loci to the nuclear periphery can be accompanied by downregulation of the loci (Bartova et al., 2000b). The position of the chromosome territory in the cell nucleus depends quite strongly on the content of euchromatin: a higher proportion of euchromatin and consequently more pronounced expression of chromosome genes is related to the more inner location of the chromosome (Croft et al., 1999).

8.2 Clinical Applications

It is well known that the induction of malignant diseases is connected with a deregulation of cell division at the so-called restriction point (R point) occurring in the G_1 phase of the cell cycle, in which the subsequent development of the cell is determined (Strauss et al., 1995). The mechanism of regulation of cell proliferation involves a number of genes of two types: proliferation-activating genes (oncogenes) and proliferation-suppressing genes (tumor suppressors). Deregulation of these genes can be caused by various defects, such as mutations, chromosomal aberrations (mainly translocations), deletions, integration of oncoviruses, etc. It has been found that many translocations are specific for a defined type of malignancy and that they often represent a single chromosomal change in malignant cells (Rabbitts, 1994). Thus, the t(9;22) translocation, giving rise to the so-called Philadelphia (Ph) chromosome, is present in 95% of patients with chronic myeloid leukemia (CML) (Kurzrock et al., 1988). Some other types of translocations in hematological malignancies are, for example, t(15;17) in acute promyelocytic leukemia (APL), t(9;22), t(4;11), or t(8;14) in acute lymphoblastic leukemia (ALL), and t(14;18), t(8;14), and t(11;14) in non-Hodgkin's lymphomas (NHL).

For the induction of translocations, and consequently for the induction of malignancies, the physical distance between critical genes is obviously important (Kozubek S. et al., 1999a). Therefore, studies of the morphology of critical genes, comparing the morphology for healthy individuals and that for patients suffering from a particular disease, and tracing the dynamics of critical genes can lead to better understanding of the oncogenic transformation of cells.

If a large number of cells are analyzed, it is possible to monitor residual disease. The detection of malignant cells on a background of normal cells requires statistically significant datasets for the oncogene–breakpoint region distribution in the range 0–0.5 μm. If thousands of nuclei are analyzed, then it is possible to distinguish between control and contaminated data (Kozubek M. et al., 1999). The greater the number of analyzed nuclei, the better the detection capabilities. Therefore, automated analysis of FISH-stained interphase nuclei is highly desirable for this purpose. It is important to note, however, that this type of study requires very good calibration of the imaging system – especially good correction of chromatic aberrations, because distances of 0–0.5 μm cannot be measured if chromatic aberrations of the optical system are of the order of several hundred nanometers. The correction of chromatic aberrations has been discussed earlier and shown in Figure 3.

8.3 Applications in Radiation Biology

It is well known that different types of malignancies (e.g., CML or thyroid cancer) are frequently induced by ionizing radiation (see, e.g., Preston et al., 1994). Therefore, radiation-induced carcinogenesis has been frequently used as a model of cancer induction. Either induction of chromosome exchange aberrations in mitotic cells or changes of chromatin structure induced by radiation are investigated.

Exchange aberrations are usually scored in mitotic cells. In this type of study, the re-allocation feature of the imaging system is very important, because it enables re-hybridization and re-acquisition of exactly the same metaphase spreads stained with different probes (see Fig. 7). In this way, it is possible to investigate, in principle, exchanges between all pairs of chromosomes. It is also possible to apply FISH after conventional staining, and thus benefit from the advantages of both techniques.

The structure of the cell nucleus is an important factor in the induction of chromosome aberrations (Kreth et al., 1998; Kozubek S. et al., 1999a). Exchange aberrations between two genetic loci are not proportional to the molecular weight of the chromosomes (Lukasova et al., 1999; Cafourkova et al., 2001) and they are obviously less probable if the genetic loci are well separated from each other in the nucleus; if two genetic loci are superimposed in a small nuclear volume, the probability of exchange aberrations will be markedly increased. The induction of aberrations is related to the higher-order chromatin structure, and consequently to the appearance of malignant diseases. Especially important are so-called "tethered" states when critical loci (e.g., *ABL* and *BCR* in chronic myeloid leukemia or *RET* and *H4* in thyroid cancer) are close to each other. Such states have been found in human blood cells (Kozubek S. et al., 1999a) and in thyroid cells (Nikiforova et al., 2000).

All methods of chromatin structure investigation can be applied also to the investigation of the influence of physical and chemical agents on the cell genome. The chromatin structure can be strongly affected by ionizing radiation. Genetic loci in the nuclei of irradiated cells move towards the center of the nucleus with subsequent recovery (Dolling et al., 1997; Jirsova et al., 2001). Radiation, beside the induction of DNA lesions, may decrease the distances between critical genes (oncogene–breakpoint region) and consequently increase the probability of the induction of translocation.

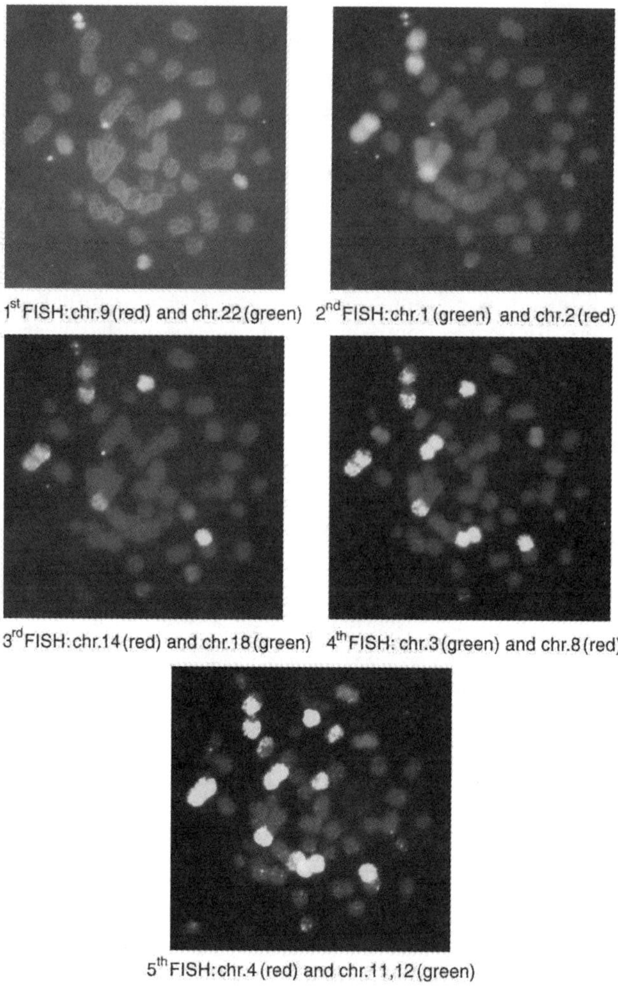

FIGURE 7 See ftp site for color figure. Repeated FISH on metaphase spreads: chromosomal aberration scoring after irradiation. Human lymphocytes were irradiated with fast neutrons from the d→Be reaction with a dose of 1.5 Gy and were repeatedly stained by FISH. Chromosome-specific probes were purchased from Cambio and Oncor. Biotin-labeled probes were stained with FITC (green) and digoxigenin-labeled probes were stained with rhodamine (red). The counterstain was DAPI (blue). The same mitosis is shown for five subsequent hybridizations. Note that whereas digoxigenin (red) can be washed out without problems, biotin (green) remains on the chromosomes also in the next staining. In the second hybridization, a dicentric between chromosomes 1 and 2 with two common acentric fragments is clearly visible.

9 CONCLUSIONS

Different aspects of FISH imaging have been discussed, including the basic principles of the FISH technique, optical systems for FISH imaging, optimization and calibration of the imaging process, selection of hardware as well as software components, optimization of the acquisition process, and automation of both the acquisition and analysis process. The use of the FISH technique has been demonstrated with several examples.

Although the FISH technique is already about 20 years old, it is still being used and has spread into further fields of cell research. Owing to its unique capability to visualize genetic material inside cells (at the level of chromosomes as well as the level of short unique sequences), FISH is an ideal tool for studying the composition and topography of cells. Therefore, it can be expected that FISH will continue to be intensively used in the future.

The evaluation of FISH-stained preparations is a very complex process, and the quality of the visualization result depends on many parameters. Some of these can be changed by the user, while some of them are determined a priori by the optical set-up. The purpose of this chapter has been to help the user with the optimization of the imaging process and with correct selection of the acquisition parameters and analysis techniques. This should help one to obtain better images with less effort. Another point of this chapter was that the automation or semiautomation of the acquisition and/or analysis process can substantially reduce the amount of human work involved. Present-day computer power is more than sufficient for most tasks of FISH imaging. What is necessary is to utilize this computer power for our purposes.

ACKNOWLEDGMENTS

I thank Stanislav Kozubek and Emilie Lukasova (Institute of Biophysics, Brno, Czech Republic) as well as Petr Matula and Pavel Matula (Faculty of Informatics, Masaryk University, Brno, Czech Republic) for careful reading of the manuscript and a number of useful comments and suggestions. This work has been supported by the Ministry of Education of the Czech Republic (Grant MSM-143300002) and the Grant Agency of the Czech Republic (Grant 202/99/P008).

REFERENCES

Adiga, P. S. U. and Chaudhuri, B. B. (1999). Deformable models for segmentation of CLSM tissue images and its application in FISH signal analysis. *Anal. Cell. Pathol.* **18**, 211–225.

Bartova, E., Kozubek, S., Kozubek, M., Jirsova, P., Lukasova, E., Skalnikova, M., and Buchnickova, K. (2000a). The influence of the cell cycle, differentiation and irradiation on the nuclear location of the abl, bcr and c-myc genes in human leukemic cells. *Leuk. Res.* **24**, 233–241.

Bartova, E., Kozubek, S., Kozubek, M., Jirsova, P., Lukasova, E., Skalnikova, M., Cafourkova, A., and Koutna, I. (2000b). Nuclear topography of the c-myc gene in human leukemic cells. *Gene* **244**, 1–11.

Boddeke, F. R., Van Vliet, L. J., Netten, H., and Young, I. T. (1994). Autofocusing in microscopy based on the OTF and sampling. *Bioimaging* **2**, 193–203.

Bubendorf, L., Kononen, J., Koivisto, P., Schraml, P., Moch, H., Gasser, T. C., Willi, N., Mihatsch, M. J., Sauter, G., and Kallioniemi, O. P. (1999). Survey of gene amplifications during prostate cancer progression by high throughput fluorescence in situ hybridization on tissue microarrays. *Cancer Res.* **59**, 803–806.

Burke, D. T., Carle, G. F., and Olson, M. V. (1987). Cloning of large segments of exogenous DNA into yeast by means of artificial chromosome vectors. *Science* **236**, 806–812.

Burke, D. T. and Olson, M. V. (1991). Preparation of clone libraries in yeast artificial chromosome vectors. *Meth. Enzymol.* **194**, 251–270.

Cafourkova, A., Lukasova, E., Kozubek, S., Kozubek, M., Govorun, R. D., Kourna, I., Bartova, E., Skalnikova, M., Jirsova, P., Pasekova, R., and Krasavin, E. A. (2001). Exchange aberrations induced with gamma-rays among 11 chromosomes of human lymphocytes. *Int. J. Radiat. Biol.* **77**, 419–429.

Castleman, K. R. (2002). Sampling, resolution, and digital image processing in the spatial and Fourier domains. In: *Confocal and Two-Photon Microscopy: Foundations, Applications, and Advances* (ed. A. Diaspro). New York: Wiley-Liss, Chap. 11.

Cheng, S., Fockler, C., Barnes, W. M., and Higuchi, R. (1994). Effective amplification of long targets from cloned inserts and human genomic DNA. *Proc. Natl. Acad. Sci. USA* **91**, 5694–5699.

Collins, J. and Hohn, B. (1978). Cosmids: a type of plasmid gene-cloning vector that is packagable in vitro in bacteriophage λ heads. *Proc. Natl. Acad. Sci. USA* **75**, 4242–4246.

Craig, A. G., Nizetic, D., Hoheisel, J. D., Zehetner, G., and Lehrach, H. (1990). Ordering of cosmid clones covering the herpes simplex virus type 1 (HSV-1) genome: a test case for fingerprinting by hybridisation. *Nucleic Acids Res.* **18**, 2653–2660.

Cremer, T., Cremer, C., Schneider, T., Baumann, H., Hens, L., and Kirsch-Volders, M. (1982). Analysis of chromosome positions in the interphase nucleus of Chinese hamster cells by laser-UV-microirradiation experiments. *Hum. Genet.* **62**, 201–209.

Cremer, T., Lichter, P., Borden, J., Ward, D. C., and Manuelidis, L. (1988). Detection of chromosome aberrations in metaphase and interphase tumor cells by in situ hybridization using chromosome-specific library probes. *Hum. Genet.* **80**, 235–246.

Cremer, T., Kurz, A., Zirbel, R., Dietzel, S., Rinke, B., Schröck, E., Speicher, M. R., Mathieu, U., Jauch, A., Emmerich, P., Scherthan, H., Ried, T., Cremer, C., and Lichter, P. (1993). Role of chromosome territories in the functional compartmentalization of the cell nucleus. *Cold Spring Harbor Symp. Quant. Biol.* **58**, 777–792.

Croft, J. A., Bridger, J. M., Boyle, S., Perry, P., Teague, P., and Bickmore, W. A. (1999). Differences in the localization and morphology of chromosomes in the human nucleus. *J. Cell Biol.* **145**, 1119–1131.

Diaspro, A. and Sheppard, C. J. R. (2002). Two-photon microscopy: basic principles and architectures. In: *Confocal and Two-Photon Microscopy: Foundations, Applications, and Advances* (ed. A. Diaspro). New York: Wiley-Liss, Chap. 3.

Dolling, J. A., Boreham, D. R., Brown, D. L., Raaphorst, G. P., and Mitchel, R. E. J. (1997). Rearrangement of human cell homologous chromosome domains in response to ionizing radiation. *Int. J. Radiat. Biol.* **72**, 303–311.

Eils, R., Bertin, E., Saracoglu, K., Rinke, B., Schröck, E., Parazza, F., Usson, Y., Robert-Nicoud, M., Stelzer, E. H. K., Chassery, J. M., Cremer, T., and Cremer, C. (1995). Application of confocal laser microscopy and three-dimensional Voronoi diagrams for volume and surface estimates of interphase chromosomes. *J. Microsc.* **177**, 150–161.

Ferreira, J., Paolella, G., Ramos, C., and Lamond, A. I. (1997). Spatial organisation of large-scale chromatin domains in the nucleus: a magnified view of single chromosome territories. *J. Cell Biol.* **139**, 1597–1610.

Firestone, L., Cook, K., Culp, K., Talsania, N., and Preston, K. (1991). Comparison of autofocus methods for automated microscopy. *Cytometry* **12**, 195–206.

Freeman, H. (1961). On the encoding of arbitrary geometric configuration. *IRE Trans. Electron. Comput.* **EC-10**, 260–268.

Gray, J. W., Caranno, A. V., Steinmetz, L. L., Van Dilla, M. A., Moore, D. H. II, Mayall, B. H., and Mendelsohn, M. L. (1975). Chromosome measurement and sorting by flow system. *Proc. Natl. Acad. Sci. USA* **72**, 1231–1234.

Groen, F. C. A., Young, I. T., and Ligthart, G. (1985). Comparison of different focus functions for use in autofocus algorithms. *Cytometry* **6**, 81–91.

Hausmann, M., Dölle, J., Arnold, A., Stepanow, B., Wickert, B., Boscher, J., Popescu, P. C., and Cremer, C. (1992). Development of a two-parameter slit-scan flow cytometer for screening of normal and aberrant chromosomes: application to a karyotype of *Sus scrofa domestica* (pig). *Opt. Eng.* **31**, 1463–1469.

Höfers, Ch., Baumann, P., Hummer, G., Jovin, T. M., and Arndt-Jovin, D. J. (1993). The localization of chromosome domains in human interphase nuclei. Three-dimensional distance determination of fluorescence in situ hybridization signals from confocal laser scanning microscopy. *Bioimaging* **1**, 96–106.

Hopman, A. H. N., Moesker, O., Smeets, A. W. G. B., Pauwels, R. P. E., Vooijs, G. P., and Ramaekers, F. C. S. (1991). Numerical chromosome 1, 7, 9 and 11 aberrations in bladder cancer detected by in situ hybridization. *Cancer Res.* **51**, 644–651.

Hosoda, F., Nishimura, S., Uchida, H., and Ohki, M. (1990). An F-factor based cloning system for large DNA fragments. *Nucleic Acids Res.* **11**, 3863–3869.

Hulspas, R., Houtsmuller, A. B., Krijtenburg, P. J., Bauman, J. G. J., and Nanninga, A. (1994). The nuclear position of pericentromeric DNA of chromosome 11 appears to be random in G0 and non-random in G1 human lymphocytes. *Chromosoma* **103**, 286–292.

Ioannou, P. A., Amemiya, C. T., Garnes, J., Kroisel, P. M., Shizuya, H., Chen, C., Batzer, M. A., and de Jong, P. J. (1994). A new bacteriophage P1-derived vector for the propagation of large human DNA fragments. *Nature Genet.* **6**, 84–89.

Jirsova, P., Kozubek, S., Bartova, E., Kozubek, M., Lukasova, E., Skalnikova, M., Cafourkova, A., Koutna, I., and Buchnickova, K. (2001). Spatial distribution of selected genetic loci in nuclei of human leukemic cells after irradiation. *Radiat. Res.* **155**, 311–319.

Kamentsky, L. A., Burger, D. E., Gershman, R. J., Kamentsky, L. D., and Luther, E. (1997a). Slide-based laser scanning cytometry. *Acta Cytol.* **41**, 123–143.

Kamentsky, L. A., Kamentsky, L. D., Fletcher, J. A., Kurose, A., and Sasaki, K. (1997b). Methods for automatic multiparameter analysis of fluorescence in situ hybridized specimens with a laser scanning cytometer. *Cytometry* **27**, 117–125.

Koch, J., Kolvaa, S., Peetersen, K., Gregersen, N., and Boland, L. (1989). Oligonucleotide priming methods for the chromosome specific labelling of alpha satelite DNA in situ. *Chromosoma* **98**, 259–265.

Koch, J. (1992). Non-radioactive labelling and detection of nucleic acids in situ by "primed in situ labelling" (PRINS). In: *Non-Radioactive In Situ Hybridization. Application Manual.* Boehringer Mannheim Biochemica, pp 41–33.

Koutna, I., Kozubek, S., Zaloudik, J., Kozubek, M., Lukasova, E., Matula, P., Bartova, E., Skalnikova, M., Cafourkova, A., and Jirsova, P. (2000). Topography of genetic loci in tissue samples: towards new diagnostic tool using interphase FISH and high-resolution image analysis techniques. *Anal. Cell. Pathol.* **20**, 173–185.

Kozubek, M. (2001). Theoretical versus experimental resolution in optical microscopy. *Microsc. Res. Tech.* **53**, 157–166.

Kozubek, M. and Matula, P. (2000). An efficient algorithm for measurement and correction of chromatic aberrations in fluorescence microscopy. *J. Microsc.* **200**, 206–217.

Kozubek, M., Kozubek, S., Lukasova, E., Mareckova, A., Bartova, E., Skalnikova, M., and Jergova, A. (1999). High-resolution cytometry of FISH dots in interphase cell nuclei. *Cytometry* **36**, 279–293.

Kozubek, M., Kozubek, S., Lukasova, E., Bartova, E., Skalnikova, M., Matula, Pa., Matula, Pe., Jirsova, P., Ganova, A., and Koutna, I. (2001). Combined confocal and wide-field high-resolution cytometry of FISH stained cells. *Cytometry* **45**, 1–12.

Kozubek, S., Lukasova, E., Ryznar, L., Kozubek, M., Liskova, A., Govorun, R. D., Krasavin, E. A., and Horneck, G. (1997). Distribution of ABL and BCR genes in cell nuclei of normal and irradiated lymphocytes. *Blood* **89**, 4537–4545.

Kozubek, S., Lukasova, E., Mareckova, A., Skalnikova, M., Kozubek, M., Bartova, E., Kroha, V., Krahulcova, E., and Slotova, J. (1999a). The topological organization of chromosomes 9 and 22 in cell nuclei has a determinatiove role in the induction of t(9,22) translocations and in the pathogenesis of t(9,22) leukemias. *Chromosoma* **108**, 426–435.

Kozubek, S., Lukasova, E., Ryznar, L., Kozubek, M., Liskova, A., and Kroha, V. (1999b). Chromatin structure and its cell cycle kinetics in intact and irradiated cell nuclei. In: *Fundamentals for the Assessment of Risks from Environmental Radiation*. (eds. C. Baumstark-Khan, S. Kozubek, and G. Horneck). Dordrecht: Kluwer, pp. 231–242.

Kreth, G., Munkel, C., Langowski, J., Cremer, T., and Cremer, C. (1998). Chromatin structure and chromosome aberrations: modeling of damage induced by isotropic and localized irradiation. *Mutat. Res.* **404**, 77–88.

Kurz, A., Lampel, S., Nickolenko, J. E., Bradl, J., Benner, A., Zirbel, R. M., Cremer, T., and Lichter, P. (1996). Active and inactive genes localize preferentially in the periphery of chromosome territories. *J. Cell Biol.* **135**, 1195–1205.

Kurzrock, R., Gutterman, J. L., and Talkpaz, M. (1988). The molecular genetics of Philadelphia chromosome-positive leukemias. *N. Engl. J. Med.* **319**, 990–998.

Lucas, J. N. and Gray, J. W. (1987). Centromeric index versus DNA content flow karyotypes of human chromosomes measured by means of slit-scan flow cytometry. *Cytometry* **8**, 273–279.

Lukasova, E., Kozubek, S., Kozubek, M., Kroha, V., Mareckova, A., Skalnikova, M., Bartova, E., and Slotova, J. (1999). Chromosomes participating in translocations typical of malignant haemoblastoses are also involved in exchange aberrations induced by fast neutrons. *Radiat. Res.* **151**, 375–384.

Munkel, C., Eils, R., Dietzel, S., Zink, D., Mehring, C., Wedemann, G., Cremer, T., and Langowski, J. (1999). Compartmentalization of interphase chromosomes observed in simulation and experiment. *J. Mol. Biol.* **285**, 1053–1065.

Netten, H., Van Vliet, L. J., Vrolijk, H., Sloos, W. C. R., Tanke, H. J., and Young, I. T. (1996). Fluorescent dot counting in interphase cell nuclei. *Bioimaging* **4**, 93–106.

Netten, H., Young, I. T., Van Vliet, L. J., Tanke, H. J., Vrolijk, H., and Sloos, W. C. R. (1997). FISH and chips: Automation of fluorescent dot counting in interphase cell nuclei. *Cytometry* **28**, 1–10.

Nikiforova, M. N., Stringer, J. R., Blough, R., Medvedovic, M., Fagin, J. A., and Nikiforov, Y. E. (2000). Proximity of chromosomal loci that participate in radiation-induced rearrangements in human cells. *Science* **290**, 138–141.

Nipkow, P. (1884). Elektrisches teleskop. *Kaiserliches Patentamt, Patentschrift* **30**, 105–105.

Ormerod, M. G. (1994). *Flow Cytometry: A Practical Approach*, 2nd edn. Oxford: IRL Press.

Ortiz de Solórzano, C., Garcia Rodrigues, E., Jones, A., Pinkel, D., Gray, J. W., Sudar, D., and Lockett, S. J. (1998a). Segmentation of confocal microscope images of cell nuclei in thick tissue sections. *J. Microsc.* **193**, 212–226.

Ortiz de Solórzano, C., Santos, A., Vallcorba, I., García-Sagredo, J. M., and del Pozo, F. (1998b). Automated FISH spot counting in interphase nuclei: Statistical validation and data correction. *Cytometry* **31**, 93–99.

Parreira, L., Telhada, M., Ramos, C., Hernandez, R., Neves, H., and CarmoFonseca, M. (1997). The spatial distribution of human immunoglobulin genes within the nucleus: evidence for gene topography independent of cell type and transcriptional activity. *Hum. Genet.* **100**, 588–594.

Petran, M., Hadravsky, M., Egger, M. D., and Galambos, R. (1968). Tandem-scanning reflected-light microscope. *J. Opt. Soc. Am.* **58**, 661–664.

Pierce, J. C., Sternberg, A., and Sauer, B. (1992). A mouse genomic library in the bacteriophage P1cloning system. Organisation and characterisation. *Mamm. Genome* **3**, 550–558.

Pinkel, D., Landegent, J., Collins, C., Fuscoe, J., Segraves, R., Lucas, J., and Gray, J. (1988). Fluorescence in situ hybridization with human chromosome specific libraries: detection of trisomy 21 and translocation of chromosome 4. *Proc. Natl. Acad. Sci. USA* **85**, 9138–9142.

Pinkel, D., Seagraves, R., Sudar, D., Clark, S., Poole, I., Kowbel, D., Collins, C., Kuo, W. L., Chen, C., Zhai, Y., Dairkee, S. H., Ljung, B. M., Gray, J. W., and Albertson, D. G. (1998). High resolution analysis of DNA copy number variation using comparative genomic hybridization to microarrays. *Nature Genet* **20**, 207–211.

Popp, S., Scholl, H. P., Loos, P., Jauch, A., Stelzer, E., Cremer, Ch., and Cremer, T. (1990). Distribution of chromosome 18 and X centric heterochromatin in the interphase nucleus of cultured human cells. *Exp. Cell Res.* **189**, 1–12.

Pratt, W. K. (1991). *Digital Image Processing*. New York: Wiley.

Preston, D. L., Kusumi, S., Tomonage, M., Izumi, S., Ron, E., Kuramoto, A., Kamada, N., Dohy, H., Matui, T., Nonaka, H., Thompson, E. E., Soda, M., and Mabuchi, K. (1994). Cancer incidence in atomic bomb survivors. Part III: Leukemia, lymphoma and multiple myeloma, 1950–1987. *Radiat. Res.* **137**, S68–S97.

Rabbitts, T. H. (1994). Chromosomal translocations in human cancer. *Nature* **372**, 143–150.

Sadoni, N., Langer, S., Fauth, C., Bernardi, G., Cremer, T., Turner, B. M., and Zink, D. (1999). Nuclear organization of mammalian genomes: polar chromosome territories build up functionally distinct higher order compartments. *J. Cell Biol.* **146**, 1211–1226.

Serra, J. (1982). *Image Analysis and Mathematical Morphology*. London: Academic Press.

Skalnikova, M., Kozubek, S., Lukasova, E., Bartova, E., Jirsova, P., Cafourkova, A., Koutna, I., and Kozubek, M. (2000). Spatial arrangement of genes, centromeres and chromosomes in human blood cell nuclei and its changes during the cell cycle, differentiation and after irradiation. *Chromosome Res.* **8**, 487–499.

Strauss, M., Lukas, J., and Bartek, J. (1995). Unrestricted cell cycling and cancer. *Nature Med.* **1**, 1245–1246.

Sun, H. B. and Yokota, H. (1999). Correlated positioning of homologous chromosomes in daughter fibroblast cells. *Chromosome Res.* **7**, 603–610.

Tanke, H. J., Florijn, R. J., Wiegant, J., Raap, A. K., and Vrolijk, J. (1995). CCD microscopy and image analysis of cells and chromosomes stained by fluorescence in situ hybridization. *Histochem. J.* **27**, 4–14.

Tkatchuk, D., Westbrook, C., Andreef, M., Donlon, T. A., Cleary, M. L., Suryanarayan, K., Homge, M., Redner, A., Gray, J. W., and Pinkel, D. (1990). Detection of BCR–ABL fusion in chronic myelogeneous leukemia by two colour fluorescence in situ hybridization. *Science* **220**, 559–562.

Török, P. and Sheppard, C. J. R. (2002). The role of pinhole size in high-aperture two- and three-photon microscopy. In: *Confocal and Two-Photon Microscopy: Foundations, Applications, and Advances* (ed. A. Diaspro). New York, Wiley-Liss, Chap. 6.

van den Engh, G., Sachs, R., and Trask, B. J. (1992). Estimating genomic distance from DNA sequence location in cell nuclei by a random walk model. *Science* **257**, 1410–1411.

Vincent, L. and Soille, P. (1991). Watersheds in digital spaces: an efficient algorithm based on immersion simulations. *IEEE Trans. Pattern Anal.* **13**, 583–598.

Weiss, M. M., Hermsen, M. A., Meijer, G. A., van Grieken, N. C., Baak, J. P., Kuipers, E. J., and van Diest, P. J. (1999). Comparative genomic hybridisation. *Mol. Pathol.* **52**, 243–251.

Wilkens, L., Tchinda, J., Burkhardt, D., Nolte, M., Werner, M., and Georgii, A. (1998). Analysis of hematologic diseases using conventional karyotyping, fluorescence in situ hybridisation and comparative genomic hybridisation. *Hum. Pathol.* **29**, 833–839.

Wilson, T. (2002). Confocal microscopy: basic principles and architectures. In: *Confocal and Two-Photon Microscopy: Foundations, Applications, and Advances* (ed. A. Diaspro). New York: Wiley-Liss, Chap. 2.

Young, J. Z. and Roberts, F. (1951). A flying spot microscope. *Nature* **167**, 231.

Zink, D. and Cremer, T. (1998). Cell nucleus: Chromosome dynamics in nuclei of living cells. *Curr. Biol.* **8**, R321–R324.

CHAPTER 20

Two-Photon Imaging of Tissue Physiology Based on Endogenous Fluorophores

LILY HSU,[1] KI H. KIM,[1] CHEN Y. DONG,[2] PETER KAPLAN,[3] THOMAS HANCEWICZ,[3] CHRISTOF BUEHLER,[4] KEITH M. BERLAND,[5] BARRY R. MASTERS,[6] and PETER T. C. SO[1]

[1] Department of Mechanical Engineering, Massachusetts Institute of Technology, 77 Massachusetts Avenue, Cambridge, MA 02139, U.S.A.; [2] Department of Physics, National University of Taiwan, Taipei, Taiwan; [3] Unilever Edgewater Laboratory, Edgewater, NJ, U.S.A.; [4] Paul Scherrer Insitut, Villigen, Switzerland; [5] Department of Physics, Emory University, Atlanta, GA, U.S.A.; and [6] Department of Ophthalmology, University of Bern, Bern, Switzerland

1 INTRODUCTION

The presence of a variety of endogenous fluorophores contributes to the ubiquitous nature of tissue autofluorescence. Microscopic imaging based on tissue autofluorescence has recently emerged as a powerful method to characterize tissue structures and functions on the submicrometer scale. Optical imaging based on autofluorescence is inherently minimally invasive, since tissue structures can be visualized in animals or humans without fixation or staining. Further, since the spectra of endogenous fluorophores are indicative of tissue biochemical composition and metabolic state, autofluorescence imaging also provide functional information about the specimen.

The use of endogenous fluorescence as an intrinsic monitor of tissue physiological states has long been recognized (MacMunn, 1880, 1914; Keilin, 1966). The redox metabolism of biological specimens can be monitored based on the fluorescence of reduced pyridine nucleotides, NAD(P)H, and the oxidized flavoproteins. Work in this area has been thoroughly covered in many previous reviews (Chance 1989, 1991; Masters, 1990; Masters and Chance, 1999). In addition to cellular proteins, extracellular matrix components, such as collagen and elastin, are also strongly fluorescent (LaBella, 1961; Thomas et al., 1963; Dabbous, 1996; Hoerman and Balekjian, 1966; Liotta, 1984; Dayan et al., 1994; Banerjee 1998, 1999; Izuishi et al., 1999; Marcu et al., 1999).

Confocal and Two-Photon Microscopy: Foundations, Applications, and Advances, Edited by Alberto Diaspro.
ISBN 0-471-40920-0

The universal presence of endogenous fluorophores allows minimally invasive two-photon imaging of tissue structures. Equally importantly, the advent of two-photon microscopy has provided an opportunity to better understand tissue biochemistry. Although the spectroscopic properties of many endogenous fluorophores have been characterized in vitro, the quantification of tissue autofluorescence in vivo remains a formidable problem. For many tissues, we do not even have a complete account of all the constituent endogenous fluorophores; their relative abundance and microscopic distribution are even less well quantified. The majority of work in this area has examined the overall spectrum from a macroscopic sample volume. This spectrum results from the emission of a mixture of many different fluorophores, and does not allow easy decomposition into individual components. The recent advent of confocal and two-photon microscopy promises to partially resolve this difficulty (Denk et al., 1990, 1991, 1995a; So et al., 2000). By optically isolating spectroscopic signal from a sufficiently small tissue volume where there are few distinct species, a better categorization of tissue fluorophores inside different tissues becomes possible. Correlative studies of tissue structures with the distributions of tissue fluorophores and their spectroscopic states also may provide new insight into tissue physiology.

In this chapter, the methodology of using two-photon endogenous fluorescence tissue imaging will be examined. We shall provide a survey of some endogenous fluorophores that are accessible using two-photon microscopy and shall review their known spectroscopic properties. We shall further present two recent technological advances in two-photon tissue imaging: video-rate two-photon microscopy and high-spatial-resolution spectroscopic analysis.

2 TWO-PHOTON TISSUE IMAGING BASED ON ENDOGENOUS FLUOROPHORES

Two-photon excitation is well suited for autofluorescence tissue imaging. The principle of two-photon excitation was described by Maria Göppert-Mayer (Göppert-Mayer, 1931). A detailed discussion of two-photon processes is presented in other chapters of this book, and will not be thoroughly examined here. In brief, two-photon excitation occurs when an electron is excited to a higher-energy electronic state by the simultaneous absorption of two photons in the infrared (IR) wavelength region. Each of the two IR photons has half of the energy required for the excitation. Fluorescence occurs when the excited electron returns to the ground state and emits a photon. Since the cross-section of the two-photon process is significantly less than that of the one-photon process, and the two-photon fluorescence has a quadratic dependence upon excitation power, two-photon excitation mainly occurs inside a localized subfemtoliter excitation volume. Outside this focal volume, there is insufficient photon density to produce appreciable two-photon fluorescence. The localization of excitation provides depth discrimination and allows three-dimensional imaging in a laser scanning microscope system as developed by Denk and Webb (Denk et al., 1990).

In addition to two-photon systems, confocal microscopy has also been applied for tissue fluorescence imaging (Corcuff et al., 1993; Bertrand and Corcuff, 1994; Masters and Thaer, 1994; Rajadhyaksha et al., 1995; Masters 1996, 1998; Masters et al., 1998a, 1999). However, two-photon excitation has a number of important advantages over the confocal approach for imaging tissue autofluorescence:

1. Since two-photon excited fluorescence only occurs in a small region, tissue photodamage and photobleaching are reduced. This results in enhanced specimen viability.
2. Endogenous tissue fluorophores have absorption in the ultraviolet (UV) spectral range. Since UV radiation is strongly absorbed and scattered by tissue, UV microscopic imaging of tissue typically has a very short penetration depth. The use of two-photon methods allows the excitation of UV fluorophores in the IR wavelength region, where tissue scattering and absorption are greatly reduced. Tissue autofluorescence imaging based on two-photon excitation typically has a better penetration depth.
3. Two-photon microscopy achieves volume localization based on the excitation process. In contrast, confocal microscopy reduces the observation volume based on the presence of a conjugated pinhole aperture in the detection light path. The unavoidable scattering of the emitted photons by optically opaque tissue reduces the fluorescence signal that would be transmitted through the pinhole aperture (Fig. 1). Two-photon microscopy typically provides superior signal-to-noise ratio inside strongly scattering samples.
4. The wide spectral separation between excitation and emission light allows accurate determination of the emission spectrum.

Although the use of IR wavelengths in two-photon microscopy reduces the effect of scattering and absorption, these effects may not be completely negligible in thick specimens. Both the absorption and scattering coefficients of many tissue types have been characterized. In the red/near-IR range, the absorption, scattering, and reduced scattering coefficients are around 0.01–1 mm^{-1}, 1–100 mm^{-1}, and 0.1–5 mm^{-1}, respectively (Flock et al., 1987; Cheong et al., 1990; Wilson and Jacques, 1990). There have been a number of reports on the imaging characteristics of two-photon microscopy in turbid media (Blanca and Saloma, 1998; Centonze and White, 1998; Daria et al., 1998; de Grauw et al., 1999; Gerritsen and de Grauw, 1999; Dunn et al., 2000). Tissue absorption and scattering not

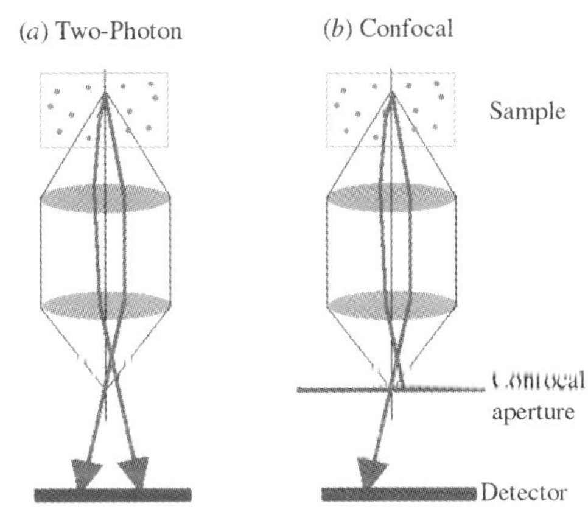

FIGURE 1 See ftp site for color figure. The effect of a confocal pinhole aperture on photon-collection efficiency in the presence of a strongly scattering specimen. (*a*) In the two-photon case, there is no pinhole and the large detector collects both deflected and undeflected photons. (*b*) In the confocal case, the deflected photons are blocked by the pinhole.

only may limit the penetration depth but may also affect image resolution. Previous studies in a few specimens have indicated that the presence of multiple scattering has negligible effects on the point spread function (PSF) (Centonze and White, 1998; Dunn et al., 2000). To further confirm these results, a thorough characterization of the two-photon PSF has been performed on a number of tissue phantoms with scattering coefficients that span typical tissue values. In these tissue phantoms, sub-diffraction-limited microspheres are immobilized in a medium with known scattering coefficients. A direct measurement of the PSF is accomplished by analyzing the images of these microspheres. These experiments have shown that the PSF is not dependent upon sample scattering coefficients over the range of scattering coefficients tested, confirming previous results (Fig. 2). However, one should also note that there exists contradictory data showing PSF degradation in the presence of multiple scattering (Schilders and Gu, 2000). Tissue scattering, absorption, and refractive-index variation all play roles in affecting microscopic resolution for tissue imaging. It is clear that additional experiments are required to further quantify these effects.

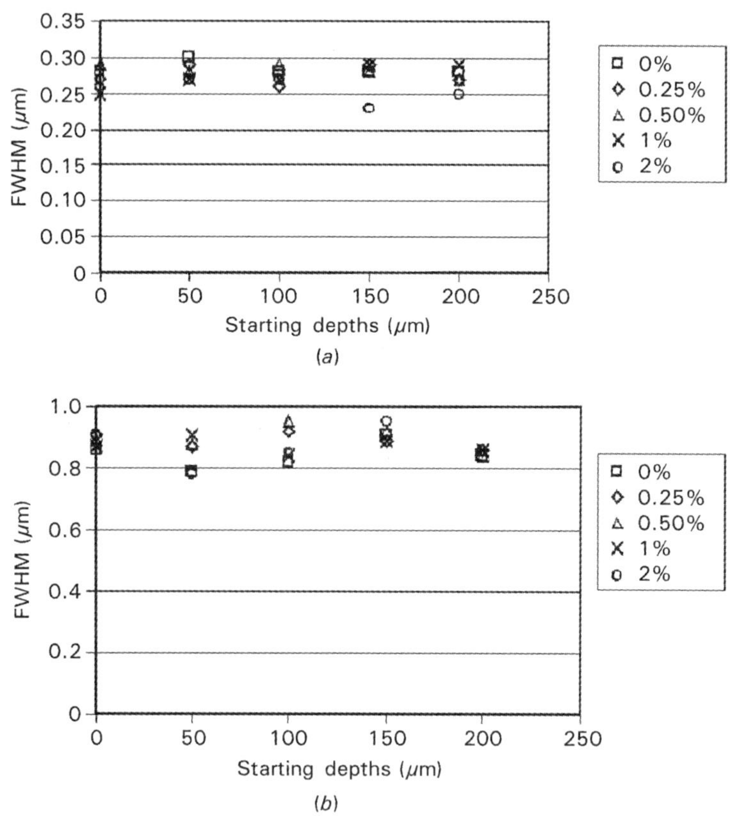

FIGURE 2 Point spread functions (PSFs) measured in tissue phantoms with different scattering coefficients. (*a*) The full-width at half-maximum (FWHM) of the PSF along the radial direction as a function of the specimen scattering coefficient. (*b*) The FWHM of the PSF along the axial direction as a function of the specimen scattering coefficient. Scattering coefficients of the specimens were varied by the volume fraction of intralipid (Liposyn III, Abbott Laboratories, IL).

3 A SURVEY OF TISSUE ENDOGENOUS FLUOROPHORES

As discussed previously, the different endogenous fluorescence species present in tissues are not completely categorized. We shall survey some of the best studied tissue endogenous fluorophores, including tryptophan, NAD(P)H, flavoproteins, collagen, elastin, and protoporphyin IX.

Tryptophan, the amino acid with the highest extinction coefficient and quantum yield, is an important contributor to native protein fluorescence under UV excitation (Kierdaszuk et al., 1995; Gryczynski et al., 1996; Banerjee et al., 1998; Brancaleon et al., 1999; Lakowicz, 1999). Tryptophan has a one-photon absorption maximum of 280 nm and an emission maximum at about 340 nm. The exact two-photon absorption maximum for tryptophan has not been measured, but is known to be below 700 nm (Xu et al., 1995, 1996). Tissue imaging based on tryptophan fluorescence provides information on protein content. Although pilot studies of tryptophan imaging based on two-photon excitation using a femtosecond colliding-beam dye-laser system have been performed, tissue imaging based on tryptophan fluorescence is relatively rare (Guo et al., 1998, 1999). The major difficulty lies in the difficulty in maintaining colliding-beam dye-laser systems. The more stable titanium : sapphire (Ti : sapphire) lasers used in most of the current multiphoton systems are only tunable from 700 to 1000 nm. These wavelengths are too long to excite tryptophan based on the two-photon process. However, there is evidence that tryptophan fluorescence may be accessible by three-photon excitation at about 900 nm (Masters et al., 1997).

NAD(P)H is the best studied endogenous fluorophore (Chance and Thorell, 1959; Chance, 1976, 1989, 1991; Chance et al., 1979; Masters, 1990; Masters and Chance, 1999). NAD(P)H emission is a major component in intracellular autofluorescence. The one-photon absorption maximum for NAD(P)H is at about 360 nm, with a typical emission maximum in the range between 400 and 500 nm. The two-photon absorption maximum of NAD(P)H is known to be below 700 nm (Xu et al., 1995, 1996; Xu and Webb, 1996). NAD(P)H distributes throughout the cytosol in the cells, but is concentrated in the mitochondria. Most importantly, NAD(P)H fluorescence is a sensitive indicator of cellular metabolism. The oxidized form, NAD, is virtually not fluorescent. However, the reduced forms of this nucleotide, NADH and NADPH, are highly fluorescent. NAD(P)H fluorescence intensity and lifetime are useful indicators of cellular redox state. The first two-photon tissue-imaging study was based on NAD(P)H fluorescence in the cornea system (Piston et al., 1995).

Another class of endogenous redox indicators comprises flavoproteins (Chance et al., 1979; Anidjar et al., 1996; Masters et al., 1997; Hockberger et al., 1999). Typical flavoproteins have one-photon excitation wavelengths of about 450 nm and emission maxima in the ranges 520–590 nm. The two-photon excitation maxima of flavin mononucleotides are below 700 nm (Xu et al., 1995, 1996; Xu and Webb, 1996). Flavoproteins are only present in the mitochondria, and are not found in the cytoplasm. Unlike NAD(P)H, flavoproteins are fluorescent in their oxidised form, but are minimally fluorescent when reduced. Since flavoprotein fluorescence is significantly lower than that of NAD(P)H in typical cells, it has not been as widely used in tissue redox spectroscopy and imaging. However, tissue imaging using both NAD(P)H and flavoprotein signals may be a very promising approach to better quantify cellular redox activities (Huang et al., 2000).

In addition to intracellular proteins, the major components of the tissue extracellular matrix are also autofluorescent. Two-photon microscopy is capable of imaging collagen/elastin bundles in excised tissues based only on their native fluorescence (LaBella, 1961; Thomas et al., 1963; Dabbous, 1966; Hoerman and Balekjian, 1966; Liotta, 1984; Dayan et al., 1994; Banerjee et al., 1998, 1999; Izuishi et al., 1999; Marcu et al., 1999). In contrast to intracellular autofluorescent proteins, the fluorescence properties of extracellular-matrix components have received relatively little attention. Nevertheless, basic spectroscopic characterizations of these tissue components have been performed. Tropocollagen and collagen fibers typically have one-photon absorption in the range of 280–300 nm and emission in the range of 350–400 nm. Elastin has been observed to have a redder emission spectrum at 400–450 nm with one-photon excitation wavelengths of 340–370 nm. The two-photon excitation and emission spectra of these tissue components are not well characterized. There is also recent evidence that there is significant second-harmonic generation in collagen structures.

Protoporphyrin IX (PpIX) is an important naturally occurring fluorophore that is typically produced in very small quantities in most tissues (Onizawa et al., 1996; Johansson et al., 1997; Svanberg et al., 1998; af Klinteberg et al., 1999; Inaguma and Hashimoto, 1999; Mehlmann et al., 1999; Madsen et al., 2000). However, the normal production of PpIX can be upregulated by the application of 5-aminolevulinic acid. PpIX can be excited by one-photon excitation at approximately 375–425 nm with fluorescence emission peaks at 635 and 705 nm. The two-photon absorption spectrum of PpIX remains to be fully characterized. Some recent studies have shown that PpIX preferentially accumulates in cancerous tissues and offers the potential of using PpIX as an endogenous marker for tissue malignancy.

TISSUE IMAGING APPLICATIONS BASED ON TWO-PHOTON-EXCITED ENDOGENOUS FLUOROPHORES

The power of two-photon microscopy for tissue imaging based on exogenous labels has been amply demonstrated in neurobiology (Denk, 1994; Denk et al., 1995b; Yuste and Denk, 1995; Svoboda et al., 1996, 1997, 1999; Kleinfeld et al., 1998; Ricken et al., 1998; Furuta et al., 1999; Horch et al., 1999; Ricken et al., 1998; Yuste et al., 1999) and embryology (Williams et al., 1994; Summers et al., 1996; Piston et al., 1998; Squirrell et al., 1999). The successes of two-photon imaging in these areas are partially due to the relative ease of introducing exogenous probes or creating transgenic animals expressing fluorescent proteins. Since exogenous probes typically have high fluorescent absorption coefficient and quantum yield, high-contrast images can be produced. Further, since many of these exogenous indicators, like calcium probes, have precisely designed spectral responses to environmental factors, it is relatively straightforward to make definitive assessments of the tissue biochemical environment. In contrast, the use of endogenous fluorophores for tissue imaging and spectroscopy is more difficult. Since typical endogenous fluorophores have low quantum yield and are more prone to photobleaching, it is often more difficult to obtain images or spectra with high signal-to-noise ratio. Further, the fluorescence properties of many endogenous fluorophores have complex relationships with their environment. For example, the binding of NAD(P)H with other proteins significantly modifies its emission intensity and spectrum. Nevertheless, the use of endogenous fluorophores has resulted in a number of interesting studies. This technique has been

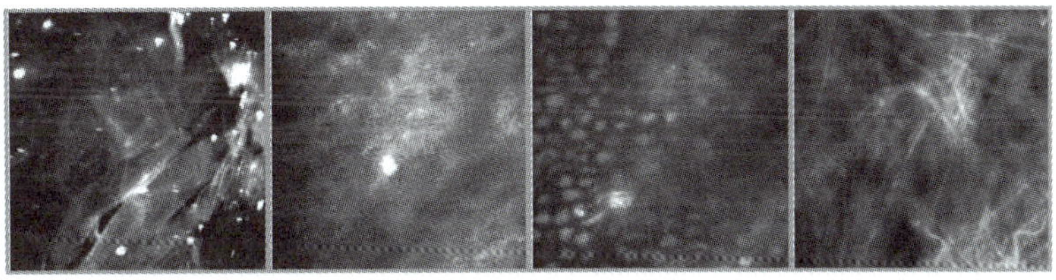

FIGURE 3 Two-photon images of ex vivo human skin at planes corresponding to (*a*) the stratum corneum, (*b*) the keratinocyte layer, (*c*) the basal cell layer, and (*d*) the dermal layer.

successfully applied to study the physiology of corneal structures in the eye (Piston et al., 1995; Buehler et al., 1999), human and mouse dermal and subcutaneous structures (Masters et al. 1997, 1999; Teuchner et al., 2000), and the metabolic processes of pancreatic islets (Bennett et al., 1996; Piston and Knobel, 1999; Piston et al., 1999).

A promising direction for two-photon tissue imaging based on endogenous fluorescence is in clinical diagnosis. Traditional biopsy involves the removal, fixation, and imaging of tissues. Most patients would prefer to avoid these invasive procedures. Further, the histological processing steps do not completely preserve the inherent biochemical information of the tissue that may be useful in clinical diagnosis. The potential of optical biopsy based on two-photon excitation has been demonstrated. Two-photon microscopy has successfully imaged skin structure of human volunteers down to a depth of 150 μm (Masters et al., 1997; Kim et al., 1998). Four distinct structural layers in the epidermis and the dermis can typically be resolved, including the stratum corneum, the keratinocyte layer, the basal cell layer and the dermis (Fig. 3). Current experiments are focusing on identifying pathological states in tissues based on the three-dimensional images obtained from two-photon microscopy.

5 FRONTIERS IN TWO-PHOTON TISSUE IMAGING BASED ON ENDOGENOUS FLUOROPHORES

Two areas of technology development are posed to greatly expand the utility of two-photon autofluorescence tissue imaging. One of these is two-photon video-rate microscopy. Since one of the main applications of two-photon autofluorescence imaging is in clinical diagnosis, where efficiency is critical, video-rate imaging capability has been developed for the rapid sampling of a large tissue area. Another application is to high-spatial resolution spectroscopic analysis. Advances in spectroscopic analysis are vital to identify all the tissue endogenous fluorophores and their relationships to tissue physiological states.

5.1 Two-Photon Video-Rate Microscopy

A number of video-rate two-photon microscope designs have been produced. Video-rate two photon microscopy can be divided into two classes. The first class of two-photon video-rate microscope achieves a high frame rate by simultaneously exciting fluorophores at multiple spatial locations (Guild and Webb, 1995; Brakenhoff et al., 1996; Bewersdorf et al., 1998; Buist et al., 1998). This type of microscope includes line-scanning and multifocal-scanning systems. The second class of video-rate system achieves the required speed by scanning a single excitation point using high-speed scanners (Fan et al., 1999; Kim et al., 1999). This type of microscope has employed high-speed resonance mirrors and rotating polygonal mirrors. A design of video-rate two-photon system based on a high-speed rotating polygonal mirror is shown in Figure 4. A 50-facet rotating polygonal mirror (Lincoln Laser, Phoenix, AZ) is used for scanning along the *x* axis (the fast axis). Because it rotates at a constant speed, it can achieve a higher scanning speed compared with a galvanometer-driven scanner. The rotation speed of the polygonal mirror can be selected from 3600 rpm to 30,000 rpm. Assuming that one frame is composed of 256×256 pixels, the frame rate ranges from 12.7 to 100 frames s^{-1}. For the *y* axis, a slower galvanometer-driven mirror (Cambridge Technology, Watertown, MA) is used. To control imaging depth, a piezoelectrically driven objective-lens translator (P-721.00, Physik Instrumente, Waldbronn, Germany) is used. The femtosecond pulse laser beam originating from a titanium : sapphire laser (Mira 900; Coherent, Palo Alto, CA) first deflects off the polygonal scanning mirror. A pair of relay lenses projects the light from the polygonal mirror onto the galvanometer-driven scanners. The laser beam is coupled into an upright microscope (Axioscope, Zeiss, Thornwood, NY) via a modified epiluminescence light path. The microscope is placed such that its telecentric plane intersects with the stationary point at the *y*-axis scan mirror. An eyepiece lens and a tube lens form a beam expander such that the back aperture of the objective lens is overfilled. The beam is reflected by a dichroic mirror towards the objective, and is focused on the specimen.

The induced fluorescence signal is collected by the same objective, passes through the dichroic mirror, and is collected at a highly sensitive photomultiplier tube (PMT) (R3896, Hamamatsu, Bridgewater, NJ). Residual scattered light is removed by an additional barrier filter (Schott BG39, Chroma Technology, Brattleboro, NH). A transimpedance amplifier (C1053-51, Hamamatsu) then converts current output from the PMT to a voltage signal with amplification. The voltage signal is sampled with a 12-bit A/D converter (AD9220EB, Analog Device, Norwood, MA) at approximately 11 MHz, and is transferred to the custom-made interface card. The sampling rate has been chosen to ensure all the signals from the incoming photons are sampled faster than the bandwidth (5 MHz) of the transimpedance amplifier. The interface card generates the signal for each pixel by summing the digital readout during the pixel residence time (720 ns), and transfers the results to a computer. Although this analog detection method has a lower signal-to-noise ratio than a single-photon counting method, it is suitable for video-rate imaging, where the use of high laser power results in a non-negligible probability that multiple photons may

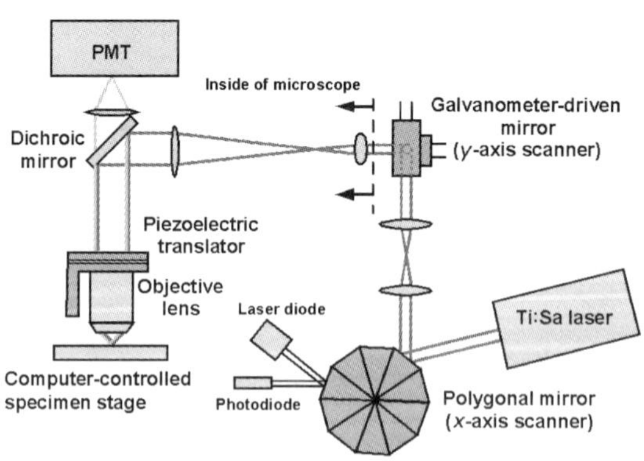

FIGURE 4 See ftp site for color figure. The schematics of a video-rate two-photon microscope based on a polygonal scanner.

arrive simultaneously at the detector. A separate laser diode (1 mW at 632 nm, Thorlab, Newton, NJ), along with a photodiode detector (Thorlab, Newton, NJ), is used to encode the polygonal mirror position and to generate a reference signal. This signal is used to synchronize the *x* and *y* scanners, the objective translator, and the PMT. A PCI-bus interface card for computer control is designed based on the reconfigurable-logic chip (XCS30, Xilinx, San Jose, CA). It receives the reference signal, and generates synchronous digital signals for actuation and acquisitions. It is composed of a PCI-bus controller board (S5933DK1, AMCC, San Jose, CA) and a custom-made piggyback board with the reconfigurable-logic chip. Most of the digital circuit for scanner control and data acquisition is implemented in the reconfigurable-logic chip. Therefore, modifications of the interface card can be easily implemented.

The ability of this system to perform high-speed tissue imaging based on endogenous fluorescence can be demonstrated in the imaging of a human skin sample. The objective lens used in this experiment has 40× magnification (Fluar, NA 1.3, oil; Zeiss, Thornwood, NY). The lateral dimension of each image is approximately 100 μm × 100 μm. Fifty layers were imaged for each section from the surface down to 67 μm with a piezoelectric objective-lens translator. Two representative layers are shown in Figure 5. The structures observed in the images correspond to collagen/elastin fibers that contribute to most of the fluorescence in the dermal layer.

5.2 High-Spatial-Resolution Spectroscopic Analysis

In addition to structural imaging of tissue, fluorescence spectroscopy provides vital complementary information. Tissue spectroscopy is a powerful method for identifying endogenous fluorescence species. Fluorescence spectroscopy has been used to characterize different tissue types, such as colon, lung, cervix, and skin (Zeng et al., 1995; Nilsson et al., 1997). In addition to tissue type characterization, spectroscopy has been used to determine the physiological state of tissue (Richards-Kortum et al., 1989, 1991; Cothren et al., 1990). The differences in tissue excitation spectra may further allow the detection and diagnosis of disease, such as distinguishing between diseased and normal tissues (Glasgold et al., 1994; Beauvoit and Chance, 1998; Ganesan et al. 1998; Licha et al., 2000).

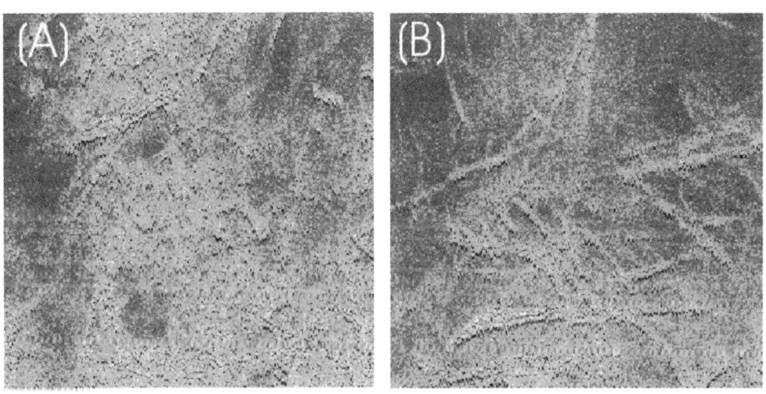

FIGURE 5 See ftp site for color figure. Video-rate images of collagen/elastin fibers in ex vivo human skin. (*a*) Image acquired at a depth of 60 μm. (*b*) Image acquired at a depth of 120 μm at the same location as (*a*).

Spectroscopy information can be extracted from the entire volume, a two-dimensional section, or even a single voxel in the sample.

We are developing high-spatial-resolution spectroscopic methods that have several advantages over the traditional approach where the spectrum is gathered from a macroscopic sample volume. As discussed previously, the spectrum obtained from macroscopic measurements can be difficult to interpret owing to the mixing of many different fluorophores. The composite fluorescence spectrum is also strongly dependent upon other tissue properties, such as scattering and absorption coefficients. The localized excitation of two-photon microscopy ensures that the fluorescence signal from a subfemtoliter sample volume that contains only a few independent fluorescent species can be acquired and analyzed. By reducing the mixing of different fluorescent species, the identity and contributions of each pure biochemical species can be assessed quantitatively.

The acquisition of spectral data from femtoliter microscopic volumes allows the extraction of "purer" spectra containing relatively fewer components. However, it is not uncommon that the autofluorescence from a single image voxel still originates from two or more endogenous fluorescent species. Therefore, a basic problem in spectroscopic image analysis is to resolve both spectral and concentration information from the spectra collected at each pixel. Multivariate curve-resolution (MCR) techniques are being developed to tackle this problem (Hancewicz and Petty, 1995; Andrew and Hancewicz, 1998).

As a demonstration of this approach, a multivariate curve-resolution approach was applied to data obtained from an ex vivo human skin specimen. The data to be analyzed are four-dimensional, including three spatial dimensions and an excitation wavelength dimension (x, y, z, λ). Ignoring possible correlation between pixels, the four-dimensional data set is converted into two dimensions (n, λ) for further analysis (two-way factor analysis), where n is the index of the voxels in a three-dimensional data set. Since no correlation between pixels is assumed in MCR analysis, the observation of a correlation between the distributions of biochemical species with the physiological structural features serves as an independent verification of the accuracy of this approach. Two-way factor analysis assumes a bilinear data structure, and attempts to separate the original data matrix of spectra, **D**, into two submatrices, **C** and **S**:

$$\mathbf{D} = \mathbf{C}\mathbf{S}^T + \mathbf{E},$$

where **C** is a matrix of coefficients related to the real concentration profiles (scores), **S** is a matrix of vectors related to the real spectra (factors), and **E** is a matrix of spectral residuals.

The above equation describes the standard abstract principal-component analysis approach. Since this solution is general, there are many possible ways to satisfy the equation. **C** and **S** are composed of linear combinations of the real concentration profiles, and linear combinations of real spectral profiles of the pure chemical species present in the original dataset. In other words, there is no unique solution to the equation; many possible solutions exist that will fit the model equally well. Since the objective is to produce pure factors (spectra) and scores (concentrations), the number of possible solutions has to be restricted in some chemically or physically meaningful way that will force the resulting concentration and spectral profiles to reflect those of the real components of the specimen. The standard way to achieve this is through a process called constrained alternating least-squares regression (ALS). ALS is also sometimes referred to as simply alternating regression (AR). The idea behind ALS is to impose realistic boundary constraints on the least-squares solution that will force the spectral profiles and concentration profiles to conform to what is known a priori about chemically significant aspects of the data structure. One of the most commonly imposed constraints is positivity, which states that spectral profiles and concentration profiles cannot be less than zero. These constraints are imposed on the data structure during the least-squares optimization process.

From the MCR analysis, at least four factors (biochemical components) contribute to most of the autofluorescence that was observed in human skin. The MCR analysis provides us with two pieces of information about each chemical component. Each piece of information alone is not sufficient to identify the different chemical components that are present. First, it provides us with the excitation spectrum of the individual component. Second, it provides us with the depth profile of the component. The depth profile shows the distribution of each of the components in the skin as a function of depth. Thus, we can understand the location and concentration of each component within the skin. Using all of this information, we can reconstruct each section of skin for each individual component. This allows us to study the structure of a particular chemical component and eventually compare it with known structures obtained by more traditional techniques such as histopathology.

One of the factors that was identified corresponds to collagen/elastin fibers present in the dermis (Fig. 6). We see that the depth profile follows the expected distribution of collagen/elastin fibers. It is concentrated in the sample at approximately 35–60 μm depth, but also exists in high concentrations at greater depths, which corresponds to the dermis. Another factor that was identified corresponds to melanin (Fig. 7). The distribution of this factor correlates very well with the known melanin distribution in skin. We found that this factor is concentrated at a depth of approximately 30–40 μm, at the position of the dermal–epidermal junction.

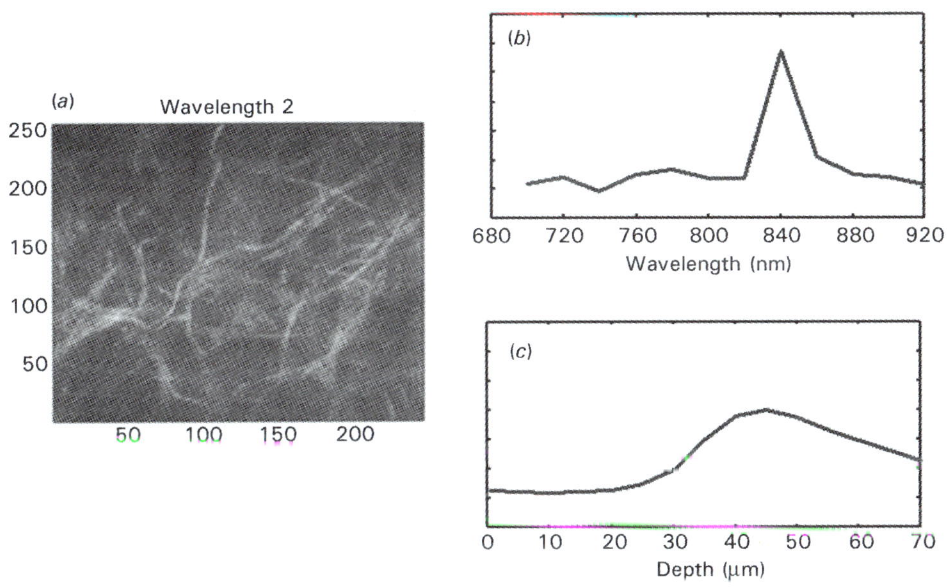

FIGURE 6 Spectral component corresponding to collagen/elastin fibers isolated by SMCR. (*a*) Reconstructed image of this spectral component. (*b*) The spectrum of this component. (*c*) The depth distribution of this component.

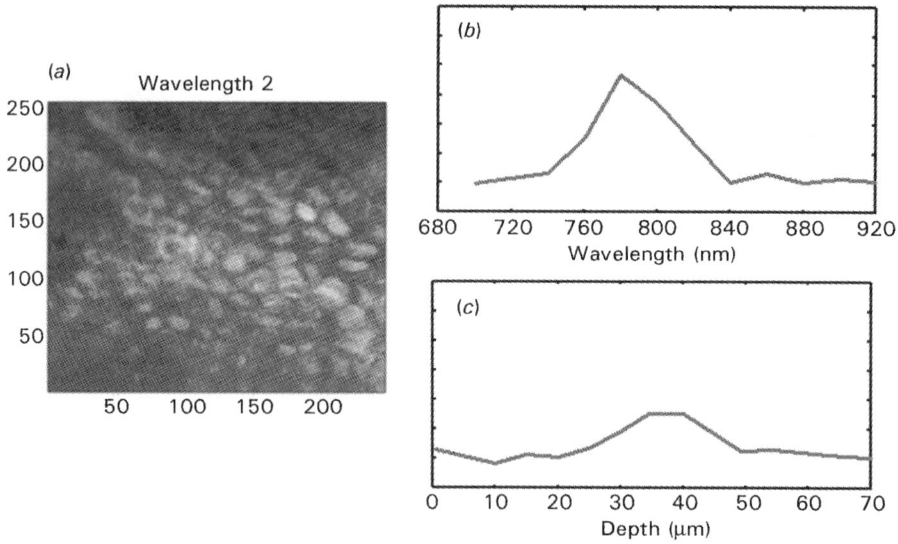

FIGURE 7 Spectral component corresponding to melanin isolated by SMCR. (*a*) Reconstructed image of this spectral component. (*b*) The spectrum of this component. (*c*) The depth distribution of this component.

6 CONCLUSIONS

The ubiquitous presence of endogenous fluorophores in tissue offers a minimally invasive approach for in situ tissue imaging and spectroscopic analysis. Although endogenous chromophores are relatively poor fluorophores, advances in microscope optics and high-sensitivity detectors have allowed tissue autofluorescence imaging up to a thickness of several hundred micrometers. With the further development of high-speed tissue imaging, two-photon autofluorescence may provide a viable complementary method to traditional histopathology for clinical diagnosis. With further advances in spectroscopic technique to identify and analyze endogenous tissue fluorophores, it may be possible to identify new endogenous markers for various tissue physiological and pathological conditions.

ACKNOWLEDGMENTS

This work is supported by Unilever Research U.S., Inc. and the American Cancer Society (RPG-98-058-01-CCE).

REFERENCES

af Klinteberg, C., Enejder, A.M., Wang, I., Andersson-Engels, S., Svanberg, S., and Svanberg, K. (1999). Kinetic fluorescence studies of 5-aminolaevulinic acid-induced protoporphyrin IX accumulation in basal cell carcinomas. *J. Photochem. Photobiol.* **B49**, 120–128.

Andrew, J.J. and Hancewicz, T.M. (1998). Rapid analysis of Raman image data using two-way multivariate curve resolution. *Appl. Spectrosc.* **52**, 797–807.

Anidjar, M., Cussenot, O., Blais, J., Bourdon, O., Avrillier, S., Ettori, D., Villette, J.M., Fiet, J., Teillac, P., and Le Duc. A. (1996). Argon laser induced autofluorescence may distinguish between normal and tumor human urothelial cells: a microspectrofluorimetric study. *J. Urol.* **155**, 1771–1774.

Banerjee, B., Miedema, B., and Chandrasekhar, H.R. (1998). Emission spectra of colonic tissue and endogenous fluorophores. *Am. J. Med. Sci.* **316**, 220–226.

Banerjee, B., Miedema, B.E., and Chandrasekhar, H.R. (1999). Role of basement membrane collagen and elastin in the autofluorescence spectra of the colon. *J. Invest. Med.* **47**, 326–332.

Beauvoit, B. and Chance, B. (1998). Time-resolved spectroscopy of mitochondria, cells and tissues under normal and pathological conditions. *Mol. Cell. Biochem.* **184**, 445–455.

Bennett, B.D., Jetton, T.L., Ying, G., Magnuson, MA., and Piston, D.W. (1996). Quantitative subcellular imaging of glucose metabolism within intact pancreatic islets. *J. Biol. Chem.* **271**, 3647–3651.

Bertrand, C. and Corcuff, P. (1994). In vivo spatio-temporal visualization of the human skin by real-time confocal microscopy. *Scanning* **16**, 150–154.

Bewersdorf, J., Pick, R., and Hell, S.W. (1998). Mulitfocal multiphoton microscopy. *Opt. Lett.* **23**, 655–657.

Blanca, C.M. and Saloma, C. (1998). Monte Carlo analysis of two-photon fluorescence imaging through a scattering medium biological microstructures application. *Appl. Opt.* **37**, 8092–8102.

Brakenhoff, G.J., Squier, J., Norris, T., Bliton, A.C., Wade, M.H., and Athey, B. (1996). Real-time two-photon confocal microscopy using a femtosecond, amplified Ti : sapphire system. *J. Microsc.* **181**, 253–259.

Brancaleon, L., Lin, G., and Kollias, N. (1999). The in vivo fluorescence of tryptophan moieties in human skin increases with UV exposure and is a marker for epidermal proliferation. *J. Invest. Dermatol.* **113**, 977–982.

Buehler, C., Kim, K.H., Dong, C.Y., Masters, B.R., and So, P.T.C. (1999). Innovations in two-photon deep tissue microscopy. *IEEE Eng. Med. Biol. Mag.* **18**, 23–30.

Buist, A.H., Muller, M., Squier, J., and Brakenhoff, G.J. (1998). Real time two-photon absorption microscopy using multipoint excitation. *J. Microsc.* **192**, 217–226.

Centonze, V.E. and White, J.G. (1998). Multiphoton excitation provides optical sections from deeper within scattering specimens than confocal imaging. *Biophys. J.* **75**, 2015–2024.

Chance, B. (1976). Pyridine nucleotide as an indicator of the oxygen requirements for energy-linked functions of mitochondria. *Circ. Res. Suppl.* **38**, I31–I38.

Chance, B. (1989). *Cell Structure and Function by Microspectrofluorometry.* New York: Academic Press, pp. 53–69.

Chance, B. (1991). *Annu. Rev. Biophys. Biophys. Chem.* 1–28.

Chance, B. and Thorell, B. (1959). Localization and kinetics of reduced pyridine nucleotide in living cells by microfluorometry. *J. Biol. Chem.* **234**, 3044–3050.

Chance, B., Schoener, B., Oshino, R., Itshak, F., and Nakase, Y. (1979). Oxidation–reduction ratio studies of mitochondria in freeze-trapped samples. NADH and flavoprotein fluorescence signals. *J. Biol. Chem.* **254**, 4764–4771.

Cheong, W.-F., Prahl, S.A., and Welch, A.J. (1990). A review of the optical properties of biological tissues. *IEEE J. Quantum Electron.* **26**, 2166–2185.

Corcuff, P., Bertrand, C., and Leveque, J.L. (1993). Morphometry of human epidermis in vivo by real-time confocal microscopy. *Arch. Dermatol. Res.* **285**, 475–481.

Cothron, R.M., Richards-Kortum, R., Sivak, M., Fitzmaurice, M., Rava, R., Boyce, G., Doxtader, M., Blackman, R., Ivanc, T., and Hayes, G. (1990). Gastrointentinal tissue diagnostic by laser-induced fluorescence spectroscopy at endoscopy. *Gastrointest. Endosc.* **36**, 105–111.

Dabbous, M.K. (1966). Inter- and intramolecular cross-linking in tyrosinase-treated tropocollagen. *J. Biol. Chem.* **241**, 5307–5312.

Daria, V., Blanca, C.M., Nakamura, O., Kawata, S., and Saloma, C. (1998). Image contrast enhancement for two-photon fluorescence microscopy in a turbid medium. *Appl Opt.* **37**, 7960–7967.

Dayan, D., Wolman, M., and Hammel, I. (1994). Histochemical study of the blue autofluorescence of collagen in oral irritation fibroma: effects of age of patients and of the duration of lesions. *Histol. Histopathol.* **9**, 11–13.

de Grauw, G.J., Vroom, J.M., Van der Voort, H.T.M., and Gerritsen, H.C. (1999). Imaging properties in two-photon excitation microscopy and effects of refractive-index mismatch in thick specimens. *Appl. Opt.* **38**, 5995–6003.

Denk, W. (1994). Two-photon scanning photochemical microscopy: mapping ligand-gated ion channel distributions. *Proc. Natl Acad. Sci. USA* **91**, 6629–6633.

Denk, W., Strickler, J.H., and Webb, W.W. (1990). Two-photon laser scanning fluorescence microscopy. *Science* **248**, 73–76.

Denk, W.J., Strickler, J.P., and Webb, W.W. (1991). Two-photon laser microscopy. United States Patent 5,034,613 (July 23, 1991).

Denk, W.J., Piston, D.W., and Webb, W.W. (1995a). Two-photon molecular excitation laser-scanning microscopy. In: *Handbook of Biological Confocal Microscopy*, 2nd edn (ed. J.B. Pawley). New York: Plenum Press, pp. 445–458.

Denk, W., Holt, J.R., Shepherd, G.M., and Corey, D.P. (1995b). Calcium imaging of single stereocilia in hair cells: localization of transduction channels at both ends of tip links. *Neuron* **15**, 1311–1321.

Dunn, A.K., Wallace, V.P., Coleno, M., Berns, M.W., and Tromberg, B.J. (2000). Influence of optical properties on two-photon fluorescence imaging in turbid samples. *Appl. Opt.* **39**, 1194–1201.

Fan, G.Y., Fujisaki, H., Miyawaki, A., Tsay, R.K., Tsien, R.Y., and Ellisman, M.H. (1999). Video-rate scanning two-photon excitation fluorescence microscopy and ratio imaging with cameleons. *Biophys. J.* **76**, 2412–2420.

Flock, S.T., Wilson, B.C., and Patterson, M.S. (1987). Total attenuation coefficients and scattering phase functions of tissues and phantom materials at 633 nm. *Med. Phys.* **14**, 835–841.

Furuta, T., Wang, S.S., Dantzker, J.L., Dore, T.M., Bybee, W.J., Callaway, E.M., Denk, W., and Tsien, R.Y. (1999). Brominated 7-hydroxycoumarin-4-ylmethyls: photolabile protecting groups with biologically useful cross-sections for two photon photolysis. *Proc. Natl Acad. Sci. USA* **96**, 1193–1200.

Ganesan, S., Sacks, P.G., Yang, Y., Katz, A., Al-Rawi, M., Savage, H.E., Schantz, S.P., and Alfano, R.R. (1998). Native fluorescence spectroscopy of normal and malignant epithelial cells. *Cancer Biochem. Biophys.* **16**, 365–373.

Gerritsen, H.C. and de Grauw, C.J. (1999). Imaging of optically thick specimen using two-photon excitation microscopy. *Microsc. Res. Tech.* **47**, 206–209.

Glasgold, R., Glasgold, M., Savage, H., Pinto, J., Alfano, R., and Schantz, S. (1994). Tissue autofluorescence as an intermediate endpoint in NMBA-induced esophageal carcinogenesis. *Cancer Lett.* **82**, 33–41.

Göppert-Mayer, M. (1931). Über Elementarakte mit zwei Quantensprungen. *Annln Phys.* **5**, 273–294.

Gryczynski, I., Malak, H., Lakowicz, J.R., Cheung, H.C., Robinson, J., and Umeda, P.K. (1996). Fluorescence spectral properties of troponin C mutant F22W with one-, two-, and three-photon excitation. *Biophys. J.* **71**, 3448–3453.

Guild, J.B. and Webb, W.W. (1995). Line scanning microscopy with two-photon fluorescence excitation. *Biophys. J.* **68**, 290a.

Guo, Y., Ho, P.P., Liu, F., Wang, Q.Z., and Alfano, R.R. (1998). Noninvasive two-photon-excitation imaging of tryptophan distribution in highly scattering biological tissues. *Opt. Commun.* **154**, 383–389.

Guo, Y., Savage, H.E., Liu, F., Schantz, S.P., Ho, P.P., and Alfano, R.R. (1999). Subsurface tumor progression investigated by noninvasive optical second harmonic tomography. *Proc. Natl Acad. Sci. USA* **96**, 10854–10856.

Hancewicz, T.M. and Petty, C. (1995). Quantitative analysis of vitamin A using Fourier transform Raman spectroscopy. *Spectrochim. Acta Part A: Mol. Spectrosc.* **12**, 2193–2198.

Hockberger, P.E., Skimina, T.A., Centonze, V.E., Lavin, C., Chu, S., Dadras, S., Reddy, J.K., and White, J.G. (1999). Activation of flavin-containing oxidases underlies light-induced production of H_2O_2 in mammalian cells. *Proc. Natl Acad. Sci. USA* **96**, 6255–6260.

Hoerman, K.C. and Balekjian, A.Y. (1966). Some quantum aspects of collagen. *Fed. Proc.* **25**, 1016–1021.

Horch, H.W., Kruttgen, A., Portbury, S.D., and Katz, L.C. (1999). Destabilization of cortical dendrites and spines by BDNF. *Neuron* **23**, 353–364.

Huang, S., Ahmed, A.H., and Webb, W.W. (2000). Multiphoton fluorescence spectroscopy of LipDH, FAD, and NAD(P)H. *Biophys. J.* **78**, 443A.

Inaguma, M. and Hashimoto, K. (1999). Porphyrin-like fluorescence in oral cancer: in vivo fluorescence spectral characterization of lesions by use of a near-ultraviolet excited autofluorescence diagnosis system and separation of fluorescent extracts by capillary electrophoresis. *Cancer* **86**, 2201–2211.

Izuishi, K., Tajiri, H., Fujii, T., Boku, N., Ohtsu, A., Ohnishi, T., Ryu, M., Kinoshita, T., and Yoshida, S. (1999). The histological basis of detection of adenoma and cancer in the colon by autofluorescence endoscopic imaging. *Endoscopy* **31**, 511–516.

Johansson, J., Berg, R., Svanberg, K., and Svanberg, S. (1997). Laser-induced fluorescence studies of normal and malignant tumour tissue of rat following intravenous injection of delta-amino levulinic acid. *Lasers Surg. Med.* **20**, 272–279.

Keilin, D. (1966). *The History of Cell Respiration and Cytochrome*. London: Cambridge University Press.

Kierdaszuk, B., Gryczynski, I., Modrak-Wojcik, A., Bzowska, A., Shugar, D., and Lakowicz, J.R. (1995). Fluorescence of tyrosine and tryptophan in proteins using one- and two-photon excitation. *Photochem. Photobiol.* **61**, 319–324.

Kim, K.H., Buehler, C., and So, P.T.C. (1999). High-speed, two-photon scanning microscope. *Appl. Opt.* **38**, 6004–6009.

Kim, K.H., So, P.T.C., Kochevar, I.E., Masters, B.R., and Gratton, E. (1998). Two-photon fluroescence and confocal reflected light imaging of thick tissue structures. *Proc. SPIE* **3260**, 46–57.

Kleinfeld, D., Mitra, P.P., Helmchen, F., and Denk, W. (1998). Fluctuations and stimulus-induced changes in blood flow observed in individual capillaries in layers 2 through 4 of rat neocortex. *Proc. Natl Acad. Sci. USA* **95**, 15741–15746.

LaBella, F.S. (1961). Studies on the soluble products released from purified elastin fibers by pancreatic elastase. *Arch. Biochem. Biophys.* **93**, 72–79.

Lakowicz, J.R. (1999). *Principles of Fluorescence Spectroscopy*. New York: Plenum Press.

Licha, K., Riefke, B., Ntziachristos, V., Becker, A., Chance, B., and Semmler, W. (2000). Hydrophilic cyanine dyes as contrast agents for near-infrared tumor imaging: synthesis, photophysical properties and spectroscopic in vivo characterization. *Photochem. Photobiol.* **72**, 392–398.

Liotta, L.A. (1984). Tumor invasion and metastases: role of the basement membrane. Warner-Lambert Parke-Davis Award Lecture. *Am. J. Pathol.* **117**, 339–348.

MacMunn, C.A. (1880). *The Spectroscope in Medicine*. London: Churchill.

MacMunn, C.A. (1914). *Spectrum Analysis Applied to Biology and Medicine*. London: Green.

Madsen, S.J., Sun, C.H., Tromberg, B.J., Wallace, V.P., and Hirschberg, H. (2000). Photodynamic therapy of human glioma spheroids using 5-aminolevulinic acid. *Photochem. Photobiol.* **72**, 128–134.

Mainen, Z.F., Maletic-Savatic, M., Shi, S.H., Hayashi, Y., Malinow, R., and Svoboda, K. (1999). Two-photon imaging in living brain slices. *Methods* **18**, 231–239, 181.

Marcu, L., Grundfest, W.S., and Maarek, J.M. (1999). Photobleaching of arterial fluorescent compounds: characterization of elastin, collagen and cholesterol time-resolved spectra during prolonged ultraviolet irradiation. *Photochem. Photobiol.* **69**, 713–721.

Masters, B.R. (1990). In vivo corneal redox fluorometry. In: *Non-Invasive Diagostic Techniques in Ophthalmology* (ed. B.R. Masters). New York: Springer-Verlag, pp. 223–247.

Masters, B.R. (1996). Three-dimensional confocal microscopy of human skin in vivo: autofluorescence of normal skin. *Bioimages* **4**, 13–19.

Masters, B.R. (1998). Three-dimensional confocal microscopy of the living in situ rabbit cornea. *Opt Express* **3**.

Masters, B.R. and Chance, B. (1999). Redox confocal imaging: intrinsic fluorescent probes of cellular metabolism. In: *Fluorescent and Luminescent Probes for Biological Activity* (ed. W.T. Mason). London: Academic Press, pp. 361–374.

Masters, B.R. and Thaer, A.A. (1994). Real-time scanning slit confocal microscopy for the in vivo human cornea. *Appl. Opt.* **33**, 695–701.

Masters, B.R. and So, P.T.C. (1999). Multi-photon excitation microscopy and confocal microscopy imaging of in vivo human skin: a comparison. *Microsc. Microanal.* **5**, 282–289.

Masters, B.R., So, P.T., and Gratton, E. (1997). Multiphoton excitation fluorescence microscopy and spectroscopy of in vivo human skin. *Biophys. J.* **72**, 2405–2412.

Masters, B.R., So, P.T., and Gratton, E. (1998a). Multiphoton excitation microscopy of in vivo human skin. Functional and morphological optical biopsy based on three-dimensional imaging, lifetime measurements and fluorescence spectroscopy. *Ann. NY Acad. Sci.* **838**, 58–67.

Masters, B.R., So, P.T.C., and Gratton, E. (1998b). Optical biopsy of in vivo human skin: multiphoton excitation microscopy. *Lasers Med. Sci.* **13**, 196–203.

Masters, B.R., So, P.T., Kim, K.H., Buehler, C., and Gratton, E. (1999). Multiphoton excitation microscopy, confocal microscopy, and spectroscopy of living cells and tissues; functional metabolic imaging of human skin in vivo. *Meth. Enzymol.* **307**, 513–536.

Mehlmann, M., Betz, C.S., Stepp, H., Arbogast, S., Baumgartner, R., Grevers, G., and Leunig, A. (1999). Fluorescence staining of laryngeal neoplasms after topical application of 5-aminolevulinic acid: preliminary results. *Lasers Surg. Med.* **25**, 414–420.

Nilsson, A.M., Heinrich, D., Olajos, J., and Andersson-Engels, S. (1997). Near infrared diffuse reflection and laser-induced fluorescence spectroscopy for myocardial tissue characterisation. *Spectrochim. Acta A: Mol. Biomol. Spectrosc.* **53A**, 1901–1912.

Onizawa, K., Saginoya, H., Furuya, Y., and Yoshida, H. (1996). Fluorescence photography as a diagnostic method for oral cancer. *Cancer Lett.* **108**, 61–66.

Piston, D.W. and Knobel, S.M. (1999). Quantitative imaging of metabolism by two-photon excitation microscopy. *Meth. Enzymol.* **307**, 351–368.

Piston, D.W., Masters, B.R., and Webb, W.W. (1995). Three-dimensionally resolved NAD(P)H cellular metabolic redox imaging of the in situ cornea with two-photon excitation laser scanning microscopy. *J. Microsc.* **178**, 20–27.

Piston, D.W., Summers, R.G., Knobel, S.M., and Morrill, J.B. (1998). Characterization of involution during sea urchin gastrulation using two-photon excited photorelease and confocal microscopy. *Microsc. Microanal.* **4**, 404–414.

Piston, D.W, Knobel, S.M., Postic, C., Shelton, K.D., and Magnuson, M.A. (1999). Adenovirus-mediated knockout of a conditional glucokinase gene in isolated pancreatic islets reveals an essential role for proximal metabolic coupling events in glucose-stimulated insulin secretion. *J. Biol. Chem.* **274**, 1000–1004.

Rajadhyaksha, M., Grossman, M., Esterowitz, D., Webb, R.H., and Anderson, R.R. (1995). In vivo confocal scanning laser microscopy of human skin: melanin provides strong contrast. *J. Invest. Dermatol.* **104**, 946–952.

Richards-Kortum, R., Rava, R.P., Fitzmaurice, M., Tong, L.L., Ratliff, N.B., Kramer, J.R., and Feld, M.S. (1989). A one-layer model of laser-induced fluorescence for diagnosis of disease in human tissue: applications to atherosclerosis. *IEEE Trans. Biomed. Eng.* **36**, 1222–1232.

Richards-Kortum, R., Rava, R.P., Petras, R.E., Fitzmaurice, M., Sivak, M., and Feld, M.S. (1991). Spectroscopic diagnosis of colonic dysplasia. *Photochem. Photobiol.* **53**, 777–786.

Ricken, S., Leipziger, J., Greger, R., and Nitschke, R. (1998). Simultaneous measurements of cytosolic and mitochondrial Ca^{2+} transients in HT29 cells. *J. Biol. Chem.* **273**, 34961–34969.

Schilders, S.P. and Gu, M. (2000). Limiting factors on image quality in imaging through turbid media under single-photon and two-photon excitation. *Microsc. Microanal.* **6**, 156–160.

So, P.T.C., Dong, C.Y., Masters, B.R., and Berland, K.M. (2000). Two-photon excitation fluorescence microscopy. *Annu. Rev. Biomed. Eng.* **2**, 399–429.

Squirrell, J.M., Wokosin, D.L., White, J.G., and Bavister, B.D. (1999). Long-term two-photon fluorescence imaging of mammalian embryos without compromising viability. *Nature Biotechnol.* **17**, 763–767.

Summers, R.G., Piston, D.W., Harris, K.M., and Morrill, J.B. (1996). The orientation of first cleavage in the sea urchin embryo, *Lytechinus variegatus*, does not specify the axes of bilateral symmetry. *Dev. Biol.* **175**, 177–183.

Svanberg, K., af Klinteberg, C., Nilsson, A., Wang, I., Andersson-Engels, S., and Svanberg, S. (1998). Laser-based spectroscopic methods in tissue characterization. *Ann. NY Acad. Sci.* **838**, 123–129.

Svoboda, K., Denk, W., Kleinfeld, D., and Tank, D.W. (1997). In vivo dendritic calcium dynamics in neocortical pyramidal neurons. *Nature* **385**, 161–165.

Svoboda, K., Helmchen, F., Denk, W., and Tank, D.W. (1999). Spread of dendritic excitation in layer 2/3 pyramidal neurons in rat barrel cortex in vivo. *Nature Neurosci.* **2**, 65–73.

Svoboda, K., Tank, D.W., and Denk, W. (1996). Direct measurement of coupling between dendritic spines and shafts. *Science* **272**, 716–719.

Teuchner, K., Ehlert, J., Freyer, W., Leupold, D., Altmeyer, P., Stucker, M., and Hoffmann, K. (2000). Fluorescence studies of melanin by stepwise two-photon femtosecond laser excitation. *J. Fluorescence* **10**, 275–281.

Thomas, J., Elsden, D.F., and Partridge, S.M. (1963). Degradation products from elastin. *Nature* **200**, 651–652.

Williams, R.M., Piston, D.W., and Webb, W.W. (1994). Two-photon molecular excitation provides intrinsic 3-dimensional resolution for laser-based microscopy and microphotochemistry. *Faseb J.* **8**, 804–813.

Wilson, B.C. and Jacques, S. (1990). Optical reflectance and transmittance of tissues: principles and applications. *IEEE J. Quantum Electron.* **26**, 2186–2199.

Xu, C. and Webb, W.W. (1996). Measurement of two-photon excitation cross sections of molecular fluorophores with data from 690 to 1050 nm. *J. Opt. Soc. Am.* **B13**, 481–491.

Xu, C., Guild, J., Webb, W.W., and Denk, W. (1995). Determination of absolute two-photon excitation cross sections by in situ second-order autocorrelation. *Opt. Lett.* **20**, 2372–2374.

Xu, C., Zipfel, W., Shear, J.B., Williams, R.M., and Webb, W.W. (1996). Multiphoton fluorescence excitation: new spectral windows for biological nonlinear microscopy. *Proc. Natl Acad. Sci. USA* **93**, 10763–10768.

Yuste, R. and Denk, W. (1995). Dendritic spines as basic functional units of neuronal integration. *Nature* **375**, 682–684.

Yuste, R., Majewska, A., Cash, S.S., and Denk, W. (1999). Mechanisms of calcium influx into hippocampal spines: heterogeneity among spines, coincidence detection by NMDA receptors, and optical quantal analysis. *J. Neurosci.* **19**, 1976–1987.

Zeng, H., MacAulay, C., McLean, D.I., and Palcic, B. (1995). Spectroscopic and microscopic characteristics of human skin autofluorescence emission. *Photochem. Photobiol.* **61**, 639–645.

CHAPTER 21

Two-Photon Near-Infrared Femtosecond Laser Scanning Microscopy in Plant Biology

UDAY K. TIRLAPUR and KARSTEN KÖNIG

Laser Microscopy Division, Institute of Anatomy II, Friedrich Schiller University Jena, Teichgraben 7, D-07743 Jena, Germany

1 INTRODUCTION

Light microscopy based on conventional light sources as well as various laser sources has to date been used extensively for spatial and temporal analysis of structural and other cellular components in plant biology (Robards, 1985; Lloyd, 1987; Gilroy, 1997; Hepler and Gunning, 1997; Buschmann and Lichtenthaler, 1998).

Our knowledge of intricate events of plant cell function has been greatly advanced in recent years with significant developments in three-dimensional (3D) fluorescence imaging technologies. Two-photon excitation laser scanning microscopy (TPLSM) using near-infrared (NIR) femtosecond laser pulses is emerging as a promising technique to observe living biological specimens in 3D over extended periods of time (Piston, 1999; König, 2000). Unlike conventional confocal laser scanning microscopy (CLSM), TPLSM has a number of inherent advantages, such as 3D resolution without a confocal pinhole, less photodamage and photobleaching above and below the plane of observation, and (more importantly) greater depth penetration with NIR laser excitation (Denk et al., 1990; Diaspro, 1999; König, 2000). Additionally the fixed wavelengths of the laser sources employed in conventional CLSM (Pawley, 1995) often constrain the possibility of using multiple fluorochromes for 3D fluorescence imaging. This limitation has been circumvented in part by the advent of multiple-laser CLSM (Fricker and White, 1992), with accompanying corrections for chromatic aberrations for different laser wavelengths. In this regard, TPLSM has an interesting advantage because the NIR radiation at a fixed wavelength can simultaneously induce visible fluorescence of any number of fluorophores without requiring additional laser sources. In recent years, a number of studies have demonstrated the use of TPLSM using different animal systems (Squirrell et al., 1999; Koester et al., 1999; Rose et al., 1999). However, comparable studies on plants have barely

Confocal and Two-Photon Microscopy: Foundations, Applications, and Advances, Edited by Alberto Diaspro.
ISBN 0-471-40920-0

begun. In this chapter, we briefly describe the TPLSM set-up and discuss some recent applications of TPLSM in plant biology.

2 LASER SOURCES AND TWO-PHOTON IMAGING SYSTEM

Femtosecond laser pulsed NIR light for TPLSM can be obtained from either CrLiSAF, frequency-doubled fiber lasers or, CrLiASGAF lasers (Wise, 1999), but the most commonly used sources are Ti : sapphire lasers. The commercially available Ti : sapphire lasers include tunable (between 700 and 1200 nm) sources and the more user-friendly compact mode-locked lasers with limited wavelength range (König, 2000). Furthermore, less expensive laser diodes have also been employed to obtain two-photon excitation (Hänninen et al., 1995; Gu and Day, 1999). Figure 1 depicts the set-up of our modified two-photon microscope.

The NIR laser beams at $\lambda = 740$, 760, and 780 nm are obtained from a tunable 76 MHz femtosecond Ti : sapphire laser source (Mira 900-F, Coherent, Santa Clara, CA) pumped by a 7 W multiline argon-ion laser. A typical width of 170 fs was determined. The 800 nm laser beam is obtained from a compact mode-locked femtosecond Ti : sapphire laser (Vitesse Coherent: operating at 80 MHz pulse frequency, 1 W output power, and 90 fs output pulse duration). The laser beams are expanded by a 1 : 3 Galilean telescope, and after transmission through two-piece neutral-density attenuators, the beams are coupled to a modified inverted Zeiss CLSM model 410 as shown in Figure 1. The beams are focused to a diffraction-limited spot by a high-numerical-aperture (NA 1.3) 40× Zeiss Neofluar Objective.

The microscope also includes a built-in He–Ne laser (633 nm) that can be used for conventional one-photon confocal imaging. In addition, it also contains a detector for NIR transmission microscopy (phase contrast, brightfield, etc.) and a host of other detectors at the baseport. The deployment of detectors at the baseport has the advantage of efficient detection of photons without transmission through the scanner and spatial filters (pinholes). We use different types of photomultipliers, cameras (video camera, slow scan camera etc.) and a time-correlated single-photon counting unit (SPC 730, Becker and Hickel, GmbH, Berlin). The SPC unit has a channel resolution of 800 fs, enabling very rapid uptake (registration) and storage of 16384 (128 × 128) decay curves at the scanning frequency. Besides measurement of fluorescence kinetics, τ-mapping can also be performed in conjunction with the recently developed SPC-Image software (Becker and Hickel, GmbH, Berlin).

For most botanical specimens, two-photon fluorescence excitation of endogenous and exogenous fluorophores require less than 10 mW of mean laser power. The appropriate mean laser powers can be attained by using neutral-density filters, and can routinely be measured at the object plane using the Fieldmaster FM power meter (Coherent, Santa Clara, CA) before and after fluorescence imaging (Tirlapur and König, 1999).

3 TWO-PHOTON IMAGING OF CELLULAR STRUCTURES, Ca^{2+}, AND BIOMOLECULES

Most of the fluorescent dyes excitable by ultraviolet (UV) light (e.g., at laser wavelengths of $\lambda = 351$ and 364 nm) and blue light ($\lambda = 488$ nm) can be used for two-photon imaging of any number of cellular structures and components in plants.

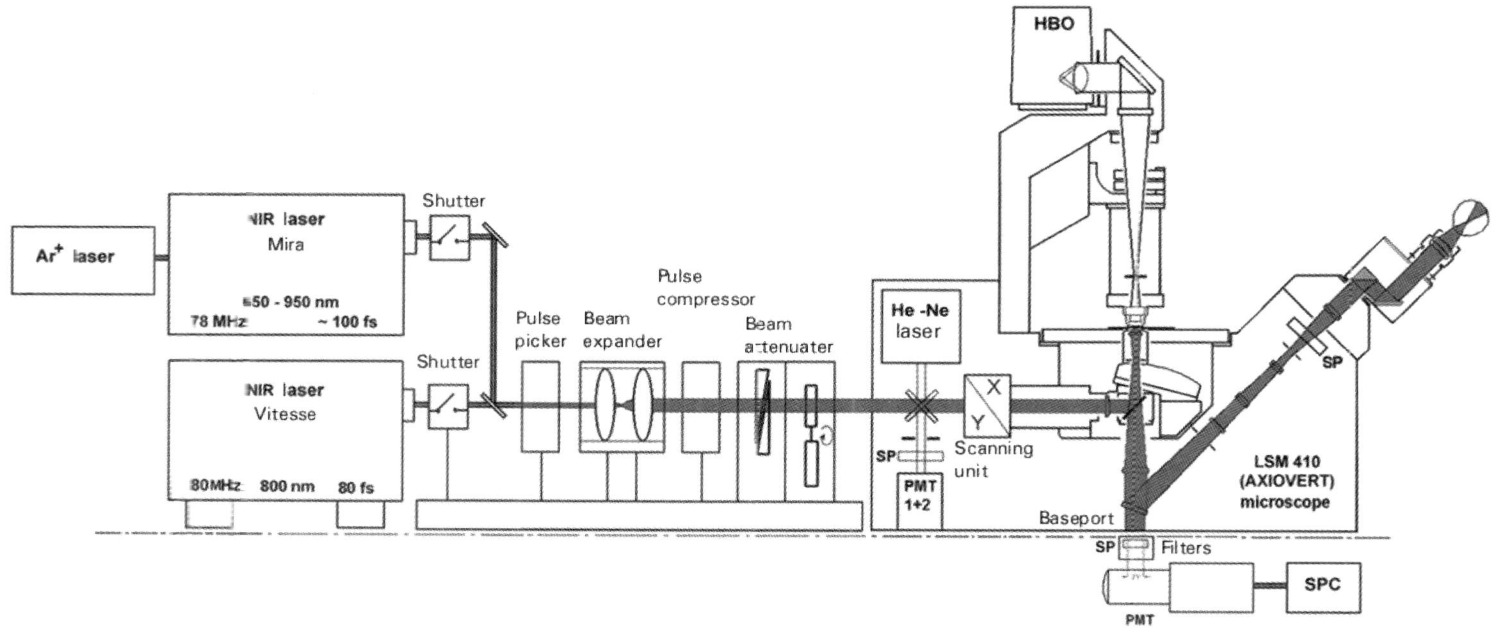

FIGURE 1 Schematic diagram of the experimental arrangement of the two- and one-photon confocal microscope with femtosecond NIR laser sources. Interchangeable detectors at the baseport include the Hamamatsu PMT for 3D black and white imaging and the SPC 730 for fluorescence-lifetime imaging.

3.1 Chromosomes/DNA

A number of fluorophores that specifically bind to nucleic acids have been extensively used for visualisation of chromosomes or DNA (Suzuki et al., 1997). Dyes such as 4′,6-diamidino-2-phenylindole hydrochloride (DAPI) and Hoechst 33342 are known to bind to the minor grove of DNA by hydrogen bonding and/or electrostatic interaction (Steiner and Kubota, 1983), whereas others such as ethidium bromide (EB) and acridine are known to show intercalative binding. *In vitro* studies have demonstrated multiphoton excitation of both DAPI and Hoechst with NIR excitation from 810 to 910 nm (Gryczynski et al., 1996).

Using a fixed wavelength of 800 nm and a mean laser power of 25 μW, we have successfully obtained 3D serial sections of 33342 Hoechst-labeled nuclei and chromosomes (Fig. 2) during stomatal differentiation in living epidermal cells of *Allium cepa* L. Similarly, the generative cell and the vegetative nuclei of maize and *Lilium* microspores labelled with DAPI (Fig. 3) and EB have also been imaged at the same NIR wavelength, even at a mean laser power of 20 μW.

Based on earlier *in vitro* studies by Gryczynski et al. (1996) and the *in vivo* studies reported, here it is conceivable that the DNA probes DAPI and Hoechst 33342 can be employed for either two- or three-photon imaging of chromosomes in plant cells as well.

In the case of microspores of maize and *Lilium* pollen, the autofluorescence of the wall was also discernible following excitation at 800 nm. This autofluorescence is most probably due to the sporopollinin and other wall components. Such NIR-induced autofluorescence from the cell walls can provide useful information about the cell shape with respect to the position of the nuclei/chromosomes.

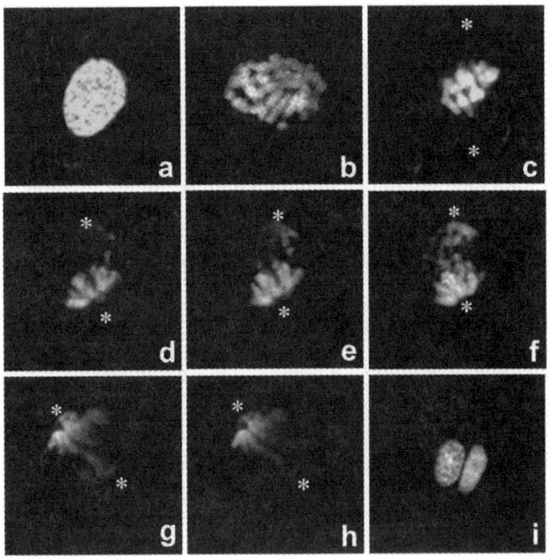

FIGURE 2 Two-photon excitation imaging of Hoechst 33342-labeled nuclei and chromosomes in living epidermal cells of *Allium* during stomatal differentiation. Maximum projection of 25 optical sections (collected at 0.5 μm intervals) of nuclei in (*a*) interphase, (*b*) prophase and (*c*) metaphase chromosomes. (*d*)–(*h*) represent median optical sections of chromosomes at anaphase. (*i*) shows the maximum projection of 25 optical sections at telophase. All images were acquired at a mean laser power of 25 μW with 800 nm NIR wavelength.

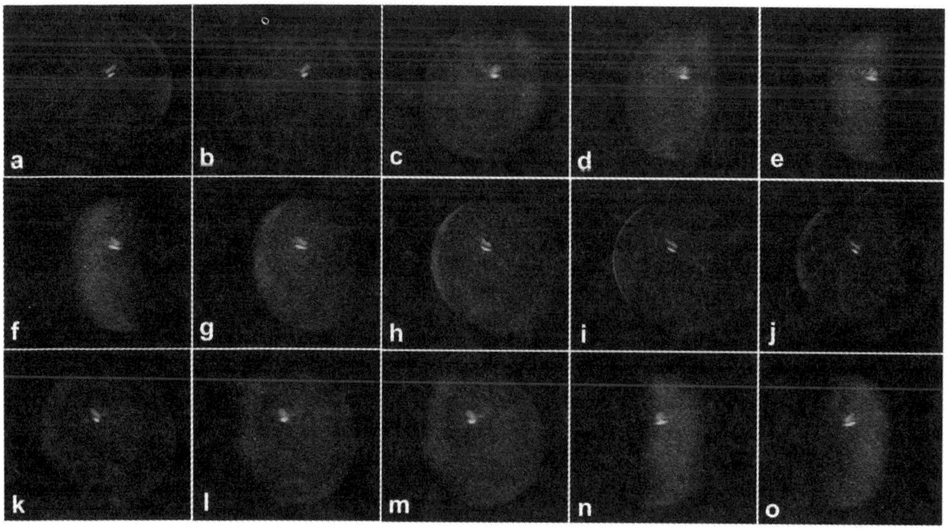

FIGURE 3 Two-photon imaging of DAPI-stained microspore of maize. (*a*)–(*o*) are 3D rotational images obtained from a stack of 28 serial optical sections, showing the vegetative nucleus and the generative cell. Note the relatively intense labeling of the nuclei in the generative cell and the autofluorescence of the microspore wall. All images were obtained with a mean laser power of 20 μW at 800 nm NIR excitation.

In addition, it is pertinent to note that two-photon excitation (at 800 nm) of various fluorophores such as spectrum blue, spectrum green, SYBR green, rhodamine 110, spectrum orange, Texas red, Cy 5, and Cy 5.5 often used in fluorescence in situ hybridization (FISH) has recently been demonstrated (König et al., 2000). Thus, the technique of multiphoton multicolor FISH can well be used for detection of multiple genes and chromosomes in plants.

3.2 Two-Photon Imaging of Plant Cell Wall Components

In general, plant cell walls are know to consist of cellulose, callose, lignin, suberin, cutin, and sporopollinin (in the case of microspores and pollen). Some, if not all, of these cell wall components tend to manifest some degree of autofluorescence when excited with UV or blue light. The UV-inducible autofluorescence of the cell wall is generally believed to be due to blue–green fluorescing ferulic acid covalently bound to cell walls (Buschmann and Lichtenthaler, 1998). Plant biologists have long been using specific fluorescent dyes such as calcofluor white and aniline blue to visualize cellulose (cellulose-like) and callose respectively in conjunction with conventional fluorescence and also with CLSM (Fricker and White, 1992). While the specificity of these fluorescent probes is somewhat debatable, they have nevertheless provided interesting information pertaining to certain underlying events of pollen pistil interactions as well as host–pathogen relationships. Using NIR laser line at 800 nm, we have been able to image callose (after staining with aniline blue) associated with the walls of pollen tubes of *Lilium* (Fig. 4) and those of *Nicotiana* at mean laser powers of 20 μW or less. In addition, the cell walls labeled with calcofluor (for cellulose) have also been imaged (see Section 7) at the same mean laser power. More

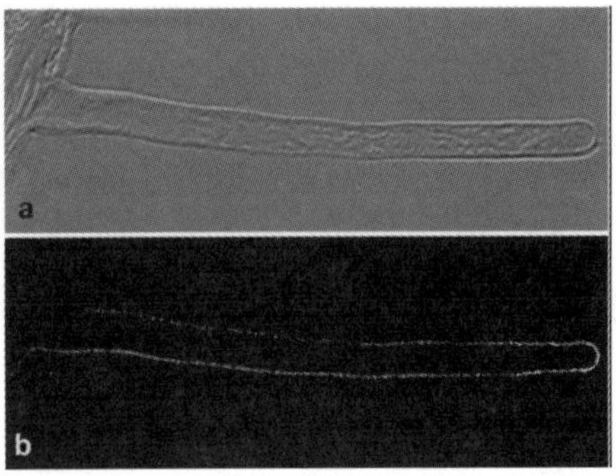

FIGURE 4 Two-photon NIR transmission (*a*) and fluorescence (callose) (*b*) images in the median optical plane of a pollen tube of *Lilium* labelled with 0.1% aniline blue in 50 mM phosphate buffer.

recently, it has been shown that propidium iodide (PI) also outlines the cell walls, particularly those of *Arabidopsis* roots, and can be visualized by excitation with the 488 nm laser from a conventional CLSM (Tirlapur and König, 1999). Interestingly, the cell walls of *Arabidopsis* as well as garden cress labeled with PI have been successfully imaged with a 800 nm NIR laser at mean powers of 5 mW or less using the internal detector of the Zeiss LSM 410 (Tirlapur and König, 1999). However, these PI-labeled cell walls can more conveniently be also imaged at as low as 75 μW mean laser power using a baseport PMT (Tirlapur and König, 1999). The latter observation reiterates the advantages of having a baseport detector over an internal one, which requires the photons to be transmitted through the scanning optics. Taken together, two-photon-induced autofluorescence as well as that imaged following labeling of cell wall components can prove very useful in obtaining important information related to structural aspects of growth and development associated with specific cells and cell types in plants.

3.3 Endoplasmic Reticulum and Mitochondria

A number of fluorescent dyes have been used for labeling of membrane components in living plant cells (see Hepler and Gunning, 1998). Of these, the carbocyanine dye $DiOC_6(3)$ has been commonly used to label endoplasmic reticulum (ER) as well as mitochondria (Quader et al., 1989; Knebel et al., 1990; Menzel,1994; Liebe and Menzel, 1995). A very interesting feature of $DiOC_6(3)$ reported by Menzel (1994) is that, at certain suboptimal concentrations, it does not label ER while the mitochondria *are* labeled. We have recently confirmed this observation, and have been able to label the mitochondria alone with $DiOC_6(3)$ in living trichomes of *Arabidopsis* (Fig. 5*a*, *b*) and *Nicotiana*, and subsequently image these structures by TPLSM at an excitation wavelength of 800 nm and a mean power of 60 μW. Furthermore, we have also confirmed the specificity of mitochondrial staining (Fig. 5*c*, *d*) by TPLSM (at 800 nm) of trichomes labeled with Mito Tracker Green FM (Molecular Probes Inc., Eugene, OR) as well as rhodamine 123. Hence, most of the routinely used fluorescent probes for labeling of mitochondria can be exited by a single NIR wavelength at 800 nm with a minimum laser power of 1 mW or less.

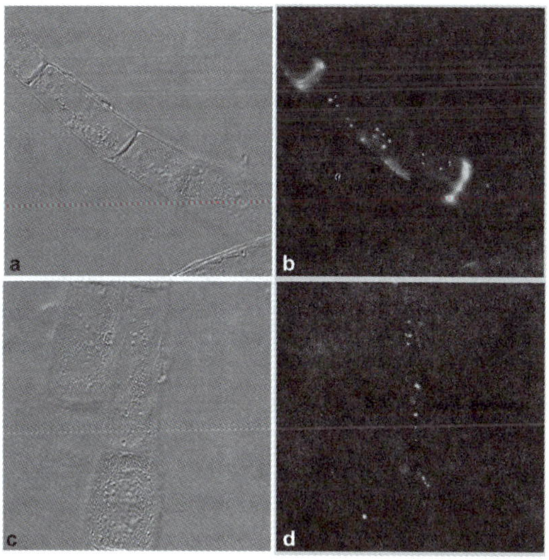

FIGURE 5 Two-photon transmission (*a*, *c*) and fluorescence (*b*, *d*) images of trichome cells of *Arabidopsis* labeled with $DiOC_6(3)$ and Mito Tracker Green FM, respectively, at a concentration of 0.5 µg/ml. Note the similarity between the punctate fluorescence labeling of the mitochondria (*b*) stained with $DiOC_6(3)$ and that labeled with Mito Tracker Green FM (*d*). In the $DiOC_6(3)$-labeled trichome, besides the punctate staining, the regions near the older transverse walls also show diffuse fluorescence, which might correspond to the ER associated with the plasmodesmatal regions.

3.4 Two-Photon Imaging of Free Cytosolic Calcium in Plant Cells

A number of fluorescent dyes have been developed to visualize changes in the free concentrations of ions such as Ca^{2+}, H^+, Mg^{2+}, K^+, Cl^-, and Zn^{2+} (Haugland, 1996). Among these, both Ca^{2+} and H^+ have been extensively studied in plants – mainly because of their important role in signal transduction and ion homeostasis (Fricker et al., 1994). In particular, changes in free cytosolic Ca^{2+} has been monitored in a wide range of cell types, including pollen tubes, root hair cells, and stomatal guard cells (Fricker et al., 1999). Quite often, measurement and imaging of intracellular Ca^{2+} are performed by using fluorescent dyes that manifest shifts in their excitation or emission spectra in response to Ca^{2+} (Tsien, 1989). Such wavelength-ratiometric methods enable measurement and/or imaging of Ca^{2+} that are independent of local probe concentration, cell thickness, photobleaching, and dye leakage. This is in contrast to methods using indicators that show quantitative change in intensity upon binding to Ca^{2+}.

The ratiometric Ca^{2+}-indicating probe indo-1 has been used elegantly for intracellular calcium imaging in root hairs of *Arabidopsis* by one-photon excitation with a conventional CLSM (Gilroy, 1997). It has, however, been reported that indo-1 displays a relatively high cross-section for two-photon excitation as well (Szmacinski et al., 1993). To the best of our knowledge, such a realization of two-photon excitation of indo-1 in living plant cells for imaging of free cytosolic Ca^{2+} has not been ascertained. We have more recently been able to image free cytoplasmic Ca^{2+} in living root hair cells of *Arabidopsis thaliana* (Fig. 6) loaded with indo-1 by two-photon excitation at 800 nm with a mean laser power of 1 mW or less. Our two-photon NIR fluorescence images reveal a highly localized Ca^{2+} gradient at the tip of the root hair, which is highly consistent with that reported by Gilroy (1997).

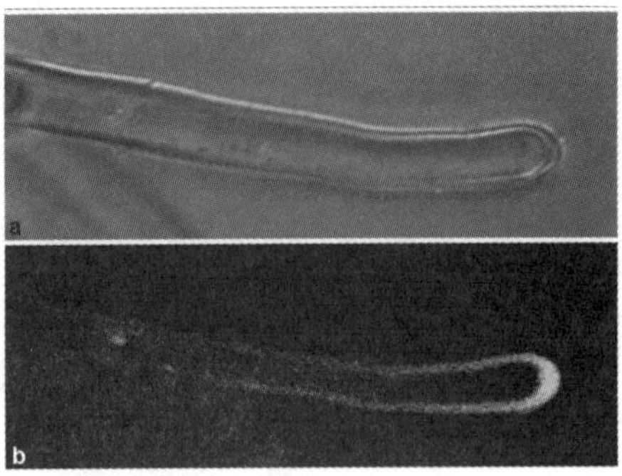

FIGURE 6 Transmission (*a*) and fluorescence (*b*) images of a living root hair cell of *Arabidopsis* loaded with indo-1 obtained by two-photon NIR (800 nm) laser scanning microscopy. Note the highly localized Ca^{2+} gradient at the tip of the root hair. The fluorescence image was obtained at a mean power of 1 mW.

Because indo-1 can provide significant emission using two-photon excitation, it could well be of immense value for fluorescence-lifetime imaging of Ca^{2+} gradients in various tip-growing plant cells, including pollen tubes.

3.5 TPLSM of the Tripeptide Glutathione in Plant Cells

In most organisms, glutathione (GSH) is known to be the most abundant low-molecular-weight thiol and plays a intricate role in protective responses against stress from adverse environmental conditions as well as in detoxification of xenobiotics and heavy metals. It can also act as a reservoir of reduced sulfur that can potentially be mobilised between various organs (Noctor et al., 1998). The levels of the tripeptide glutathione (γ-glutamyl-cysteinyl-glycine) and its fluorescent glutathione S-bimane conjugate (GSB) have recently been measured in developing trichoblasts (root hair cells) and atrichoblasts (non-root hair cells) of intact *Arabidopsis* roots using monobromobimane and TPLSM at 770 nm (Meyer and Fricker, 2000). Interestingly, these authors have noted improved contrast when images were collected from deep within the root tissue using two-photon excitation at 770 nm in comparison with images obtained by conventional CLSM at 442 nm. This observation is highly consistent with the general expectation that NIR laser wavelengths have relatively higher penetration depth than UV and visible laser lines (often used in conventional CLSM) and are thus more efficient for in-depth imaging of cellular components in thick botanical specimens. The possibility of measuring metabolites such as GSH on a per-cell basis by TPLSM circumvents certain of the limitations of conventional biochemical techniques, which require extraction of tissue with the consequence that only the average level of the metabolite in the tissue is estimated. It is therefore possible that TPLSM can bring important new insights to the study of biomolecules and metabolism at cellular level in intact plants as well.

4 IMAGING OF GREEN FLUORESCENT PROTEIN BY TWO-PHOTON EXCITATION

In recent years, the green fluorescent protein (GFP) from the jellyfish *Aequorea victoria* has proven to be an excellent endogenous fluorescent marker for spatially monitoring gene expression, protein folding, and trafficking (Tsien, 1998). In addition, a number of mutant forms have been engineered (Chiu et al., 1996) and widely employed to unravel structure–function relationships in living cells. In plants, GFP has been targeted to different cellular components, including the nucleus (Chytilova et al., 1999), plastids (Tirlapur et al., 1999), endoplasmic reticulum, Golgi and mitochondria (Haseloff and Siemering, 1998), cytoskeleton (Marc et al., 1998), and phragmoplastin (Gu and Verma, 1997), and has also been successfully used to track viral movement proteins (Blackman et al., 1998). Wild-type GFP is known to absorb violet/blue ($\lambda_{max} = 395$ nm along with a minor peak at $\lambda = 470$ nm) and has an emission maxima at $\lambda = 508$ nm, while GFP mutants have different spectral properties (see Haseloff, 1999). Recently, the one- and two-photon-induced fluorescence of wild-type as well as other recombinant GFPs has been examined (Xia et al., 1999; Volkmer et al., 2000) in vitro. Periasamy et al. (1999) have imaged GFP in single living cells using a 880 nm NIR femtosecond laser. We have instead used an 800 nm NIR Ti : sapphire femtosecond laser for imaging of plastid-targeted (König, 2000) as well as ER-targeted GFP (Fig. 7) in *Arabidopsis* at a mean laser power of less than 1 mW. It is interesting to note that with the 800 nm NIR laser, it is possible to simultaneously induce autofluorescence of the chloroplast as well as that of the ER-targeted GFP in living epidermal cells (Fig. 7) of *Arabidopsis* by the two-photon absorption process. Serial

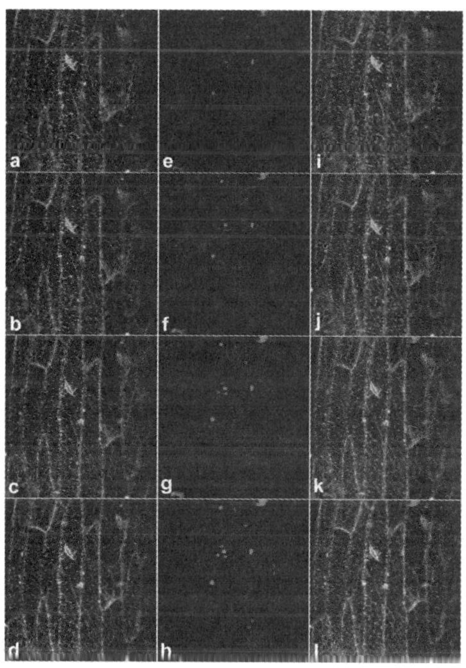

FIGURE 7 See ftp site for color figure. Simultaneous detection of ER-targeted GFP and autofluorescence of chloroplasts in the cortical region of the hypocotyl of *Arabidopsis* with two-photon femtosecond NIR excitation at 800 nm. (*a*)–(*d*) are serial optical sections (obtained at intervals of 0.25 μm) of the cortical ER, while (*e*)–(*h*) are the corresponding serial sections showing chlorophyll autofluorescence associated with the chloroplasts. (*i*)–(*l*) are electronically merged images of corresponding GFP and chlorophyll fluorescence. Note the distinct presence/association iation of chloroplasts in the vicinity of the nucleus (arrows), and the reticulate organisation of the cortical ER. The bright green spots associated with the lateral walls correspond to the plasmodesmatal regions.

optical sectioning of epidermal cells has enabled us to gain information regarding the 3D distribution of GFP-tagged ER as well as the precise positioning of the chloroplasts in these cell types. Similarly, using two-photon microscopy, Köhler et al. (1997) have imaged chloroplast tubules in tobacco mesophyll cells, and have shown by photobleaching and recovery of GFP that movement of proteins (Tirlapur et al., 1999) can occur between individual chloroplasts through the interconnecting tubules. More recently, we have also been able to demonstrate two-photon excitation of mutant GFPs (Pepperkok et al., 1999) such as cyan fluorescent protein (CFP) and yellow fluorescent protein (YFP) at an NIR excitation wavelength of 800 nm (Tirlapur and König, unpublished results). Thus, TPLSM can invariably offer more interesting possibilities for simultaneous detection of multiple GFPs in living plant cells, and hence extends the number of protein processes and/or organelles that can be imaged simultaneously.

5 NIR TWO-PHOTON IMMUNOFLUORESCENCE IMAGING OF F-ACTIN AND MICROTUBULES

The organization of the cytoskeleton is quite crucial in plant cell signaling and associated events of morphogenesis. To date, conventional CLSM has proven to be an important imaging technique to unravel spatial relationships between the microfilaments and microtubules in plant cells. This has been facilitated by the use of fluorophore-coupled secondary antibodies as well as fungal toxins such as phalloidin and phallacidin (for visualization of F-actin). The most commonly used fluorophores have been rhodamine, FITC, and Texas red. Some of the most remarkable CLSM images describing F-actin localization include those associated with the plasmodesmata (White et al., 1994), focal projections in protonemata of *Ceratodon* (Walker and Sack, 1995), and those obtained during cell division (see Hepler and Gunning, 1998). Similarly impressive and compelling CLSM images of microtubules have also been documented during cellular morphogenesis in other plant cells (Cyr, 1994; Marc et al., 1998). In a recent collaboration (Baluska and Volkmann, University of Bonn), we have acquired high-resolution serial optical sections as well as generating 3D immunofluorescence images of F-actin and microtubules in maize root tip sections using two-photon excitation at different NIR wavelengths (760–800 nm) at mean laser powers of 50–80 μW. Unlike the organization of these cytoskeletal components documented by conventional fluorescence microscopy (see Baluska and Volkmann, 2000), our TPLSM images, particularly at an NIR excitation wavelength of 780 nm, show a more elaborate organization of the F-actin as well as that of microtubules in the cells of the elongation region of maize root tips (Fig. 8).

6 NONINVASIVE LOADING OF FLUORESCENT PROBES INTO SINGLE CELLS USING FEMTOSECOND NIR LASER PULSES

Symplastic communication (intercellular transport through plasmodesmata between cells) in plants has been investigated by monitoring the cell-to-cell movement of plasma-membrane-impermeable fluorescent dyes of different sizes and characteristics (Oparka, 1991).

Techniques such as phloem loading (Oparka et al., 1994), microinjection (Goodwin et al., 1990), and noninvasive approaches such as ester loading (Duckett et al., 1994, Zhu et

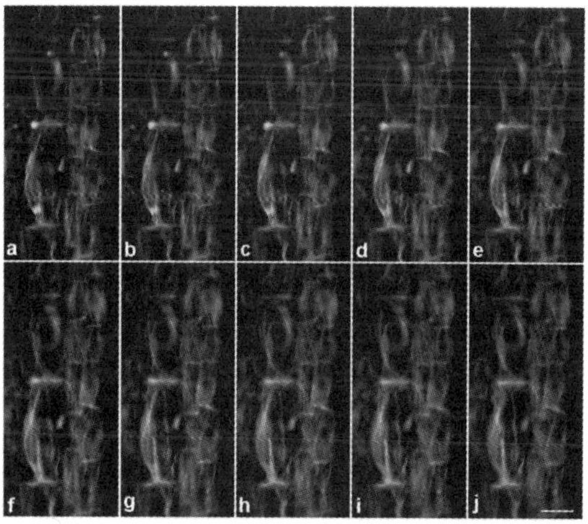

FIGURE 8 Two-photon immunofluorescence imaging of actin in maize root. (*a*)–(*j*) are serial optical sections of maize root tip (immunolabeled with FITC as secondary antibody) obtained at space steps of 0.25 μm with a 780 nm NIR TPLSM. Note the presence of fine filamentous actin localization around the nucleus and thick actin cables extending across the cells in the elongation region.

al., 1998) have been employed to label cells with fluorescent probes such as lucifer yellow and carboxyfluorescein.

The movement of fluorescent dyes (dye coupling) between individual living plant cells has been studied in shoot apical meristem (van der Schoot and Lucas, 1993), epidermis of leaves (Erwee and Goodwin, 1983; Palevitz and Hepler, 1985; Erwee et al., 1985), staminal hairs (Tucker, 1988) and roots (Duckett, 1994) by microinjection of membrane-impermeable fluorescent dyes. Because single epidermal cells are more amenable to microinjection of fluorescent dyes, our knowledge of symplastic coupling in plants has largely been limited to these cell types. Furthermore, because of the limitation of the microinjection technique, dye-coupling studies between individual cells or a group of cells that are deeply seated in the meristematic tissue have so far not been forthcoming.

We have recently reported the successful application of NIR femtosecond laser beams as a novel noninvasive means for dye permeation/loading and simultaneous excitation of the fluorochrome PI in meristematic cells of a higher plant, *Arabidopsis thaliana*, and also to monitor local dye coupling between cells during development by 3D imaging (Tirlapur and König, 1999). Representative time-lapse images of the dye-coupling pattern in the root tip of *Arabidopsis* are shown in Figure 9.

7 NANOSURGERY OF PLANT CELL WALLS WITH FEMTOSECOND NIR LASER BEAMS

To date, several laboratories have used high-energy UV laser microbeams either to destroy the walls of plant cells (Greulich and Weber, 1992) or to ablate a small area of the extracellular matrix (Henriksen and Assmann, 1997; Berger, 1998). These laser-mediated cell-ablation studies have been of great value in gaining non-enzymatic access to the plasma membrane, and understanding root development as well as cell polarization.

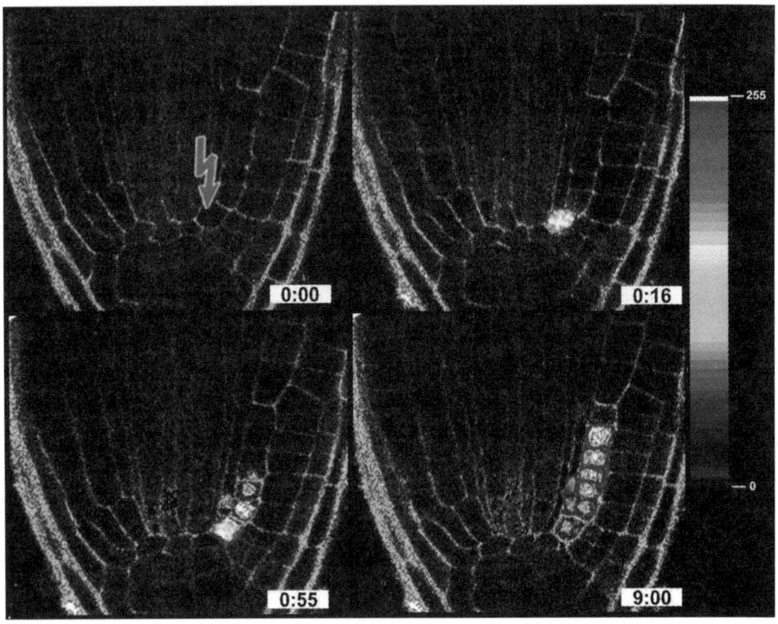

FIGURE 9 Noninvasive dye permeation and imaging of the dye-coupling pattern in *Arabidopsis* root. Time-lapse (0–9 minute) optical images of the *Arabidopsis* root meristem demonstrate the dye-coupling pattern between the constituent cells following noncontact NIR femtosecond laser-mediated dye (propidium iodide) permeation. The dye was loaded into a single cortex/endodermal initial (lightning symbol at 0 : 00). Note that the dye moves preferentially into the derivative cell files of the cortex and the endodermis rather than to the lateral pericycle and stelar initials. The inset wedge shows pixel intensities from 0 to 255.

Nevertheless, the use of high-intensity UV laser light has major disadvantages. First, even millisecond exposure of plant cells to UV irradiation has been shown to drastically affect them by altering calcium homeostasis (Frohnmeyer et al., 1999) and also inducing cell death (Danon and Gallois, 1998). Second, the penetration depth of UV light in highly turbid and refractive plant tissue is low, and thus it is extremely difficult to perform intratissue nanoprocessing (Berger, 1998). Third, the pulse duration is longer (often in nanoseconds) and the beam quality is poor.

As described in the preceding sections, safe nondestructive two-photon fluorescence imaging of botanical specimens can be performed using an 80 MHz Ti : sapphire laser with NIR femtosecond (170 fs) excitation intensities in the range of megawatts to gigawatts per square centimeter at mean powers of 10 mW or less. Nevertheless, above a certain higher mean power threshold where the laser intensities are in the range of terawatts per square centimeter, nonlinearly induced instantaneous destructive effects based on multiphoton ionization of molecules (Oraevsky et al., 1996) and plasma formation are known occur in the subfemtoliter excitation volume (König et al., 1999; König, 2000). We have used these spatially confined highly destructive nonlinear multiphoton effects of intense NIR femtosecond laser pulses and have performed nanodissection of cell walls (Fig. 10) as well as noninvasive intratissue nanoprocessing. For nanodissection of the walls, the same experimental set-up as described for two-photon fluorescence imaging was used, but at higher mean laser powers of either 10–15 mW or 30–50 mW, and a 1 μm thick region of

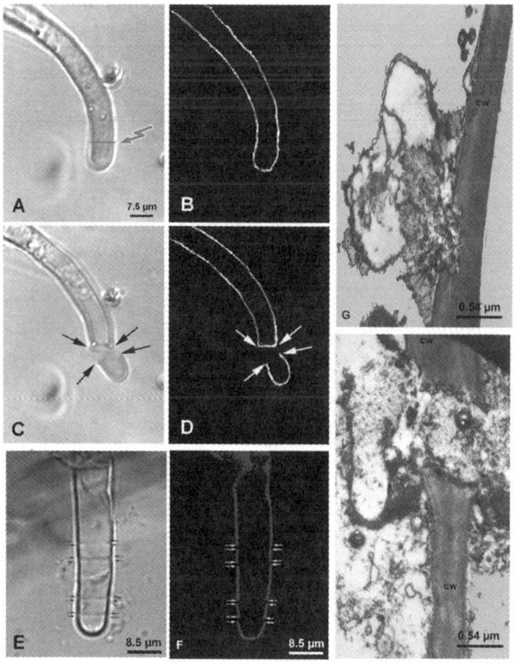

FIGURE 10 Nanodissection of cell walls and intratissue nanoprocessing with intense near infrared femtosecond laser pulses. Two-photon transmission (A, C, and D) and fluorescence images (B, D, and F) of root hair cells of *Arabidopsis* stained with calcofluor white (that labels the cell wall) before (A and B) and after dissection/selective separation (C and D) of the root hair tip with intense near infrared femtosecond laser beam at mean laser power of 10–15 mW. A single root hair cell can also be nanodissected into as many as 4 to 5 separate parts (E and F). The two electron micrographs (G and H) show partial ejection of the cell wall by point exposure (G) and intratissue nanodissection (H) of the cell wall of *Elodea densa* after scanning a small region of interest with NIR laser beam at a mean point of 30 to 50 mW.

interest was scanned for 0.64 s at the focused wall region of a cell in the lower epidermis of the leaf. Hence, the use of an NIR femtosecond laser-based technique can potentially become the method of choice for noninvasive intratissue nanosurgery and nanoprocessing of intracellular organelles in different cell types (König et al., 2001). This technique can also be extended to understand the fundamental role of specific statocytes and the precise number of statoliths required for gravitropic response (Blancaflor et al., 1998) of roots and other organs (Chen et al., 1999).

8 TWO-PHOTON EXCITATION OF CHLOROPHYLL AUTOFLUORESCENCE

In plants, light quanta are mainly absorbed by photosynthetic pigments such as chlorophyll and carotenoids associated with the chloroplasts. The UV (355 nm)-induced 690 nm and 740 nm chlorophyll *a* fluorescence emissions (F690 and F740) known to emanate from the antenna and reaction centre of the photosynthetic photosystem II have been elegantly exploited (Buschmann and Lichtenthaler, 1998) as a diagnostic tool for the evaluation of plant stress (see Lichtenthaler and Miehe, 1997).

In higher plants, in vivo chlorophyll fluorescence to date has been induced using a range of laser sources such as a frequency-tripled Nd : YAG UV laser (at $\lambda = 355$ nm), a frequency-doubled Nd : YAG laser (at $\lambda = 532.5$ nm), a He–Ne laser ($\lambda = 632.8$ nm), and a blue laser (see Buschmann and Lichtenthaler, 1998). In particular, the Nd : YAG laser at $\lambda = 355$ nm has been extensively used (Lichtenthaler and Miehe, 1997) for excitation of endogenous fluorophores in leaves of different species and the changes in chlorophyll autofluorescence at maxima near 690 nm and near 730–740 nm have been employed as diagnostic signatures to detect physiological stress.

We have recently observed that excitation with ultrashort 90 and 170 fs NIR laser pulses at $\lambda = 740$, 760, 780, 800, 820, 840, 860, 880, and 900 nm (at mean powers of 1 mW or less) invariably induces red chlorophyll autofluorescence of the chloroplasts present in the mesophyll cells (Fig. 11) of *Arabidopsis* leaf.

9 FLUORESCENCE-LIFETIME IMAGING OF CHLOROPHYLL FLUORESCENCE EVOKED BY TWO-PHOTON ABSORPTION

Time-resolved fluorescence-lifetime imaging (FLIM) is a powerful noninvasive technique for vital measurements of fluorescence lifetimes in biological specimens in two as well as in three dimensions (see Schneckenburger et al., 1998). FLIM provides a number of advantages, among which is the fact that such measurements are independent of the concentration of the chromophore and the local intensity, thus enabling visualization of molecular environs in a single living cell (Bastiaens and Squire, 1999) and an organelle (this study). These remarkable properties of FLIM have been elegantly adopted for spatially resolving specific physiological parameters such as calcium-ion concentrations (Lakowicz et al., 1992, 1994), changes in pH (Sanders et al., 1995; Lin et al., 1999), oxygen levels (Gerritsen et al., 1997), molecular associations (Gadella and Jovin, 1995), and proteolytic processing (Bastiaens and Jovin, 1996) in single living cells.

To date, FLIM has been accomplished using wide-field microscopes (see Bugiel et al., 1989; Lakowicz et al., 1992) and conventional confocal microscopes (Buurman et al., 1992; König et al., 1996) in both the frequency domain (Szmacinski et al., 1993; König et al., 1996) and the time domain (O'Connor and Phillips, 1984; Cubeddu et al., 1993; König, 2000).

With the use of FLIM and conventional CLSM, it has become possible to correlate time-dependent photophysical as well as biochemical details with the spatial distribution of cellular structures and molecules in a microscopic image (Buurman et al., 1992).

The elegant combination of two-photon excitation (TPE) with fluorescence-lifetime imaging (FLIM) has thus far been implemented only using frequency-domain methods (Piston et al., 1992; So et al., 1995, 1998; König et al., 1996).

Recently, we have demonstrated the first successful use of a combination of two-photon excitation with FLIM in the time domain for detailed in vivo characterization of chloroplasts and their division profiles in different cell types of *Arabidopsis* leaves (Tirlapur and König, 2000). Figure 12 shows the fluorescence intensity and the corresponding TPFLIM image of chloroplasts in mesophyll cells of *Arabidopsis* leaf.

10 CONCLUSIONS AND FUTURE PERSPECTIVES

Perusal of the pertinent literature on applications of TPLSM in biology reveals that to date there have been only a handful of published studies demonstrating the unique capabilities

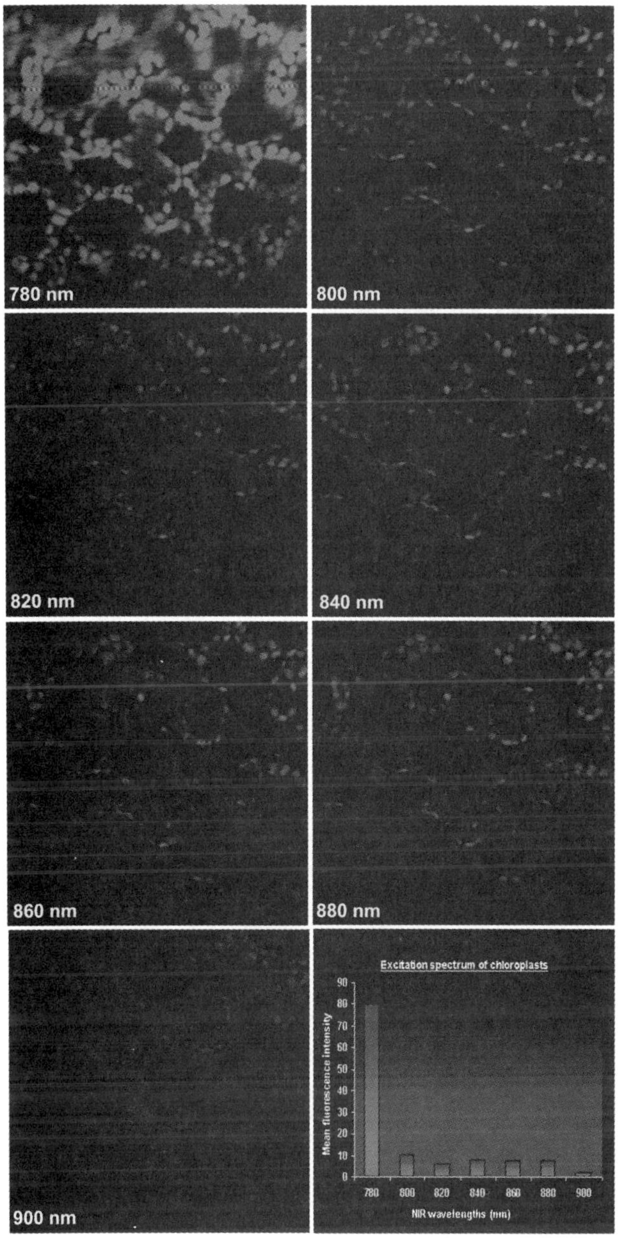

FIGURE 11 See ftp site for color figure. Two-photon imaging of chlorophyll fluorescence in mesophyll cells following excitation with different NIR wavelengths at 780, 800, 820, 840, 880, and 900 nm. As depicted in the spectrum.

and the use of this powerful technology in plant sciences. In this chapter, we have highlighted the efficacy with which TPLSM can be realized for 3D and 4D fluorescence imaging of exogenous and endogenous fluorophores in living and fixed vegetative and reproductive plant cells and structures. In addition, we have also demonstrated the interesting novel applications of TPLSM for vital noninvasive dye permeation, nanosur-

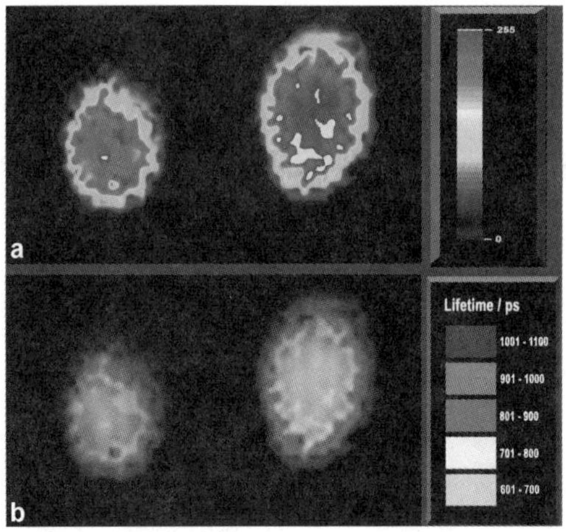

FIGURE 12 See ftp site for color figure. Two-photon fluorescence intensity (*a*) and τ-image (*b*) of chloroplasts.

gery, and nanoprocessing of botanical specimens. More importantly, we have in most instances provided the relevant details and other parameters, including the safe mean laser powers that would eventually be required to minimize NIR femtosecond laser-induced cellular and subcellular perturbations in plant cells.

The progress made so far in realizing TPLSM in plant biology indicates that it will certainly find even broader and interesting applications, enabling plant biologists to unravel some of the most intricate sequential series of spatio-temporal events underlying (1) signal transduction and ion dynamics, (2) protein–protein interactions, (3) symplastic communication, (4) basic aspects of organelle and cell division, (5) tip growth, and (6) plant morphogenesis as a whole. Hence, TPLSM in plants is likely to make an enormous impact on revolutionizing our basic thinking of structure–function relationships in three as well as in four dimensions.

REFERENCES

Baluska, F., Barlow, P.W., and Volkmann, D. (2000). Actin and myosin VIII in developing root apex cells. Development- and tissue-specific distributions with possible relevance for diverse root cell functions. In: *Actin: A Dynamic Framework for Multiple Plant Cell Functions* (ed. C.J. Staiger, F. Baluska, D. Volkmann, and P.W. Barlow). Dordrecht: Kluwer, pp. 457–476.

Bastiaens, P.I. and Jovin, T.M. (1996). Microspectroscopic imaging tracks the intracellular processing of a signal transduction protein: fluorescent labelled protein kinase CβI. *Proc. Natl Acad. Sci. USA* **93**, 8407–8412.

Bastiaens, P.I. and Squire, A. (1999). Fluorescence lifetime imaging microscopy: spatial resolution of biochemical processes in the cell. *Trends Cell Biol.* **9**, 48–52.

Berger, F. (1998). Cell ablation studies in plant development. *Cell Mol. Biol.* **44**, 711–719.

Blackman, L.M., Boevink, P., Santa Cruz, S., Palukaitis, P., and Oparka, K.J. (1998). The movement protein of cucumber mosaic virus traffics into sieve elements in minor veins of *Nicotiana clevelandii*. *Plant Cell* **10**, 525–537.

Blancaflor, E.B., Fasano, J.M., and Gilroy, S. (1998). Mapping the functional roles of cap cells in the response of *Arabidopsis* primary roots to gravity. *Plant Physiol.* **116**, 213–222.

Bugiel, I., König, K., and Wabnitz, H. (1989). Investigation of cells by fluorescence laser scanning microscopy with subnanosecond time resolution. *Lasers Life Sci.* **3**, 47–53.

Buschmann, C. and Lichtenthaler, H.K. (1998). Principles and characteristics of multi-colour fluorescence imaging in plants. *J. Plant Physiol.* **152**, 297–314.

Buurman, E.P., Sanders, R., Draaijer, A., Gerritsen, H.C., van Veen, J.J.F., Houpt, P.M., and Levine, Y.K. (1992). Fluorescence lifetime imaging using confocal laser scanning microscope. *Scanning* **14**, 155–159.

Chen, R., Rosen, E., and Masson, P.H. (1999). Gravitropism in higher plants. *Plant Physiol.* **120**, 343–350.

Chiu, W.-L., Niwa, Y., Zeng, W., Hirano, T., Kobayashi, H., and Sheen, J. (1996). Engineered GFP as vital reporter in plants. *Curr. Biol.* **6**, 325–330.

Chytilova, E., Macas, J., and Galbraith, D.W. (1999). Green fluorescent protein targeted to the nucleus, a transgenic phenotype useful for studies in plant biology. *Ann. Bot.* **83**, 645–654.

Cubeddu, R., Canti, G., Taroni, P., and Valentini, G. (1993). Time-gated fluorescence imaging for diagnosis of tumers in a murine model. *Photochem. Photobiol.* **57**, 480–483.

Cyr, R.J. (1994). Microtubules in plant morphogenesis: role of cortical array. *Annu. Rev. Cell Biol.* **10**, 153–180.

Danon, A. and Gallois, P. (1998). UV radiation induces apoptotic-like changes in *Arabidopsis thaliana*. *FEBS Lett.* **437**, 131–136.

Denk, W., Strickler, J.H., and Webb, W.W. (1990). Two-photon laser scanning fluorescence microscope. *Science* **248**, 73–76.

Diaspro, A. (1999). Introduction to two-photon microscopy. *Microsc. Res. Tech.* **47**, 163–164.

Duckett, C.M., Oparka, K.J., Prior, D.A.M., Dolan, L., and Roberts, K. (1994). Dye coupling in the root epidermis of *Arabidopsis* is progressively reduced during development. *Development* **120**, 3247–3255.

Erwee, M.G. and Goodwin, P.B. (1983). Characterization of *Egeria densa* Planch. Leaf symplast. Inhibition of intercellular movement of fluorescent probes by group II ions. *Planta* **158**, 320–328.

Erwee, M.G., Goodwin, P.B., and Van Bel, J.E. (1985). Cell–cell communication in the leaves of *Commelina cyanea* and other plants. *Plant Cell Environ.* **8**, 173–178.

Frohnmeyer, H., Loyall, L., Blatt, M.R., and Grabov, A. (1999). Millisecond UV-B irradiation evokes prolonged elevation of cytosolic-free Ca^{2+} and stimulates gene expression in transgenic parsley cell cultures. *Plant J.* **20**, 109–117.

Fricker, M.D. and White, N.S. (1992). Wavelength considerations in confocal microscopy of botanical specimens. *J. Microsc.* **166**, 29–42.

Fricker, M.D., Tlalka, M., Ermantraut, J., Obermeyer, G., Dewey, M., Gurr, S., Patrick, J., and White, N.S. (1994). Confocal fluorescence ratio imaging of ion activities in plant cells. *Scanning Microsc. Suppl.* **8**, 391–405.

Fricker, M.D., Plieth, C., Knight, H., Blancaflor, E., Knight, M.R., White, N.S., and Gilroy, S. (1999). Fluorescence and luminescence techniques to probe ion activities in living plant cells. In: *Fluorescent and Luminescent Probes*, 2nd edn (ed. T.W. Mason). New York: Academic Press, pp. 569–596.

Gadella, T.W. Jr and Jovin, T.M. (1995). Oligomerization of epidermal growth factor receptors on A431 cells studied by time-resolved fluorescence imaging microscopy. A stereochemical model for tyrosine kinase receptor activation. *J. Cell Biol.* **129**, 1543–1558.

Gerritsen, H.C., Sanders, R., Draaijer, A., Ince, C., and Levine, Y.K. (1997). Fluorescence lifetime imaging of oxygen in living cells. *J. Fluoresc.* **7**, 11–15.

Gilroy, S. (1997). Fluorescence microscopy of living plant cells. *Annu. Rev. Plant Physiol. Plant Mol. Biol.* **48**, 165–190.

Goodwin, P.B., Sheperd, V., and Erwee, M.G. (1990). Compartmentation of fluorescent tracers injected into the epidermal cells of *Egeria densa* leaves. *Planta* **181**, 129–136.

Greulich, K.O. and Weber, G. (1992). The light microscope on its way from an analytical to preparative tool. *J. Microsc.* **167**, 127–151.

Gryczynski, I., Malak, H., and Lakowicz, J.R. (1996). Multiphoton excitation of the DNA stains DAPI and Hoechst. *Bioimaging* **4**, 138–148.

Gu, M. and Day, D. (1999): Two-photon multilayer bit data storage by use of continuous wave illumination. *Proc. SPIE* **3749**, 444–445.

Gu, X. and Verma, D.P.S. (1997): Dynamics of phragmoplastin in living cells during cell plate formation and uncoupling of cell elongation from the plane of cell division. *Plant Cell* **9**, 157–169.

Hänninen, P.E., Schrader, M., Soini, E., and Hell, S.W. (1995). Two-photon excitation fluorescence microscopy using a semiconductor laser. *Bioimaging* **3**, 70–75.

Haseloff, J. (1999). GFP variants for multispectral imaging of living cells. *Meth. Cell Biol.* **58**, 139–151.

Haseloff, J. and Siemering, K.R. (1998). The uses of GFP in plants. In: *Green Fluorescent Protein: Strategies, Applications and Protocols* (ed. M. Chalfie and S. Kain). New York: Wiley, pp. 191–220.

Haugland, R.P. (1996). *Handbook of Fluorescent Probes and Research Chemicals*, 6th edn. Eugene, OR: Molecular Probes.

Henriksen, G.H. and Assmann, S.M. (1997). Laser-assisted patch clamping: a methodology. *Eur. J. Physiol.* **433**, 832–841.

Hepler, P.K. and Gunning, B.E.S. (1998). Confocal fluorescence microscopy of plant cells. *Protoplasma* **201**, 121–157.

Knebel, W., Quader, H., and Schnepf, E. (1990). Mobile and immobile endoplasmic reticulum in onion bulb epidermal cells: short and long-term observations with a confocal laser scanning microscope. *Eur. J. Cell Biol.* **52**, 328–340.

Koester, H.J., Baur, D., Uhl, R., and Hell, S.W. (1999). Ca^{2+} fluorescence imaging with pico- and femtosecond two-photon excitation: signal and photodamage. *Biophys. J.* **77**, 2226–2236.

Köhler, R.H., Cao, J., Zipfel, W.R., Webb, W.W., and Hanson, M.R. (1997). Exchange of protein molecules through connections between higher plant plastids. *Science* **276**, 2039–2042.

König, K. (2000). Multiphoton microscopy in life sciences. *J. Microsc.* **200**, 83–104.

König, K., So, P.T.C., Mantulin, W.W., Tromberg, B.J., and Gratton, E. (1996). Two-photon excited lifetime imaging of autofluorescence in cells during UVA and NIR photostress. *J. Microsc.* **183**, 197–204.

König, K., Göhlert, A., Liehr, T., Loncarevic, I.F., and Riemann, I. (2000). Two-photon multicolour FISH: a versatile technique to detect specific sequences within single DNA molecules in cells and tissues. *Single Mol.* **1**, 41–51.

König, K., Riemann, I., Fischer, P., and Halbhuber, K.-J. (1999). Intracellular nanosurgery with near infrared femtosecond laser pulses. *Cell. Mol. Biol.* **45**, 195–201.

König, K., Riemann, I., Fritzsche, W. (2001). Nanodissection of human chromosomes with near-infrared laser pulses. *Opt. Lett.* **26**, 819–821.

Lakowicz, J.R., Szmacinski, H., Nowaczyk, K., and Johnson, M.L. (1992). Fluorescence lifetime imaging of free and protein bound NADH. *Proc. Natl Acad. Sci. USA* **89**, 1271–1275.

Lakowicz, J.R., Szmacinski, H., Nowaczyk, K., Lederem W.J., Kirby, M.S., and Johnson, M.L. (1994). Fluorescence lifetime imaging of intracellular calcium in COS cells using Quin-2. *Cell. Calcium* **15**, 7–27.

Lichtenthaler, H.K. and Miehe, J.A. (1997). Fluorescence imaging as a diagnostic tool for plant stress. *Trends Plant Sci.* **2**, 316–320.

Liebe, S. and Menzel, D. (1985). Actomyosin-based motility of endoplasmic reticulum and chloroplast in *Vallisneria* mesophyll cells. *Biol. Cell* **85**, 207–222.

Lin, H.J., Szmacinski, H., and Lakowicz, J.R. (1999). Lifetime-based pH sensors: indicators for acidic environments. *Anal. Biochem.* **269**, 162–167.

Lloyd, C.W. (1987). The plant cytoskeleton: impact of fluorescence microscopy. *Annu. Rev. Plant Physiol.* **38**, 119–139.

Marc, J., Granger, C., Brincat, J., Fisher, D.D., Kao Teh-hui, McCubbin, A.G., and Cyr, R.J. (1998). A *GFP-MAP4* reporter gene for visualising cortical microtubule rearrangements in living epidermal cells. *Plant Cell* **10**, 1927–1939.

Menzel, D. (1994). An interconnected plastidom in *Acetabularia*: implications for the mechanism of chloroplast motility. *Protoplasma* **179**, 166–171.

Meyer, A.J. and Fricker, M.D. (2000). Direct measurement of glutathione in epidermal cells of intact *Arabidopsis* roots by two-photon laser scanning microscopy. *J. Microsc.* **198**, 174–181.

Noctor, G., Strohm, M., Jouanin, L., Kunert, K.J., Foyer, C.H., and Rennenberg, H. (1998). Direct measurement of glutathione in epidermal cells of intact *Arabidopsis* roots by two-photon laser scanning microscopy. *J. Microsc.* **198**, 174–181.

O'Connor, D.V. and Phillips, D. (1984). *Time-Correlated Single Photon Counting*. London: Academic Press.

Oparka, K.J. (1991). Uptake and compartmentation of fluorescent probes by plant cells. *J. Exp. Bot.* **42**, 565–567.

Oparka, K.J., Duckett, C.M., Prior, D.A.M., and Fisher, D.B. (1994). Real-time imaging of phloem unloading in the root tip of *Arabidopsis*. *Plant J.* **6**, 759–766.

Oraevsky, A.A., Da Silva, L.B., Rubenchik, A.M., Feit, M.D., Glinsky, M.E., Perry, M.D., Mammini, B.M., Small, W., and Stuart, B.C. (1996). Plasma mediated ablation of biological tissues with nanosecond-to-femtosecond laser pulses: relative role of linear and nonlinear absorption. *IEEE J. Select. Top. Quantum Electron.* **2**, 801–809.

Palevitz, B.A. and Hepler, P.K. (1985). Changes in dye coupling of stomatal cells of *Allium* and *Commelina* demonstrated by microinjection. *Planta* **164**, 473–479.

Pawley, J.B. (Ed.) (1995). *Handbook of Confocal Microscopy*, 2nd edn. New York: Plenum.

Periasamy, A., Sharman, K.K., Ahuja, R., Eto, M., and Brautigan, D.L. (1999). Fluorescence lifetime imaging of green fluorescent protein in a single living cell. *Proc. SPIE* **3604**, 6–12.

Pepperkok, R., Squire, A., Geley, S., and Bastiaens, P.I.H. (1999). Simultaneous detection of multiple green fluorescent proteins in live cells by fluorescence lifetime imaging microscopy. *Curr. Biol.* **9**, 269–272.

Piston, D.W. (1999). Imaging living cells and tissues by two-photon excitation microscopy. *Trends Cell Biol.* **9**, 66–69.

Piston, D.W., Sandison, D.R., and Webb, W.W. (1992). Time-resolved fluorescence imaging and background rejection by two-photon excitation in laser scanning microscopy. *Proc. SPIE* **1640**, 379–387.

Quader, H., Hofmann, A., and Schnepf, E. (1989). Reorganization of the endoplasmic reticulum in epidermal cells of onion bulb scales after cold stress: involvement of cytoskeletal elements. *Planta* **177**, 273–280.

Robards, A.W. (1985). *Botanical Microscopy*. Oxford: Oxford University Press.

Rose, C.R., Kovalchuk, Y., Eilers, J., and Konnerth, A. (1999). Two-photon Na^+ imaging in spines and fine dendrites of central neurons. *Eur. J. Physiol.* **439**, 201–207.

Sanders, R., Draaijer, A., Gerritsen, H.C., Houpt, P., and Levine, Y.K. (1995). Quantitative lifetime imaging microscopy. *Anal. Biochem.* **227**, 302–308.

Schneckenburger, H., Gschwend, M.H., Sailer, R., Mock, H.P., and Strauss, W.S.L. (1998). Time-gated fluorescence microscopy in cellular and molecular biology. *Cell. Mol. Biol.* **44**, 795–805.

So, P.T.C., French, T., Yu, W., Berland, K., and Dong, C. (1995). Time-resolved fluorescence microscopy using two-photon excitation. *Bioimaging* **3**, 49–63.

So, P.T.C., König, K., Berland, K., Dong, C.Y., French, T., Bühler, C., Ragan, T., and Gratton, E. (1998). New time-resolved techniques in two-photon microscopy. *Cell. Mol. Biol.* **44**, 771–793.

Squirrell, J.M., Wokosin, D.L., White, J.G., and Bavister, B.D. (1999): Long-term two-photon fluorescence imaging of mammalian embryos without compromising viability. *Nature Biotechnol.* **17**, 763–767.

Steiner, R.F. and Kubota, Y. (1983). Fluorescent dyes–nucleic acid complexes. In: *Excited States of Biopolymers*. New York: Plenum, pp. 203–254.

Suzuki, T., Fujikura, K., Higashiyama, T., and Takata, K. (1997). DNA staining for fluorescence and laser confocal microscopy. *J. Histochem. Cytochem.* **45**, 49–53.

Szmacinski, H., Gryczynski, I., and Lakowicz, J.R. (1993). Calcium-dependent fluorescence lifetimes of indo-1 for one- and two-photon excitation of fluorescence. *Photochem. Photobiol.* **58**, 341–345.

Tirlapur, U.K. and König, K. (1999). Near-infrared femtosecond laser pulses as a novel non-invasive means for dye-permeation and 3D imaging of localised dye-coupling in the *Arabidopsis* root meristem. *Plant J.* **20**, 363–370.

Tirlapur, U.K. and König, K. (2001). Femtosecond near infrared lasers as a novel tool for non-invasive real-time high resolution time-lapse imaging of chloroplast division in living bundle sheath cells of *Arabidopsis*. *Planta* DOI 10.1007/S004250100597.

Tirlapur, U.K., Dhase, B., Riess, I., and Olemiller, P. (1999). Characterization of the activity of a plastid-targeted green fluorescent protein in *Arabidopsis*. *Eur. J. Cell Biol.* **78**, 233–240.

Tsien, R.Y. (1989). Fluorescent indicators of ion concentrations. *Meth. Cell Biol.* **30**, 127–157.

Tsien, R.Y. (1998). The green fluorescent protein. *Annu. Rev. Biochem.* **67**, 509–544.

Tucker, E.B. (1988). Inositol biphosphate and inositol triphosphate inhibit cell-to-cell passage of carboxyfluorescein in staminal hairs of *Setcreasea purpurea*. *Planta* **174**, 358–363.

Van der Schoot, C. and Lucas, W.J. (1993). Microinjection and the study of tissue patterning in plant apices. In: *Methods in Plant Molecular Biology* (ed. P. Maliga, D. Klessig, A. Cashmore, W. Gruissein, and W. Varner). Cold Spring Harbor Press, NY, Cold Spring Harbor Press, pp. 173–189.

Volkmer, A., Subramaniam, V., Birch, D.J.S., and Jovin, T.M. (2000). One- and two-photon excited fluorescence lifetimes and anisotropy decays of green fluorescent proteins. *Biophys. J.* **78**, 1589–1598.

Walker, L.M. and Sack, F. (1995). Microfilament distribution in protonemata of moss *Ceratodon*. *Protoplasma* **189**, 229–237.

White, R.G., Badelt, K., Overall, R.L., and Vesk, M. (1994). Actin associated with plasmodesmata. *Protoplasma* **180**, 169–184.

Wise, F. (1999). Lasers for two-photon microscopy. In: *Imaging: A Laboratory Manual* (ed. R. Yuste, F. Lanni, and A. Konnerth). Cold Spring Harbor, NY: Cold Spring Harbor Press, pp. 18.1–18.9.

Xia, A.D., Wada, S., Tashiro, H., and Huang, W.H. (1999). One- and two-photon-induced fluorescence from recombinant green fluorescent protein. *Arch. Biochem. Biophys.* **372**, 280–284.

Zhu, T., Lucas, W.J., and Rost, T.L. (1998). Directional cell-to-cell communication in the *Arabidopsis* root apical meristem I. An ultrastructural and functional analysis. *Protoplasma* **203**, 35–47.

CHAPTER 22

Two-Photon Excitation Microscopy for Image Spectroscopy and Biochemistry of Tissues, Cells, Organelles, and Lipid Vesicles Under Physiological Conditions*

TIZIANA PARASASSI,[1] LUIS BAGATOLLI,[2] ENRICO GRATTON,[3] MOSHE LEVI,[4] FULVIO URSINI,[5] WEIMING YU,[6] and HUBERT K. ZAJICEK[4]

[1] Istituto di Neurobiologia e Medicina Molecolare, CNR, Viale Marx 15-43, 00137 Roma, Italy; [2] Departamento de Química Biológica, Facultad de Ciencias Química, UNC, Pabellón Argentina, Cuidad Universitaria, 5000 Cordoba Argentina; [3] Laboratory for Fluorescence Dynamics, University of Illinois at Urbana-Champaign, 1110 West Green Street, Urbana, IL 61801 U.S.A.; [4] Department of Internal Medicine, The University of Texas Southwestern Medical Center at Dallas and VA Medical Center, 4500 South Lancaster Road, Dallas, TX 75216, U.S.A; [5] Dipartimento di Chimica Biologica, Universita' di Padova, Viale G. Colombo 3, 35121 Padova Italy; and [6] Department of Cell and Molecular Biology, Northwestern University, School of Medicine, 303 East Chicago Avenue, Chicago, IL 60611, U.S.A.

1 INTRODUCTION

The exclusive occurrence of two-photon excitation at the focal point of the microscope (Denk et al., 1990) offers the valuable advantage of minimizing out-of-focus photodamage and photobleaching (König et al., 1996) – decisive limiting factors in imaging living cells. Cell and tissue autofluorescence is generally accompanied by a rapid fading due to photobleaching, which hinders the observation of biologically relevant molecules such as NAD(P)H, currently considered as a monitor of the cell redox state (Patterson et al., 2000; Piston, 1999; Bennett et al., 1996). In addition, the near-infrared (NIR) wavelength used in two-photon excitation can penetrate deeper into tissues than ultraviolet (UV) can (Masters et al., 1997), thus allowing observation of the interior of thick living samples with no need for slicing and fixation (Parasassi et al., 2000). Several probes that absorb in the UV also fade rapidly, and their imaging is only possible by using NIR two-photon excitation. These include, for instance, membrane probes such as Laurdan (Yu et al., 1996; Parasassi et al., 1997; Bagatolli and Gratton, 1999, 2000a,b) and Prodan (Bagatolli and Gratton, 2000a),

* The experiments reported in this chapter were performed at the Laboratory for Fluorescence Dynamics of the Department of Physics, University of Illinois at Urbana-Champaign.

Confocal and Two-Photon Microscopy: Foundations, Applications, and Advances, Edited by Alberto Diaspro. ISBN 0-471-40920-0 which have been used in imaging natural membranes and phospholipid vesicles since the introduction of two-photon excitation. We present in this chapter some examples related to biochemical processes imaged and quantified in fresh tissues, as well as examples of the biochemical and biophysical characterization of localized environments in cells and in phospholipid vesicles. Fluorescence parameters such as polarization, lifetime, and generalized polarization that require repeated scanning of the same microscopic area can be easily achieved from the images themselves, without compromising the viability of the sample or the emission properties of the fluorophores.

2 MATERIALS AND METHODS

2.1 Preparation of Aorta Samples and Incubation with LDL

Aorta cross-section rings, about 0.5 mm thick, were sectioned from the ascending aorta of six- to eight-week-old Long–Evans male rats using a tissue chopper, immediately after the animals were sacrificed. Samples were maintained in Earle's balanced salt solution (EBSS) containing 24.6 mM glucose and 26.3 mM $NaHCO_3$, pH 7.4, saturated with 95% O_2 and 5% CO_2. Samples were imaged within a few hours by mounting in a hanging drop slide in fresh EBSS buffer. Minimally oxidized human low-density lipoprotein (LDL) (obtained by storage at 4°C for one to two weeks of LDL isolated from healthy fasting volunteers by routine ultracentrifugation; Chapman et al., 1988), at a protein concentration of 3 mg/ml, or 50 μM *tert*-butyl hydroperoxide (TBH) were added to aorta rings, and they were incubated for 30 minutes at 37°C.

2.2 Labeling of Vascular Cells and of LDL

A lipophilic probe, 2-dimethylamino-6-lauroylnaphthalene (Laurdan) (Molecular Probes Inc., Eugene, OR) was used to visualize cells in the space between matrix fibers. Three microliters of a 2 mM solution of Laurdan in dimethylsulfoxide were added to 3 ml of EBSS, containing the aorta ring and minimally oxidized LDL, and incubated for 30 minutes at 37°C. Labeled samples were then washed and mounted in fresh EBSS medium for microscopic observations.

2.3 Cell Growth and Labeling

(*a*) Normal human epidermal keratinocytes (NHEK, Clonetics Corp., San Diego, CA), plated at a density of $8 \times 10^3\ cm^{-2}$ on culture dishes with #1.5 coverglass bottom (Bioptechs, Butler, PA), were grown in KGM medium (Clonetics Corp.), supplemented with 0.1 ng ml human recombinant epidermal growth factor (EGF), 5.0 $\mu g\ ml^{-1}$ insulin, 0.5 $\mu g\ ml^{-1}$ hydrocortisone, 50 $\mu g\ ml^{-1}$ gentamicin, 50 $ng\ ml^{-1}$ amphotericin B, 0.15 mM Ca^{2+}, 7.5 $mg\ ml^{-1}$ bovine pituitary extract. *N*-acetyl-L-cysteine (NAC; Sigma Chem. Co., St Louis, MO) stock solution (20 mM in KGM) was used within one week of preparation and sterilized by filtration before addition to the cell culture. An aliquot of NAC stock solution, to a final concentration of 2 mM, was added to the cell culture 24 hours after seeding. Cells were then allowed to grow for one to two more days, until microscopy observations performed directly in the culture dishes.

(*b*) OK cells, a renal proximal tubular cell line derived from the opossum kidney (Arar et al., 1995), were plated on coverglass-bottom dishes (Bioptechs), and grown in Dulbecco's modified Eagle's high-glucose medium (DMEM) supplemented with 10% fetal bovine serum (FBS), 100 IU penicillin G and 100 μg ml^{-1} streptomycin for 24 hours. Cells where then (1) grown on regular DMEM medium in the presence of 10% FBS, penicillin, and streptomycin, or (2) grown on DMEM plus 5% lipoprotein-deficient serum (LPDS, Intracel, Rockville, MD) for up to two days to deplete the cells of cholesterol, or (3) grown on DMEM plus 5% LPDS for six hours, to upregulate their LDL receptors, and then grown on DMEM plus 400 μg ml^{-1} of human LDL (Calbiochem, San Diego, CA) for up to two days. Cell-culture supplies were obtained from Life Technologies (Rockville, MD) and Sigma Chemical Co. Laurdan labeling of OK cells was obtained by adding 1 μl of a 2 mM solution of Laurdan in dimethylsulfoxide directly to 2 ml of the medium in the growth dishes and incubated for 30 minutes at 37°C. Labeled cells were then washed with fresh medium and imaged directly in the culture dishes.

2.4 Preparation of Giant Unilamellar Vesicles (GUVs)

Stock chloroform solutions of equimolar dilauroyl- and dipalmitoyl-phosphatidylcholine (DLPC and DPPC, respectively), with or without 30 mol% cholesterol, at a final lipid concentration of 0.2 mg ml^{-1}, were used. The electroformation method developed by Angelova and Dimitrov (1986) was followed, and a previously described (Bagatolli and Gratton, 1999) home-made temperature-controlled chamber was used. About 3 μl of the lipid stock solution were spread on each platinum wire under a stream of N_2. Residues of organic solvent were removed by lyophilization for about two hours. To add the aqueous solvent (Millipore water, 17.5 MΩ cm^{-1}) inside the chamber, the bottom part of the chamber was sealed with a coverslip. The water was previously heated to 50°C. Sufficient water was added to the chamber to cover the platinum wires (about 4 ml). The platinum wires were then connected to a function generator (Hewlett-Packard, Santa Clara, CA), and a low-frequency AC field (sinusoidal wave function with a frequency of 10 Hz and an amplitude of 2 V) was applied for 90 minutes. After vesicle formation, the AC field was turned off and the temperature scan (from high to low temperature) initiated. The experiments were carried out in the same chamber after vesicle formation using an inverted microscope (Axiovert 35, Zeiss, Thornwood, NY) as previously reported (Bagatolli and Gratton, 1999). The temperature was measured inside the sample chamber using a digital thermocouple (model 400B, Omega Inc., Stamford, CT) with a precision of 0.1°C. To label the vesicles, a small amount ($<$ 1 μl) of Prodan in DMSO was added after vesicle formation (final Prodan : lipid ratio = 1 : 700 mol/mol). The GUV yield was approximately 95% and their mean diameter was about 30 μm.

2.5 Microscopy Measurements

The 770 nm excitation beam was from a titanium : sapphire laser (Mira 900, Coherent, Palo Alto CA) pumped by an argon-ion laser (Innova 310, Coherent). The microscope configuration and scanning routine have been described previously (So et al., 1995). A frame-scanning rate of 20 s was used. The average laser power on the sample was of the order of 3 mW. A quarter-wave plate was used to change the laser light polarization from linear to circular. Autofluorescence was collected through a broadband BG39 barrier filter. For polarization measurements, vertically polarized two-photon excitation was obtained by inserting a polarizer after the quarter-wave plate in the excitation path. The autofluorescence emission was collected after an analyzer inserted into the optical path. Images of parallel and perpendicular emission were used to calculate the images of autofluorescence polarization. The polarization was not corrected for the instrument geometry. Images of Laurdan-labeled samples were acquired using two different optical bandpass filters (Ealing Electro Optics, New Englander Industrial Park, Holliston, MA) centered in the blue (440 nm) and red (490 nm) regions of the Laurdan emission spectrum. The two filters were exchanged each time a full frame was scanned. From the images at the two emissions, the generalized polarization (GP) (Parasassi et al., 1990) image was calculated as previously described (Yu et al., 1996). The lifetime images were obtained using the microscope configuration described in So et al. (1995) and the harmonic-response technique. The Ti : sapphire laser was operated with a repetition rate of 80 MHz and a pulsewidth of 150 fs. The autofluorescence emission was detected by a gain-modulated photomultiplier. The laser source, detection system, and computer comprise a phase-locked loop. The images of the demodulation and of the phase shift were used to calculate the modulation-lifetime and the phase-lifetime images using an image of a fluorescein solution in water as the reference, as described in So et al. (1995). Section images of OK cells were achieved by using an objective nanopositioner (p-720 PIFOC, Physik Instrumente, Waldbronn, Germany). Two-photon excitation fluorescence microscopy images of GUVs were obtained by employing a LD-Acroplan 20× long-working-distance air objective (Zeiss, Homldale, NJ) with a numerical aperture (NA) of 0.4. The excitation light at 780 nm was attenuated to 50 mW. The fluorescence emission was observed through a broad-bandpass filter, from 350 nm to 600 nm (BG39, Chroma Technology Inc., Brattleboro, VT). The diameter of the vesicles was measured by size-calibrated fluorescent spheres (latex FluoSpheres, polystyrene, blue fluorescent 360/415, diameter 15.5 μm, Molecular Probes Inc.).

3 RESULTS AND DISCUSSION

3.1 Autofluorescence Imaging

3.1.1 Tissues The pyridinoline and desmosine crosslinks in the extracellular matrix proteins elastin and collagen, respectively, show a bright fluorescence after UV absorption (Baraga et al., 1991). By using two-photon excitation (770 nm), long, straight, parallel, thick extracellular matrix proteins have been imaged in fresh cross-section rings of rat aorta maintained in a glucose- and oxygen-supplemented buffer (Figure 1*A*) (Parasassi et al., 2000). The visualization of this morphology of matrix fibers was only possible due to the minimum manipulation of the tissue. Indeed, we used rat aorta slices, about 0.5 mm thick, sectioned by a tissue chopper immediately after the animals were sacrificed, and maintained and imaged in a physiological buffer. In contrast, images obtained by other microscopies from slices thinner by at least one order of magnitude and following freezing–thawing, inclusion in matrices, fixation, and staining, showed thinner and curly fibers (Baraga et al., 1991; de Carvalho and Taboga, 1996; Young et al., 1998). Given the regular pattern of extracellular fibers observable in fresh aorta samples, we were able to appreciate the relevant alteration of their ordered structure after short-term incubation with minimally modified LDL (Parasassi et al., 2000). We used LDL separated by human plasma and minimally oxidized by storage at 4°C for one to two weeks – a rather short storage period, after which only lipid hydroperoxides are reported to increase (Berliner et al., 1990), with virtually no other oxidative damage (i.e., production of aldehydes, consumption of vitamin E, or modification of apo-B charge). In fact, we measured (Parasassi et al., 2000) a hydroperoxide content of 30 nmol or less of lipid hydroperoxides per milligram of cholesterol. After a 30-minute incubation of aorta, at 37°C, with these minimally modified LDLs, the aorta matrix proteins mainly constituted curved and broken fibers (Figure 1*B*). Similar and even stronger alterations were observed after incubation of the aorta with TBH, supporting the specific involvement of lipid hydroperoxide (not shown). We were able to prevent this alteration of the fibers by co-incubation of the aorta with LDL (or with TBH) and with antioxidants or with antiproteases (Parasassi et al., 2000), thus probing an oxidation-sensitive mechanism, triggered by lipid hydroperoxides in LDL, leading to enhanced activity of proteases on the extracellular matrix proteins. The decreased polarization and the slightly shorter lifetime measured directly at every pixel of the autofluorescence image of the fibers after treatment with minimally oxidized LDL or TBH fits the morphological fragmentation, indicating an increased mobility and a more hydrophilic environment of the fluorescent crosslink in fibers. After incubation with minimally oxidized LDL for 30 minutes, the average polarization of matrix protein autofluorescence decreased from 0.38 ± 0.01 to 0.29 ± 0.01 (Fig. 2), and the average lifetime decreased by about 6% (Fig. 3) (Parasassi et al., 2000). An even lower average polarization was observed after a longer (45-minute) incubation with minimally oxidized LDL, resulting in a value of 0.24 ± 0.01. Owing to this direct relationship with the time of aorta incubation with LDL, the decrease in autofluorescence polarization seems particularly appropriate for the quantitative evaluation of fiber proteolysis.

3.1.2 Cells Among the first examples of the features of two-photon excitation microscopy was the direct observation of cell autofluorescence (Piston et al., 1995; König et al., 1996). When using NIR excitation in the wavelength range 700–800 nm,

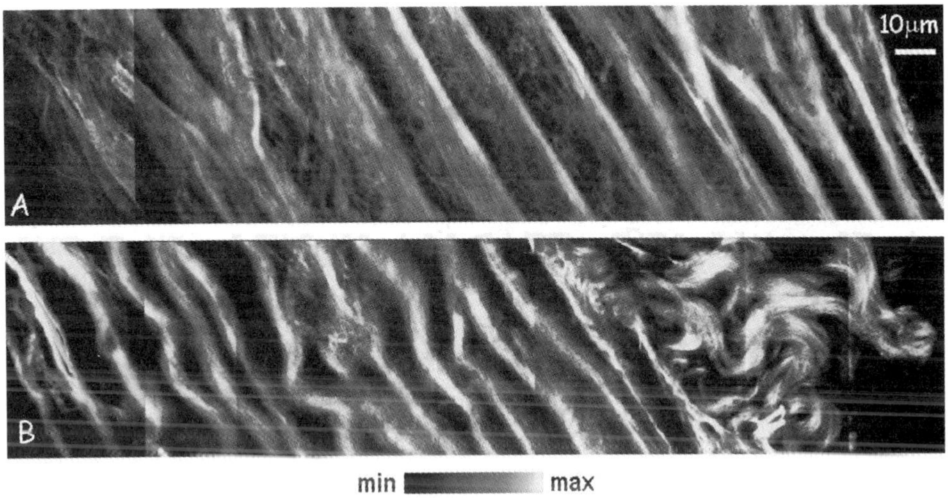

FIGURE 1 See ftp site for color figure. Pseudocolor representation of a composite image of several adjacent frames of deep optical sections showing the autofluorescence intensity of the matrix fibers: (*A*) in a control sample of a rat aorta cross-section ring from the lumen (right) to the ring exterior (left); (*B*) following treatment with minimally oxidized LDL.

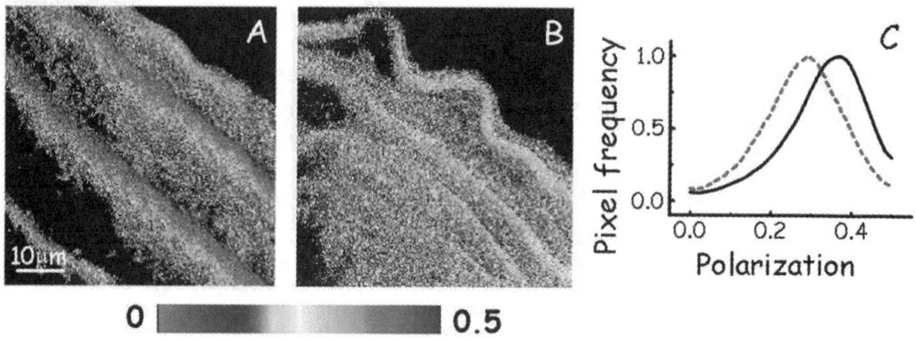

FIGURE 2 See color insert. Pseudocolor representation of autofluorescence polarization values of matrix fibers in the control aorta ring (*A*) and after incubation with minimally oxidized LDL (*B*). The lumen is on the top right. (*C*) Normalized histograms of the pixel frequency versus autofluorescence polarization were obtained from images of the control (continuous black line) and of LDL-treated samples (dotted red line). Each histogram represents the average of three images.

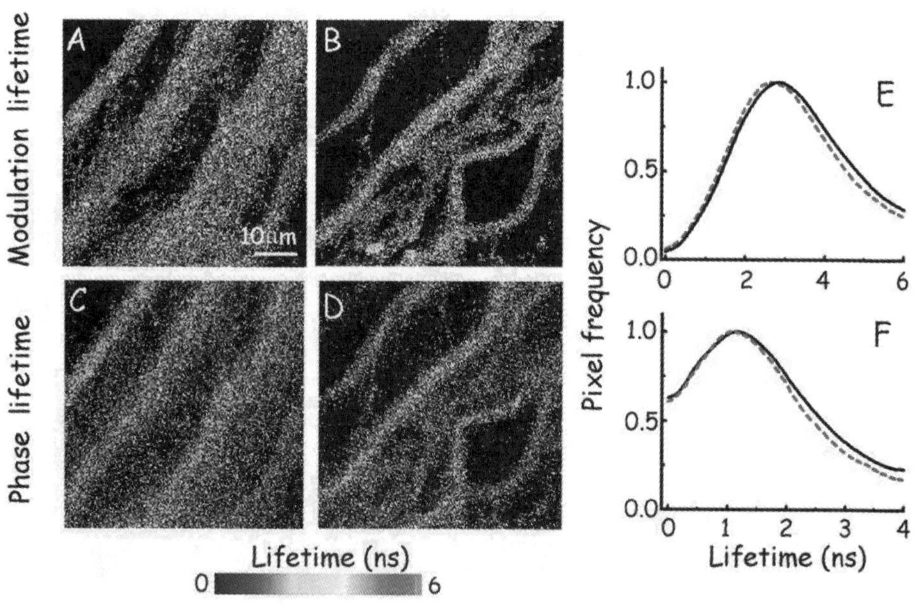

FIGURE 3 See color insert. Pseudocolor representation of autofluorescence lifetime values (*A*, *B*: modulation lifetimes; *C*, *D*: phase lifetimes) of matrix fibers in the control aorta ring (*A*, *C*) and after incubation with minimally oxidized LDL (*B*, *D*). The lumen is on the top right. (*E*, *F*) Normalized histograms of the pixel frequency versus autofluorescence lifetime were obtained from images of the control (continuous black line) and of LDL-treated samples (dotted red line). Each histogram represents the average of four images.

reduced pyridine nucleotides (both reduced nicotinamide–adenine dinucleotide and reduced nicotinamide–adenine dinucleotide phosphate, NAD(P)H) are the compounds showing two-photon absorption. Although diffuse also in the cytoplasm, a high concentration of NAD(P)H is present in mitochondria in an oxidized or reduced state that depends on the cell metabolism, in particular on the oxidative metabolism, and with fluorescence intensity directly related to their reduced state (Patterson et al., 2000). Indeed, cells observed by two-photon excitation microscopy show bright organelles in the vicinity of the dark nucleus that can be identified as mitochondria (Fig. 4). Oxidative stress of various origins can alter the cell redox status with a variety of responses – for instance, the induction of apoptosis by reactive oxygen species, which, in turn, can be inhibited by antioxidants (Meyer et al., 1993). Taking advantage of the exclusive occurrence of two-photon excitation at the focal point of the microscope and of the optimum background rejection, we observed cells directly in their culture medium. Indeed, cells may be grown on thin glass-bottomed dishes and imaged during their growth and during treatments. We treated normal human epidermal keratinocytes with a cytoplasmic thiol reductant compound, *N*-acetyl-L-cysteine (NAC), which has been used to inhibit the induction of apoptosis by a variety of stimuli (Wedi et al., 1999). The aim of our experiment was to monitor the cell redox state by the fluorescence emission of the mitochondria (i.e., of NAD(P)H) after treatment with an antioxidant compound not directly involved in the respiratory chain reactions. Indeed, after treatment with NAC, we observed a dramatic increase in mitochondrial brightness (Fig. 4), thus showing that this cytoplasmic

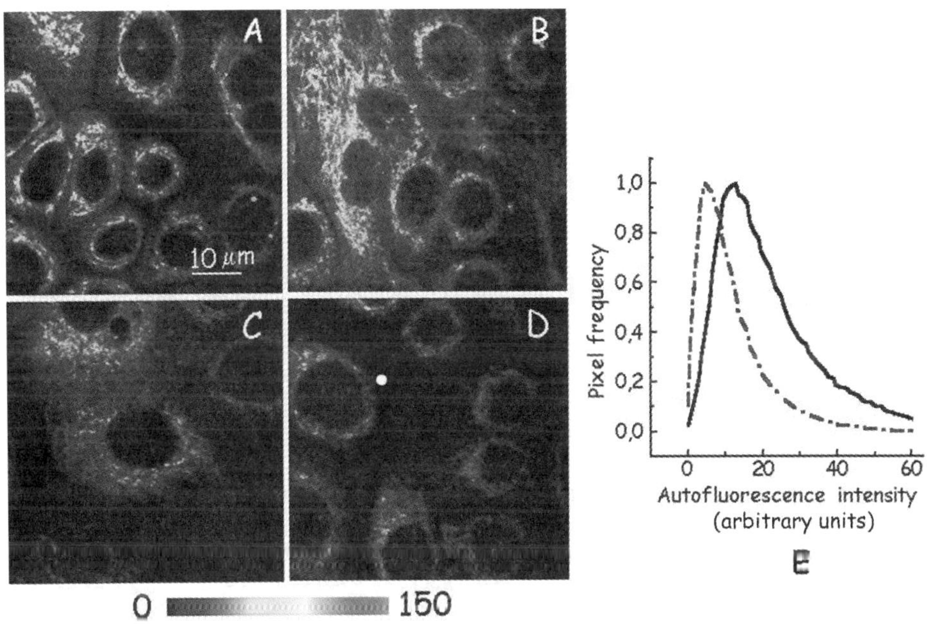

FIGURE 4 See ftp site for color figure. Pseudocolor representation of the autofluorescence intensity in normal human epidermal keratinocytes. The cells imaged in (*A*) and (*B*) were grown for two days after supplementation with 2 mM NAC. (*E*) Normalized histograms of the autofluorescence intensity for control samples (blue dashed line) and for NAC-treated samples (black continuous line). Each histogram represents the average intensity of four different images.

compound can influence the overall cell redox state as monitored by the autofluorescence of mitochondria.

3.2 Rapidly Fading UV Probes

3.2.1 Tissues Besides problems related to sample photodamage when using UV excitation, the use of some probes absorbing in the UV region also poses problems in one-photon excitation microscopy – mainly because of their rapid photobleaching. This is the case with the lipophilic probes 6-lauroyl- and 6-propionyl-2-dimethylaminolauroyl-naphthalene (Laurdan and Prodan, respectively), used in fluorescence spectroscopy for characterizing membrane properties by simple steady-state measurements (Parasassi et al., 1990; Krasnowska et al., 1998). Indeed, these probes show an emission redshift in polar lipid environments, the polarity of lipids being related to their phase state and to their composition. For the study of lipid properties we developed the generalized polarization function (GP) (Parasassi et al., 1990), which is inversely related to the lipid polarity (or to the lipid "fluidity"), which offers several advantages with respect to standard ratiometric measurements, including the possibility of avoiding simultaneous calibration. We used Laurdan to label the aorta tissue and to visualize the adhesion of LDL particles to endothelial cells and their internalization. The Laurdan probe diffuses into tissue lipids with no need for fixation or permeabilization, and, owing to its virtually complete partitioning into lipids with respect to the aqueous environment, its labeling persists while the tissue is imaged in its buffer. The lipid components of LDL particles are also labeled by Laurdan. In our images, we analyzed Laurdan emission by calculating the GP value at each pixel (Parasassi et al., 2000), thus producing a pseudocolor GP image. Cells between matrix fibers were imaged in the aorta tissue, showing higher GP values (i.e., lower "fluidity") in endothelial cell membranes with respect to smooth muscle-like cell membranes in the tunica media, in agreement with the higher cholesterol content in endothelial cell membranes. In Figure 5*A*, red (high GP) endothelial cells are facing the lumen and yellow-green (low GP) smooth muscle-like cells are visible in the space between the matrix fibers. Of course, in this figure, the autofluorescence of extracellular matrix proteins is also visible, with a blue color, due to the typical low GP value of their autofluorescence in our experimental conditions. When the aorta sample is incubated with LDL, these particles can be identified by their color, which is redder than that of the endothelial cell membranes (Fig. 5*B*, *C*). This high GP value is accounted for by the high cholesterol content in LDL. Indeed, our spectroscopic measurements of Laurdan-labeled minimally oxidized LDL gave relatively high GP values, of about 0.46 ± 0.01 (Brunelli et al., 2000). In Figure 5*B*, several LDL particles are adhering to the endothelium, while in Figure 5*C*, they are compartmentalized in the subendothelial space as large aggregates. Adhering and compartmentalized LDL were simultaneously present in different areas of all aorta samples after incubation with LDL, thus demonstrating the possibility of probing both the adhesion and the compartmentalization of these particles in vessels maintained in their physiological buffer, with no need for specific labeled antibodies. Also in these images, after incubation with LDL, extracellular fibers, seen in blue, were curved and with some fragmentation.

3.2.2 Cells The Laurdan probe has also been used to achieve spatially resolved biochemical and biophysical information in living cells. In OK cells, GP values for Laurdan show marked heterogeneity (Fig. 6*A*), consistent with our earlier study that

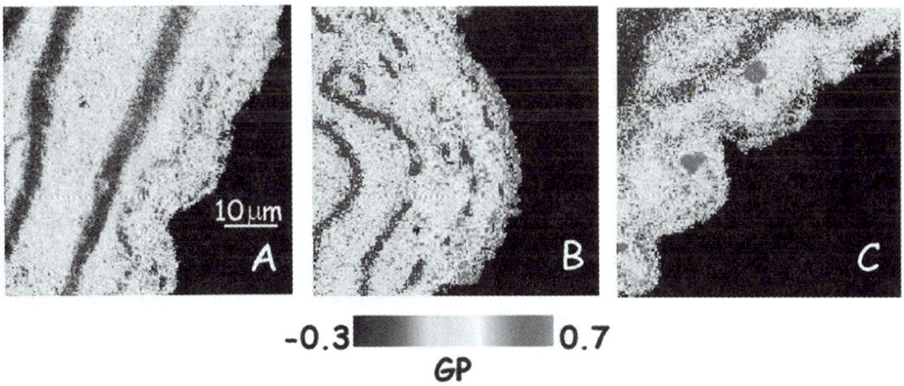

FIGURE 5 See color insert. Pseudocolor representation of Laurdan GP values in rat aorta sections. In (*A*), a control samples is imaged; in (*B*) and (*C*), the aorta samples were incubated for 30 minutes at 37°C with minimally oxidized LDL. In all panels, the lumen is on the right.

suggested the presence of lipid domains in OK cell membranes (Parasassi et al., 1997). The GP values are higher on the apical membranes of OK cells, which is definitely consistent with the higher cholesterol content of the apical membrane when compared with the basolateral membrane and intracellular organelle membranes (Levi et al., 1987; Loghman-Adham et al., 1995). Treatment of OK cells with LDL, which caused a further increase in cholesterol content, resulted in significantly higher Laurdan GP values, especially at the level of the apical membrane (Fig. 6*C*). This can be shown even more dramatically when low GP values are excluded and only GP values higher than 0.4 are shown (Fig. 6*D*). In contrast, treatment of OK cells with LPDS, which resulted in a decrease in cholesterol content, resulted in markedly lower Laurdan GP values, especially in the apical membrane (Fig. 6*E*). These changes are illustrated in an even more dramatic way when low GP values are excluded and only GP values above 0.4 are shown (Fig. 6*F*). In conclusion, these images of the Laurdan GP value allow the biophysical characterization of control OK cell membranes, with a lower "fluidity" of the apical with respect to the basolateral membrane, directly from the images and with no need for separation and purification of the different membranes. Similarly, the effect of cell incubation with LDL, rich in cholesterol, is directly imaged as an impressive increase in the Laurdan GP in the apical membrane, related to a decreased "fluidity"; while, conversely, an increased "fluidity" of the apical membranes is imaged after the cells are grown in a lipid-depleted medium. Apparently, the basolateral membranes were not affected by either treatment.

3.2.3 *Vesicles* Lipid vesicles of a variety of dimensions and lamellarities have been widely characterized by two-photon excitation microscopy, taking advantage of the inherent sectioning capability of this microscopy (Denk et al., 1990; Bagatolli et al., 2000), of the photoselection effect when using polarized incident light (Parasassi et al., 1997; Bagatolli and Gratton, 1999, 2000a, b), and of the negligible photobleaching of UV-absorbing probes when excited by NIR wavelengths (Piston et al., 1995). Here, we only report an example related to the visualization of the partition coefficient of the membrane probe Prodan in giant unilamellar vesicles with or without cholesterol, also demonstrating

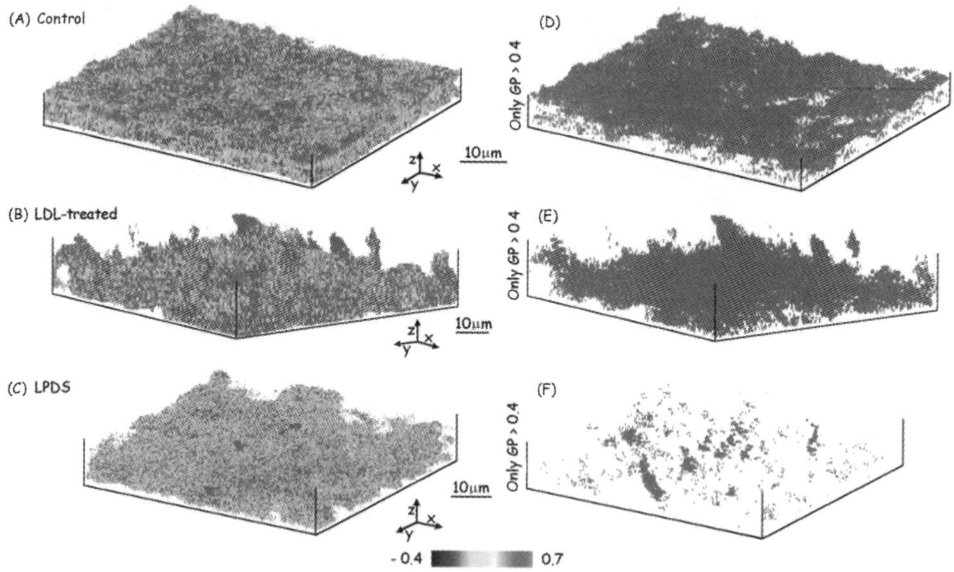

FIGURE 6 See ftp site for color figure. Three-dimensional pseudocolor reconstruction of Laurdan GP values in membranes of OK cell layers. (*A*), (*C*), (*E*) are obtained from a confluent monolayer of OK cells in control, cholesterol-enriched (LDL-treated), and cholesterol-depleted (LPDS) conditions, respectively. (*B*), (*D*), and (*F*) display the same cell monolayers, viewed as GP values above 0.4 only. Cholesterol enrichment causes a significant increase in GP, with most GP > 0.4, while cholesterol depletion causes a marked decrease in GP, with most of GP < 0.4.

a relevant modification of surface properties of the phospholipid bilayer when cholesterol is present.

Similarly to Laurdan (Parasassi et al., 1990), Prodan is a membrane probe sensitive to the polarity properties of the bilayer (MacGregor and Weber, 1981; Parasassi et al., 1994a; Massey and Pownall, 1998; Krasnowska et al., 1998). Mainly due to the different lengths of their acyl residues, partitioning of Laurdan and Prodan between the membrane and water differs. Indeed, while Laurdan partitioning in the membrane is virtually complete, an appreciable partitioning of Prodan in water is observed. In addition, owing to its short propionic acid tail, the fluorescent moiety of Prodan is located closer to the aqueous bilayer surface than that of Laurdan, thus offering sensitive monitoring of the properties of this portion of the bilayer (Krasnowska et al., 1998). In particular, depending on the bilayer surface packing, Prodan partitioning can show relevant variation. For instance, the Prodan partition coefficient between the membrane and water is higher in the liquid-crystalline phase with respect to the gel, tightly packed phase by a factor of about 30 (Krasnowska et al., 1998). These properties of Prodan can be directly visualized by using giant unilamellar vesicles (GUVs) composed of coexisting phospholipid phases, imaged with two-photon excitation microscopy (Bagatolli and Gratton, 2000a). In addition to a different phase state, other factors can modify the bilayer packing, the most biologically relevant being cholesterol. Indeed, in the presence of cholesterol, the packing properties of the gel phase phospholipids are modified in such a way that Prodan partitioning in the membrane increases and Prodan labels the whole membrane. As described previously, in GUVs composed of mixed gel and liquid-crystalline phase in the absence of cholesterol, the fluorescence intensity is not uniform but shows dark areas corresponding to unlabeled gel phase domains (Fig. 7*A*). By increasing the temperature, i.e., when the membrane is composed of a homogeneous liquid-crystalline phase without gel phase domains, dark areas are no longer observable and the vesicle is uniformly labeled (Fig. 7*B*). Instead, in the presence of cholesterol, both at low and at high temperatures, we could not observe dark areas (Figs. 7*C*, *D*), thus directly showing the effect of cholesterol in decreasing the surface packing of the bilayer, by means of an increased partitioning of Prodan in favor of the membrane. For the photoselection effect (Parasassi et al., 1997, 1999; Bagatolli and Gratton, 2000a), these images also clearly show that at low temperature and in the presence of 30 mol% cholesterol (Fig. 7*C*), although the gel phase domains are no longer present, the overall lipid dynamics is reduced, as expected in a homogeneous liquid-ordered phase (Ipsen et al., 1987; Mouritsen, 1991). Indeed, at the top of the vesicle, the lipids are oriented perpendicularly to the plane of the membrane, and the Prodan fluorescent moiety is aligned with the lipids (Parasassi et al., 1997, Bagatolli and Gratton, 2000). Since the direction of propagation of the excitation light is parallel to the plane of the membrane, a partial photoselection effect occurs, so that at the top of the vesicle only those Prodan molecules that can rotate in a small cone around the membrane normal can be excited (Parasassi et al., 1997; Bagatolli and Gratton, 2000). This explains the low fluorescence intensity observed at the top of the vesicle in Figure 7*C*, and is in agreement with the reported physical properties of mixed-phase phospholipids in the presence of 30 mol% cholesterol (Ipsen et al., 1987; Mouritsen, 1991; Parasassi et al., 1994b).

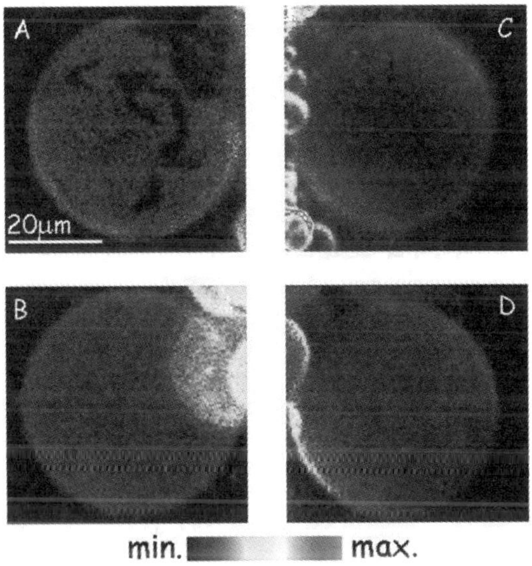

FIGURE 7 See ftp site for color figure. Two-photon excitation fluorescence microscopy images of GUVs, labeled with Prodan, composed of an equimolar mixture of dilauroyl- and dipalmitoyl-phosphatidylcholine in the absence (*A*, *B*) and presence (*C*, *D*) of 30 mol% cholesterol.

4 CONCLUSIONS

We have reported some examples of the attractiveness of using two-photon excitation microscopy for biochemical and biophysical investigations on biological samples. We believe that the main advantages of two-photon excitation reside in the deeper penetration of NIR compared with shorter wavelengths and in the reduced out-of-focus photodamage and photobleaching. These properties allow, on the one hand, direct observation of tissues and cells with no need for manipulation that can alter their properties, and, on the other hand, the imaging of both autofluorescence, with limited sample photodamage, and of probes whose use in microscopy has been limited by their rapid photobleaching. Under these conditions, the biological response of tissues and cells to biochemical treatments is directly imaged with no need for complicated and time-consuming procedures that, besides the isolation of the compound or of the structure of interest from its environment, may also affect its properties. Indeed, we have shown that richer information is obtained when a biological sample is observed under near-physiological conditions, as, for instance, in the case of the effect of incubation of fresh aorta tissue with LDL, which revealed the activation of a redox-sensitive proteolytic mechanism occurring on the extracellular matrix fibers.

REFERENCES

Angelova, M.I. and Dimitrov, D.S. (1986). Liposome electroformation. *Faraday Discuss. Chem. Soc.* **81**, 303–311.

Arar, M., Baum, M., Biber, J., Murer, H., and Levi, M. (1995). Epidermal growth factor inhibits Na–Pi cotransport and mRNA in OKP cells. *Am. J. Physiol.* **268**, F309–F314.

Bagatolli, L.A. and Gratton, E. (1999). Two-photon fluorescence microscopy observation of shape changes at the phase transition in phospholipid's giant unilamellar vesicles. *Biophys. J.* **77**, 2090–2101.

Bagatolli, L.A. and Gratton, E. (2000a). Two-photon fluorescence microscopy of coexisting lipid domains in giant unilamellar vesicles of binary phospholipid mixtures. *Biophys. J.* **78**, 290–305.

Bagatolli, L.A. and Gratton, E. (2000b). A correlation between lipid domain shape and binary phospholipid mixture composition in free standing bilayers. A two-photon fluorescence microscopy study. *Biophys. J.* **79**, 434–447.

Bagatolli, L.A., Parasassi, T., and Gratton, E. (2000). Giant phospholipid vesicles: comparison among the whole lipid sample characteristics using different preparation methods. A two photon fluorescence microscopy study. *Chem. Phys. Lipids* **105**, 135–147.

Baraga, J.J., Rava, R.P., Fitzmaurice, M., Tong, L.L., Taroni, P., Kittrell, C., and Feld, M.S. (1991). Characterization of the fluorescent morphological structures in human arterial wall using ultraviolet-excited microspectrofluorimetry. *Atherosclerosis* **88**, 1–14.

Bennett, B.D., Jetton, T.L., Ying, G., Magnuson, M.A., and Piston, D.W. (1996). Quantitative subcellular imaging of glucose metabolism within intact pancreatic islets. *J. Biol. Chem.* **271**, 3647–3651.

Berliner, J.A., Territo, M.C., Sevanian, A., Ramin, S., Kim, J.A., Bamshad, B., Esterson, M., and Fogelman, A.M. (1990). Minimally modified low density lipoprotein stimulates monocyte endothelial interactions. *J. Clin. Invest.* **85**, 1260–1266.

Brunelli, R., Mei, G., Krasnowska, E.K., Pierucci, F., Zichella, L., Ursini, F., and Parasassi, T. (2000). Estradiol enhances the resistance of LDL to oxidation by stabilizing apoB-100 conformation. *Biochemistry* **39**, 13897–13903.

Chapman, M.J., Laplaud, P.M., Luc, G., Forgez, P., Bruckert, E., Goulinet, S., and Lagrange, D. (1988). Further resolution of the low density lipoprotein spectrum in normal human plasma: physicochemical characteristics of discrete subspecies separated by density gradient ultracentrifugation. *J. Lipid Res.* **29**, 442–458.

de Carvalho, H.F. and Taboga, S.R. (1996). Fluorescence and confocal laser scanning microscopy imaging of elastic fibers in hematoxylin-eosin stained sections. *Histochem. Cell Biol.* **106**, 587–592.

Denk, W., Strickler, J.H., and Webb, W.W. (1990). Two-photon laser scanning fluorescence microscopy. *Science* **248**, 73–76.

Ipsen, J.H., Karlstrom, G., Mouritsen, O.G., Wennerstrom, H., and Zuckermann, M.H. (1987). Phase equilibria in the phosphatidylcholine-cholesterol system. *Biochim. Biophys. Acta* **983**, 162–172.

König, K., So, P.T.C., Mantulin, W.W., Tromberg, B.J., and Gratton, E. (1996). Two-photon excited lifetime imaging of autofluorescence in cells during UVA and NIR photostress. *J. Microsc.* **183**, 197–204.

Krasnowska, E.K., Gratton, E., and Parasassi, T. (1998). Prodan as a membrane surface fluorescent probe: partitioning between water and phospholipid phases. *Biophys. J.* **74**, 1984–1993.

Levi, M., Molitoris, B.A., Burke, T.J., Schrier, R.W., and Simon, F.R. (1987). Effects of vitamin D-induced hypercalcemia on rat renal cortical brush border and basolateral membranes and mitochondria. *Am. J. Physiol.* **256**, F85–F94.

Loghman-Adham, M., Motock, G.T., Wilson, P., and Levi, M. (1995). Characterization of Na–phosphate cotransporters in renal cortical endosomes. *Am. J. Physiol.* **269**, F93–F102.

MacGregor, R.B. and Weber, G. (1981). Fluorophores in polar media: spectral effects of the Langevin distribution of electrostatic interactions. *Ann. NY Acad. Sci.* **366**, 140–154.

Massey, J.B. and Pownall, H.J. (1998). Surface properties of native human plasma lipoproteins and lipoprotein models. *Biophys. J.* **74**, 869–878.

Masters, B.R., So, P.T., and Gratton, E. (1997). Multiphoton excitation fluorescence microscopy and spectroscopy of in vivo human skin. *Biophys. J.* **72**, 2405–2412.

Meyer, M., Schreck, R., and Baeuerle, P.A. (1993). H_2O_2 and antioxidants have opposite effects on activation of NF-κB and AP-1 in intact cells: AP-1 as secondary antioxidant responsive factor. *EMBO J.* **12**, 2005–2015.

Mouritsen, O.G. (1991). Theoretical models of phospholipid phase transitions. *Chem. Phys. Lipids* **57**, 179–194.

Parasassi, T., De Stasio, G., d'Ubaldo, A., and Gratton, E. (1990). Phase fluctuation in phospholipid membranes revealed by Laurdan fluorescence. *Biophys. J.* **57**, 1179–1186.

Parasassi, T., Giusti, A.M., Gratton, E., Monaco, E., Raimondi, M., Ravagnan, R., and Sapora, O. (1994a). Evidence for an increase in water concentration in bilayers after oxidative damage of phospholipids induced by ionizing radiation. *Int. J. Radiat. Biol.* **65**, 329–334.

Parasassi, T., Di Stefano, M., Loiero, M., Ravagnan, G., and Gratton, E. (1994b). Influence of cholesterol on phospholipid bilayers phase domains as detected by Laurdan fluorescence. *Biophys. J.* **66**, 120–132.

Parasassi, T., Gratton, E., Yu, W.M., Wilson, P., and Levi, M. (1997). Two-photon fluorescence microscopy of Laurdan GP-domains in model and natural membranes. *Biophys. J.* **72**, 2413–2429.

Parasassi, T., Gratton, E., Zajicek, H., Levi, M., and Yu, W.M. (1999). Detecting membrane lipid microdomains by two-photon fluorescence microscopy. *IEEE Eng. Med. Biol.* **18**, 92–99.

Parasassi, T., Yu, W.M., Durbin, D., Kuriashkina, L., Gratton, E., Maeda, N., and Ursini, F. (2000). Two-photon fluorescence of aorta fibers shows proteolysis induced by LDL hydroperoxides. *Free Radicals Biol. Med.* **28**, 1589–1597.

Patterson, G.H., Knobel, S.M., Arkhammar, P., Thastrup, O., and Piston, D.W. (2000). Separation of the glucose-stimulated cytoplasmic and mitochondrial NAD(P)H responses in pancreatic islet beta cells. *Proc. Natl Acad. Sci. USA* **97**, 5203–5207.

Piston, D.W. (1999). Imaging living cells and tissues by two-photon excitation microscopy. *Trends Cell Biol.* **9**, 66–69.

Piston, D.W., Masters, B.R., and Webb, W.W. (1995). Three-dimensionally resolved NAD(P)H cellular metabolic redox imaging of the in situ cornea with two-photon excitation laser scanning microscopy. *J. Microsc.* **178**, 20–27.

So, P.T.C., French, T., Yu, W.M., Berland, K.M., Dong, C.Y., and Gratton, E. (1995). Time-resolved fluorescence microscopy using two-photon excitation. *Bioimaging* **3**, 49–63.

Wedi, B., Straede, J., Wieland, B., and Kapp, A. (1999). Eosinophil apoptosis is mediated by stimulators of cellular oxidative metabolisms and inhibited by antioxidants: involvement of a thiol-sensitive redox regulation in eosinophil cell death. *Blood* **94**, 2365–2373.

Young, A.A., Legrice, I.J., Young, M.A., and Smaill, B.H. (1998). Extended confocal microscopy of myocardial laminae and collagen network. *J. Microsc.* **192**, 139–150.

Yu, W.M., So, P.T., French, T., and Gratton, E. (1996). Fluorescence generalized polarization of cell membranes: a two-photon scanning microscopy approach. *Biophys. J.* **70**, 626–636.

CHAPTER 23

Real-Time In Situ Calcium Imaging with Single- and Two-Photon Confocal Microscopy

KATSUMASA FUJITA and TETSURO TAKAMATSU

Department of Pathology and Cell Regulation, Kyoto Prefectural University of Medicine, 465 Kajii-cho, Kawaramachi-Hirokoji, Kamigyo-ku, Kyoto 602-0841, Japan

1 INTRODUCTION

The recent development of confocal microscopes and accompanying techniques has enabled us to observe activities of living cells. This advance in microscopy has important implications in biology and medicine if subcellular phenomena can be observed in situ (Takamatsu, 1998). It is known that the behavior of cells depends strongly on surrounding conditions (Fig. 1). For example, heart muscle cells change their beat frequency with maturation of surrounding cells (Oyamada, 1997). There must be intercellular communications for their synchronization. However, intercellular communication has previously been observed in cultured cells that have neither three-dimensional structure nor interstitial tissue such as vessels.

In this chapter, we introduce fundamental techniques and applications of fluorescence microscopy for real-time in situ imaging that is applicable to the investigation of subcellular activities and intercellular communications in living tissue. The techniques, including scanning methods, and excitation methods in both single- and two-photon fluorescence microscopes, are introduced from the viewpoint of the instrumentation. Experimental results on the monitoring of Ca^{2+} concentration in living cells of a perfused rat whole heart, where heart muscle cells and interstitial tissues are present in three dimensions, are shown as examples of the investigations that can be performed with both single- and two-photon confocal fluorescence microscopes.

Confocal and Two-Photon Microscopy: Foundations, Applications, and Advances, Edited by Alberto Diaspro.
ISBN 0-471-40920-0

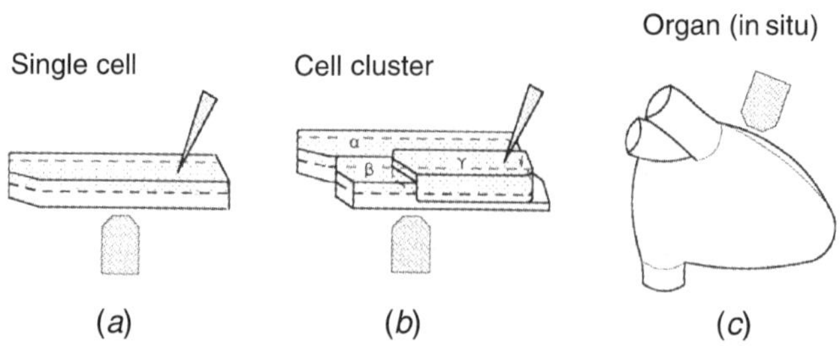

FIGURE 1 Three different conditions for cell observation.

2 REAL-TIME IN SITU IMAGING WITH SINGLE-PHOTON EXCITATION

2.1 Real-Time Scanning Techniques

After the development of the confocal microscope, several researchers attempted to realize real-time or faster scanning techniques in the microscope (Kino and Corle, 1989). Shack et al. (1982) used a polygonal mirror (Fig. 2) to change the angle of incident light to an objective lens illuminating a specimen. The focal point of the objective lens moves on the focal plane in a specimen with a change in the path direction of the incident light. The polygonal mirror is used for scanning the direction that requires a faster scanning rate in the *x*–*y* plane, and a galvanometer mirror scans the other direction. A scanning system using a polygonal mirror can be achieved with a relatively simple set-up compared with other techniques, such as a system using a resonant galvanometer mirror. However, users have to take care about noise caused by the rotation and vibration of the mirror, and dead time when the laser beam hits the edges of the mirror. It is also possible to use a resonant galvanometer mirror (Fig. 3) for real-time scanning. However, such a mirror scans the specimen at a constant rate by being driven at the resonant frequency of the mirror and the torsion rod, and the frequency cannot be changed. The time variation of the scanning angle follows a sinusoidal curve and the scanning distance on the specimen also. This makes development of the system more complex, and allows less adaptability of the system to a variety of applications.

In another approach, Petran et al. employed a Nipkow disc (which was originally developed for television sets and TV cameras), for real-time scanning in a confocal

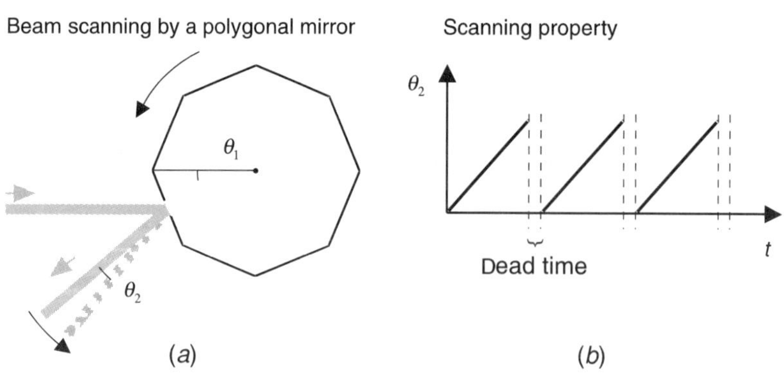

FIGURE 2 Rotation of a polygonal mirror changes the angle between the mirror surface and the reflected beam. The number of mirrors determines the number of scanning lines per rotation.

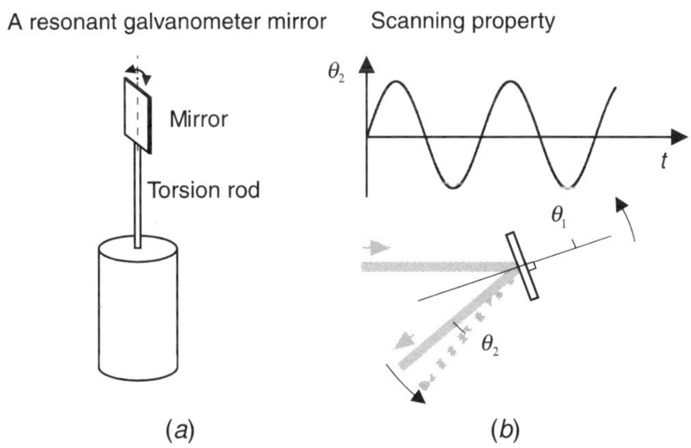

FIGURE 3 Driving a galvanometer mirror at the resonant frequency of the mirror and a torsion rod allows fast scanning.

microscope (Egger and Petran, 1967). This remarkable scanning technique allows one to observe a specimen even faster than video rate with confocal optics. However, this technique did not become popular for confocal microscopes because of its complex configuration and low detection efficiency of light. Later, Xiao and co-workers proposed a simplified scanning system, which became a popular style for this scanning method (Xiao and Kino, 1987). Figure 4 shows the scanning system proposed by them. In this system, light transmitted through the Nipkow disc illuminates a specimen with multiple foci. The multiple foci scan a view area in the specimen with rotation of the disc. Reflected light or fluorescence from the specimen is collected by the same objective lens, and traces back on the same optical path to be detected by a two-dimensional detector, such as a CCD camera.

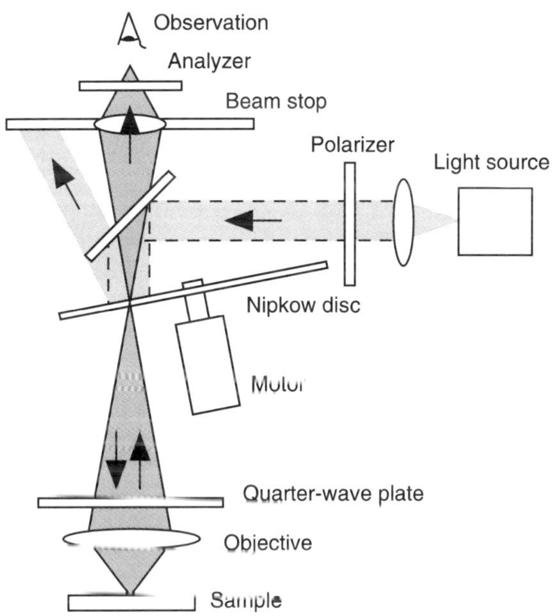

FIGURE 4 Real-time confocal scanning microscope using a Nipkow disc.

Although the Nipkow disc enables one to scan a specimen fast enough for real-time observation, it is still difficult to apply this system to fluorescence since the transmittance of light through the disc is too small. Recently, Ichihara et al (1996) solved this problem by introducing a microlens-array disc in front of the Nipkow disc to make the transmittance higher.

The scanning techniques mentioned above allow us to observe a specimen at video rate with a confocal microscope. In the former two systems using polygonal or resonant mirrors, the maximum scanning rate is determined by the rotation speed of the polygonal mirror or the resonant frequency of the galvanometer mirror, respectively. With the polygonal mirror, the scanning rate is usually no higher than the video rate, since noise caused by vibration can be a big problem. The use of the resonant galvanometer allows one to scan a specimen at more than video rate, depending on the number of pixels scanned on one image. Currently, a 130 frames s^{-1} scanning rate is available with a commercial system from BioRad Laboratories, U.S.A. On the other hand, scanning systems using a Nipkow disc allows an image acquisition rate much higher than video rate. For example, the Yokogawa Electric Company, Japan has commercialized a system with a microlens-array disc that can scan a specimen within 1 ms.

For applications of fluorescence imaging, the fluorescence flux that can be obtained through the optical system is one of the important factors when selecting a scanning system. A system with a polygonal or a resonant galvanometer mirror uses a single focus to scan the specimen. In this case, the amount of fluorescence detected per unit time decreases proportionally to the increase in scanning speed. The increase in excitation intensity allows one to enhance the fluorescence flux. However, it also brings more phototoxicity and faster photobleaching. Therefore, in the use of single-focus scanning, there is a trade-off between scanning rate and fluorescence intensity (signal-to-noise ratio of images). On the other hand, in a system using multiple foci for illuminating the specimen, there is no such trade-off, because the total amount of fluorescence can be increased proportionally to the number of excitation spots. This enables one to maintain signal-to-noise ratio without increasing the intensity of excitation, even in fast-scanning cases.

The principles for constructing fluorescence images are also different, depending on the scanning system. As shown in Figure 5, scanning with a single focus imposes a time difference between the starting point and the end point of scanning. Even though the image acquisition is at video rate, there is 33 ms of time difference in an image. In contrast, the multifocus scanning technique scans a specimen uniformly without a large time difference at different positions on the specimen. This property is useful for observing responses of specimens for which the time scale is close to the scanning rate.

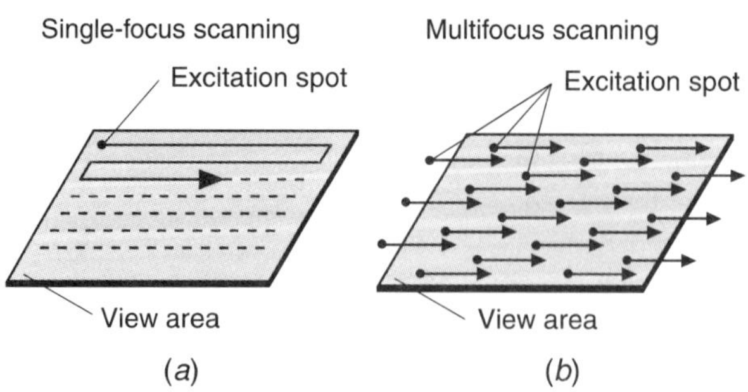

FIGURE 5 Comparison of (*a*) single-focus scanning and (*b*) multifocus scanning.

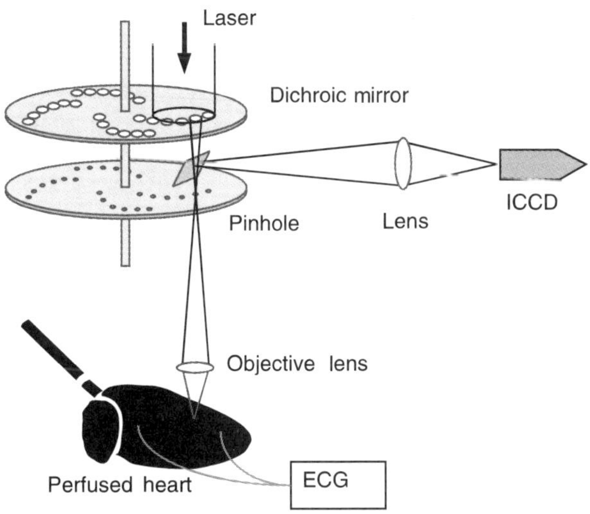

FIGURE 6 Optical set-up for in situ imaging of a perfused rat whole heart.

2.2 Real-Time In Situ Imaging of a Whole Heart

Takamatsu and co-workers demonstrated the first example of in situ imaging of organs using real-time confocal fluorescence microscope (Hama et al., 1998; Kaneko et al., 2000). Figure 6 shows the set-up of the microscope that they developed for observing Ca^{2+} dynamics in a rat whole heart. In this microscope, a multifocus scanning system using microlens- and pinhole-array discs was adopted because of its scanning rate and fluorescence-detection efficiency. In their demonstrations, a whole heart extracted from a rat was placed under a microscope objective lens after being loaded with a fluorescent Ca^{2+} indicator (fluo-3/AM). The heart was perfused even during the observation with a Tyrode solution to keep the heart alive for as long as possible.

In Figure 6, the heart was immersed in a water bath to protect it, and was observed by a water-immersion objective lens. Taking account of the fluorescence-detection efficiency, the use of an oil-immersion lens seems to be reliable for those applications with weak fluorescence. However, when using an oil-immersion lens with an inverted microscope, mismatched refractive indices between coverslip and solution cause distortion of images. The use of the water-immersion lens was the best choice in this situation – and probably under other in situ conditions also.

Figure 7*a* shows sequential in situ images of intracellular Ca^{2+} concentration represented as the intensity of the fluorescent Ca^{2+} indicator. The brightness of the images is proportional to Ca^{2+} concentration in the heart. Sudden increases in Ca^{2+} concentration (Ca^{2+} transients) were observed to be synchronized with heartbeat signals on a simultaneously obtained electrocardiogram (Fig. 7*b*). Figure 8 shows another example of Ca^{2+} dynamics in a rat whole heart. Propagation of highly concentrated Ca^{2+} in a single cell can be seen. This propagation is known as a Ca^{2+} wave, and is an abnormal behavior of Ca^{2+} in muscle cells (Takamatsu and Wier, 1990).

Techniques for introducing fluorescent dyes into specimens are among the significant aspects of in situ imaging. In the case of cultured cells, specimens are immersed in a

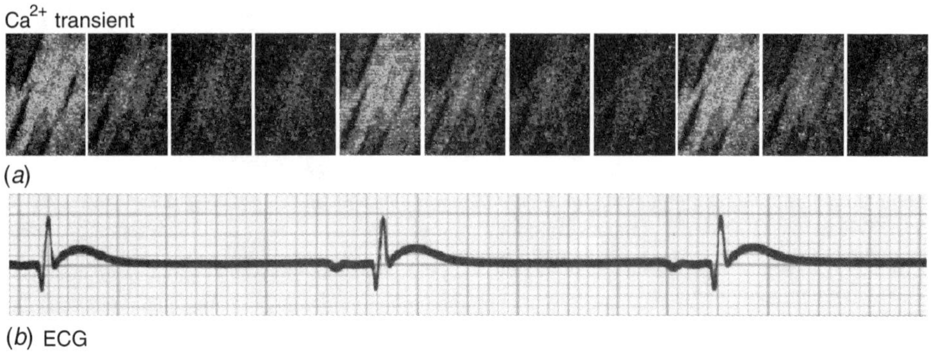

FIGURE 7 (*a*) Ca^{2+} transient observed in a perfused rat whole heart. (*b*) The electrocardiogram (ECG) obtained with the fluorescence images.

solution that contains a membrane-permeable fluorescent dye. Obviously, in cases of in situ imaging, specimens are too thick for a fluorescent dye to permeate. In the demonstration described above, the fluorescent indicator was loaded into the rat whole heart by perfusion of the solution containing the indicator through the coronary artery (Fig. 9) (Hama et al., 1998).

2.3 Applications in Pathology

Myocardial infarction is characterized by insufficiency of oxygen and nutrient substances and inadequate removal of metabolites. Most cases have conduction disturbance of

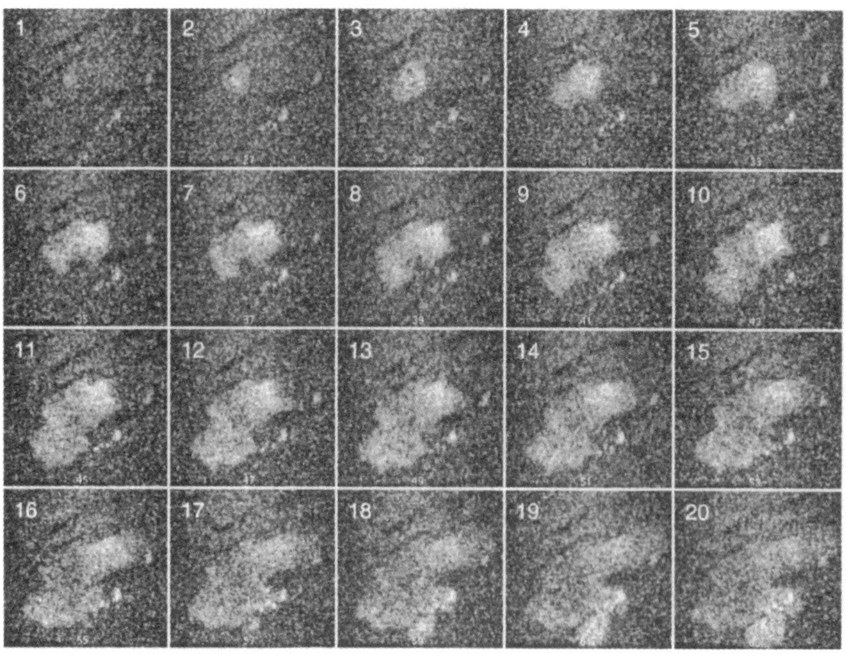

FIGURE 8 Ca^{2+} wave observed in a perfused rat whole heart.

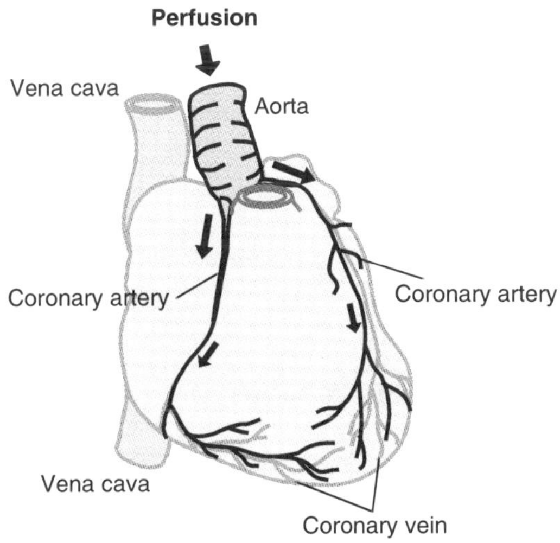

FIGURE 9 Perfusion through the coronary artery for loading a fluorescent Ca^{2+} indicator.

excitation or myocardial irritability, namely arrhythmia including fetal ventricular tachycardia and fibrillation. Although extensive electrophysiological and histological studies show that such tachyarrhythmia originate at the border of the myocardial infarction (Matsushita et al., 1999), the precise pathogenesis is still unknown. To clarify intracellular calcium dynamics in heart muscle cells at the border of the myocardial infarction, a rat whole heart was observed after coronary ligation of the left coronary artery. Figure 10 shows the calcium dynamics at the border of the myocardial infarction. A Ca^{2+} transient

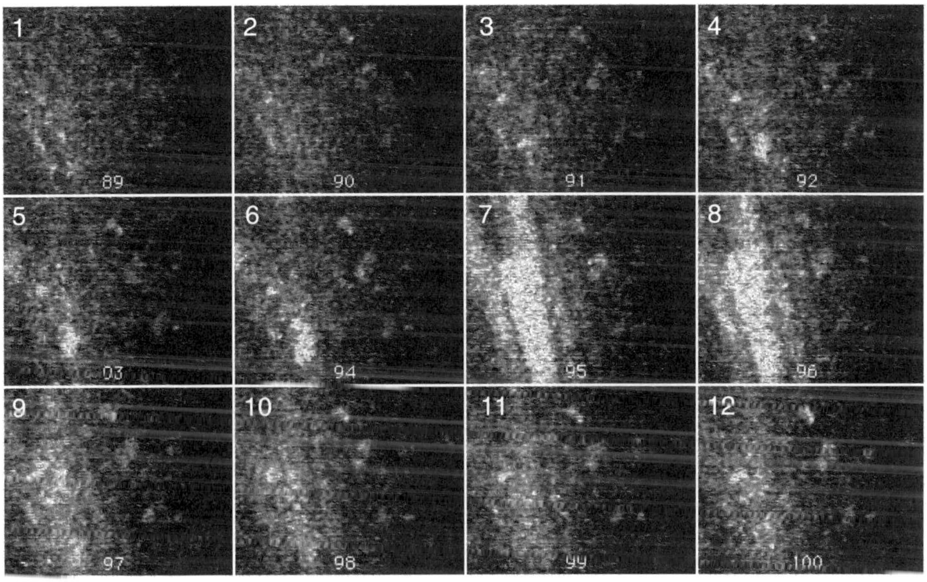

FIGURE 10 Ca^{2+} transient at the border of a myocardial infarction. The right halves of the images show the infarcted parts of the heart.

does not appear at the right half of the images, and a clear border exists between infarcted and noninfarcted areas.

However, it is important to observe a larger view area for understanding how this calcium dynamic acts to cause heart disease. In a confocal microscope with fast scanning, a high-numerical-aperture (NA) objective lens is often used to acquire as much fluorescence as possible. Usually, a higher-NA objective lens has a greater magnification. The development of new objective lenses that have both low magnification and high NA is required for applying in situ imaging techniques to pathology more effectively.

3 REAL-TIME IMAGING WITH TWO-PHOTON EXCITATION

3.1 Scanning Techniques for Real-Time Two-Photon Fluorescence Microscopy

Several techniques have also been proposed for two-photon fluorescence microscopes. These techniques are based on those for a confocal single-photon fluorescence microscope. Xie et al. (1998) developed a real-time imaging system by fast raster scanning of a single focus with a resonant galvanometer mirror, and observed single green algae cells swimming in an aqueous solution. Fan et al. (1999) used a similar configuration for observing Ca^{2+} concentration in cultured rat cardiomyocytes.

Since these two-photon fluorescence microscopes are based on a single-focus scanning technique, the total fluorescence intensity that can be obtained in an image decreases with increasing scanning rate, as with single-photon excitation. This problem is more serious in two-photon fluorescence microscopes, because introducing a higher-intensity excitation to enhance the fluorescence flux reduces cell viability or brings about nonlinear optical phenomena such as self-focusing, continuum generation, and ablation. In addition, high-intensity excitation easily induces saturation in fluorescence intensity (König et al., 1999; Smith et al., 2000; 2001). Figure 11 shows the fluorescence intensity from a rhodamine B solution as a function of excitation power. The circles show the experimental results obtained with a rhodamine B solution excited by a mode-locked Ti : sapphire laser with wavelength 800 nm, repetition rate 80 MHz, pulsewidth 100 fs, and an NA 1.3 objective

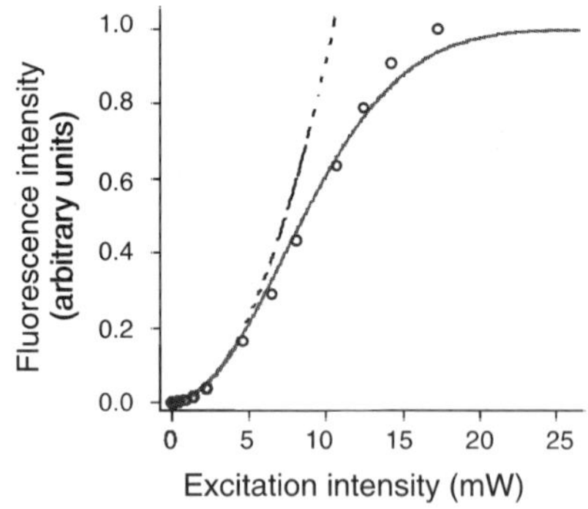

FIGURE 11 Variation of two-photon fluorescence intensity from a rhodamine B solution with excitation intensity. The solid line and the circles show calculated and measured intensities from the solution, respectively. The dotted line is the quadratic curve fitted to the results calculated at several low excitation intensities.

lens. The solid curve shows the result calculated using the two-photon absorption cross-section of rhodamine B under the same conditions as the experiment. It can be seen that the fluorescence intensity is saturated completely at excitation intensities above 20 mW in the calculated results. In both experimental and calculated results, there is a saturation effect at excitation intensities of more than about 5 mW. This saturation effect is an important factor that determines the scanning rate in two-photon fluorescence microscopes. If it is possible to obtain a fluorescence intensity strong enough for detection with the required scanning rate, this problem can be ignored. However, in cases of observing ion dynamics in living bodies, for instance, the saturation effect limits the availability of the two-photon fluorescence microscope.

The use of a multifocus scanning technique can solve this problem also in cases of two-photon excitation. Brewersdorf et al. (1998) proposed a multifocus two-photon excitation microscope, where fluorescence was excited with several tens of foci produced on the focal plane of an objective lens by using a microlens array. Figure 12*a* shows the optical set-up of the proposed microscope. In the case of two-photon excitation, a pinhole-array disc is not required to obtain three-dimensional resolution. Later, Fujita et al. (2000a) introduced

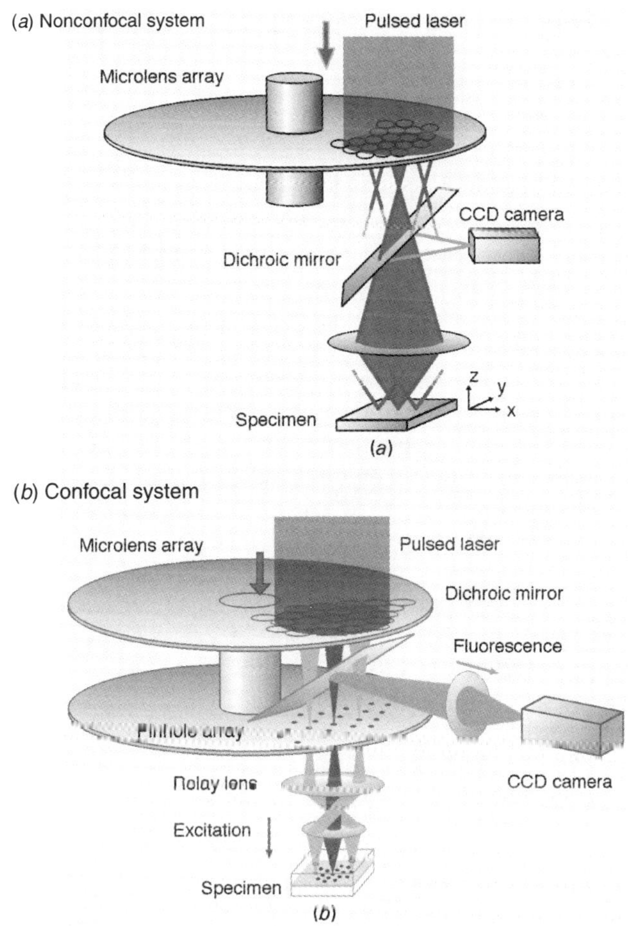

FIGURE 12 Optical set-up of (*a*) a nonconfocal multifocus two-photon microscope and (*b*) a confocal multifocus two-photon microscope.

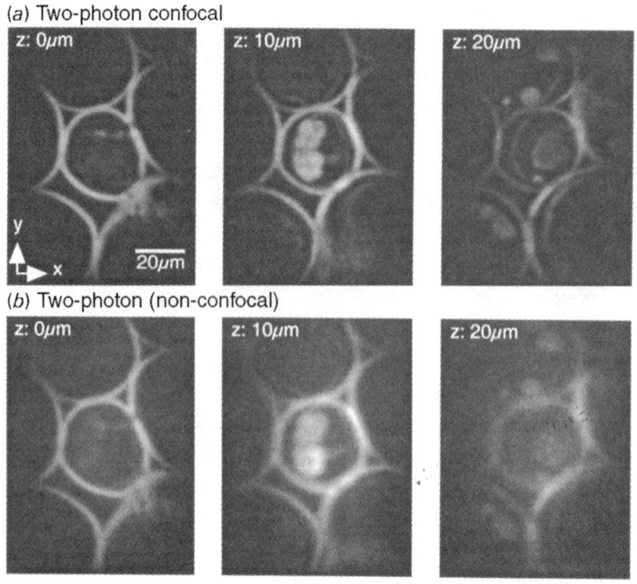

FIGURE 13 Comparisons of the images obtained by (*a*) a confocal system and (*b*) a nonconfocal system.

a pinhole-array disc to introduce confocal detection in the multifocus two-photon microscope as shown in Figure 12*b*. In principle, a confocal pinhole in a two-photon microscope does not work as effectively as one in a single-photon confocal fluorescence microscope. However, in multifocus scanning, it works well to enhance the contrast of images and penetration depth. Figure 13 shows a comparison of two-photon fluorescence images obtained by a confocal and by a nonconfocal system using multifocus scanning. Figure 14 shows x–z cross-sectional images of the same dataset as in the previous figure. These comparisons show that scattering of fluorescence in a specimen reduces the contrast of images, and this becomes more prominent with increasing observation depth. It also can be seen that the pinhole-array disc works effectively even in observing the shallow parts of the specimen, because backward scattering from the deeper parts of the specimen can be eliminated by the pinhole array. In a typical single-focus scanning two-photon fluorescence microscope, the use of a pinhole decreases the amount of fluorescence intensity at the detector. Scattering of fluorescence does not have much effect in this point-to-point detection system. For this reason, confocal detection is not used in many cases, especially when observing deep parts of a specimen. On the other hand, in the case of multifocus scanning, two-dimensional detectors such as CCD cameras pick up the scattered fluorescence as a background noise, which reduces the contrast of images and degrades the depth-penetration benefit of two-photon excitation.

3.2 Efficiency of Two-Photon Excitation in Multifocus Scanning

Since the excitation intensity required to induce two-photon fluorescence is much higher compared with the single-photon case, the number of foci used for multifocus two-photon

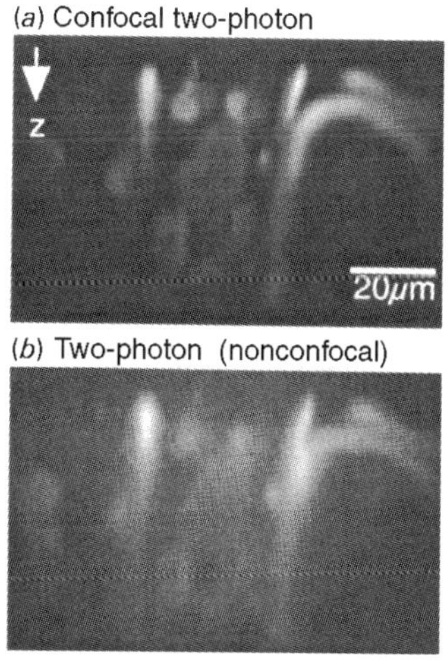

FIGURE 14 The x–z cross-sectional images obtained by (a) a confocal system and (b) a nonconfocal system.

excitation has to be much smaller than with single-photon excitation (Xu and Webb, 1996). Usually, several hundreds (or even more) foci are used in a single-photon multifocus fluorescence microscope. It is difficult to achieve this number of foci in two-photon excitation with the limited intensity of a laser source. Therefore, in order to apply a two-photon microscope system to real-time in situ imaging it is necessary to optimize the system, including the excitation power, repetition rate, pulsewidth of the laser, and the number of foci, by estimating the efficiency of two-photon excitation of fluorescence.

For estimating the efficiency of two-photon excitation, it is convenient to consider the case of a single focus first. The amount of fluorescence emitted from a focus in a specimen can be shown to be

$$F(t) \propto C\eta R(t), \tag{1}$$

where C is the concentration of fluorescent molecules, η is the quantum yield, and $R(t)$ is the excitation rate of the molecules. The excitation rate of the molecules with a short-pulsed laser can be obtained by considering a simple two-level system representing the molecular status. Figure 15 shows the simple model of the two-level system used for estimation. The number of molecules in the excited state and those in the ground state can be described by the rate equations

$$\frac{dn_{\text{ex}}}{dt} = n_g(t)r_{\text{ex}}(t) - \frac{n_{\text{ex}}(t)}{\tau_{\text{ex}}} \tag{2}$$

and

$$\frac{dn_g}{dt} = -n_g(t)r_{\text{ex}}(t) + \frac{n_{\text{ex}}(t)}{\tau_{\text{ex}}}, \tag{3}$$

where n_{ex} is the population of the excited molecules, n_g is the population of molecules in the ground state, r_{ex} is the excitation probability, and τ_{ex} is the lifetime of the excited

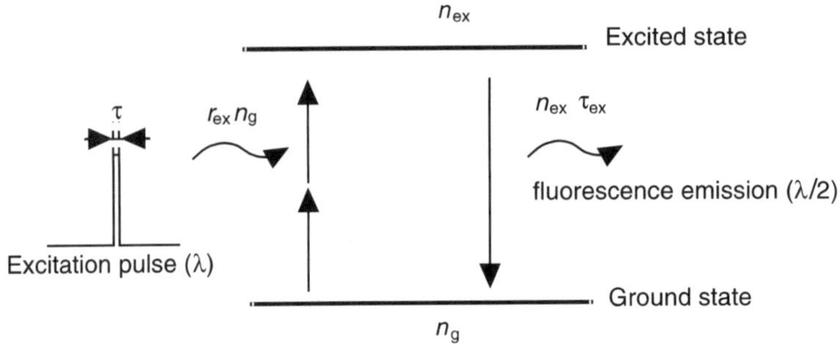

FIGURE 15 A model of a two-level system used for the calculation of two-photon excitation efficiency.

molecules (depending on the relaxation time). The excitation rate of the molecules, r_{ex}, is found by considering the photon density at the focus to be given by

$$r_{ex}(t) = \sigma_{2p}\Phi^2_{peak}(t), \tag{4}$$

where σ_{2p} is the two-photon absorption cross-section of the molecule and Φ_{peak} is the photon density at the focus, given by

$$\Phi_{peak}(t) \approx \frac{NA^2}{4\hbar c\lambda}\frac{I_{ave}(t)}{f\tau}, \tag{5}$$

where λ is the wavelength of the excitation light, c is the speed of light in vacuum, NA is the numerical aperture of the objective lens used to focus the light into the sample, I_{ave} is the average power of the incident light, and τ and f are the pulsewidth and the repetition rate of the pulse laser, respectively. Since the relaxation time from the excited state can be assumed to about 10 ns for many fluorescent molecules, the terms including τ_{ex} can be omitted. So, by solving these equations, assuming a rectangle laser pulse and that no molecule is in the excited state before the incidence of the pulse, we obtain

$$F(t) \propto C\eta ft\left\{1 - \exp\left[-\sigma_{2p}\left(\frac{NA^2}{4\hbar c\lambda}\frac{I_{ave}}{f\tau}\right)^2\tau\right]\right\}, \tag{6}$$

which was used to obtain the theoretical curve shown in Figure 11.

In a multifocus fluorescence microscope, a laser beam is divided into a number of foci. Under this condition, the fluorescence excitation is somewhat different from that with a single focus, where the intensity of the excitation laser used can be high enough to reach the saturation of the fluorescence emission. When the average power $I_{total}(t)$ reaches the focal plane of an objective lens, we obtain

$$F(t) \propto C\eta ftN\left\{1 - \exp\left[-\sigma_{2p}\left(\frac{NA^2}{4\hbar c\lambda}\frac{I_{total}(t)}{f\tau N}\right)^2\right]\right\}, \tag{7}$$

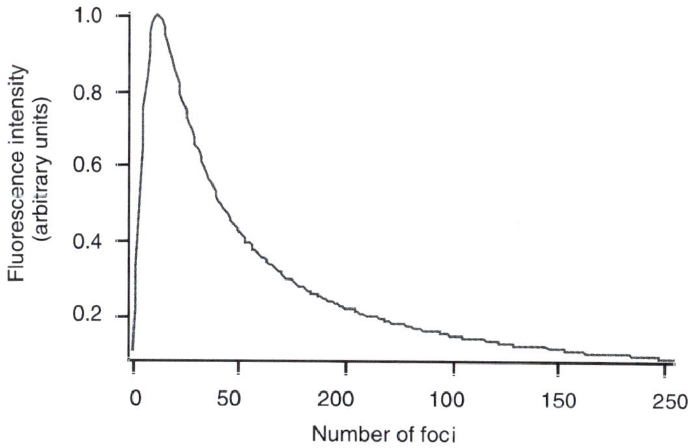

FIGURE 16 Total intensity of fluorescence induced with a different number of foci.

where N is the number of foci in use. From Eq. (7), the optimized number of foci depending on the laser source used can be calculated. If we assume that a mode-locked Ti : sapphire laser with a wavelength of 800 nm, a repetition rate of 80 MHz, and a pulsewidth of 100 fs is used to induce two-photon excitation in rhodamine B by an NA 1.3 objective lens, the amount of fluorescence emission at a different number of foci can be calculated as shown in Figure 16. The calculated result shows that the maximum fluorescence intensity can be obtained with 12 foci for the excitation. When the number of foci is smaller than 12, the amount of the fluorescence is almost proportional to the number of foci. This is because the fluorescence excitation rate is saturated, and the increase in the excitation intensity for each focus does not help one to obtain more fluorescence. On the other hand, when the number of foci is larger than 12, the amount of the fluorescence is almost inversely proportional to the square of the number of foci. In this situation, the increase in the number of foci brings about a lack of excitation intensity at each focus.

3.3 Real-Time Two-Photon Fluorescence Imaging

Since the first development of a real-time two-photon microscope in 1998, there have not been many examples of imaging living specimens in real time using two-photon excitation. However, as mentioned above, a system using a resonant galvanometer mirror has already been commercialized by Biorad Laboratories.

For in situ imaging, Fujita et al. (2000b) showed the first images of Ca^{2+} concentration in a perfused rat whole heart using their real-time confocal two-photon fluorescence microscope (Fig. 17). The Ca^{2+} concentration in the heart muscle cells is visualized using a fluorescent Ca^{2+} indicator (fluo-4/AM). Other conditions for the observation are the same as in the case of single-photon excitation mentioned above. Figures 17*a* and 17*b* show a Ca^{2+} wave and a Ca^{2+} transient recorded with about 66.6 ms and 33.3 ms of temporal resolution, respectively. Although the view area in the specimen is smaller than with single-photon excitation, the contrast of the images is high enough (or even higher than the single-photon case) for observing the Ca^{2+} concentration in the cells. This is

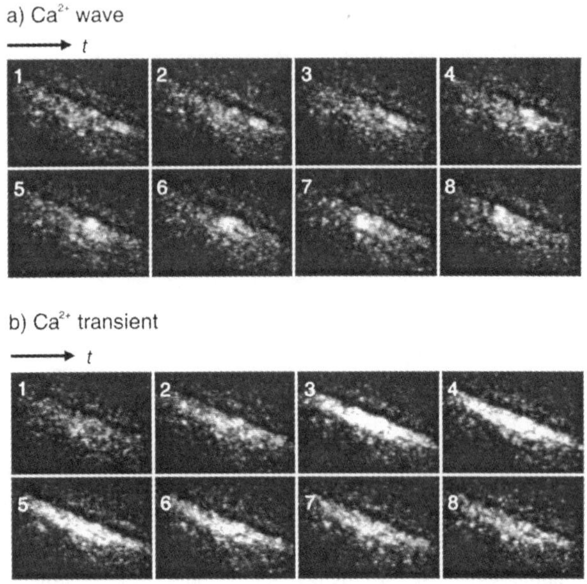

FIGURE 17 Sequential images of Ca^{2+} concentration obtained by a confocal multifocus two-photon microscope in a perfused rat whole-heart. (*a*) Ca^{2+} wave obtained with about 66 ms of temporal resolution. (*b*) Ca^{2+} transient obtained with about 33 ms temporal resolution.

probably because the localization of two-photon fluorescence excitation works effectively to reduce background fluorescence coming from out-of-focus planes.

When using two-photon excitation, it is expected that a greater penetration depth can be obtained, especially with thick specimens. Figure 18 shows our preliminary data on fluorescence intensity measured from different depths in a specimen using a typical two-photon microscope. The specimen was a rat whole heart loaded with fluo-4/AM, with the other conditions of the experiment being the same as those of in situ single-photon imaging mentioned above. It can be seen that the intensity of fluorescence decreases rapidly with observation depth. The reasons for this rapid decay of fluorescence intensity with observation depth are probably scattering of excitation laser and fluorescence,

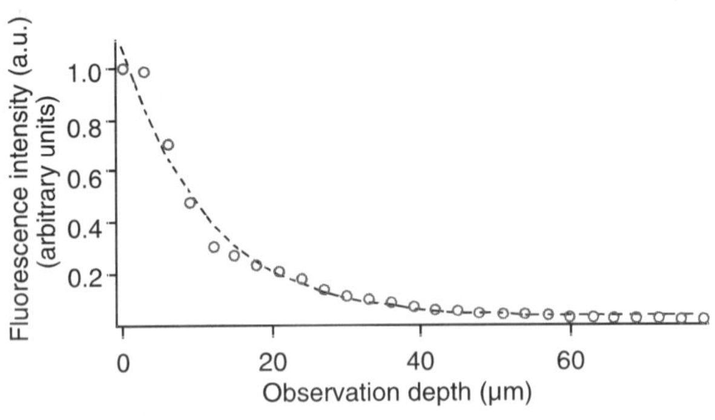

FIGURE 18 Fluorescence intensity measured at different observation depths in a perfused rat whole heart.

reabsorption of fluorescence, and broadening of the excitation pulse during laser propagation in the specimen. The weight of each factor in determining the fluorescence intensity is still under investigation. However, individual differences among specimens make these experiments more difficult, especially under in situ conditions.

4 CONCLUSIONS

In this chapter, techniques and applications of in situ imaging using single- and two-photon fluorescence microscopes have been described. Even with currently available systems, in situ imaging techniques have the potential to become popular and powerful tools in biology and medicine. Advances in accompanying techniques are required to allow the use of two-photon microscopes for in situ imaging. Low fluorescence excitation efficiency is a main disadvantage of two-photon microscopes when compared with the case of single-photon excitation. At present, researchers are using fluorescent dyes that were designed for single-photon excitation, and whose two-photon absorption cross-sections are, of course, much smaller than those of single-photon absorption. For full exploitation of the advantages of two-photon excitation in various applications in biology and medicine, it is desirable to develop fluorescent dyes that have two-photon absorption cross-sections large enough to give greater fluorescence with lower excitation intensity.

ACKNOWLEDGMENT

The authors thank Professor Satoshi Kawata and Dr. Osamu Nakamura in Osaka University, Japan, for useful discussions on composing the manuscript. The authors also thank Takeo Tanaami in Yokogawa Electric Company, Japan, for the technical information about the microlens array scanner.

REFERENCES

Brewersdorf, J., Pick, R., and Hell, S. W. (1998). Multifocus multiphoton microscopy. *Opt. Lett.* **23**, 65–67.

Egger, M. D. and Petran, M. (1967). New reflected-light microscope for viewing unstained brain and ganglion cells. *Science* **157**, 305.

Fan, G. Y., Fujisaki, H., Miyawaki, A., Tsay, R. K., Tsien, R. Y., and and Ellisman, M. H. (1999). Video-rate scanning two-photon excitation fluorescence microscopy and ratio imaging with cameleons. *Biophys. J.* **76**, 2412–2420.

Fujita, K., Nakamura, O., Kaneko, T., Oyamada, M., Takamatsu, T., and Kawata, S. (2000a). Confocal multipoint multiphoton excitation microscope with microlens and pinhole arrays. *Opt. Commun.* **174**, 7–12.

Fujita, K., Kaneko, T., Nakamura, O., Oyamada, M., Takamatsu, T., and Kawata, S. (2000b). Real-time intracellular Ca ion observation by using a real-time two-photon confocal fluorescence microscope. In: *Extended Abstracts, 40th Annual Meeting of the American Society for Cell Biology.*

Hama, T., Takahashi, A., Ichihara, A., and Takamatsu, T. (1998). Real-time in situ confocal imaging of calcium wave in the perfused whole heart of rats. *Cell. Signal.* **5**, 331–337.

Ichihara, A., Tanaami, T., Isozaki, K., Sugiyama, Y., Kosugi, Y. Mikuriya, K., Abe, M., and Uemura, I. (1996). High-speed confocal fluorescence microscopy using a nipkow scanner with microlenses for 3-D imaging of single fluorescent molecule in real time. *Bioimaging* **4**, 57–62.

Kaneko, T., Tanaka, H., Oyamada, M., Kawata, S., and Takamatsu, T. (2000). Three distinct types of Ca^{2+} waves in Langendorff-perfused rat heart revealed by real-time confocal microscopy. *Circ. Res.* **86**, 1093–1099.

Kino, S. G. and Corle, T.R (1989). Confocal scanning optical microscopy. *Phys. Today*, September, 55–62.

König, K., Riemann, I., Fischer, P., and Halbhuber, K. J. (1999). Intracellular nanosurgery with near infrared femtosecond laser pulses. *Cell. Mol. Biol.* **45**, 195–201.

Matsushita, T., Oyamada, M., Fujimoto, K., Yasuda, Y., Masuda, S., Wada, Y., Oka, T., and Takamatsu, T. (1999). Remodeling of cell–cell and cell–extracellular matrix interactions at the border zone of rat myocardial infarcts. *Circ. Res.* **85**, 1046–1055.

Oyamada, M., Oyamada, Y., Komatsu, K., Takamatsu, T., and Mori, M. (1997) Connexin expression during in vitro cardiac differentiation of mouse embryonic stem cells. In: *Gap Junctions* (ed. R. Wesner). ISO Press. 352–356.

Shack, R., Bell, B., Hillman, D., Kingston, R., Landesman, A., Shoemaker, R., Vukobratovich, D., and Bartels, P. H. (1982). Ultrafast laser scanner microscope – first performance test (biological application). In: *Proceedings of International Workshop on Physics and Engineering in Medical Imaging*, pp. 49–57.

Smith, N., Fujita, K., Nakamura, O., and Kawata, S. (2001a). Subsurface microprocessing of collagen in 3D by ultrashort laser pulses. *Appl. Phys. Lett.* **78**, 999–1001.

Smith, N., Fujita, K., Kaneko, T., Kato, K., Nakamura, O., Kawata, S., and Takamatsu, T. (2001b). Generation of calcium waves in living cells by pulsed-laser-induced photodisruption. *App. Phys. Lett.* **79**, 1208–1210.

Takamatsu, T. (1998). Confocal microscopy: applications in research and practice of pathology. *Anal. Quant. Cytol. Histol.* **20**, 529–532.

Takamatsu, T. and Wier, W. G. (1990). Calcium waves in mammalian heart: quantification of origin, magnitude waveform, and velocity. *FASEB J.* **4**, 1519–1525.

Xiao, G. Q. and Kino, G. S. (1987). A real-time confocal scanning optical microscope. *Proc. SPIE* **809**, 107–13.

Xie, S., Holtom, G., Sanchez, E., Novotony, L., and Arnett, D. (1998). New developments of two-photon microscopy. In: *Proceedings of Focus on Microscopy, Sydney*, Abstract 73.

Xu, C. and Webb, W. W. (1996). Measurement of two-photon excitation cross sections of molecular fluorophores with data from 690 to 1050 nm. *J. Opt. Soc. Am.* **B13**, 481–491.

CHAPTER 24

Superresolution in Fluorescence Confocal Microscopy and in DVD Optical Data Storage

ROY PIKE

Physics Department, King's College London, Strand, London, WC2R 2LS, U.K.

1 INTRODUCTION AND BACKGROUND

As is described in detail in other chapters of this book, in a confocal scanning microscope [1], the object is illuminated by a focused beam, either the beam or the specimen is raster-scanned, and the transmitted, reflected, or fluorescent light that passes through a pinhole at the center of the image plane is then detected. The optical imaging principles of the compact disc are the same as those of a confocal scanning microscope using coherent illumination.

In a series of earlier theoretical papers on optical imaging [2–5], which neglected high-aperture corrections, it was shown that when focused beams are used for illumination with either coherent or incoherent light and at all numerical apertures (NA), the conventional theory of optical resolution underestimates the correct diffraction limit and that a suitably designed system can robustly achieve an increase in linear resolution approaching twice the Rayleigh criterion. To achieve this extra "superresolution," a somewhat new approach to optical imaging was needed, namely, that the observed image must be regarded only as data that must be processed further numerically or optically to yield the final output.

Two novel microscopes, which implement the necessary processing instantaneously by optical means, were proposed [6, 7]. In the coherent case, the extra resolution can be achieved by performing predetection optical processing using a special holographic mask placed in the image plane, a Fourier lens, a pinhole, and a single detector. In the incoherent case, the processing is performed by the use of an image-plane intensity transmittance mask, two integrating detectors, and an electronic subtractor.

Confocal and Two-Photon Microscopy: Foundations, Applications, and Advances, Edited by Alberto Diaspro.
ISBN 0-471-40920-0

The holographic image-plane mask may be conveniently fabricated as a "binary-coded" version of the calculated continuous mask. The binary ring pattern of the mask is devised to emulate the continuous profile with sufficient accuracy to preserve the resolving power of the microscope while still being easy to manufacture [8].

In this early work, the neglect of high-aperture corrections allowed a number of the calculations to be performed analytically, and solutions could therefore be found. Experimental verification in the coherent case was achieved in 1992 [9], where a gain in linear resolutions of a factor of 1.7 was demonstrated. Experimental verification in the incoherent case proved to be more difficult, but was achieved in 1993 [10], again only for low NA.

More recently, calculation of the mask for the high-aperture incoherent microscope, which is important for fluorescence microscopy, particularly in the biological sciences, has been completed. The results have been confirmed satisfactorily in the laboratory. In the case of NA = 1.3, a resolution limit of 70 nm using a 488 nm laser was established. All these results, including the forward numerical simulation of the superresolving imaging, were published in reference [11].

Since at high NA, the numerical demands of the direct sampling of the imaging kernel grow rapidly, in [11] a new semi-analytical approach was devised using circular symmetry given by the use of circularly polarized light. A new numerical framework was developed, which led to a fast algorithm, and was easy to optimize and to implement. We shall describe these results below.

Application of the superresolution theory to the coherent illumination used in compact disc read-out, with the aim of improving the density of optical data storage by a factor of at least two, requires full vector analysis of the problem of high-aperture coherent imaging; this is, not unexpectedly, much more difficult. However, this vector imaging problem has now also been solved successfully [12, 13] and we shall describe briefly below the computation of the requisite superresolving masks to achieve the gain in storage capacity. Such a gain would allow full-length motion pictures to be viewed on an uncomplicated DVD reader, rather than the two-layer or double-sided reading systems now being promoted.

Experimentally, as the full vector masks are not yet available, we have built a suitable test reader, have created test disc surfaces, and have carried out encouraging tests with masks calculated in the scalar approximation.

The descriptions of the techniques and results for superresolution in coherent and incoherent scanning imaging to follow will lean heavily on a number of our recent papers, particularly [11–13].

2 THEORY OF THE HIGH-APERTURE CONFOCAL FLUORESCENCE MICROSCOPE

Since, in a confocal system, the illumination and collector lenses have a common focal point at the specimen, this point will be chosen as the origin of the coordinate system. The optical axis will be adopted as the z axis, such that $z = 0$ coincides with the confocal plane. Throughout, we shall use the mixed cylindrical–polar, three-dimensional, coordinate system ($\boldsymbol{\rho} = (\rho, \theta), z)$), with the dimensionless system of optical units [14] defined by

$$x = \frac{2n \sin\alpha}{\lambda} x_0, \qquad y = \frac{2n \sin\alpha}{\lambda} y_0, \qquad z = \frac{2n \sin^2\alpha}{\lambda} z_0, \tag{2.1}$$

where x_0, y_0, and z_0 are the standard geometrical coordinates,

$$\mathrm{NA} = n \sin \alpha. \tag{2.2}$$

is the numerical aperture of the microscope objective lens, and λ is the wavelength of the excitation light. Here, α denotes the semi-angle of acceptance of the objective lens and n denotes the refractive index of the objective immersion oil. In this paper, we shall take $\mathrm{NA} = 1.3$, which for standard immersion oil of $n = 1.518$ gives $\alpha = 58.9°$. Without narrowing our scope too much, for practical purposes we assume that:

(i) the optical system is rotationally invariant around the optical axis;
(ii) the lenses and immersion system are aberration-free;
(iii) the excitation light is circularly polarized;
(iv) the fluorescent material is weakly absorbing, so that incident light is not appreciably depleted in passing through the specimen;
(v) the fluorescent radiation is completely incoherent and randomly polarized;
(vi) the slight difference between the wavelength of the excitation radiation and the wavelength of the fluorescence is negligible.

Let us consider a three-dimensional fluorescent object and denote the distribution function of its fluorescent centers by $f(\boldsymbol{\rho}; z)$. Under the assumptions listed above, the basic imaging equation of the scanning confocal microscope gives the intensity distribution in the image plane as

$$g(\boldsymbol{\rho}) = \int W_2(|\boldsymbol{\rho} - \boldsymbol{\rho}'|; z') W_1(|\boldsymbol{\rho}'|; z') f(\boldsymbol{\rho}'; z') d\boldsymbol{\rho}' dz', \tag{2.3}$$

where $W_1(|\boldsymbol{\rho}|; z)$ and $W_2(|\boldsymbol{\rho}|; z)$ are the rotationally symmetric point spread functions (PSFs; i.e., the time-averaged energy distributions in the focal region), respectively, of the illuminating lens and the imaging lens. We shall consider here the confocal microscope working in the epifluorescence mode, with a single lens used both for illumination and imaging, so that

$$W_1(\rho; z) = W_2(\rho; z) \equiv W(\rho; z). \tag{2.4}$$

Concerning the illumination defined by the assumptions above and for lenses with high numerical apertures, we use the full expression for the PSF $W(\rho; z)$, derived first by Ignatowsky [14] (see also Richards and Wolf [15]) as

$$W(\rho; z) = |I_0(\rho; z)|^2 + 2|I_1(\rho; z)|^2 + |I_2(\rho; z)|^2, \tag{2.5}$$

where

$$I_0(\rho; z) = \int_0^\alpha \sqrt{\cos\theta} \sin\theta\, (1 + \cos\theta) J_0\left(\frac{\sin\theta}{\sin\alpha}\rho\right) \exp\left(i \frac{\cos\theta}{\sin^2\alpha} z\right) d\theta, \tag{2.6}$$

$$I_1(\rho; z) = \int_0^\alpha \sqrt{\cos\theta} \sin^2\theta\, J_1\left(\frac{\sin\theta}{\sin\alpha}\rho\right) \exp\left(i \frac{\cos\theta}{\sin^2\alpha} z\right) d\theta, \tag{2.7}$$

$$I_2(\rho; z) = \int_0^\alpha \sqrt{\cos\theta} \sin\theta\, (1 - \cos\theta) J_2\left(\frac{\sin\theta}{\sin\alpha}\rho\right) \exp\left(i \frac{\cos\theta}{\sin^2\alpha} z\right) d\theta. \tag{2.8}$$

Our problem is to find the function $f(\boldsymbol{\rho}'; z')$ for a given g. In fact, since scanning is involved, it is sufficient to recover only $f(0; 0)$, that is, the value of the object at the confocal point. Complete reconstruction of f can then be achieved by repeating this procedure at each scanning position. In order to solve the Fredholm equation (2.3), we consider its singular system, say $\{\alpha_k; u_k, v_k\}$, $k = 0, 1, 2, \ldots$, which is defined as a set of solutions of the coupled integral equations

$$\mathbf{A}u_k = \alpha_k v_k, \qquad \mathbf{A}^* v_k = \alpha_k u_k, \tag{2.9}$$

where $\mathbf{A}$ is defined by the integral operator of Eq. (2.3) and where $\mathbf{A}^*$ denotes its adjoint, given by

$$(\mathbf{A}^* g)(\boldsymbol{\rho}; z) = W(\boldsymbol{\rho}; z) \int W(|\boldsymbol{\rho} - \boldsymbol{\rho}'|; z) g(\boldsymbol{\rho}')\, d\boldsymbol{\rho}'. \tag{2.10}$$

In Eq. (2.9), α_k stands for the kth singular value of the system, while $u_k(\boldsymbol{\rho}; z)$ and $v_k(\boldsymbol{\rho})$ are the singular functions spanning the object and image domains, respectively. Notice that these functions provide complete orthonormal bases in the corresponding domains.

An approximate solution (truncated singular-value decomposition) of the inverse problem stated above is given by

$$f(0; 0) = \sum_{k=0}^{K-1} \frac{1}{\alpha_k} (g, v_k) u_k(0; 0). \tag{2.11}$$

The number of terms, K, in the summation is limited by the noise affecting the data (cf. [16]). In Eq. (2.11), (g, v_k) is the scalar product of g and v_k; that is,

$$(g, v_k) = \int g(\boldsymbol{\rho}) v_k(\boldsymbol{\rho})\, d\boldsymbol{\rho}. \tag{2.12}$$

By substituting this expression into Eq. (2.11) and then changing the order of summation and integration, one gets

$$f(0; 0) = \int M(\boldsymbol{\rho}') g(\boldsymbol{\rho}')\, d\boldsymbol{\rho}', \tag{2.13}$$

where

$$M(\boldsymbol{\rho}) = \sum_{k=0}^{K-1} \frac{1}{\alpha_k} u_k(0; 0) v_k(\boldsymbol{\rho}). \tag{2.14}$$

As follows from Eq. (2.13), in order to solve the problem, one first multiplies the image g with a known function $M(\boldsymbol{\rho})$ that is determined by the singular system, and then spatially integrates the result. The function $M(\boldsymbol{\rho})$ will be called the "optical mask". The multiplication and spatial integration are both analog operations that can be implemented optically, which allows the processing to be carried out prior to detection.

The most important property of $W(\boldsymbol{\rho}; z)$ is that its Fourier transform

$$
\begin{aligned}
\hat{W}(\boldsymbol{\omega}; \eta) &= \int_{\mathbf{R}^3} W(\boldsymbol{\rho}; z) e^{-i(\boldsymbol{\rho}, \boldsymbol{\omega}) - i\eta z} d\boldsymbol{\rho}\, dz \\
&= 2\pi \int_0^{+\infty} \rho\, d\rho \int_{-\infty}^{+\infty} dz\, J_0(\omega\rho) e^{-i\eta z} W(\rho z),
\end{aligned}
\tag{2.15}
$$

where $\boldsymbol{\omega} = (\omega_1, \omega_2)$, is bounded, with support contained within a cylinder

$$|\boldsymbol{\omega}| \leqslant \Omega_\perp, \qquad |\eta| \leqslant \Omega_\parallel, \tag{2.16}$$

where

$$\Omega_\perp = 2\pi, \qquad \Omega_\parallel = \frac{\pi}{1 + \cos\alpha}. \tag{2.17}$$

It can be noted here that, in view of the above, the use of higher-NA lenses implies the enhancement of the ratio between axial and lateral resolution. It can also be shown that

(i) $\hat{g}(\boldsymbol{\omega})$ has its support in the circle $|\boldsymbol{\omega}| \leqslant \Omega_\perp$;
(ii) $\hat{f}(\boldsymbol{\omega}; \eta)$ has its support in the cylinder $|\boldsymbol{\omega}| \leqslant 2\Omega_\perp$, $|\eta| \leqslant 2\Omega_\parallel$.

Using these properties, it can be demonstrated that the integral operator (2.3) is of the Hilbert–Schmidt class, and therefore it is compact. The line of the proof follows that of [17].

In the following section, we shall analyze the circularly symmetric case, and then discretize the resulting integral equation using sampling theorems in order to find the singular system.

3 SAMPLING OF THE IMAGING KERNEL

The singular system for this case can be determined only numerically. This is done by means of discretization of the operator **A** using sampling theory [18]. Using these components, the mask function can be calculated.

Using circularly polarized light, although the instantaneous illumination pattern at the object is that due to a given direction of linear polarization, the detector averages the fluorescent emission over a time that is long compared with an optical cycle, and the optical system shows cylindrical symmetry. Since we are interested in recovering the object only at the axial point, the solution of this problem is thus invariant with respect to rotations about the optical axis. This implies that we can simplify our inverse problem by replacing both the full image g and the full object f by their axially averaged projections

$$g_0(\rho) = \frac{1}{2\pi} \int_0^{2\pi} g(\rho, \phi)\, d\phi \tag{3.1}$$

and

$$f_0(\rho'; z') = \frac{1}{2\pi} \int_0^{2\pi} f(\rho', \phi'; z')\, d\phi' \tag{3.2}$$

on the subspace of functions with circular symmetry. In the above, (ρ, ϕ) and (ρ', ϕ') denote the polar coordinates describing $\boldsymbol{\rho}$ and $\boldsymbol{\rho}'$, respectively. As the singular functions u_k and v_k are now circularly symmetric, so is the resulting optical mask (cf. Eq. (2.14)). Consequently, it must be noted that only the rotationally invariant component of the image function g now contributes to the scalar product (2.13); the angular integration on the right-hand side of (2.13) can be carried out, and as a result g_0 replaces g. As the identity $f(0, 0) \equiv f_0(0, 0)$ is obvious, we can limit ourselves to the radial inverse problem, and thus Eq. (2.3) can now be rewritten as

$$g_0(\rho) = (\mathbf{A}_0 f_0)(\rho) = \int_0^{\infty} \int_{-\infty}^{+\infty} W_0(\rho, \rho'; z') W(\rho'; z') f(\rho'; z') \rho' d\rho' dz', \tag{3.3}$$

where

$$W_0(\rho, \rho'; z') = \int_0^{2\pi} W(\sqrt{\rho^2 + \rho'^2 - 2\rho\rho' \cos\beta}; z')\, d\beta. \tag{3.4}$$

Evidently, $g_0(\rho)$ is the mean value of $g(\rho, \phi)$ over the circle of radius ρ, and $f_0(\rho'; z')$ has a similar meaning. Also, because of (2.16), it can be demonstrated that

(i) the Hankel transform of $g_0(\rho)$ has support in $\leqslant \omega \leqslant \Omega_\perp$;
(ii) the Hankel transform of $f_0(\rho'; z')$ with z' fixed has support in $0 \leqslant \omega \leqslant 2\Omega_\perp$, while its Fourier transform with respect to z' with ρ' fixed has support in $|\eta| \leqslant 2\Omega_\parallel$.

In the above, by a Hankel transform of function h, we understand

$$\tilde{h}(\omega) = \int_0^{\infty} \rho J_0(\omega\rho) h(\rho)\, d\rho, \tag{3.5}$$

where J_0 is the zeroth-order Bessel function of the first kind. Defined in this way, for rotationally symmetric functions, this transform is equivalent to the ordinary Fourier transform.

In terms of the projected functions, the new singular system of the problem, say $\{\alpha_{0,k}; u_{0,k}, v_{0,k}\}$, will be the solutions of the coupled homogeneous equations

$$\mathbf{A}_0 u_{0,k} = \alpha_{0,k} u_{0,k}, \qquad \mathbf{A}_0^* v_{0,k} = \alpha_{0,k} u_{0,k}, \tag{3.6}$$

where $\mathbf{A}_0^*$ is the adjoint of $\mathbf{A}_0$, given by

$$(\mathbf{A}_0^* g_0)(\rho) = W(\rho; z) \int_0^{\infty} W_0(\rho, \rho'; z) g_0(\rho') \rho' d\rho'. \tag{3.7}$$

In order to tackle this system, we shall develop below a practical method of discretizing this problem.

If the function h is Hankel band-limited, with bandwidth Ω, it can be expressed by means of the so-called sampling expression [19] as

$$h(\rho) = \sum_{n=1}^{\infty} \frac{2x_n}{J_1(x_n)} h\left(\frac{x_n}{\Omega}\right) \frac{J_0(\Omega\rho)}{x_n^2 - (\Omega)^2}. \tag{3.8}$$

In other words, $h(\rho)$ can be expressed in terms of its values at the sampling points $\rho_n = x_n/\Omega$, where x_n is the nth zero of $J_0(x)$. If we introduce the functions

$$S_n(\Omega;\, \rho) = \Omega\sqrt{2}\, \frac{x_n J_0(\Omega\rho)}{x_n^2 - (\Omega\rho)^2}, \qquad n = 1, 2, \ldots, \tag{3.9}$$

then Eq. (3.8) can be rephrased as

$$h(\rho) = \sum_{n=1}^{\infty} \frac{\sqrt{2}}{\Omega\, J_1(x_n)} h\left(\frac{x_n}{\Omega}\right) S_n(\Omega;\, \rho). \tag{3.10}$$

In fact, it can be demonstrated that in the space of circularly symmetric and band-limited functions, the functions $S_n(\Omega;\, \rho)$ form an orthonormal basis; that is,

$$\int_0^{\infty} S_n(\Omega;\, \rho) S_m(\Omega;\, \rho) \rho\, d\rho = \delta_{nm}, \tag{3.11}$$

and also

$$\int_0^{\infty} \rho S_n(\Omega;\, \rho) h(\rho) d\rho = \frac{\sqrt{2}}{\Omega J_1(x_n)} h\left(\frac{x_n}{\Omega}\right). \tag{3.12}$$

Since the Hankel transforms of the projected functions $g_0(\rho)$ and $f_0(\rho';\, z')$ have bandwidths $\Omega_\perp$ and $2\Omega_\perp$, respectively, the sampling expansion in their case yields

$$f_0(\rho';\, z') = \sum_{m=1}^{\infty} \frac{\sqrt{2}}{2\Omega_\perp J_1(x_m)} f_0\left(\frac{x_m}{2\Omega_\perp};\, z'\right) S_M(2\Omega_\perp;\, \rho'). \tag{3.13}$$

On substituting Eq. (3.13) into Eq. (3.3) and then using the resulting equation for the sampled values of $g_0(\rho)$ at the points $\rho_n = x_n/\Omega_\perp$, by invoking the property (3.12) one gets

$$g_0\left(\frac{x_n}{\Omega_\perp}\right) = \sum_{m=1}^{\infty} \frac{2}{(2\Omega_\perp)^2 J_1^2(x_m)} \int_{-\infty}^{+\infty} W_0\left(\frac{x_n}{\Omega_\perp}, \frac{x_m}{2\Omega_\perp};\, z'\right) W\left(\frac{x_m}{2\Omega_\perp};\, z'\right) f_0\left(\frac{x_m}{2\Omega_\perp};\, z'\right) dz'. \tag{3.14}$$

On the other hand, since the functions $f_0(x_m/2\Omega_\perp;\, z')$, as functions of the variable z', are band-limited with bandwidth $2\Omega_\parallel$, from the usual one-dimensional Whittaker–Shannon sampling theorem we obtain

$$f_0\left(\frac{x_m}{2\Omega_\perp};\, z'\right) = \sum_{l=-\infty}^{+\infty} f_0\left(\frac{x_m}{2\Omega_\perp};\, z_l\right) \operatorname{sinc}\left[\frac{2\Omega_\parallel}{\pi}(z' - z_l)\right], \qquad m = 1, 2, \ldots, \tag{3.15}$$

where we have introduced

$$z_l = \frac{\pi}{2\Omega_\parallel} l, \qquad l = 0, \pm 1, \pm 2, \ldots, \tag{3.16}$$

and

$$\operatorname{sinc}\left[\frac{2\Omega_\parallel}{\pi}(z' - z_l)\right] = \frac{\pi}{2\Omega_\parallel} \frac{\sin[2\Omega_\parallel(z' - z_l)]}{\pi(z' - z_l)}. \tag{3.17}$$

Since the product $W_0 W$ appearing in Eq. (3.14) is itself band-limited with bandwidth $2\Omega_\perp$, one also has

$$\begin{aligned}\int_{-\infty}^{+\infty} & W_0\left(\frac{x_n}{\Omega_\perp}, \frac{x_m}{2\Omega_\perp};\, z'\right) W\left(\frac{x_m}{2\Omega_\perp};\, z'\right) \operatorname{sinc}\left[\frac{2\Omega_\parallel}{\pi}(z' - z_l)\right] dz' \\ &= \frac{\pi}{2\Omega_\parallel} W_0\left(\frac{x_n}{\Omega_\perp}, \frac{x_m}{2\Omega_\perp};\, z_l\right) W\left(\frac{x_m}{2\Omega_\perp};\, z_l\right).\end{aligned} \tag{3.18}$$

Finally, substitution of Eq. (3.15) into Eq. (3.14) with the help of Eq. (3.18) results in the linear system

$$b_n = \sum_{m=1}^{\infty} \sum_{l=-\infty}^{+\infty} \mathbf{A}_{n;m,l}\, a_{m,l}, \qquad n = 1, 2, \ldots, \tag{3.19}$$

where

$$b_n = \frac{\sqrt{2}}{\Omega_\perp J_1(x_n)} g_0\left(\frac{x_n}{\Omega_\perp}\right), \qquad a_{m,l} = \sqrt{\frac{\pi}{2\Omega_\parallel}} \frac{\sqrt{2}}{2\Omega_\perp J_1(x_m)} f_0\left(\frac{x_m}{2\Omega_\perp};\, z_l\right), \tag{3.20}$$

$$\mathbf{A}_{n;m,l} = \sqrt{\frac{\pi i}{2\Omega_\parallel}} \frac{W_0\left(\frac{x_n}{\Omega_\perp}, \frac{x_m}{2\Omega_\perp};\, z_l\right) W\left(\frac{x_m}{2\Omega_\perp};\, z_l\right)}{\Omega_\perp^2 J_1(x_n) J_1(x_m)}. \tag{3.21}$$

The singular values of the infinite-dimensional matrix (3.21) coincide with the singular values of the integral operator (3.3). Approximations of these singular values can be computed by considering finite sections of the matrix (3.21) obtained by limiting the values of the indices to

$$n = 1, 2, \ldots, N_0, \qquad m = 1, 2, \ldots, M_0, \qquad l = 0, \pm 1, \pm 2, \ldots, m \pm L_0. \tag{3.22}$$

As $N_0, M_0, L_0 \to \infty$, the singular values of the finite-dimensional matrix will converge to the singular values of the continuous operator. If we now denote by $U_{m,l}^{(k)}$ and $V_n^{(k)}$ the components of the singular vectors of the finite-dimensional matrix (normalized to unity with respect to Euclidean norm), from Eqs. (3.20) together with similarly truncated sampling expansions, the approximations to the singular functions of the integral operator (3.3) can be expressed as

$$v_k(\rho) = 2\pi\sqrt{2} \sum_{n=1}^{N_0} V_n^{(k)} \frac{x_n J_0(2\pi\rho)}{x_n^2 - (2\pi\rho)^2} \tag{3.23}$$

and

$$u_k(\rho, z) = 4\sqrt{\frac{2\pi}{1+\cos\alpha}} \sum_{m=1}^{M_0} \sum_{l=-L_0}^{L_0} U_{m,l}^{(k)} \frac{x_m J_0(4\pi\rho)}{x_m^2 - (4\pi\rho)^2} \operatorname{sinc}\left[\frac{2(z' - z_l)}{1+\cos\alpha}\right]. \tag{3.24}$$

Using these formulas in Eq. (2.14), we obtain the superresolving mask.

4 COMPUTATION OF THE DISCRETIZED KERNEL

Calculation of the matrix kernel of Eq. (3.21) requires computation of the function $W(\rho; z)$ and its projection $W_0(\rho, \rho'; z)$ onto the subspace of circularly symmetric functions (cf. Eq. (3.4)). Owing to the complicated nature of the PSF in the high-NA case (the expressions (2.6)–(2.8)), calculation of the integral appearing in Eq. (3.4) is the most difficult part of the whole computation. The straightforward method, where one evaluates the double integral numerically, is painstakingly slow and its accuracy is rather difficult to control.

In order to find a more suitable method, let us return to the function W. Since its Hankel transform has bandwidth $\Omega_\perp = 2\pi$, according to the sampling expansion (3.10) we can write

$$W(\rho; z) = \sum_{k=1}^{\infty} W_k(z) S_k(2\pi; \rho), \tag{4.1}$$

where

$$W_k(z) = \frac{\sqrt{2}}{2\pi J_1(x_n)} W\left(\frac{x_k}{2\pi}; z\right). \tag{4.2}$$

In the following, we shall also need to remember the identity (see Chapter 5 of [19])

$$\int_0^{2\pi} S_k(2\pi; \sqrt{\rho^2 + \rho'^2 - 2\rho\rho'\cos\theta'})\, d\theta = \frac{1}{\rho'} S_k(2\pi; \rho) * \delta(\rho - \rho'), \tag{4.3}$$

where $*$ denotes the convolution and $\delta(\rho)$ is the Dirac delta distribution. This convolution can be evaluated by using Hankel-transform techniques to obtain

$$S_k(2\pi;\rho) * \delta(\rho-\rho') = \sqrt{2}\,\frac{4\pi^2\rho'}{x_k J_1(x_k)} I\left(\rho,\rho',\frac{x_k}{2\pi}\right), \tag{4.4}$$

where we have introduced the symbol

$$I(a,b,c) = \int_0^{2\pi} J_0(ax)J_0(bx)J_0(cx)x\,dx. \tag{4.5}$$

Then, substituting Eq. (4.1) into Eq. (3.4) and taking into account Eqs. (4.3) and (4.4), we obtain

$$W_0(\rho,\rho';z) = 4\pi\sum_{k=1}^{\infty}\frac{W\left(\frac{x_k}{2\pi};z\right)}{[J_1(x_k)]^2} I\left(\rho,\rho',\frac{x_k}{2\pi}\right) \tag{4.6}$$

To get $\mathbf{A}_{n;m,l}$, we have to take $\rho = x_n/2\pi$ and $\rho' = x_m/4\pi$ and calculate the sequence of integrals (4.5) for these values, namely

$$I\left(\frac{x_n}{2\pi},\frac{x_m}{4\pi},\frac{x_k}{2\pi}\right), \qquad k = 1,2,\ldots. \tag{4.7}$$

Fortunately, integrals of this kind were already considered by Fettis in 1957 [20]. The expressions below are derived directly from his formulas. If we denote

$$q \equiv \frac{x_n^2 + \frac{1}{4}x_m^2 - x_k^2}{x_n x_m}, \tag{4.8}$$

then, depending on whether $|q|$ is greater, equal to, or smaller than 1, the result assumes the following forms:

(i) In the case $|q| > 1$, we have $|q| = \cosh\theta$, and, for $q > 0$,

$$I\left(\frac{x_n}{2\pi},\frac{x_m}{4\pi},\frac{x_k}{2\pi}\right) = 8\pi^2\frac{x_k J_1(x_k)}{x_n x_m}\sum_{p=1}^{\infty} J_p(x_n)J_p(\tfrac{1}{2}x_m)\frac{e^{-p\theta}}{\sinh\theta}. \tag{4.9}$$

For $q < 0$, and alternating sign factor $(-1)^{p+1}$ has to be included in the sum.

(ii) In the case $|q| < 1$, we have $q = \cos\theta$, and

$$I\left(\frac{x_n}{2\pi},\frac{x_m}{4\pi},\frac{x_k}{2\pi}\right) = 8\pi^2\frac{x_k J_1(x_k)}{x_n x_m}\sum_{p=1}^{\infty} J_p(x_n)J_p(\tfrac{1}{2}x_m)\frac{\sin p\theta}{\sin\theta}. \tag{4.10}$$

(iii) The case $|q| = 1$ is a natural limit of the previous one. We have $q = \cos\theta$, where either $\theta = 0$ or $\theta = \pi$. In both cases in Eq. (4.10), we have to replace $\sin p\theta/\sin\theta$ by its limiting value as $\theta \to 0$ or $\theta \to \pi$, respectively. In the first case it is p, while in the second it is $p(-1)^{p+1}$.

Crucial for this calculation of the discretized operator matrix $\mathbf{A}_{n;m,l}$ is the fact that we can use precalculated tables of $\mathbf{J}_p(x_n)$ and $J_p(\frac{1}{2}x_m)$ as well as the array of

$W(x_k/2\pi, z_l)/[J_1(x_k)]^2$. For each pair (n, m), a series of coefficients (4.7) is calculated, which is then used unchanged for all values of l. Convergence of the final series is dynamically controlled, further reducing unnecessary computation. The simplicity of the resulting algorithm makes it very easy to implement, and also susceptible to compiler optimization.

5 THEORY OF HIGH-APERTURE COHERENT IMAGING

To construct a superresolving mask for a compact disc reader, our initial intention was to calculate a mask specifically designed for an NA of 0.6 for circularly polarized light, as we had implemented in the incoherent case. It turned out, however, that it is theoretically impossible to cast the equations into the hoped-for circularly symmetric first-kind Fredholm linear form. This is possible in the fluorescence microscope only because the fluorescence lifetime is long compared with the period of the optical wave, which allows the nonsymmetric vector illumination pattern to average out to a circularly symmetric one.

The implication of this fact is that a superresolving coherent-light system, in so far as the scalar approximation is inadequate for the higher numerical apertures, must use linearly polarized light and a uniaxial image plane mask. The theoretical and computational requirements are thus, unfortunately, enormously increased over those for a circularly symmetric calculation.

The calculation of the vector case was eventually achieved by application of the approach of Sheppard and Wilson [21] and, following Huygens' principle, treating analytically each point of the illuminated surface as a re-radiating dipole. An integral is then taken over all the image-plane amplitudes, resulting in a vector imaging formula with a matrix imaging kernel. Below, we present the derivation of this kernel, starting with a thorough presentation of Sheppard and Wilson's result, since we need more detail than is presented in their published paper. To a large extent, we shall follow their notation.

Let us consider a radiating electric dipole $\mathbf{p}$ placed at a point O close to the focus of the thick lens. The spherical waves radiated by the dipole originating in O will then, after passing through the lens, converge at the conjugate point O'. Let O and O' be the centers of two sets of spherical coordinates (r, θ, ϕ) and (r', θ', ϕ'), defined such that the optical axis z is common to both of them, with opposite orientations, and with the remaining two Cartesian coordinates parallel. In this way, we always have $\phi = \phi'$. If R and R' are the distances of the points O and O' from the lens, then the constant ratio R'/R defines the magnification M of the lens. We shall assume that M is large. It is then customary to introduce two reference spheres with radii R and R', respectively, at O and O'. We shall consider only the sections $\mathbf{S}$ and $\mathbf{S}'$ of these spheres covering the corresponding opening and closing angles of the lens, α and α'. When O coincides exactly with the focus of the lens, α defines its numerical aperture as

$$\mathrm{NA} = n \sin \alpha, \tag{5.1}$$

where n is the refractive index of the lens immersion oil.

The far field of our radiating dipole is usually represented as an angular spectrum of plane waves ("rays"). Let us follow a ray traveling in the direction of the unit vector $\mathbf{r}(\theta, \phi)$. The lens will then refract into the direction $\mathbf{k}(\theta', \phi)$, almost parallel to the optical axis for sufficiently large magnification M. Let us denote the meridional plane, spanned by

r and **k**, by P. Let us note that P depends only on ϕ. The vector $\mathbf{n} = (\mathbf{r} \times \mathbf{k})/\sin\theta$ (of unit length) is normal to P. If we define two additional vectors $\mathbf{g} = \mathbf{r} \times \mathbf{n}$ and $\mathbf{h} = \mathbf{k} \times \mathbf{n}$, then we can introduce two orthonormal bases that will be very helpful in describing the effect of refraction in the lens on the electric field of the ray. The base "before" consists of vectors $(\mathbf{r}, \mathbf{n}, \mathbf{g})$, while the base "after" is defined by $(\mathbf{k}, \mathbf{n}, \mathbf{h})$.

Up to a constant factor, the far field of our dipole, which is "carried" by our ray, is given as [22]

$$\mathbf{E}_1 = \mathbf{p} - (\mathbf{p} \cdot \mathbf{r})\mathbf{r}, \tag{5.2}$$

and has no component along the propagation vector **r**. The refracted field in turn must have no component along the new wave vector **k**. The component of $\mathbf{E}_1$ normal to P will pass through the lens unchanged, while the component parallel to **g** will change its direction into **h**. That is, if the field before the refraction is

$$\mathbf{E}_1 = (\mathbf{p} \cdot \mathbf{n})\mathbf{n} + (\mathbf{p} \cdot \mathbf{g})\mathbf{g}, \tag{5.3}$$

then the refracted field is given by

$$\mathbf{E}_2 = (\mathbf{p} \cdot \mathbf{n})\mathbf{n} + (\mathbf{p} \cdot \mathbf{g})\mathbf{h}. \tag{5.3}$$

The above formula then has to be further normalized, by dividing its right-hand side by $\sqrt{\cos\theta}$ to account for the direction change in the energy stream (cf. [14, 15, 23].

In spherical coordinates, we have

$$\mathbf{r} = (\sin\theta\cos\phi, \sin\theta\sin\phi, \cos\theta), \tag{5.4}$$

$$\mathbf{k} = (0, 0, 1) \tag{5.5}$$

$$\mathbf{n} = (\sin\phi, -\cos\phi, 0), \tag{5.6}$$

$$\mathbf{g} = (\cos\theta\cos\phi, \cos\theta\sin\phi, -\sin\theta), \tag{5.7}$$

$$\mathbf{h} = (\cos\phi, \sin\phi, 0). \tag{5.8}$$

Strictly speaking, **k** should depend on θ' and ϕ' in the same way as **r** depends on θ and ϕ, but when the source is located almost at the focal point of the lens and when the magnification is large), we can limit ourselves to the leading-order approximation and assume that the refracted rays propagate essentially along the optical axis. Using these formulas we can easily calculate $\mathbf{p} \cdot \mathbf{n}$ and $\mathbf{p} \cdot \mathbf{g}$ and insert them into Eq. (5.3). The resulting $\mathbf{E}_2$ has only x and y components:

$$(E_2)_x = p_x(\cos\theta\cos^2\phi + \sin^2\phi) + p_y(\cos\theta - 1)\sin\phi\cos\phi - p_z\sin\theta\cos\phi, \tag{5.9}$$

$$(E_2)_y = p_x(\cos\theta - 1)\sin\phi\cos\phi + p_y(\cos\theta\sin^2\phi + \cos^2\phi) - p_z\sin\theta\sin\phi. \tag{5.10}$$

After refraction, the component rays are propagating almost parallel to the optical axis until they converge in the image plane, that is, at the back focal plane of the lens. The electric field there is given by the vector analog of the Debye integral (cf. [24]). This will be valid for the Fresnel numbers of interest here:

$$\mathbf{E}_3(\boldsymbol{\rho}_i) \propto \int_{S'} \frac{\mathbf{E}_2(\theta, \phi)}{\sqrt{\cos\theta}} e^{-ik\mathbf{n}_\perp \cdot \boldsymbol{\rho}_i} \, d\mathbf{s}', \tag{5.11}$$

where $d\mathbf{s}'$ is a surface element of the fictitious reference sphere S', $\mathbf{n}_\perp$ is normal to that surface, and $\boldsymbol{\rho}_i$ is the coordinate vector in the image plane. Expressing $\boldsymbol{\rho}_i$ in circular coordinates as $(\rho_i \cos\phi_i, \rho_i \sin\phi_i)$, we get

$$\mathbf{n}_\perp \cdot \boldsymbol{\rho}_i = \rho_i \cos(\phi' - \phi_i)\sin\theta'. \tag{5.12}$$

Then, remembering that according to the lens law, we have $\sin\theta' = M^{-1}\sin\theta$, we can perform the change of variables in the integral (5.11) to the unprimed ones to obtain

$$\mathbf{E}_3(v_i, \phi_i) \propto \int_0^{2\pi} d\phi \int_0^{\alpha} d\theta \mathbf{E}_2(\theta, \phi)\sin\theta\sqrt{\cos\theta}e^{-iv_i\cos(\phi-\phi_i)\sin\theta/\sin\alpha}, \tag{5.13}$$

where $v_i = k\rho_i \sin\alpha/M$ defines the standard optical coordinates in the image plane. This formula can be interpreted as a Fourier transform that has been "refraction-corrected." The next step consist in inserting Eqs. (5.9) and (5.10) into Eq. (5.13) and carrying out an integration over ϕ. To achieve that, we shall need the following well-known formulas from the theory of Bessel functions (cf., e.g., [23]):

$$\int_0^{2\pi} d\beta \cos\beta e^{it\cos(\beta-\gamma)} = 2\pi i J_1(t)\cos\gamma, \tag{5.14}$$

$$\int_0^{2\pi} d\beta \sin\beta e^{it\cos(\beta-\gamma)} = 2\pi i J_1(t)\sin\gamma, \tag{5.15}$$

$$\int_0^{2\pi} d\beta \sin\beta\cos\beta e^{it\cos(\beta-\gamma)} = -\pi J_2(t)\sin 2\gamma, \tag{5.16}$$

$$\int_0^{2\pi} d\beta \cos^2\beta e^{it\cos(\beta-\gamma)} = \pi J_0(t) - \pi J_2(t)\cos 2\gamma, \tag{5.17}$$

$$\int_0^{2\pi} d\beta \sin^2\beta e^{it\cos(\beta-\gamma)} = \pi J_0(t) + \pi J_2(t)\cos 2\gamma. \tag{5.18}$$

The remaining integration over θ will now involve Bessel functions and cannot be carried out analytically. It is customary to use here a set of functions for shorthand notation:

$$I_0(q;\alpha) = \int_0^{\alpha} d\theta\sqrt{\cos\theta}\sin\theta(1+\cos\theta)J_0\left(q\frac{\sin\theta}{\sin\alpha}\right), \tag{5.19}$$

$$I_1(q;\alpha) = \int_0^{\alpha} d\theta\sqrt{\cos\theta}\sin^2\theta J_1\left(q\frac{\sin\theta}{\sin\alpha}\right), \tag{5.20}$$

$$I_2(q;\alpha) = \int_0^{\alpha} d\theta\sqrt{\cos\theta}\sin\theta(1-\cos\theta)J_2\left(q\frac{\sin\theta}{\sin\alpha}\right) \tag{5.21}$$

(first introduced by Ignatowsky [14]). In this notation, we have

$$(E_3)_x = p_x(I_0 + \cos 2\phi_i I_2) + p_y \sin 2\phi_i I_2 - 2ip_z\cos\phi_i I_1, \tag{5.22}$$

$$(E_3)_x = p_x \sin 2\phi_i I_2 + p_y(I_0 - \cos 2\phi_i I_2) - 2ip_z \sin\phi_i I_1; \tag{5.23}$$

where we have put $I_m = I_m(v_i; \alpha)$ and α as before is the opening angle of the objective. We shall use this notation also below. If we define a matrix $\mathbf{K} = \mathbf{K}(v_i, \phi_i)$ as

$$\mathbf{K} = \begin{bmatrix} I_0 + \cos 2\phi_i I_2 & \sin 2\phi_i I_2 & -2i \cos \phi_i I_1 \\ \sin 2\phi_i I_2 & I_0 - \cos 2\phi_i I_2 & -2i \sin \phi_i I_1 \end{bmatrix}, \tag{5.24}$$

the electric field of the radiating dipole in the image plane can be described simply as

$$\mathbf{E}_3(v_i, \phi_i) = \mathbf{K}(v_i, \phi_i)\mathbf{p}. \tag{5.25}$$

Because of the linearity of the formula (5.25), we can make use of the superposition principle as well as Huygens' principle. The surface of the sample (or the compact disc) is illuminated by a tiny spot of a focused laser beam. Every atom of the sample within the illumination range will become the re-radiating dipole, with the dipole moment being a product of the illuminating field $\mathbf{E}_0$ and a general (perhaps complex) response function of the sample f (we shall assume that f is a scalar). In the case of the compact disc, the response function would be a pure phase function. As the size of the illuminating spot is small, we can invoke the aplanarity of the lens and assume that the image of a point removed slightly from the focus will be the same as if it were exactly at the focus, only appropriately shifted. Contributions of all the secondary dipoles are then added together. As a result, the full image becomes a vector convolution:

$$g = \mathbf{K} * (\mathbf{E}_0 \cdot f), \tag{5.26}$$

which is a close vector analogue of the familiar Fredholm equation of imaging, with $\mathbf{E}_0$ and $\mathbf{K}$ corresponding to the point spread functions of the objective and illuminating lens, respectively. In spite of using the same lens for both purposes, because of the vector nature of imaging, these two functions are now different. The imaging system is assumed to be perfect, but the analysis could be extended to account for aberrations – in fact, such modifications, implemented dynamically for example by a programmable liquid crystal mask, would be a good way to compensate for spherical aberration due to penetration into the recording medium of a multilayer disc. $\mathbf{E}_0$ is the field of a plane wave focused by the illuminating lens. As such, it is given by the well-known Richards and Wolf formulas [15] first derived by Ignatowsky [14]. If we denote the electric field amplitude of the incident plane wave as $\mathbf{E} = (E_x, E_y, 0)$, then $\mathbf{E}_0$ will result as

$$\mathbf{E}_0 = \begin{bmatrix} E_x(I_0 + \cos 2\phi_o I_2) + E_y \sin 2\phi_o I_2 \\ E_x \sin 2\phi_o I_2 + E_y(I_0 - \cos 2\phi_o I_2) \\ -2iI_1(E_x \cos \phi_o + E_y \sin \phi_o) \end{bmatrix}, \tag{5.27}$$

where $I_m = I_m(u_o; \alpha)$ and (u_o, ϕ_o) are the circular coordinates of the object space in the optical units as before. It is worth noting that because the z component of Eq. (5.27) is purely imaginary, after inserting Eq. (5.27) into our imaging equation (3.26), with $\mathbf{K}$ being given by Eq. (5.24), the resulting operator will be purely real.

6 CALCULATION OF THE MASK IN THE INCOHERENT CASE

The singular system for the kernel matrix whose elements were calculated by the technique given in the previous section have been computed for various values of the parameters N_0, M_0, and L_0 using a NAG library SVD routine. By increasing these parameters, we have observed that the convergence of the singular values is rather slow. However, for $N_0 = 70$, $M_0 = 140$, and $L_0 = 70$, we have found that increasing them further changes the resulting singular values only very slightly (after the fourth significant digit), implying that a satisfactory degree of convergence has been achieved. In Table 1, we present the first nine singular values of this system. If we assume that the signal-to-noise ratio in our system does not exceed 100, then only the first seven of them will be significant, owing to the regularizing cutoff (cf. Eq. (2.11). Singular functions corresponding to the singular values given in Table 1 can be found in [11].

Knowledge of these singular-system components allows evaluation of the mask function of Eq. (2.14). Figure 1 shows the form of the mask calculated for seven singular values (i.e., $K = 6$). It has a continuous profile, and changes sign. These features make it rather difficult to implement the mask for the case of incoherent light, because it is not possible to have a negative light intensity. However, if two detectors are used (with separate light processing) and the final readout is formed by subtraction of their signals, one can easily obtain the correct result. The practical difficulty, however, of how to implement the continuous profile accurately remains.

A practically feasible solution to this problem is to use a partially reflecting and partially transparent plate inserted at 45° to the optical axis in the image plane (see [8, 10]). The light that passes through the uncoated parts of such a mask (and is measured by one detector) corresponds to the positive components of the mask profile, while the light that is reflected at 90° (and is measured by a separate detector) corresponds to the negative components. The mirror design consists of prolate elliptical annuli (forming circles when orthoprojected onto the image plane). Their positions and widths are chosen to emulate as closely as possible the action of the computed mask.

A precise algorithm is described in [8]. Speaking in mathematical terms, it attempts to reproduce the radial weight of the continuous mask profile by a binary-valued function (with two values +1 and −1), which is, in a sense, "close" to the original in the function space while also retaining its original zero-crossings. Although our choice of a function-

TABLE 1 The first 9 singular values of the NA = 1.3 fluorescence microscope

k	α_k	α_k/α_0 (%)
0	152.0312	100
1	33.18858	21.83
2	15.01833	9.88
3	6.817549	4.48
4	3.572776	2.35
5	2.158534	1.42
6	1.295880	0.85
7	0.752640	0.50
8	0.499181	0.33

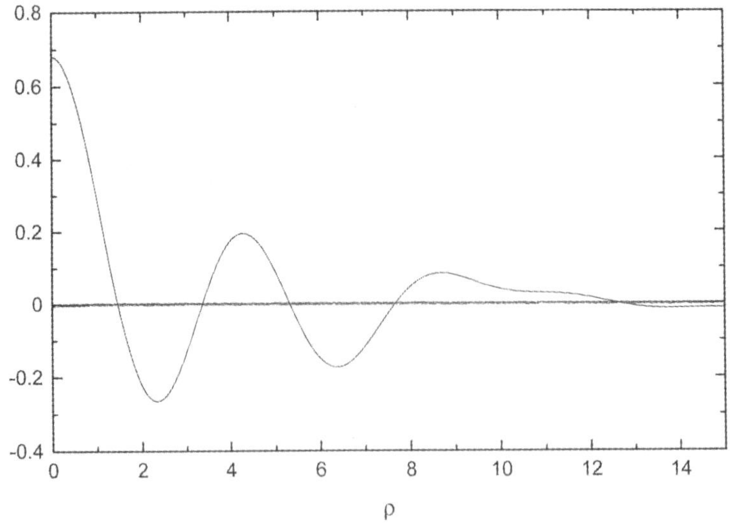

FIGURE 1 The continuous mask function for the superresolving microscope calculated using the parameters in the text.

space measure is quite arbitrary, its justification is satisfactory performance of the resulting mask. A "binary" mask, constructed according to this prescription, is shown in Figures 2*a* and 2*b*.

7 CALCULATION OF THE MASK IN THE COHERENT CASE

In order to solve the Fredholm equation (5.26), we again consider its singular system, $\{\alpha_k; u_k, v_k\}$, $k = 0, 1, 2, \ldots$, as in Eq. (2.9), where **A** is defined by its integral operator and where **A*** is its operator adjoint.

The truncated singular-value decomposition solution of the inverse problem stated above is given, as in Eq. (2.11), by

$$f(0) = \sum_{k=0}^{L-1} \frac{1}{\alpha_k} (g, v_k) u_k(0). \tag{7.1}$$

The analog of Eq. (2.11) is

$$f(0) = \int M(v) g(v)\, dv, \tag{7.2}$$

where, in the same way, the mask is given by

$$M(v) = \sum_{k=0}^{L-1} \frac{1}{\alpha_k} v_k^*(v) u_k(0). \tag{7.3}$$

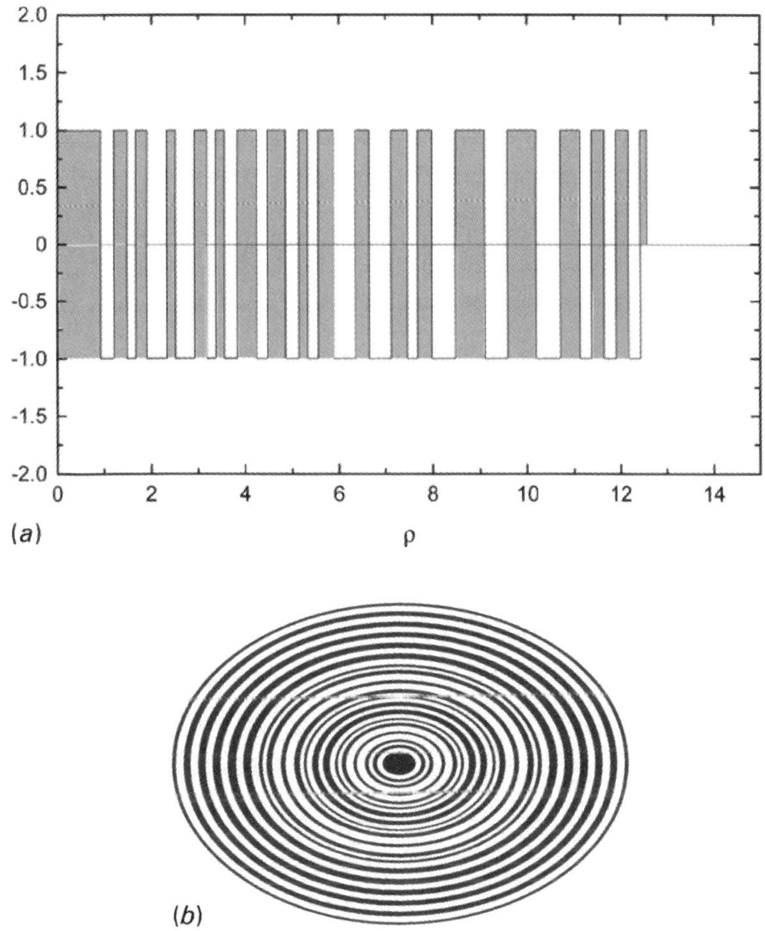

FIGURE 2 (*a*) The approximate binary form of the superresolving mask of Figure 1. (*b*) The design of the binary image-plane mask to be placed at 45° to the optical axis at the focal point.

For the program outlined above to work in our case, the fundamental imaging equation (5.26) has to be cast in a scalar form, otherwise the concept of the optical mask will not be feasible. We can achieve that by restricting our problem to measuring only one component (say x) of the image. This is achieved most easily by a linear polarizer. In this case, Eq. (5.26) can be rewritten as

$$g = \sum_{m=1}^{3} K_m * (E_m f), \tag{7.1}$$

where E_m denotes the three components of the illuminating field and K_m denotes the elements of the first row of **K**. For the computations below, we have assumed that the illuminating field of Eq. (5.27) results from a plane wave polarized along the y axis. In contrast to the incoherent case the resulting kernel is not rotationally symmetric (no matter what relative orientation of the polarizing filter and the illuminating light is adopted).

In order to find the singular system associated with the above equation, we first have to discretize its integral operator. As we can infer by analogy with the incoherent case [11], the bandwidth of the kernel **K** is π. Because g in Eq. (7.4) appears through convolution with **K**, it is obvious that the same bandwidth bound also holds. Then, it can be easily shown that the frequency components of f exceeding 2π are annihilated by the operator of the form (7.4) (cf. [5]). For this reason, we can limit the bandwidth of f to 2π, since the frequency components cannot be recovered anyway. Knowing the above bandwidth limits, we can use the two-dimensional Whittaker–Shannon sampling theorem to discretize our imaging kernel.

We shall need two results of the sampling theory (for a complete account see the book by Papoulis [19]). The first states that any function of two variables, with bandwidth limited by Ω in both dimensions, can be represented as

$$h(x, y) = \sum_{n,m} h(n\Delta, m\Delta)\,\mathrm{sinc}[\Omega(x - n\Delta)]\,\mathrm{sinc}[\|\Omega(y - m\Delta)], \tag{7.5}$$

where $\Delta = \pi/\Omega$, $\mathrm{sinc}\, x = \sin x/x$, and the summation extends from $-\infty$ to $+\infty$. We shall refer below to the arguments $(n\Delta, m\Delta)$ in the above formula as sampling points and to the corresponding values of sampled functions as sampling coefficients. In practical applications, the summation above will be limited by a cutoff, say N. It is worth noting here that the size of the double-sinc basis will grow very rapidly, as $(2N)^2$.

According to the second result, the value of a convolution of two functions, both of which are band-limited by Ω at the sampling points, can be expressed as a discrete convolution of their sampling coefficients. Adopting one-dimensional notation for simplicity, if $g = k * f$, then

$$g(m\Delta) = \Delta \sum_{n} f(n\Delta)g((m - n)\Delta), \tag{7.6}$$

where $\Delta = \pi/\Omega$ as before.

The discretization scheme then proceeds as follows. We start by considering the double-sinc sampling basis $f_{i,j}$ corresponding to the decomposition of f. If each of them is inserted into Eq. (7.4), because the products $E_m f_{i,j}$ are band-limited by $\Omega = 3\pi$, the resulting convolutions can be performed according to the formula (7.6). We only have to sample the *product* $E_m f_{i,j}$ and the kernels K_m at relevant points and convolve the resulting sampling coefficients using the discrete fast Fourier transform. Some care is only necessary to avoid artifacts connected to circularity of such convolution (careful zero-padding does the trick). The result determines at once all the sampling coefficients of decomposition of the image of $f_{i,j}$ into the sampling basis in the image space. These coefficients build the discretized imaging kernel. From this moment on, the only remaining obstacle is the size of the resulting array, growing as $(2N)^4$. Next, the numerical SVD of the discrete kernel is performed. Then, using the sampling bases again, we recreate the "continuous" singular functions of Eq. (5.26) and use them to calculate the mask (7.3). To complete these calculations, Columbus, the Engineering and Physical Sciences Research Council's supercomputing facility at Rutherford Appleton Laboratory, was used.

We should like to remark that although the above computation was carried out using only real functions, the entire analysis can be adapted to the complex case by routine complexification. This will not bring anything new, however, since the mask form will remain unchanged. The mask that we have calculated, can thus be used for general

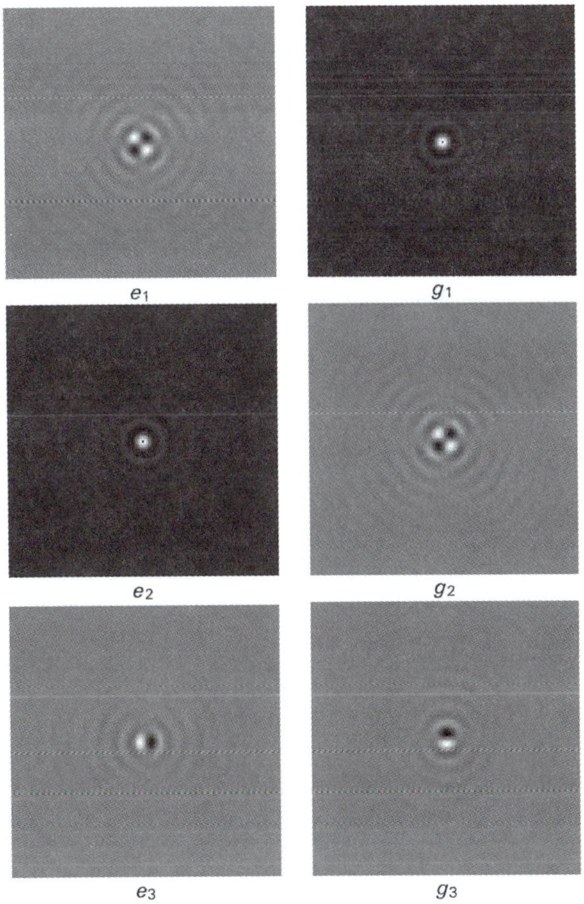

FIGURE 3 The kernel functions for an NA of 0.6.

complex response functions of the specimen, which of course includes the purely phase-like response of the optical disc.

Figure 3 shows the three pairs of components of the vector kernel function for an NA of 0.6. The singular spectrum of this kernel is shown in Figure 4. The form of the functions can be found in [13]. Masks using various numbers of functions are shown in Figure 5. The number of functions chosen in a practical case will depend on the signal-to-noise ratio. More functions will increase the superresolution, but the singular value of the highest index function, as given in Figure 4, must not be smaller than the inverse of this ratio.

8 EXPERIMENTAL CONFIRMATION IN THE INCOHERENT CASE

An incoherent-imaging, superresolving, high-aperture, laser-scanning microscope has been built based on the above theoretical analysis of incoherent imaging [11]. The microscope is modeled on the optical dimensions of the well-known Bio-Rad MRC1024 instrument. A binary mask in the final image plane, calculated using the

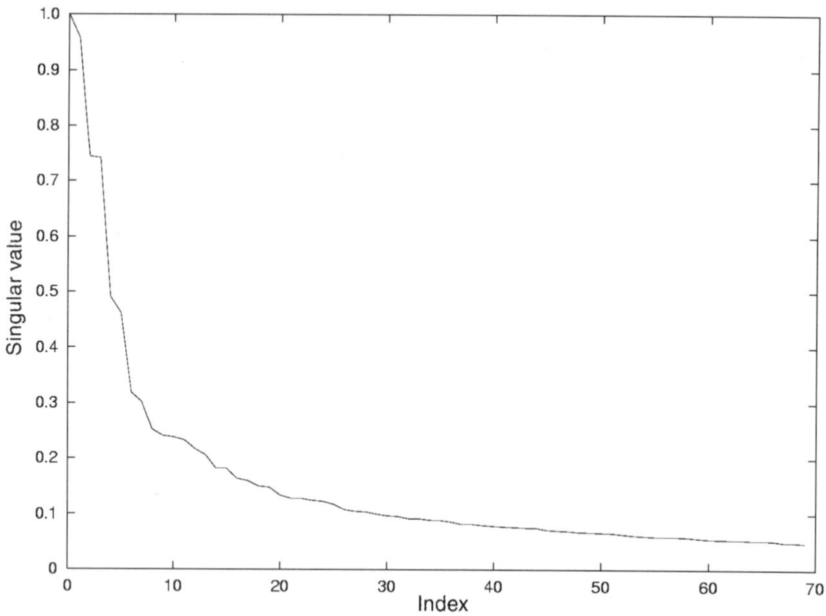

FIGURE 4 The singular spectrum for an NA of 0.6.

above results, performs the singular-value decomposition of the image intensity, and superresolves images by instantaneous analog optical processing.

The experimental set-up is shown in Figure 6. The laser is focused in the object plane by a set of lenses, the eyepiece ($EP1$), the tube lens ($TL1$), and the objective lens (OL). At $\lambda = 488$ nm and NA $= 1.3$, the radius of the central lobe of the Airy pattern amounts to just under 230 nm. The objective lens gathers the fluorescent light from an FITC-stained sample at the central wavelength $\lambda_c = 520$ nm. A narrow-bandpass filter with a window of 25 nm around λ_c is fitted to eliminate any undesirable stray light in the system.

Along the detection path, real images are created at two points before the final image plane at M_{im}, where the overall magnification is set to $m = 3300$, as in the BioRad instrument. The use of such large magnification also allows a simple photo-etching technique to be used for fabrication of the mask.

At M_{im}, the mask splits the image beam into two parts: the reflected light is measured by one detector, PMT_2, while the complementary transmitted part is registered by PMT_1. The difference of the signals from the two detectors yields the superresolved pixel value, while the sum gives the signal value of the ordinary type 1 microscope at the same instant. A confocal image can also be produced by reducing the size of the diaphragm in front of the reflection-arm detector, although it cannot be obtained simultaneously in our present system.

Because of its insertion at 45°, the mask pattern consists of concentric fully reflective and transmissive elliptic rings. The maximum dimensions correspond to a projected circular diameter of 5.0 mm.

After careful alignment, the instrument is calibrated for the relative sensitivity of the two detectors. A zero level is chosen in order to suppress the impact of autofluorescence produced by the immersion oil and the mounting fluid as well as of constant stray light.

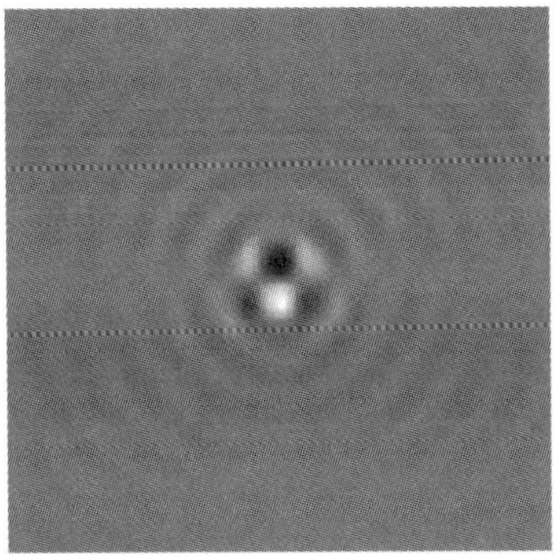

$k = 4$

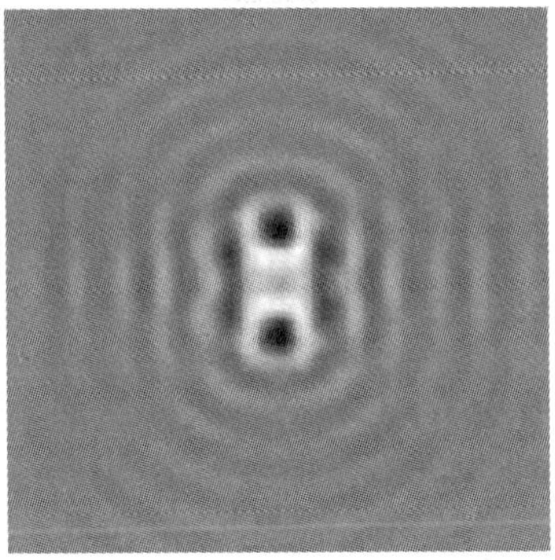

$k = 10$

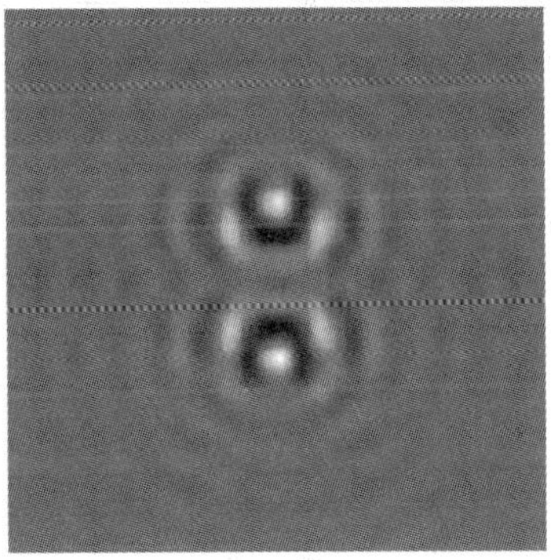

$k = 20$

FIGURE 5 Masks calculated for higher and higher superresolution: from 4, 10, and 20 singular functions.

The microscope is an object scanning device, so a 3D piezo-scanning stage shifts the object. The optical axis of the system remains stationary. Interactive software running on a PC controls the movements of the scanning stage through a D/A board while reading out both the detector single-photon counting outputs at the same time. The software also renders the final image as the calibrated difference of the two measured signals. The experiments were performed using a test pattern supplied by the Cornell Nanofabrication Facility, consisting of a periodic set of lines and spacings both 125 nm wide (resulting in a 250 nm period).

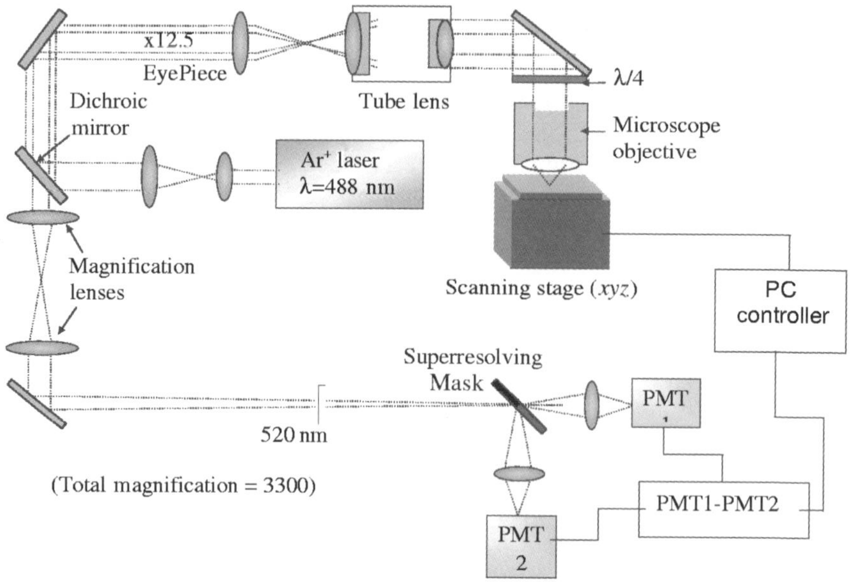

FIGURE 6 Schematic diagram of the experimental microscope. (See text for the legend.)

The theoretical performance expected from the superresolving mask system is shown in Figure 7, against that of a type 1 microscope. Figure 8 shows the experimental results. In both of these figures part (*a*) shows a type 1 scan, while (*b*) is the scan in the superresolving mode.

For the calculation of Figure 7, an object of parallel lines with 250 nm period has been formulated as $f(\rho, z=0)$, and its angular average $f_0(\rho, z=0)$ derived for any scanning position; see Eq. (3.1). Then, the angular-averaged image $g_0(\rho)$ could be computed by sampling $f_0(\rho)$ and applying the operator $\mathbf{A}_{n,m;l}$ to it, Eq. (3.2). $g_0(\rho)$ is the blurred image

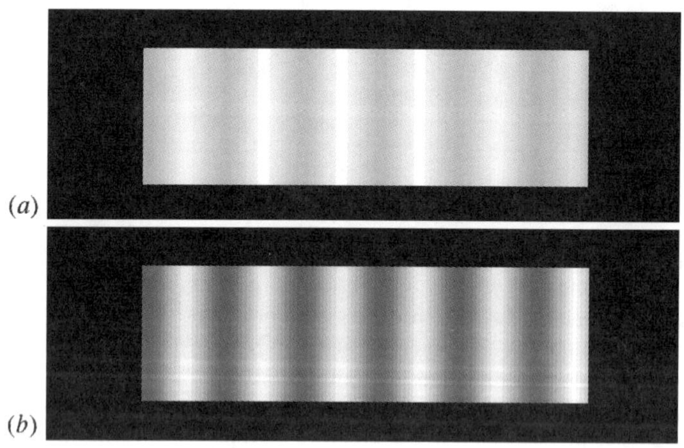

FIGURE 7 Calculated image of a set of parallel lines. The lines are 125 nm wide, separated by another 125 nm. Part (*a*) corresponds to a type 1 microscope and (*b*) to a superresolving microscope.

(*a*)

(*b*)

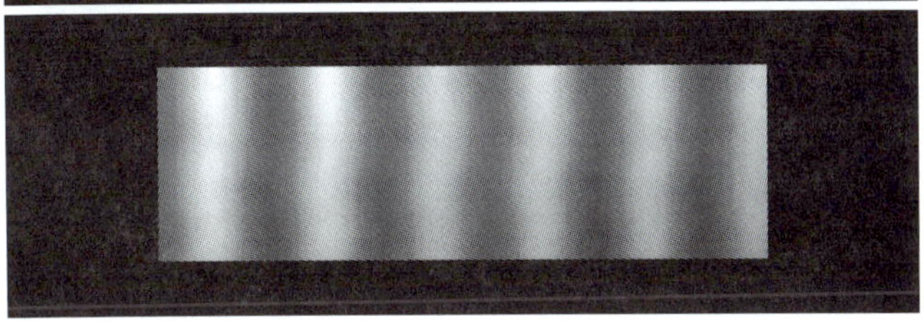

FIGURE 8 Experimental images of the set of parallel lines of Figure 7 using our superresolving microscope. Part (*a*) shows pixel intensities in type 1 mode. Part (*b*) is the in-band superresolved image.

at a specific scanning position. So, Eq. (2.13) gives the reconstructed superresolved pixel value for this position. A mask function has been chosen that represents the mask used in the experiment with seven singular functions. In the case of the type 1 image, the pixel intensity arises from the same process by setting the mask function to unity for all ρ in Eq. (2.13).

No noise is present in the theoretical results, and the experimental results (Fig. 8) were band-limited to the seven singular functions used in the make-up of the mask by projecting the image onto them (offline), thus removing the out-of-band photon and dark-current noise. The modulations between lines (i.e., the calculated ratios of low intensity in the space to the high intensity at a line) are 0.83 and 0.44 for the type 1 and superresolving microscopes, respectively. The experimental results gave modulations of 0.80 for the type 1 and 0.43 for the superresolving microscope. For the values of wavelength and numerical aperture used, a period of 250 nm represents just about the Rayleigh limit. As in the previous work at low aperture, these experimental images at high working aperture show close agreement with the full superresolving gain in microscopy that was predicted theoretically.

9 FORWARD CALCULATIONS IN THE COHERENT CASE

We have carried out numerical calculations on the results to be expected from a compact disc reader both with and without the superresolving mask. These are full solutions of the equations for the forward-imaging problem for the test specimens used at 488 nm wavelength and an NA of 0.6. The test specimens were designed and made by electron-beam lithography by Dr. P. Charalambous in the Department of Physics at King's College, London. They had lines of square pits at a 1 : 1 mark-space ratio ranging from 1.5 μm down to 122 nm spacing. Micrographs of a test-specimen surface are shown in Figure 9. Exact calculations of scattering from rectangular grooves have been carried out in [25]; these show that the apparent phase depth of a nominal $\frac{1}{2}\pi$ pit is greater than or less than $\frac{1}{2}\pi$,

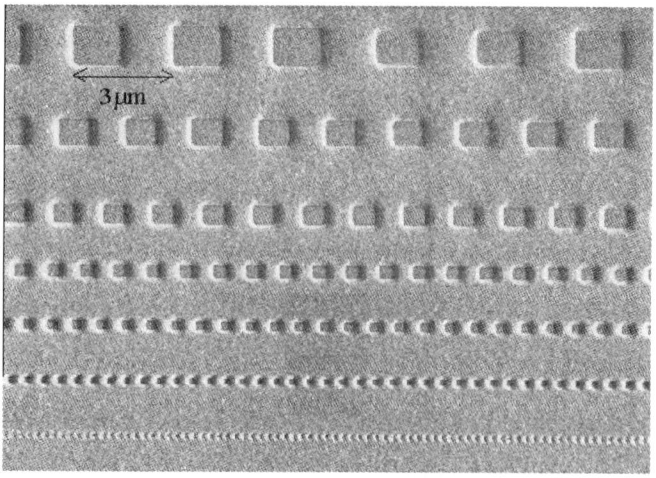

FIGURE 9 Micrographs of a test-specimen surface.

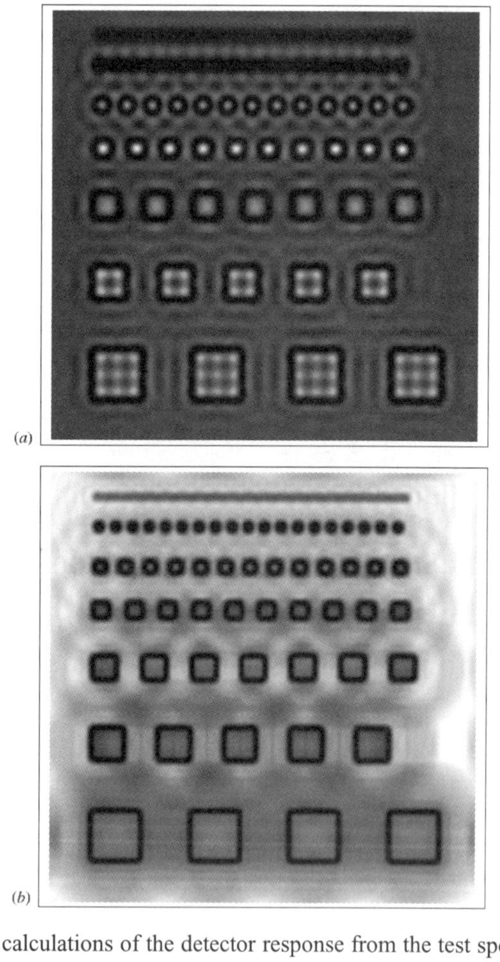

FIGURE 10 Forward calculations of the detector response from the test specimen: (*a*) without the superresolving mask; (*b*) with the superresolving mask.

depending on the incident polarisation. For our present calculations for square pits, we have taken the pit depth to be $\frac{1}{4}\lambda$. Typical results are shown in Figure 10. It can be seen that the results of scanning a line of pits of the larger sizes show a dependence upon the exact trajectory of the laser spot with respect to the centerline of the row. Along the centerline, the frequency of the response is equal to twice that obtained at the edges. This is because the modulation is caused by the transit of the spot across an edge, so that both in-phase and out-of-phase components are present simultaneously at the detector from the top and bottom of the pit. Along the edge of the row, both phases are present for the whole length of the pit transit, while at the center line, both phases are only present during the crossing of the spot into and out of the pit. In the middle of the pit, the whole beam is reflected from the bottom surface and gives the same signal strength as the top disc surface. The work in the coherent case is continuing under a new collaboration within the European Union, led by the Philips company in the Netherlands, with the aim of designing a possible future generation of DVD systems.

ACKNOWLEDGMENTS

The author wishes to thank all his coauthors named in the references below, a number of whom were postgraduate students, postdoctoral fellows, academic visitors, and staff at King's College London. Special thanks are due to Mario Bertero and his colleagues at the University of Genoa, and Christine De Mol at the Université Libre de Bruxelles, who have been long-term collaborators in various inverse problems, including especially optical imaging. Thanks are also due to Fred Brakenhof and Hans van der Voort for their help and joint work with us at the University of Amsterdam in the early days. This work has been supported in its various phases by the NATO Scientific Affairs Division, the European Union, the U.S. Army, and the U.K. Engineering and Physical Sciences Research Council; we have also had generous support from Bio-Rad Sciences Ltd.

REFERENCES

1. Wilson, T, and Sheppard C., *Theory and Practice of Scanning Optical Microscopy*. London: Academic Press, 1984.
2. Bertero, M. and Pike, E. R. (1982). Resolution in diffraction-limited imaging, a singular value analysis: I The case of coherent illumination. *Optical Acta* **29**, 727–746.
3. Bertero, M., Boccacci, P., and Pike, E. R. (1982). Resolution in diffraction-limited imaging, a singular value analysis: II The case of incoherent illumination. *Optica Acta* **29**, 1599–1611.
4. Bertero, M., Brianzi, P., Parker, P., and Pike, E. R. (1984). Resolution in diffraction-limited imaging, a singular value analysis: III The effect of sampling and truncation of the data. *Optica Acta* **31**, 181–201.
5. Bertero, M., De Mol, C., Pike, E. R., and Walker, J. G. (1984). Resolution in diffraction-limited imaging, a singular value analysis: IV The case of uncertain localisation or non-uniform illumination of the object. *Optica Acta*, **31**, 923–946.
6. Walker, J. G., Pike, E. R., and Bertero, M. (1989). Scanning optical microscope. British Patent Application No. 89/13129.
7. Davies, R. E., Inverse problems in confocal scanning microscopy. PhD Thesis, University of London, 1990.

8. Grochmalicki, J., Pike, E. R., Walker, J. G. Bertero, M., Boccacci, P., and Davies, R. E. (1993). Supcrresolving masks for incoherent scanning microscopy. *J. Opt. Soc. Am.* **A10**, 1074–1077.

9. Young, M. R., Jiang, S. H., Davies, R. E., Walker, J. G., Pike, E. R., and Bertero, M. (1992). Experimental confirmation of superresolution in confocal scanning microscopy using holographic masks. *J. Microsc.* **165**, 131.

10. Grochmalicki, J., Pike, E. R., and Walker, J. G. (1993). Experimental confirmation of superresolution in incoherent scanning microscopy. *J. Eur. Opt. Soc. A, Pure Appl. Opt.* **2**, 565–568.

11. Akduman, I., Brand, U., Grochmalicki, J., Hester, G., and Pike, E. R. (1998). Superresolving masks for incoherent high-numerical-aperture scanning microscopy in three dimensions. *J. Opt. Soc. Am.* **A15**, 2275–2287.

12. Brand, U., Hester, G., Grochmalicki, J., and Pike, E. R. (1999). Superresolution in optical data storage. *J. Opt. A: Pure Appl. Opt.* **1**, 794–800.

13. Grochmalicki, J. and Pike, E. R. (2000). Superresolution for digital versatile discs. *Appl. Opt.* **39**, 6341–6349.

14. Ignatowsky, V. S. (1921). Diffraction by an objective lens with arbitrary aperture (in Russian). *Trans. Opt. Inst. Petrograd* Vol. I, Issue IV, pp 1–36.

15. Richards, B. and Wolf, E. (1959). Electromagnetic diffraction in optical systems: II. Structure of the image field in an aplanatic system. *Proc. R. Soc. Lond.* **A253**, 358–379.

16. Bertero, M., Boccacci, P., Defrise, M., De Mol, C., and Pike, E. R. (1989). Superresolution in confocal scanning microscopy: II. The incoherent case. *Inverse Problems* **5**, 441–461.

17. Bertero, M., Boccacci, P., Malfanti, F., and Pike, E. R. (1994). Superresolution in confocal scanning microscopy: V. Axial superresolution in the incoherent case. *Inverse Problems* **10**, 1059–1077.

18. Bertero, M., and Boccacci, P. (1989). Computation of the singular system for a class of integral operators related to data inversion in confocal microscopy. *Inverse Problems* **5**, 935–957.

19. Papoulis, A., *Systems and Transforms with Applications in Optics*. New York: McGraw-Hill, 1968.

20. Fettis, H. E. (1957). Lommel-type integrals involving three Bessel functions. *J. Math. and Phys.* **36**, 88–95.

21. Sheppard, C. and Wilson, T. (1982). The image of a single point in microscopes of large numerical aperture. *Proc. R. Soc. Lond.* **A379**, 145–158.

22. Stratton, J. A., *Electromagnetic Theory*. New York: McGraw-Hill, 1941.

23. Stamnes, J. J., *Waves in Focal Regions*. Bristol: Adam Hilger, 1986.

24. Sheppard, C. J. R., Choudhury, A., and Gannaway, J. (1977). Electromagnetic field near the focus of wide-angular lens and mirror systems. *Microwaves, Opt. Acoust.* **1**, 129–132.

25. Dil, J. G. and Jacobs, B. A. J. (1979). Apparent size of reflecting obstacles of the order of one wavelength. *J. Opt. Soc. Am.* **69**, 950–960.

CHAPTER 25

Photobleaching by Confocal Microscopy

JAMES G. McNALLY[1] and CAROLYN L. SMITH[2]

[1]Laboratory of Receptor Biology and Gene Expression, National Cancer Institute, Bethesda, MD 20892, U.S.A.
[2]National Institute for Neurological Disorders and Stroke, Light Imaging Facility, Bethesda, MD 20892, U.S.A.

1 INTRODUCTION

In this chapter, we review photobleaching techniques with an emphasis on how to implement them with modern laser scanning confocal microscopes. By photobleaching, we mean the irreversible destruction of fluorescence in a defined region within a sample, and then the subsequent observation of the redistribution of fluorescence by exchange of bleached and/or unbleached fluorescent molecules. If the molecules are mobile, then the time course and spatial characteristics of the redistribution can provide information about the cellular processes involved (Fig. 1). Movement of the labeled molecules may be due to diffusion or directed transport, or a combination of these mechanisms. Which mechanism dominates will depend on their relative time scales and the distances over which they operate. The recovery process also may reflect structural characteristics of the sample, such as compartment boundaries. Finally, the labeled molecule may be engaged in binding reactions within the sample. When this holds, the redistribution of bleached and unbleached molecules will reflect not only the mechanisms of movement but also the rates of binding. Thus, photobleaching techniques have the potential to assay a wide variety of cellular processes.

Photobleaching techniques were first developed and applied to biological and molecular specimens in the mid 1970s (Poo and Cone, 1974; Liebman and Entine, 1974; Edidin et al., 1976; Schlessinger et al., 1976). For many years thereafter, these methods thrived in the hands of a relatively small group of biophysicists who constructed their own photobleaching systems and derived sophisticated quantitative analyses to manipulate the resulting data. With the advent of the confocal microscope, use of photobleaching methods has proliferated. The reasons for this growth are twofold. First, the technique is easily implemented on a laser scanning confocal microscope. Second, the introduction of green fluorescent protein (GFP) as an endogenous protein marker has permitted ready fluorescent tagging of a number of important biological molecules in live cells. With this

Confocal and Two-Photon Microscopy: Foundations, Applications, and Advances, Edited by Alberto Diaspro.
ISBN 0-471-40920-0

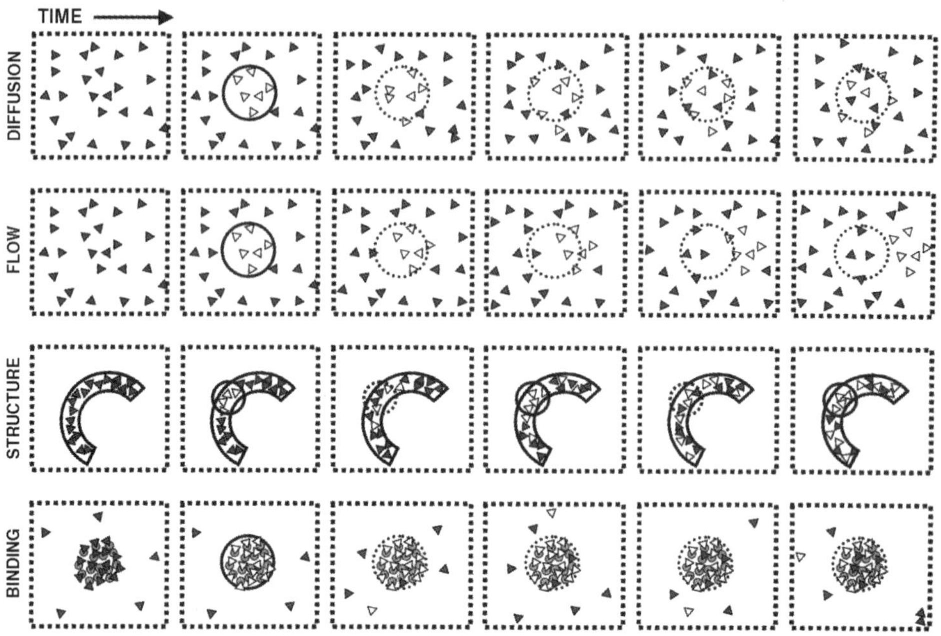

FIGURE 1 Schematic illustrating how photobleaching can be used to assess different cellular processes. The black dotted outer square represents a subregion of a cell. Black triangles are fluorescent molecules, and white triangles are their bleached counterparts. A photobleach is represented by a solid gray circle. After the bleach, the dotted gray circle indicates where the bleach was applied. Each row of the figure represents how different cellular processes affect photobleach recovery, with time increasing to the right. In the first row, reflecting pure diffusion, each molecule executes a random walk. Thus, bleached molecules exit the bleach zone at random and unbleached molecules enter it at random. The rate of fluorescent recovery in the bleach zone can be used to determine the diffusion coefficient. In the second case – flow – the fluorescent molecules are moving directionally to the right, owing, for example to active transport. After the bleach, fluorescent molecules enter the bleach zone from the left, and bleached molecules exit from the right, providing a characteristic diagnostic for directed transport. (New fluorescent molecules also flow into the subregion from the left.) In the third row, photobleaching is used to detect whether a cellular structure (the half-circle) is interconnected. Bleached molecules at the third time point interchange by diffusion, in this case, with unbleached neighboring molecules. By the fourth time point, the original bleached molecules have left the bleach zone and have been replaced by unbleached molecules, which are then themselves bleached by a second bleach pulse (solid gray circle). This process is iterated to gradually deplete fluorescence throughout the structure, demonstrating its continuity. The fourth row represents a binding reaction that occurs at a particular site in the cell. Solid gray circles correspond to immobile binding sites for the molecules represented by triangles. In this example, the off rate of binding is presumed to be the rate-limiting step. Diffusion and the on rate are therefore fast, so that once a site becomes unbound, it quickly binds another molecule. Thus, the rate of fluorescent recovery will show a very fast rate (perhaps undetectable) due to diffusion, and a slower rate due to the binding off rate.

increased interest in photobleaching, several reviews have appeared in recent years (Lippincot-Schwartz et al. 1999; Meyvis et al., 1999; White and Stelzer, 1999). The reader should consult these for additional information and viewpoints.

2 PHOTOBLEACHING STRATEGIES: FRAP, FLIP, AND IFRAP

Here we give an overview of these three photobleaching strategies, before discussing them in detail in separate sections below. The conventional photobleaching set-up is composed of a laser focused by an objective to a spot in the specimen. The spot is centered on the microscope's optical axis, and so the region to be bleached must be positioned by a manual or motor-driven stage and aligned to the optical axis, typically with the aid of a cross-hair in the oculars. The bleach is accomplished using a short pulse (microsecond to millisecond range) of high-intensity laser light, and then the recovery of fluorescence within the bleached area is recorded over time with a photomultiplier tube also centered to record from the focal plane spot. Rapid data collection, as provided by a photomultiplier tube (PMT), is important for cases of a freely diffusing molecule, where recovery times can be fast (e.g., of the order of 100 ms for GFP in cytoplasm (Swaminathan et al., 1997)).

On a laser scanning confocal microscope, the FRAP ("fluorescence recovery after photobleaching") strategy is essentially the same but the scanning capability provides greater flexibility in defining the bleach and imaging regions. The strategy in a typical experiment is to collect a prebleach image in order to identify a region of interest, bleach the selected area by scanning it with high laser power, and then monitor fluorescence recovery by scanning with low laser power. Analysis of the time course of fluorescence recovery can show the fraction of fluorescent molecules that are mobile/immobile and provide an estimate of both the diffusion coefficient and potentially the exchange rates of binding interactions that may be occurring within the bleached region.

The confocal microscope's ability to monitor fluorescence recovery at arbitrary sites enables an alternative photobleaching strategy, namely FLIP, ("fluorescence loss in photobleaching"; Cole et al., 1996). This approach is unique to confocal microscopy, because it uses the scanning capabilities of the microscope to alternate between bleaching at one site and monitoring of fluorescence at another site. If there is exchange of the fluorescent molecule between these two sites, then there will be recovery of fluoresence at the bleach site and loss of fluorescence at the monitoring site. Iteration of the bleach at a time interval sufficiently long to allow exchange to occur leads to progressive depletion of the fluorescent pool. The rate at which fluorescence is lost at the monitoring site is a measure of the exchange rate between the two sites. If the labeled molecules are immobile, or if a structural barrier exists between the region of interest and the bleach site, then no exchange will occur. FLIP can provide structural information, such as whether fluorescent molecules are contained in a single, continuous structure or within a series of discrete, contiguous structures (see below). Another advantage of FLIP is that it monitors the redistribution of fluorescence in regions of the cell that are not themselves subjected to bleaching, and so provides a control to test if the mobility of fluorescence molecules observed in a standard FRAP experiment is an artifact caused by light damage at the bleach site.

The confocal microscope's ability to bleach arbitrary patterns enables yet another alternative photobleaching strategy. Here we describe this approach for the first time, and call it IFRAP ("inverse-FRAP"). In IFRAP, all regions of a cell except the region of interest are bleached, and then loss of fluorescence from the region of interest is monitored. This approach simplifies the analysis of the recovery data in cases where a binding reaction occurs, because it reveals the off rate of binding without complicating contributions from the on rate. In conventional FRAP, the on rate plays an important role, since influx of fluorescence to a site is measured. But the off rate is also involved in a FRAP recovery, since bleached molecules that remain at a binding site prevent unbleached molecules from replacing them. In IFRAP, only the loss of fluorescent molecules is detected, and so this approach provides a more direct and simpler assay for the off rate.

3 TECHNICAL REQUIREMENTS FOR PHOTOBLEACHING ON A CONFOCAL MICROSCOPE

In order to carry out photobleach experiments on a laser scanning confocal microscope, the microscope must have the ability to: (1) rapidly change the laser illumination intensity; (2) scan a specified region of interest in a specimen; (3) collect a time-lapse series of images in order to monitor fluorescence recovery. We discuss below the strengths and limitations of the strategies by which these capabilities are achieved in current commercial confocal microscopes.

3.1 Controlling Laser Illumination Intensity

Rapid bleaching of a robust fluorophore such as GFP requires illumination intensities on the order of several milliwatts. In contrast, monitoring fluorescent recovery requires significantly lower illumination levels, typically about 1000-fold less than the photobleach intensity. These widely different intensities can be achieved in several different ways. First, it may be possible to vary the output of the laser, either manually or electronically. However, the range of laser outputs is unlikely to be sufficiently large to achieve optimal bleaching and monitoring intensities. Moreover, it is preferable to run lasers at high output levels, because they are then more stable than at low output levels. A second approach to produce a large change in intensity is to insert a neutral-density filter in front of the laser to attenuate the beam during monitoring. Some confocal microscopes, such as the Zeiss 410, have a series of neutral-density filters in a motorized filter wheel that vary the illumination intensity by a factor of up to 1000 on a timescale of about 0.5 s. A third and much faster way to change illumination intensity is with an acousto-optic modulator (AOM) or acousto-optic tunable filter (AOTF). These devices allow the illumination intensity to be varied by 1000-fold within a few microseconds. Moreover, they allow graded control over the laser illumination, thus providing much more flexibility than can be obtained with filters. Acousto-optical devices are used in the current generation of confocal microscopes from Zeiss, Leica, and Biorad. The rapid switch in intensity levels afforded by these microscopes is critical for applications where the mobility processes under study are themselves quite rapid (see Section 3.3). Laser illumination intensity and extent of bleaching will also depend on the transmission of the optical elements (objectives and filters), as well as on the pixel dwell time, the number of scans, and the scan zoom (see below).

3.2 Defining the Scan Region

Different confocal microscopes provide varying degrees of flexibility with regard to the size, shape and location of the scan region. Nearly all spot-scanning confocal microscopes allow the scan zoom to be varied over a range of 1–8, or even more. The scan area at zoom

8 is $\frac{1}{64}$ the area at zoom 1 and the rate of photobleaching is 64 times faster, since photobleach increases with the square of the zoom factor (Centonze and Pawley, 1995). In addition, some confocal microscopes have the capablity to limit the extent of the scan in the y axis, or both the x and y axes, making it possible to define rectangular regions of interest of varying dimensions. Some microscopes can also scan a single line or single spot. The most sophisticated microscopes allow the user to define regions of interest that can be essentially any shape or size. This is achieved by using an AOM or AOTF to blank the beam at defined times during the scan.

3.3 Monitoring Fluorescence Recovery

In a typical FRAP experiment, the goal is to measure fluorescence recovery, and often the process under study, such as diffusion, is very rapid. When this is the case, there are two key parameters. First, the time lag between the bleach and the first postbleach image must be minimized, because short lags permit more accurate assessment of fast recovery processes. This lag time is significantly shorter in systems with an AOM or AOTF (see Section 3.1). A second factor critical for speed is the time required to collect monitor images. Collection of a full-field image on a scanning confocal microscope typically takes about 1 s, or longer if averaging is needed to reduce noise. This is too slow for analysis of a fast process such as diffusion. Faster image collection can be achieved if the confocal microscope enables collection of images from a smaller region of interest. Even on systems that do not explicitly allow scanning in a small region of interest, it may be possible to achieve the same effect by scanning fewer points (e.g. 124×124 rather than 512×512) and increasing the zoom factor to deliver the needed resolution.

4 PHOTOBLEACHING STRATEGIES

Having considered the general instrumental features important for photobleaching with a confocal microscope, we now discuss in more detail three basic photobleaching strategies: FRAP, FLIP, and IFRAP.

4.1 FRAP on a Confocal Microscope

The protocol for a typical FRAP experiment is to collect a prebleach image, then perform the bleach with appropriate bleaching parameters applied to a specified area, and then finally monitor recovery of the bleach with appropriate imaging parameters in a defined imaging region (Fig. 2). Bleaching should be performed as rapidly as possible to minimize movement of molecules into and out of the bleach area during the bleach process. Rapid bleaching is achieved by using high laser power and a high-numerical-aperture (NA), high-transmission objective (e.g., a Fluar or Planapo objective) and by selecting a small bleach area. For determination of a diffusion coefficient, bleach intensity and duration should be set to reduce the fluorescence intensity by 30–70% (Meyvis et al., 1999). Images should be acquired rapidly during the initial period of recovery, when the intensity is changing rapidly and the process should be monitored until a plateau is reached. The latter part of the recovery process can be monitored at longer time intervals. The area monitored during recovery can be the same as the bleach area, or can include a larger area. An advantage of imaging a larger area is that this makes it possible to verify that the specimen remains stationary during the experiment.

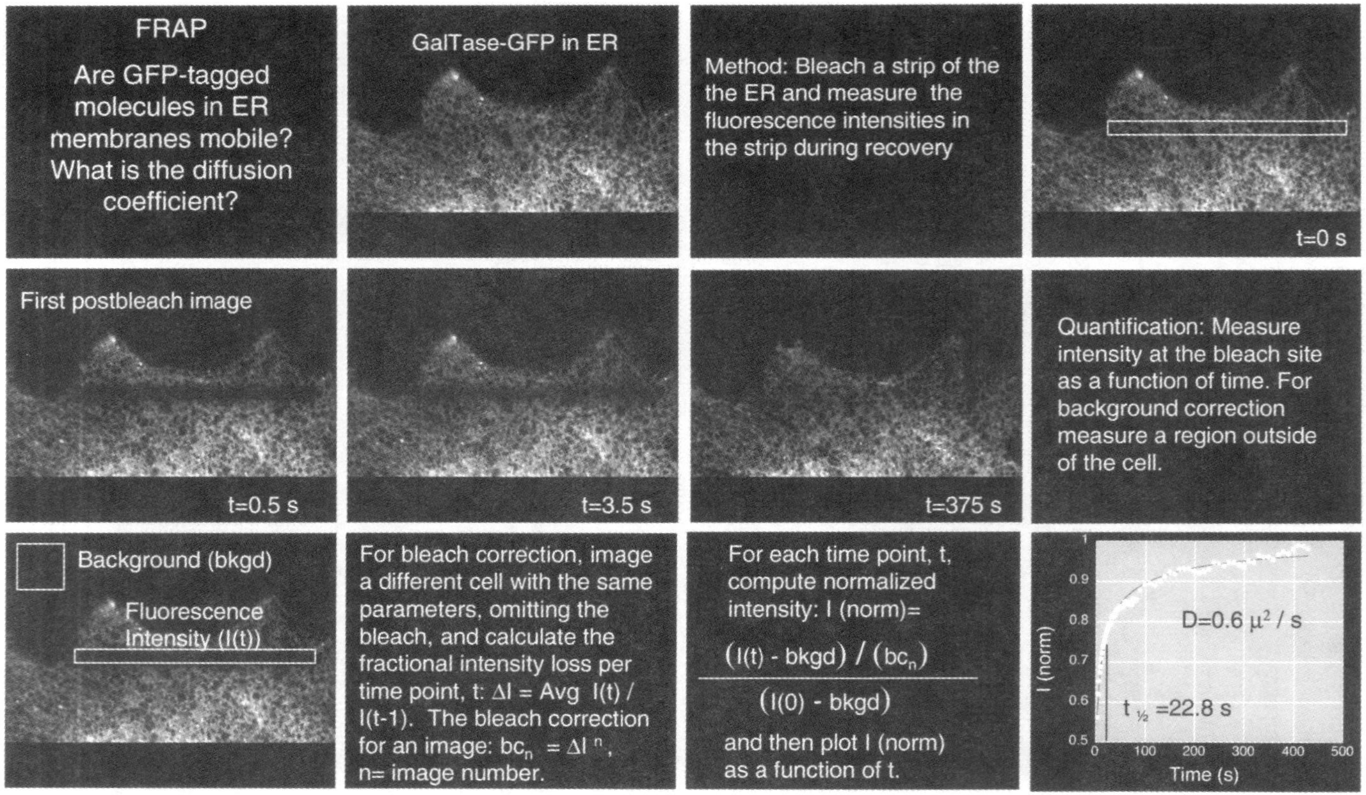

FIGURE 2 FRAP. Cos 7 cells expressing a GFP fusion protein (GalTase–GFP; Cole et al., 1996) that targets to the Golgi complex were treated with brefeldin A to redistribute Golgi membrane proteins to the endoplasmic reticulum (ER). The FRAP experiment was carried out on a LSM 510 (Carl Zeiss) using a 40× NA 1.2 Fluar objective. The bleach was performed by scanning the area outlined in white six times with the 488 nm line from a 40 mW argon laser at 75% output and no attenuation. Images were collected using the same laser output, but with the beam attenuated to 0.1%. The first 50 images were collected at 1.5 s intervals and the remaining images at 10 s intervals.

The dimensions of the bleach region will depend on the point spread function (PSF) of the objective. A simple way to check the dimensions of the bleach region is to bleach a specimen in which the fluorescence has been immobilized by fixation and then collect a series of optical sections through the specimen. The dimensions of the bleach region are critical for quantitative analysis of a FRAP experiment. The parameters used for imaging the recovery of fluorescence are also important. In particular, the PMT gain and offset should be set such that all intensity values are within range (no saturation), and when possible images should be collected at better than 8-bit resolution. The choice of pinhole diameter requires a tradeoff between optimizing signal to noise ratio and restricting measurement to the region most completely bleached. A larger pinhole improves signal to noise, but also increases the optical section thickness, which should not exceed the z depth of the bleach volume. The actual depth of the bleach volume for the optics in use can be determined by bleaching a fixed specimen.

In any photobleach experiment, there is both desirable and undesirable bleaching. The desirable bleaching destroys fluorescence in the area of interest, but the undesirable bleaching occurs during monitoring of the bleach recovery. To reduce this undesirable bleaching, it is always advisable to monitor bleach recovery at settings that minimize sample bleaching. However, the recorded signal must be sufficiently large compared with background and noise levels that recovery can be measured accurately. As a consequence, laser power during monitoring is often sufficient to induce some bleaching. For a quantitative analysis and even for qualitative comparisons, this bleaching must be accounted for.

The bleaching rate due to the monitoring process can be estimated in at least three ways. First, a different but comparable specimen (or set of specimens) can be used to estimate the rate of bleaching under the conditions used to monitor recovery in the actual experiment. Second, the same specimen and region of interest to be bleached can be first measured using the monitoring conditions to estimate the rate of bleaching due to monitoring there. This approach has the advantage that it accounts for any site-specific bleaching properties in the region of interest. Third, and most commonly, a different region from the experimental specimen can be used in parallel to estimate bleaching during the monitor phase. In this approach, a region much larger than the bleached region is acquired during recovery, and then loss of fluorescence in this larger area during the monitor phase is determined. The principal disadvantage of this approach is that it takes longer to acquire these larger images; therefore, for very rapid processes, the first or second method may be preferable. A compromise for the third method is afforded by AOTF-based scanning systems in which fluorescence can be monitored in two small areas – one encompassing the bleach zone and the other sufficiently far away to be unaffected by the bleach. The scan time will depend on the size of the rectangular region that encompasses the two regions.

Once the basic data have been acquired, a recovery curve can be generated. The first step is to measure intensities as a function of time both for the region of interest and for the region to be used for bleaching correction due to monitoring. Most confocal microscope software packages enable such measurement on a stack of time-lapse data, and then import the results into a spreadsheet program. Once these data are tabulated, the next step is to subtract the average background level from all of these measurements. The background measurement should be taken on the same sample used in the experiment, since it includes any autofluorescence in the medium as well as the PMT offset. If a large enough image has been acquired during the monitor phase, then it is often possible to make the background measurement directly from this image. The background-corrected data can now be normalized and corrected for bleaching due to monitoring. The percentage decrease in fluorescence due to monitoring should be a constant value. The bleach correction for each image can be computed from the average percentage change and the image number, as shown by the equation in Figure 2. In the experiment shown in Figure 2, the average percentage decline per image was 0.001%, so bleach correction had little impact. An alternative approach for bleach correction is to divide the intensity measured at each time point in the FRAP experiment by the value measured at the matching time point in the sample that was not bleached (see Figure 4 below). The resultant data can then be plotted to yield a recovery curve. An advantage of this latter approach is that it also corrects for intensity changes due to changes in focus, but a disadvantage is that it incorporates into the bleach correction fluctuations due to noise.

Several features of the recovery curve in Figure 2 should be noted. First, the bleach reduces the fluorescence intensity by about 50%. This extent of bleaching is sufficient to adequately measure fluorescence recovery, but not so large that the recovering signal is lost in the noise. Second, the recovery reaches a 50% value at 22.8 s. The half-time for recovery is an easily calculated value that can be used to make comparisons among different specimens. Third, the recovery curve plateaus slightly below 100% of its original value by the end of the observation period (430 s). The failure to recover to 100% of the prebleach intensity could indicate that some of the bleached fluorescent molecules were immobile. A more likely explanation in this instance is that the difference between the fluorescence intensity after recovery and the prebleach intensity represents the reduction in total cellular fluorescence caused by the bleach.

A diffusion coefficient D can be calculated by fitting a function to the recovery curve. In Figure 1, D was calculated from an empirical formula for diffusion in one dimension (into the slit), as was described by Ellenberg et al. (1997) and Lippincott-Schwartz et al. (1999). The calculated D does not take into account the complex geometry of the endoplasmic reticulum, and probably underestimates the actual diffusion rate.

4.2 FLIP by Confocal Microscopy

The standard FLIP experiment is performed by iteration of two successive steps: (1) bleach a site at some distance from the region of interest; then (2) acquire a monitor image that includes the region of interest. The time interval between successive bleaches should be sufficiently long to allow exchange of labeled molecules between the bleach and monitoring sites. If this condition holds, then fluorescent molecules will move into the bleach site and subsequently be bleached, leading to the progressive decline of fluorescence at the monitoring site.

The principal use of FLIP is to probe the structure of intracellular compartments, as illustrated in Figure 3. In this example, the fluorescent molecules are distributed within an intracellular membrane network (the ER). Bleaching the fluorescent molecules in one region of the cell leads to loss of fluorescence throughout the entire network, indicating not only that the fluorescent molecules are free to move (as also could be seen in a FRAP experiment), but in addition that ER throughout the entire cell is interconnected.

The measurements needed for analysis of a FLIP experiment are comparable to those needed for a FRAP experiment. Bleaching due to monitoring can be measured in a separate sample, or in an unconnected region in the same sample, as was described above for FRAP. The background, measured in an area outside of any cell, must be subtracted

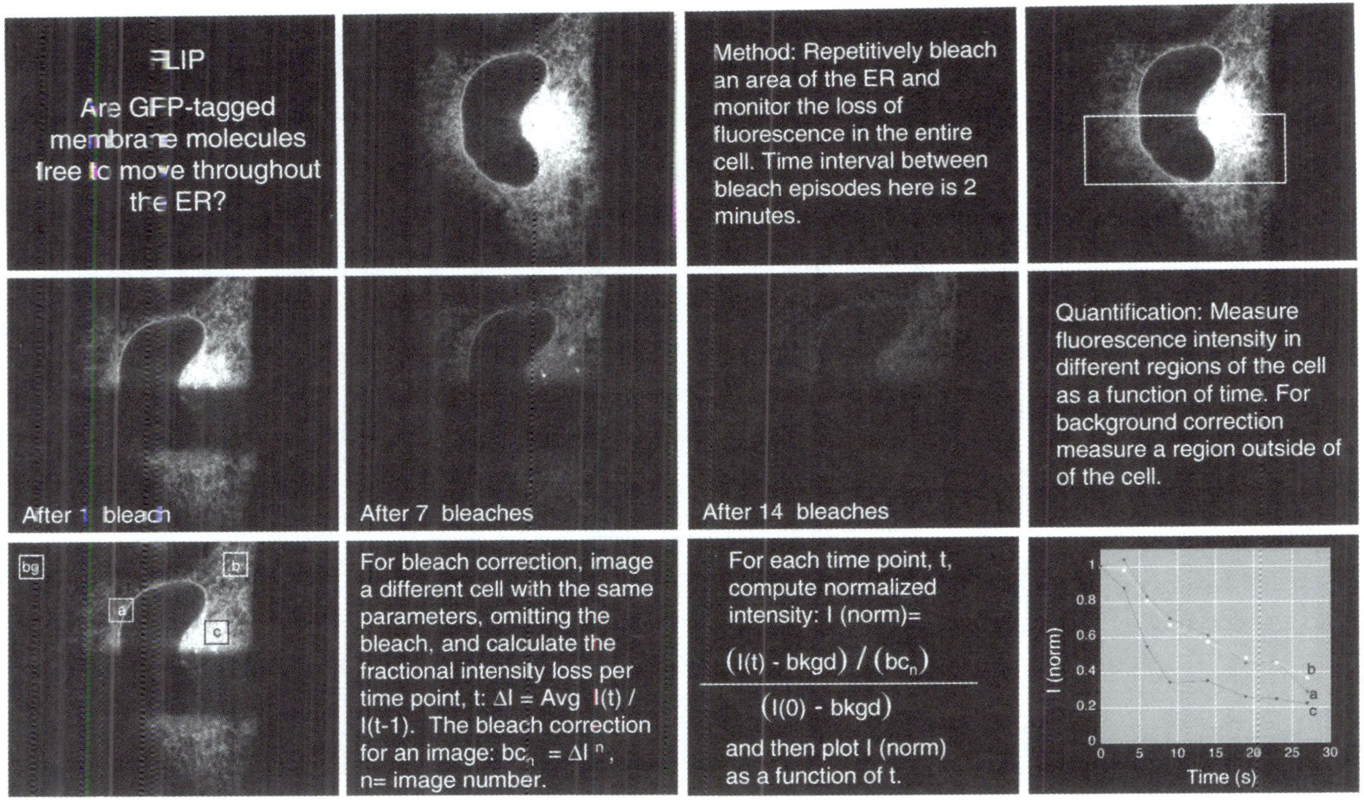

FIGURE 3 FLIP. Cos 7 cells expressing a GFP fusion protein (GalTase–GFP; Cole et al., 1996) that targets to the Golgi complex were treated with brefeldin A to redistribute Golgi-membrane proteins to the ER. The experiment was carried out on a LSM 510 (Carl Zeiss) with a 40× NA 1.2 Fluar objective. Bleaching was performed by scanning the outlined area 12 times with the 488 nm line from a 40 mW argon laser at 75% output and no attentuation. Images were collected using the same laser output, but with the beam attenuated to 2%. The time interval between successive bleaches was 2 minutes. Note that after 14 bleaches, much of the fluorescence is lost in the cell, demonstrating connectedness of the ER.

from all measurements. Bleaching correction can also be performed as described for the FRAP experiment.

The resultant plot of a FLIP experiment in its simplest form reveals two features: first, the rate of decay due to loss of fluorescence from the region of interest; second, the presence or absence of an immobile fraction. If all of the fluorescent molecules in the monitoring site are free to move into the bleach site, then repeated bleaches will cause the fluorescence in the monitoring site to decline toward zero. However, if some of the fluorescent molecules in the monitoring site were immobile, they would never be replaced by bleached molecules. The fraction of the fluorescent pool that is immobile can be determined by bleaching until the fluorescence in the monitoring region falls to a stable value.

4.3 IFRAP by Confocal Microscopy

The principal of IFRAP is to bleach everything except the region of interest – hence the name "inverse FRAP."

IFRAP has advantages over both FRAP and FLIP for analysis of binding-site kinetics. As already noted, FRAP data incorporate information about both on and off rates at a binding site, whereas IFRAP under reasonable assumptions provides a direct estimate of off rates only. These assumptions are that the number of unbleached molecules at the site of interest is small compared with the total number of bleached molecules in the system. Under this hypothesis, unbleached molecules do not rebind, because they are lost in a large pool of bleached molecules.

IFRAP can also be viewed as an instantaneous FLIP that eliminates contributions from diffusion or transport. Ordinarily in FLIP, the fluorescent pool is gradually depleted. The time required to sense this loss in the measurement region is a measure of how rapidly bleached molecules move into this region. As a result, FLIP is useful for measurement of diffusion or transport. In contrast, it can be a complicating factor when binding reactions are under study in the region of interest. Then the loss of fluorescence within the region is due to a combination of processes: the diffusion of bleached molecules to the binding sites and the loss of fluorescent molecules from the binding sites. This is not a serious problem if diffusion is very rapid compared with the rate of exchange at the binding sites, because then the binding kinetics dominate. However, when diffusion or transport rates are comparable to or slower than exchange rates, interpretation of FLIP data is complex. In IFRAP, these complications are mitigated because depletion of the fluorescent pool is essentially instantaneous, and so loss of fluorescence reflects only binding reactions and diffusion at the site. When the proportion of bound molecules at the site is much higher than the proportion of freely diffusing molecules there, the loss of fluorescence in IFRAP reflects predominantly binding kinetics.

The IFRAP protocol is to first acquire an image of the specimen, and then use it to define the region to be bleached. This region is outlined, leaving only the region of interest untouched. Then the bleach is performed – ideally as quickly as possible to permit measurement of the recovery as soon as possible. High laser intensity is needed so that ideally a complete bleach can be performed in a single iteration. Following the bleach, monitor images are then acquired until the region of interest loses all of its fluorescence (Fig. 4).

Analysis of the IFRAP data follows similar principles as outlined above for FRAP and FLIP. Bleaching correction is also critical here, and comparable strategies to those

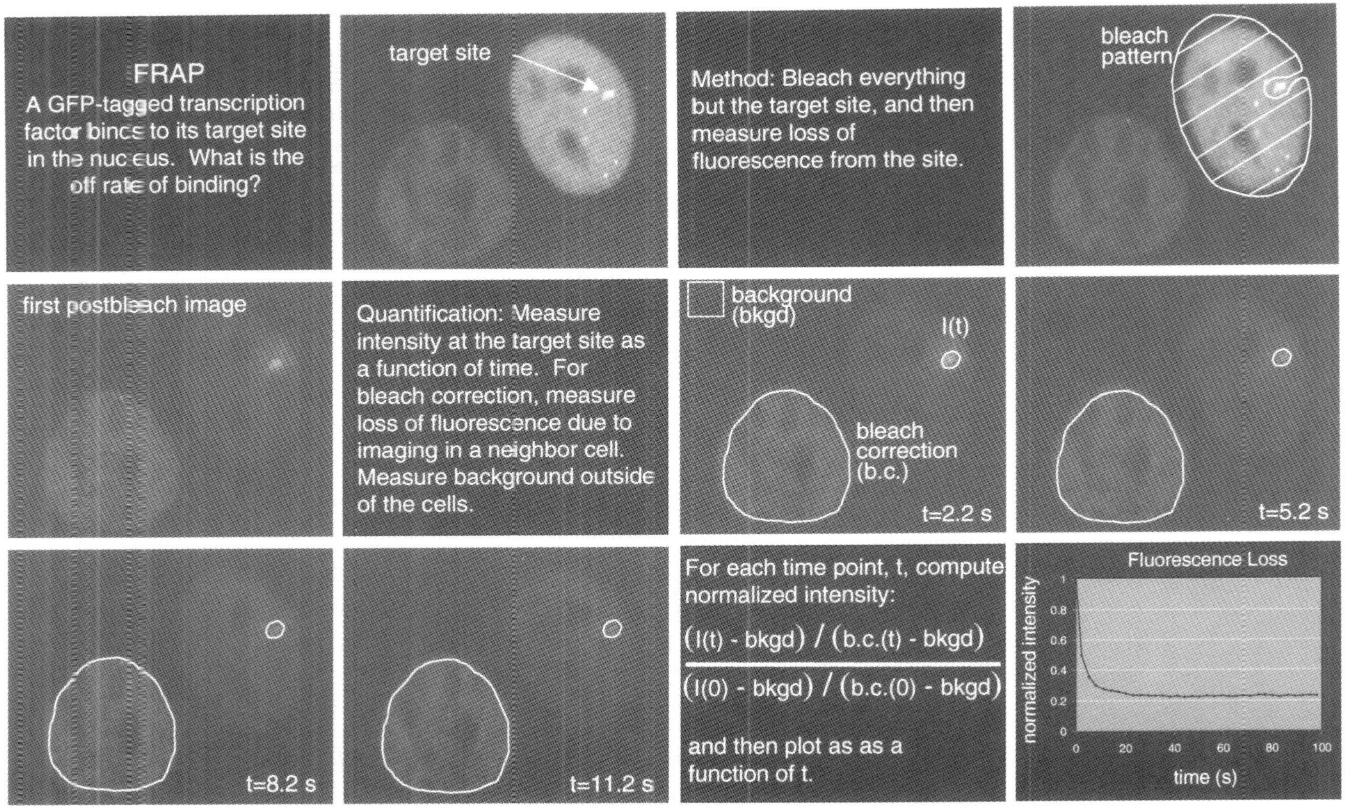

FIGURE 4 IFRAP. Images are from a mouse cell line (3617) expressing a GFP fusion with the glucocorticoid receptor (GFP–GR) and also containing a tandem array of promoter binding sites for this receptor (McNally et al., 2000). Binding of the GFP-tagged receptor to the tandem array yields a distinctive punctate spot in the nucleoplasm (arrow labeled target site). The experiment was carried out on a LSM 510 (Carl Zeiss) with a 40× NA 1.2 Fluar objective. Bleaching was performed by scanning the hatched area two times with the 488 nm and 514 nm lines from a 40 mW argon laser at 100% output and no attentuation. Images were collected using the same laser output, but with the beam attenuated to 0.1%. Following the bleach, the time interval between monitor images was 3 seconds

described above can be employed. Background must be subtracted from all measurements, and then bleaching correction performed by dividing intensities in the region of interest by the bleach correction intensity at that time point. A practical advantage of IFRAP is that the region of interest for measurement need not be so precisely defined, since intensities are low beyond the unbleached zone. This is helpful for many specimens in which the object of interest moves somewhat during the recovery phase.

The intensity curve for IFRAP is similar to that obtained by FLIP, but compressed in time. The loss of fluorescence from the region is a measure of movement out of the region. An immobile fraction appears as any residual fluorescence. In practice (as for FLIP as well), not all fluorescence can be bleached, and so there is an equilibration between the site of interest and the residual fluorescence in the specimen.

5 QUANTITATIVE ANALYSIS OF PHOTOBLEACHING DATA

First, it should be emphasized that a considerable amount of information can be extracted from photobleaching data without recourse to a detailed quantitative analysis. With the techniques described above, it is possible to generate recovery or loss curves that in themselves are informative. In many cases, it is of interest to merely demonstrate that a specific protein exchanges rapidly, or conversely that it is nearly immobile, or that there is some combination of these two aspects (e.g., a rapid recovery with a 25% immobile fraction). In other cases, it is of interest to compare mobilities under different experimental conditions. For example, it may be important to demonstrate that fluorescence recovery occurs much faster in FRAP in the presence of a certain drug or in a mutant background. These qualitative comparisons often supply sufficient information.

More quantitative information can be gleaned from a photobleaching experiment, although more effort and sophistication are required to retrieve it. In the simplest case of FRAP with a focused beam and a system exhibiting pure diffusion, the diffusion constant can be measured by well established methods (for a review, see Wolf, 1989). In this scenario, the process of recovery is described by the diffusion equation, which can be solved based on the initial conditions operative in the FRAP experiment. The result is a series expansion that can be fitted to the fluorescence-recovery data to yield an estimate of the diffusion constant. The initial conditions for the bleach involve the laser intensity and beam profile, which can be measured from a fixed specimen. Equations for deriving a diffusion coefficient from FRAP with circular or strip geometry also have been described – by Soumpasis (1983) for a uniform circle and by Ellenberg et al. (1997) and Lippincott-Schwartz et al. (1999) for a strip. A common feature of these approaches is that they take into account only the fluorescence in the bleached zone. The intensity values are low in the early part of the recovery process, resulting in a low signal-to-noise ratio. Other approaches calculate a diffusion coefficient by analyzing the entire distribution of bleached molecules during the recovery process (Kubitscheck et al., 1994; Wedekind et al., 1996).

When binding reactions are suspected in addition to pure diffusion, the first step is to fit the pure-diffusion model and demonstrate that it cannot account for the observed recovery data. At this point, a more complex model involving reaction and diffusion components is necessary (Icenogle and Elson, 1983; Elson and Qian, 1989). Depending on the relative rates of diffusion and reaction components, there are various simplifications that can be considered (Elson and Reidler, 1979; Meyvis et al., 1999).

FLIP data can be quantitatively analyzed by kinetic analysis to yield rate constants governing exchange of molecules between different cellular compartments (Phair and Misteli, 2000). Quantitative analysis of IFRAP has not been performed to date, but, as discussed above, it should be simplified in that diffusion and transport into the volume are eliminated. To evaluate a chemical reaction in the region of interest, the measured loss of fluorescence should reflect a combination of the off rate and diffusive loss from the region.

6 CONCLUSIONS

Confocal microscopy has provided wider access to photobleaching methods, and at the same time has enabled alternative bleaching strategies. These benefits do not depend on confocal optics per se, but rather on the laser scanning capabilities of most confocal microscopes. Indeed, in many photobleaching applications, it may be desirable to open the confocal pinhole somewhat to keep the region of interest in focus. Laser scanning, however, enables both conventional spot bleaching for FRAP as well as new approaches such as FLIP. Incorporation of AOTFs or AOMs into the light path permit arbitrary bleaching patterns to be defined, as in IFRAP. These latest innovations make the confocal microscope a versatile tool in photobleaching studies.

ACKNOWLEDGMENTS

We thank Kristien Zaal for assistance in performing and analyzing experiments in Figures 2 and 3. We thank Gordon Hager, Waltraud Müller, and Dawn Walker for the cell line used in Figure 4. We are also grateful to Kristien Zaal and Eliot Elson for comments on the manuscript.

REFERENCES

Centonze, V. and Pawley, J.B. (1995). Tutorial on practical confocal microscopy and use of the confocal test specimen. In: *Handbook of Biological Confocal Microscopy.* (ed. J.B. Pawley). New York: Plenum Press, pp. 549–569.

Cole, N.B., Smith C.L., Sciaky, N., Terasaki, M., Edidin, M., and Lippincott Schwartz, J. (1996). Diffusional mobility of Golgi proteins in membranes of living cells. *Science* **273**: 797–801.

Edidin, M., Zagyansky, Y., and Lardner, T.J. (1976). Measurement of membrane protein lateral diffusion in single cells. *Science* **191**, 466–468.

Ellenberg, J., Siggia, E.D., Moreira, J.E., Smith, C.L., Presley, J.F., Worman, H.J., and Lippincott-Schwartz, J. (1997). Nuclear membrane dynamics and reassembly in living cells: targeting of an inner nuclear membrane protein in interphase and mitosis. *J. Cell. Biol.* **138**, 1193–1206.

Elson, E.L. and Qian, H. (1989). Interpretation of fluorescence correlation spectroscopy and photobleaching recovery in terms of molecular interactions. *Meth. Cell Biol.* **30**, 307–332.

Elson, E.L. and Reidler, J.A. (1979). Analysis of cell surface interactions by measurements of lateral mobility. *J. Supramol. Struct.* **12**, 481–489.

Icenogle, R.D. and Elson, E.L. (1983). Fluorescence correlation spectroscopy and photobleaching recovery of multiple binding reactions. I. Theory and FCS measurements. *Biopolymers* **22**, 1919–1948.

Kubitscheck, U., Wedekind, P., and Peters, R. (1994). Lateral diffusion measurement at high spatial resolution by scanning microphotolysis in a confocal microscope. *Biophys. J.* **67**, 948–956.

Liebman, P.A. and Entine, G. (1974). Lateral diffusion of visual pigment in photoreceptor disk membranes. *Science* **185**, 457–459.

Lippincott-Schwartz, J., Presley, J.F., Zaal, K.J., Hirschberg, K., Miller, C.D., J., and Ellenberg, J (1999). Monitoring the dynamics and mobility of membrane proteins tagged with green fluorescent protein. *Meth. Cell Biol.* **58**, 261–281.

McNally, J.G., Müller, W.G., Walker, D., Wolford, R., and Hager, G.L. (2000). The glucocorticoid receptor: rapid exchange with regulatory sites in living cells. *Science* **287**, 1262–1265.

Meyvis, T.K., De Smedt, S.C., Van Oostveldt, P., and Demeester, J. (1999). Fluorescence recovery after photobleaching: a versatile tool for mobility and interaction measurements in pharmaceutical research. *Pharm. Res.* **16**, 1153–1162.

Phair, R.D. and Misteli, T. (2000). High mobility of proteins in the mammalian cell nucleus. *Nature* **404**, 604–609.

Poo, M. and Cone, R.A. (1974). Lateral diffusion of rhodopsin in the photoreceptor membrane. *Nature* **247**, 438–441.

Schlessinger, J., Koppel, D.E., Axelrod, D., Jacobson, K., Webb, W.W., and Elson, E.L. (1976). Lateral transport on cell membranes: mobility of concanavalin A receptors on myoblasts. *Proc. Natl Acad. Sci. USA* **73**, 2409–2413.

Soumpasis, D.M. (1983). Theoretical analysis of fluorescence photobleaching recovery experiments. *Biophys. J.* **41**, 95–97.

Swaminathan, R., Hoang, C.P., and Verkman, A.S. (1997). Photobleaching recovery and anisotropy decay of green fluorescent protein GFP-S65T in solution and cells: cytoplasmic viscosity probed by green fluorescent protein translational and rotational diffusion. *Biophys. J.* **72**, 1900–1907.

Wedekind, P., Kubitscheck, U., Heinrich, O., and Peters, R. (1996). Line-scanning microphotolysis for diffraction-limited measurements of lateral diffusion. *Biophys. J.* **71**, 1621–1632.

White, J. and Stelzer, E. (1999). Photobleaching GFP reveals protein dynamics inside live cells. *Trends Cell Biol.* **9**, 61–65.

Wolf, D.E. (1989). Designing, building, and using a fluorescence recovery after photobleaching instrument. *Meth. Cell Biol.* **30**, 271–306.

CHAPTER 26

Two-Photon Laser Scanning Microscopy for Characterization of Integrated Circuits and Optoelectronics

CHRIS XU

Bell Laboratories, Lucent Technologies, Holmdel, NJ 07733, U.S.A.

1 INTRODUCTION

Laser scanning optical-beam-induced current (OBIC) imaging has become an increasingly useful tool in the failure analysis of integrated circuits (ICs) (Wilson and Sheppard 1984; Richards and Footner, 1992; Komoda and Shimizu, 1994; Cole, 1996). In combination with confocal reflected-light imaging, a variety of failure modes can be identified nondestructively by OBIC. However, front-surface OBIC is limited by nonuniform light transmission through the multiple levels of metallic interconnects and power distribution planes used in modern IC technology. In other situations, such as in the case of flip-chip packaging, the chip's front surface is entirely inaccessible. The infrared (IR) laser scanning microscope (Ziegler, 1990; Bossmann et al., 1992) takes advantage of the transparency of silicon in the near infrared and allows imaging through the polished backsides of ICs, circumventing the difficulties in front-surface imaging. Using excitation wavelengths longer than the silicon band gap, "through-the-backside" reflected-light images can be easily obtained. Backside one-photon OBIC (1P-OBIC) faces the conflicting requirements that illumination light not be too strongly attenuated by the silicon substrate (with a typical thickness of hundreds of microns) and an efficient generation of electron–hole pairs. These conditions are only approximately met, even at relatively long wavelengths near the silicon band edge. The spatial resolution of 1P-OBIC is therefore reduced and the usefulness of OBIC for the analysis of chips fabricated with submicrometer lithography is limited.

Two-photon absorption excited OBIC (2P-OBIC) imaging through the backsides of integrated circuits (Xu and Denk 1997, 1999) remedies some of the problems faced by 1P-OBIC. As in the case of multiphoton laser scanning fluorescence microscopy (Denk et al., 1990), 2P-OBIC gains several important advantages by using nonlinear optical excitation. Because the energies of two photons are combined to generate each electron–hole pair,

Confocal and Two-Photon Microscopy: Foundations, Applications, and Advances, Edited by Alberto Diaspro.
ISBN 0-471-40920-0

single-photon energies below the silicon band gap can be used to generate free carriers. This almost completely avoids attenuation of the excitation light by the substrate. Because the absorption rate increases quadratically with the excitation intensity, free-carrier generation is confined almost completely to the vicinity of the focal point. Such spatial localization provides intrinsic three-dimensional resolution and eliminates out-of-focus background. Furthermore, the spatial-frequency limit of the optical transfer function is doubled for two-photon excitation compared with that of one-photon excitation at the same excitation wavelength (Nakamura, 1993). Because the excitation wavelength only needs to be slightly longer than what is usable for backside 1P-OBIC, a resolution enhancement of a factor of almost 2 results.

The extension of 2P-OBIC to optoelectronic devices, such as vertical cavity surface-emitting lasers (VCSELs), is natural. Driven by a wide variety of applications in optical communications such as optical interconnects, high-speed local area networks, and optical signal processing, VCSEL technology has made significant progress. Thus far, VCSEL structures with the highest efficiencies have been devices with oxide-defined current apertures. However, VCSELs with implant-defined current apertures require only photolithographic processing steps. The basic theory of ion implantation is understood and ion implants have been used in VCSEL processing since the early 1990s (D'Avanzo, 1982; Lee et al., 1990). These were deep implants (several micrometers) into the top mirror. Recently, the use of shallow implants has been introduced (Chirovsky et al., 1999). The effectiveness of these implants is still a concern with regard to VCSEL performance and reliability.

To aid VCSEL design and fabrication, we employ a two-photon photocurrent testing method to quantitatively study VCSELs on a microscopic level. This study focuses mainly on the current aperture formation by the implantation process, an issue of critical importance to VCSEL performance. Similar to two-photon backside imaging of ICs, two-photon excitation provides a convenient way of injecting localized carriers in the P–I–N diodes of the laser through the GaAs substrate. Because the photon energy of the excitation beam is below the substrate band gap, two-photon photocurrent generation circumvents the problem caused by the opacity of the substrate and the top metal layer. Backside illumination allows the use of ordinary microprobes to contact the VCSELs on the front side, without the restriction of the probe movement and the requirement of long-working-distance lenses. Thus, high-numerical-aperture (NA) objectives can be used to achieve high spatial resolution ($\leqslant 1$ μm). Furthermore, the technique requires no packaging, is non-destructive, and can be implemented as a wafer-scale tool.

In this chapter, we shall review our experiments on two-photon laser scanning microscopy in the characterization of ICs and VCSELs. We will study the transition from 1P- to 2P-OBIC and explore the different contrast mechanisms in 1P- and 2P-OBIC. Signal-to-background issues and the spatial resolution will also be discussed.

2 TWO-PHOTON LASER SCANNING MICROSCOPY FOR IC ANALYSIS

2.1 Experimental Set-up

We used an optical parametric oscillator (OPO) (Opal, Spectra Physics), providing wavelength tunable pulses ($\lambda = 1.15$–1.35 μm about 120 fs pulsewidth, and 80 MHz repetition rate), as the illumination source (Fig. 1). The excitation beam was raster-

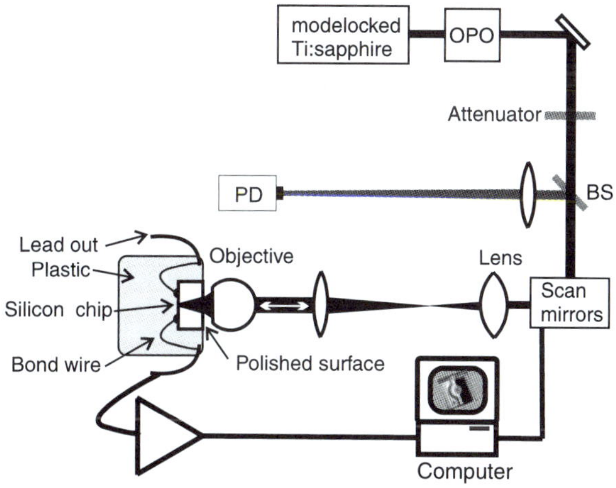

FIGURE 1 Experimental set-up of 1P/2P-OBIC of ICs and confocal reflected light-imaging.

scanned by two galvanometer-driven mirrors (Cambridge Technology 6800) and focused by a microscope objective (Zeiss, 0.7–1.3 NA/40× oil, Iris). A pair of relay lenses between the scan mirrors and the objective also formed a beam expander, allowing us to fill the back aperture of the objective lens (approximately 5 mm in diameter), so we could achieve the smallest possible focus at the sample.

The backside of the silicon chip was exposed by grinding away the plastic packaging material and die-attach, and was then polished to a mirror finish. As a contrast signal we used either the variation in supply current for the memory chip (HM628128BLFP-7, Hitachi) or the change in the inverting input bias current for the operational amplifier (μA741; Texas Instruments). Current was recorded and displayed as a function of the beam position, thus generating an OBIC map of the IC. The average power at the sample varied between 0.5 mW and 10 mW.

For comparison and registration purposes, confocal reflected-light images were also obtained. The back-reflected light was collected by the objective lens and descanned by the scan mirrors. A beam splitter (BS, 70% reflectivity) was used to redirect the majority of the reflected light, which was then focused by a lens ($f = 600$ mm) onto a germanium photodiode (PD, Thorlab). The small active area of the photodiode (about 0.2 mm^2) served as the confocal aperture.

2.2 Results Of IC Imaging

2.2.1 Power and Wavelength Dependence of the OBIC signal Sufficient substrate transmission requires the excitation wavelength for backside 1P-OBIC to be right around the band gap. Although 2P-OBIC could be performed with any wavelength larger than but less than twice the band-edge wavelength, the quest for the highest spatial resolution and, in some substrates, the presence of free-carrier absorption, which increases toward longer wavelengths, places the optimum wavelength region for 2P-OBIC close to the band edge as well. A detailed study of the nature of the OBIC signal in the transition region between one- and two-photon excitation can thus be achieved by tuning the excitation wavelength through only a small range around the silicon band edge. The dominant excitation mode can be determined by using the incident-power dependence of the photocurrent, which is linear and quadratic for one- and two-photon absorption, respectively.

The power dependences of the OBIC signal of the SRAM in the wavelength region of 1.15–1.25 μm are shown in Figure 2. The average power at the sample was varied from 0.5 mW to 10 mW, a range suitable for backside 1P- or 2P-OBIC imaging. At 1.15 μm, the OBIC varies linearly with the incident power in this power range, indicating that one-photon generation of free carriers is dominant. The transition from one-photon to two-photon OBIC occurs between 1.19 μm and 1.21 μm, where the two-photon contribution starts to become significant at high incident powers. At 1.23 μm and 1.25 μm, two-photon absorption dominates throughout the whole power range.

These data define the wavelength region where the transition between 1P- and 2P-OBIC occurs, using the total photocurrent as a measure. However, as will be discussed in Section 4 (Eqs. (2) and (3)), the ratio of 2P to 1P excitation depends also on the sample thickness, the depth of effective charge collection, and the temporal structure of the excitation source (i.e., pulselength and repetition frequency). Furthermore, it is possible that in parts of the transition region the average photocurrent is still dominated by one-photon excitation while the image contrast is already two-photon-dominated (see Section 4.1).

Contrast (i.e., the spatial variation in the detected current) is determined by variations in both the rate of free-carrier generation and the free-carrier collection efficiency. The generation rate depends on not only the absorption cross-section and the incident power but also the light pathlength (in 1P-OBIC) or the focus profile (in 2P-OBIC). Once a carrier is generated, there are several factors that influence the collection efficiency, for

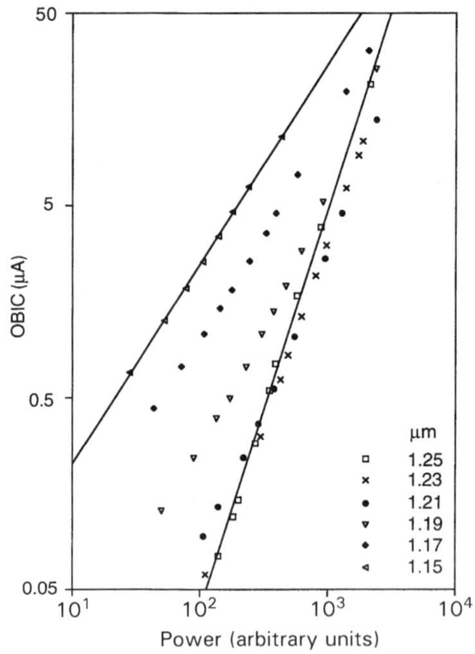

FIGURE 2 Power dependence of the OBIC signal using the static RAM chip. The best-fit slopes at 1.15 μm and 1.25 μm are 1.04 and 2.09, respectively.

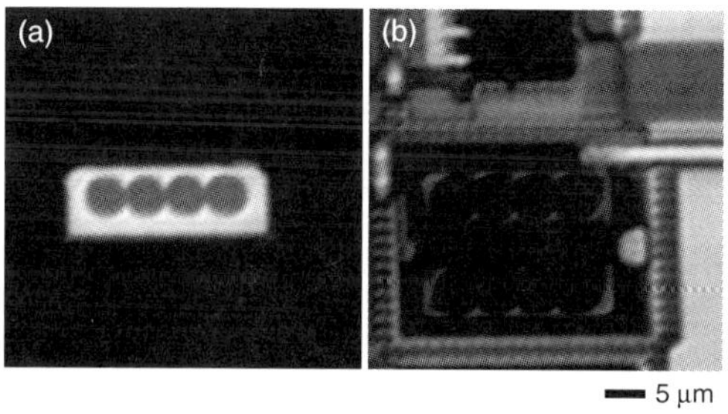

FIGURE 3 2P-OBIC images of the same SRAM using the current in (*a*) one of the address pins (A15) and (*b*) the power supply under otherwise identical conditions. The brightness of the image represents the current values.

example the presence of an electric field (intrinsic or caused by an external voltage) and the density of recombination centers. The signal also depends on the circuit used to detect the current. Figure 3 demonstrates this by showing images that were taken using, respectively, the power-supply current and the current in one of the address pins of the same IC under otherwise-identical experimental condition. Most areas of the chip showed some contrast in the power-supply mode, but only a very small patch was visible when the address pin current was measured.

2.2.2 Comparison of 1P- and 2P-OBIC Images Striking differences between 1P- and 2P-OBIC images, obtained at 1.15 μm and 1.25 μm, respectively, suggest that the contrast mechanisms for 2P-OBIC and 1P-OBIC are substantially different. Figure 4, for example, compares the 1P- and 2P-OBIC images of the 741 Op-amp. The polarity of the induced current, as indicated by an image intensity lighter or darker than the background, abruptly changed at the transistor boundary in the 2P-OBIC image (Fig. 4*b*). The 1P-OBIC image (Fig. 4*a*), on the other hand, showed a much more gradual transition from light to

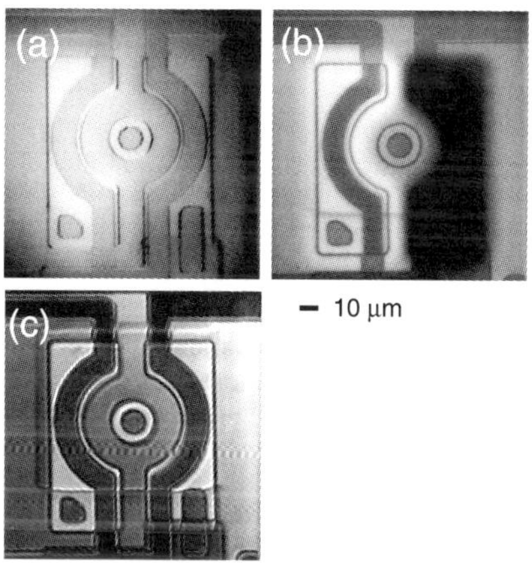

FIGURE 4 One-photon (*a*) and two-photon (*b*) OBIC and confocal reflected-light (*c*) images of the same region in a μA741 operational amplifier. (*a*) and (*c*) were obtained at 1.15 μm and (*b*) at 1.26 μm. (The confocal images at 1.15 μm and 1.26 μm look identical.) The excitation powers at the sample was about 5 mW for (*b*) and 0.4 mW for (*a*) and (*c*).

dark. Such behavior is expected from differences in the distribution of the free carriers generated, which for one-photon absorption is smeared out because absorption occurs throughout the substrate along the illumination path, but for two-photon absorption is confined to the focal volume. (See also Section 4.1.)

Figure 5*a* compares 1P- and 2P-OBIC images of the SRAM. To test whether a simple shift in focus could be responsible for some of the differences seen, 1P-OBIC and 2P-OBIC (again at 1.15 μm and 1.25 μm excitation wavelengths) images of a SRAM chip were obtained at closely spaced focus intervals. Figure 5*b* shows that 1P- and 2P-OBIC images match each other at no focal position. To facilitate the comparison, the average currents were subtracted before displaying the images. The averages were then plotted as a function of focus position in Figure 5*c* to conform, in particular, that the 1P current is rather insensitive to the focal position.

The most prominent feature of 1P-OBIC images is that edges of metal interconnect lines showed strong enhancement, which was entirely missing from the 2P-OBIC counterpart. Edge enhancement had been reported in previous studies of backside 1P-OBIC (Ziegler, 1990), but the mechanism was not fully understood. Comparison of 1P and 2P images rules out both of the most commonly cited explanations – Schottky effects and enhanced reflection – because both mechanisms would lead to an enhanced signal in 2P-OBIC as well, but we never saw any such enhancement in 2P-OBIC images. Our interpretation is that the enhancement in 1P-OBIC is instead caused by the wide-angle light scattering that is expected at edges. To understand how this affects the current, one has to appreciate that even with high-NA illumination, the opening angle of the illumination (and of the specularly reflected light) cone *inside* the Si substrate is rather small (the half opening angle is only about 17° for NA = 1.0). Scattered light, on the other hand, is distributed over a large solid angle, and at large deviations from the surface normal travels a much longer distance in the Si substrate, and might even be totally internally reflected and so entirely prevented from leaving the substrate. The difference in the resulting current is then due to the fact that scattered light is readily absorbed by one-photon processes, as long as the quantum energy is large enough, but two-photon absorption of the scattered light is minimal explaining the observed complete lack of edge enhancement in 2P-OBIC. In fact, scattering effectively reduces the reflection from the top surface, leading to a reduction in the signal level at edges, in confocal reflected-light and 2P-OBIC images.

The similarities between 2P-OBIC and confocal reflected-light images are thus not as surprising as one might initially think. If the collection efficiency of the free carriers is uniform throughout the IC, then variations in the 2P-OBIC result mainly from changes in the intensity of the reflected light which is, of course, the source of contrast for the reflected light image as well. In fact, to pinpoint variations in the free-carrier collection efficiencies, one would best look for differences between 2P-OBIC and confocal reflected-light images.

One such difference is another striking feature often seen in 2P-OBIC images, but in neither 1P-OBIC nor confocal images, which comprises an intensity step that is much sharper than the optical resolution (Fig. 5*a*, indicated by an arrow). We have investigated this step at various incident optical power levels under both 2P and 1P conditions, and have found that, as long as the supply voltage is above a certain value, the step is always *present* in 2P-OBIC and always *absent* in 1P-OBIC. In the absorption-mode transition region around 1.19 μm, the step is only visible at high incident power, where two-photon induced current is dominant. Although these steps were visible in 2P-OBIC even at the lowest laser

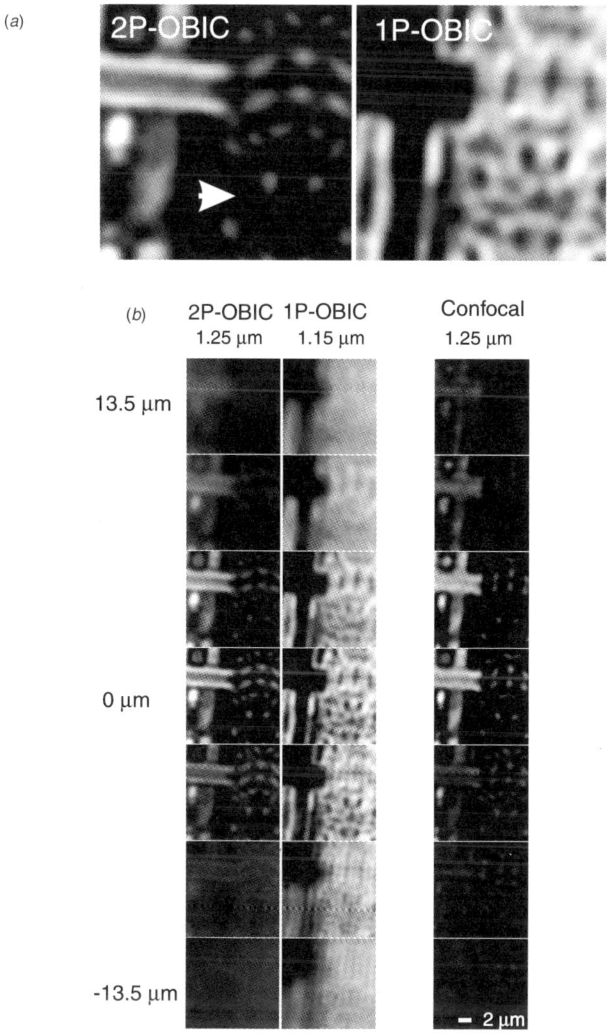

FIGURE 5 (*a*) 2P- and 1P-OBIC images of the SRAM. (*a*) Focal series of 2P-OBIC, 1P-OBIC, and confocal reflected-light images of the SRAM. The excitation powers at the sample were about 10 mW for 2P-OBIC and confocal images and 2.5 mW for 1P-OBIC. Images displayed correspond to objective positions spaced by 2 μm intervals (corresponding to shifts in the focal plane inside silicon by 4.5 μm. DC components of the images have been removed. At the best focus position ($z = 0$ μm), the range of the current variations before background subtraction are approximately 5–15 μA, 6.5–15 μA, and 0.4–18 μA for 2P-OBIC, 1P-OBIC, and confocal reflected light, respectively. (*c*) DC component (i.e., current averaged over the image) of 2P-OBIC, 1P-OBIC, and confocal reflected-light images of the same area as displayed in (*b*). Data were taken by stopping the objective lens at 100 nm intervals.

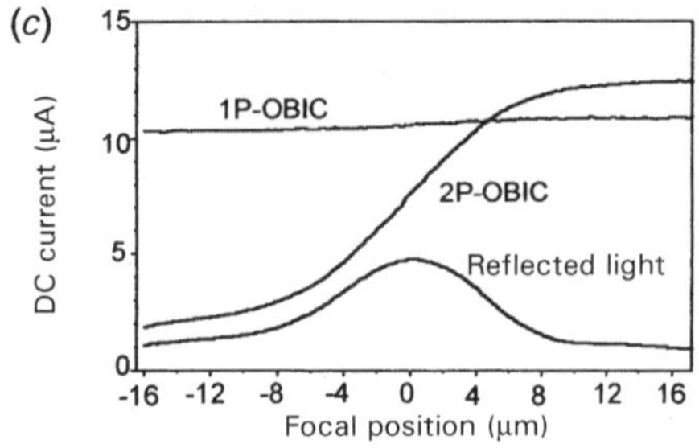

FIGURE 5 (*continued*)

powers that we used (approximately 0.3 mW at the sample at 1250 nm), the exact location of the lines did depend on the incident power, and also shifted slightly when the scan direction was changed. Figure 6 shows how these superresolution structures appear and shift, depending on the supply voltage. We observe little if any voltage hysteresis in the location of the lines.

While we do not completely understand the origin of these structures, which we call "switch lines," one possible explanation is that an internal circuit element is switching between two states that have different collection efficiencies, leading to a step in the current signal. Switching itself would be induced by a shift in the balance of carrier generation in different regions as the beam is scanned. Such a mechanism would obviously require a fairly localized charge generation as provided only by 2P but not by 1P excitations, making this model consistent with the fact that switch lines are absent in 1P-OBIC in our chips. In highly doped substrates, where carrier collection is confined to a thin layer because of very high recombination rates, switch lines are occasionally seen under 1P conditions (C. Xu and W. Denk, unpublished data). The exact location of the

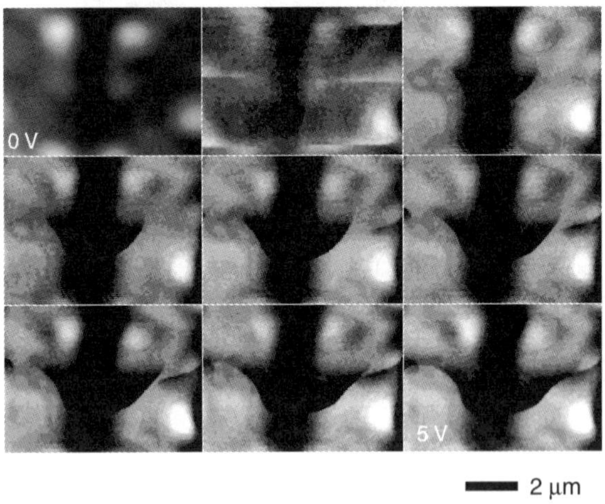

FIGURE 6 Dependence of the "switch" lines on the power-supply voltage in 2P-OBIC. The power-supply voltage was increased from 0 to 5 V in 0.625 V intervals. The sequence of the images are left to right and then top to bottom. Differences (scaled up 5× in intensity) to the zero-supply-voltage image are shown for nonzero supply voltage.

switch line most likely depends sensitively on abnormalities in the local circuit parameters and could, therefore, be very useful for failure analysis.

3 TWO-PHOTON LASER SCANNING MICROSCOPY FOR VCSEL ANALYSIS

3.1 Experimental Set-up

The experimental layout is illustrated in Figure 7. In essence, the set-up combines an inverted laser scanning microscope (the same as described in Section 2.1) with a standard probe station. A compact mode-locked fiber laser (Femtolite B-35, IMRA America) is used as the illumination source. The laser provides an average power of 40 mW and $\leqslant$ 300 fs pulses with repetition rate of about 48 MHz at 1.55 μm. Special microscope objectives (either 100×/0.85 NA, Hamamatsu A3717 or 100×/0.9 IR, Zeiss) featuring correction collars for imaging through thick substrates were used. By carefully adjusting the collar, spherical aberration caused by the index mismatch of the air/substrate interface can be completely compensated (performances of the lenses are confirmed by measuring the axial response of confocal imaging). The backsides of the VCSEL wafers were chemically/mechanically polished to a mirror finish, the only sample preparation step required before testing. Ordinary microprobes were used to contact the VCSELs. An IR-sensitive camera was used to guide the probes and locate the incident beam. Two-photon-generated current was recorded as a function of the beam position, and displayed as photocurrent maps of the VCSELs. Although results reported here are for VCSELs under zero bias, images have also been taken for VCSELs under reverse bias, for example, to measure the spatial variations of the breakdown voltage. Detailed VCSEL structure and processing steps have been reported elsewhere (Chirovsky et al., 1999).

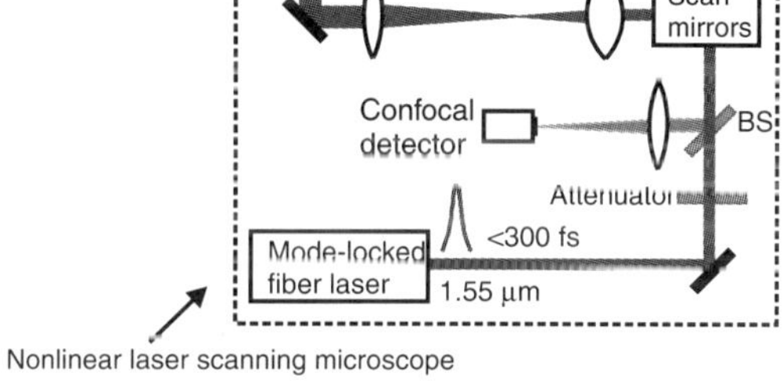

FIGURE 7 Experimental set-up for the two-photon photocurrent imaging of VCSELs.

3.2 Results of VCSEL Imaging

The effectiveness of the two-photon photocurrent method is demonstrated in Figure 8. Two 980 VCSELs – one a normal device with a 10 μm current aperture and the other a test device with implant throughout the active layer – were imaged. While the confocal reflected-light images showed virtually no difference, the contrast is apparent in the two-photon photocurrent images. Figure 9 compares quantitatively the photocurrent levels of a normal VCSEL and two fully implanted devices. Within the implant-defined current aperture, the photocurrent measured in the normal device is more than 15 times larger than that of the fully implanted devices, showing clearly the current-blocking effect of the implant process. Outside of the current aperture, the measured photocurrent levels are about the same for the normal and fully implanted devices; that is, the large difference in conductivity within the current aperture has no effect on the region outside. Combined with the observation of the sharp aperture boundary, we can conclude that lateral carrier diffusion is negligible in the detected photocurrent, and the detected photocurrent represents the local property of the device. The devices shown in Figures 8 and 9 are identical except for the implant layer. Thus, Figures 8 and 9 also show that the two-photon photocurrent image represents the properties of the implant layer, despite the fact that free carriers are generated throughout the confocal length (about 6 μm inside the substrate), which is much larger than the active layer of the VCSEL (< 0.1 μm) (Chirovsky et al., 1999). Although only results for 980 nm VCSELs have been reported here, we have also successfully used the same technique to study 850 nm VCSELs.

The study of the failure mechanisms of VCSELs can also be aided by the two-photon photocurrent technique, particularly in locating the defect sites. As an example, the VCSEL shown in Figure 10 was intentionally subjected to extreme conditions (24 mA and 6 V, approximately 12× threshold or 1.2× roll-over current level) in order to speed up the

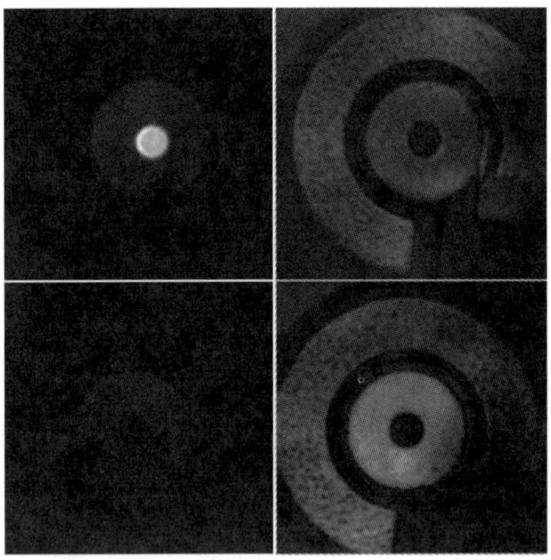

(a) (b)

FIGURE 8 Simultaneous two-photon photocurrent and confocal reflected-light images of 980 VCSELs with implant-defined current aperture. The gray scales represent the photocurrent and reflected-light levels. The top row shows a normal VCSEL and the bottom row a fully implanted test VCSEL. The top and bottom images were obtained under identical conditions. The size of the current aperture is 10 μm. For detailed VCSEL structure, see Chirovsky et al. (1999).

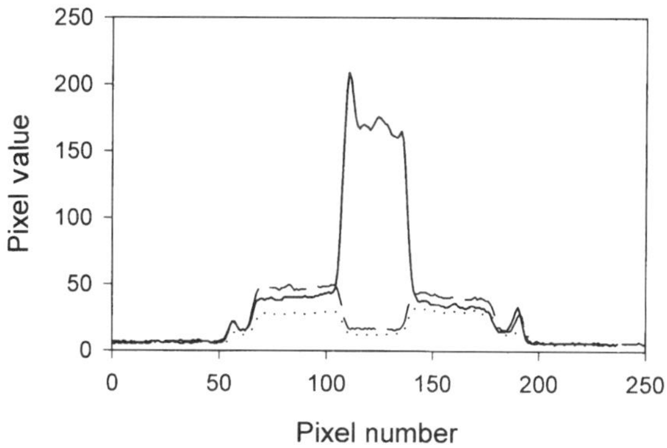

FIGURE 9 Line plots of the photocurrent levels. The lines were taken through the middle of the 10 μm current apertures. The solid line is data taken with a normal VCSEL. The dashed and dotted lines are data taken with two fully implanted devices. A pixel value of 150 represents a photocurrent of 10 μA.

degradation process. The uniform degradation of the mesa (i.e., the lack of localized defects in the mesa region) is likely caused by the thermal annealing of the implant; while the localized breakdown at the P-contact possibly indicates regions of high field strength. The degraded VCSEL still lased, but only at about 20% of its normal power level. The bias voltage to achieve the same current also dropped significantly, consistent with the observation of a leaky aperture. We also note that the backside confocal reflected-light images before and after degradation showed little difference, indicating no significant change in the optical properties of the device.

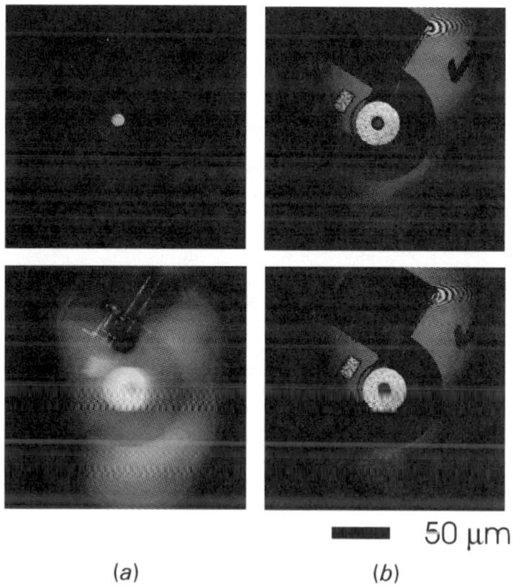

FIGURE 10 Simultaneous two-photon photocurrent (*a*) and confocal reflected-light (*b*) images of 980 VCSELs before (top) and after (bottom) operating at 24 mA for several minutes.

4 DISCUSSION

4.1 Signal Fraction in One- and Two-Photon Backside OBIC

In general, the *total* induced current at scan position **r** can be written as

$$N(\mathbf{r}) = \int_V d\mathbf{r}'[f(\mathbf{r}')\alpha(\mathbf{r})I(\mathbf{r}-\mathbf{r}') + f(\mathbf{r}')\beta(\mathbf{r}')I^2(\mathbf{r}-\mathbf{r}')], \tag{1}$$

where the integration is over the entire substrate, $\alpha(\mathbf{r}')$ and $\beta(\mathbf{r}')$ are the one- and two-photon absorption coefficients, respectively, $I(\mathbf{r}-\mathbf{r}')$ is the light-intensity distribution, which includes contributions by scattering and back-reflection, and $f(\mathbf{r}')$ is the collection efficiency, at location $\mathbf{r}'$. The first term in Eq. (1) represents the current contributed by one-photon excitation; while the second term represents the contribution from two-photon excitation. Although Eq. (1) describes the total current generated, only the fraction of the current that is spatially modulated with high resolution should be considered signal. We have to distinguish further between modulations that are due to variations in scattering or reflectivity, and that are therefore not specific to the OBIC signal but can for example be seen in the reflected-light image, and modulations that are due to variations in generation ($\alpha(\mathbf{r}')$ and $\beta(\mathbf{r}')$) or collection ($f(\mathbf{r}')$) efficiencies, which are dependent on optically invisible device and material properties.

Because only that fraction of carriers that is generated close (within about one confocal length, z_0) to the device surface can possibly contribute to the OBIC-specific modulation (sensing local variations in $\alpha(\mathbf{r}')$, $\beta(\mathbf{r}')$, and $f(\mathbf{r}')$), it is most informative to calculate that fraction for both 1P- and 2P-OBIC. We further assume that the collection efficiency varies mostly with distance z from the device layer. The signal fractions for 1P- and 2P-OBIC are then, respectively,

$$S_1 = \frac{\int_0^{z_0} \alpha P e^{-\alpha(d-z)} f(z)\, dz}{\int_0^{d} \alpha P E^{-\alpha(d-z)} f(z)\, dz} \tag{2}$$

and

$$S_2 = \frac{\int_0^{z_0} dz \int_0^{\infty} 2\pi\rho\, d\rho\, \beta I^2(\rho, z) f(z)}{\int_0^{d} dz \int_0^{\infty} 2\pi\rho\, d\rho\, \beta I^2(\rho, z) f(z),} \tag{3}$$

where d is the substrate thickness. Since $z_0 \leqslant 10\ \mu\text{m}$ for a high-NA focus, $d \gg z_0$.

S_2 is always of order unity, because two-photon excitation is strongly confined to within the confocal length, and S_2 is therefore not dependent much on the variation of the collection efficiency with z. The signal fraction in 1P-OBIC, on the other hand, varies very much with the z dependence of the collection efficiency. For example, in the case of uniform collection, the highest signal fraction for 1P-OBIC is obtained by choosing the excitation wavelength where $\alpha = 1/d$, and $S_1 \approx e^{-1} z_0/d \ll 1$. In the other extreme, where the collection efficiency is localized within z_0, $S_1 \approx 1$, and S_1 and S_2 will become comparable, as may be expected for heavily doped substrates (for which no data are shown here, however). When the collection range is much greater than the confocal depth, on the other hand, 2P-OBIC has far higher signal fraction.

The total current originating from a layer of thickness can be estimated by using the numerators in Eqs. (2) and (3). For the powers used in Figure 5 and assuming a collection efficiency of 1, the expected 1P and 2P currents originating from a layer z_0 thick are, respectively, about 21 μA and 200 μA using the two-photon absorption coefficient $\beta \sim 1$ cm GW^{-1} at 1.26 μm, which we measured (C. Xu and W. Denk, unpublished data), using the in situ autocorrelation technique (Xu et al., 1995), because the published values of the two-photon absorption coefficient of silicon vary widely (Chase and Van Stryland, 1995) (from 1.5 to 7300 cm GW^{-1}). These current values are much higher than the experimental values shown in Figure 5, possibly indicating a much smaller collection efficiency. Nevertheless, these values mean that for these conditions, which lead to comparable levels of image contrast, the current originating in the device layers is only about 10% of the current modulation visible in the 1P-OBIC image. This is quite consistent with the view that most of the apparent contrast in 1P-OBIC images is due to scattering modulation, which would by far mask any contribution from charges generated near the device layer.

4.2 Spatial Resolution

The ability to resolve the smallest design features is critical in the examination of modern ICs. The large difference in image appearance caused by the difference in contrast mechanisms for 1P- and 2P-OBIC imaging makes a direct experimental comparison of the spatial resolution difficult, but the improvement in the spatial resolution can be estimated theoretically.

Compared with frontside OBIC imaging, where visible light ($\lambda \approx 0.5$ μm) can be used, the spatial resolution of one-photon backside IR imaging is reduced because of the longer excitation wavelength. Two-photon absorption, because it depends quadratically on the incident power, promises an improvement in the lateral optical resolution. In the spatial-frequency domain, the cutoff in the optical transfer function of 2P-OBIC is twice that of 1P-OBIC for the same excitation wavelength. In practice the excitation wavelength used in backside 1P-OBIC, which needs to be around the silicon band gap (i.e., about 1.1 μm) in order to avoid complete absorption by the substrate, is very similar to the shortest useful 2P-OBIC wavelength (1.23 μm) (Fig. 2). An improvement in the spatial frequency cutoff by a factor of almost 2 can, therefore, be expected. To actually achieve such an improvement in spatial resolution, deconvolution techniques need to be used, since a much faster fall-off in the optical transfer function, as in confocal imaging, towards the limiting frequencies has to be compensated. A reduction in the fall-off might also be achievable instrumentally by using, for example, annular apertures.

Substrate absorption also limits the attainable spatial resolution in 1P-OBIC. The reason is that high-NA rays travel for a longer distance inside the substrate compared with paraxial rays, leading to greater attenuation (Kempe, et al., 1994), effectively reducing the numerical aperture. Such a reduction is not substantial for low-NA focusing, but becomes relevant when the use of solid-immersion objective lenses is considered.

Spherical aberration caused by the refractive-index mismatch between the sample and the immersion medium degrades resolution in high-NA imaging (Sheppard and Gu, 1992; Hell et al., 1993). As shown in the VCSEL study, special objective lenses already exist for imaging through thick semiconductor substrates. Furthermore, special optical schemes could be employed outside the objective to compensate for the aberration (Sheppard, 1977; Sheppard and Gu, 1992). Ultimately, the absence of linear absorption by silicon in 2P-

OBIC permits the use of objective lenses made of silicon. Silicon lenses could further dramatically increase the spatial resolution, putting 2P-OBIC on a par with deep UV lithography.

5 SUMMARY

Backside 2P-OBIC eliminates the difficult problems that limit the usefulness of conventional IR 1P-OBIC. There is no attenuation of the excitation beam, even by thick substrates. The three-dimensional confinement of two-photon excitation eliminates background due to out-of-focus generation of free carriers. 2P-OBIC also significantly improves the spatial resolution of backside OBIC. The combination of two-photon backside imaging and the probe station is well suited for internal and full-wafer-scale testing. The fact that the image is obtained through the backside of the devices makes the technique particularly useful for characterizing optoelectronic devices designed for flip-chip bonding, even before the removal of the substrates. The required peak and average power levels for testing can be easily satisfied by available compact ultrafast laser sources, making the technique practical and user-friendly.

Future possibilities are the exploitation of the pulsed nature of the excitation light to provide an additional contrast mechanism, particularly in high-speed circuits. The fact that imaging occurs through the polished backside of the chip holds the prospect of further resolution improvements through the use of high-NA liquid-immersion or even solid-immersion techniques.

ACKNOWLEDGMENTS

Two-photon imaging of ICs was carried out in Dr. W. Denk's laboratories at Murray Hill, Lucent Technologies. The author thank G. W. Kammlott for polishing the samples. The author is grateful to M. Kelly and J. Cargo for helpful discussions.

REFERENCES

Bossmann, B., Baurschmidt, P., et al. (1992). Failure analysis techniques with the confocal laser scanning microscope. *Proc. ISTFA*, 351–361.

Chase, L.L. and Van Stryland, E.W. (1995). Nonlinear optical properties. In: *CRC Handbook of Laser Science and Technology: Optical Materials* (ed. J.M. Weber). Boca Raton, FL: CRC Press, Supplement 2.

Chirovsky, L.M.F., Hobson, W.S., and R.E. Leibenfuth et al. (1999). Implant-aperture and index-guided vertical-cavity surface-emitting laser. *IEEE Photon. Technol. Lett.* **11**, 500–502.

Cole, E.I. Jr. (1996). Electron and optical beam testing of integrated circuits using CIVA, LIVA, and LECIVA. *Microelectron. Eng.* **31**, 13–24.

D'Avanzo, D.C. (1982). Proton isolation for GaAs integrated circuits. *IEEE Trans. Electron Devices* **ED-29**, 1051–1059.

Denk, W., Strickler, J.H., et al. (1990). Two-photon laser scanning fluorescence microscopy. *Science* **248**, 73–76.

Hell, S., Reiner, G., et al. (1993). Aberrations in confocal microscopy induced by mismatches in refractive index. *J. Microsc.* **169**, 391–405.

Kempe, M., Thon, A., et al. (1994). Resolution limits of microscopy through scattering layers. *Opt. Commun.* **110**, 492–496.

Komoda, H. and Shimizu, K. (1994). Optical beam induced current techniques for failure analysis of very large scale integrated circuits devices. *Jpn. J. Appl. Phys., Part 1* **33**, 3393–3401.

Lee, Y.H., Tell, B. et al. (1990). Top-surface-emitting GaAs four-quantum-well lasers emitting at 850 nm. *Electron. Lett.* **26**: 710–711.

Nakamura, O. (1993). Three-dimensional imaging characteristics of laser scan fluorescence microscopy: two-photon excitation vs. single-photon excitation. *Optik* **93**, 39–42.

Richards, B.P. and Footner, P.K. (1992). *The Role of Microscopy in Semiconductor Failure Analysis.* New York: Oxford University Press.

Sheppard, C.J.R. (1977). The use of lenses with annular aperture in scanning optical microscopy. *Optik* **48**, 329–334.

Sheppard, C.J.R. and Gu, M. (1992). Axial imaging through an aberrating layer of water in confocal microscopy. *Opt. Commun.* **88**, 180–190.

Wilson, T. and Sheppard, C.J.R. (1984). *Theory and Practice of Scanning Optical Microscopy.* New York: Academic Press.

Xu, C. and Denk, W. (1997). Two-photon optical beam induced current imaging through the backside of integrated circuit. *Appl. Phys. Lett.* **71**: 2578–2580.

Xu, C. and Denk, W. (1999). Near infrared two- and one-photon optical beam induced current imaging through the backside of integrated circuits. *J. Appl. Phys.* **86**, 2226–2231.

Xu, C., Guild, J., et al. (1995). Determination of absolute two-photon excitation cross-sections by in situ second-order autocorrelation. *Opt. Lett.* **20**: 2372–2374.

Ziegler, E. (1990). Applications of an infrared laser scan microscope in device testing. *Microelectron. Eng.* **12**, 73–80.

INDEX